NATURAL SCIENCES IN AMERICA

NATURAL SCIENCES IN AMERICA

HISTORY

OF

NORTH AMERICAN

PINNIPEDS

BY JOEL ASAPH ALLEN

ARNO PRESS
A New York Times Company
New York, N. Y. • 1974

Reprint Edition 1974 by Arno Press Inc.

Reprinted from a copy in the University
  of Illinois Library

NATURAL SCIENCES IN AMERICA
ISBN for complete set: 0-405-05700-8
See last pages of this volume for titles.

Manufactured in the United States of America

———◆———

Library of Congress Cataloging in Publication Data

Allen, Joel Asaph, 1838-1921.
    History of the North American pinnipeds.

    (Natural sciences in America)
    Reprint of the 1880 ed. published by the U. S. Govt.
Print. Off., Washington, which was issued as Miscella-
neous publications no. 12 of the U. S. Geological and
Geographical Survey of the Territories.
    1. Pinnipedia--North America. 2. Mammals--North
America. I. Title. II. Series. III. Series:
United States. Geological and Geographical Survey of
the Territories. Miscellaneous publications, no. 12.
QL737.P6A4 1974   599'.745   73-17792
ISBN 0-405-05702-4

QL
737
.P6
A4
1974

DEPARTMENT OF THE INTERIOR
U. S. GEOLOGICAL AND GEOGRAPHICAL SURVEY OF THE TERRITORIES
F. V. HAYDEN, GEOLOGIST-IN-CHARGE

MISCELLANEOUS PUBLICATIONS—No. 12

HISTORY

OF

NORTH AMERICAN

PINNIPEDS

A MONOGRAPH OF THE

WALRUSES, SEA-LIONS, SEA-BEARS

AND

SEALS

OF

NORTH AMERICA

BY JOEL ASAPH ALLEN

Assistant in the Museum of Comparative Zoölogy at Cambridge
Special Collaborator of the Survey

WASHINGTON
GOVERNMENT PRINTING OFFICE
1880

PREFATORY NOTE.

UNITED STATES GEOLOGICAL AND
GEOGRAPHICAL SURVEY OF THE TERRITORIES,
Washington, D. C., July 1, 1880.

The present series of monographs of the North American
Pinnipedia, by Mr. J. A. Allen, may be considered as a second
installment of the systematic History of North American Mammals, of which the Fur-Bearing Animals by Dr. Elliott Coues,
U. S. A., forming No. 8 of the Miscellaneous Publications of
the Survey, was published as a specimen fasciculus. The first
monograph of this series, treating of the Walruses, was prepared
nearly three years since for publication in the Bulletin of the
Survey, but before it was quite ready for the press, Dr. Coues,
owing to his pressing engagements in other directions, invited
Mr. Allen to extend his treatise to embrace the entire suborder
of the Pinnipeds, to which he had already given special attention, with a view to its incorporation as a part of the proposed
general History of North American Mammals. Since, however, considerable time must elapse before the whole work can
be completed, it has been thought best not to delay the publication of the part already prepared relating to the Pennipeds.

As nearly all of the species belonging to this group found in
the northern hemisphere are members of the North American
fauna, the present treatise is virtually a monograph of all the
species occurring north of the equator, and includes incidentally
a revision of those of other seas. The literature of the whole
group is not only reviewed at length, but the economic phase
of the subject is treated in detail, embracing, in fact, a general
history of the Sealing industries of the world. The technical
treatment of the subject is based mainly on the rich material
of the National Museum, supplemented at many important
points by that of the Museum of Comparative Zoölogy of Cambridge, which, through the kindness of the directors of these
institutions, was generously placed at the author's disposal.
That contained in the other principal museums of the country

was also examined, so that so far as the species of the northern
hemisphere is concerned the amount of material consulted
doubtless far exceeds that ever before studied by any single
investigator of the group. For the biographical part, to which
much space has been allotted, matter has been freely gathered
from all available sources. In addition to the results here first
published, the work may be considered as a compendium of
our present knowledge of the subject.

In regard to the need of a work like the present, it may be
stated that with the exception of Dr. Theodore Gill's important
"Prodrome" of a proposed monograph of North American
Pinnipeds, published in 1866, there has been no general treatment of the species since the excellent compilations of Drs.
Harlan and Godman appeared, now more than half a century
ago. Respecting foreign works, nothing has been recently
published covering the ground here taken beyond a very general synopsis of the technical phases of the subject. The best
accounts of the species occurring along the shores of Europe
are in other languages than English, while no general history
of the economic relations of the subject exists. In relation to
the important Fur Seal Fisheries of Alaska, the author has been
able to present *in extenso* the results of Captain Charles Bryant's
long experience at the Fur Seal Islands, where for nearly ten
years he was the government agent in charge of the islands.
Although not received until the article on this species had
been transmitted to the printer, it proves to be, to only a small
degree, a repetition of the account given by Mr. Elliott, also
reproduced at length. The history Captain Bryant gives of
the changes in the numbers and relations of the different
classes of these animals at the rookeries, under the present
system of management of the Fur Seal business, forms a valuable basis for generalization in regard to the future regulation
of this industry, and is also an important contribution to the
life-history of the species.

The cuts, some thirty in number, illustrating the cranial characters of the Walruses, were drawn for the present work by Mr. J.
H. Blake, of Cambridge, and engraved by Messrs. Russell and
Richardson, of Boston. The Survey is indebted to Professor
Baird, Secretary of the Smithsonian Institution, for a series of sixteen original figures, engraved by Mr. H. H. Nichols, of Washington, from photographs on wood, illustrating the skulls of *Callorhinus ursinus*, Peale's "*Halichœrus antarcticus*," *Cystophora cristata*,

and *Macrorhinus angustirostris;* also to the Zoölogical Society of London for electros of Gray's "*Halicyon richardsi,*" and of a series of historic figures of the walrus published in the Society's "Proceedings," by the late Dr. Gray, and to the proprietors of "Science Gossip," for electros of the full-length figures of seals. These were received through Dr. Coues, who also furnished the full-length views of *Eumetopias stelleri* and *Callorhinus ursinus.*

Mr. Allen desires me to express, in this connection, acknowledgments of his indebtedness to Prof. Spencer F. Baird, Secretary of the Smithsonian Institution, and to Prof. Alexander Agassiz, Director of the Museum of Comparative Zoölogy, for the liberality with which they have placed at his service the rich material relating to this group of animals contained in the museums respectively under their charge; to Prof. Henry A. Ward, of Rochester, N. Y., for the use of much valuable material relating to the Walruses that he would not otherwise have seen; and to Captain Charles Bryant, late special agent of the United States Treasury Department, for his report, kindly prepared at the author's request, for the present work. Also to Dr. Elliott Coues, Secretary of the Survey, for the use of many of the cuts, for valuable suggestions during the prepation and printing of the monograph, and revision of the proof-sheets.

F. V. HAYDEN,
United States Geologist.

LETTER OF TRANSMITTAL.

———

CAMBRIDGE, MASS.,
May 25, 1880.

SIR: I have the honor to transmit herewith for approval
and for publication the "History of North American Pinnipeds,"
being a monograph of the Walruses, Sea-Lions, Sea-Bears, and
Seals of North America.

Very respectfully, your obedient servant,

J. A. ALLEN.

F. V. HAYDEN,
  Washington, D. C.

TABLE OF CONTENTS.

LIST OF ILLUSTRATIONS.

LIST OF ILLUSTRATIONS.

HISTORY OF THE NORTH AMERICAN PINNIPEDS.

The Pinnipeds, or *Pinnipedia*, embracing the Seals and Walruses, are commonly recognized by recent systematic writers as constituting a suborder of the order *Feræ*, or Carnivorous Mammals. They are, in short, true *Carnivora*, modified for an aquatic existence, and have consequently been sometimes termed "*Amphibious Carnivora*." Their whole form is modified for life in the water, which element is their true home. Here they display extreme activity, but on land their movements are confined and labored. They consequently rarely leave the water, and generally only for short periods, and are never found to move voluntarily more than a few yards from the shore. Like the other marine Mammalia, the *Cetacea* and *Sirenia* (Whales, Dolphins, Porpoises, Manatees, etc.), their bodies are more or less fish-like in general form, and their limbs are transformed into swimming organs. As their name implies, they are *fin-footed*. Generally speaking, the body may be compared to two cones joined basally. Unlike the other marine mammals, the Pinnipeds are all well clothed with hair, while several of them have, underneath the exterior coarser hair, a thick, soft, silky under-fur. In contrasting them with the ordinary or terrestrial mammals, we note that the body is only exceptionally raised, and the limbs are confined within the common integument to beyond the knees and elbows, and are hence to only a slight degree serviceable for terrestrial locomotion. The first digit of the manus is generally lengthened and enlarged, as are both the outer digits of the pes. As compared with other *Feræ*, they present, in osteological characters, many obvious points of difference, especially in relation to the structure of the skull, limbs, and pelvis, and in dentition. The skull is distinctively characterized by great compression or constriction of the interorbital portion, the large size of the orbital fossæ, in the lachrymal bone being imperforate (without a lachrymal canal) and contained within the orbit, and in the presence (generally) of considerable vacuities between the palatine and frontal bones and the tympanic and exoccipital bones. The deciduous dentition is rudimentary, never to any great extent functional, and frequently does not persist beyond the fœtal life of the animal. In the permanent dentition, the canines are greatly developed, sometimes enormously so; the lower incisors are never more than four in number, and sometimes only two; the upper incisors usually number six, but sometimes only four, or even two; the grinding teeth (premolars and molars) are generally simple in structure, and usually differ from each other merely in respect to size, or the number of roots by which they are inserted. The pelvis differs from that of the terrestrial *Feræ* in the shortness of the iliac portion and the eversion of its anterior border; the ischiac bones barely meet for a short distance in the male, and are usually widely separated in the female, the pelvic arch thus remaining in the latter permanently open ventrally.

The existing Pinnipeds constitute three very distinct minor groups or families, differing quite widely from each other in important characters: these are the Walruses, or *Odobænidæ*, the Eared Seals, or *Otariidæ*, and the Earless Seals, or *Phocidæ*. The first two are far more nearly allied than are either of these with the third, so that the *Odobænidæ* and *Otariidæ* may be together contrasted with the *Phocidæ*. The last named is the lowest or most generalized group, while the others appear to stand on nearly the same plane, and about equally remote from the *Phocidæ*. The Walruses are really little more than thick, clumsy, obese forms of the Otarian type, with the canines enormously developed, and the whole skull correlatively modified. The limb-structure, the mode of life, and the whole economy are essentially the same in the two groups, and, aside from the cranial modifications presented by the *Odobænidæ*, which are obviously related to the development of the canines as huge tusks, the Walruses are merely elephantine Otariids, the absence or presence of an external ear being in reality a feature of minor importance.

The characters of the suborder and its three families may be more formally stated as follows: *

---

* The characters here given are in part those collated by Dr. Theodore N. Gill in 1873 ("Arrangement of the Families of the Mammals." Smithsonian Miscellaneous Collections, No. 230, pp. 56, 68, 69), by whom the distinctive features of these groups were first formulated. They have, however, been carefully verified and further elaborated by the present writer, while the families are here quite differently associated.

Limbs pinniform, or modified into swimming organs, and enclosed to or beyond the elbows and knees within the common integument. Digits of the manus decreasing in length and size from the first to the fifth; of those of the pes, the first and fifth largest and longest, the three middle ones shorter and subequal. Pelvis with the iliac portion very short, and the anterior border much everted; ischia barely meeting by a short symphysis (never anchylosed) and in the female usually widely separated. Skull generally greatly compressed interorbitally; facial portion usually short and rather broad, and the brain-case abruptly expanded. Lachrymal bone imperforate and joined to the maxillary, enclosed wholly within the orbit. Palatines usually separated by a vacuity, often of considerable size, from the frontals. Tympanic bones separated also by a vacuity from the exoccipitals. Dentition simple, generally unspecialized, the molars all similar in structure. Deciduous dentition rudimentary, never truly functional, and generally not persistent beyond the foetal stage of the animal. Permanent incisors usually $\frac{6}{4}$ or $\frac{4}{4}$, sometimes $\frac{4}{2}$ (*Cystophora* and *Macrorhinus*), or even $\frac{2}{2}$ (*Odobænus*); canines $\frac{2}{2}$; molars $*$ $\frac{5}{5}$, $\frac{6}{5}$, or $\frac{5}{3}$ .... PINNIPEDIA.

A. Hind legs capable of being turned forward and used in terrestrial locomotion. Neck lengthened (especially in family II). Skull with the mastoid processes large and salient (especially in the males), and with distinct alisphenoid canals. Anterior feet nearly as large as the posterior, their digits rapidly decreasing in length from the first to the fifth, without distinct claws, and with a broad cartilaginous border extending beyond the digits. Hind feet suceptible of great expansion, the three middle digits only with claws, and all the digits terminating in long, narrow, cartilaginous flaps, united basally. Femur with the trochanter minor well developed .... GRESSIGRADA.

   I. Without external ears. Form thick and heavy. Anterior portion of the skull greatly swollen, giving support to the enormously developed canines, which form long, protruding tusks. Incisors of deciduous (foetal) dentition $\frac{6}{6}$; of permanent dentition $\frac{2}{0}$. No postorbital processes, and the surface of the mastoid processes continuous with the auditory bullæ ...... *Odobænidæ.*

   II. With small external ears. Form slender and elongated. Anterior portion of the skull not unusually swollen, and the canines not highly specialized. Incisors of deciduous dentition $\frac{6}{4}$, only the outer on either side cutting the gum; of permanent dentition $\frac{6}{4}$, the two central pairs of the upper with a transverse groove. Postorbital processes strongly developed. Surface of the mastoid processes not continuous with the auditory bullæ... *Otariidæ.*

B. Hind legs not capable of being turned forward, and not serviceable for terrestrial locomotion. Neck short. Skull with the mastoid

---

$*$ In view of the uncertainty respecting the proper notation of the grinding teeth, they will in the present work be designated simply as molars, with no attempt at distinguishing "premolars" from "molars."

processes swollen, but not salient, and without distinct alisphenoid canals. Anterior limbs smaller than the posterior, the first
digit little, if any, longer than the next succeeding ones, all
armed with strong claws, which are terminal. Hind feet capable of moderate expansion, short; digits (usually) all armed
with strong claws, and without terminal cartilaginous flaps.
Femur with no trace of the trochanter minor...REPTIGRADA.*

III. Without external ears. Postorbital processes wanting, or very
small. Incisors variable ($\frac{6}{4}$, $\frac{4}{4}$, or $\frac{4}{2}$). Deciduous dentition not
persistent beyond fœtal life....... .. ..*Phocidæ.*

The Pinnipeds present a high degree of cerebral development, and are easily domesticated under favorable conditions.
They manifest strong social and parental affection, and defend
their young with great persistency and courage. They are carnivorous (almost without exception), subsisting upon fishes,
mollusks, and crustaceans, of which they consume enormous
quantities. The Walruses and Eared Seals are polygamous,
and the males greatly exceed the females in size. The ordinary
or Earless Seals are commonly supposed to be monogamous, and
there is generally little difference in the size of the sexes. The
Walruses and Eared Seals usually resort in large numbers to
certain favorite breeding grounds, and during the season of reproduction leave the water, and pass a considerable period upon
land. The Earless Seals, on the other hand, with the exception
of the Sea Elephants, do not so uniformly resort to particular
breeding grounds on land, and leave the water only for very
short intervals. They usually bring forth their young on the
ice, most of the species being confined to the colder latitudes.
Only one of the various species of the *Pinnipedia* appears to
be strictly tropical, and very few of them range into tropical
waters. As a group, the Pinnipeds are distinctively characteristic of the Arctic, Antarctic, and Temperate portions of the
globe, several of the genera being strictly Arctic or Subarctic
in their distribution. The Walruses are at present confined
mainly within the Arctic Circle, and have no representatives
south of the colder portions of the Northern Hemisphere. The
Otariidæ and *Phocidæ*, on the other hand, are abundantly
represented on both sides of the equator, as will be noticed
more in detail later.

---

* For the suggestion of the terms *Gressigrada* and *Reptigrada* I am indebted
to my friend Dr. Elliott Coues.

FAMILY ODOBÆNIDÆ.

Walruses.

"*Trichecidæ*, GRAY, London Med. Repos., 1821, 303" (family). *Apud* Gray.

Trichechidæ, GRAY, Ann. of Philos., 1825, 340; Ann. and Mag. Nat. Hist., 3d
  ser., xviii, 1866, 229; *ibid.*, 4th ser., iv, 1869, 268; Suppl. Cat. Seals
  and Whales, 1871, 5 (family).

Trichecina, GRAY, Loudon's Mag. Nat. Hist., i, 1837, 538; "Zoöl. Erebus and
  Terror, 3" (subfamily). In part only, or exclusive of *Halichœrus*.

Trichechina, GRAY, Cat. Mam. Brit. Mus., pt. ii, 1850, 29; Cat. Seals and
  Whales, 1866, 33 (subfamily). In part only = *Trichecina* Gray, 1837.

"*Trichechidæ* seu *Campodontia*, BROOKES, Cat. Anat. and Zoöl. Mus. 1828, 37."

Trichechoidea, GIEBEL, Fauna der Vorwelt, i, 1847, 221: Säugeth., 1855, 127
  (family).

Trichecina, TURNER, Proc. Zoöl. Soc. Lond., 1848, 85, 88 (subfamily).

Rosmaridæ, GILL, Proc. Essex Institute, v, 1866, 7, 11; Families of Mam.,
  1872, 27, 69, 70 (= "*Trichechidæ* Brookes, Gervais").—ALLEN, Bull.
  Mus. Comp. Zoöl., ii, 1870, 21.

Rosmaroidea, GILL, Fam. Mam., 1872, 70 ("superfamily" = *Rosmaridæ* Gill).

Broca, LATREILLE, Fam. Règ. Anim., 1825, 51 (family).

Les Morses, F. CUVIER, Dents des Mam., 1825, 233; Dict. Sci. Nat., lix, 1829,
  465 (family).

GENERAL OBSERVATIONS.

Among the distinctive features of the *Odobænidæ* are the
enormous development of the upper canines, and the consequent
great enlargement of the anterior portion of the skull for their
reception and support, the early loss of all the incisors except
the outer pair of the upper jaw, the caducous character of the
posterior molars, and the molariform lower canines. The Walruses share with the Eared Seals the ability to turn the hind
feet forward, and consequently have considerable power of locomotion on land. This is further aided by a greater freedom of
movement of the fore feet than is possessed by the Earless
Seals. The Walruses differ from the Eared Seals by their
much thicker bodies, shorter necks, and longer caudal vertebræ,
the dorsal and lumbar vertebræ remaining of proportionately
the same length. In consequence of their obesity, the ribs and
the proximal segments of the limbs are longer in the Walruses
than in the Eared Seals, while the distal segments of the limbs
are relatively shorter. The scapula is long and narrow, instead
of short and broad, as in the *Otariidæ*, and its crest is placed

5

more anteriorly. Accordingly, in respect to general form, we
have slenderness of both body and limbs in the one contrasted
with great thickness of body, and distally a disproportionate
reduction of the extremities in the other. The most striking
differences, however, exist in the cranial characters, resulting
from the great development of the upper canines in the Walruses, and the consequent modifications of the facial portion of
the skull. In the *Otariidæ*, the general contour of the skull is
strongly Ursine; in the *Odobænidæ*, it is unique, owing to its
great expansion anteriorly. In respect to other cranial features,
the Walruses differ from the Eared Seals in having no post-orb-
ital processes, and in the mastoid processes being not separated
from the auditory bullæ. The teeth are all single-rooted, and
have in the permanent dentition no distinct crowns.

On comparing the *Odobænidæ* with the *Phocidæ*, the differences in general structure are found to be far greater than obtain between the Walruses and Eared Seals, especially in regard
to the hind extremities; these in the *Phocidæ* being directed
backward, and useless as organs of terrestrial locomotion.
Hence, in so far as the *Odobænidæ* and *Otariidæ* agree in limb-
and skull-structure, they both similarly depart from the Phocine
type. As already indicated in the synopsis of the suborder *Pinnipedia*, the *Phocidæ* differ far more from either the *Odobænidæ*
and *Otariidæ* than do these latter from each other. This difference is especially emphasized in the skull; for while the *Odobænidæ* and *Otariidæ* agree in all important cranial characters,
aside from the special features correlated with the immense
enlargement of the upper canines in the former, they widely
differ from the *Phocidæ*. This is especially seen in the absence
in the latter of an alisphenoid canal, in the greatly swollen auditory bullæ, the position of the carotid foramen, and the non-
salient character of the mastoid processes.

The few points in which the Walruses differ in myology from
other Pinnipeds, Dr. Murie states to be "the presence of a co-
raco-brachialis, a flexor brevis manus, a pronator quadratus, an
opponens pollicis, and a palmaris brevis," in the possession of
which it differs both from *Otaria* and *Phoca*, but that in other
respects they "muscularly present general agreement." "Compared with the Seals [*Phoca?*] there are two extra peronei and
a flexor brevis hallucis." "Though deficient in concha, the auricular muscles are remarkably large."*

* Proc. Zoöl. Soc. Lond., 1870, p. 545.

"Considering the very different attitudes assumed by the *Trichechidæ* and *Otariidæ* as compared with the *Phocidæ*," he further adds, "it is remarkable how very little deviation follows in the muscular development. The two former, as might be anticipated, present a general agreement, especially in the mode of implantation of the muscles of the hind leg, and in this respect recede from the Seal, yet but slightly."*

In respect to the position and character of the viscera, a general agreement has beeen noted with those of the other Pinnipeds, and they present nothing that calls for special notice in the present connection. As Dr. Murie has stated, there is little appreciable difference exhibited throughout the Pinnipeds in the construction of the alimentary canal.. "It is simply that of a Carnivore, with, however, a moderate-sized cæcum. The great glandular superficies and correlated large lymphatics point to means of speedy and frequent digestion; and in the Walrus these apparatus are extraordinarily developed."†

In accordance with the characters already given (p. 3), if any subdivision of the Pinnipeds into groups of higher rank than families is to be made, it seems evident that the *Odobænidæ* and *Otariidæ* are to be collectively contrasted with the *Phocidæ*; in other words, that to unite the *Otariidæ* and *Phocidæ* as a group of co-ordinate rank with the *Odobænidæ* is to lose sight of the wide differences that separate the two first-named families, as well as of the many important features shared in common by the *Odobænidæ* and *Otariidæ*, by which both are trenchantly separated from the *Phocidæ*.

Although the Walruses are now very generally recognized as constituting a natural family of the Pinnipeds, ranking co-ordinately with the Eared Seals on the one hand and with the Earless Seals on the other, the affinities of few groups have been more diversely interpreted. As early as the thirteenth century, the author of the "Speculum Regale",—one of the earliest works relating to natural history, in which the Walrus is mentioned,—stated distinctly that the Walrus was an animal closely related to the Seals; and we find that nearly all natural-history writers prior to the middle of the eighteenth century who referred to the Walruses, gave them the same association. It was the technical systematists of the last half of the eighteenth century who broke up this natural juxtaposition, and variously grouped

---

* Trans. Zoöl. Soc. Lond., vol. vii, 1872, p. 459.
† Trans. Zoöl. Soc. Lond., vol. vii, 1872, p. 461.

them with forms with which they had no relationship. In the infancy of science, nothing was perhaps more natural than that animals should be classified in accordance with their mode of life, their habitat, or their external form, and we are hence not surprised to find that Rondelet, Gesner, Aldrovandus, Jonston, and other pre-Linnæan writers, arranged the Pinnipeds, as well as the Sirenians and Cetaceans, with the fishes, or that other early writers should term all four-footed creatures "Quadrupeds," and divide them into "Land Quadrupeds" and "Quadrupeds of the Sea." While all marine animals were by some early writers classified as "fishes,"* the Pinnipeds were much sooner disassociated from the true fishes than were the Cetaceans and Sirenians, the mammalian affinities of which were not at first recognized by even the great Linné himself, who, as late as the tenth edition of his "Systema Naturæ" (1758), still left them in the class "*Pisces.*"

In view of the several excellent descriptions and very creditable figures of the Atlantic Walrus that appeared as early as the sixteenth and seventeenth centuries (a detailed account of which will be given later), it is surprising that the early systematic writers should display such complete ignorance of some of the most obvious external characters of this animal, as was notably the case with Linné, Klein, Brisson, Erxleben, and Gmelin, who strangely associated the Walrus and the Manatee as members of the *same genus*, and grouped them with such diverse creatures as Sloths and Elephants. Linné, it is true, in the earlier editions of the "Systema Naturæ," placed the Walrus with the Seals in the genus *Phoca*, in his order *Feræ*,—a near hit at their true affinities. Later, however, following probably Klein and Brisson, he fell into the grave error of removing them to nearly the most unnatural association possible. In this connection, it may prove not uninteresting to sketch, in brief outline, the strange history of the classification of this singular group of fin-footed Carnivores.

As already stated, Linné's first allocation of the group was the natural one. Brisson,† in 1756, led in the long rôle of error by forming his third "order" of mammals of the Elephant, the

<hr/>

*Most modern languages still retain relics of this ancient custom, as evinced, for example, in such English words as *shell-fish, cray-fish, whale-fishery, seal-fishery*, etc., while *hvalfish* (Swedish), *walrisch* (Danish), *wallfisch* (German), etc., are common vernacular names applied to Cetaceans.

†Règne Animal, 1756, p. 48.

Walrus, and the Manatee, the two last named constituting his "genus *Odobenus*." This was a marked retrocession from even the system of Klein,* of a few years' earlier date, who brought together as one family the Seals, Otters, Beaver, Walrus, and Manatee. Linné, in 1766,† not only removed the Walrus from the genus *Phoca*, in which he had previously placed it, to *Trichechus*, but also transferred it from his order *Feræ* to *Bruta*, which thus contained not only the Walrus, but such a diverse assemblage as Elephants, Sloths, and Anteaters. Linné's genus *Trichechus*, as at this time constituted, was equivalent to Brisson's genus "*Odobenus*." Erxleben,‡ who recognized no higher groups than genera, placed the Walruses and Seals together immediately after the Carnivores. Schreber,§ at about the same date (1777), adopted a similar classification, the Walrus standing next after the Elephant and preceding the Seals. Schreber's genus *Trichechus* contained also the Dugong and the Manatee. Gmelin,‖ in 1788, followed the Linnæan arrangement of 1766; the Walrus, as usual from the time of Brisson to Gmelin, standing next to the Elephant, and associated generically with the Sirenians. Blumenbach,¶ from 1788 till as late even as 1825, still arranged the Walrus and the Sirenians in the genus *Trichechus*. In other respects, the Walrus appears with new associates, the genus *Trichechus* being united with *Ornithorhynchus* to form a "family"(!) of his "order" *Palmata*. The order *Palmata*, as the name implies, was composed of the web-footed mammals, and divided into three "families," namely, "*A. Glires*" (consisting of the genus *Castor*); "*B. Feræ*" (*Phoca* and *Lutra*); and "*C. Bruta*" (*Ornithorhynchus* and *Trichechus*). This is essentially also the arrangement proposed by Klein in 1751.

The first step toward dismembering the unnatural conglomeration known previously under the names *Trichechus* and *Odobenus* was made by Retzius** in 1794, who divided the genus *Trichechus* of former authors into three genera, namely, *Manatus*, for the Manatee; *Hydromalis*, for Steller's Sea-Cow (= *Rhytina* Illiger, 1811); and *Trichechus*, the last embrac-

---

* Quad. Disp. Brev. Hist. Nat., 1751, pp. 40, 92.
† Syst. Nat., ed. 12, 1766, p. 49.
‡ Syst. Reg. Anim., 1777, p. 593.
§ Säugeth., ii, [1776?], p. 260.
‖ Syst. Nat., i, 59.
¶ Handb. d. Naturgesch., 1788, p. 142, and later editions.
** Kongl. Vetensk. Acad. nya Handling., xv, 1794, pp. 286–300.

ing both the Walrus and the Dugong. While this was in the main a most important and progressive innovation, Retzius seems to have labored, like several still earlier writers, under the impression that the Walrus, like the Dugong, *had no hind feet*. Ozeretskovsky,[*] about a year later, and probably ignorant of Retzius's paper, also placed, as curiously happened, the Walrus and the Dugong together in the genus *Trichechus*, because he supposed the Dugong *had hind feet*, like the Walrus! These curious antithetical mistakes indicate how little was known by systematic writers about the structure of these animals as late as the close of the last century.

The elder Cuvier, [†] in 1798, while retaining the Walrus and the Sirenians in the genus *Trichechus*, separated them from some of their former unnatural entanglements by again associating *Trichechus* and *Phoca* in his group "Mammifères Amphibies," which he placed between the "Solipèdes" and "Mammifères Cétacés." He divided this group into "I. Les Phoques (*Phoca*)" and "II. Les Morses (*Trichecus*, L.)"; the latter including "1. *Trichecus rosmarus*"; "2. *Trichecus dugong*"; "3. *Trichecus manatus*."

As already shown, Retzius nearly disentangled the Walrus from the Sirenians, leaving of the latter only the Dugong in the genus *Trichechus*. G. Fischer, [‡] in 1803, completed the separation by removing the Dugong and the Manatee, to which he gave the generic names respectively of *Platystomus* (=*Halicore*, Illiger, 1811) and *Oxystomus* (=*Manatus*, Retzius, 1794), leaving only the Walrus in *Trichechus*. The genus *Trichechus*, however, as first instituted by Artedi (1738) and Linné (1758), as will be shown later, did not relate in any way to the Walrus, being applied exclusively to the Manatee. It was not till 1766 that the term was first made to cover both the then known Sirenians and the Walrus, although the embroilment of the two groups began with Brisson, ten years earlier.

The Pinnipeds and Sirenians, collectively considered, were first separated as distinct groups by Illiger [§] in 1811, who raised them to the rank of orders, they forming respectively his orders *Pinnipedia* and *Natantia*. The former consisted of two genera, *Phoca*, embracing all the Seals, and *Trichechus*, containing only the Walruses. They were regarded as forming a single family,

---

[*] Nova Act. Acad. Petrop., xiii, 1796, pp. 371–375.

[†] Tabl. Élément., p. 172.

[‡] Das National-Museum der Naturgeschichte, ii, 1803, pp. 344–358.

[§] Prodromus Systematis Mammalium et Avium, 1811, pp. 138, 139; Abhandl. der Akad. Wissensch. zu Berlin, 1804–1811, (1815), pp. 39–159, *passim*.

equivalent in extent with the order *Pinnipedia*. The propriety
of the changes introduced by Illiger was not speedily recognized
by contemporary writers; Cuvier, and many subsequent systematists for half a century, placing the Pinnipeds among the
Carnivora and the Sirenians among the *Cetacea*, with the rank
respectively of families, the family *Phocidæ* embracing all the
Pinnipeds. Dr. J. E. Gray, in 1821,* and again in 1825,† widely
separated the Walruses from the Seals as a family, *Trichechidæ*,
which he most strangely placed (together with the Sirenians) in
the order *Cete*. Later, however, in 1837, ‡ he reunited the Walruses and the Seals into the single family *Phocidæ*, which he
divided into five subfamilies, *Trichechina* being the third and
central group, and embracing the genera *Halichœrus* and *Trichechus*. This highly artificial classification he retained till 1866,
when, following other systematists, he again raised the Walruses to the rank of a distinct family.

Latreille, § in 1825, not only treated the Pinnipeds as an
order (*Amphibia*), but separated the Walruses from the others
as a distinct family (*Broca*), the Seals forming his family *Cyno-
morpha*.

In 1829, F. Cuvier‖ divided the Pinnipeds into the Seals proper
("les Phoques proprement dits"), and the Walruses ("les Morses").
Brookes, ¶ in 1828, again recognized the Walruses as forming
a family ("*Trichechidæ* seu *Campodontia*") distinct from the
other Pinnipeds. Wagler,** in 1830, made the Walruses merely
a genus of his order *Ursi*. Nilsson, †† in 1837, divided the Pinnipeds into two sections, the second of which embraced not only
Trichechus, but also *Halichœrus*, *Cystophora*, and *Otaria*. Turner, ‡‡ in 1848, from a study of the skulls, separated the Pinnipeds into three natural groups, considered by him to hold the
rank of subfamilies, namely : *Arctocephalina*, embracing *Otaria*
and *Arctocephalus ;* *Trichecina*, consisting of the genus "*Trichecus*"; and *Phocina*, embracing all the other Seals. Gill, §§ in 1866,

* "London Med. Repos., 1821, p. 302," *apud* Gray.
† Annals of Philosophy, 2d ser., vol. x, 1825, p. 340.
‡ Loudon's Mag. Nat. Hist., vol. i, p. 583.
§ Fam. Règ. Anim., p. 51.
‖ Dict. Sci. Nat., t. lix, p. 367.
¶ "Cat. of his Anatom. and Zoöl. Mus., p. 36," *apud auct.*
** Naturl. Syst. Amph., p. 27.
†† Vetensk. Akad. Handl., 1837, 235; Wiegmann's Arch. f. Naturg., 1841,
p. 306 (transl.).
‡‡ Proc. Zoöl. Soc. Lond., 1848, pp. 85, 88.
§§ Proc. Essex Institute, vol. v, p. 7.

was the next author who recognized the Walruses as forming a distinct family, which he termed *Rosmaridæ*. In this step, he was immediately followed by Gray,* and by the present writer† in 1870. Lilljeborg,‡ in 1874, also accorded them family rank, as has been the custom of late with various other writers. Gill, § in 1872, raised them to the rank of a "superfamily" (*Rosmaroidea*), treating them as a group co-ordinate in rank with his "*Phocoidea*," consisting of the *Phocidæ* and *Otariidæ*.

Their final resting-place in the natural system has now probably been at last reached, the majority of modern systematists agreeing in according to them the position and rank of a family of the *Pinnipedia*. To Illiger seems due the credit of first distinctly recognizing the real affinities of both the Pinnipeds and Sirenians to other mammals, and with him originated the names by which these groups are now commonly recognized, the chief modification of Illiger's arrangement being the reduction of the *Pinnipedia* from a distinct order to the rank of a suborder of the *Feræ*.

GENERA.

The family *Odobænidæ* (*Trichechidæ* Gray and Brookes = *Rosmaridæ* Gill) includes, so far as at present known, only the existing genus *Odobænus* (= *Trichechus* of many authors, not of Artedi nor of Linné) and the two extinct genera *Trichechodon* and *Alactherium*, recently described from fossil remains found in Belgium. *Alactherium*,‖ while evidently referable to the *Odobænidæ*, differs quite strikingly from the existing Walruses. The parts known are the left ramus of the lower jaw, the greater portion of the cranium (the facial portion and teeth only wanting), several cervical vertebræ, a portion of the pelvis, and various bones of the extremities. The rami of the lower jaw are not anchylosed as in the Walrus, and the dentition is quite different from that of *Odobænus*, that of the lower jaw being I. 2, C. 1, M. 4. The symphysis occupies nearly half of the length of the jaw. Van Beneden describes the skull as resembling in some characters the skull of the Otaries, and in others those of the Morses. The molar teeth he says could not be easily distinguished from those of the Morse if they were found isolated. No

---

* Ann. and Mag. Nat. Hist., 3d ser., vol. xviii, 1836, p. 229.
† Bull. Mus. Comp. Zoöl., vol. ii, p. 21.
‡ Fauna öfer Sveriges och Norges Ryggr., p. 674.
§ Arrangement of Families of Mammals, 1872, p. 69.
‖ Van Beneden, Ann. Mus. d'Hist. Nat. de Belgique, i, 1877, p. 50.

canines were found at Anvers, but Van Beneden is strongly of
the opinion that the teeth described by Ray Lankester,* from
the Red Crag of England, in 1865, and named *Trichechodon huxleyi*, are those of his *Alactherium cretsii*. The other bones referred to *Alactherium* bear a general resemblance to the corresponding bones of the existing Walrus, but indicate an animal of
much larger size. The femur and some of the other bones bear
also a resemblance to corresponding parts of the Otaries. A cast
of the cerebral cavity shows that the brain was not much unlike
that of the existing Walruses and Otaries, but with the cerebel-
lum smaller. *Alactherium* thus proves to have been a Pinniped
of great size, closely related in general features to the Walruses
of to-day, but presenting features also characterizing the Eared
Seals as well as others common to no other Pinniped. The genus
Trichechodon of Van Beneden (probably not = *Trichechodon*,
Lankester, 1865) is much less well known, the only portion of
the skull referred to it being part of a right ramus. The other
bones believed to represent it are nine vertebræ, part of a pelvis, a humerus, a femur, several metatarsal, metacarpal, and
phalangeal bones, etc., and part of a tusk. Says Van Beneden:
" Une branche de maxillaire est tout ce que nous possédons de la
tête. Les dents manquent, mais le bord est assez complet
pour qu'on puisse bien juger de leurs caractères par les alvéoles.
Nous pouvons, du reste, fort bien aussi apprécier la forme de
cet os, distinguer sa symphyse et sa brièveté.

" L'os est brisé à son extrémité antérieure, la symphyse est
fort courte et l'os n'a pas plus d'épaisseur sur la ligne médiane
que sur le côté. Les alvéoles sont comparativement fort grandes :
les trois dernières sont à peu prè semblables, l'antérieure est la
plus petite. C'est l'inverse dans le Morse. La canine devait être
fort grande. Il n'y a qu'une seule alvéole pour une dent incisive.

" Le *corps* du maxillaire est remarquable pour sa courbure.
Toute la partie postérieure qui constitue la *branche* du maxillaire
manque. On voit sur la face externe trois trous mentonniers.

" En comparant ce maxillaire à celui du Morse vivant, on voit
que la symphyse est toute différente, qu'il existe une grande
alvéole pour la dent canine et des traces d'une petite alvéole
pour une incisive qui restait probablement cachée sous les gen-
cives. Dans le Morse vivant, il n'y a pas de place pour une
canine [grande] au maxillaire inférieur."

---

* See beyond, p. 62.

The other bones are described as more or less resembling those of the Walrus, and do not much exceed them in size. Some of them are also said to closely resemble corresponding parts of *Alactherium*.

Van Beneden's descriptions and figures of the lower jaw fragment indicate features widely different from those of the corresponding part in the Walrus, especially in the shortness of the symphysis and in the curvature of the part represented, but above all in the number, relative size, and form of the alveoli, and particularly in the large size of that of the canine, which must have been almost as highly specialized as in the Sea Lions. That the tusks referred to it by Van Beneden (those described by Lankester especially, as well as the fragment he himself figures) belong here, there seems to be at least room for reasonable doubt.* The differences presented by the jaw fragment of *Trichechodon* as compared with the corresponding part of *Alactherium* are even still more marked.

The more obvious characters distinctive of the three genera of the *Odobænidæ*, as at present known, may be briefly indicated as follows :

Synopsis of the Genera.†

1. ODOBÆNUS.—Rami of lower jaw firmly anchylosed, even in early life; symphysis short. Incisors (in adult) 0; canines 1—1; molars 3—3, the last much smaller than the others.

2. ALACTHERIUM.—Rami of lower jaw not anchylosed; symphysis very long. Incisors (in adult) 2—2; canines 1—1; molars 4—4, the last smaller than the preceding ones.

3. "TRICHECHODON" (Van Beneden).—Rami of lower jaw (apparently) unanchylosed. Incisors 1—1?, very small; canines 1—1, highly specialized; molars 4—4, the first small, the last three much larger and subequal.

GENUS ODOBÆNUS, *Linné.*

Odobenus, LINNÉ, Syst. Nat., i. 1735 (ed. Fée), 59 (applied exclusively to the Walrus in a generic sense).—BRISSON, Règne Anim., 1756, 48 (used strictly in a generic sense, but embracing "1. La Vache marine— *Odobenus*"=Walrus; "2. Le Lamantine—*Manatus*." The characters given apply almost exclusively to the Walrus).

---

* Van Beneden himself says: "M. Ray Lankaster avait vu en Angleterre différentes grandes dents, provenant du crag et qui différaient surtout entre elles par leurs dimensions. Nous croyons devoir rapporter ces dents au genre *Alactherium*." Yet he cites "*Trichechodon huxleyi* Ray Lankaster" as a synonym of *Trichechodon konninckii,* described by himself much later! In view of the uncertainties of the case, it is to be regretted that he did not propose a new generic as well as specific name for his *Trichechodon konninckii.*

†With reference only to the lower jaw, the only known part, in case of the extinct types, readily susceptible of comparison.

Odobœnus, MALMGREN, Öfvers. K. Vet. Akad. Forh. 1863, (1864), 130.
Rosmarus, KLEIN, Quad. Disp. Brev. Hist. Nat., 1751, 40, 92 (applied in a generic sense exclusively to the Walrus).—"SCOPOLI, Introd. Hist.
  Nat., 1777, —."—GILL ("ex Scopoli"), Proc. Essex Inst., v, 1866, 7.
Phoca, LINNÉ, Syst. Nat., i, 1758 (in part only).
Trichechus, LINNÉ, Syst. Nat., 1766, 49 (in part only; not of Linné, 1758, nor
  Artedi, 1738; based exclusively in both cases on Sirenians).
Trichechus (in part only), ERXLEBEN, SCHREBER, GMELIN, BLUMENBACH,
  RETZIUS, and other early writers.
Trichechus, G. FISCHER, Nat. Mus. Naturgesch. zu Paris, 1803, 344.—ILLI
  GER, Syst. Mam. et Av., 1811, 139.—Also of GRAY, and most writers
  of the present century.
Odobenotherium, GRATIOLET, Bull. Soc. Géol. de France, 2ᵉ sér., xv, 1858,
  624 (= "*Trichechus rosmarus*" auct.—founded on a supposed fossil).
Odontobœnus, SUNDEVALL, Öfvers. K. Vet. Akad. För'a., 1859, 441.
? *Trichechodon*, LANKESTER, Quarter. Journ. Geol. Soc. Lond., xxi, 1865, 226,
  pl. x, xi (based on fossil tusks from the Red Crag, England).

The name *Trichechus*, for so long a time in general use for
the Walruses, proves not, as long ago shown by Wiegmann,* von
Baer, Müller, Stannius, and later by other writers, to belong at
all to these animals, but to the Manatee. The name *Trichechus*
originated with Artedi in 1738, in a posthumous work † edited
by Linné. The characters given were "*Dentes* plani in utraque
maxilla. *Dorsum* impenne. *Fistula* . . . ." The citations under *Trichechus* embrace no allusion to the Walrus, but
relate wholly to Sirenians, or to the Manatee, as the latter was
then known.‡ Artedi's description of the Manatee is quite full
and explicit, but includes also characters and references belonging to the Dugong. § *Trichechus* forms Artedi's "genus LI,"
and is placed in his "order V, *Plagiuri*" (embracing the Cetaceans and Sirenians, the other genera of this order being *Physeter, Delphinus, Balœna, Monodon*, and *Catodon*), and is hence

---

* Respecting the proper generic name of the Walruses, Wiegmann, in 1838,
thus forcibly expressed his views: "Die Gattung *Odobenus* [von Brisson, 1756]
hätte beibehalten werden müssen, da der ganz abgeschmackte Name *Trichechus* gar nicht dem Walrosse, sondern ursprünglich dem Manati angehört, und
von Artedi für diesen gebildet war, um die bei einem Fische oder vielmehr
Wallfische auffallende Behaarung zu bezeichnen."—*Archiv für Naturgeschichte*, v. Jahrg., Band i, 1838, p. 116.

† Ichthyologia, 1738, pars i, p. 74; pars iii, p. 79; pars iv, p. 109. In Artedi's work the name is twice written *Trichechus* and twice *Thrichechus*. On p.
74 of pars i, where it first occurs, its derivation is given, namely: "*Trichechus* a ϑρίξ *crinis* & ἰχθος *piscis* quia solus inter pisces fere hirsutus sit."

‡ The references in a general way appear to include all the Sirenians then
known.

§ E. g., "Dentium duo utrinque eminent, longitudine spithamæ crassitu
pollicis."

equivalent to the *Cete* of Linné (Syst. Nat., ed. x, 1758). Linné, in 1758, first introduced Artedi's genus *Trichechus*, at which time he placed in it only the Manatee, Dugong, and Steller's Sea Cow, leaving the Walruses still in *Phoca*. His diagnosis of the genus* embraced none of the distinctive characters of the Walrus. In 1766 (12th ed., Syst. Nat.), he transferred the Walrus from *Phoca* to *Trichechus*, making it the first species of the genus. The diagnosis, though slightly changed verbally, has still little, if any, reference to the characters of the Walruses, unless it be the phrase " Laniarii superiores solitarii,"† which is equally applicable to the Dugong, and is not at all the equivalent of "Phoca dentibus canines exsertis," previously ascribed to the Walrus in former editions, when the Walruses were placed under *Phoca*. Hence, to whatever the generic name *Trichechus* may be referable, it certainly is not pertinent to the Walrus. This being settled, the question arises, What generic name is of unquestionable applicability to the Walruses? Here the real difficulty in the case begins, for authors who admit the inapplicability of *Trichechus* to this group are not agreed as to what shall be substituted for it. Scandinavian writers, as Malmgren (1864) and Lilljeborg (1874), and Peters (1864) among German authorities, have for some years employed *Odobænus*, a name apparently originating with Linné (as *Odobenus*) in 1735, and adopted in a generic sense by Brisson in 1756. A modified form of it (*Odontobænus*) was also employed by Sundevall in 1859. Gill, in 1866, and other recent American writers, have brought into some prominence the name *Rosmarus*, first used in a generic sense by Klein in 1751, by Scopoli in 1777, by Pallas‡ in 1831, and by Lamont§ in 1861; while the great mass of English and Continental writers still cling to *Trichechus*.

The genera *Odobenotherium* and *Trichechodon*, based on fossil remains of the Walrus, have also been recently introduced into the literature of the subject, the former by Gratiolet in 1858, and the latter by Lankester in 1865; but these (especially the first)

---

* "Dentes primores nulla, laniarii superiores solitarii, molares ex osse rugoso utrinque inferius duo. Labia gemiata. Pedes posteriores coadunati in pinnam."—*Syst. Nat.*, ed. x, i, 1758, p. 34.

† The second diagnosis of *Trichechus* is, in full, as follows: "Dentes primores nulli utrinque. Laniarii superiores solitarii. Molares ex osse rugoso utrinque ; inferius duo. Labia geminata. Pedes posteriores compedes coadunati in pinnam."—*Syst. Nat.*, ed. xii, 1766, i, p. 48.

‡ Zool. Rosso-Asiat., vol. i, 269.

§ Seasons with the Sea-horses, pp. 141, 167.

appear to be referable to the existing Walruses, and of course become merely synonyms of earlier names. Consequently the choice evidently lies between *Odobœnus* and *Rosmarus*. *Odobœnus* has sixteen years' priority over *Rosmarus*, if we go back to the earliest introduction of these names into systematic nomenclature.* It is true that *Rosmarus* was the earliest Latin name applied to the Walrus, its use dating back to the middle of the sixteenth century, when it was employed interchangeably with *Mors* and *Morsus* by Olaus Magnus, Gesner, Herberstain, and others, *but only in a vernacular sense.* Although used by Klein systematically in 1751, Gill adopted it from Scopoli, 1777, probably because Klein was not a "binomialist." Linné used *Odobœnus* generically in 1735, as did also Brisson in 1756. The whole question turns on what shall be considered as the proper starting-point for generic nomenclature, about which opinion is still divided. If the early generic names of Artedi, Klein, Brisson, and Linné (prior to 1758) are admissible, as many high authorities believe, then *Odobœnus* is unquestionably the only tenable generic name for the group in question, of which *Rosmarus* is a synonym. †

SPECIES.

The existing Walruses have been commonly considered as belonging to a single circumpolar species. A few authors have recognized two, or deemed the existence of two probable, while one appears to have admitted three. Altogether, however, not less than six or seven specific names have been given to the existing species, besides several based on fossil remains of the Atlantic Walrus. In the present paper, the attempt will be made to establish the existence of two; but before entering further upon the discussion, it may not be out of place to glance briefly at the views previous authors have held respecting the point in question.

Pennant appears to have been the first to call attention to the probable existence of more than a single species of Walrus, who, in 1792, in speaking of the Walruses of the Alaskan coast, says: "I entertain doubts whether these animals [of "Unalascha, Sandwich Sound, and Turnagain River"] are of the same species

---

* *Odobœnus*, Linné, "Digiti ant., post. 5, palmipes. Ross *Morsus*. Dentes intermedii superiores longissimi."—*Syst. Nat.*, 1735 (ed. Fée), 59.—*Rosmarus*, Klein, Quad. Disp. Brev. Hist. Nat., 1751, 40, 92.

† In accordance with custom in similar cases, the name of the family becomes *Odobœnidæ*,—neither *Rosmaridæ* nor *Trichechidæ* being tenable.

with those of the Gulph of St. *Laurence*. The tusks of those of the Frozen Sea are much longer, more slender, and have a twist and inward curvature."* Shaw, a few years later, thought that the Walrus described and figured in the account of Captain Cook's last voyage, though perhaps not specifically distinct from those of the Arctic shores of Europe, should be regarded as belonging to a different variety.† He appears, however, to have based his opinion wholly on figures of the animals, and particularly on those given by Cook and Jonston (the latter a copy of Gerrard's, at second-hand from De Laët). Illiger, in 1811, formally recognized two species in his "Ueberblick der Säugthiere nach ihrer Vertheilung über die Welttheile,"‡ namely, *Trichechus rosmarus*, occurring on the northern shores of (Western?) Asia, Europe, and North America, and *T. obesus*, occurring on the northwestern shores of North America and the adjoining northeastern shores of Asia. While I do not find that he has anywhere given the distinctive characters of those two species, he, in the above-cited paper, also named the animal described and figured by Cook, *T. divergens*. F. Cuvier, in 1825, in describing the dentition of the "Morses," says: "Ces dents ont été décrites d'après plusieurs têtes qui semblent avoir appartenu à deux espèces, à en juger du moins par les proportions de quelques unes de leurs parties, et non seulement par l'étendue de leurs défenses, caractère qui avait déjà fait soupçonner à Shaw

---

* Arctic Zoölogy, vol. i, 1792, pp. 170, 171.

† He says: "An excellent representation is also given in pl. 52 of the last voyage of our illustrious navigator, Captain Cook. It is easy to see, however, a remarkable difference between the tusks of this last, and those of the former kind figured in Jonston, and it clearly appears, that though this difference is not such as to justify our considering them as two distinct species, yet it obliges us to remark them as varieties; and it should seem, that, in the regions then visited by Captain Cook, viz. the icy coasts of the American continent, in lat. 70, the Walrus is found with tusks much longer, thinner, and far more sharp-pointed, in proportion, than the common Walrus; and they have a slight inclination to a subspiral twist: there is also a difference in the position of the tusks in the two animals; those of the variety figured in Captain Cook's voyage curving inwards in such a manner as nearly to meet at the points, while those of the former divaricate. These differences appear very striking on collating different heads of these animals. Something may, however, be allowed to the different stages of growth as well as to the difference of sex. In order that these differences may be the more clearly understood, we have figured both varieties on the annexed plates . . . . ."—*General Zoology*, vol. i, 1800, pp. 236, 237, pls. 68, 68*.

‡ Abhandl. der Akad. der Wissensch. zu Berlin, 1804–1811, p. 64. Read before the Academy Feb. 28, 1811, but apparently not published till 1815.

l'existence de deux espèces de morses."[*] Fremery, in 1831,
having before him a series of eleven skulls, distinguished three
species, namely, *Trichechus rosmarus, T. longidens*, and *T. cooki*.
The first (*T. rosmarus*) was principally characterized by having
diverging tusks, about as long as the length of the whole head,
faintly grooved on the outside, and with two distinct grooves on
the inside; by the possession of five back teeth, the last two
very small; by the lower edge of the nasal opening being but
little produced; by the occipital crest being strongly developed;
and by the great specific gravity of the bones of the skull. The
second (*T. longidens*) was principally characterized by the tusks
equalling or exceeding in length two-thirds of the length of the
skull, with a single deep groove on the inner side; by having
only four back teeth, the last one small; a smaller development of the occipital crest (*except in old animals!*); and a lighter
specific gravity of the bones. The third (*T. cooki*), considered
as a doubtful species, was based wholly on Shaw's plate 68 (from
Cook), already noticed, and hence is the same as Illiger's *T.
divergens*. Wiegmann, von Baer, Stannius, and most subsequent
writers, have properly regarded Fremery's characters of his *T.
rosmarus* and *T. longidens* as based merely on ordinary individual or sexual differences. Wiegmann, and also Temminck,
according to Fremery, believed the female to be distinguishable
from the male by its longer and thinner tusks, with the crests
and ridges of the skull less developed, while other differences,
as the relative prominence of the bony lower edge of the nasal
opening, were differences characteristic merely of different individuals.[†] Stannius, however, in 1842, after passing in review

---

[*] Dents des Mammifères, p. 235.

[†] Wiegmann, in commenting upon Fremery's supposed specific differences,
observes as follows respecting probable sexual and individual differences
in the tusks and skulls of Walruses: "Hr. Fremery führt an, dass Hr.
Temminck einen (nach Deutlichkeit der Nähte) noch jungen Schädel des
Reichsmuseums mit ausgezeichnet langen dünnen Stosszähnen für den
eines Weibchens gehalten habe. Ich erinnere mich auch von Grönlands-
fahrern gehört zu haben, dass sich das Weibchen durch längere, dünnere,
dass Männchen durch kürzere, aber viel dickere Stosszähne auszeichne." The
alleged difference in the specific gravity of the bones of the skull he believes also to be a sexual feature, as possibly also the difference in the number of molar teeth. Respecting the prominence of the lower border of the
nasal opening he says: "Die mehr oder minder starke Hervorragung des
unteren Randes der Nasenöffnung kann ich dagegen nur für eine individuelle
Verschiedenheit halten, da ich sie bei einem Schädel mit kurzen Stosszähnen, der die übrigen vom Verf. hervorgehobenen Merkmale besitzt, sehr
stark, und umgekehrt bei einem alten Schädel mit langen Stosszähnen kaum
merklich finde."—*Archiv für Naturgesch.*, 1838, pp. 128, 129.

the characters assigned by Fremery as distinctive of several species, and after mentioning at length other features of variation observed by him in a considerable series of skulls, describing several of his specimens in detail, and arriving at the conclusion that up to that time all the supposed species of Walrus constituted really but a single species, added another, under the appropriate name *Trichechus dubius*. This with subsequent authors has shared the fate of Fremery's species,* being considered as based merely on individual variation.

As will be more fully noticed later, two nominal species have been founded on the fossil remains of the Walrus, namely, *Trichechus virginianus*, DeKay, 1842, and *Odobenotherium lartetianum* of Gratiolet, the former based on remains from Accomac County, Virginia, and the latter on remains from near Paris, France. Lankester, in 1865, added still another, based on tusks from the Red Crag of England, under the name *Trichechodon huxleyi*.

Dr. Leidy, in 1860, in a paper on fossil remains of the Walrus from the eastern coast of the United States, again noticed the differences in the size, length, and curvature of the tusks in specimens from the northwest coast of North America and the common Walrus of the North Atlantic. He says: "In the course of the preceding investigations [referring to previous portions of his paper], I was led to examine a specimen, in the cabinet of the Academy of Natural Sciences [of Philadelphia], consisting of the stuffed skin of a portion of the head enveloping the jaws of a species of Walrus apparently differing from the true *Trichecus rosmarus*, of which, as characteristic, I have viewed the figures of the skull and skeletons as given by Daubenton, Cuvier, and De Blainville. The specimen was presented by Sandwith Drinker, Esq., of Canton, China, and was probably derived from the Asiatic shore of the Arctic Ocean. From the worn condition of the upper incisor and molars, it appears to have belonged to an old individual; and in the case of the lower jaw, the teeth appear to have been entirely worn out. The tusks are very much larger and are narrower than in

---

* Giebel, in 1855, referred to Fremery's and Stannius's species as still needing confirmation: "Die von Fremery nach der Beschaffenheit der Zähne unterschiedenen Arten, *Tr. longidens* und *Tr. Cooki;* sind längst als unhaltbar erkannt worden und auch die von Stannius auf Schädeldifferenzen begründete Art, *Tr. dubius*, entbehrt noch der weitern Bestätigung."—*Säugethiere*, p. 128, footnote.

the *T. rosmarus*, and they curve downward, outward, and inward, instead of continually diverging as in this species.   At their emergence from the alveoli the tusks are two and three-quarter inches apart, near the middle five and a quarter inches, and at their tips only one inch.   Their length is twenty-two inches and their diameter at the alveolar border antero-posteriorly two and a quarter inches, and transversely one and a half inches.   Towards their lower part they are twisted from within, forwards and outwardly." After quoting Pennant's remark (already given, see p. 17) about similar differences noted by him, he adds that "the superior incisor and molar teeth are also very much smaller than in the fossils of *T. rosmarus*," and he gives measurements showing this difference. He then says : "The hairs of the upper lip of the *T. rosmarus* are stated by Shaw, to be about three inches long, and almost equal to a straw in diameter.* In the specimen under consideration, the hairs of the moustache are stiff-pointed spines, not more than one line long at the upper part of the lip, and they gradually increase in size until at the lower and outer part of the lip they are about one inch in length." He further adds, in the same connection : " Since presenting the above communication to the Society, the Academy has received from Mr. Drinker, of Canton, an entire specimen of the Walrus from Northern Asia.   In this individual, which measures in a straight line eight feet from the nose to the tail, the tusks are ten inches long, and diverge from their alveoli to the tips, where they are five and a half inches apart, but they are slender, as in the stuffed head above mentioned, and appear as if they would ultimately have obtained the same length and direction.   Perhaps the peculiarities noticed may prove to be of a sexual character."†

As will be shown later, we have here the more prominent external differences characterizing the two species of Walrus for the first time explicitly stated from direct observation of specimens.   If Dr. Leidy had had at that time good skulls of the two species for comparison, the other important cranial differences (noted beyond) could not have escaped him, and he perhaps would have been led to formally recognize the Pacific Walrus as a species distinct from the Atlantic Walrus.

I have met with nothing further touching this subject prior to Mr. H. W. Elliott's report on the Seal Islands of Alaska,

---

* "Shaw's Zoology, vol. i, pt. i, p. 234."
†Trans. Amer. Phil. Soc. Phila., vol. xi, pp. 85, 86.

published in 1873, in which, under the heading "The Walrus of Bering Sea, (Rosmarus arcticus)" he says:—"I write 'the Walrus of *Bering Sea*', because this animal is quite distinct from the Walrus of the North Atlantic and Greenland, differing from it specifically in a very striking manner, by its greater size and semi-hairless skin."* This is all he says, however, respecting their differences, no reference being made to the really distinctive features. Thus the matter rested till, in 1876, Gill formally recognized two species in his "List of the Principal Useful or Injurious Mammals," † in a catalogue of a "Collection to Illustrate the Animal Resources of the United States" in the exhibit of the National Museum at the International Exhibition of 1876, held in Philadelphia. This is merely a nominal list, in which appears, under "*Rosmaridæ*," the following, which I here fully and literally transcribe:

> ROSMARUS OBESUS, (Illiger,) Gill.
> The [Atlantic] Walrus.
> Atlantic Coast.

> ROSMARUS COOKII, (Fremery,) Gill.
> The [Pacific] Walrus.
> Pacific Coast.

Here is simply a nominal recognition of two species without expressed reasons therefor. In an article on the *Rosmaridæ*, published in 1877, Dr. Gill again says: "Two species appear to exist—one (*R. obesus*) inhabiting the northern Atlantic, and the other (*R. Cookii*) the northern Pacific." ‡

Van Beneden, on the other hand, in 1877, distinctly affirms his disbelief in the existence of two species. In referring to the subject he says: "Nous ne croyons pas que les Morses du détroit de Behring diffèrent spécifiquement de ceux de la mer de Baffin ou de la Nouvelle-Zemble, et c'est à tout, à notre avis, que Fremery a essayé de les répartir en espèces distinctes d'après les modifications de leurs dents." He further adds the testimony of von Baer as follows: "Von Baer s'est occupé en 1835 de cette question à l'Académie de St. Pétersbourg et l'illustre naturaliste m'écrivait, peu de temps avant sa mort, au sujet de la différence légère des Morses, à l'Est et à l'Ouest de

---

* Report on the Prybilov Group or Seal Islands of Alaska, 1873 (not paged). Also, Report on the Condition of Affairs in Alaska, 1875, p. 160.

† This "List" is anonymous, and is hence, perhaps, not properly quotable in this connection, although its authorship is known to the present writer.

‡ Johnson's New Universal Cyclopædia, vol. iii, 1877, p. 1725.

la mer Glaciale, qu'il regardait les différences comme des modifications locales*. Ce n'est pas l'avis de Henry W. Elliot, qui considère le Morse du nord du Pacifique comme un animal distinct."† In another connection he refers to the subject as follows : " Il y a des auteurs qui pensent que le Morse du Nord Pacifique est assez différent de celui du Groënland, pour en faire une espèce distincte. Nous ne partageons pas cet avis. Les modifications sont assez peu importantes et nous croyons pouvoir le mettre sur le compte de variations locales."‡

ODOBÆNUS ROSMARUS, *Malmgren.*

Atlantic Walrus.

"*Rosmarus, seu Morsus Norvegicus,* OLAUS MAGNUS, Hist. de Gent. Sept. 1555, 757 (figure)" ; also later editions.

Rosmarus, GESNER, Hist. Anim. Aquat., 1558, 249 ; also later editions.

Rosmarus, Wallross, JONSTON, Hist. Nat. de Piscibus et Cetis, 1649, 727, pl. xliv (two lower figures ; upper one from Gesner, the lower from De Laët) ; also later editions.—"KLEIN, Reg. Anim., 1754, 67."—"SCOPOLI, Hist. Nat., 1777, —."—ZIMMERMANN, Spec. Zool. Geograph. Quad. etc., 1777, 330.

Equus marinus et Hippopotamus falso dictus, Morse or Sea Horse, RAY, Syn., 1695, 191.

Wallross, MARTENS, Spitzb., 1675, 78, pl. P, fig. b.—EGEDE, Beschr. und Natur-Gesch. Grönland, 1742, 54 ; 1763, 106 ; Descrip. et Hist. Nat. du Grœnl., 1765, 61 (with a figure).—CRANTZ, Hist. von Grönl., 1765, 165 ; English ed., 1768, 125.—GOETHE, "Morphol., 1, 1817, 211" ; Act. Acad. Cæs. Leop. Carol., xv, i, 1831, 8, pl. iv (dentition, etc.).—VON BAER, Mém. Acad. St. Pétersb. Math. etc., viᵉ sér., ii, 1835, 199 (blood-vessels of limbs).—JAEGER, Müller's Arch. für Anat., 1844, 70 (dentition—Labrador specimens).

Walross, MARTENS, Zoologische Garten, xi, 1870, 283 (etymology).

Wallrus seu Mors, RUYSCH, Theatr. Animal., 1718, 159, pl. xliv (figure same as Jonston's).

Walrus, WORM, Mus. Worm., 1655, 289 (fig. from De Laët).—WYMAN, Proc. Bost. Soc. Nat. Hist., iii, 1850, 242 (relation to Pachyderms).—LEA, Proc. Acad. Nat. Sci. Phila., 1854, 265 (use of the skin).—WHEAT-LAND, Proc. Essex Institute, 1, 1854, 62 (remarks on a skull).—SONNTAG, Nar. Grinnell Expl. Exp. 1857, 113 (woodcut—group of Walruses).—MURRAY, Geogr. Distr. Mam., 1866, 128, map, xxviii * (distribution ; in part).—HAYES, Open Polar Sea, 1867, 404 (hunting).—PACKARD, Bull. Essex Institute, i, 1869, 137 (former existence in Gulf of St. Lawrence).—ATWOOD, Proc. Bost. Soc. Nat.

---

* " Les Morses des côtes de Sibérie ou de l'est de l'Asie ont les dents canines plus fortes que les Morses de Spitzberg et de Grœnland, me disait-il dans une lettre . . . . . "

† Ann. du Mus. Roy. d'Hist. Nat. Belgique, pt. 1, 1877, 45.

‡ Ibid., p. 17.

Hist., xiii, 1870, 220 (remarks on a skull from the Gulf of St. Lawrence).—TURNER, Journ. Anat. and Phys., v, 1870, 115 (relations of pericardium).—RINK, Danish Greenland, 1877, 126 (distribution), 248, 252, 272 (chase).

Arctic Walrus, PENNANT, Synop. Quad., 1781, 335; Arctic Zoöl., 2d ed., i, 1792, 168 (in part).

Fossil Walrus, BARTON, London Phil. Mag., xxxii, 1805, 98 (no locality).—MITCHELL, SMITH, & COOPER, Ann. New York Lyc. Nat. Hist., ii, 1828, 271 (fossil, Accomac Co., Va.—doubtfully referred to the existing species); Edinb. New Phil. Journ., v, 1828, 325 (abstract of the last).—HARLAN, Edinb. New Phil. Journ., xvii, 1834, 360; Trans. Geol. Soc. Penn., i, 1835, 75; Med. and Phys. Researches, 1835, 277 (same specimen).—LYELL & OWEN, Proc. Lond. Geol. Soc., iv, 1843, 32 (fossil, Martha's Vineyard, Mass.); Amer. Journ. Sci. and Arts, xlvi, 1843, 319 (same).

La Vache marine, BRISSON, Règ. Anim., 1756, 48.

"Morsch, GMELIN, Reise durch Russland, iii, 1751, 165.

Morse ou Vache marine, BUFFON, Hist. Nat., xiii, 1765, 358, pl. liv (animal).—DAUBENTON, Buffon's Hist. Nat., xiii, 1765, 415, pl. lv (skull).—HOLLANDRE, Abrégé d'Hist. Nat. des Quad. Vivip., i, 1790, pl. xii, fig. 3.—F. CUVIER, Dict. des Sci. Nat., xxxiii, 1816, 27; Dents des Mam., 1825, 233, pl. xcv.

Morse, HUET, Coll. de Mam. du Mus. d'Hist. Nat., 1808, 59, pl. liii (fig. from Buffon).

Sea Cow, SHULDHAM, Phil. Trans., lxv, 1775, 249.

Phoca, BONNANIO, Rerum Nat. Hist., i, (no date), 159, pl. xxxix, fig. 27 (a poor representation of De Laët's figure, with the young one omitted).

Phoca rosmarus, LINNÉ, Syst. Nat., i, 1758, 38.

Trichechus rosmarus, LINNÉ, Syst. Nat., i, 1766, 49.—MÜLLER, Prod. Zool. Dan., i, 1776, 1.—SCHREBER, Säugeth., ii, 1775, 262, pl. lxxix (from Buffon).—ZIMMERMANN, Geogr. Geschichte, i, 1778, 299; ii, 1780, 424.—FABRICIUS, Fauna Grœnl., 1780, 4.—ERXLEBEN, Syst. Reg. Anim., 1787, 593.—GMELIN, Syst. Nat., i, 1788, 59.—SHAW, Nat. Miscel., 1791, pl. cclxxvi; Gen. Zoöl., i, 1800, 234 (in part), fig. 68, (from Jonston).—BLUMENBACH, Handb. der Naturgesch., 1788, 142; 1821, 136; 1825, 112; Abbild. natur. Gegenst., 1796–1810, No. 15, text and plate (from Jonston).—DONNDORFF, Zool. Beyträge, 1792, 124.—RETZIUS, Kong. Vet. Akad. Nya Handl., xv, 1794, 391; Fauna Suecicæ, 1800, 48.—OZERETSKOVSKY, Nov. Act. Acad. Sci. Imp. Petrop., xiii, (1796), 1802, 371.—BARTON, Phil. Mag., xxxii, 1805, 98 (fossil; locality not stated).—G. CUVIER, Tableau élément., 1798, 172; Leçons d'Anat. Comp., 1800–1805, — ; 2ᵉ éd., ——, — ; 3ᵉ éd., 1837, 207, 257, 293, 329, 398, 472; Règ. Anim., i, 1817, 168; i, 1829, 171; Ossem. Foss., iv, 280; 3ᵉ éd., v, 1ᵉ ptie., 1825, 234; v, 2ᵉ ptie., 521, pl. xxxiii (osteology).—ILLIGER, Prod. Syst. Mam. et Av., 1811, 139; Abhandl. der Berliner Akad. (1804–1811), 1815, 56, 61, 64, 68, 75 (distribution).—DESMAREST, Nouv. Dict. d'Hist. Nat., xxi, 1818, 390; Mam., 1820, 253.—SCORESBY, Account Arct. Regions, i, 1820, 502 (general history).—"KERSTERN, Capitis Trichechi Rosmari Descrip. Ost., 1821,—."—SCARTH, Edinb. Phil. Journ., ii, 1825, 283 (Orkney); Jardine's Nat. Library, Mam., vii, 1838, 219, pl. xx (original figure of animal).—HARLAN, Faun. Amer., 1825, 114; Edinb. New Phil. Journ., xvii, 1834, 360 (fossil); Trans. Geol. Soc. Penn., i, 1835, 72 (same); Med. and Phys. Repos., 1835, 277 (same).—GODMAN, Amer. Nat. Hist., i, 1826, 354.—SCHINZ, Naturg. der Säugeth., 1827, 169, pl. lxv (two figures—"Abbildung nach Blumenbach und Schmid").— LESSON, Man. de Mam., 1827, 208.—ROSS, App. Parry's Fourth Voy., 1828, 192; App. Ross's Second Voy., 1835, xxi.—FLEMING, Brit. Anim., 1828, 18.—RAPP, Naturw. Abhandl. Würtemb., ii, 1828, 107 (dentition); "Bull. Sci. Nat., xvii, 1829, 280" (abstract).—FISCHER, Synop. Mam., 1829, 243.—GUÉRIN-MÉNEVILLE, Icon. du Règne Anim. de G. Cuvier, Mam., 1829–1838, 19, pl. xix, fig. 5 (animal).—FREMERY, Bijdrag tot de natuurk. Wetensch., vii, 1831, 384.—DELONGCHAMPS, Mém. Soc. Linn. de Normandie, v, 1835, 101 (dentition).—WILSON, Nat. Hist. Quad. and Whales, 1837, 145, pl. cccxxxiv, fig. 2 (animal); Encycl. Brit., 7th ed., xiv, 125.—BELL, Brit. Quad., 1837, 258 (animal and skull; original figures).—VON BAER, Mém. Acad. St. Pétersb. Math., etc., 6e sér., iv, 1838, 97, pl. xlvii (distribution).—WIEGMANN, Arch. für Naturgesch., 1838, 113 (dentition).— HAMILTON, Jard. Nat. Libr., Mam., viii, 1839, 103, pl. i (animal, and woodcut of skull,—original figure).—RICHARDSON, Zoöl. Beechey's Voy., 1839, 6.—BLAINVILLE, Ostéographie, Des Phoques, 1840–51, 19, pl. i (skeleton), pl. iv (skull).—DEKAY, Nat. Hist. New York, Zoöl., i, 1842, 56.—ZIMMERMANN, Jahrb. für Mineral., 1845, 73.— WAGNER, Schreber's Säugeth., vii, 1846, 84, pl. lxxix.—GIEBEL, Fauna der Vorwelt, 1847, 222 (fossil); Säugeth., 1855, 128; Odontog., 1855, 82, pl. xxxvi, fig. 5 (dentition).—NILSSON, Skand. Faun., 1847, 318.—GERVAIS, Zool. et Pal. Français, i, 1848–52, 140.—GRAY, Cat. Seals in Brit. Mus., 1850, 32; Proc. Zoöl. Soc. Lond., 1853, 112 (on attitudes and figures); Cat. Seals and Whales, 1836, 36, 367.—OWEN, Proc. Zoöl. Soc. Lond., 1853, 103 (anat. and dentition); Ann. and Mag. Nat. Hist., xv, 1855, 226 (from the foregoing); Cat. Osteol. Coll. Mus. College Surg., 1853, 631 (skeleton); Encycl. Brit., xvi, 1854, 463, fig. 112 (skull); Odontography, 1854, 510, pl. cxxxii, fig. 8 (dentition); Orr's Circle of the Sciences, Zoöl., i, 1854, 230, fig. 27 (skeleton); Comp. Anat. and Phys. Vertebrates, ii, 1866, 490, 498, 507; iii, 1868, 338, 524, 780.—BLASIUS, Faun. Wirb. Deutschl., i, 1857, 262, figs. 148–150 (skull).—VAN DER HOEVEN, Handb. Zoöl. Engl. Ed. ii, 1858, 697.—VON SCHRENCK, Reisen im Amur-Lande, i, 1859, 179 (in part only).—LEIDY, Trans. Amer. Phil. Soc. Phila., xi, 1860, pls. iv, v (in part); Journ. Acad. Nat. Sci. Phila., 2d ser., vii, 1869, 416.—WOLF & SCLATER, Zoöl. Sketches, i, 1861, No. 16.—GERRARD, Cat. Bones Mam. Brit. Mus., 1862, 145.—NEWTON, Proc. Zoöl. Soc. Lond., 1864, 499.—SCLATER & BARTLETT, Proc. Zoöl. Soc. Lond., 1867, 818, 819.—VON MIDDENDORFF, Sibirische Reise, iv, 1867, 934 (in part only).—BROWN, Proc. Zoöl. Soc. Lond., 1868, 335, 427 (habits and distribution); Man. Nat. Hist. Greenland, 1875, 35.—MURIE, Proc. Zoöl. Soc. Lond., 1868, 67 (report on cause of death of specimen in Zoöl. Gard., Lond.); 1870, 581; Trans. Zoöl. Soc. Lond., vii, pt. vi, 1871, 411, pls. li–lv (anatomy).—GILPIN, Proc. and Trans. Nova Scotia Inst. Nat. Sci., ii, pt. 3, 1870, 123 (with a plate).—REEKS, Zoölogist, 1871, 2550 (St. George's Bay, Newfoundland).—HEUGLIN, Reisen nach dem Nordpolarmeer, iii, 1874, 43 (habits and distribution).—DEFRANCE, Bull. Soc. Géol., 3e sér., ii, 1874, 164 (fossil, France).—GULLIVER, Proc. Zoöl. Soc. Lond., 1874, 580 (size of blood-corpuscles).—FEILDEN, Zoölogist, 3d ser., i, 1877, 360 (distribution and food).—VAN BENEDEN, Ann. Mus. Roy. d'Hist. Nat. Belgique, i, 1877, 39 (distribution, general habits, and fossil remains).—RINK, Danish Greenland, its People and its Products, 1877, 126.

Trichechus longidens, FREMERY, Bijdrag tot de Natuurk. Wetensch., vi, 1831, 384.

Trichechus virginianus, DEKAY, Nat. Hist. New York, Zoöl., i, 1842, 56, pl. xix, figs. 1, *a*, *b* (fossil, Accomac Co., Va.).

? *Trichechus dubius*, STANNIUS, Müller's Arch. für Anat., 1842, 407 (without locality).

Rosmarus arcticus, LILLJEBORG, Fauna öfvers Sveriges och Norges Ryggr., 1874, 674.

Rosmarus trichechus, LAMONT, Seasons with the Sea-horses, 1861, 141, 167 (two plates).—GILL, Johnson's New Univ. Cyclop., iii, 1877, 633.

Rosmarus'obesus, GILL, Proc. Essex Inst., v, 1866, 13 (in part only); International Exhib. 1876, Anim. Resources U. S., No. 2, 1876, 4 (Atlantic Walrus; no description); Johnson's New Univ. Cycl., iii, 1877, 1725.—PACKARD, Proc. Bost. Soc. Nat. Hist., x, 1866, 271; Mem. Bost. Soc. Nat. Hist., i, 1867, 246 (fossil).—LEIDY, Journ. Acad. Nat. Sci. Phila., viii, 1877, 214, pl. xxx, fig. 6 (fossil, South Carolina).

Odobenotherium lartetianum, GRATIOLET, Bull. Soc. Géol. de France, 2e sér., xv, 1858, 624, pl. v (fossil, near Paris, France).

Odontobænus rosmarus, SUNDEVALL, Öfver. K. Vet. Akad. Forh., 1859, 441; Zeitsch. Gesammt. Naturw. Halle, xv, 1860, 270.

Odobænus rosmarus, MALMGREN, Öfver. K. Vet. Akad. Forh. 1863, (1864), 130 (food and habits), 505, pl. vii (dentition); Wiegmann's Archiv f. Naturgeschichte, 1864, 67 (translated from Öfvers. K. Vet. Akad. Forh., 1863, 130 et seq.).—PETERS, Monatsb. Akad. Wiss. Berlin, 1864, 685, pl. (dentition); Ann. and Mag. Nat. Hist., (3), xv, 1865, 355 (abstract).—RINK, Danish Greenland, 1877, 430.

? *Trichecodon huxleyi*, LANKESTER, Quarter. Journ. Geol. Soc. Lond., xxi, 1865, 226, pls. x, xi (fossil; Red Crag, England).

? *Trichechus manatus*, FABRICIUS, Fauna Grœnl., not *Rhytina gigas;* see BROWN, Proc. Zoöl. Soc. Lond., 1868, 357, 358.

? *Phoca ursina*, FABRICIUS, Fauna Grœnl., not *Callorhinus ursinus;* see BROWN, Proc. Zoöl. Soc. Lond., 1868, 357, 348.

Morse; Vache marine; Cheval marine; Bête à la grande dent (French).

Bos marinus, RUYSCH, l. c.

Hvalross (Swedish and Danish).

Havhest; Hvalruus (Norwegian).

Morsk (Lapp).

Wallross; Meerpferd (German).

Walrus; Sea Cow; Sea Horse (English).

EXTERNAL CHARACTERS.—As regards general form, the head, in comparison with the size of the body, is rather small, squarish in outline, but much longer than broad, with the muzzle abruptly truncated and somewhat bilobed by the depression surrounding the nasal opening. The lower jaw is pointed and narrow anteriorly. The upper lip is heavily armed with thick, strong, pellucid bristles. The nostrils are somewhat crescentic in shape, placed vertically, with the upper part more expanded than the lower, and hence bear some resemblance to two commas placed with their convex surfaces toward each other. The eyes are situated rather high up, about midway between the muzzle and the occiput. The ear is wholly destitute of a pinna, forming merely an orifice on the side of the head in a deep fold of the skin. The most prominent facial character in the adults is, of course, the long protruding upper canines, which extend 12 to 15 or more inches beyond the rictus. The neck is short, being only about as long as the head; it gradually thickens toward the body, into which it insensibly merges. The body is exceedingly thick and heavy, presenting everywhere a rounded outline, and attaining its greatest circumference at the shoulders, whence it gradually tapers posteriorly. The tail is scarcely, if at all, visible, being enclosed within the teguments of the body. The fore limbs are free only from the elbow; as in the Pinnipeds generally, they are greatly expanded, flat, and somewhat fin-like, but with much more freedom of motion than is the case in the *Phocidæ.* They are armed with five small flat nails, placed at considerable distance from the end of the cartilaginous toe-flap. The first or inner digit is slightly the longest, the others being each successively a little shorter till the fifth, which nearly equals the first. The hind limb is enclosed within the teguments of the body nearly to the heel; the free portion when expanded is fan-shaped, but when closed the sides are nearly parallel. The first and fifth digits are considerably longer and larger than the middle ones, the fifth being also rather larger than the first. They are all provided with small nails, placed at some distance from the end of the toe-flap. The soles of both fore and hind extremities are bare, rough, and "warty," and the dorsal surface of the digits as far as the proximal phalanges is also devoid of hair. In the young and middle-aged, the body is rather thickly covered with short hair, which, however, is thinner and shorter on the ventral surface of the neck and body and on the limbs than elsewhere. It is everywhere of a yellowish-brown color, except on the belly and at the base of the flippers, where it passes into dark reddish-brown or chestnut. The bristles are pale yellow or light yellowish horn-color. In old animals, the hair becomes more scanty, and often gives place to nearly bare scarred patches, frequently of considerable area. Very old individuals sometimes become almost naked, presenting the same appearance that has been so often observed among very old males of the Alaskan Walrus. The skin is everywhere more or less wrinkled and thrown into folds, especially over the shoulders, where the folds are deep and heavy. The average length of four adult males examined is about 10½ feet, varying from 9½ to 11 feet. Authors, however, commonly give rather larger dimensions, and a length of twelve feet is said to be not infrequently attained. The largest bristles vary in length from 2.25 to 2.75 inches.

From Dr. Murie's paper on the general anatomy of a young individual I add a few further details. Dr. Murie describes the muzzle as capable of great mobility, and the mystacial bristles as curving in different directions according to the muscular tension of the parts to which they are attached. "When the nostrils are relaxed they drop forwards and the bristles inwards. At such times the nares are apart fully 1½ inch; but when they are contracted a septum 0.6 of an inch wide only divides them. Occasionally, when alive, I observed the animal retract its upper lip, as a dog would in snarling; and this caused a deep furrow in the facial region. This change in the features gives quite a different expression to the physiognomy . . . . . When seen in front and from above, the face has a most curious expression, recalling to mind that of the cranium of an Elephant rather than the Walrus's ally *Otaria*. The auricular region then acquires a prominent aspect, as do the orbits. The great breadth of the muzzle also comes out better. The face is entirely hairy to the roots of the bristles . . . . . On the lower surface of the muzzle and chin, the upper lip passes one inch beyond the lower lip, and the snout, with its adpressed bristles, one or two inches beyond that. A portion of the upper rosy lip, in this view, is seen thrust upwards or puckered outside the canines. These upper canine teeth, which grow to massive tusks in the adult and aged Walruses, in ours had little more than protruded beyond the mandibular lips. The chin and anterior portion of the throat are very hairy; this diminishes backwards; and on the throat the almost hairless skin is thrown into longitudinal and parallel narrowish flat-topped rugæ."*

* Trans. Zoöl. Soc. Lond., 1872, vol. vii, p. 419.

In respect to the mystacial bristles, Dr. Murie's figures of the head and muzzle of the young specimen described by him (drawn from photographs, some from the living animal) represent them as quite long, the longest being said to be from 4 to 5 inches in length, and those of the sides of the muzzle as curving inward and nearly meeting beneath the chin. Lamont also speaks of them as being in the adult 6 inches in length. Hamilton describes the Orkney specimen as having the largest nearly 5 inches in length, "and as thick as a Thrush's quill." Dr. Kane says: "The cheeks and lips are completely masked by the heavy quill-like bristles." The authors of the history of the Swedish expedition to Spitzbergen and Bear Island in 1861 state that they are 4 inches long and nearly a line thick.* In the four or five adult male specimens I have had the opportunity of examining, the exserted portions of the longest bristles were less than 3 inches in length, and when extracted measured scarcely more than $4\frac{1}{2}$, the shortest being mere points projecting through the skin. From Dr. Murie's figures and description of the young, and from other accounts, it would seem that the bristles become shorter in adult life, being perhaps worn off by constant friction. The bristles in the specimens I have seen bore no resemblance to the long curving bristles figured and described by Dr. Murie as existing in the young animal. They were considerably (one-third) longer, however, in the youngest of four specimens in Professor Ward's collection than in the oldest, giving support to the opinion already stated that they become shorter as the animal advances in age.†

As already noted, the fore feet are formed much as in other Pinnipeds, more nearly agreeing, however, with those of the *Otariidæ* than with those of the *Phocidæ*, especially with respect to freedom of movement, having the power of pronation and supination to a considerable degree. "In the Walrus," says Dr. Murie, "the humerus, radius, and ulna can be so placed that they meet at an acute angle, the lower limb of which is in a great measure free. The digits, on the other hand, can together be turned backwards at a sharp angle with the radius and ulna, so that the bones of the limbs altogether form an S-shaped figure. In the Seal the antibrachium and digits bend on each

---

* See Passarge's German translation, p. 132.

† In Pallas's figure (in his "Icones") of a young example of the Pacific Walrus, the mystacial bristles are represented as very long, as in the young of the Atlantic species.

other more angularly, thus <. .... In the act of swimming
the Walrus evidently can use its fore limb as far as the elbow,
with a kind of rotary movement of the manus and antibrachium;
but in the Seal the rotary action takes place only at the wrist,
and above that a sort of ginglymoid or back and forward movement."

"The palmar surface or sole of the manus is not unlike a par-
lor shovel in figure. There is a great callous, roughened and
warty pad at the proximal end or ball of the hand; and this,
from discoloration incident to use, is of an intense dark brown
or almost black colour. From the radial margin, where it is
stoutest and roughest, it trends towards the base of the fifth
digit. Circumscribed digital pads, as in Carnivora, there are
none; but furrows and ridges traverse obliquely forwards the
policial to the opposite side." This "remarkable rough and
warty palmar surface," continues Dr. Murie, "affords above
everything a stay and firm leverage on slippery ground; no
stocking or wisp of straw used by man to bind round the foot
when on smooth ice can equal nature's provision of coarse tegu-
mentary papillæ." Also, "The angle at which the carpo-meta-
carpal joint is set, and the very odd manner of foot-implanta-
tion on the ground, namely, semiretroverted, evidently make it
an easier task to go forwards or upwards on a smooth surface
than to retrograde."* The hind foot (pes) is similarly roughened and furrowed. The notion advanced by Sir Everard
Home,† that the feet of the Walrus were provided with suc-
torial power, like that of the disk of a fly's foot, by which they
were enabled to maintain firm footing on smooth ice and rocks,
Dr. Murie considers untenable. No one who has ever seen
a Walrus walk, says Dr. Murie, could for a moment suppose
that its massive weight was sustained by a pedal vacuum, as in
a fly's foot.

As regards the proportionate size of the limbs, the fore limbs,
in an animal 8 to 10 feet long, are stated by Edwards,‡ to measure from the "shoulder joint to the finger ends, two feet; expansion, one foot; the hind limbs measuring twenty-two inches, and
extending, when outstretched, eighteen inches beyond the body,
with an expansion of two feet." Scoresby says the fore feet are
"from two to two and a half feet in length, and being expansive

* Trans. Zoöl. Soc. Lond., vol. vii, 1872, pp. 420, 421.
†Phil. Trans., 1824, pp. 233–235, pl. iv.
‡MSS. as quoted by Richardson, Suppl. Parry's Sec. Voy., p. 340.

may be stretched to the breadth of fifteen to eighteen inches." The hind feet, he says, have a length of "about two to two and a half feet," the breadth, when fully extended being "two and a half to three feet."[*]

Dr. Gilpin[†] gives about the same dimensions for a specimen 12 feet long, namely, fore-flippers, length 2 feet; breadth 13 inches; hind flippers, length 22 inches, breadth (when stretched) 2 feet 6 inches. Dr. Murie gives for a specimen about 7¾ feet long: from shoulder-joint to extreme end of first digit, 23½ inches; extreme length from os calcis to tip of fifth digit, 17½ inches; extreme breadth, when forcibly distended, 13 inches. My own measurements, taken from three unmounted skins of adult males preserved in salt in the collection of Prof. Henry A. Ward of Rochester, are as follows: manus, from carpal joint to end of digits, 14 to 15 inches; transverse diameter at base, 9½ to 10 inches; pes, from tarsal joint to end of longest digit, 15 to 18 inches; transverse diameter at tarsus, about 7 inches. The rigidity of the feet did not permit of ready expansion.

In respect to the tail, Dr. Murie says: "Strictly speaking, the Walrus possesses no free tail, as do the *Phocidæ* and *Otariidæ;* for a broad web of skin stretches across from os calcis to os calcis, enveloping the caudal representative. This remarkable elastic membrano-tegumentary expansion, reminding one of the more delicate web similarly situated in Bats, has posteriorly, when the legs are outspread, a wide semilunar border with little if any medio-caudal projection. What appears as a tail when the limbs are approximated is in reality fibroid tissue and skin; for the caudal vertebræ stop short about an inch from the free margin."[‡]

The number of mammæ is stated by various writers to be four. According to Edwards (as quoted by Richardson[§]), these are placed, in the adult, 15 inches apart, in the corners of a quadrangle having the umbilicus in the centre. Owen and Murie give them as "two abdominal and two inguinal."

In respect to general size, authors vary greatly in their statements, the length ranging for adults from about 10 to 12 and even 15 or 16 feet, while the weight given ranges from 1,500 to 5,000 pounds! Among what may be termed recent writers, Parry

---

*Account of Arctic Regions, vol. i, p. 503.

†Proc. and Trans. Nova Scotia Inst. Nat. Sci., vol. ii, pt. 3, p. 123.

‡ Trans. Zoöl. Soc. Lond., vol. vii, p. 425.

§ Suppl. to Parry's Sec. Voyage, p. 340.

gives the weight of a "moderate-sized female," but evidently from his account quite young, as 1,550 pounds. Scoresby says: "The Walrus is found on the shores of Spitzbergen twelve to fifteen feet in length and eight to ten in circumference."* Dr. Gilpin gives the weight of a full-grown male as 2,250, while Lamont says a full-grown old male will weigh at least 3,000 pounds.† Aside from Dr. Murie's measurements of a young specimen, I have met with no detailed measurements of the Atlantic Walrus, except those given by Dr. Gilpin,‡ which are as follows:

	Ft.	In.
Extreme length	12	3
Length of head	1	5
Breadth of muzzle	1	0
Distance from nose to eye	0	8
Distance between eyes	0	9½
Extension of tusk beyond the mouth	1	0
Distance of tusks apart at base	0	4
Distance of tusks apart at tips	0	11
Length of fore-flipper	2	0
Breadth of fore-flipper	1	1
Length of hind-flipper	1	10
Breadth of hind-flipper, distended	2	6
Thickness of skin	0	1
Thickness of blubber	0	1½

Weight said to be 22 cwt.

Fleming § gives the length of the Walrus as 15 feet, with a circumference at the shoulders of 10 feet; and the length of the tusks as 20 inches. Hamilton ‖ says an individual killed in Orkney, in 1825, which he saw, "was about ten feet in length," with the head 13½ inches in length. From the size of the tusks (exserted 8½ inches) it appears to have been far from fully grown. Daubenton gives the length of the specimen he described as 11½ feet, with a circumference at the shoulders of 8 feet. Lamont

---

* Account of the Arct. Reg., vol. i, p. 502.

† Mr. Lamont, in his "Seasons with the Sea-horses" (p. —), gives the weight of an old male as 3,000 pounds, but in his "Yachting in the Arctic Seas" (p. 89), he says, "A full-sized old bull Walrus must weigh at least 5,000 lbs., and such a Walrus, if very fat, will produce 650 lbs. of blubber, but seldom more than 500 lbs., which is I think the *average* amount yielded by the most obese of our victims." He speaks, however, in another place (p. 183), of one that "yielded between 700 and 800 pounds of fat." The weight of the entire animal, as last estimated by Mr. Lamont, is probably much too great.

‡ Proc. & Trans. Nova Scotia Inst. Nat. Sci., vol. ii, pt. 3, pp. 123, 124.

§ Hist. Brit. Mam., 1828, p. 18.

‖ British Quad., p. 223.

speaks of having got one day "a very large and fat cow," the length of which he gives as 11 feet 5 inches.* My own measurements of three adult males from unstuffed (salted) skins are as follows: (1) length (from nose to tail), 10 feet 5 inches; (2) 9 feet 6 inches; (3) 10 feet 10 inches; (4) 8 feet 5 inches. The first three were fully adult, while one of them, to judge from its broken, worn tusks and partly naked, scarred skin, was very old; the other was not more than two-thirds grown. These may all have been specimens of less than the average size. Adding, however, 15 to 18 inches for the length of the hind limb (not here included), would give a length of about 12 feet for the larger individuals.†

Most of the old writers were content with stating it to be as large as an ox and as thick as a hogshead. The accounts of the color are also discrepant; Fabricius's statement that the color varies with age, the young being black, then dusky, later paler, and finally in old age white, having been quoted by most subsequent compilers. Writers who have given the color from actual observation have never, however, confirmed Fabricius's account, they usually describing the color of the hair as "yellowish-brown," "yellowish-gray," "tawny," "very light yellowish-gray," etc., some of whom explicitly state that after extended observations they have never met with the changes of color with age noted by Fabricius. Thus, Mr. Robert Brown says that although he has seen Walruses of all stages, from birth until nearly mature age, he never saw any of a black color, all being of "the ordinary brown color, though, like most animals, they get lighter as they grow old."‡ Scoresby says that the skin of the Walrus is covered "with a short yellowish-brown colored hair."§

Dr. Gilpin states that his Labrador specimen was thinly covered with "adpressed light yellowish-green hair," about an inch in length. He adds that the surface of the whole skin was

---

* Yachting in the Arctic Seas, p. 77.

† I find it to be a nearly universal custom with writers (especially with non-scientific writers), in giving the length of Pinnipeds to measure from the point of the nose to the end of the outstretched hind flippers, so that "length" must generally be understood as the total length from "point to point," and not merely that of the head and body. Taking, for example, Dr. Gilpin's specimen, and deducting the length of the hind flipper from the "extreme length," would leave 10 feet 5 inches.

‡ Proc. Zoöl. Soc. Lond., 1868, p. 428.

§ Account of the Arctic Regions, vol. i, p. 503.

covered by " scars and bald warty patches," and that the skin itself was thrown into " welts and folds " on the neck and shoulders.

Mr. Brown further says that " the very circumstantial account of the number of mystacial bristles given in some accounts is most erroneous; they vary in the number of rows and in the number in each row in almost every specimen. They are elevated on a minute tubercle, and the spaces between these bristles are covered with downy whitish hairs."*

Many other writers also note the scars and warty patches and partial absence of hair referred to above by Dr. Gilpin. Mr. Brown, in speaking of those he met with in Davis Straits, says: " I have seen an old Walrus quite spotted with leprous-looking marks consisting of irregular tubercular-looking white cartilaginous hairless blotches; they appeared to be the cicatrices of wounds inflicted at different times by ice, the claws of the Polar Bear, or met with in the wear and tear of the rough-and-tumble life a Sea-horse must lead in N. lat. 74°."* Mr. Lamont further adds that in the Spitzbergen seas the "old bulls are always very light-colored, from being nearly devoid of hair; their skins are rough and rugose, like that of a Rhinoceros, and they are generally quite covered with scars and wounds, inflicted by harpoons, lances, and bullets which they have escaped from, as well as by the tusks of one another in fights among themselves."† From these reports, especially that of Mr. Brown, Dr. Murie‡ has inferred that the Walrus is subject to skin diseases, and that the " glandular spots " thus produced are mistaken " for healed cutaneous wounds." However this may be, it is pretty well established that many of these marks are really scars of wounds.

Respecting other external characters, especially the tusks, and their variations with age, sex, and accidental causes, I transcribe the following from Mr. Lamont's entertaining book, which will be found so freely quoted in subsequent pages: " Old bulls," he observes, " very frequently have one or both of their tusks broken, which may arise from using them to assist in clambering up the ice and rocks. . . . . The calf has no tusks the first year, but the second year, when he has attained to about the size of a large Seal, he has a pair about as large as

---

* Proc. Zoöl. Soc. Lond., 1868, p. 428.
† Seasons with the Sea-horses, p. 137.
‡ Trans. Zoöl. Soc. Lond., vol. vii, 1872, p. 422.

the canine teeth of a lion; the third year they are about six inches long.

"Tusks vary very much in size and shape according to the age and sex of the animal. A *good pair* of bull's tusks may be stated as twenty-four inches long,* and four pounds apiece in weight; but we ᴐbtained several pairs above these dimensions, and in particular one pair, which measured thirty-one inches in length when taken out of the head, and weighed eight pounds each. Such a pair of tusks, however, is extremely rare, and I never, to the best of my belief, saw a pair nearly equal to them among more than one thousand Walruses, although we took the utmost pains to secure the best, and always inspected the tusks carefully with a glass before we fired a shot or threw a harpoon.

"Cows' tusks will *average* fully as long as bulls', from being less liable to be broken, but they are seldom *more* than twenty inches long and three pounds each in weight. They are generally set much closer together than the bull's tusks, sometimes overlapping one another at the points, as in the case with the stuffed specimen at the British Museum. The tusks of old bulls, on the contrary, generally diverge from one another, being sometimes as much as fifteen inches apart at the points." †

Mr. Brown observes: "The whalers declare that the female Walrus is without tusks; I have certainly seen females without them, but, again, others with both well developed. In this respect it may be similar to the female Narwhal, which has occasionly no 'horn' developed." ‡

Captain Parry states that Captain Lyon obtained the head of a small Walrus, remarkable on account of its having *three* tusks, all very short, but two of them close together on the right side of the jaw, and placed one behind the other. §

Scoresby gives the length of the tusks externally as from "ten to fifteen inches," and their full length when cut from the skull as from "fifteen to twenty, sometimes almost thirty," and their weight as from "five to ten pounds each, or upward." ‖

The sexual differences described by Lamont were long since

---

* This probably includes their whole length when removed from the sockets, of which probably not more than eighteen to twenty inches were exposed in life.

† Loc. cit., pp. 137–140.

‡ Proc. Zoöl. Soc. Lond., 1868, p. 429.

§ Narrative of Parry's Second Voyage, p. 415.

‖ Account of the Arctic Regions, vol. i, p. 502.

suspected by Wiegmann and Stannius (see *anteà*, p. 19), who believed that the female had longer, slenderer, and more converging tusks than the male. There is also a specimen in the collection of the Museum of Comparative Zoölogy, Cambridge, in which the tusks are very long and slender, and converge to such a degree that their points actually overlap.

In concluding this rather rambling notice of the external characters and aspect of the Atlantic Walrus, I append the quaint and very correct description of this animal, written by the missionary Egede as early as 1740. I give it from Krünitz's German translation from the original Danish:

"Der Wallross, oder das Meerpferd, ist eine Art von Fisch, dessen Gestalt einem Seehunde gleichkömmt: jedoch ist es weit grösser und stärker. Seine Pfoten sind mit fünf Klauen versehen, wie die Pfoten des Seehundes; doch kürzer von Nägeln; und der Kopf is dicker, runder und stärker. Die Haut dieses Thieres ist, vornehmlich am Halse, einen Daumen dick, und aller Orten faltig, und runzlig. Es hat ein dickes und braunes Haar. In dem obern Kinnbacken sitzen zwey krumme Zähne, welche aus dem Munde über der Unterlippe hervorragen; und einen oder zwey Fuss lang, und bisweilen auch wohl noch länger sind. Die Wallrosszähne sind in eben solchem Werth, als die Elephantenzähne. Inwendig sind sie dicht und fest, an der Wurzel aber hohl. Sein Maul ist wie ein Ochsenmaul; unten und oben mit stachlichten Borsten, in der Dicke eines Strohhalms, besetzt, und diese dienen ihm anstatt eines Bartes. Oberhalb des Mundes sind zwey Naselöcher, wie bey dem Seehunde. Seine rothe Augen sehen ganz feurig aus; und weil sein Hals ganz ausserordentlich dick ist, kann er nicht leicht um sich herum sehen; und dieserhalb dreht er die Augen im Kopfe herum, wann er etwas ansehen will. Er hat, gleich dem Seehunde, einen sehr kurzen Schwanz. Sein Fleisch hat eine Aehnlichkeit mit dem Schweinenfleische. Es pflegt sich dieses Thier mehrentheils auf dem Eise aufzuhalten. Indessen kann es so lange auf dem Lande bleiben, bis es der Hunger nöthigt, in die See zu gehen; indem es sich von denen Fischen und Meer-Insekten unterhält. Wann es im Zorne ist, brüllt es wie ein Ochs. Die Meerpferde sind beherzt, und stehen sich einander bis in den Tod bey. Sie leben in beständigem Kriege mit denen Bären, denen sie mit ihren grossen und starken Zähnen genug zu schaffen machen. Oefters tragen sie den Sieg davon; und

wenigstens kämpfen sie so lange, bis sie todt zur Erde nieder-
fallen."*

Another account of the Walrus, from its being one of the
earliest extant, is also of especial interest in the present connection. Though repeatedly copied, in part or wholly, by the
earlier authors, and also by von Baer, I think it deserving of
reproduction here. It was written by Prof. A. E. Vorst, and was
based on the young specimen taken to Holland in 1613. It is
here copied from De Laët (Descrip. Indiæ Occident.), by whom it
was published in 1633:

"Belluam hanc marinam vidi, magnitudine vituli, aut canis
Britannici majoris, Phocæ non dissimilem; capite rotundo, ocu-
lis bovillis, naribus depressis ac patulis, quos modo contrahe-
bat, modo diducebat, aurium loco utrinque foramina; rictus
oris rotundo nec ita vasto, superiori parte aut labro mystaca
gestabat setis cartilagineis, crassis ac rigidis constantem. Inferior maxilla trigona erat, lingua crassa brevisque, atque os interius
dentibus planis utrimque munitum, pedibus anterioribus posteri-
oribusque latis, atque extrema corporis parte Phocam nostratem
plane referebat. Pedes anteriores antrorsum, posteriores retrorsum spectabant cum ingrederetur. Digiti quinque membrana in-
tersepiente distincti, eaque crassa, posterioribus digitis ungues
impositi, non prioribus, cauda plane carebat. Postica parte
repebat magis quam incedebat. Cute crassa, coreacea, pilisque
brevibus ac tenuisibus obsita vestiebatur, colore cinereo. Grun-
nitum apri instar edebat, seu crocitabat voce gravi et valida.
Repebat per aream extra aquam, quotidie per semihoram aut
amplius dolio aqua pleno immittebant, ut se ibi oblectaret. Ca-
tulus erat, ut ferebant qui attulerant ex nova Zembla, decem
hebdomadarum, dentes seu cornua exerta, ut adultiores, non-
dum habens, tubercula tamen in superiori labro percipieban-
tur, unde brevi proditura facile apparebat. Ferum et validum
animal calebat ad tactum, validique per nares spiribat. Pul-
mentarium ex avena miliove comedebat lente et suctu magis,
quam deglutiendo, herumque gestantem cibum ac offerentem
magno nisu ac grunnitu accedebat, sequebaturque, nidore ejus
allectus. Lardum ejus gustantibus haud insuave visum est.

---

* Herrn Hans Egede, Missionärs und Bischofes in Grönland, Beschreibung
und Natur-Geschichte von Grönland, übersetzet von D. Joh. Ge. Krünitz.
Mit Kupfern. Berlin, verlegts August Mylius, 1763. pp. 106–108.—Since
transcribing the above I have met with an early (1768) English translation
of this work, in which an English rendering of the above description may
be found at p. 125.

Conspiciebantur ibidem duo majorum capita, dentibus duobus exertis Elephantorum instar, longis ac crassis et albicantibus munita, qui deorsum versus pectus spectabant. Eorum coria CCCC aut IC pondo pendisse ferebant Angli qui attulerant. Hisce dentibus rupes ascendere seque sustinere ajebant, et prodeunt in continentem seu terram ut sommum ibi capiant gregatim. Pabulum ajebant illis esse folia oblonga ac magna, herbæ cujusdam e fundo maris nascentis. Nec piscibus vivere aut carnivorum esse. Vidi ibidem penem ejusdem animalis osseum, rotundum, cubitum et amplius longum, crassum, ponderosum ac solidum, in fine prope glandem longe crassiorem ac rotundiorem. Hujus pulvere ad calculum pellendum Moscovitæ retuntur." *

A still earlier description of the Walrus is given by Purchas †
in his account of the first voyage "into the North Seas," by William Barents, a Dutch navigator, who met with Walruses on Orange Island, in 1594, translated from the Dutch by W. Philip. The account says they "went to one of those Islands [of Orange], where they found about two hundred Walrushen, or Sea-horses, lying upon the shore to bast themselves in the Sunne. This Sea-horse is a wonderful strong Monster of the Sea, much bigger than an Oxe, which keeps continually in the Seas, having a skin like a Sea-calfe or Seale, with very short hayre, mouthed like a Lion, and many times they lye upon the Ice; they are hardly killed unlesse you strike them just upon the forehead, it hath foure Feet, but no Eares, and commonly it hath one or two young ones at a time. And when the Fishermen chance to find them upon a flake of Ice with their young ones, shee casteth her young ones before her into the water, and then takes them in her Armes and so plungeth up and downe with them, and when shee will revenge her-selfe upon the Boates, or make resistance against them, then shee casts her young ones from her againe, and with all her force goeth towards the Boate . . . . . . thinking to overthrow it. . . . . . They have two teeth sticking out of their mouthes, on each side one, each being about half an Ell long, and are esteemed to bee good as any Ivory or Elephants teeth, especially in *Muscouvia, Tartaria,* and thereabouts where they are knowne, for they are as white, hard, and even as Ivorie."

SEXUAL DIFFERENCES.—The subject of sexual differences in the Walruses has received very little attention at the hands of

---

* Novus Orbis seu Descriptio Indiæ Occidentalis, pp. 38, 39, 1633.
† His Pilgrimes, vol. iii, p. 476.

systematic writers, who have, indeed, no positive information to offer, and very little can be gleaned from other sources. All that I have met with, after pretty extensive research, has already been incidentally given in the foregoing account of the external characters. All that can be gathered is that in the female the tusks are smaller and thinner, and the general size of the animal may be inferred to be somewhat smaller than in the male. In fact, the external characters in the adult animal of the species under consideration have never as yet been given with much detail, the few naturalists who have met with it in life seeming to take it for granted that an animal so long known, and so familiar to them, must be well known, thereby rendering a careful and detailed description unnecessary. The very good description given by Dr. Gilpin (see *anteà*, pp. 31, 32, 33) of an adult is about all that I have met with in the way of detailed descriptions of the adults of either sex.

The figures and descriptions given of the young, especially those recently published by Dr. J. Murie,[*] leave little to be desired as regards the external characters in early life. The absence of references to any strongly marked sexual differences in the adult might perhaps be taken as negative evidence that none exist; but on the basis of analogy with the other Pinnipeds, especially with the *Otariidæ*, we should hardly expect their absence. Even in the case of the skulls, few sexed specimens appear to have come under the observation of specialists. We here and there, however, meet with references to supposed sexual differences in the size and character of the tusks, and also in respect to the size of the skull and the density and weight of the bones in those of supposed females as compared with those of supposed males. Thus, Wiegmann, in 1832, in referring to the species described by Fremery, in 1831, says, in remarking upon Fremery's "*Trichechus Cookii*," that he remembers having heard from a Greenland traveller that the female Walrus has longer and slenderer tusks than the male, and states, on the authority of Fremery, that a young specimen in the Royal Museum of Holland, having long, slender tusks, was regarded by Temminck as a female. He also considers, on the ground of analogy, that the greater or less development of the occipital and other crests of the skull, as well as the relative weight of the bones,

---

[*] "Researches upon the Anatomy of the Pinnipedia.—Part I. On the Walrus (*Trichechus rosmarus*, Linn.)."—Trans. Zoöl. Soc. Lond., vol. vii, 1872, pp. 411–464, with woodcuts, and plates li–lv.

to be only differences of a sexual character.* Stannius,† ten years later, cited the views of Temminck and Wiegmann (as above given) respecting sexual differences in Walruses, but adds nothing new to the subject. Lamont (see *anteà*, p. 35) states that the "tusks vary very much in size and shape according to the age and sex of the animal." "Cows' tusks," he says, "will *average* fully as long as bulls', from being less liable to be broken, but they are seldom *more* than twenty inches long and three pounds each in weight. They are generally set much closer together than the bull's tusks, sometime overlapping at the points, as in the case with the stuffed specimen at the British Museum." He gives the length of tusks in the male as 24 inches, and the weight as 4 pounds each.

A skeleton, marked as that of a female, in the Museum of Comparative Zoölogy, collected in the Greenland seas by Dr. Kane, has the bones very light, soft, and porous, as compared with those of male specimens. The skull (see figg. 1–3) is much smaller, with the crests and ridges very slightly developed, and the tusks long and slender, and overlapping at the points. This skull, though of a rather aged individual, is 2 to $2\frac{3}{4}$ inches shorter than male skulls of corresponding age, and about 2 inches narrower; but these figures scarcely express the real difference between them, owing to the very much weaker development and slighter structure of all parts of the skull, which certainly has not one-half the weight of average adult male skulls. The weaker structure is especially marked in the lower jaw. The tusks, on the other hand, are several inches longer than in any male skulls of the Atlantic species I have yet examined, but they are so much weaker and slenderer that their weight is more than one-half less. The same difference of lightness and smaller size extends throughout all the bones of the skeleton, indicating that the size of the animal in life was far less than that of ordinary males. The very great length of the

---

* Says Wiegmann : "Hr. Fremery führt an, dass Hr. Temminck einen (nach Deutlichkeit der Nähte) noch jungen Schädel des Reichsmuseums mit ausgezeichnet langen dünnen Stosszähnen für den eines Weibchens gehalten habe. Ich erinnere mich auch von Grönlandsfahren gehört zu haben, dass sich das Weibchen durch längere, dünnere, das Männchen durch kürzere, aber viel dickere Stosszähne auszeichne. Die geringere Entwicklung der Hinterhauptleiste, die geringere Schwere der Knochen, selbst das Zurückbleiben des hintersten Backenzahnes im Oberkiefer könnte, wenn es wirklich nur sexuelle Verschiedenheit sein sollte, mit Analogien belegt werden."—*Arch. für Naturgesch.*, 1832, pp. 128, 129.

† Müller's Arch. für. Anat., 1844, p. 392.

tusks (see fig. 1) is doubtless abnormal, and is doubtless owing to their unsymmetrical development and overlapping at the points, which must have interfered to some extent with their use, and hence have preserved them from wearing.

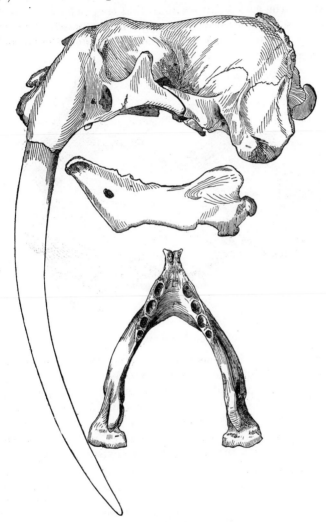

Fig. 1.—*Odobœnus rosmarus*, ♀.

In the National Museum at Washington are also four skulls, which, though unmarked as to sex, are unquestionably those of females.* They agree with the one already described as to small size, the absence of well-developed crests and ridges for muscular attachment, small, slender tusks, and general weakness of structure, as compared with male skulls of corresponding age.† The closed sutures show that they belonged to aged individuals, but in other respects might be presumed to be skulls of young animals, for which such skulls are doubtless usually mistaken.

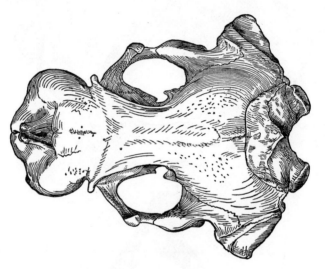

FIG. 2.—*Odobænus rosmarus,* ♀ .

From these data it seems fair to conclude that there are well-marked sexual differences among Walruses, manifested especially in the inferiority of size of the female, in the comparatively weak development of the bones of the skull, the smaller size of the bones of the general skeleton, and in the size and form of the tusks. These differences are, in short, just such as, from analogy, one would naturally expect to exist, and confirm the

* This I inferred from their small size and light structure, and was pleased to have my determination confirmed by so competent an authority as Dr. Emil Bessels, who pronounced them to be unquestionably those of females. Dr. Bessels's judgment, it is perhaps needless to say, is based on personal experience while on the Polaris Expedition, during which he secured and prepared numerous specimens of both sexes, which were lost with the ill-fated vessel.

† In the National Museum there is also a female skull of the Pacific Walrus that presents corresponding differences as compared with male skulls of the same species.

conjectures of Wiegmann and Temminck. What other differences obtain, especially in external characters, can as yet be only conjectured. It is to be hoped, however, that we shall not have long to wait for detailed accounts of the external characters of the adults of both sexes.

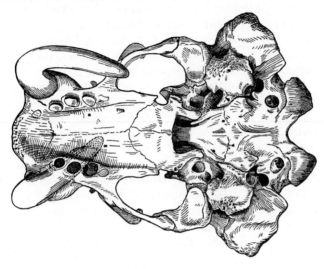

FIG. 3.—*Odobœnus rosmarus*, ♀.

INDIVIDUAL VARIATIONS, AND VARIATIONS DEPENDENT UPON AGE.—That a wide range of individual variation obtains in this species is sufficiently evident from an examination of even a limited series of skulls. These differences have been noted in considerable detail by Fremery, Wiegmann, Stannius, and Jaeger, as will be presently noticed more in detail in presenting the general history of the species. Still greater differences, of course, result from differences of age. These collectively, as will be noted later, have formed the basis of several nominal species. All the Pinnipeds appear to be subject to a wide range of variations of this character, and none more so than the Walruses. These affect to a considerable extent the general proportions of the skull, and especially the form and relative development of different bones. These latter differences are best seen in comparatively young skulls, since most of the sutures close at a rather early age. Among these variations are especially noteworthy those of the nasal bones, the intermaxillaries, and the frontals, and to a less degree those of the base of the skull. The crests and ridges for muscular attachment of course greatly increase with age, and vary considerably in respect to direction, position, and relative development in different individuals. The bony crests at the junction of the intermaxillaries below the anterior nasal opening are especially variable with age, becoming gradually obliterated in adult life by the general thickening of the bones of the skull. They are certainly less prominent in old age than in youth, and the same is true of the incisive border of the intermaxillaries. The intermaxillaries, as a rule, only meet the nasals in their upward extension, but in occasional specimens there is a narrow extension of them posteriorly between the nasals and maxillaries, reaching for one-half to two-thirds the length of the nasals. This variation is seen in the skulls figured by Goethe* and by Blainville,† and has been noted in two skulls by Stannius.‡ In other cases the intermaxillary rises to the surface between the nasals and maxillaries only in the form of narrow isolated areas, as is seen in a skull figured by Goethe,§ and in two skulls I have myself examined. Hence Blainville, when he says, " . . . . et le prémaxillaire, épais, remonte jusque entre le nasal et le maxillaire, de manière à circonscrire avec le premier l'orifice nasal . . . . ,"‖ describes the exceptional instead of the normal condition.

The nasals vary greatly in breadth and in length in different specimens, and even in the same specimen one is sometimes much wider than the other. The concavity and width of the bony palate is also subject to much variation, in this respect hardly two specimens being found to agree. In some, the concavity is nearly one-fourth greater than in others.

---

*Act. Acad. Cæs. Leop. Carol., xi, pt. i, pl. iv, fig. 2.

† Ostéographie, Des Phoques, pl. i.

‡ Says Stannius: "Bisweilen aber, wie bei den Kieler Schädeln a und c, tritt noch eine dünne Leiste dieses Fortsatzes zwischen die das Oberkieferbein und das Nasenbein verbindende Längsnaht und trennt eine Strecke weit diese Knochen. So sieht man es auch auf der in dem Blainville'schen Werke befindlichen Abbildung. Indem diese Leiste an einigen Stellen stärker, an andern Stellen weniger stark oder gar nicht nach aussen hervortritt und zu Tage kömmt, hat es bisweilen den Anschein, als fänden sich isolirte Knochenstückchen in der eben genannten Naht. Wirklich erwähnt de Fremery eines zwischen Nasenbein und Oberkieferbein vorkommenden Ossiculum Wormianum bei seinem aus Labrador stammenden Walross-Schädel."—*Müller's Archiv für Anat.*, 1842, p. 401.

§ Act. Acad. Cæs. Leop. Carol., Bd. xv, pt. i, 1831, pl. iv, fig. 1.

‖ Ostéographie, Des Phoques, p. 20.

The frontals vary greatly in form at their posterior border, especially in respect to their interparietal extension. This portion has sometimes a breadth equal to that of the nasal bones, and terminates quite squarely; at other times, it has less than half this breadth, and is rapidly narrowed posteriorly.

The tusks vary considerably in length, size, and form, and more especially in direction, in specimens of the same age and sex. They become much larger in old age than in middle life, but are then more or less abraded and broken at the points. The grooving on the sides varies more or less with each individual, and even in the two tusks of the same animal. The tusks generally widely diverge, but are sometimes nearly parallel, but appear to be very rarely convergent, while in the female they are frequently more or less convergent, and sometimes touch at the points, or even overlap.

In regard to external characters, considerable changes result from age, especially in respect to the size and amount of abrasion of the tusks, and through the loss of the hair incident to old age, and the shortening of the mystacial bristles.

The following table of measurements (given in millimeters) shows to some extent the variations that occur in the general size and form of the skull.

Measurements of Seven Skulls of ODOBÆNUS ROSMARUS.

Catalogue number.	Locality.	Sex.	Length.	Breadth at zygomata.	Breadth at mastoid processes.	Least breadth between temporal fossæ.	Nasal bones, length.	Nasal bones, width posteriorly.	Nasal bones, width anteriorly.	Anterior border of intermaxillæ to posterior end of palate.	Canines, length (from plane of molars).	Canines, circumference at base.	Canines, distance between external edges at base.	Canines, distance apart at tips.	Length of upper molariform series.	Width of palate at last molar.	Upper incisors, distance apart.	Lower jaw, length.	Lower jaw, height at coronoid process.	Age.
*7156	North Greenland	♀	373	204	237	60	72	38	57	173	254	126	158		65	38	72	216	78	Adult.
*4645	Greenland	♀		184	230	57				170	237	95	144		63	39	63	223	70	Adult.
*9570	......do	♀	313	178	234	63	59	44	52	163	96	70	120		59	32	59	200	72	Young.
†1721	......do	♂	387	250	332	90				228	330	197	198	273	98	28	78	334	98	Old.
†	......do	♂		220	293	78				235	250	177	178	248	87	26	72			Middle-aged.
†1720	......do	♀	322	195	262					173	418	70	133	63				226	76	Adult.
‡	......do	♂	349	185	243	80	70	62	68	197	267	120	156	193	74	37	73	243	73	Young.

* In National Museum, Washington, D. C.
† In Museum of Comparative Zoölogy, Cambridge, Mass.
‡ In Museum of Boston Society of Natural History, Boston, Mass.

DENTITION.—The dentition of Walruses, for various reasons, has been a perplexing subject, and has engaged the attention of many eminent writers. In the adult stage it presents many abnormalities, and, besides, is subject to much individual variation, both in the temporary and adult series. For a long time its deviations from the normal type were not well understood, and even now leading authorities do not appear to be quite in harmony in respect to the proper notation. As previously stated, the incisors of both jaws, except the outer pair in the upper, disappear soon after birth, and before middle life is reached the last tooth of the molar series on each side in both jaws also usually disappears. A brief history of the principal investigations, and opinions held at different times respecting the dentition of the Atlantic Walrus (for the investigations respecting the dentition of this group appear to have been based almost wholly upon this species), is herewith appended as forming a highly interesting chapter in the technical history of the species. In this historical sketch will be found noted many facts relating to the general subject, given by the authors whose papers are here briefly summarized.

The dentition of the Atlantic Walrus has been discussed in greater or less detail by Rapp, von Baer, Wiegmann, Fremery, Stannius, Jaeger, Owen, Malmgren, Peters, and various other writers. Wiegmann, in 1838, pretty fully presented the early history of the subject, noting the almost total lack of information respecting the matter shown by Linné, who evidently paid little attention to the references to the subject made by previous writers. The credit of first giving any definite statement respecting the number of the teeth and their character is due to Anderson, who, in 1734, gave the number of molars as four above and three below. Brisson, in 1756, gave the number as four both above and below; while Crantz, in 1765, again gives four above and three below, and quite fairly describes the normal dentition of the adult.* In the same year, Daubenton gave also again four below on each side as well as four above.

---

* I append in full Crantz's description:

"It had no sharp *incisores* in its mouth, and none at all before, but only four teeth on each side; on the right side of the under-jaw three pretty broad concave grinders. . . . . The two long tusks or horns growing out of its face above the nose, and bending down over its mouth, so as almost to barricade it up, seem to be more an impediment than a help to it. . . . . The right tusk is about an inch longer than the left, and its whole length

No author prior to Schreber (1775) appears to have met with deciduous incisors, who found two such upper incisors on each side in a young skull in the Museum of Erlangen. These he correctly conjectured were temporary, disappearing at a later stage of life.*

To Goethe, however, is given the credit of recognizing the true character of the first tooth of the upper molariform series. Says Camper (as quoted by Wiegmann): "Es ist der Hr. Göthe, sachsenweimarscher Geheimer Rath, der mir zuerst die *ossa intermaxillaria* des Wallrosses und der Schneidezähne desselben hat kennen lernen, indem er mir eine vortreffliche Abhandlung mit schönen Zeichnungen dieser Knochen verschiedener Thiere zugeschickt hatte." Camper, in criticising Linné's errors regarding the Walrus,† gives four incisors $\left(\frac{2-2}{0-0}\right)$, and four molars above and five below $\left(\frac{4-4}{5-5}\right)$ (or sometimes only four below). The observations of Schreber, Goethe, and Camper appear to have been generally overlooked by subsequent writers, so that it was left for G. Cuvier to discover anew the presence of deciduous incisors in the young Walrus. Between the canines he recognized two incisors similar to the molars, which he says the majority of observers had overlooked, because they are not fixed in the intermaxillary, and between these again two pointed small ones in young individuals. He gave the number of molars as four on each side, above and below, and stated that there are neither incisors nor canines in the

---

is 27 inches, 7 of which are grafted into the scull; its circumference is 8 inches. They stand about three inches asunder in the head, and at their extremities 9 inches apart, bent a little downwards."—*History of Greenland*, etc., English translation, London, 1767, p. 126.

*Schreber's account is as follows: " . . . . Die ERSTE Gattung, das insonderheit sogennante WALLROSS, hat zwar, ob gleich kein Schriftsteller etwas davon sagt, *zween Vorderzähne in der obern Kinnlade;* sie sind aber sehr klein, ragen wenig aus ihren Hölen hervor, und werden allem Ansehen nach auserhalb dem Zahnfleische nicht zu bemerken seyn, zumal da sie nicht am Rande der obern Kinnlade, sondern mehr hineinwärts stehen. Ich finde sie an einem zur Naturaliensammlung hiesiger Universität gehörigen Wallrossschädel; und da derselbe, besage seiner Grösse, von einem jungen Thiere ist; so glaube ich beynahe gar, dass sie bey zunehmendem Alter des Thieres ausfallen und nicht wieder wachsen. Sie kommen also hier in keine weitere Betrachtung, als dass sie dem Systematiker einen Wink geben, dis Thier nicht zu weit von dem Robbengeschlechte zu entfernen."—*Säugethiere*, Th. ii, p. 260.

†I quote the French edition of Camper's works (Œuvres, tom. ii, p. 480, Paris, 1803), the only one accessible to me.

lower jaw.* F. Cuvier gave later also the same dental formula. He deemed that the peculiar or anomalous dentition of the Walruses indicated that they were an isolated group, having affinities, on the one hand, with the Carnivora, and, on the other, with the Ruminants!†

According to Wiegmann, Rudolphi‡ (in 1802) recognized the first of the series of *lower* grinding teeth as a canine. §

Thus, as Wiegmann long since observed, the subject remained till Rapp was so fortunate, in 1828, as to have opportunity to examine a fœtal specimen. In this example, he found six incisors in the upper jaw and five in the lower $\left(\frac{3-3}{3-2}\right)$. He also expressed it as his belief that the first lower molar should be regarded as a canine, because (1) it was somewhat further removed from the rest than the others were from each other; because (2) of its greater length and thickness in the adult animal; because (3) it stands close to the temporary or milk incisors, and shuts against the outermost of the upper incisors; and because (4) it lacks the transverse depression seen on the inner side of the crown of the back-teeth. The dental formula recognized by him for the Walrus' may be considered as I. $\frac{3-3}{3-3}$; C. $\frac{1-1}{1-1}$; M. $\frac{4-4}{3-3}$.

Fremery, in 1831, also made reference to the dentition of the Walruses; but his paper bears mainly upon the question of whether there are one or more species of these animals, and will be further noticed in another connection. He notes particularly the presence, in some of his skulls, of two small molars above, behind the large ones.

Wiegmann,‖ in 1838, contributed facts additional to those already recorded, but his memoir is largely devoted to a discussion of the observations of preceding writers. He assents to Rudolphi's and Rapp's interpretation of the homological relation of the first large tooth of the lower jaw; refers to finding

---

* Règne Animal, tom. i, 1817, p. 168.

† Dents des Mam., p. 234.

‡ Anatomisch-physiologische Abhandlungen, p. 145.

§ Wiegmann says: " . . . . Ueberdies ist Rudolphi der erste, der die unteren Eckzähne erkennt. Er bemerkt nämlich, dass der erste Backenzahn des Unterkiefers sich von den übrigen durch seine Grösse auszeichne, und wenn auch der Form nach einem Backenzahne ähnlich, doch seiner Grösse nach beinahe für einen Eckzahn zu halten wäre, was später durch Rapp, dem indessen diese Notiz unbekannt blieb, ausser Zweifel gesetzt ist."—*Arch. für Naturg.*, 1838, pp. 119, 120.

‖ Archiv für Naturgesch., 1838, pp. 113–130.

traces of alveoli of six incisors both above and below, and to the early deciduous character of the last (fifth) upper molar, and the frequent disappearance of the fourth. He concludes that the normal number of the back-teeth is $\frac{5-5}{4-4}$, and that in early life the dentition of the Walrus is not widely different from that of other Pinnipeds.

The same year (1838), Macgillivray[*] considered the normal dentition of the Walrus to be I. $\frac{3-3}{2-2}$; C. $\frac{1-1}{0-0}$; Pm. + M. $\frac{5-5}{5-5}$ $=\frac{18}{14}=32$. His conclusion was based on the examination of a quite young specimen, of which he speaks as follows: "The normal dentition of the Walrus is shown by the skull of a young individual in the Museum of the Edinburgh College of Surgeons. In the upper jaw there are on each side three incisors, the first or inner extremely small, the second a little larger, the third or outer disproportionately large, being equal to the larger grinders. The socket of this tooth is placed in the intermaxillary bone, but towards its mouth it is partly formed by the maxillary. The small incisors have deep conical sockets. The canine tooth is displaced, being thrust outwards beyond the line of the other teeth, and causing the peculiar bulging of the head. The lateral incisor is on the level of its anterior margin, and the first grinder is opposite to its middle. There are five grinders, having conical obtuse sockets, and consequently single roots; the first smaller than the last incisor, the second and third largest, the fourth much smaller, the fifth very small, all shortly conical, and blunt, with enamel on the tip only. The canine tooth is also at first enamelled at its extremity. In the lower jaw there are two very small conical incisors on each side; the canine tooth is wanting; five grinders, with single conical compressed roots, and short compressed conical crowns, enamelled at the point; the first, second, and third nearly equal, the latter being a little larger, the fourth much smaller, and the fifth very small. The tusks, or enormously developed canine teeth of the upper jaw, are compressed, conical, a little curved backward, directed downwards and a little forwards, and somewhat diverging, but in some individuals, when very long, they again converge towards the points. In adults, the incisors are obliterated, excepting the lateral pair of the upper jaw; the fifth grinder in both jaws has also disappeared, and sometimes the fourth in one or both jaws."

---

[*] British Quadrupeds, 1838, pp. 220, 221.

Stannius, in 1842, further contributed to the subject by adding observations respecting variation in the number of the teeth resulting from age, describing in detail the incisive dentition of a series of four skulls of different ages. In two young skulls, the outer temporary incisor of the upper jaw on either side remained; the alveoli of the second pair were still distinct, while the alveoli of the middle pair were nearly obliterated. In another, the alveoli of the inner pair of incisors were wholly obliterated; those of the second pair were barely recognizable, while those of the outer pair were distinct, the teeth having fallen later. He was also able to recognize the alveoli of six incisors in the lower jaws of the skulls just mentioned, and states that he thought Rapp's view of the homology of the first lower back-tooth (considered as a canine) was probably correct. He further takes exceptions to the value of the characters assumed by Fremery as the basis of several species of Walruses.* He adds, in respect to the tusks, that in old age they become wholly solid to the base. In regard to the upper molars, he notes the presence of five in several instances, and finds that, as a rule, the fourth disappears before the fifth, or, at least, that its alveolus becomes sooner obliterated. He also confirms the statement previously made •by Wiegmann, that the alveoli become filled by depositions of bony matter in concentric layers at the bottom and on the sides.

---

*These alleged specific characters he notices in detail, and considers them as dependent upon age. He says: "Das Missliche dieser Charaktere erhellt schon aus dem Umstande, dass dieselben nur für völlig ausgewachsene Thiere, nicht aber für junge anwendbar sind, denn das stärkste Wachsthum der Eckzähne fällt erst offenbar in eine spätere Lebensperiode in welcher namentlich die beiden innersten Schneidezähne und die beiden letzten Backzähne jeder Seite der oberen Kinnlade schon geschwunden sind. Hierzu kömmt noch der Umstand, dass auch die Eckzähne bei sehr alten Thieren an der Spitze bedeutend abgenutzt sind, demnach in späteren Lebensstadien an Länge wahrscheinlich wieder abnehmen. Endlich scheint es ja selbst, als ob die Länge dieser Zähne je nach den Geschlechtern verschieden wäre.

"Eben so wenig Gewicht möchte ich auf die Furchungen dieser Zähne legen. Ihrer Zahl, wie ihrer Stärke nach sind sie bei verschiedenen übrigens nicht von einander abweichenden Individuen verschieden, wie ich mich durch Vergleichung einer grossen Anzahl von Walrosszähnen überzeugt habe; ja diese Furchen sind bisweilen an beiden, bisweilen nur an Einem dieser Zähne spurlos verschwunden."

Respecting *Trichechus cooki*, he adds: "Auch an einem Schädelfragmente des Kieler Museums finde ich etwas convergirende Eckzähne, möchte aber zweifeln, ob dieser Umstand eine Artunterscheidung rechtfertigt."—*Müller's Archiv für Anatomie*, etc., 1842, pp. 398, 399.

Jaeger,[*] in 1844, described the dentition of three rather young Walrus skulls from Labrador. In the youngest skull ($8\frac{1}{3}$ inches long, $6\frac{1}{2}$ inches broad, Paris measure), which had the canines about two inches long, he notes that the last upper molar had already fallen from the right side, but still remained on the left, behind which, as well as behind the alveolus of the fourth molar on the right side, was a little shallow pit, in which, during fœtal life, a tooth had perhaps stood. In front of these beforementioned teeth were three molars on each side, and in front of these a conical incisor, and the alveoli of the others were traceable, although already filled with a spongy substance. In the lower jaw, there were five teeth on each side, with traces of three already fallen fœtal incisors on one side and of two on the other. The second skull ($9\frac{3}{4}$ inches by $7\frac{2}{3}$) was somewhat older, the canines being about five inches long. There were present in this skull three upper molars on each side, and a filled-up alveolus behind them. Of these teeth (as also in the other skull), the middle one was the largest and most worn.[†] In front of these, and somewhat distant from them ($5'''$), was an incisor on each side, and in addition to these another pair of small conical incisors. In the lower jaw of this skull were, on each side, four teeth homologized as C. 1—1, M. 3—3; Jaeger thus recognizing, as had Rudolphi, Rapp, Wiegmann, Fremery, and Stannius, the first of the lower-jaw series as a canine. Behind these were traces of the alveoli of the fourth pair of molars. In the third skull (length $12\frac{1}{2}$ inches, breadth $10\frac{1}{2}$), still older, with tusks a foot long, were three upper "back-teeth" on each side, close together, the middle one being the largest, and in front of these a cutting-tooth. The lower jaw had also four teeth on each side, homologized as before. In this skull, there remained no trace of the middle incisors. Another still older skull had the same dental formula as the last.

Owen,[‡] in 1853, gave the following formula for the deciduous dentition of the Walrus: I. $\frac{2-2}{2-2}$; C. $\frac{1-1}{0-0}$; M. $\frac{2-2}{2-2}$=18. This was based on the examination of a young animal, which had died in the Zoölogical Gardens of London. He, at the same time, proposed the following as the formula of the normal or functional dentition of the Walrus: I. $\frac{1-1}{1-1}$; C. $\frac{1-1}{0-0}$; Pm. $\frac{3-3}{3-3}$=18.

---

* Müller's Arch., 1844, pp. 70–75.

† In the young skulls described by Stannius, the middle molar is mentioned as being uniformly the largest and most worn.

‡ Proc. Zoöl. Soc. Lond., 1853, pp. 105, 106.

Professor Owen, in referring to instances of deviation from this formula, dependent on differences of age and sex, stated "that occasionally a small tooth was found anterior to the normal series of four, and more commonly in the upper than the lower jaw; and that, more rarely, a small tooth was superadded behind the normal four, in the upper jaw, and still more rarely in the lower jaw; the formula of the dentition of such varieties, in excess, being,—I. $\frac{2-2}{2-2}$, C. $\frac{1-1}{0-0}$, Pm. $\frac{3-3}{3-3}$, M. $\frac{1-1}{1-1}=26$." Owen here makes no reference to the literature of the subject, and evidently gave a very erroneous interpretation of the dental formula. In his later references to the subject he gives an entirely different interpretation, and one more nearly agreeing with that now commonly accepted. In his latest reference to the subject,[*] he writes: "In the Walrus (*Trichechus rosmarus*) the normal incisive formula is transitorily represented in the very young animal, which has three teeth in each premaxillary and two on each side of the fore part of the lower jaw; they soon disappear except the outer pair above, which remain close to the maxillary suture, on the inner side of the sockets of the enormous canines, and commence the series of small and simple molars which they resemble in size and form. In the adult there are usually three such molars on each side, behind the permanent incisor, and four similar teeth on each side of the lower jaw; the anterior one passing into the interspace between the upper incisor and the first molar. . . . . The canines are of enormous size. . . . . Their homotype below retains the size and shape of the succeeding molars." The formula of the normal dentition apparently here recognized is: I. $\frac{3-3}{2-2}$; C. $\frac{1-1}{1-1}$; M. $\frac{3-3}{3-3}=\frac{14}{12}=26$.

Giebel,[†] in 1855, gave six incisors both above and below as the number existing in the young before and for a short time after birth. Of these, the lower are said to soon fall out, their alveoli then becoming filled with a bony deposit. Of the upper incisors, the inner pair first disappear, and soon after them also the middle pair, leaving only the outer pair, which begin the molariform series, and to which they are often referred, this outer pair persisting till late in life. The upper canines, he says, are never cast.[‡] In the lower jaw, the first permanent tooth is regarded as a canine, because it is thicker and rounder than the posterior teeth, and lacks the cross-furrow that marks the oth-

---

[*] Anatomy of Vertebrates, vol. iii, p. 338.

[†] Odontog., p. 82, pl. 36, fig. 7; Säugeth., p. 129.

[‡] They are, however, as shown by Malmgren (see beyond), preceded in the embryo by temporary teeth.

ers. The young animal has five upper molariform teeth ("Back-zähne"), the last two of which are smallest and early disappear, and also later the third, leaving only two behind the canine, and an anterior molariform incisor. In the lower jaw there are only four "Backzähne" on each side, of which the last and smallest very soon falls away. The dental formula given is as follows: "$\left(\frac{6-1}{6-0}\right) + \frac{1}{1} + \left(\frac{5-2}{4-2}\right)$" = temporary dentition: I. $\frac{3-3}{3-3}$, C. $\frac{1-1}{1-1}$, M. $\frac{5-5}{4-4}$; adult dentition: I. $\frac{1-1}{0-0}$, C. $\frac{1-1}{1-1}$, M. $\frac{2-2}{2-2}$. While Giebel accepts the first permanent tooth of the molariform series of the upper jaw as an incisor, and the first in the lower jaw as a canine, he recognizes only two persistent molars on each side, above and below.

Malmgren,[*] in 1864, figured the dentition from a fœtal speci-men, and published an elaborate paper on the dentition of the Walrus, in which he reviewed at some length the history of the subject, noticing quite fully the writings of the early authors, from Crantz to the Cuviers, and the papers of Rapp, Owen, Wiegmann, Nilsson, and other later writers. The formula he presents as that of the permanent dentition is: I. $\frac{1-1}{0-0}$, C. $\frac{1-1}{1-1}$, M. $\frac{3-3}{3-3}=\frac{10}{8}=18$; and for the deciduous dentition: I. $\frac{3-3}{3-3}$, C. $\frac{1-1}{1-1}$, M. $\frac{4-4}{4-4}=\frac{16}{16}=32$. [†]

The specimen figured shows both the permanent and decidu-ous dentition. The deciduous teeth are most of them separately figured, of natural size, as minute, slender, spindle-rooted teeth, with short, thickened crowns. The permanent teeth are already in place, although even the upper canines had probably not pierced the gum. The middle pair of incisors of both jaws had already disappeared, leaving only their distinctly recognizable alveoli. His specimen appears to have had but a single cadu-cous molar behind the permanent series, from which he assumes the number of upper molars to be 4—4 instead of 5—5.

The following year, Peters [‡] referred to Malmgren's paper, publishing a plate illustrating the dentition as existing in a some-what older skull (received from Labrador) than that figured by Malmgren. Peters here takes exception to Malmgren's assumed number of back-teeth, which, in accordance with the views of Rapp and Wiegmann, Peters believed should be $\frac{5-5}{5-5}$, instead of $\frac{4-4}{4-4}$.

---

[*] Öfversigt af Kongl. Vet.-Akad. Förhandl., 1863, pp. 505–522, pl. vii.

[†] The paper being published in Swedish, I am unable to follow him in his discussion of the subject.

[‡] Monatsb. K. P. Akad., 1865, pp. 685–687, pl. facing p. 685.

According to Peters, Malmgren, from not finding more than four upper back-teeth in any of the many skulls of various ages he had examined, concluded that when a fifth is present it is abnormal. The young skull figured and described by Peters, however, has in the upper jaw the fourth and fifth back-teeth still in place on the right side, and the fourth on the left side, with an alveolus of a fifth. This Peters considered as affording new proof of the correctness of Wiegmann's formula. As already noticed, five molars have been recognized by Fremery, Rapp, Giebel, and Owen, and, though perhaps not always present, are frequently to be met with.

The dental formula of the Walrus, as determined by Rapp and Wiegmann, has been adopted by Van der Hoeven [*] and Blasius, [†] as well as by Peters, and essentially by Giebel. Giebel, however, gives only four deciduous lower incisors, instead of six. Owen, in his later works, agrees in this point with Giebel, but takes apparently no cognizance of the deciduous fourth and fifth molars, to which he refers, however, in his earlier papers.

Gray, [‡] in 1866, although quoting the formula given by Rapp, adopts the following: "Cutting teeth $\frac{4}{2}$ in young, $\frac{2}{0}$ in adult; grinders $\frac{5-5}{4-4}$ in adult, truncated, all single-rooted; canines, upper very large, exserted." He, however, quotes Rapp's formula, and also that given by Owen in his "Catalogue of the Osteological Series of the Museum of the Royal College of Surgeons" (1853, p. 630).

Professor Flower, [§] in 1869, gave a diagram of the dentition of the Walrus based on many observations made by himself and on "those of others, especially Professor Malmgren," in which both the temporary and permanent dentition is indicated as follows: Milk dentition: I. $\frac{2-2}{2-2}$, C. $\frac{1-1}{1-1}$, M. $\frac{4-4}{4-4}$; permanent dentition: I. $\frac{1-1}{0-0}$, C. $\frac{1-1}{1-1}$, M. $\frac{3-3}{3-3}$. He adds that "it is probable that an anterior rudimentary incisor is developed in the upper if not in the lower jaw," making the temporary incisors hypothetically $\frac{3-3}{3-3}$. "I believe," he says, "that the rudimentary milk teeth never cut the gum, and are absorbed rather than shed. This process commences before birth, ... The rudimentary teeth, however, in front of and behind the large teeth . . . . . are not

---

* Lehrbuch der Zoologie, 1856, p. 738, English ed.
† Säugethiere Deutschlands, 1857, pp. 261, 262.
‡ Cat. Seals and Whales, p. 35.
§ Journ. Anat. and Phys., iii, p. 272.

unfrequently persistent to extreme old age, although commonly lost in macerated skulls. These rudimentary teeth are usually described as 'milk-teeth'; even the posterior ones are sometimes so called, but it appears to me an open question whether they do not rather represent permanent teeth in a rudimentary or aborted condition."

Huxley, in his "Anatomy of Vertebrated Animals" (pp. 360, 361), published in 1872, adopts the following as the dental formula of the Walruses: "I. $\frac{1-1}{0-0}$, C. $\frac{1-1}{1-1}$, p. m. m. $\frac{3-3}{3-3} + \frac{2-2}{1-1}$." He says: "The dentition of the Walrus is extremely peculiar. In the adult, there is one simple conical tooth in the outer part of the premaxilla, followed by a huge tusk-like canine, and three, short, simple-fanged teeth. Sometimes, two other teeth, which soon fall out, lie behind these, on each side of the upper jaw. In the mandible there are no incisors, but a single short canine is followed by three similar, simple teeth, and by one other, which is caducous."* Both here and in the formula no reference is made to the deciduous incisors, although the caducous molars are recognized.

In the foregoing résumé, we have seen how vague was the information bearing on this subject possessed by all writers prior to about the beginning of the present century; how the earlier notices of the existence of incisors in the young were overlooked and rediscovered by later writers, as well as how slowly the first permanent tooth of the molariform series of the upper jaw came to be generally recognized as a true incisor and not a molar; how, later, the number of incisors in the young was found to be six in the upper jaw and six in the lower jaw, with, as a rule, two small caducous molars on each side in the upper jaw, and one on each side in the lower behind the permanent grinding teeth; that the first permanent molariform tooth of the lower jaw was a canine and not a molar; and that by different writers the number of incisors recognized in the lower jaw has been sometimes four and sometimes six, and the caducous upper molars regarded sometimes as one and sometimes as two. Finally, that the true formula of the full dentition of the Walrus is I. $\frac{3-3}{3-3}$; C. $\frac{1-1}{1-1}$; Pm. M. $\frac{5-5}{4-4} = \frac{18}{16} = 34$. It hence appears that the dentition of the Walruses is peculiar and somewhat abnormal in four features, namely, (1) the early disappearance of all the incisors except the outer pair of the upper

---

* Anat. Vertebr. Anim., pp. 360, 361.

jaw, (2) the enormous development of the upper canines, (3) the slight specialization of the lower canines, and (4) the caducous character of the two posterior pairs of molars of the upper jaw and the posterior pair in the lower jaw. The early dentition of the Walrus differs mainly from that of most other Pinnipeds in having six lower incisors instead of four, the incisive formula of other Pinnipeds, as generally recognized, being usually $\frac{3-3}{2-2}$, frequently $\frac{2-2}{2-2}$, and sometimes (as in *Macrorhinus* and *Cystophora*) $\frac{2-2}{1-1}$,—never, at least in the permanent dentition, $\frac{3-3}{3-3}$, but I am far from sure this number may not sometimes appear in the deciduous dentition. In the Sea Otter (*Enhydris*), there are said to be six lower incisors in the young, while only four are present in adult life. The middle pair of lower incisors so early disappear that even in very young specimens they are sometimes wanting. Rapp found in a fœtal specimen three on one side and only two on the other, and quite a number of prominent writers on the subject have recognized two pairs of lower incisors as the normal number. In many specimens, the alveoli of three pairs have been found, and, in addition to the instances already given, I may add that there is a young skull in the Museum of Comparative Zoölogy that shows decided traces of three pairs, the outer incisor on one side being still in place.

In view of all that is at present known respecting the subject, I adopt the following formulæ as being well-established,— premising, however, that they are substantially in accord with the view of the case presented by Professor Flower in 1869 :—

Temporary dentition: I. $\frac{3-3}{3-3}$; C. $\frac{1-1}{1-1}$; M. $\frac{4-4}{4-4} = \frac{16}{16} = 32$.

Permanent dentition: I. $\frac{1-1}{0-0}$; C. $\frac{1-1}{1-1}$; M. $\frac{5-5}{5-5} = \frac{14}{12} = 26$; the last two upper molars and the last lower one on each side being rudimentary and often absent.

FOSSIL REMAINS.—Remains of the Atlantic Walrus, in a fossil state, have been found at various points along the Atlantic coast from Maine to South Carolina, and in Europe as far south as England and France. The first noticed from American localities was thus mentioned by Barton in 1805, but the locality is not given. He says: "The bones of one of these large animals have been found. These appear to have belonged to a species of *trichechus;* perhaps to the *trichechus rosmarus* or *morse.*"* Messrs. Mitchill, Smith, and Cooper described, in

---

* London Phil. Mag., vol. xxxii, 1805, p. 98.

1828,* a specimen consisting of the anterior portion of a skull, found on the sea-beach in Accomac County, Virginia. The same specimen was also described later by Harlan.† These writers all considered it as bearing the closest resemblance to the corresponding portion of the skull of the existing Walrus, to which they doubtfully referred it; but later it was regarded by DeKay as representing a distinct species, to which he gave the name *Trichechus virginianus.*‡ In 1844, Lyell described a tusk obtained from the Tertiary Clays of Gardiner, Maine, which Owen regarded as probably belonging to an extinct species.§ Lyell‖ also refers to a skull he obtained at Martha's Vineyard, Massachusetts. He describes this skull as "differing from skulls of the existing species (*Trichechus rosmarus*, Linn.), with which it was compared by Professor Owen, in having only six molars and two tusks, whereas those of the recent have four molars on each side, besides occasionally a rudimentary one. The front tusk is rounder than that of the recent walrus."¶

In 1857, Dr. Leidy** described and figured a skull found on the sea-beach at Long Branch, Monmouth County, New Jersey, where it was obtained by Prof. J. F. Frazer in 1853, and refers to another specimen (consisting of the facial portion of a skull) discovered at the same locality by Prof. Geo. H. Cook. The first-named specimen, says Dr. Leidy, " has lost a portion of the cranium proper, and the exserted portion of one tusk, but otherwise, except being a little water-worn, is in a good state of preservation. It is unchanged in texture, and nearly so in colour; and it belonged to an old individual, as all the sutures are completely obliterated. The form of the facial portion of this specimen corresponds with that of the specimen from Virginia, [described by DeKay and preceding writers,] above mentioned; and the entire skull closely resembles that of the recent Walrus, *Trichechus rosmarus*, as represented in the figures of Daubenton, Cuvier, and De Blainville; and its measurements are also sufficiently near those given by the first-named author to recognize it as the same species.

---

* Ann. Lyc. Nat. Hist. New York, vol. ii, 1828, p. 271.
† Edinb. New Phil. Journ., vol. xvii, 1834, p. 360.
‡ Nat. Hist. New York, Zoölogy, pt. i, 1842, p. 56, pl. xix, figs. 1, *a*, *b*.
§ See Packard, Mem. Bost. Soc. Nat. Hist., vol. i, 1867, p. 246.
‖ Amer. Journ. Sci. and Arts, vol. xlvi, 1844, p. 319.
¶ As is well known, the existing Walrus has occasionally only the number of teeth found in the Martha's Vineyard specimen.
** Trans. Am. Phil. Soc., vol. xi, 1857, p. 83, pls. iv, v, fig. 1.

" The tusks in the fossil curved downwardly in a diverging manner, and were about four inches distant from each other at their emergence from the alveoli, and ten inches at their tips. The remaining tusk in the specimen is thirteen inches long from its alveolar border, and in this latter position it is three inches in diameter antero-posteriorly and one and three quarters inches transversely . . . ." The other specimen, from New Jersey, mentioned above, he says is also "unchanged from its original texture, but is brown from the infiltration of oxide of iron. It also belonged to an old individual, as all the sutures are obliterated, and the third molars together with the greater part of their alveoli are gone. In its anatomical details the specimen agrees with the corresponding portion of Professor Frazer's specimen, except it is an inch and a half broader in the position of the canine alveoli, and the antero-posterior diameter of the tusk is rather less."* Of both these specimens, Dr. Leidy gives figures, and they agree entirely with corresponding parts of the existing North Atlantic species. Dr. Leidy, however, notes differences between these specimens and those of the Walrus of the North Pacific.

Dr. Leidy adds : " An important question now arises in relation to the age or geological period to which the three Walrus skulls, thus discovered on the coast of New Jersey and Virginia, belong. As they appear to be of the same species as the recent *Trichechus rosmarus*, which once lived in great numbers in the Gulf of St. Lawrence, they are most probably the remains of individuals that were once floated upon fields of ice southerly, and left on the present United States coast. Or, perhaps they may be the remains of the same species which probably during the glacial period extended its habitation very far south of the latitude in which it has been found in the historic period."† In view of the now well-known former extension of the habitat of the Moose, Caribou, Reindeer, Musk Ox, and other northern mammals, southward to Kentucky, the latter hypothesis seems the more probable one, and that the species in glacial times inhabited the eastern coast of the United States southward to Virginia, if not even beyond this point.

More recently, Dr. Leidy has announced the occurrence of Walrus remains in the phosphate beds of Ashley River, South Carolina, and has described and figured a tusk from that locality.

---

* Trans Amer. Phil. Soc., vol. xi, pp. 83, 84.
† Ibid., p. 84.

"This specimen," he says, "is as black as ebony, dense, heavy, and brittle, and is nearly complete, except at the thin border of the pulp cavity. The curvature is slight, and it indicates the tooth to be of the left side." He gives its dimensions or length externally, following the curvature, thirteen inches; near the root it has an antero-posterior diameter of three and five-eighths inches, and a transverse diameter of one and three-fourths inches, and at the middle the transverse diameter is two and one-eighth inches, while the antero-posterior diameter is about the same as at the base. "In robust character," he adds, "the tusk quite equals those of the largest mature recent skulls which have come under my observation, but is much shorter and more abruptly tapering. The specimen looks like what we might suppose the tusks of the living animal would be were they broken off near the middle and then worn away little more than one-fourth the length in a curved line deflected from the course of the anterior longitudinal convexity to the tip. The comparative brevity of the tusk and its worn condition at the end may perhaps have depended upon just such an accident and subsequent wear. In a mature skull from the shore of Sable Island, and preserved in the Museum of the Academy, the tusks, which are of the usual size, are worn in the same manner as the Ashley specimen for more than half their length."

After describing in detail the fluting of the tusks, and the variation noticeable in this respect in different skulls of the living Walrus, he concludes that, while the fluting differs somewhat in the fossil tusks from that usually seen in the tusks of the existing animal, these differences cannot be considered as having specific value. In referring to DeKay's "Trichechus virginianus," he says: "No remains of an undoubtedly extinct species known to me have been discovered anywhere." He finally adds, respecting the Ashley fossil, that "it is an interesting fact to have learned that this [the living] or a closely related species formerly existed so far south as the Ashley River, South Carolina."*

The discovery of the greater part of the skeleton of a Walrus, including the skull, with the tusks over five inches long, and all the teeth except two, in the Quaternary Clays at Portland, Me., was made during July of the present year (1878). It was found in excavating for the foundation of the new "Boston & Maine" transfer station, at about seven feet from the surface.

---

* Journ. Acad. Nat. Sci. Phila., 2d ser., vol. viii, 1877, pp. 214–216, pl. xxx, fig. 6.

" It was partially imbedded in a layer of blue clay a foot in thickness, overlaid by a layer of lighter clay two feet two inches thick, containing casts and shells of *Mya arenaria, Macoma subulosa, Mytylus edulis, Cardium (Serripes) grœnlandicum, Astarte truncata, Saxicava distorta, Nucula antiqua, Leda tenuisulcata, L. truncata, Natica clausa* and *pusilla,* and *Balanus.* The skeleton is in the Museum of the Portland Society of Natural History."*

In Europe, Walrus remains were reported by Cuvier† as found at Angers, France, but Gervais‡ found later that the only portion of those remains accessible to him belonged not to the Walrus, but to the *Halitherium.*

In 1858, however, a part of a cranium was described by Gratiolet, from the diluvial deposits of Montrouge. near Paris. He, however, considered it as distinct from the existing species, even generically, and gave it the name *Odobenotherium lartetianum.*§ In 1874, a nearly entire skull was described by Defrance, from similar deposits near the village of Sainte-Ménehould, Marne, which he not only considered as identical with the living species, but also referred the fragment previously described by Gratiolet to the same species. Respecting these specimens he says :

" En comparant entre elles les têtes du *Trichechus rosmarus* de nos mers, de l'*Odobenotherium Lartetianum* et du *Trichechus de Sainte-Ménehould,* on leur trouve une ressemblance aussi complète que possible, sauf en ce qui concerne la forme et le volume de l'apophyse mastoïde, point qui présente des différences assez sensible. On sait que dans le *T. rosmarus* cette apophyse est très-grande, presque verticale, et saillante la partie inférieure du crâne; celle de l'*Odobenotherium,* également très-volumineuse, se prolonge presque horizontalement en arrière, sans dépasser le crâne inférieurement; celle du *Trichechus de Sainte-Ménehould* présente un volume plus considérable encore que dans les deux autres, sans se prolonger en arrière comme dans l'*Odobenotherium,* mais inférieurement comme dans le *Trichechus* actuel. Ces nuances légères indiquent évidemment une étroite parenté entre ces trois individus; aussi est-il difficile de comprendre que Gratiolet ait voulu établir un nouveau genre sur des particularités peu accentuée que celles que lui présentait la portion de crâne dont il était possesseur, et qui ne

* American Naturalist, vol. xii, p. 633, Sept., 1878; see also *Portland* (Maine) *Argus,* of July —, 1878.

† Ossem. Foss.

‡ Zool. et Paléont. Françaises, 1859, p. 88.

§ Bull. Soc. Géol. de France, 2e sér., xv, 1858, p. 624.

sont d'ailleurs que des particularités relatives pour la plupart à l'âge et au sexe, ainsi que l'a établi M. Gervais."*

Van Beneden† refers to Gratiolet's specimen at some length, giving its full history and exposing its true character. He says: "On a trouvé à Montrouge, près de Paris, il y a quelques années, un crâne dont on s'est beaucoup occupé et que Gratiolet a décrit sous le nom d'*Odobénothère*. Lartet l'avait remis à Gratiolet. Nous avons examiné cette tête avec tout le soin nécessaire et nous partageons complètement l'avis que M. Paul Gervais a exprimé à son sujet dans la *Zoologie et la Paléontologie françaises* (p. 88), c'est-à-dire, que ce crâne fracturé et qui a subi l'action du feu, n'est autre chose qu'un crâne de Morse vivant qui été rapporté du Nord.

"Nous avons étudié cette pièce avec M. Paul Gervais, ayant devant nous tous les éléments de comparaison que possède le Muséum et c'est après avoir sérieusement hésité si l'Odobénothére est un Morse ou non, que nous nous somme rangé de l'avis de notre savant confrère.

"Cet Odobénothère repose sur un fragment de crâne dont la cavité cérébrale a été utilisée pour un usage quelconque et qui aura été apporté dans cet état par quelque pêcheur du Nord. C'est le côté droit et non le côté gauche qui est conservé.

"Celui qui se trouve devant ce fragment de crâne et qui a devant lui un choix de sections des diverses régions de la tête, comprend aisément comment a pu se tromper.

"L'importance que Le Hon a attachée à la présence de cette tête dans le Diluvium rouge, tombe ainsi complètement; à propos de la période glaciaire, Le Hon avait accordé une grande valeur à cette prétendue découverte de gratiolet."‡

Lankester, in 1865, described fossil tusks, from the Red Crag of England, of an animal evidently closely allied to the Walrus. He enumerates no less than twelve or fifteen specimens of these remains, mostly fragments, collected from various localities, all from the so-called "Red Crag" formation of England, or its equivalent. The principal localities are Sutton, Felixstow, and Bawdsey, in England, but he refers also to their occurrence at different points in Belgium. The majority of the specimens of the tusk obtained, writes Mr. Lankester, "are its pointed

---

*Bull. Soc. Géol. de France, 3e sér., ii, 1874, pp. 169, 170.

†Descrip. des Ossements Fossiles des Environs d'Anvers, Ann. Mus. d'Hist. Nat. de Belgigue, i, 1877, pp. 40, 41.

‡"LE HON, *L'homme fossile*, 1867, p. 304.—ID., *Mouvement des mers* . . . . , p. 48, 1870."

terminations; but other specimens, of the base and intermediate portions, have come to light. Throughout its length," Mr. Lankester continues, " which in some examples must have been fully three feet, the tusk is slightly curved; but in those which appear to be fully grown the curve is considerably greater towards the terminal point, the direction of the curve probably giving the tusk, if its Pinnigrade affinities be established, a retroflected position, as in the *Dinotherium.* The Crag tusk is very much compressed laterally, so that its transverse section has an elliptical outline, whilst that of the *Dinotherium*-tusk is nearly circular. The amount of lateral compression is, however, extremely variable, as it is also in the living Walruses; the amount also of the lateral as well as the antero-posterior flection of the tusk appears to vary, as in the recent *Trichecus,* the variability of which in the size and form of its tusks is well known. A single large furrow on the outer surface, two on the inner, and one on the inner curved margin, extend along the whole length of the tusk in many specimens, exactly similar to those on some tusks of Walrus; but in both the recent and fossil specimens they are subject to much variation, in their major or minor development. No appearance of any wearing of the point of the tusks by use during life is observable; and indeed the greater backward curvature of that part seems to result from its freedom from usage, since in the Walrus the point of the tusk is rapidly worn away, which of course checks any tendency to curvature which might become apparent if the tusk were not used against such hard substances as rocks and blocks of ice.

" From an examination of the general contour and form of the tusks, without regard to their substance or structure, one would unquestionably be led to regard them as belonging to an animal similar to the existing Walrus, inasmuch as it is in this animal alone that this form of tusk, with its longitudinal furrows, great length, and gentle curvature, is found."

After describing in detail the structure of these fossil tusks, as shown in sections and as revealed by the microscope, Mr. Lankester further observes: " In its microscopical structure, the dentine of the fossil tusks presents a complete resemblance to that of the Walrus.* . . . . The dentine tubes are very

---

* Their microscopical structure, as well as external form, are illustrated by numerous figures, forming plates x and xi, accompanying Mr. Lankester's paper.

nearly of the same size, and equally closely packed, and are connected with stellate lacunæ in some numbers near the periphery of the tooth. This structure, which is not peculiar to the Walrus, is, nevertheless, a test of affinity, inasmuch as the form of the lacunæ varies in different animals. They are not met with in the tusks of the Proboscidea or the Hippopotamus, but occur in the curious incisors of the Dugong. The ' dentinal cells' of the Crag tusks also resemble those of the Walrus. . . . . In structure the cement exactly resembles that of the Walrus, displaying vascular canals, bone-lacunæ, and canaliculi, of the same form and disposition; but the proportion which it bears to the thickness of the other tooth-tissues appears to be larger in the Walrus than in the fossil.

" From the foregoing remarks it will be apparent that we have in these fossil tusks characters which ally them most closely to the large canines of the genus *Trichecus*." After enumerating the points of form and structure which distinguish these tusks from those of other animals, and those which assimilate them to those of the Walrus, he thus generalizes the results of his investigations: " Lastly, they resemble the large canine tusks of the living *Trichecus* in their curvature, varying lateral compression, large surface-furrows, short and wide pulp-cavity, globular ' osseo-dentine', and every detail of minute structure. They differ from them in their greater curvature at the point of the tusk, their greater lateral compression, and minor development of cement.

" I accordingly propose to establish the genus *Trichecodon* to receive the animal thus indicated. The justification of a generic separation must be sought in the fact of the great antiquity of the Red Crag, and the consequent probability of the association of other and more distinctive attributes with those of the tusks."

As regards its geological position and associations, Mr. Lan·kester adds: " It appears that the *Trichecodon Huxleyi*, like the Cetacean remains of the Crag and large Sharks' teeth, ... is a derived fossil in the Red Crag, belonging properly to the Middle Crag, which is not now observable in this country [England], but is well developed at and near Antwerp."*

It thus appears that Mr. Lankester was as much, or more, influenced in his generic differentiation of these fossils from their

---

* Quarterly Journal of the Geological Society of London, xxi, 1865, pp. 226–231.

nearest living allies, by the geological evidence of their antiquity as by the actually observed and admittedly slight differences of form and structure. Mr. Lankester does not inform us respecting the locality whence came his specimens of the tusks of the living Walrus with which he compared the fossil tusks. In this connection it may be added (see further on this point the account of *Odobœnus obesus* given beyond) that the tusks of the Pacific species (*Odobœnus obesus*) are not only longer and slenderer than those of the Atlantic species (*O. rosmarus*), but are sharper-pointed and more incurved, and do not present the worn and broken appearance so often (indeed, usually) seen in the tusks of old individuals of the latter. Whether or not they present differences of structure has not, so far as known to me, been microscopically determined. The tusks of the Pacific species, furthermore, sometimes attain the size indicated for the tusks of "*Trichecodon huxleyi.*" For the present I must consider Lankester's *Trichecodon huxleyi* as certainly not generically separable from the existing Walruses, although it may have differed from the existing Atlantic species in larger size and possibly in other characters, as so often happens among the immediate progenitors of existing species in other groups of mammals.

Van Beneden has recently reviewed at considerable length the history of the supposed and actual fossil remains of the Walrus,* showing that most of those reported as found in different parts of France and Germany were really those of different species of extinct Sirenians or other animals than the Walrus. Van Beneden, however, describes and figures a dorsal vertebra he considers as that of the Walrus, found near Deurne, and a scaphoid bone from Anvers.

GEOGRAPHICAL DISTRIBUTION, PRESENT AND PAST.—1. *Coast of North America.*—As already shown (*anteà*, pp. 57–61), the Walrus, like the Musk Ox, the Caribou, and the Moose, ranged during the great Ice Period much beyond the southern limit of its boundary at the time the eastern coast of North America was first visited by Europeans. While its remains have been found as far south as New Jersey, Virginia, and even South Carolina, there is no evidence of its existence on the New England coast within historic time, or during the last three hundred and fifty years. During the last half of the sixteenth century they are known to have frequented the southern coast of

---

* Ann. Mus. d'Hist. Nat. de Belgique, i, 1877, pp. 39–42.

Nova Scotia, as well as the shores and islands to the northward; but this appears to have been at that time their southern limit of distribution.

In May, 1534, they were met with by James Cartier, about the island of "Ramea" (probably Sable Island), who thus refers to them: "About the said Island [Ramea] are very greate beastes as great as oxen, which have two great teeth in their mouths like unto Elephants teeth, & live also in the Sea. We saw them sleeping upon the banke of the water: wee think-ing to take it, went with our boates, but so soone as he heard us, he cast himselfe into the sea."* They were afterward hunted here for their tusks and oil. Thus Richard Fischer, in speaking of the same island, says: "On which Isle [of Ramea] are so great abundance of the huge and mightie Sea Oxen with great teeth in moneths of April, May and June, that there have bene fifteene hundreth killed there by one small barke, in the year 1591."† The same writer tells us that George Drake, two years later, "found a shippe of Saint Malo three parts freighted with these fishes." Another writer says that he had seen a "dry flat full at once" of their teeth, "which are a foote and sometimes more in length." They also, at about the same time, frequented the so-called "Bird Islands" off Cape Breton. Says Charles Leigh: "Upon the lesse of these Islands of Birds we saw greate store of Morsses or Sea Oxen, which were a sleepe upon the rockes: but when we approached nere unto them with our boate they cast themselves into the sea and pursued us with such furie as that we were glad to flee from them." It is later said that the number of these "Sea Oxen" was "about thirty or forty."‡ From the accounts of other writers we learn that these "Sea Oxen" were accustomed to resort to these various islands during April, May, and June, for the purpose of bringing forth their young. Thus, "Thomas James of Bristoll," in speaking of the "Isle of Ramea," says it was situated "in 47 degrees, some fiftie leagues from the Grand Bay, neere Newfoundland: and is about twentie leagues about, and some part of the Island is flat Sands and shoulds: and the fish commeth on banke (to do their kinde) in April, May & June, by numbers of thousands, which fish is very big: and hath two great teeth: and the skinne of them is like Buffes leather: and they will not away from their yong ones. The yong ones are

---

* Hakluyt, Voyages, vol. iii, p. 254.   † Ibid., p. 238.   ‡ Ibid., pp. 242, 249.

as good meat as Veale. And with the bellies of five of the saide fishes they make a hogshead of Traine, which Traine is very sweet, which if it will make sope, the king of Spaine may burne some of his Olive trees."* Charlevoix also alludes to the Walrus fishery at Sable Island, which the English at one time established there, but says it was soon abandoned, being found unprofitable.†

Molineux Shuldham has left us quite a full account (and one that has been often quoted) of the habits of these animals, and of the wholesale destruction by which they were speedily extirpated from the Atlantic coast south of Labrador. This account, written in 1775, says: "The sea-cow is a native of the Magdalen Islands, St. John's, and Anticosti in the Gulph of St. Lawrence. They resort very early in the spring to the former of these places, which seems to be by nature particularly adapted to the wants of these animals, abounding with clams of a very large size, and the most convenient landing-places, called Echouries. Here they crawl up in great numbers, and sometimes remain for fourteen days together without food, when the weather is fair; but on the first appearance of rain, they immediately retreat to the water with great precipitation. They are, when out of the water, very unwieldy, and move with great difficulty. They weigh from 1500 to 2000 pounds, producing, according to their size, from one to two barrels of oil, which is boiled out of a fat substance that lies between the skin and the flesh. Immediately on their arrival they calf, and engender again about two months after; so that they carry their young about nine months. They never have more than two at a time, and seldom more than one.

"The echouries are formed principally by nature, being a gradual slope of soft rock, with which the Magdalen Islands abound; about 80 to 100 yards wide at the water side, and spreading so as to contain, near the summit, a very considerable number. Here they are suffered to come and amuse themselves for a considerable time, till they acquire a boldness, being at their first landing so exceedingly timid as to make it impossible for any person to approach them. In a few weeks they assemble in great numbers; formerly, when undisturbed by the Americans, to the amount of seven or eight thousand; and the form of the echourie not allowing them to remain contiguous to the water, the foremost ones are insensibly pushed above the slope. When

---

* Hakluyt, Voyages, vol. iii, p. 237. † Charlevoix, vol. v, p. 216.

they are arrived to a convenient distance the fishermen, having provided the necessary apparatus, take the advantage of a sea wind, or a breeze blowing rather obliquely on the shore, to prevent the smelling of these animals (who have that sense in great perfection, contributing to their safety), and with the assistance of very good dogs, endeavour in the night time to separate those that are the farthest advanced from those next the water, driving them different ways. This they call making a cut, and is generally looked upon to be a most dangerous process, it being impossible to drive them in any particular direction, and difficult to avoid them; but as they are advanced above the slope of the echourie, the darkness of the night deprives them of every direction to the water, so that they stray about and are killed at leisure, those that are nearest the shore being the first victims. In this manner there has been killed fifteen or sixteen hundred at one cut. They then skin them, and take off a coat of fat that always surrounds them, which they dissolve by heat into oil. The skin is cut into slices of two or three inches wide, and exported to America for carriage traces, and to England for glue. The teeth is an inferior sort of ivory, and is manufactured for the same purposes, but soon turns yellow."*

According to Dr. A. S. Packard, jr., its bones are still found at the localities mentioned by Shuldham. "According to tradition," he further says, "it also inhabited some of the harbors of Cape Breton; and I have been informed by a fisherman in Maine, whose word I do not doubt, that on an islet near Cape Sable, Nova Scotia [probably the "Isle of Ramea" of the early voyagers already quoted], its bones are found abundantly on the sandy shore, fifteen to twenty feet above the sea. In the St. Lawrence Gulf they were exterminated during the middle of the last century. The last one seen or heard of in the Gulf, so far as I can ascertain, was killed at St. Augustine, Labrador, twenty-five years since. One was seen at Square Island fifteen years since, and two shortly before that, and another was killed at the same place about eight years since. I saw the head of a young Walrus, which was found floating, dead, having been killed, apparently by a harpoon, in the drift ice north of Belle Isle."†

Dr. J. Bernard Gilpin, writing a few years later (in 1869), in referring to the former occurrence of the Walrus on the shores

---

* Phil. Trans., vol. lxv, p. 249.
† Proc. Bost. Soc. Nat. Hist., vol. x, 1866, p. 271.

and islands of the Gulf of St. Lawrence, says: "At Miscou, Bay Chaleur, Perley found only their bones, but in such numbers as to form artificial sea beaches. These were doubtless victims of 'The Royal Company of Miscou', founded during the earlier part of the seventeenth century, by the King of France, and whose ephemeral city of New Rochelle, numbering at one time some thousands, has passed away leaving no sign. The murdered Sea-horses have left a more enduring monument than the murderers." He further adds: "Though we have no accounts later than the seventeenth and eighteenth centuries of their inhabiting Sable Island, yet it is very probable that they continued to resort there until they entirely left these latitudes. Its difficulty of access; its being uninhabited, and its sandy bars fringed with a ceaseless surf, point it out as their last hold."*

Dr. Gilpin also records the capture of a Walrus in the Straits of Belle Isle, Labrador, in March, 1869, which was dragged on the ice for five miles, and then taken by ship to St. John's, Newfoundland, and thence to Halifax, Nova Scotia, where it was described and figured by Dr. Gilpin.† Mr. Reeks‡ states that a "specimen was driven ashore in St. George's Bay," Newfoundland, about 1868, and alludes to the frequent occurrence of their bones along the Newfoundland coast.

It is still an inhabitant of the shores of Hudson's Bay, Davis Strait, and Greenland, where, however, its numbers are annually decreasing. In Greenland, according to Mr. Robert Brown (writing of its distribution in 1867), " it is found all the year round, but not south of Rifkol, in lat. 65°. In an inlet called Irsortok it collects in considerable numbers, to the terror of the natives, who have to pass that way. . . . It has been found as far north as the Eskimo live, or explorers have gone. On the western shores of Davis's Strait it is not uncommon about Pond's, Scott's, and Home Bays, and is killed in considerable numbers by the natives. It is not now found in such numbers as it once was; and no reasonable man who sees the slaughter to which it is subject in Spitzbergen and elsewhere can doubt that its days are numbered. It has already become extinct in several places where it was once common. Its utter extinction is a foregone conclusion."§

* Proc. and Trans. Nova Scotia Inst. Nat. Sci., vol. ii, pt. 3, pp. 126–127.
† Ibid., pp. 123–127, with a plate.
‡ Zoölogist, 1871, 2550.
§ Proc. Zoöl. Soc. Lond., 1868, p. 433.

70   ODOBÆNUS ROSMARUS—ATLANTIC WALRUS.

Kane and Hayes, during the years 1853 to 1855, found the Walrus very abundant about Port Foulke, on the western coast of Greenland, in latitude 79°, but they seem to have, since that date, greatly decreased in numbers along the whole of the Greenland coast. Captain Feilden, in his paper on the "Mammalia of North Greenland and Grinnell Land," observed in 1875 by the British Arctic Expedition, after alluding to their former abundance about Port Foulke, as observed by Kane and Hayes, says: "Curiously enough, we did not see one of these animals in the vicinity of Port Foulke nor in Smith Sound, until we reached Franklin Pierce Bay. There, in the vicinity of Norman Lockyer Island, we saw several Walruses, and killed two or three. . . . Near Cape Fraser I saw a single Walrus; but as far as my observation goes, it does not proceed further north than the meeting of the Baffin Bay and Polar tides near the above mentioned Cape."*

Mr. Ludwig Kumlien, naturalist of the Howgate Polar Expedition of 1877, states:† "The Walrus is quite common about Cape Mercy and the southern waters of Cumberland Sound, but at the present day rarely strays up the Sound. Their remains, however, are by no means rare, even in the greater Kingwah, and many of the old Eskimo hut foundations contain the remains of this animal. The Eskimo say they got mad and left. Certain it is, they are found around Annanactook only as stragglers at the present day. Considerable numbers were observed on pieces of floating ice near Cape Mercy, in July. About Nugumeute they are largely hunted by the Eskimo living there."

Respecting their occurrence more to the southward, on the Greenland coast, Dr. Rink states: "The Walrus is only rarely met with along the coast, with the exception of the tract between 66° and 68° N. lat., where it occurs pretty numerously at times. The daring task of entering into contest with this animal from the kayak on the open sea forms a regular sport to the natives of Kangamiut in 66° N. lat. The number yearly killed has not been separately calculated, . . . but they can hardly exceed 200."‡

The westernmost point at which it has been observed is said to be the western shore of Hudson's Bay. Mr. J. C. Ross states it to be an inhabitant of the west coast of Baffin's Bay and

* The Zoölogist, 3d ser., vol. i, p. 360, September, 1877.
† In MSS. notes he has kindly placed at my disposal.
‡ Danish Greenland, its People and its Products, pp. 126–127, 1877.

Repulse Bay, and to be occasionally met with in the northern part of Prince Regent's Inlet, but says it is unknown to the natives of Boothia.* Dr. Richardson says: "The Walruses were very numerous at Igloolik and on the other parts of the coast to the eastward of the Fury and Hecla's Strait. They are not found, however, at the mouth of the Copper Mine River, although the black whale had been sometimes drifted thither."† He also refers to its being unknown to the Eskimos of the Coppermine and Mackenzie Rivers.‡ No species of Walrus appears to have ever been seen on the Arctic coast of America between the 97th and 158th meridians, or for a distance of about sixty degrees of longitude.

2. *Coast of Europe.*—On the western shores of Europe the Walrus has been taken at no remote date as far south as Scotland,§ and Mr. Robert Brown, in 1868, stated that he suspected it to be a "not unfrequent visitor" to the less frequented portions of the Scottish shores, he considering it probable that "not a few of the 'Sea-horses' and 'Sea-cows' which every now and again terrify the fishermen on the shores of the wild western Scottish lochs, and get embalmed among their folk-lore, may be the Walrus."‖ Fleming states that one was killed in the Sound of Stockness, on the east coast of Harris, in December, 1817,¶ while another, according to Macgillivray and others, was killed in Orkney in June, 1825.** Mr. R. Brown adds that one was seen in Orkney in 1857, and another in Nor' Isles about the same time.†† It appears to have never occurred in Iceland, except as a rare straggler. Many years ago they are said to have lived on the shores of Finmark, and at a much later date to have abounded on some of the islands off this coast. Mr. Lamont says: "We learn from the voyage of Ohthere, which was performed about a thousand years ago, that the Walrus then abounded on the coast of Finmarken itself; they have, however, abandoned that coast for some centuries, although individual stragglers have been occasionally captured there up to within

* Ross's 2d Voy., App., 1835, p. xxi.
† Suppl. Parry's 2d Voy., p. 338.
‡ Zoölogy of Beechey's Voyage, Mam., 1839, p. 6.
§ Hector Boece's History of Scotland, as quoted by British zoölogists.
‖ Proc. Zoöl. Soc. Lond., 1868, p. 433.
¶ British Animals, p. 19.
** Edinb. New Phil. Journ., vol. ii, p. 389; British Quad., Jard. Nat. Libr., Mam., vol. vii, p. 223. See also Bell, Hamilton, etc., l. c.
†† Proc. Zoöl. Soc. Lond., 1868, p. 433.

the last thirty [now about forty-six] years. [*]   After their
desertion of the Finmarken coast, Bear Island [or Cherie Island,
lying about two hundred and eighty miles north of the North
Cape] became the principal scene of their destruction; and next
the Thousand Islands [southeast of Spitzbergen], Hope Island
[a little further north, but still in the southeast corner], and
Ryk Yse Island, which in their turn are now very inferior
hunting-ground to the banks and skerries lying to the north of
Spitzbergen.

" Fortunately for the persecuted Walruses, however, these latter districts are only accessible in open seasons, or perhaps once
in three or four summers, so that they get a little breathing
time there to breed and replenish their numbers, or undoubtedly the next twenty or thirty years would witness the total
extinction of *Rosmarus trichecus* on the coasts of the islands of
Northern Europe.

" The Walrus is also found all round the coasts of Nova Zembla, but not in such numbers as at Spitzbergen; and he undergoes, if possible, more persecution in those islands from some
colonies of Russians or Samoïedes, who, I am told, regularly
winter in Nova Zembla for the purpose of hunting and fishing."†

" The war of extermination," says Mr. Lamont, in his later
work, " which has been carried on for many years in Spitzbergen and Novaya Zemlya has driven all the Arctic fauna [mam-
mals] from their old haunts, and, in seeking retreats more inaccessible to man, it is probable that they have had in some
degree to alter their habits.   For example, up to about twenty
years ago it was customary for all Walrus-hunters to entertain
a reasonable hope that by waiting till late in the season all former ill-luck might be compensated in a few fortunate hours by
killing some hundreds on shore; in fact, favorite haunts were
well-known to the fishers, and were visited successively before
finally leaving the hunting-grounds.   Now, although the Arctic
seas are explored by steamers and visited annually by as bold
and enterprising hunters as formerly, such a windfall as a herd
of Walruses ashore is seldom heard of.

" Each year better found vessels and more elaborate weapons

---

* Mr. Lamont has since reported the capture of a large bull " in Magerö
Sound near the North Cape about 1868."—*Yachting in the Arctic Seas*, p. 58,
footnote.

† Seasons with the Sea-horses, pp. 167, 168.

are sent out to harry the Walrus; as a consequence every season there is greater difficulty in obtaining a cargo—for two reasons, those animals which have ventured into what was safe feeding-ground last year meet their enemy, and half are killed, while the other half escaping will be found next year a step farther away. This intelligent retreating of the Walrus before a superior enemy will, I believe, preserve the species after its scarcity in accessible waters renders it no longer an object of sport and commerce. That the Walrus, . . . is being driven from every district where the hand of man is felt, is certain."*

Mr. Alfred Newton, writing in 1864, respecting their former presence on the coasts of Finmark, and their distribution at that date, observes : " I see no reason to doubt the assertion, or perhaps it would be safer to say the inference, that in former days Walruses habitually frequented the coasts of Finmark. In the sixteenth and seventeenth centuries they were certainly abundant about Bear Island; they are spoken of there as 'lying like hogges upon heaps,' [†] . . .; yet for the last thirty years probably not one has been seen there. Now they are hemmed in by the packed ice of the Polar Sea on the one side and their merciless enemies on the other. The result cannot admit of any doubt. . . . . Its numbers are apparently decreasing with woful rapidity. The time is certainly not very far distant when the *Trichechus rosmarus* will be as extinct in the Spitzbergen seas as *Rhytina gigas* is in those of Behring's Straits." ‡

In Richard Chancellor's account of his " discoverie of Moscovia," in 1553–1554, we read : " To the North part of that Countrey are the places where they have their Furres, as Sables, Marterns, greese Bevers, Foxes white, blacke, and red, Minkes, Ermines, Minivers, and Harts. There are also a fishes teeth, which fish is called a Morsse. The takers thereof dwell in a place called Postesora, which bring them upon Harts to *Lampas* to sell, and from *Lampas* carrie them to a place called Colmogro, where the high Market is holden on Saint *Nicolas* day."§ On Hondius's map of Russia accompanying this account Lampas is placed on the White Sea, near the mouth of the Dwina River.

---

* Yachting in the Arctic Seas, 1876, pp. 59, 60.

† [" It seemed very strange to us," says Jonas Poole, in his account of his visit to Cherie Island in 1604, " to see such a multitude of monsters of the Sea, lye like hogges upon heapes."—*Purchas his Pilgrimes*, vol. iii, p. 557.]

‡ Proc. Zoöl. Soc. Lond., 1864, p. 500.

§ Purchas his Pilgrimes, vol. iii, pp. 213, 214.

The Walruses appear to have been first met with on Cherie Island in 1603, and to have become nearly exterminated there within a very few years. The history of their destruction there and at Spitzbergen during the early part of the seventeenth century is given in the following excerpts : "In the yeare 1603. Stephen Bennet was imployed by the Companie,* in a Ship called the *Grace*, to those parts Northwards of the Cape ["of Norway"], and was at Cherie *Iland* and killed some Sea-horses, and brought home Lead Oare from thence . . .

" Heere it is to bee understood, that the Companie having by often resort and imployment to those parts, observed the great number of Sea-horses at *Cherie Iland*, and likewise the multitude of Whales, that shewed themselves upon the coast of Greenland [now Spitzbergen]; They first applyed themselves to the killing of Morces, which they continued from yeere to yeere with a Ship or two yeerely, in which Ships the Companie appointed *Thomas Welden* Commander, and in the yeere 1609. the Companie imployed one *Thomas Edge* their Apprentice, for their Northern Voyage, and joyned him in Commission with the foresayd *Welden*. Now the often using of Cherie Iland, did make the Sea-horse grow scarce and decay, which made the Companie looke out for further Discoveries."†

During the expedition of 1604, Jonas Poole, who has left an account of the "Divers Voyages to Cherie Iland in the yeeres. 1604, 1605, 1606, 1608, 1609," says that as they approached Cherie Island, "We had not furled our Sayles, but we saw many Morses swimming by our ship, and heard withall so huge a noyse of roaring, as if there had beene an hundred Lions. Immediately wee manned our Boate. . . . wee landed, and saw abundance of Morses on thes hoare, close by the Sea-side," etc. They attacked them with muskets, "not knowing whither they could runne swiftly or seize upon us or no." Owing to inexperience, they succeeded in killing only fifteen out of "above a thousand," but secured a hogshead of teeth, which they picked up on the shore. Two days later they found, on another part of the island, "neere a thousand Morses," of which they killed "thirtie or thereabouts, and when wee had taken off their heads, we went aboard." The next day they went on shore again and

---

* Incorporated some time prior to the year 1556, under the name "The Merchants of England," and called also the "Muscovia Merchants" and the "Muscovia Companie."

† Purchas his Pilgrimes, vol. iii, p. 464.

" fell a killing of the beasts. . . . We killed that day six-ty Morses, all the heads whereof were very principall." They departed soon after for England.

The next year (1605) they returned to Cherie Island. On the 8th of July, says the account, " we entred into a Cove, having all our men on shoare with shot and javelins, and slue abundance of Morses. The yeere before we slue all with shot, not think-ing that a javelin could pierce their skinnes : which we found now contrarie, if they be well handled, for otherwise, a man may thrust with all his force and not enter : or if he doe enter, he shall spoyle his Lance upon their bones; for they will strike with their fore-feet and bend a Lance round and breake it, if it bee not all the better plated. They will also strike with their Teeth at him that is next them : but because their Teeth grow downward, their strokes are of small force and danger." They took in " eleven tunnes of Oyle, and the teeth of all the beasts aforesaid."

The following year (1606) they again set out for Cherie Island, arriving there July 3. They found the ice still about the island, and the Walruses not yet on shore; " For their nature is such, that they will not come on land as long as any Ice is about the land." On the 14th they perceived on shore " of the beasts sufficient to make our voyage, wee prepared to goe killing. Master *Welden* and Master *Bennet* appointed mee to take eleven men with mee, and to goe beyond the beasts where they lay; that they and wee might meet at the middest of them, and so enclose them, that none of them should get into the Sea, ... and before six houres were ended, we had slayne about seven or eight hundred Beasts. . . . For ten dayes space we plyed our businesse very hard, and brought it almost to an end." They took in " two and twentie tuns of the Oyle of the Morses, and three hogsheads of their Teeth."

In 1608 they again reached Cherie Island toward the end of June, and on the 22d " came into a Cove where the Morses were, and slew about 900. or 1000. of them in less than seven houres : and then we plied our business untill the second of July : at what time we had taken into our ship 22. tunnes and three hogs-heads of Oyle." On their return they took with them two live young Walruses, one of which lived till they reached London.*

The voyage in 1609 was less successful. They slew at one time eighty, at another one hundred and fifty, and at still an-

---

* Purchas his Pilgrimes, vol. iii, pp. 557–560.

other time forty-five; but they lost most of them in consequence of bad weather. "In the yeefe 1610. the Companie set out two Ships, viz. the *Lionesse* for *Cherie Iland, Thomas Edge* Commander; and the *Amitie*, for a Northerne Discoverie, the Master of which ship was *Jonas Poole:* who in the moneth of May fell with a Land, and called it *Greenland*, this is the Land that was discovered by Sir *Hugh Willoughby* long before [*Spetsberg* of the Hollanders], which Ship *Amitie* continued upon the coast of *Greenland*, discovering the Harbours and killing of Morces [the first killed by the English on Spitzbergen], untill the moneth of August, and so returned for *England*, having gotten about some twelve Tunnes of goods, and an Unicornes horne.

"In the yeere 1611. the Companie set foorth two Ships, the *Marie Margaret* Admirall, burthen one hundred and sixtie tunnes, *Thomas Edge* Commander; and the *Elizabeth*, burthen sixtie tunnes, *Jonas Poole* Master, well manned and furnished with all necessarie Provisions, they departed from Blackwall the twentieth of Aprill, and arrived at the Foreland in *Greenland* in the Latitude of 79. degrees, the twentieth of May following, the Admirall had in her six Biskayners expert men for the killing of the Whale: this was the first yeere the Companie set out for the killing of Whales in *Greenland*, and about the twelfth of June the Biskayners killed a small Whale, which yeelded twelve Tunnes of Oyle, being the first Oyle that ever was made in *Greenland*. The Companies two Shalops looking about the Harbour for Whales, about the five and twentieth of June rowing into Sir *Thomas Smith his* Bay, on the East side of the Sound saw on the shoare great store of Sea-horses: after they had found the Morses they presently rowed unto the ship, being in crosse Road seven leagues off, and acquainted the Captayne what they had found. The Captayne understanding of it, gave order to the Master, *Stephen Bennet*, that he should take into his Ship fiftie tunnes of emptie Caske, and set sayle with the Ship to goe into *Foule Sound*. The Captayne went presently away in one Shallop with sixe men unto the Seamorse, and tooke with him Lances, and comming to them they set on them and killed five hundred Morses, and kept one thousand Morses living on shoare, because it was not profitable to kill them all at one time. The next day the Ship being gone unto the place & well mored where the Morse were killed, all the men belonging to the Ship went on shoare, to worke and make Oyle of the Morses; and when they had wrought two or three dayes, it fortuned that a small quantitie of Ice come out of *Foule Sound*, and put the Ship from her Moring. . . . The Ship being cast away without hope of recoverie, the Commander *Thomas Edge* gave order, that all the Morse living on shoare shoold be let goe into the Sea, and so gave over making of Oyle. . . ." Fitting up their boats as well as they could they soon after abandoned the coast of Spitzbergen ("Greenland"), and set sail for Cherie Island, where they found the "Elizabeth" and returned to Spitzbergen "to take in such Goods as the sayd *Edge* had left in *Foule Sound*, woorth fifteene hundred pounds."*

As early as the year 1611, the previous persecutions of the Walruses at Cherie Island had made them very wary. Thomas Finch, in his account of a visit to this island by William Gourdon in August of that year, says: "At our comming to the Iland, wee had three or foure dayes together very fine weather: in which time came in reasonable store of Morses, . . . yet by no meanes would they go on those beaches and places, that formerly they have been killed on. But fortie or fiftie of them together, went into little holes within the Rocke, which were so little, steepe and slipperie, that as soone as wee did approach towards them, they would tumble all into the sea. The like whereof by the Masters and *William Gourdons* report, was never done."†

During the years 1612, 1613, 1614, and 1615, numerous vessels were sent out from England to Spitzbergen for the products of the Walruses and Whales, but generally met with indifferent success, being much troubled with Spanish, Dutch, and Danish "interlopers."

" In the yeere 1616, the Companie set out for Greenland eight Sayle of great ships, and two Pinnasses under the command of *Thomas Edge*, who following his course, arrived in *Greenland* about the fourth of June, having formerly appointed all his ships for their severall Harbours, for their making of their Voyage upon the Whale, and having in every Harbour a sufficient number of expert men, and all provisions fitting for such a Voyage. This yeere it pleased God to blesse them by their labours, that they full laded all their ships with Oyle, and left an overplus in the Countrey, which their ships could not take in. They imployed this yeere a small Pinnasse unto the East-ward part of *Greenland*, Namely, the Iland called now *Edges* Iland,

---

* Purchas his Pilgrimes, vol. iii, pp. 464, 465.   † Ibid., p. 536.

and other Ilands lying to the North-wards as farre as seventie
eight degrees, this Pinnasse was some twentie tunnes, and had
twelve men in her, who killed one thousand Sea-horses on
Edges Iland, and brought all their Teeth home for *London*."

In 1617, they "employed a ship of sixtie tunnes, with twenty
men in her, who discovered to the Eastward of *Greenland*, as
farre North-wards as seventy-nine degrees, and an Iland which
he named *Witches* Iland, and divers other Ilands as by the
Map appeareth, and killed store of Sea-horses there . . ."*

The Dutch, Danes, and Spaniards began, in 1612, also to
visit Spitzbergen in pursuit of Whales and Sea-horses, but are
reported by the English to have made indifferent voyages. The
company soon also had rivals in the "Hull-men," who, as well
as the Dutch, did them much "ill service."†

About the years 1611 and 1612, the Whale-fishery was found
to be more profitable than Walrus-hunting, and subsequently
became the main pursuit, not only by the English, but by the
Dutch and Danes. Yet the Walruses were by no means left
wholly unmolested, having been constantly hunted, with more
or less persistency, down to the present day, and, as already
shown, were long since exterminated from Cherie Island and
other smaller islands more to the northward, and greatly reduced in numbers on the shores of Spitzbergen.

Walruses have been recently reported as occurring on the
outer or northwestern coast of Nova Zembla, but as not existing on the inner or southeastern coast. Von Baer, on the authority of S. G. Gmelin and others, gave the eastern limit of the
distribution of the Atlantic Walruses as the mouth of the Jenesei River, though very rarely single individuals wandered as
far eastward as the Piasina River. He even regarded the Gulf
of Obi as almost beyond their true home.‡ Von Middendorff,
however, considers von Baer's eastern limit as incorrect, and
cites old Russian manuscript log-books ("handschriftliche
Schiffsbücher") in proof of their occurrence in numbers in August, 1736, as far east as the eastern Taimyr Peninsula, and of
their being met with in August, 1739, as far east as Chatanga
Bay. Still further eastward, in the vicinity of the mouth of the
Lena River, he gives similar authority for their occurrence in
August, 1735, and says that Dr. Figurin attests their presence

---

*Purchas his Pilgrimes, vol. iii, p. 467.   †Ibid., pp. 472, 473.
‡Mém. de l'Acad. des Sci. de St. Pétersb., vi° sér., Sci. math., phys. et nat.,
tome iv, 2ᵈᵉ pars, pp. 174, 184.

on the shores of the delta-islands of the Lena. Respecting the
more easterly coast of the Siberian Ice Sea, he says it is certainly known that the Walruses of Behring's Sea extend westward in great numbers to Koljutschin Island. Only the males,
however, reach this limit, the females not extending beyond the
vicinity of the mouth of the Kolyma River.[*]

It hence appears that about 1735 to 1739 Walruses were
met with as far eastward as the mouth of the Lena River; but
Wrangell, nearly a century later, explored quite thoroughly
this whole region without meeting with them, and I have found
only one reference to their existence on the Siberian coast between the Kolyma and Jenesei Rivers later than those cited by
von Middendorff.

According to a recent letter[†] from Professor Nordenskjöld, of
the Swedish Northeast Passage Expedition, "two Walruses"
were seen in August, 1878, a little to the eastward of the Jenesei
River, and that open water was found as far as the mouth of
the Lena. From this it would seem that there is nothing to
prevent, at least in favorable years, the Walruses from passing
eastward to the mouth of the Lena. There still remains, however, a breadth of some thirty degrees of longitude (between
130° and 160°) where as yet no Walruses have been seen. They
appear to have been only very rarely met with to the eastward
of the Jenesei (longitude 82° E.), and to be uncommon east of
the Gulf of Obi.

At present the Atlantic Walrus ranges along the northeastern coast of North America from Labrador northward to Repulse Bay and Prince Regent's Inlet, and along the shores of
Greenland; in the Old World only about the islands and in
the icy seas to the northward of Eastern Europe and the neighboring portions of Western Asia, where it rarely, if ever, now
visits the shores of the continent.

On the eastern coast of North America, Walruses have been
met with as far north as explorers have penetrated, and as far
as the Esquimaux live. They winter as far north as they can
find open water, retiring southward in autumn before the advance of the unbroken ice-sheet. Kane speaks of their remaining in Renssellaer Harbor (latitude 78° 37') in 1853, till the second week of September, when the temperature reached zero
of Fahrenheit.[‡]

---

[*] Von Middendorff's Sibirische Reise, Bd. iv, 1867, pp. 935, 936.
[†] See Nature, vol. xix, p. 102, December 5, 1878.
[‡] Arctic Exploration, vol. i, p. 140.

NOMENCLATURE.—Several specific names have been in more
or less current use for the Atlantic Walrus, or rather for the
Atlantic and Pacific species collectively. Accepting *Odobænus*
as the proper generic name of the group, there is nothing to
prevent the adoption of *rosmarus* for the specific name of the
Atlantic species. It was used for this species exclusively by
Linné, Erxleben, and other early systematic writers, the Pacific
Walrus being at that time unknown to the systematists. If *Rosmarus* be used as the generic name of the group, as it has been
by a few late writers, as a substitute for the wholly untenable
one of *Trichechus*, it will be, of course, necessary to adopt some
other name for the species. Dr. Gill has used *obesus* of Illiger ;
but as this was applied by Illiger exclusively to the Pacific Walrus, it cannot properly be used for the Atlantic species. It
would be difficult to select a subsequent name that would not
be open to objection, if one should stop short of *trichechus*, used
(inadvertently?) in a specific sense ("*Rosmarus trichechus*") by
Lamont in 1861. The name *longidens* of Fremery, 1831, was
based on what subsequent writers have considered as probably
the female, but the name is highly inappropriate, inasmuch as
it is the Pacific species, and not the Atlantic, that has the longer
tusks. There are left *virginianus* of DeKay and *dubius* of
Stannius : the first is objectionable on account of its geographical significance ; the other is only doubtfully referable to the
Atlantic species. Adopting *Odobænus* for the genus, leaves
rosmarus available for the species, thus settling the whole difficulty.

As already noticed (*anteà*, p. 20), two species besides *virginianus* have been based on fossil remains, and have been made
the basis of new genera. The first of these is the *Odobenotherium lartetianum* of Gratiolet, since referred by Defrance to the
existing species ; the other is the *Trichecodon huxleyi* of Lankester, which there is perhaps reason for regarding as the large
extinct progenitor of the existing Walruses.

ETYMOLOGY.—The term *rosmarus* was originally used by
Olaus Magnus, about the middle of the sixteenth century, in a
vernacular sense, interchangeably with *morsus*, the Latinized
form of the Russian word *morsz* (or *morss*). It was used in the
same way by Gesner a few years later, as well as by numerous
other pre-Linnæan authors. Respecting the etymology of the
word, von Baer gives the following : " In dem historisch-topo-

graphischen Werke: *De gentium septentrionalium conditionibus cet.* *Romae* 1555 heisst es: *Norvagium littus maximos ac grandes pisces elephantis habet, qui morsi seu rosmari vocantur, forsitan ob asperitate mordendi sic appellati,* (Eine recht witzige Etymologie!) *quia, si quem hominem in maris littore viderint apprehendereque poterint, in eum celerime insiliunt, ac dente lacerant et in momento interimunt.*" *

The same author also gives the following from Herberstain (1567): "Under andern ist auch ein thier, so grösse vvie ein ochs, und von den einwonern Mors oder der Tod geheissen wird." † Hence, either from superstitious notions of the terrible character of this animal, or from the resemblance of the Russian word *morss* to the Latin word *mors*, these terms became early confounded, and rendered by the German word *Tod*, or death. ‡

In the account of the exploits of the Norman Othere, where the Walrus first finds its place in literature, it is termed *Horsewael*. As noted by Martens § and other writers, equivalent words in other languages have become current for this animal, as *Walross* or *Wallross* of the Germans, *Wallrus* of the Dutch,

---

* In an early (1658) English version of Olaus Magnus's work ("A Compendious History of the Goths, Svvedes, & Vandals and other Northern Nations. Written by Olaus Magnus, Arch-Bishop of Upsal, and Metropolitan of Svveden", p. 231), this passage is rendered as follows: "The *Norway* Coast, toward the more Northern parts, hath huge great Fish as big as Elephants, which are called *Morsi*, or *Rosmari*, may be they are so from their sharp biting; for if they see any man on the Sea-shore, and can catch him, they come suddenly upon him, and rend him with their Teeth, that they will kill him in a trice." From this it would appear that *Morsus*, as used by Olaus Magnus, *might* be simply the Latin word *morsus*, from *mordere*, to bite.

† See von Baer, Mém. de Acad. des Sci. de St. Pétersb., viᵉ sér., Sc. math., phys. et nat., tome iv, 2ᵈᵉ pars, pp. 112, 113.

‡ Von Baer quotes a passage from the "Rerum Moscoviticarum auctores varii," originally published early in the sixteenth century, in which occurs the phrase "scandut ex mari pisces morss nuncupati," which he regards as the first introduction into Latin of the Sclavic name Моржь. In Western Europe it a little later became current in the form of *Mors*, which was soon written *Morss* or *Mors*, from which Buffon later formed the name *Morse*, which has since been the common appellation of this animal among French writers. Von Baer further observes that the accidental resemblance in sound of this word to that of the Latin word for death (*mors*) appears to have contributed not a little to the strange conception of the terribleness of this animal which was early entertained and even still prevails in Western Europe, although the Russian accounts do not speak of it.

§ Zoolog. Garten, Jahrg. xi, 1870, p. 283, where the etymology of the names of the Walrus is briefly discussed.

Misc. Pub. No. 12——6

and *Walrus* of the English. By the early Scandinavian writers
it was termed *Rosmhvalr*, which later became resolved into *Ros-
mul*, from which, perhaps, originated the Latin term *Rosmarus*,
which has the same significance, introduced by Olaus Magnus
and Gesner, and the Norwegian word *Rostungr*. Gesner and
several subsequent writers also used the word *Meerross*, and we
have in English the equivalent term *Sea-horse*, as one of the appellations of the Walrus, and also, but more rarely, *Meerpferd*
in German, and *Cheval marin* in French.

The current French term *Morse* appears, as already stated, to
have been introduced by Buffon as a modification of the Russian word *morss*, used by Michow (1517) and Herberstain (1549).
Among other old vernacular names we find in English *Sea
Cow*, in French *Vache marine*, in Latin *Bos marinus*, etc., while
by the early French settlers in America it was commonly termed
Bête à la grande dent.

LITERATURE.—1. *General History.*—Passing over the by some
supposed allusions to the Walrus by Pliny as too vague and
uncertain for positive identification, * we meet, according to
von Baer, with the first positive reference to the present
species in the account of the exploits of the famous Norman ex-
plorer Othere, or Octher, who, about the year 871 (890 according to some authorities), made a voyage to some point beyond
the North Cape, where he met with large herds of Walruses,
some of the tusks of which he is said to have taken to England
as a present to King Alfred. † Walruses appear to have been

---

* See K. E. von Baer, Mém. de l'Acad. Imp. des Sci. de St. Pétersb., vime
sér., Sci. math., phys. et nat., tome iv, 3me livr., 1836, (1837), pp. 101, 102. To
this admirable monograph I am greatly indebted for information respecting
the earlier publications bearing upon the history of the Walruses. To this
exhaustive memoir the reader is referred for a full exposition of this part of
the subject. The following short summary is based, so far as the early history of the subject is concerned, mainly upon von Baer's monograph, an
analysis of which will be presented at a subsequent page. (See *posteà*, p. 88,
footnote.)

† Hakluyt's rendering of this account is as follows: "The principall
purpose of his [Othere's] traveile this way, was to encrease the knowledge
and discoverie of these coasts and countreyes, for the more comoditie of fishing of horsewhales, which have in their teeth bones of great price and ex-
cellencie: whereof he brought some at his returne unto the king. Their
skinnes are also very good to make cables for shippes, and so used. This
kind of whale is much lesse in quantitie then other kindes, having not in
length above seven elles."—HAKLUYT'S *Voyages*, vol. i, p. 5.

an object of chase on the coast of Finmark as early as 980, and must have been met with by the Norsemen when they visited Greenland about the end of the tenth century. Their tusks were an article of commercial value among the Mongolian and Tartar tribes as early as the twelfth to the fifteenth centuries. Aside from the various notices by Scandinavian writers, the earliest unmistakable reference to the Walrus, other than that connected with Othere, as above mentioned, was, according to von Baer (l. c., p. 108), by Albertus Magnus, in the first half of the thirteenth century.

Says this writer (as quoted by von Baer), whose account is here paraphrased: The hairy Cetaceans have very long tusks, by which they suspend themselves to the rocks in order to sleep. Then comes the fisherman and separates near the tail as much skin as he can from the underlying fat, and then attaches a cord, which has at the other end a large ring, which he makes fast to a post or tree. Then when the fish awakens (by all of these operations he was not yet awakened), they cast a huge sling-stone upon his head. Being aroused, he attempts to get away, and is held by the tail near to the place and captured, either swimming in the water or half alive on the shore. This ludicrous description von Baer believes had for its foundation misunderstood reports of the Walruses' habit of reposing upon the shore or upon ice-bergs, the use of their tusks in climbing up to these places of rest, and their deep sleep, and that the account of the mode of capture was based on an incorrect knowledge of the use of the harpoon; and that the account shows that as early as the thirteenth century the Walrus was harpooned on the coast extending from the White Sea northwards. *

---

* This curious legend is quoted by Gesner in his Historia Animalia Aquatilia, 1558, p. 254. The following rendering appears also in the above-cited English version of Olaus Magnus: "Therefore, these Fish called *Rosmari*, or *Morsi*, have heads fashioned like to an Oxes, and a hairy Skin, and hair growing as thick as straw or corn-reeds, that lye loose very largely. They will raise themselves with their Teeth as by Ladders to the very tops of Rocks, that they may feed on the Dewio Grasse, or fresh Water, and role themselves in it, and then go to the Sea again, unless in the mean while they fall very fast asleep, and rest upon the Rocks, for then Fisher-men make all the haste they can, and begin at the Tail, and part the Skin from the Fat; and into this that is parted, they put most strong cords, and fasten them on the rugged Rocks, or Trees, that are near; then they throw stones at his head, out of a Sling, to raise him, and they compel him to descend, spoiled of the greatest part of his Skin which is fastened to the Ropes : he being thereby

The Walrus is also referred to by Hector Boëthius in 1526, in his History of Scotland;* by Herberstain (or Herberstein, as also written) in 1549; by Paré about the year 1600; and by Aldrovandus in 1642.

Herberstain also very correctly indicates the habits of these animals, which, he says, repair to the shore in large herds to repose, and that while the herd sleeps one of their number keeps watch. He compares their feet to those of the Beaver, and refers to the value of their tusks to the Russians, Turks, and Tartars, and observes that they called them fish-teeth. †

Even before the middle of the sixteenth century, Walruses had been met with on the eastern shore of North America. In May, 1534, they were seen by Cartier, and later in the same century by Fischer, Drake, and others, on the coast of Nova Scotia and adjacent islands, and later still by other explorers on the islands in the Gulf of Saint Lawrence (see anteà, p. 66), in the accounts of whose voyages‡ occur interesting notices of these animals.

In the year 1553, Edward VI of England sent an expedition under Willoughby and Chancellor to the White Sea, which resulted in still further increasing our knowledge of the Walruses, especially of their distribution eastward along the Arctic coast of Europe and Asia. Chancellor's short account§ refers especially to the uses made of the skins and tusks.

The earliest delineations of the Walrus appear to have been made by Olaus Magnus in his "Tabula Terrarum Septentrionalium" (1555), where he has portrayed many strange and fabulous animal forms which there is reason to believe were based upon this animal. ‖ Gesner a few years later (1558), in his "His-

---

debilitated, fearful, and half dead, he is made a rich prey, especially for his Teeth, that are very pretious amongst the *Scythians*, the *Moscovites*, *Russians*, and *Tartars*, (as Ivory amongst the *Indians*) by reason of its hardness, whiteness, and ponderousnesse. For which cause, by excellent industry of Artificers, they are made fit for handles for Javelins: And this is also testified by *Mechovita*, an Historian of *Poland*, in his double *Sarmatia*, and *Paulus Jovius* after him, relates it by the Relation of one *Demetrius*, that was sent from the great Duke of *Moscovy*, to Pope *Clement* the 7th."—Loc. cit., pp. 231, 232.

* "Scotorum Regni Descriptio, p. 90," as cited by various writers.
† Herberstain, as cited by von Baer, l. c., p. 111.
‡ See Hakluyt's Voyages, vol. iii, ed. 1810, pp. 237, 238, 242, 249, 254, etc.
§ See Hakluyt's Voyages, vol. i, ed. 1599, p. 237.
‖ Olaus Magnus's figures will be noticed later under the section devoted to the figures of the Walrus (posteà, p. 92 et seq.).

torium Animalium" (in the volume devoted to the "Animalia
Aquatilia"), faithfully copied all of Olaus Magnus's figures under
the heading "De Cetis," and then presents, under the name
Rosmarus, the figure of the Walrus from Olaus Magnus. This
figure, however, he judiciously criticises, stating that the tusks
should be in the upper jaw, and not in the lower, as they were
represented by Olaus Magnus. This last-named author, in the
later editions of his work "De Gentium Septentrionalium Conditionibus," etc. (as in that of 1563), rightly places, according to
von Baer, the tusks in the upper jaw. Gesner (continues von
Baer) knew only the first edition of this work, and took his
figure from the above-mentioned "Tabula Terrarum Septentrionalium." Also were unknown to him the accounts of the Walrus given by "Herberstain, Chancellor, and Othere," so that he
made extracts from only Michovius and Albertus Magnus. He
also knew no better than to offer, as a figure of the Walrus, a
drawing he had received from Strassburg, representing, pretty
fairly, the head and tusks, while the rest was purely a fabrica-
tion. Some rhymes, which he further inflicts upon his readers,
show clearly how "awful" the conceptions of the Walrus then
were (or, as von Baer puts it, "Wie schauerlich noch die Vor-
stellungen vom Wallrosse waren").*

In 1608, a young living Walrus was taken to England, having
been captured on Bear or Cherie Island off the coast of Norway,† while four years later (1612) another young Walrus, with
the stuffed skin of its mother, was taken to Holland. The first
appears to have been very intelligently described by Ælius Ever-
hard Vorstius, whose description is quoted by De Laet.‡ The
specimen taken to Holland was well figured by Hessel Gerard,
the young one doubtless from life, the figures being published
by him in 1613,§ and subsequently repeatedly copied (as will
be more fully noticed later).

In 1625, Purchas, in his history of the voyages of the English
to Cherie Island and Spitzbergen (then called "Greenland"),
gives much interesting information respecting the chase of the

* To show what these conceptions were, von Baer cites the passages
already quoted (*anteà*, p. 81), in reference to the singular misinterpretations
given in Western Europe to the Russian name *Morss*. See von Baer, l. c.,
p. 113.

† Recueil de Voy. au Nord, 2ᵉ éd., tome ii, p. 368.

‡ Nov. Orb. s. Descrip. Ind. Occ., 1633, p. 41.

§ See von Baer, l. c., p. 128; Gray, Proc. Zoöl. Soc. Lond., 1853, p. 115.

Walrus at these islands, and in one place a quaint description and some very curious figures of the animal.*

In 1675, the Walrus was again described and wretchedly figured by Martens, † who is said to have been the first "naturalist" who ever saw the Walrus in its native haunts. Zorgdrager, ‡ in 1720, supplied by far the fullest account of these animals, as observed by him in Spitzbergen, that had appeared up to that date. He gives not only a quite detailed and truthful account of their habits, especially under persecution, but also of their wholesale destruction at that early time in the Spitzbergen seas, and of their extermination at some of the points at which they had formerly been accustomed to land in immense herds. He also notes the increasing difficulties of their capture owing to the great shyness of man they had acquired in consequence of persecution, and describes the manner in which they were captured, and also their products. Copious extracts from Zorgdrager's account of the Walrus are given by Buffon (translated into French from a German edition), and he has also been extensively quoted by even much later writers.

The Greenland Walrus was described by Egede § in 1741, by Anderson ‖ in 1747, by Ellis ¶ in 1748, by Cranz ** in 1765, and by Fabricius †† in 1780, some of whom added much information respecting its habits and distribution, its usefulness to the natives and their ways of hunting it, as well as respecting its external characters.

The above-cited accounts of the Walrus formed the basis of numerous subsequent compilations, and most of those last given are cited by the early systematic writers, few of whom, as previously shown (see *anteà*, pp. 8–11), had any just appreciation of even its most obvious external characters. Linné, as already noted (*anteà*, p. 8), profited little by what had been written by preceding authors, while Brisson, Erxleben, and Gmelin manifest a scarcely better acquaintance with this badly misrepresented and poorly understood creature. No little confusion has hence arisen in systematic works respecting its posi-

---

* See *anteà*, p. 74–78, and *posteà*.

† Spitzbergen, pp. 78–83, pl. P, fig. *b*.

‡ Bloeyende Opkomst der Aloude en Hedendaagsche Groenlandsche Visschery, etc., ed. 1720, pp. 165–172.

§ Det gamle Grønlands nye Perlustration, etc., 1741, p. 45.

‖ Nachrichten von Island, Grönland und der Strasse Davis, p. 258.

¶ Voyage to Hudson's Bay, p. 134.

** Historie von Grönland, pp. 165, 167.

†† Fauna Groenl., p. 4.

tion and affinities (see *anteà*, pp. 7–12). The accounts by Houttuyn, Buffon, Pennant, P. S. L. Müller, and Schreber are excellent for their time. These authors all recognized the close relationship of the Walrus to the Seals, and quite correctly indicated its external characters and habits. Some of these accounts, however, include references to both species.

Daubenton, in Buffon's "Histoire Naturelle,"* gave a description and figure of a Walrus's skull, and made the first contribution to our knowledge of its internal anatomy, based on the dissection of a fœtal specimen.

Since the beginning of the present century, the Walrus has been the subject of almost numberless notices, as well as of several elaborate papers, devoted in most cases to special points in its anatomy, very few of which need be here enumerated. †

The elder Cuvier, beginning with his "Leçons d'Anatomie comparée" (1800–1805), and ending with the third edition of his "Ossemens fossiles" (1825), contributed considerably to our general knowledge of its structure and affinities, especially of its osteology; he in 1825 ‡ first figuring and describing its skeleton. A paper by Sir Everard Home,§ in 1824, figured and described the stomach and feet from specimens taken to England from Hudson's Bay, preserved in salt. This paper is noteworthy mainly on account of the singularly erroneous interpretation there made of the structure and functions of the feet, Home supposing that these organs were provided with sucking discs, by means of which the creature was enabled to adhere firmly to the ice in climbing. The skeleton of the Walrus was again figured and described by Pander and d'Alton || in 1826, and still later by Blainville ¶ about 1840. Von Baer, ** in 1835, published some account of the arterial system of the Walrus, based on a dissection of a young specimen. Its general anatomy, especially its limb-structure, myology, vascular and respiratory systems, viscera and generative organs, and external cha-

---

* Tome xiii, 1765; pp. 415–424, pll. liv, lv. The skull had been previously figured by Houttuyn (in 1761), as will be noticed later.

| Those relating to its dentition have been already noticed in detail (see *anteà*, pp. 47–57); several others have also been specially referred to, and nearly all are cited in the references given at pp. 23–26.

‡ Ossem. Foss., 3e éd., tome v, iime pt., pp. 521–523, pl. xxxiii.

§ Phil. Trans., 1824, pp. 233–241, pl. iv.

|| Skelete der Robben und Lamantine, pll. i, ii.

¶ Osteographie, Des Phoques, pll. i and iv.

** Mém. de l'Acad. St. Pétersb., vime sér., Sci. math., phys. et nat., tome iime, 1835, pp. 199–212.

racters, were quite fully and satisfaetorily treated by Dr. J. Murie* in 1872.

Illiger, in 1811, in a paper on the geographical distribution of the mammals of the Northern Hemisphere (see *anteà*, p. 18), first nominally recognized the Pacific Walrus as a species distinct from the Atlantic animal, while Fremery, in 1831, recognized three species, and Stannius, in 1842, admitted two,† but, as already noticed, only one species of Walrus has been commonly recognized. The matter of variation dependent upon sex, age, and individual peculiarities, has received, as already noticed (see *anteà*, pp. 38–43), special attention at the hands of Wiegmann, Stannius, Jaeger, and other writers.

Unquestionably, the most important paper relating to the literature, geographical distribution, and habits of the Walruses is the well-known and justly celebrated memoir by von Baer,‡

---

* Trans. Lond. Zoöl. Soc., vol. vii, pp. 411–462, 8 woodcuts, and pll. li–lv.

† For a notice of the literature of this part of the subject see *anteà*, pp. 17–23.

‡ Anatomische und zoologische Untersuchungen über das Wallross (*Trichechus rosmarus*) und Vergleichung dieses Thiers mit andern See-Säugethieren. Von Dr. K. E. v. Baer. Gelesen den 6. Nov. 1835. <Mém. de l'Acad. Impér. des Sciences de Saint-Pétersbourg, vime sér., Sc. math., phys. et nat., tome ivme, pp. 96–236. [Mit einer Tafel.] Publié par ordre de l'Académie. En Février, 1837.

The paper has the following contents:

I. Zoologische Abtheilung.

Cap. I. Veranlassung und Inhalt dieser Untersuchungen (pp. 97–100). § 1. Veranlassung. § 2. Anatomische Untersuchung. § 3. Zoologische Nachforschungen. § 4. Alter des lebend beobachteten Thiers.

Cap. II. Geschichte der Kenntniss des Wallrosses und kritische Musterung der bisher gelieferten Abbildungen (pp. 100–130). § 1. Urzeit und Alterthum. § 2. Mittelalter. § 3. Vom Schlusse des fünfzehnten Jahrhunderts bis auf Linné und Buffon. § 4. Von Linné und Buffon bis jetzt. § 5. Uebersicht der bisher gelieferten Abbildungen vom Wallrosse.

Cap. III. Beobachtungen an dem lebenden Thiere (pp. 130–148). § 1. Frühere Fälle von der Anwesenheit lebender Wallrosse in mittleren Breiten. § 2. Allgemeines Ansehen des Thiers. § 3. Der Kopf. § 4. Die Bewegungen. § 5. Blasen oder Ausspritzen von Wasser. § 5 [*bis*]. Wartung des jungen Wallrosses. § 6. Geistiges Naturel des Thiers. § 7. Bildsamkeit und Anhänglichkeit.

Cap. IV. Allgemeine Betrachtungen über die Bildsamkeit der See-Säugethiere und über die Anhänglichkeit der Individuen Einer Art unter einander (pp. 148–171). § 1. Aufgabe. § 2. Gezähmte Wallrosse. § 3. Gezähmte Robben. § 4. Wahre Cetaceen. § 5. Gesellschaftliches Leben. § 6. Liebe der Aeltern zu den Jungen und der Jungen gegen die Aeltern. § 7. Gatten-Liebe. § 8. Allgemeine Begründung dieser Verhältnisse.

Cap. V. Verbreitung der Wallrosse (pp. 172–204). § 1. Sie wohnen in zwei getrennten Verbreitungs-Bezirken. § 2. Oestlicher Verbreitungs-Bezirk.

published in 1837. This elaborate memoir, so often already cited in the present article, gives a general summary of nearly all papers, references, and figures relating to the Walruses that appeared prior to 1835, the date of its presentation to the Imperial Academy of Sciences of Saint Petersburg for publication. It also contains many original biological and anatomical observations, based on a young living specimen brought to Saint Petersburg in 1828, which, surviving for only a week after its arrival, soon fell into his hands for dissection. *

Von Baer, after a few preliminary remarks respecting the occasion and objects of his paper, and a few words on the anatomy of the Walrus, devotes some thirty pages to a critical and exhaustive historical *résumé* of the literature relating to the general subject. Then follow some eighteen pages detailing his observations on the living animal, in which he gives some account of the few young individuals that had, up to that time, been taken alive to Middle Europe; also a detailed account of the external appearance of the specimen he had examined in life. He notes especially its attitudes, movements, and limb-structure, and compares it in these points with the Seals. After describing the position and character of the limbs in the Seals, and the restriction of their movements on land to a wriggling movement, with the belly lying on the ground, he refers to the freer use of the extremities possessed by the Walrus, which he found was able to truly stand upon its four feet, and says that,

* He seems, however, to have never published in full the results of his observations upon its anatomy, he apparently reserving the anatomical part of his memoir in the hope of perfecting it through the study of additional material.

in respect to the use of its limbs, it occupies an intermediate place between the Pinnipeds and the ordinary four-footed Mammals, among which latter its less pliant feet give it the appearance of a cripple. If we should call, he says, the Seal a crawler or slider, we should have to term the Walrus a waddler, since in walking it throws its plump body to the right and left. Here we have fairly described, for the first time, the flexibility of the extremities,—the bending of the hind feet sometimes forward, sometimes backward, and the free turning of the fore feet,— although an allusion was made to this by Vorstius * two centuries before, yet the fact of flexibility remained generally unrecognized till 1853, when a young living specimen reached London. Von Baer points out the fallacy of Sir Everard Home's notion that the feet of the Walrus are provided with suction-discs, and the "blowing" of the Walrus mentioned by Martens, who described it as throwing water from its nostrils like a whale.

Following this chapter on its external features, movements, temperament, behavior, etc., is an interesting dissertation of some twenty or more pages on the domesticability of the marine mammals in general, which is devoted largely to a history of the behavior of the Seals in captivity, with a short notice of the different examples of the Walrus, the Sirenians, and the smaller Cetaceans that had been observed in confinement. The next thirty pages are given to a discussion of the geographical distribution of the Walruses, the treatment of which subject is marked by the same pains-taking research that characterizes. the other parts of this learned monograph. He shows that Walruses are confined to two widely separated habitats, and not, as previously supposed, found all along the Arctic coasts. He describes them as limited to two regions, an eastern and a western, the first including the northwestern coast of North America from the Peninsula of Aliaska northward, and the corresponding parts of the neighboring Asiatic coast. To the eastward he could trace them only to the vicinity of Point Barrow, and to the westward only to a few degrees beyond East Cape.

The western region, he affirmed, embraces only the Arctic coast of Europe eastward to the mouth of the Jenesei River,

---

* "Pedes anteriores antrorsum, posteriores retrorsum spectabant cum ingrederetur," says Vorstius as quoted by De Laët (see *anteà*, p. 37). The hind feet are also represented as turned forward in Hessel Gerard's figure, published in 1613 (see *posteà*).

and, on the other side of the Atlantic, the shores of Greenland and Arctic America westward to the western shores of Hudson's Bay and Fox Channel. There is thus left between these two regions nearly the whole of the coast of Asia bordering on the Polar Sea on the one hand, and almost the whole of the coast of North America formed by the Arctic Sea on the other.

In the later portion of his chapter on the distribution of the Walruses he devotes a few pages to a consideration of their migrations, and the physical causes which limit their distribution. Their migrations, he believes, are very imperfectly known, but he inclines to the opinion that they only periodically visited such points in their former range as Sable Island and other southerly lying islands. The causes which limit their range he considers to be mainly temperature, since he finds the southern boundary of their distribution is deflected northward and southward in accordance with the curves of isothermal lines.

The former range of the Walruses is also considered at length, to which subject are devoted nearly twenty-five pages. A short account is given of their reproduction and food, the paper closing with an inquiry into their systematic relationship to other animals. The map accompanying his memoir shows not only the distribution of the Walruses as at that time known, but indicates also the region over which they are known to have formerly occurred, and also the habitat of the *Rhytina*, or Sea-cow of Steller.

The reception in London, in 1853, of a young living Walrus gave rise to a paper by Owen[*] on its anatomy and dentition, and another by Gray,[†] " On the Attitudes and Figures of the Morse." A short paper was contributed by Sundevall[‡] in 1859 on its general history.

Leidy, in 1860, published an important paper on the fossil remains of Walruses found on the eastern coast of the United States, while Gratiolet, Defrance, Lankester, and Van Beneden have also written about those that have been met with in France, England, and Belgium.[§]

Malmgren, in 1864, in a paper on the Mammalian Fauna of

---

[*] Proc. Zoöl. Soc. Lond., 1853, pp. 103–106.

[†] Ibid., pp. 112–116, figs. 1–10.

[‡] Om Walrossen, Ofversigt K. Vet. Akad. Forh. (Stockh.), xvi, 1859, pp. 441–447; also translated in Zeitschr. gesammt. Naturw. Halle, xv, 1860, pp. 270–275.

[§] See *anteà*, pp. 61–65.

Finmark and Spitzbergen,* published many interesting notes relating to its habits and food, and later a special paper on its dentition (noticed *anteà*, p. 54.) Malmgren's observations on their habits, distribution, etc., also appear in the history of the Swedish Expedition to Spitzbergen and Bear Island in the year 1861,† together with a somewhat detailed and very interesting general history of the animal, with several illustrations.

Brown, in 1868, in his "Notes on the History and Geographical Relations of thePinnipedia frequenting the Spitzbergen and Greenland Seas,"‡ devotes several pages to the Walruses (pp. 427–435), in which he considers especially their habits and food, geographical distribution, and economic value.

In addition to the special papers cited in the foregoing pages, their general history has been more or less fully presented in several general works treating of the mammalia, and in several faunal publications.§ Much information respecting their general history may also be found in the narratives of various Arctic explorers, as Parry, Wrangell, Keilhau, Kane, Hayes, Lamont, and others, whose contributions will be more fully noticed in the following pages relating to the habits of the Walruses.

2. *Figures.*—As von Baer facetiously remarks, no animal has had the honor of being depicted in such strange and widely diverse representations as the Walrus. These, as has been previously stated, began with Olaus Magnus, about the middle of the sixteenth century, who opened the series with half a dozen phantastic figures, based apparently upon this animal, only one of which, however, bore the name *Rosmarus* (*Rosma-*

---

* Iakttagelser och anteckningar till Finmarkens och Spetsbergens Dägg-djursfauna. Öfversigt af Kongl. Vetensk.-Akad. Förhandl. 1863, (1864), pp. 127–155. [Walruses noticed, pp. 130–134.] Also republished in German in Wiegmann's Arch. für Naturgesch., 1864, pp. 63–97.

† I have seen only Passarge's German translation, entitled "Die schwedischen Expeditionen nach Spitzbergen und Bären-Eiland ausgeführt in den Jahren 1861, 1864, und 1868, unter Leitung von O. Torell und A. E. Nordenskiöld. Aus dem Schwedischen übersetzt von L. Passarge. Nebst 9 grossen Ansichten in Tondruck, 27 Illustrationen in Holzschnitt und einer Karte von Spitzbergen in Farbendruck. Jena, Hermann Costenoble, 1869." See pp. 131–143 (general history), 147, 151, etc.

‡ Proc. Zoöl. Soc. Lond., 1868, pp. 405–440.

§ See, among others, Macgillivray, British Quad., 1838, pp. 219–224; Hamilton, Amphib. Carnivora, etc., 1839, pp. 103–123; Nilsson, Skand. Faun., i, 1847, pp. 318–325; Giebel, Säugethiere, 1855, pp. 127–129; Lilljeborg, Fauna Sveriges och Norges Däggdj., 1874, pp. 654–667; etc., etc.

rus seu Morsus Norvegicus).* This figure † represents an animal standing half erect, resting against a rock, having four feet,

Fig. 4.—"*Rosmarus seu Morsus Norvegicus.* Olaus Magnus, 1568, p. 789."

a long, thick, cylindrical tail, terminating abruptly in an irregular expansion, with a low dorsal spiny crest, and two rather long porcine tusks projecting downward from about the middle of the mouth. Another,‡ a prone figure, called *Porcus mon-*

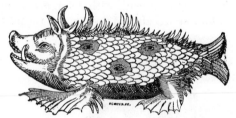

Fig. 5.—"*Porcus monstrosus Oceani Germanici.* Olaus Magnus, 1568, p. 788."

strosus Oceani Germanici, has a thick, short body, a fish-like tail, and a swine's head, ears, and tusks; the latter placed only in the lower jaw and directed upward. Behind the prominent pointed ears are two horns. The body is covered with heavy scales, among which are placed three eyes. The back is crested with large, somewhat recurved, spines of irregular size, and the feet are webbed and fin-like, especially the anterior. Another, called *Vacca marina*, represents the head of an ox, with a long beard on the chin. A fourth represents a dolphin-like body, with four feet, a fish's tail, a pair of long, ascending,

*My remarks respecting Olaus Magnus's figures are based on Gesner's (Hist. Animal. Aquat., 1558, pp. 247–249), and Gray's copies of them (Proc. Zoöl. Soc. Lond., 1853, p. 113), Olaus Magnus's work not being accessible to me. The figures herewith given (Figs. 4–12) are from electros of Gray's figures.

† See Fig. 4, copied by Gray from Olaus Magnus.

‡ See Fig. 5, copied by Gray from Olaus Magnus.

curved tusks near the posterior angle of the mouth, and long spines from the chin, throat, head, and back.*   A fifth † is a

FIG. 6.—"*Rosmarus.* Gesner, Addenda, 368, 16, 1560. (Reduced one-ninth.)"

creature having a swinish head, with long, ascending tusks in the lower jaw, four short, clawed feet, and a rather long, cylin-

FIG. 7.—"*Vacca marina.* Gesner, Addenda, p. 369, 1560."

drical body wrapped in armor!   Another is a monstrous sea animal, with a circle of long spines around the head, and tufts of spines from the nostrils and chin, four feet, the anterior only with claws, a forked tail with the points laterally recurved, and

* See Fig. 6, copied by Gray from Gesner.
† See Fig. 7, copied by Gray from Gesner.

two great tusks in the upper jaw, but no other resemblance to the Walrus.*

The figures published by Olaus Magnus were, according to von Baer, all faithfully copied by Gesner,† who added to them

FIG. 8.—"*Rosmarus.* Gesner, Icones Animalium, 1560, p. 178. De Cetis, Ord. xii. (Reduced two-thirds.)"

another, which he received from Strassburg. This‡ represents, as von Baer terms it, the "morphological paradox" of a vertebrate with two pairs of feet, a pair of wings or floats ("Flossen"), and a fish's tail. The head has considerable resemblance to that of a Walrus, with the large tusks properly situated in the upper jaw, and the eyes and nose are passably represented. The feet are all directed backward, in a swimming posture, and armed with strong claws. The Seal-like body has engrafted upon it the tail of a fish, while at the shoulder is seen a sort of wing-like appendage. The figure of the head is said to have been drawn in Strassburg from an actual specimen afterward sent by the bishop of Drontheim to the Pope, but to this was added a wholly imaginary figure of the body. Gesner's figure was several times copied, among others by Ambrosinus in 1642, in his Addenda to Aldrovandus's work,§ and also by Jonston‖ in 1657.

In 1598, De Veer, in a work entitled "The Navigation into the North-Seas,"¶ gave an illustration entitled "The Portraiture of our boats and how we nearly got into difficulty with the Seahorses." In this picture are depicted several Seal-like ani-

---

* This figure is not included in Gray's series.

† Icones Animalium, 1553, and Historia Animalium, 1558.

‡ See Fig. 8, copied by Gray from Gesner.

§ "Paralipomena, etc., adnexa ad Aldrovandi Historiam Monstrorum, p. 106," according to von Baer.

‖ De Piscibus, pl. xliv.

¶ Amsterdam, 1598; translated and republished in London in 1609, and reprinted from the London edition by the Hakluyt Society in 1853, the last-named being the edition here quoted.

mals standing on the ice, with long tusks and an arched body, supported on a decurvate bifurcate tail and the fore limbs, while the heads of several others are seen in the water. They are represented as having distinct pointed ears and no hind feet, unless the tail-like ending of the body may be supposed to represent the hind limbs.* An explanation of the inaccuracy of

FIG. 9.—*"Sea Horse*, 1609."

these figures is evidently afforded by the context (pp. 218–219 of the Hakluyt edition), in which we find the following: "And passing along by it ["Admiralty Island"], we saw about two hundred seahorses lying upon a flake of ice, and we sailed close by them and draue them from there, which had almost cast us doun; for they being mighty strong fishes [Zee-monsters, the editor says is the term used in the original Dutch], and of great force swam towards us (as if they would revenge on us for the despight that we had done them) round about our scuts [boats] with a great noyse, as if they would have devoured us; but we escaped from them by reason of a good gale of wind; yet it was not wise of us to wake sleeping wolves."

FIG. 10.—*" Walruss.* Ad vivum delineatum ab Hesselo G. A. 1613. (Reduced four-sevenths.)"

In 1613 a very correct and in many ways admirable representation of the Walrus was published by Hessel Gerard† (or

---

* One of these figures has been copied by Gray (Proc. Zoöl. Soc. London, 1853, p. 114, fig. 6), but omitting the ears and somewhat reduced in size. Gray's figure is here reproduced (see Fig. 9).

† "Histoire de Spitsberghe," as cited by Gray. Blumenbach and von Baer cite doubtfully "Descriptio ac delineatio geographica detectionis freti, s. transitus ad occasum supra terras Americanas in China atque Japonem ducturi, etc." Von Hessel Gerard. Amsterdam, 1613. 4°.

Gerrard, as also written), drawn from life from a young animal, which, with the stuffed skin of its mother, arrived in Holland in 1612. This representation consists of two figures, one of a full-grown animal, the other of a young one a few months old.* The hind portion of the larger animal is partly hidden by the figure of the smaller one. The general form of the body, the tusks, and extremities are all faithfully portrayed, the hind limbs being turned forward in their natural position,—the first figure, and the only one for the next two hundred and fifty years, in which the hind limbs are placed in a natural position. This figure has been many times copied,—first by De Laët† in 1633, and subsequently from De Laët, by Wormius‡ in 1665, by Jonston,§ Shaw,|| Schinz,¶ Gray,** and doubtless by many others. Most of the early authors, as Wormius, Jonston, and others, copied, not directly from Gerard, but from De Laët, while Shaw copied from Jonston, and Schinz from Blumenbach, in several cases these second and third hand representations doing great injustice to Gerard's original figure. Blumenbach,†† through the kindness of his friend Forster, was enabled to take his from Gerard's original imprint, and it is a much finer illustration than that afforded by De Laët, the one usually copied. Von Baer‡‡ also refers to a colored copy of Gerard's figure, which he obtained, with a collection of natural-history illustrations, from a bookseller in Leipsic, in which the coloring was truthfully executed, agreeing closely with the color of the young animal he saw alive in St. Petersburg.§§ Gerard's often-copied drawing

---

* See Fig. 10, copied by Gray, and here reproduced.
† Novus Orbis, seu Descrip. Ind. Occident., 1633, p. 38.
‡ Mus. Worm., p. 289.
§ De Piscibus et Cetis, 1649, pl. xliv (also in subsequent editions).
|| General Zoöl., i, 1800, pl. lxviii*.
¶ Naturgesch. und Abbild. der Säuget., pl. lxv, lower figure.
** Proc. Zoöl. Soc. Lond., 1853, p. 115, fig. 9. Gray's figure is here given (see Fig. 10).
†† Abbild. naturhist. Gegenstände, 1796–1810, No. 15 (plate and text).
‡‡ Loc. cit., p. 129.
§§ Von Baer's account of this important early figure is as follows: "Diese vortreffliche Zeichnung wurde in Kupfer gestochen und einigen Exemplaren von dem Abdrucke der *Descriptio ac delineatio geographica detectionis freti, s. transitus ad occasum supra terras Americanas in Chinam atque Japonem ducturi* etc., der von Hessel Gerard in Amsterdam 1613, 4°. besorgt ist, beigegeben. In diesem Buche wurde der Originalkupferstich von Forster gefunden und Blumenbach mitgetheilt. Da er sich, wie Blumenbach sagt, in keiner andern Ausgabe desselben Werkes und auch in dieser nur in den wenigsten Exemplaren findet, so ist wohl wahrscheinlich, dass er gar nicht zu dem

was well worthy of repetition, being incomparably better than
any other made prior to those taken from the living specimen
received at the Zoölogical Gardens in London in 1853.

Purchas, in his "Pilgrimes,"* gives some very interesting,
and in many respects excellent, representations of the Walrus,
to which I find no reference in the writings of von Baer or Gray,
or, in fact, anywhere. In the principal of these figures, the
general form of the Walrus is more correctly delineated than
in any figure, except Gerard's, that appeared prior to 1857.
Barring its facial expression, and the presence of what seems
to be a mane, it is excellent. The general outline of both the
body and the limbs are surprisingly truthful, as is likewise the
attitude. The hind feet are turned forward, and the size and
position of the tusks are correctly represented. The face, however, has a most ludicrous half-leonine, half-human expression,
which is heightened by the addition of an ear having the general form of a human ear. In addition to this, the creature is

---

Werke gehört und nur von dem Herausgeber oder von den Käufern einigen
Exemplaren beigebunden ist. Ich habe nicht Gelegenheit gehabt, das hier
genannte Werke zu sehen und darnach zu bestimmen, ob das Kupfer zu dem
Buche gehört, vermuthe aber eines Theils aus der angegebenen Seltenheit
seines Vorkommens und andern Theils aus dem Umstande, dass die Figur
in mehreren Werken des 17. Jahrhunderts wiederholt wurde, dass sie damals
bekannter war, als im 18ten. Ja, ich besitze selbst ein colorirtes Blatt, das ich
in einer Sammlung naturhistorischer Abbildungen in Leipzig aus dem Nach-
lasse eines Naturalienhändlers kaufte und welches, zwar nicht der Original-
Kupferstich, doch eine Copie desselben ist. Die Farbe, welche beide Thiere
auf meinem Blatte haben, ist ganz übereinstimmend mit der Farbe des
jungen Wallrosses, das hier zu sehen war. Da nun die erwachsenen Wallrosse in der Regel heller sind, so ist es mir wahrscheinlich, dass auch die
Colorirung damals nach dem jungen Thiere gemacht ist."—VON BAER, *l. c.*,
pp. 128, 129.

Gray says: "In a small quarto tract, called the 'Histoire du Pays nommé
Spitsberghe, écrit par H. G. A., Amsterdam, chez Hessel Gerrard A.', 1613, a
plate at page 20 contains an excellent figure of the Morse and its young, 'ad
vivum delineatum ab Hesselo G. A.' This figure was repeated in De Laët's
'Amer. Descript.', p. 28, 1633, by Jonston, 'Pisces', t. 44, in 1657, and by
Shaw, 'Zoölogy', t. 68*, from Jonston." Gray copies this figure with the
following title: "Fig. 9. *Walruss*. Ad vivum delineatum ab Hesselo G. A.
Histoire de Spitsberghe, by H. G. A., 1613. Another edition, same date.
(Reduced four-sevenths.)"—*Proc. Zoöl. Soc. Lond.*, 1853, p. 115.

It would thus appear that either Gerard's figure was published simultaneously in several different works, since that mentioned by Gray is not the
one cited by von Baer, or else that, as von Baer suspected, and as seems
more probable, the plate did not really belong to the work von Baer cites,
but merely happened to be bound with it.

* Vol. iii, pp. 472-473 bis.

represented as having a heavy mane, extending from the head to the middle of the back. The figure bears the quaint legend, "*The Seamorce is in quantity as bigg as an oxe.*" Another illustration on the opposite page shows "The manner of killing ye Seamorces," and represents a small herd of Walruses attacked by a party of hunters armed with lances. The Walruses are all headed toward the water, the men being between them and the sea. The Walruses are depicted in the attitude of walking, all having the hind feet turned forward, these figures giving apparently a correct idea of the Walrus's manner of progression on land. These two illustrations form part of a series that embellish a map of "Greenland" (Spitzbergen), the others representing different scenes in Whale-fishing and "the manner of killing Beares."

Zorgdrager,* in 1720, gave a figure of a Walrus which has a Seal-like head with two long tusks in the upper jaw, and the general body-contour of a Walrus. The posterior third of the body and hind limbs are fortunately, to judge by the rest of the figure, left to the imagination, being hidden behind the figure of a Seal ("Zee Rob"); the fore limbs bear no resemblance to those of a Walrus.

In 1741, Egede† gave a Seal-like figure of a Walrus, with its calf, confronting a Polar Bear. The open mouth displays a series of long sharp teeth, looking even less like Walrus tusks than the general form of the animal does like the outline of a Walrus. This figure of the Walrus is surprisingly poor, considering the excellent description Egede gives of the animal.

In 1748, Ellis‡ further enriched the iconographic literature of the Walrus by furnishing a figure, respecting which he says: "I shall not detain the Reader with an Account of a Creature ["Sea Horse"] so often described, but refer him to the Cut, in which he will find it very truly represented."§ The figure, however, is one of the worst imaginable, considering the opportunity Ellis evidently had for observation. In some respects it bears some resemblance to that given by De Veer. Ellis's figure combines a Lynx-like face with Lion-like fore limbs, short,

---

* Bloeyende Opkomft der Aloude en Hedendaagsche Groenlandsche Visschery, 1720, plate facing p. 162, upper left-hand figure.

† Beschreibung und Naturgeschichte von Grönland, 1763, p. 106, pl. vi, lower left-hand figure, Krunitz's German translation. The work appeared in Danish as early as 1741.

‡ Voyage to Hudson Bay, pl. facing p. 134, middle figure.

§ Loc. cit., p. 236.

round, prominent ears, small, pointed, inward-curving tusks, no hind feet, and a body tapering to a doubly emarginate fish-like tail, possibly intended to represent hind limbs.

Pontoppidon, in his Natural History of Norway, published in 1751, gave a figure of the Walrus in which the resemblance consisted mainly in the presence of two huge tusks in the upper jaw. Only the head, neck, and upper portion of the body are represented; but the general outline, as far as seen, is suggestive of the animal it was intended to represent.

Houttuyn,* in 1761, gave a very fair figure of the skull and os penis of a Walrus. As P. L. S. Müller, in 1773, used Houttuyn's plates in his "Natursystem," these figures are there again called into service, to which was added a noteworthy representation of the animal.† This represents an apparently young Walrus as lying partly on the side, with the diminutive hind feet

FIG. 11.—" *Wall-Ross*, Marten's Spitzbergen, &c. 1675, t. P, fig. *b*.   (Reduced three-tenths.)"

turned forward. The general outline of the body indicates the obese form of the Walrus; but the head, with its small, short tusks, has scarcely the faintest resemblance to the head of that animal.

---

* Natuurlyke Historie of uitvoerige Beschryving der Dieren, Planten en Mineraalen, volgens het Samenstel van den Heer Linnaeus. Met naawkeurige Afbeeldingen. Eerste Deels, tweede Stuk, 1761, pl. xi, figg. 1, 4.

† Des Ritters Carl von Linné Königlich Schwedischen Leibarztes, &c. &c. vollständiges Natursystem nach der zwölften lateinischen Ausgabe und nach Anleitung des holländischen Houttuynischen Werks mit einer ausführlichen Erklärung ausgefertiget von Philipp Ludwig Statius Müller, etc. Erster Theil. Nürnberg, 1773. Pl. xxix, fig. 2. This is one of the few original plates added by Müller to Houttuyn's series

In 1765, a most wretched and ludicrous caricature of the Walrus was contributed by Martens.* In this figure, the much-abused Walrus is represented as having an enormously large and shapeless head, in which the small tusks are set widely apart; it has small Seal-like fore feet, and no hind limbs, or, if present, they are directed backward, and look more like a fish's tail than distinct limbs. The tusks alone give the figure any suggestion of what it was intended to represent.

The next figure of which I have knowledge was published by Buffon,† also in 1765, and soon after copied by Schreber.‡ This

FIG. 12.—"*Le Morse,* Buffon, xiii, t. 548, 1765. (Reduced two-fifths.)"

was evidently drawn from a stuffed specimen, to which the taxidermist had given the attitude and general form of a common Seal. In 1827, a very fair figure of the head (the animal being supposed to be in the water, with only the head visible) was published in Griffith's Animal Kingdom (vol. ii, pl. v), which was later repeated by Hamilton,§ and also elsewhere. In 1836, a very fair, colored figure (evidently from a stuffed specimen), barring the posterior direction of the hind limbs, appeared in the "Disciples edition"‖ of Cuvier's Règne Animal, copied from Pal-

* Spitzbergische Reisebeschreibung, pl. P, fig. *b.* This fig. is also reproduced by Gray (l. c., fig. 7), and is here copied as Fig. 11.

† Histoire Naturelle, t. xiii, pl. liv.

‡ Säuget., pl. lxxix.

§ Amphibious Carnivora, p. 106, in Jard. Nat. Library, Mam., vol. viii.

‖ Le Règne Animal, etc., par Georges Cuvier. "Edition accompagnée des planches gravées, ... par une réunion de disciples de Cuvier," etc. Paris, 1836 et seq.

The Walrus is figured in "Mammifères," pl. xliv. The history of the figure is given as follows: "Figure dessinée d'après celle qu'a donnée Pallas dans la *Zoographia Rosso -Asiatica,* et réformée, pour le pose, d'après un croquis inédit de Choris; au vingtième environ de la grandeur naturelle."

The only copy of Pallas's "Icones" accessible to me is imperfect, and has not the figure here copied. There is, however, a quite different one, which will be noticed later in another connection. Whether Pallas's figure here copied represents the Atlantic or the Pacific species cannot well be determined.

las. Another much like it was published soon after in Macgillivray's British Quadrupeds,* and still another, also quite similar, in Hamilton's Marine Carnivora.† The vignette-titlepage of the last-named work also represents a "Walrus hunt," in which a boat's crew are depicted as attacking a group of five old Walruses. The plate in Hamilton's "Amphibious Carnivora" purports to have been drawn from a specimen in the Edinburgh Royal Museum, and seems to be essentially the same as that in Macgillivray's British Quadrupeds, with a somewhat altered position and different background. In each of these plates are represented two other distant figures of the Walrus. In each, the tusks are long, and seem to represent the Pacific rather than the Atlantic species, as is also the case in the "Disciples edition" of the Règne Animal. In all these last-named figures, the hind limbs are directed posteriorly, but in other respects they are fair representations.

Dr. Kane,‡ in 1856, gave several illustrations of the animal, and also of its breathing-holes, and of the implements employed by the Innuits in Walrus-hunting. In Sonntag's "Narrative of the Grinnell Exploring Expedition,"§ published in 1857, a

---

* Jardine's Nat. Library, Mam., vol. vii, 1838, pl. xx.

† Ibid., vol. viii, 1839, pl. i.

‡ Arctic Exploration, vol. i, pp. 141 ("Walrus Sporting"), 142 ("Walrus-hole"), 419 ("portrait"); vol. ii, plate facing p. 214 ("Walrus Hunt off Pikantlik"—a nearly full figure.

§ This curious and apparently little known brochure, by the eminent astronomer of the Expedition, is well worthy of attention, notwithstanding the ludicrously sensational character of the titlepage affixed by the enterprising publishers. The titlepage, transcribed in full, is as follows: "Professor Sonntag's Thrilling Narrative of the Grinnell Exploring Expedition to the Arctic Ocean, in the years 1853, 1854, and 1855, in search of Sir John Franklin, under the command of Dr. E. K. Kane, U. S. N. Containing the History of all previous explorations of the Arctic Ocean, from the year 1618 down to the present time; showing how far they advanced northward, what discoveries they made and their scientific observations. The present whereabouts of Sir John Franklin and his party, if they are still alive. A statement of the only practicable method by which the North Pole may be reached; the reasons why all exploring expeditions have hitherto failed to penetrate the icy barriers of the Polar Regions. Highly important astronomical observations, proving that there is no such thing as apparent time at the North Pole; sufferings of Dr. Kane's exploring party; how they were buried for two years in the ice, enduring a degree of cold never experienced by any human being before; their miraculous escapes and unprecedented hardships; their abandonment of the ship; and perilous journey of four hundred miles over the ice. With nearly one hundred splendid engravings. By Professor August Sonntag, Astronomer to the

group of four old Walruses is figured (full-page woodcut, p. 113). The animals are disposed in various attitudes, and represent admirably the grim visage, postures, and uncouth proportions of the Atlantic Walrus. The figure in the foreground is presented in profile, with both fore and hind limbs in a *natural position;* behind this are two old veterans seen in half-profile, and behind these a third lying on its back with the hind limbs thrust upward. This illustration, *evidently a study from life,* is by far the best representation of the adult Atlantic Walrus with which I am acquainted. In 1857, Dr. Gray reproduced, as previously detailed (*anteà,* pp. 93–100), a series of the early figures from Olaus Magnus, Gesner, Jonston, Gerard, Martens, Buffon, and Cook.

The next original figures of the Walrus with which I am acquainted were drawn from the living specimen in the Gardens of the Zoölogical Society of London by Mr. Wolf, and appear in Wolf and Sclater's "Zoölogical Sketches,"[*] published in 1861. In plate xviii is represented a group of Walruses in various attitudes. Those in the foreground are young and tuskless, with a heavy array of long mystacial bristles, and much thinner necks and shoulders than the Walrus is commonly represented as having, doubtless owing to the very emaciated condition of the living original.

At about this date (1861), some very good pictures of groups of Walruses were published by Mr. Lamont in his entertaining and instructive book entitled "Seasons with the Sea-horses." In a spirited plate (called "Chase of the Walrus"), facing the titlepage, is portrayed a group of Walruses in the sea, attacked

---

Expedition, formerly of the Royal Observatory at Vienna, and late of the U. S. National Observatory, Washington City, D. C. Philadelphia, Penn.: Jas. T. Lloyd & Co. Cincinnati, Ohio: Jas. T. Lloyd & Co." No date. Large 8vo, pp. 176, paper. Copyright dated 1857.

The publishers state: "The undersigned having purchased Professor Sonntag's Narrative of the Grinnell Expedition, some months since, have used their best judgment and abilities in preparing this thrilling narrative for the press, to make it as acceptable to the reading public as possible," etc.

The name of the author is alone sufficient guaranty of the trustworthy and instructive character of the work, which, despite the dime-novel aspect of its exterior, is a valuable contribution to the history of the Arctic Regions. Pages 80 to 85 are devoted to a general account of the Walrus. At page 83 is a sketch of a "Desperate attack of Walruses on the English Boat," based apparently on Captain Beechey's account of an adventure with these animals.

[*] Vol. i, pl. xviii.

by a boat's crew, one of the poor animals having been already harpooned. Another plate, facing page 72, entitled " Walruses on the Ice," represents a herd on the ice in various attitudes, most, but not all, of which have the hind feet extended backward, in the manner of Seals. In his later work, " Yachting in the Arctic Seas," he has given (plate opposite p. 56) a very fine side-view of the head, and on p. 221 a large vignette figure of the head seen in front.

Mr. Brown also refers to " the excellent figures of the Walrus taken by the artist of the Swedish Expedition," namely, a "chromolithograph and head, both drawn by Herr von Yhlen,"— " under the direction of such well-informed naturalists as Torell, Malmgren, Smith, Goes, Blomstrand, &c.," in which "the fore flippers are represented as rather doubled back, and the hind flippers extended." This work (" Svenska Expeditioner til Spetsbergen år 1861, pp. 168–182, pl. facing p. 169, and head p. 308 ") I have been unable to see, but presume the figures are the same as those in the German translation of this work, which appeared in 1869.* The frontispiece of this work represents a group of four old Walruses resting on the ice, with a fifth in the water in the foreground. A woodcut of the head of a young, or more probably a female, is given on p. 132, and on p. 136 a hunting-scene.

In 1867 appeared figures of the second living specimen received at the Zoölogical Society's Gardens. According to Dr. Murie† these were published in " The Field," " Land and Water," " Illustrated London News," and elsewhere. The figure originally appearing in " The Field " (drawn by Mr. Wood) is republished by Dr. Murie in his " Memoir on the Anatomy of the Walrus "‡ from the original wood-block. This is a rather more robust figure than those published by Wolf and Sclater, but is likewise tuskless (being also that of a very young animal), and shows similarly the long, descending, curved mystacial bristles.

In 1870, Dr. Gilpin figured a male Walrus killed in March, 1869, in the Straits of Belle Isle, Labrador. In this figure, the general form of the body is very well represented, but the hind

---

* Die schwedischen Expeditionen nach Spitzbergen und Bären-Eiland, ausgeführt in den Jahren 1861, 1864 und 1868, etc. (for full transcript of the titlepage see *anteà*, p. 92).

† Trans. Zoöl. Soc. Lond., vol. vii, 1872, p. 413.

‡ Loc. cit., p. 416.

limbs are turned backward, as in the common Seals. A view of the muzzle forms a second figure, and the form of one of the fore limbs is given in outline.

Wells, in his "Gateway to the Polynia," published in 1873, gives a plate (facing page 201) in illustration of the Walrus. The figure in the foreground represents an individual flat on its belly with all the limbs directed posteriorly. Other figures represent other individuals reposing in various attitudes.

The above-enumerated figures of the Walrus embrace all the original figures of the Atlantic species thus far known to me, and all to which I have seen references, so far as figures of the entire animal are concerned. In recapitulation, it may be stated that Gesner's figure, published in 1558, is the first that had an actual foundation in nature, all the preceding (the mythical ones of Olaus Magnus) being purely fictitious or based on erroneous conceptions. Gesner's, as already noticed, was a curious combination of reality with myth, the head only being drawn from nature, and a fanciful body added! The first really drawn from nature ("ad vivum") was Hessel Gerard's excellent figure published in 1613. Subsequently appeared numerous figures in the works of travellers, drawn apparently either from memory or by artists who had never seen the animal they so confidently attempted to depict.

The first representation based on a museum specimen appears to have been Buffon's, in 1765, which has been aptly described as being merely a common Seal with tusks. Other figures followed later, as those in the so-called "Disciples edition" of Cuvier's Règne Animal, and in the two already cited volumes of Jardine's Naturalist's Library, drawn also from stuffed specimens, in which the hind limbs were always placed in a wholly false attitude, though in other respects passably fair figures. Not until a living specimen reached London, in 1853, did the correct attitudes of the animal and the natural position of the hind limbs become generally known to naturalists, and not until then was the truthfulness of Gerard's early figure duly recognized and appreciated, notwithstanding that von Baer, nearly twenty years earlier, testified to its excellence, and correctly described the flexibility of the limbs. Now, through the two living specimens seen and figured in London, and through excellent recent figures of the Pacific Walrus, the attitudes and external bearing of few of our marine mammalia are better known than those of the Walruses.

In addition to the above-described figures of the general animal, representations of various anatomical details, both of the osteology and the soft parts, have been from time to time published. As early as 1761, the skull, as previously stated, was figured by Houttuyn, and again by Daubenton* in 1765, these being the earliest figures of the skull to which I find reference. Goethe, in his "Morphologie" (see *anteà*, p. 48), gave important figures illustrative of the dentition and structure of the anterior portion of the skull.

Home,† in 1824, published a series of excellent figures of the extremities and stomach. G. Cuvier,‡ in 1825, figured skulls and the skeleton, his figures of the skull being also reproduced in the "Disciples" edition of Cuvier's Règne Animal. § Pander and d'Alton, in 1826, in their "Vergleichende Osteologie,"‖ gave an excellent figure of the skeleton and detail illustrations of the skull and limbs. In the figure of the skeleton, the hind feet are turned forward in a plantigrade position, and the fore limbs are given their natural pose. Von Baer,¶ in 1835, figured the blood-vessels of the limbs, and, in 1840, De Blainville** figured the skeleton and the skull. Gray,†† in 1850, gave a view of the skull, the same figure being repeated in some of his later works.‡‡ Owen, in 1845, figured the dentition in his "Odontography" (pl. cxxxii, fig. 8), the skull and dentition in 1854,§§ and gave another figure in 1868.‖‖ In 1857, Walrus skulls were figured by Blasius,¶¶ and Leidy*** the same year figured a fossil skull from Monmouth County, New Jersey. Later, as already noticed (*anteà*, p. 54), the milk dentition was figured by Malmgren, and also by

* Buffon's Histoire Nat., tome xiii, pl. lv. An artistically much improved (but unaccredited) copy of Daubenton's figures appears in Hamilton's "Amphibious Carnivora" (Jardine's Naturalist's Library, Mam., vol. viii, pp. 100, 101).

† Phil. Trans., 1824, pp. 235–241, pll. iv–viii.

‡ Ossem. Fossiles.

§ Mam., pl. xliv.

‖ Lieferung xi, Die Robben und Lamantine, pll. i, ii.

¶ Mém. de l'Acad. St. Pétersb., Sci. Nat., vime sér., 1835, t. iii, pl. —.

** Ostéog., Des Phoques, pll. i (skeleton) and iv (skull); eight figures.

†† Cat. Mam. Brit. Mus., p. 31, fig. 11 (small woodcut).

‡‡ Cat. Seals and Whales, 1866, p. 35, fig. 12.

§§ Encycl. Brit., article Odontography, p. 463, fig. 112.

‖‖ Comp. Anat. and Phys. Vertebr., vol. iii, p. 338, fig. 265.

¶¶ Fauna Wirbelth. Deutschl., pp. 261, 262, figg. 148–150.

*** Trans. Amer. Phil. Soc. Phila., (2), vol. xi, pll. iv, v.

Peters. Dr. Murie, in 1874, gave numerous figures illustrative of its external characters, myology, dentition, generative, digestive, and vocal organs, based on a dissection of the young Walrus that died in the Garden of the Zoölogical Society of London in 1867, these being the only figures, so far as known to me, devoted to the general anatomy. Doubtless other figures of the skull, and possibly of the dentition, have appeared that are not here noted.

HABITS AND THE CHASE.—The Walruses are at all times more or less gregarious, occurring generally in large or small companies, according to their abundance. Like the Seals, they are restricted in their wanderings to the neighborhood of shores or large masses of floating ice, being rarely seen far out in the open sea. Although moving from one portion of their feeding-grounds to another, they are said to be in no true sense a migratory animal.* *They delight in huddling together on the ice-floes or on shore, to which places they resort to bask in the sun, pressing one against another like so many swine. *They are also said to repair in large herds to favorable shores or islands,† usually in May and June, to give birth to their young, at which times they sometimes remain constantly on land for two weeks together, without ever taking food.‡ They are believed to be' monogamous, and to bring forth usually but a single young at a time, and never more than two. The period of gestation is commonly believed to be about nine months. The young are born from April to June, the time probably varying with the latitude. Malmgren states that the pairing of the Walruses takes place about the end of May or the beginning of June; that the female gives birth to a single young in May or June; and that the period of pregnancy lasts probably for a year.*He states that Dr. A. von Goes found a month-old fœtus in the uterus of a female on the 8th of July, in latitude 80° N., but adds that females with mature young in the uterus have been taken as late as the end of

---

* See Brown, Proc. Zoöl. Soc. Lond., 1868, p. 433.

† Says Zorgdrager (writing in 1750), as quoted by Buffon, in referring to this habit: "Anciennement & avant d'avoir été persécutés, les morses s'avançoient fort avant dans les terres, de sorte que dans les hautes marées ils étoient assez loin de l'eau, & que dans le temps de la basse mer, la distance étant encore beaucoup plus grande, on le abordoit aisément."—*Hist. Nat.,* tome xiii, p. 366.

‡ See Shuldham, Phil. Trans., vol. lvi, 1777, p. 249, quoted *anteà,* p. 67.

June or July. The females, he believes, suckle their young for
two years, and that hence not less than three years elapse between each birth. The females with their newly-born young
are said to keep aloof from the society of other Walruses, and
that females are never found to be pregnant during the year
following the birth of their young. Females in the second year
of suckling their young collect in large herds and live apart
from the full-grown males. Of thirty full-grown Walruses killed
by Malmgren's harpooner in Henlopen Straits, in the month of
July, not one was a male. Where the full-grown males were at
this time was unknown, but they were believed by the hunters
to be "on the banks," remote from the land, while the females
with their young sought the bays and open sea near the shores,
the two sexes thus living in separate herds.*

Notwithstanding the explicitness of Malmgren's account,
who no doubt correctly details his own experience in the
matter, there is much rebutting testimony, most observers
reporting that both sexes and the young occur in the same
herds.† The only detailed account of the pairing and repro-

---

* See further Malmgren's paper, as translated in Arch. für Naturgesch.,
1864, pp. 70–72.

† Says Dr. Kane: "The early spring is the breeding season, . . . at
which time the 'female with her calf is accompanied by the grim-visaged
father, surging in loving trios from crack to crack, sporting around the berg-
water, or basking in the sun."—*Arctic Exploration*, vol. ii, p. 131.

Dr. Hayes, referring to a herd upon which he made an attack, thus observes: "Besides the old bulls, the group contained several cows and a few
calves of various sizes—some evidently yearlings, others but recently born,
and others but half or three quarters grown. Some were without tusks,
while on others they were just sprouting; and above this they were of various sizes up to those of the big bulls, which had great curved cones of ivory
nearly three feet long."—*Open Polar Sea*, p. 406.

Lamont also refers to the presence of young and old, males and females,
in the same herd, and to the custom of the Walrus-hunters of striking a
young one in order to detain the herd, which, through sympathy, join con-
certedly in its defense, thus affording the hunters opportunity for further
slaughter.—*Seasons with the Sea-horses*, pp. —.

Through the kindness of Prof. Henry A. Ward, of Rochester, N. Y., I am
in receipt, in answer to inquiries respecting the habits and reproduction of
the Walrus, of the following information from the pen of Captain Adams,
of the whaling-steamer "*Arcturus*," from Dundee to Baffin's Bay. Captain
Adams, writing from long experience in Walrus-hunting, says: "I am of
opinion that the female Walrus prefers low flats of land on which to bring
forth her young. The time is in mid-spring. In early May I have seen very
young Walruses on the ice with their mothers. I have also seen afterbirths
on the ice, but still think that low flat land is preferred when attainable. I do

duction of the Walrus is that long since given by Shuldham, based on observations made a century ago at the Magdalen Islands in the Gulf of St. Lawrence (see *anteà*, p. 67), to which he says they repair "early in spring" and immediately bring forth their young. Captain Parry states that he met with females accompanied by their young in Fox Channel, July 13, and Mr. Lamont speaks of meeting with young accompanied by their parents at the same season (July 15) in the vicinity of Spitzbergen. Captain Hayes refers to meeting with "calves newly born" as early as July 3 in Frobisher's Bay. Captain Parry says that Walruses killed by the Esquimaux in March (in the years 1822 and 1823) were observed to be with young.[*]

When repairing to the land or to the ice-floes to rest, those first arriving are described as generally composing themselves for a nap at the place where they first land, but their comrades still in the water having a strong desire to land at the same spot, the latter force those already on shore higher up, while they in turn are pushed forward by later comers, their habits in this, as well as in many other respects, resembling those of the Sea Lions and Sea Bears.

The Walrus, like the common Seals, is said to have its breathing-holes in the ice. These are described by Dr. Kane as being similar to those of the Seals, having "the same circular, cleanly-finished margin," but made in much thicker ice, with the "radiating lines of fracture round them much more marked." The ice around the holes is much discolored, while near them are numbers of broken clam-shells, and in one instance Dr. Kane found "gravel, mingled with about half a peck of coarse shingle of the beach."[†] Kane says the Walrus often sleeps in the water between the fields of drift-ice. "In this condition," he relates, "I frequently surprised the young ones whose mothers were asleep by their sides."[‡] Other writers refer to the same habit.

---

not think that the females and young live in separate herds from males, but the males herd alone in early spring. In the middle of summer both sexes herd together; then the males are very wild. I have seen many females alone in the autumn. I do not think the females nurse their young over twelve months."—*Communicated by Prof. H. A. Ward* in a letter of date March 31, 1878.

[*] Narrative of Parry's Second Voyage, p. 415.

[†] Arctic Exploration, vol. i, 1856, pp. 141, 142. On page 142 is a figure of a "Walrus hole." Mr. Robert Brown gives a similar account (Proc. Zoöl. Soc. Lond., 1868, p. 429), using, in fact, in part the *same phraseology*.

[‡] Ibid., vol. i, p. 141.

The voice of the Walrus is a loud roaring or "hucking," and can be heard to a great distance, often giving notice of the presence of a herd long before they can be seen. "Like some of the higher order of beings to which he has been compared," says Dr. Kane, he "is fond of his own music, and will lie for hours listening to himself. His vocalization is something between the mooing of a cow and the deepest baying of a mastiff: very round and full, with its bark or detached notes repeated rather quickly seven to nine times in succession."* Other writers speak of the roaring of a herd as being distinguishable at the distance of several miles.

The Walrus, unless molested, is represented as inoffensive and harmless, but as exhibiting when attacked great fierceness, and even vindictiveness, proving a powerful and often dangerous antagonist. Their strong affection for their young and their sympathy for each other in times of danger are strong traits in their character, in which qualities they are rarely exceeded by any members of the mammalian class. When one of their number is wounded, the whole herd usually join in an intelligent and concerted defense. With their enormous size and threatening tusks it is little wonder that they inspired the early voyagers with terror, and that their powers and ferocity were to some degree overestimated. Their aspect is, in short, as affirmed by recent intelligent observers, little less than terrible. That the accounts given by the early navigators of the fierce attacks made upon them by the "Sea Horses," as they commonly termed them, are not to be by any means wholly attributed to the superstitious fears so prevalent respecting sea-monsters in the early times, is evident from the trustworthy accounts given us of these creatures by the intrepid explorers of the Arctic region in our own times, as will be shown by the copious testimony presently to be given. That there is much in his aspect that is truly formidable is evident from Mr. Lamont's graphic description, who says: "The upper lip of the Walrus is thickly set with strong, transparent, bristly hairs, about six [?] inches long, and as thick as a crow-quill; and this terrific mustache, together with his long white tusks, and fierce-looking, blood-shot eyes, gives *Rosmarus trichecus* altogether a most unearthly and demoniacal appearance as he rears his head above the waves. I think it not unlikely that the old fable of the mermaid may have originated by their grim resemblance to

---

* Arctic Exploration, vol. i, 1856, p. 410.

the head of a human being when in this position."* The
confounding, in early times, of the Russian name *Morss* by the
peoples of Western Europe with the Latin word *Mors* and the
German word *Tod*, as already alluded to (*anteà*, p. 81), finds its
explanation doubtless in exaggerated accounts of its terrible
aspect and power.

The Walrus, either through confidence in its own power, or
through ignorance of the character of its human foes, is generally
not easily alarmed, and permits a near approach before manifesting uneasiness or fear, sometimes, indeed, treating its human
visitors with quiet indifference. When found reposing on land,
it is, in fact, easily dispatched, unless it has been previously
subjected to repeated attacks, when it profits by dearly-bought
experience and makes a timely retreat to the water, and thus
commonly escapes its pursuers. With due caution, however,
the Walrus-hunters succeed in cutting off their retreat to the
sea, when hundreds of the then helpless creatures fall victims
to the hunter's rapacity. Says Zorgdrager, as translated by
Buffon: "On marchoit de front vers ces animaux pour leur cou-
per la retraite du côté de la mer; ils voyoient tous ces prépara-
tifs sans aucune crainte, & souvent chaque chasseur en tuoit
un avant qu'il pût rengagner l'eau. On faisoit une barrière de
leurs cadavres & on laissoit quelques gens à l'assût pour assom-
mer ceux qui restoient. On en tuoit quelquefois trois ou quatre cents. . . . . On voit par la prodigieuse quantité d'ossemens de ces animaux dont la terre est jonchée qu'ils ont été
autrefois très nombreux."† This manner of attack was also
well described a little later by Lord Shuldham, his detailed account of their destruction at the Magdalen Islands during the
last century being fully corroborated by scores of modern observers at numerous other localities. According to Lord Shuldham, the hunters allowed them to come on shore to the number
of several hundred, and then cautiously approaching them from
the seaward, under cover of the darkness of night, would endeavor, by the aid of well-trained dogs, to cut off their retreat
to the water and drive them further inland. These attacks
were sometimes so successful that fifteen or sixteen hundred
have been killed in a single attack.‡ A similar wholesale destruction of Walruses was carried on by the English in the

---

* Seasons with the Sea-horses, pp. 141, 142.
† Buffon's Hist. Nat., tom. xiii, pp. 366, 367.
‡ For Lord Shuldham's account in full see *anteà*, p. 67.

early part of the seventeenth century at Cherie or Bear Island, as already related (*anteà*, pp. 73–78). Mr. Lamont, in his "Seasons with Sea-horses," gives a similar account of their recent destruction in the Spitzbergen Seas, where he says, by a similar mode of attack, two ships' crews killed nine hundred in a single day.*

The habits of the Walruses as met with in their native waters,—their strong affection for their young and for each other, inducing the whole herd to join in defense of a wounded comrade, and their power and courage in the water in repelling the attacks of man,—I have chosen to detail in the language of actual observers, believing the vivid portrayal of a few scenes from real life, by trustworthy eye-witnesses, to be far preferable to any epitomized account of the subject, however well and carefully elaborated. The personal incidents involved and the circumstances of pursuit are necessarily important accessories to a correct appreciation of the scenes described.

As stated in several of the earliest accounts of these animals, they are always more or less wary, and at times difficult to approach, usually keeping a sentinel on guard while the herd is asleep. Respecting their habits at such times, Mr. Robert Brown observes as follows: "On the floes, lying over soundings and shoals, the Walruses often accumulate in immense numbers, and lie huddled upon the ice. More frequently, in Davis's Strait and Baffin's Bay, they are found floating about on pieces of drift ice, in small family parties of six or seven; and I have even seen only one lying asleep on the ice. Whether in large or small parties, one is always on the watch, as was long ago observed by the sagacious Cook: the watch, on the approach of danger, will rouse those next to them; and the alarm being spread, presently the whole herd will be on the *qui vive*."†

Mr. Lamont thus describes a scene in the Spitzbergen waters: "At 3 a. m. this morning [July 13, 1859], we were aroused by the cheering cry of 'Hvalruus paa Ysen' (Walruses on the ice). We both got up immediately, and from the deck a curious and exciting spectacle met our admiring gaze. Four large flat icebergs were so densely packed with Walruses that they were sunk awash with the water, and had the appearance of being solid *islands of Walrus!*

"The monsters lay with their heads reclining on one another's

---

* Mr. Lamont's account will be given later in full. (See p. 114.)
† Robert Brown in Proc. Zoöl. Soc. Lond., 1868, p. 429.

backs and sterns, just as I have seen Rhinoceroses lying asleep in the African forests: or, to use a more familiar simile, like a lot of fat hogs in a British straw-yard. I should think there were about eighty or one hundred on the ice, and many more swam grunting and spouting around, and tried to clamber up among their friends, who, like surly people in a full omnibus, grunted at them angrily, as if to say, 'Confound you, don't you see that we are full?' There were plenty more good flat icebergs about, but they always seem to like being packed as closely as possible for mutual warmth. These four islands were several hundred yards apart, ..."*

Mr. Lamont thus refers to the number seen on another occasion, and incidentally to their watchful habits: "We had a pleasant row of four or five miles over calm water quite free of ice, and were cheered for the latter half of the distance by the sonorous bellowing and trumpeting of a vast number of Walruses. We soon came in sight of a long line of low flat icebergs crowded with Sea-horses. There were at least ten of these bergs so packed with the Walruses that in some places they lay two deep on the ice. There can not have been less than three hundred in sight at once; but they were very shy and restless, and, although we tried every troop in succession as carefully as possible, we did not succeed in getting within harpooning distance of a single Walrus. Many of them were asleep; but there were always some moving about who gave the alarm to their sleeping comrades by flapping them with their fore feet, and one troop after another manage to shuffle into the sea always just a second or so in time to avoid a deadly harpoon." †

"With reference to the Walrus," says Captain Hall, " Mr. Rogers told me that one day, when out cruising for Whales, he went, with two boats and crews, half way across Frobisher Bay, and then came to an iceberg one hundred feet above the sea, and, mounting it, with a spy-glass, took a look all around. Whales there were none; but Walrus—'Why', to use his figurative but expressive words, 'there were millions out on the pieces of ice, drifting with the tide—Walrus in every direction—millions on millions'." ‡ While these numbers are not, doubtless, to be taken literally, they certainly imply an immense number of Walruses. The context states that while the whalers

---

* Seasons with the Sea-horses, p. 72.
† Ibid., pp. 80, 81.
‡ Arctic Researches, etc., p. 234.

in Frobisher's Bay had met with no Whales, "Walrus in any numbers could be obtained, and many had been secured for their skins and tusks."

The Walruses in the Spitzbergen waters, according to Mr. Lamont, usually congregate in August in great numbers on land, " sometimes to the number of several thousands, and all lie down in some secluded bay or some rocky island, and there remain in a semi-torpid sort of state, for weeks together, without moving or feeding." They do not usually do this, he adds, till near the end of August, or some months later than they were found to do in the sixteenth and seventeenth centuries on the shores and islands of the Gulf of St. Lawrence. This is possibly owing to the difference in the climate, although it seems hardly probable that this can be the whole cause of the difference. Mr. Lamont, in this connection, makes no reference to the time of bringing forth of the young, and does not give this as one of the reasons for their visiting the land. He alludes, however, to their sudden disappearance at this time from the ice-floes. He says the Walrus-hunters consider themselves fortunate if they find one of these resorts, as then they can kill in a few hours a " small fortune's-worth of them." His account of these " trysting-places," however, is mainly at second hand, and possibly the date is not carefully given.*

Mr. Lamont's account of the great havoc the hunters often make with the then helpless beasts, destroying many hundreds in a few hours, is quite similar, so far as the destruction of life is concerned, to the account given by Lord Shuldham of their destruction a century and a half ago at the Magdalen Islands. Referring to one of the southwesternmost of the Thousand Islands, Mr. Lamont says: "It seems that this island had long been a very celebrated place for Walruses going ashore, and great numbers had been killed upon it at different times in bygone years. In August, 1852, two small sloops sailing in company approached the island, and soon discovered a herd of Walruses, numbering, as they calculated, from three to four thousand, reposing upon it. Four boats' crews, or sixteen men, proceeded to the attack with spears. One great mass of Walruses lay in a small sandy bay, with rocks enclosing it on each side, and on a little mossy flat above the bay, but to which the bay formed the only convenient access for such unwieldy animals. A great many hundreds lay on other parts of the island at a little dis-

---

* Seasons with the Sea-horses, pp. 173, 174.

tance. The boats landed a little way off, so as not to frighten them, and the sixteen men, creeping along shore, got between the sea and the bay full of Walruses before mentioned, and immediately commenced stabbing the animals next them. The Walrus, although so active and fierce in the water, is very unwieldy and helpless on shore, and those in front soon succumbed to the lances of their assailants; the passage to the shore soon got so blocked up with the dead and dying that the unfortunate wretches behind could not pass over, and were in a manner barricaded by a wall of carcasses. Considering that every thrust of a lance was worth twenty dollars, the scene must have been one of terrific excitement to men who had very few or no dollars at all; and my informant's eyes sparkled as he related it. He said the Walruses were then at their mercy, and they slew, and stabbed, and slaughtered, and butchered, and murdered until most of their lances were rendered useless, and themselves were drenched with blood and exhausted with fatigue. They went on board their vessels, ground their lances, and had their dinners, and then returned to their sanguinary work; nor did they cry 'Hold, enough!' until they had killed *nine hundred* Walruses, and yet so fearless or so lethargic were the animals, that many hundreds more remained sluggishly lying on other parts of the island at no great distance. . . . When I visited the island six years afterward, there still remained abundant testimony to corroborate the entire truth of the story. The smell of the island was perceptible at several miles' distance, and on landing we found the carcasses lying as I have described them, and in one place two and three feet deep. The skin and flesh of many remained tolerably entire, notwithstanding the ravages of Bears, Foxes and Gulls. So many Walruses have been killed on this island at different times that a ship might easily load with *bones* there. . . ."* The worst feature of this wholesale slaughter was the fact that their small vessels, already partly loaded, could carry away only a small portion of the spoil. A subsequent attempt to reach the island later in the season for the purpose of securing the rest failed, owing to its being surrounded by impenetrable ice.

Respecting the parental affection displayed by the Walruses, Mr. Lamont relates the following: "I never in my life witnessed anything more interesting and more affecting than the wonderful maternal affection displayed by this poor Walrus. After she

---

* Seasons with the Sea-horses, pp. 175–177.

was fast to the harpoon and was dragging the boat furiously through the icebergs, I was going to shoot her through the head that we might have time to follow the others; but Christian called to me not to shoot, as she had a 'junger' with her.   Although I did not understand his object, I reserved my fire, and upon looking closely at the Walrus when she came up to breathe, I then perceived that she held a very young calf under her right arm, and I saw that he wanted to harpoon it; but whenever he poised the weapon to throw, the old cow seemed to watch the direction of it, and interposed her own body, and she seemed to receive with pleasure several harpoons which were intended for the young one.   At last a well-aimed dart struck the calf, and we then shortened up the lines attached to the cow and finished her with the lances.   Christian now had time and breath to explain to me why he was so anxious to secure the calf, and he proceeded to give me a practical illustration of his meaning by gently 'stirring up' the unfortunate junger with the butt end of a harpoon shaft.   This caused the poor little animal to emit a peculiar, plaintive, grunting cry, eminently expressive of alarm and of a desire for assistance, and Christian said it would bring all the herd round about the boat immediately.   Unfortunately, however, we had been so long in getting hold of our poor decoy duck that the others had all gone out of hearing, and they abandoned their young relative to his fate, which quickly overtook him in the shape of a lance thrust from the remorseless Christian.

"I don't think I shall ever forget the faces of the old Walrus and her calf as they looked back at the boat!   The countenance of the young one, so expressive of abject terror, and yet of confidence in its mother's power of protecting it, as it swam along under her wing; and the old cow's face showing such reckless defiance for all that we could do to herself, and yet such terrible anxiety as to the safety of her calf!

"This plan of getting hold of a junger and making him grunt to attract others is a well-known 'dodge' among hunters; and, although it was not rewarded on this occasion, I have several times seen it meet with the full measure of success due to its humanity and ingenuity."*

When in the water, to again quote from Mr. Lamont, "the herd generally keep close together, and the simultaneousness with which they dive and reappear again is remarkable; one moment

---

* Seasons with the Sea-horses, pp. 70, 71.

you see a hundred grisly heads and long gleaming white tusks above the waves; they give one spout* from their blow-holes, take one breath of fresh air, and the next moment you see a hundred brown hemispherical backs, the next a hundred pair of hind flippers flourishing, and then they are all down.   On, on, goes the boat as hard as ever we can pull the oars; up come the Sea-horses again, pretty close this time, and before they can draw breath the boat rushes into the midst of them: *whish!* goes the harpoon: *birr!* goes the line over the gunwale: and a luckless junger on whom Christian has kept his eye is 'fast': his bereaved mother charges the boat instantly with flashing eyes and snorting with rage; she quickly receives a harpoon in the back and a bullet in the brains, and she hangs lifeless on the line: now the junger begins to utter his plaintive grunting bark, and fifty furious Walruses are close round the boat in a few seconds, rearing up breast high in the water, and snorting and blowing as if they would tear us all to pieces.   Two of these auxiliaries are speedily harpooned in their turn, and the rest hang back a little, when, as bad luck would have it, the junger gives up the ghost, owing to the severity of his harpooning, and the others no longer attracted by his cries, retire to a more prudent distance.   But for the 'untoward' and premature decease of the junger, the men tell me we should have had more Walruses on our hands than we could manage.   We now devote our attention to 'polishing off' the two live Walruses—well-sized young bulls—who are still towing the heavy boat, with their two dead comrades attached, as if she were behind a steam-tug, and struggling madly to drag us under the icebergs: a vigorous application of the lances soon settles the business, and we now, with some difficulty, tow our four dead victims to the nearest flat iceberg and fix the ice-anchor, by which, with the powerful aid of block and tackle, we haul them one by one on the ice and divest them of their spoils. . . .

   " While we were engaged in cutting up these Walruses, there were at least fifty more surrounding the iceberg, snorting and bellowing, and rearing up in the water as if smelling the blood

---

* It is, perhaps, almost needless to say that the "spouting" here referred to is merely the spray thrown upward by the forcibly expelled breath as they rise to the surface, although a "spouting from their blow-holes" has occasionally been attributed to them since the time of Martens, who says they "blow water from their nostrils like a whale." See on this point von Baer (l. c., pp. 139–147), who has discussed the matter at length in his above-cited memoir on the Walruses.

of their slaughtered friends, and curious to see what we were doing to them now. They were so close that I might have shot a dozen of them; but, as they would have been sure to sink before the boat could get to them, I was not so cruel as wantonly to take their lives. When the Walruses were all skinned, we followed the herd again with success; and when we left off, in consequence of dense fog suddenly coming on, we had secured nine altogether—a very fair morning's bag we thought. . . . During this morning's proceedings I realized the immense advantage of striking a junger first, when practicable. This curious clannish practice of coming to assist a calf in distress arises from their being in the habit of combining to resist the attacks of the Polar Bear, which is said often to succeed in killing a Walrus. If, however, Bruin, pressed by hunger and a tempting opportunity, is so illadvised as to snap a calf, the whole herd come upon him, drag him under water, and tear him to pieces with their long sharp tusks. I am told this has been seen to occur, and I quite believe it."*

Capt. William Edward Parry, in his narrative of his second voyage for the discovery of a northwest passage, makes frequent reference to the Walrus, and his report of encounters with them shows that serious and even fatal consequences sometimes result to the boats' crews who venture to attack them.

"In the course of this day [July 15, 1822, in Fox Channel] the Walruses," says Captain Parry, "became more and more numerous every hour, lying in large herds upon loose pieces of drift-ice; and it having fallen calm at one P. M., we despatched our boats to endeavor to kill some for the sake of the oil they afford. On approaching the ice our people found them huddled in droves of from twelve to thirty, the whole number near the boats being perhaps about two hundred. Most of them waited quietly to be fired at, and even after one or two discharges did not seem to be greatly disturbed but allowed the people to land on the ice near them, and, when approached, shewed an evident disposition to give battle. After they had got into the water, three were struck with harpoons and killed from the boats. When first wounded they became quite furious, and one, which had been struck from Captain Lyon's boat, made a resolute attack upon her, and injured several of the planks with its enormous tusks. A number of the others came round them, also repeatedly striking the wounded animals with their tusks, with

* Seasons with the Sea-horses; pp. 81–83, 84.

the intention of either getting them away or else of joining in the attack upon them. Many of these animals had young ones which, when assaulted, they either took between their fore-flippers to carry off, or bore away on their backs. Both of those killed by the Fury's boats were females, and the weight of the largest was fifteen hundred-weight and two quarters nearly; but it was by no means remarkable for the largeness of its dimensions. The peculiar barking-noise made by the Walrus, when irritated, may be heard, on a calm day, with great distinctness at the distance of two miles at least. We found musquet-balls the most certain and expeditious way of despatching them after they had been once struck with the harpoon, the thickness of the skin being such, that whale-lances generally bend without penetrating it. One of these creatures, being accidentally touched by one of the oars of Lieutenant Nias's boat, took hold of it between its flippers and forcibly twisting it out of the man's hand, snapped it in two."*

Again, says the same writer, " The Heckla's two boats had one day a very narrow escape in assaulting a herd of these animals [Walruses]; for several of them, being wounded, made so fierce an attack on the boats with their tusks, as to stave them in a number of places, by which one was immediately swamped and the other much damaged. The Fury's being fortunately in sight prevented any further danger; two of the Walruses were killed and secured, and the damaged boats lightened and towed to the shore, from which they had been several miles distant." †

In addition to the foregoing testimony respecting the power and courage of these animals when in the water, I add the following: Mr. Lamont states that " a boat belonging to a sloop from Tromsöe had been upset two or three days before, in our immediate vicinity, and one of the crew killed by a Walrus. It seemed that the Walrus, a large old bull, charged the boat, and the harpooner, as usual, received him with his lance full in the chest; but the shaft of the lance broke all to shivers, and the Walrus, getting inside of it, threw himself on the gunwale of the boat and overset it in an instant. While the men were floundering in the water among their oars and tackle, the infuriated animal rushed in among them, and, selecting the unlucky harpooner, who, I fancy, had fallen next him, he tore him nearly

---

* Narrative of Parry's Second Voyage, p. 268.   † Ibid., p. 469.

into two halves with his tusks. The rest of the men saved themselves by clambering on to the ice until the other boat came to their assistance.

"Upon another occasion I made the acquaintance of the skyppar of a sloop who had been seized by a bereaved cow Walrus, and by her dragged twice to the bottom of the sea, but without receiving any injury beyond being nearly drowned, and having a deep scar plowed in each side of his forehead by the tusks of the animal, which he thought did not wish to hurt him, but mistook him for her calf as he floundered in the water.

"Owing to the great coolness and expertness of the men following this pursuit, such mishaps are not of very frequent occurrence, but still a season seldom passes without two or three lives being lost in one way or another."*

Among the numerous writers who have described a "Walrus hunt," no accounts that I have seen more vividly portray the scene, or give more information respecting the nature and habits of the Atlantic species, than Dr. I. I. Hayes, for which reason I deem no apology is necessary for transcribing his lengthy account in full. Under date of July 3, 1861 (the scene being in Frobisher's Bay), he says:

"I have had a Walrus hunt and a most exciting day's sport. Much ice has broken adrift and come down the Sound, during the past few days; and, when the sun is out bright and hot, the Walrus come up out of the water to sleep and bask in the warmth on the pack. Being upon the hilltop this morning to select a place for building a cairn, my ear caught the hoarse bellowing of numerous Walrus; and, upon looking over the sea, I observed that the tide was carrying the pack across the outer limit of the bay, and that it was alive with the beasts, which were filling the air with such uncouth noises. Their numbers appeared to be even beyond conjecture, for they extended as far as the eye could reach, almost every piece of ice being covered. There must have been, indeed, many hundreds or even thousands.

"Hurrying from the hill, I called for volunteers, and quickly had a boat's crew ready for some sport. Putting their rifles, a harpoon, and a line into one of the whale boats, we dragged it over the ice to the open water, into which it was speedily launched.

"We had two miles to pull before the margin of the pack

---

* Seasons with the Sea-horses, pp. 84, 85.

was reached. On the cake of ice to which we first came, there were perched about two dozen animals; and these we selected for the attack. They covered the raft almost completely, lying huddled together, lounging in the sun or lazily rolling and twisting themselves about, as if to expose some fresh part of their unwieldy bodies to the warmth—great, ugly, wallowing sea-hogs, they were evidently enjoying themselves, and were without apprehension of approaching danger. We neared them slowly, with muffled oars.

"As the distance between us and the game steadily narrowed, we began to realize that we were likely to meet with rather formidable antagonists. Their aspect was forbidding in the extreme, and our sensations were perhaps not unlike those which the young soldier experiences who hears for the first time the order to charge the enemy. We should all, very possibly, have been quite willing to retreat had we dared own it. Their tough, nearly hairless hides, which are about an inch thick, had a singularly iron-plated look about them, peculiarly suggestive of defense; while their huge tusks, which they brandished with an appearance of strength that their awkwardness did not diminish, looked like very formidable weapons of offense if applied to a boat's planking or to the human ribs, if one should happen to find himself floundering in the sea among the thick-skinned brutes. To complete the hideousness of a facial expression which the tusks rendered formidable enough in appearance, Nature had endowed them with broad flat noses, which were covered all over with stiff whiskers, looking much like porcupine quills, and extending up to the edge of a pair of gaping nostrils. The use of these whiskers is as obscure as that of the tusks; though it is probable that the latter may be as well weapons of offense and defense as for the more useful purpose of grubbing up from the bottom of the sea the mollusks which constitute their principal food. There were two old bulls in the herd who appeared to be dividing their time between sleeping and jamming their tusks into each other's faces, although they appeared to treat the matter with perfect indifference, as they did not seem to make any impression on each other's thick hides. As we approached, these old fellows—neither of which could have been less than sixteen feet long, nor smaller in girth than a hogshead—raised up their heads, and, after taking a leisurely survey of us, seemed to think us unworthy of further notice; and, then punching each other again in the face, fell once more asleep. This was exhibiting a degree of coolness rather alarming. If they had showed the least timidity, we should have found some excitement in extra caution; but they seemed to make so light of our approach that it was not easy to keep up the bold front with which we had commenced the adventure. But we had come quite too far to think of backing out; so we pulled in and made ready for the fray.

"Beside the old bulls, the group contained several cows and a few calves of various sizes,—some evidently yearlings, others but recently born, and others half or three quarters grown. Some were without tusks, while on others they were just sprouting; and above this they were of all sizes up to those of big bulls, which had great curved cones of ivory, nearly three feet long. At length we were within a few boat's lengths of the ice raft, and the game had not taken alarm. They had probably never seen a boat before. Our preparations were made as we approached. The Walrus will always sink when dead, unless held by a harpoon-line; and there were therefore but two chances for us to secure our game—either to shoot the beast dead on the raft, or to get a harpoon well into him after he was wounded, and hold on to him until he was killed. As to killing the animal where he lay, that was not likely to happen, for the thick skin destroys the force of the ball before it can reach a vital part, and indeed, at a distance, actually flattens it; and the skull is so heavy that it is hard to penetrate with an ordinary bullet, unless the ball happens to strike through the eye.

"To Miller, a cool and spirited fellow, who had been after whales on the 'nor-west coast', was given the harpoon, and he took his station at the bows; while Knorr, Jensen, and myself kept our places in the stern-sheets, and held our rifles in readiness. Each selected his animal, and we fired in concert over the heads of the oarsmen. As soon as the rifles were discharged, I ordered my men to 'give way', and the boat shot right among the startled animals as they rolled off pell-mell into the sea. Jensen had fired at the head of one of the bulls, and hit him in the neck; Knorr killed a young one, which was pushed off in the hasty scramble and sank; while I planted a minie-ball somewhere in the head of the other bull and drew from him a most frightful bellow,—louder, I venture to say, than ever came from wild bull of Bashan. When he rolled over into the water, which he did with a splash that sent the spray flying all over us, he almost touched the bows of the boat and gave Miller a good opportunity to get in his harpoon, which he did in capital style.

"The alarmed herd seemed to make straight for the bottom, and the line spun out over the gunwale at a fearful pace; but, having several coils in the boat, the end was not reached before the animals began to rise, and we took in the slack and got ready for what was to follow. The strain of the line whipped the boat around among some loose fragments of ice, and the line having fouled among it, we should have been in great jeopardy had not one of the sailors promptly sprung out, cleared the line, and defended the boat.

"In a few minutes the whole herd appeared at the surface, about fifty yards away from us, the harpooned animal being among them. Miller held fast to his line, and the boat was started with a rush. The coming up of the herd was the signal for a scene which baffles description. They uttered one wild concerted shriek, as if an agonized call for help; and then the air was filled with answering shrieks. The '*huk! huk! huk!*' of the wounded bulls seemed to find an echo everywhere, as the cry was taken up and passed along from floe to floe, like the bugle-blast passed from squadron to squadron along a line of battle; and down from every piece of ice plunged the startled beasts, as quickly as the sailor drops from his hammock when the long-roll beats to quarters. With their ugly heads just above the water, and with mouths wide open, belching forth the dismal '*huk! huk! huk!*' they came tearing toward the boat.

"In a few moments we were completely surrounded, and the numbers kept multiplying with astonishing rapidity. The water soon became alive and black with them.

"They seemed at first to be frightened and irresolute, and for a time it did not seem that they meditated mischief; but this pleasing prospect was soon dissipated, and we were forced to look well to our safety.

"That they meditated an attack there could be no longer a doubt. To escape the onslaught was impossible. We had raised a hornets' nest about our ears in a most astonishingly short space of time, and we must do the best we could.

"It seemed to be the purpose of the Walrus to get their tusks over the gunwale of the boat, and it was evident that, in the event of one such monster hooking on us, that the boat would be torn in pieces, and we would be left floating in the sea helpless. We had good motive therefore to be active. Miller plied his lance from the bows, and gave many a serious wound. The men pushed back the onset with their oars, while Knorr, Jensen, and myself loaded and fired our rifles as rapidly as we could. Several times we were in great jeopardy, but the timely thrust of an oar, or the lance, or a bullet saved us. Once I thought we were surely gone. I had fired and was hastening to load; a wicked-looking brute was making at us, and it seemed probable that he would be upon us. I stopped loading, and was preparing to cram my rifle down his throat, when Knorr, who had got ready his weapon, sent a fatal shot into his head. Again, an immense animal, the largest that I had ever seen, and with tusks apparently three feet long, was observed to be making his way through the herd, with mouth wide open, bellowing dreadfully. I was now as before busy loading; Knorr and Jensen had just discharged their pieces, and the men were well engaged with their oars. It was a critical moment, but, happily, I was in time. The monster, his head high above the boat, was within two feet of the gunwale, when I raised my piece and fired into his mouth. The discharge killed him instantly and he went down like a stone.

"This ended the fray. I know not why, but the whole herd seemed suddenly to take alarm, and all dove down with a tremendous splash almost at the same instant. When they came up again, still shrieking as before, they were some distance from us, their heads all now pointed seaward, making from us as fast as they could go, their cries growing more and more faint as they retreated in the distance.

"We must have killed at least a dozen, and mortally wounded as many more. The water was in places red with blood, and several half-dead and dying animals lay floating about us. The bull to which we were made fast pulled away with all his might after the retreating herd, but his strength soon became exhausted; and, as his speed slackened, we managed to haul in the line, and finally approached him so nearly that our rifle-balls took effect and Miller at length gave him the *coup de grace* with his lance. We then drew him to the nearest piece of ice, and I had soon a fine specimen to add to my Natural History collections. Of the others we secured only one; the rest had died and sunk before we could reach them.

"I have never before regarded the Walrus as a formidable animal; but this contest convinces me that I have done their courage great injustice. They are full of fight; and, had we not been very active and self-possessed, our boat would have been torn to pieces, and we either drowned or killed. A more fierce attack than that which they made upon us could hardly be imagined, and a more formidable looking enemy than one of these huge monsters, with his immense tusks and bellowing throat, would be difficult to find. Next time I try them I will arm my boat's crew with lances. The rifle is a poor reliance, and, but for the oars, the herd would have been on top of us at any time."*

Captain Hall, in his "Arctic Researches," also thus makes reference to a Walrus-fight in Frobisher Bay: "On their way back, Mr. Lamb, in charge of the second boat, had a fight with some Walrus in the following manner. Approaching a piece of ice on which some of these creatures were basking, he attacked one of them, whereupon all the rest immediately rushed toward the boat, and vigorously set upon him and his crew. For a time it seemed necessary to fly for safety; but all hands resisted the attack, and would have got off very well, but that one of the Walrus herd pierced the boat's side with his tusks, and made the invaders retreat to repair damages. Mr. Lamb had to drag his boat upon an ice-floe near by, and stuff in oakum to stop a serious leak thus caused. Finally he succeeded, though with some difficulty, in getting back, and thus ended his encounter with a shoal of Walrus."†

Dr. Kane, in describing the Innuit method of attacking the Walrus from the ice, says: "When wounded, he rises high out of the water, plunges heavily against the ice, and strives to raise himself with his fore-flippers upon its surface. As it breaks under his weight, his countenance assumes a still more vindictive expression, his bark changes to a roar, and the foam pours from his jaws till it froths his beard. . . . He can strike a fearful blow; but prefers charging in a soldierly manner. I do not doubt the old stories of the Spitzbergen and Cherie Island fisheries, where the Walrus put to flight the crowds of European boats. Awuk [Walrus] is the lion of the Esquimaux and they always speak of him with the highest respect.

"I have heard of oomiaks being detained for days at a time at a crossing of straits and passages which he infested. Governor Flaischer told me that, in 1830, a brown Walrus, which according to the Esquimaux is the fiercest, after being lanced

* The Open Polar Sea, pp. 404–411.
† Arctic Researches and Life among the Esquimaux, pp. 334, 335.

and maimed at Upernavik, routed his numerous assailants,
and drove them in fear to seek for help from the settlement,
His movements were so violent as to jerk out the harpoons that
were stuck into him. The governor slew him with great difficulty after several rifle-shots and lance-wounds from his whale-
boat.

"On another occasion, a young and adventurous Inuit
plunged his nalegeit into a brown Walrus; but, startled by the
savage demeanor of the beast, called for help before using his
lance. The older men in vain cautioned him to desist. 'It is a
brown Walrus,' said they; '*Aúvek-Kaiok!*' 'Hold back!' Finding the caution disregarded, his only brother rowed forward
and plunged the second harpoon. Almost in an instant the
animal charged upon the kayacker, ripping him up, as the description went, after the fashion of his sylvan brother, the wild
boar. The story was told me with much animation; how the
brother remaining rescued the corpse of the brother dead; and
how, as they hauled it up on the ice-floes, the ferocious beast
plunged in foaming circles, seeking fresh victims in that part
of the sea which was discolored by his blood.

"Some idea may be formed of the ferocity of the Walrus,"
continues Dr. Kane, "from the fact that the battle which
Morton witnessed, not without sharing some of its danger,
lasted four hours; during which the animal rushed continually
at the Esquimaux as they approached, tearing off great tables
of ice with his tusks, and showing no indication of fear whatever. He received upward of seventy lance-wounds,—Morton
counted over sixty; and even then he remained hooked by
his tusks to the margin of the ice, unable or unwilling to retire.
His female fought in the same manner, but fled on receiving a
lance-wound. The Esquimaux seemed to be fully aware of the
danger of venturing too near; for at the first onset of the Walrus they jumped back far enough to be clear of the broken ice.
Morton described the last three hours as wearing, on both
sides, the aspect of an unbroken and seemingly doubtful combat."*

From the foregoing it appears that the early accounts of the
courage of the Walrus and its attacking and even destroying
boats in defense of its young, or in retaliation for an assault,
finds ample corroboration. I conclude the abundant evidence
on this subject by the following from the pen of Mr. Robert

* Arctic Exploration, vol. i, pp. 414–417.

Brown, who says: "When attacked, unlike the other Seals (unless it be the *Cystophora* [Hoodel Seal]), it [the Walrus] will not retreat but boldly meet its enemies. I was one of a party in a boat which harpooned a solitary Walrus asleep on a piece of ice. It immediately dived, but presently arose, and, notwithstanding all our exertions with lance, axe, and rifle, stove in the bows of the boat; indeed we were only too glad to cut the line adrift and save ourselves on the floe which the Walrus had left, until assistance could reach us. Luckily for us the enraged Morse was magnanimous enough not to attack its chop-fallen enemies, but made off grunting indignantly, with a gun-harpoon and a new whale-line dangling from its bleeding flanks."[*]

The foregoing pages sufficiently indicate the methods and implements commonly employed in destroying the Walrus for commercial or other purposes. To complete the account of the chase it is only necessary to note the special equipment of a Walrus-hunter, and to describe the manner of disposing of the animal when captured, with a brief account of its products and their uses. This will be given from Mr. Lamont's work, already so often quoted, who, in a chapter devoted to the subject, has furnished the only connected and detailed account known to me. From this I condense the following:

A well-appointed Walrus-boat for five men is twenty-one feet long by five feet beam, having her main breadth about one-third from the bow, and strongly built. She is *bow-shaped* at both ends, and should be light, swift, and strong, and easy to manage, and hence has the keel well depressed in the middle. She is always "carvel-built," being thus much less liable to injury from ice or the tusks of the Walruses than if "clinker-built," and easier to repair when damaged. She is braced with thick and strong stem- and stern-pieces, to resist concussions with the ice. There is a deep notch in the centre of the stem-piece, and three others in a block of hard wood on each side of it, for the lines to run through, in addition to which there is also sometimes an upright post on the bow for making fast the lines, but usually the foremost thwart is used for this purpose. Each man rows with a pair of oars hung in grummets to single stout thole-pins. The steersman rows with his face to the bow, and steers with his pair of oars instead of with a single oar or rudder; and each man rowing with a pair of oars enables the crew to turn

---

[*] Proc. Zoöl. Soc. Lond., 1868, p. 429.

the boat much quicker than it could be done otherwise, while the shortness of the oars renders them easier to handle and less in the way among the ice than longer ones would be.   The harpooner rows the bow-oars and is the commander of the boat, he alone using the weapons and the telescope.   The strongest man in the boat is placed next the harpooner, to haul in the line when a Walrus is struck and to be the assistant of the harpooner. The boats are always painted white outside to assimilate their color to that of the ice.   Each boat is provided with six harpoons, placed in racks, three on each side of the bow (inside), and protected by a painted canvas curtain.   To each harpoon is attached twelve or fifteen fathoms of line, each coiled separately in flat boxes under the front thwart, the end being firmly fastened to some strong part of the boat.   The lines should be of the finest quality of two-inch tarred rope, "very soft laid," of the best workmanship and materials.   Four shafts for the harpoons are usually carried, made of white-pine poles about twelve feet long, and about an inch and a half in thickness, fitted at one end to enter the socket of the harpoon.   The harpoons are used for either thrusting or darting, and a skillful harpooner is said to be able to secure a Walrus at a distance of four or five fathoms.   When possible, they are thrust into the victim, and a precautionary twist given in order to disengage the shaft and more securely entangle the barbs in the monster's blubber or skin.   In addition to the harpoons are usually carried four or five very large lances, with heavy, white-pine shafts about nine feet long, and increasing in thickness from an inch and a half to two and a half where it enters the socket of the lance.   This is for the double purpose of giving the necessary strength to the shaft, and to afford buoyancy enough to float the lance-head in case it becomes disengaged from the animal, the lance-head being secured to the shaft by a double thong of raw seal-skin.   Each boat is also provided with five "haak-picks," or boat-hooks, which may be used in dispatching Seals, as well as for the ordinary uses of a boat-hook; also with several axes, a large one for decapitating the dead Walruses, and a small one for cutting the line in case the Walrus proves too fierce and mischievous, or in case of accidents; five or six large, sharp "flensing" knives; an ice-anchor, with tackle for hauling the dead Walruses on to flat icebergs; lockers supplied with various smaller implements and a small outfit of provisions, to guard against the uncertainties arising from accidents and thick weather. In the way of additional weapons, heavy rifles with plenty of ammunition are considered desirable, and often prove of great service when the Walruses are too wary to permit a near approach, as often happens. Generally a mast and sail are, or should be, also carried, though by no means always needed.*

According to the same writer, the manner of "flensing," or taking off and securing the skin and blubber, is as follows: The huge beasts being drawn up on to an ice-floe, the skin, with the blubber adhering, is then removed by dividing the skin into halves† by a slit along the ventral and dorsal lines of the body. It is then loaded into the boats and taken to the ships and thrown into the hold in bulk. Afterward, as leisure or opportunity offers, the skins are drawn up, spread across an inclined platform erected on deck for the purpose, and the blubber removed. This is done by two men who act as "blubber-cutters," clad in oil-skin suits, and armed with large, sharp knives having curved edges. The blubber is then dexterously removed from the skin, cut into slabs of twenty or thirty pounds' weight, and thrown down the hatchway, where two men are stationed to receive it and slip it into the square bung-holes of the casks. From its oiliness it soon finds its own level in the casks, which, when full, are tightly closed. ‡

Captain Hall describes the Esquimaux method of taking the Walrus as follows: "The hunter has a peculiar spear, to which is attached a long line made of Walrus hide; this line is coiled, and hung about the neck; thus prepared, he hides himself among the broken drifting ice, and awaits the moment for striking his game. The spear is then thrown, and the hunter at once slips the coil of line off his head, fastens the end to the ice by driving a spear through a loop in it, and waits till the Walrus comes to the surface of the water, into which he has plunged on feeling the stroke of the harpoon; then the animal is quickly dispatched by the use of a long lance. The recklessness and cool daring of the Innuit is forcibly shown in this operation, for if he should fail to free his neck of the coil at just the right moment, he would inevitably be drawn headlong beneath the ice."§

---

* Compiled from Lamont's Seasons with the Sea-horses, pp. 43–51.

† In the case of full-grown Walruses; but in the case of "calves," the skin is left entire.

‡ Compiled from Lamont's Seasons with the Sea-horses, pp. 76, 77.

§ Arctic Researches, etc., p. 500.

"In attacking the Walrus in the water they [the Esquimaux] use the same gear [as in attacking Whales], but much more caution than with the Whale, always throwing the *katteelik* from some distance, lest the animal should attack the canoe and demolish it with his tusks. The Walrus is in fact the only animal with which they use any caution of .this kind."* This "gear," or *katteelik*, is said to be the largest-of their weapons, and to be used only in attacking Whales and Walruses. It has a shaft of light wood, about four feet in length, like those of their weapons used in killing Seals, but the shaft is much thicker than in the others, especially near the middle, where is lashed a small shoulder of ivory for the thumb to rest against, in order to give additional force in throwing or thrusting the spear. The spear-point is of ivory, fitted into the socket at the end of the shaft, where it is secured by double thongs, in such a way as to give it steadiness when a strain is put upon it in the direction of its axis, but provided with a spring that disengages it when a lateral strain endangers its breaking. To the line attached to the *katteelik* a whole Seal-skin, inflated like a bladder, is fastened, for the purpose of impeding the progress of the animal in the water when struck. †

Dr. Kane gives a graphic account of a Walrus hunt by a party of Innuits. They set off with three sledges drawn by dogs, for the open water, ten miles distant. As they neared the new ice, they would from time to time remove their hoods and listen intently for the animal's voice. Myouk, one of the party, becoming convinced, by signs or sounds, or both, that the Walruses were waiting for him, moved gently on and soon heard the characteristic bellow of a bull. The party now forming in single file followed in each other's steps, winding among hummocks and approaching in a serpentine course the recently frozen ice-spots surrounded by firmer ice. "When within half a mile of these, the line broke, and each man crawled toward a separate pool; Morton on his hands and knees following Myouk. In a few minutes the Walrus were in sight. They were five in number, rising at intervals through the ice in a body, and breaking it up with an explosive puff that might have been heard for miles. Two large grim-looking males were conspicuous as the leaders of the group.

"Now for the marvel of the craft. When the Walrus is above

* Narrative of Parry's Second Voyage, p. 510.
† See Parry's Second Voyage, pp. 507, 508, and pl. facing p. 550, figs. 20, 21

water, the hunter is flat and motionless; as he begins to sink, alert and ready for a spring. The animal's head is hardly below the water-line before every man is in a rapid run; and again, as if by instinct, before the beast returns, all are motionless behind protecting knolls of ice. They seem to know beforehand not only the time he will be absent, but the very spot at which he will reappear. In this way, hiding and advancing by turns, Myouk, with Morton at his heels, has reached a plate of thin ice, hardly strong enough to bear them, at the very brink of the water-pool the Walrus are curvetting in.

"Myouk, till now phlegmatic, seems to waken with excitement. His coil of Walrus-hide, a well-trimmed line of many fathoms' length, is lying at his side. He fixes one end of it in an iron barb, and fastens this loosely by a socket upon a shaft of Unicorn's [Narwhal's] horn: the other end is already looped, or, as sailors would say, 'doubled in a bight'. It is the work of a moment. He has grasped the harpoon: the water is in motion. Puffing with pent-up respiration, the Walrus is within a couple of fathoms, close before him. Myouk rises slowly; his right arm thrown back, the left flat at his side. The Walrus looks about him, shaking the water from his crest: Myouk throws up his left arm; and the animal, rising breast-high, fixes one look before he plunges. It has cost him all that curiosity can cost: the harpoon is buried under his left flipper.

"Though the Awuk [Innuit name of the Walrus] is down in a moment, Myouk is running at desperate speed from the scene of his victory, paying off his coil freely, but clutching the end by its loop. He seizes as he runs a small stick of bone, rudely pointed with iron, and by a sudden movement drives it into the ice: to this he secures his line, pressing it close down to the ice surface with his feet.

"Now comes the struggle. The hole is dashed in mad commotion with the struggles of the wounded beast; the line is drawn tight at one moment, the next relaxed: the hunter has not left his station. There is a crash of the ice; and rearing up through it are two Walruses, not many yards from where he stands. One of them, the male, is excited and seemingly terrified: the other, the female, collected and vengeful. Down they go again, after one grim survey of the field; and on the instant Myouk has changed his position, carrying his coil with him and fixing it anew.

"He has hardly fixed it before the pair have again risen,

breaking up an area of ten feet diameter about the very spot
he left. As they sink once more he again changes his place.
And so the conflict goes on between address and force, till the
victim, half exhausted, receives a second wound, and is played
like a trout by the angler's reel."

The method of landing the beast upon the ice is thus described: "They made two pair of incisions in the neck, where
the hide is very thick, about six inches apart and parallel to each
other, so as to form a couple of bands. A line of cut hide, about
a quarter of an inch in diameter, was passed under one of these
bands and carried up on the ice to a firm stick well secured in
the floe, where it went through a loop, and was then taken back
to the animal, made to pass under the second band, and led off
to the Esquimaux. This formed a sort of 'double purchase',
the blubber so lubricating the cord as to admit of a free movement. By this contrivance the beast, weighing some seven
hundred pounds, was hauled up and butchered at leisure." *

Referring again to the chase of the Walrus, Dr. Kane says
the manner of hunting varies considerably with the season of
the year. In the fall, when the pack is but partly closed, they
are found in numbers about the neutral region of mixed ice and
water, when the Esquimaux assail them in cracks and holes
with nalegeit and line. This fishery, as the season grows colder,
darker, and more tempestuous, is attended with great hazard,
and scarcely a year passes without a catastrophe. The spring
fishery begins in March. The Walrus is now taken in two ways.
Sometimes when he has come up by the side of an iceberg or
through a tide-crack to enjoy the sunshine, he lingers so long
that he finds his retreat cut off by the freezing-up of the opening through which he ascended. The Esquimaux, scouring the
ice-floes with keen hunter-craft, then scent him out by the aid
of their dogs and despatch him with spears. Again they are
found "surging in loving trios from crack to crack, sporting
around the berg-water or basking in the sun," when they are
attacked by their vigilant enemies with the spear and harpoon.
This mode of attack "often becomes a regular battle, the male
gallantly fronting the assault and charging the hunters with
furious bravery. Not unfrequently the entire family, mother,
calf, and bull, are killed in one of these combats." †

---

* Arctic Exploration, vol. i, pp. 407–414, 417.   † Ibid., vol. ii, pp. 131–133.

PRODUCTS.—The commercial products of the Walrus are its oil, hide, and tusks. The oil is said to be much inferior in quality to that of Seals, but is used for nearly the same purposes.* The yield is also much less in proportion to the size of the animal, in the largest specimens seldom exceeding five hundred pounds.† The hide is said to be a valuable commodity, and " sells for from two to four dollars per half skin, calves only counting for a half; it is principally exported to Russia and Sweden, where it is used to manufacture harness and sole leather; it is also twisted into tiller-ropes, and is used for protecting the rigging of ships from chafing. In former times nearly all the rigging of vessels on the north coasts of Norway and Russia used to be composed of Walrus-skin. [‡] When there is a superfluity of the article in the market I believe it is boiled into glue. It is from an inch to an inch and a half thick, very pliable in its green state, but slightly spongy, so that I should doubt the quality of the leather made from it." §

As noted in the earlier portions of this paper, the tusks were in very early times a valuable article of traffic among the barbarous tribes of Eastern Europe and Northern Asia. Brown states that "there is said to be a letter in the library of the Vatican proving that the old Norse and Icelandic colonists in Greenland paid their 'Peter's Pence' in the shape of Walrus tusks and hides." ‖ The ivory afforded by the tusks, though

---

* Lamont says it is usual to mix the Seal and Walrus oil indiscriminately together, and that "the compound is always exported into Southern Europe under the name of Seal oil."—*Yachting in the Arctic Seas*, p. 89.

† Scoresby states that he "never met with any that afforded above twenty or thirty gallons of oil."—*Account of the Arctic Regions*, vol. i, p. 503.

‡ [In the instructions given to Jonas Poole by the Muscovie Company in March, 1610, occurs the following: "And in as much as we have agreed here with a Tanner for all the Morses hides which wee kill and bring into *England*, and have sent men of purpose for the slaying, salting, and ordering of the same, whereof we have appointed one to goe in your ship: We would have you reserve the hides, and stoore your ship therewith in stead of ballast. And if you obtayne a greater quantitie then you can bring away with you, having alwaies regard to commodities of more value, which are Oyle, Teeth, and Whales finnes [whalebone], that none of them be left behind; We would have you leave the said overplus of hides in some convenient place, till the next yeere, that we send more store of shipping."—*Purchas his Pilgrimes*, vol. iii, p. 709.]

§ Seasons with the Sea-horses, p. 77.

‖ Proc. Zoöl. Soc. Lond., 1868, p. 434.

inferior in quality to Elephant ivory, is used for nearly the same purposes. It is said, however, to sooner become yellow by exposure, to be of coarser texture, and hence to have less commercial value. I have met with no statistics relating to the amount annually obtained, or the price it brings in market.[*]

The flesh of the Walrus is sometimes used as food by Arctic voyagers, and forms an important article of diet with the Esquimaux and Tschuchchis. Captain Hall states that while his party remained at Cape True they were never in want of food. "Walrus," he says, "was abundant, and was indeed almost exclusively our diet. We had Walrus brains for supper; stewed Walrus, or Walrus boiled, for dinner; but always Walrus, and no bread."[†] Richardson states that "their flesh is preferred by the Esquimaux before that of the Small Seal (*Phoca hispida*), their feet or fins are considered delicacies, and the heart and liver were pronounced by our navigators to be excellent. The tongue is said to be good when fresh, but becomes oily by keeping."[‡] In the narrative of Cook's last voyage it is stated that the fat of the Pacific Walrus "is as sweet as marrow," but that it soon grows rancid unless salted, when it will "keep much longer." The lean flesh is described as being coarse and black, and as having a rather strong taste, but the heart is said to be "nearly as well tasted as that of a bullock."[§] Captain Parry, in a passage already quoted (*anteà*, p. 119), states that the meat was not only eaten by his men, but was "eagerly sought after on this and every other occasion throughout the voyage, by all those among us who could overcome the prejudice arising chiefly from the dark color of the flesh. In no other respect that I could ever discover, is the meat of the Walrus when fresh-killed in the slightest degree offensive or unpalatable. The heart and liver are indeed excellent."[||]

FOOD.—The food of the Walruses has long been a subject of dispute, not less from the varied character of the substances

---

[*] Mr. Lamont says, respecting products of the Walrus and their value: "Curiosity led me once to weigh and value the marketable parts of a large bull Walrus, and the following results were arrived at:—Weight of Walrus blubber = 520 pounds, about one fifth of a ton, which at 40*l.* a ton is worth 8*l.* ; 300 pounds of skin at 2*d.* a pound = 2*l.* 10*s.*, and 8 pounds of ivory at 5*s.* a pound = 2*l.*, giving a value of 12*l.* 10*s.*"—*Yachting in the Arctic Seas*, p. 89.

[†] Arctic Researches, etc., p. 557.

[‡] Suppl. Parry's Second Voyage, p. 338.

[§] Cook's Last Voyage, vol. ii, p. 457.

[||] Narrative of Parry's Second Voyage, p. 268.

found in their stomachs by different observers than from the peculiar conformation of their teeth. Martens, judging from the appearance of their excrement, thought it must subsist mostly upon sea-grass. Anderson, however, correctly stated that they subsisted upon Mollusca, which they obtained from the bottom of the sea by digging with their tusks. Cranz also says its food seems to consist wholly of "muscles and such kind of shell-fish" and "sea-grass." F. Cuvier, Bell, and others, thought the dentition indicated that their diet must be mainly, if not wholly, vegetable. Most modern observers who have given attention to the matter state that they have often found vegetable matter mixed with other food in their stomachs, some claiming the food to be in small part vegetable, but mainly animal, while others think the fragments of sea-weed so frequently met with in their stomachs are only accidentally present. Mr. Brown, who appears to have had excellent opportunity of obtaining information on this point, observes: "I have generally found in its stomach various species of shelled Mollusca, chiefly *Mya truncata*, a bivalve very common in the Arctic regions on banks and shoals, and a quantity of green slimy matter which I took to be decomposed Algæ which had accidentally found their way into its stomach through being attached to the shells of the Mollusca of which the food of the Walrus chiefly consists. I cannot say that I ever saw any vegetable matter in its stomach which could be decided to have been taken in as food, or which could be distinguished as such. As for its not [*sic*] being carnivorous, if further proof were necessary, I have only to add that whenever it was killed near where a Whale's carcass had been let adrift, its stomach was invariably found *crammed* full of the *krang* or flesh of that Cetacean. As for its not being able to hold the slippery cuirass of a fish, I fear the distinguished author of 'The British Mammalia' [Bell] is in error. The Narwhal, which is even less fitted in its want of dentition for an ichthyophagous existence, lives almost entirely upon platichthyoid fishes and Cephalopoda. Finally the *experimentum crucis* has been performed, in the fact that fish have been taken out of its stomach; and a most trustworthy man, the captain of a Norwegian sealer, has assured me (without possessing any theory on the subject) that he has seen one rise out of the water with a fish in its mouth."*

That it will readily subsist on fish, as well as other animal

* Proc. Zoöl. Soc. Lond., 1868, pp. 430, 431.

food, is further proven by Mr. Bartlett, who states that the one received at the Gardens of the London Zoölogical Society in 1867 was fed on fish, mussels, whelks, clams, and the stomachs, intestines, and other soft parts of fishes, and that while on the way from the Davis's Straits to the Shetland Islands was fed on strips of boiled pork, and subsequently during the voyage on mussels. He says he is inclined to believe it would eat carrion or decomposed flesh, and raises the question whether the Walruses may not " be the scavengers of the Arctic seas, the Vultures among mammals," and suggests that the strong bristles of the muzzle may have something to do with the gathering of this kind of food, " as well as with shrimp-catching." He further states that it declined every kind of sea-weed offered.*

Mr. Lamont informs us that he has found their stomachs to contain great quantities of sand-worms, star-fish, shrimps, clams (*Tridacna*), and cockles (*Cardium*), and that he *believes* that they also eat marine algæ, or sea-weeds.

Malmgren states that he found that the Walruses of Spitzbergen subsist almost exclusively upon two species of mussel, namely, *Mya truncata* and *Saxicava rugosa*, which live buried from 3 to 7 inches deep in the mud, in 10 to 50 fathoms of water. By aid of their grinding teeth and tongue they remove the shells, and swallow usually only the soft parts of the animal. Only once among many thousands examined did Malmgren find any to which a piece of the shell adhered. The young subsist for two years almost solely upon the milk of the mother, they being unable to dig mussels from the mud until their tusks have attained a length of 3 or 4 inches, which length is not acquired till the animals have reached the age of two years.†

In common with some other Pinnipeds, the White Whale and probably other Cetaceans, the Walrus takes into its stomach small stones and gravel, but for what purpose appears as yet unknown. Mr. Brown tells us that considerable quantities of these are always seen around its *atluk*, or breathing-holes.‡

* Proc. Zoöl. Soc. Lond., 1867, p. 820.

† See Malmgren as translated in Toschel's Arch. für Naturgesch., 1864, pp. 68–72. The reasons here given to account for the long period of nursing seem reasonable, but other authorities believe that they derive nourishment from the mother for only one year.

‡ Proc. Zoöl. Soc. Lond., 1868, p. 430.

FUNCTIONS OF THE TUSKS.—The functions of the tusks have been also a matter of dispute, more especially as to whether they are to any degree organs of locomotion. References to their use in effecting a landing upon ice-bergs or upon icy or rocky shores have come down to us from the earliest times, and enter into nearly all the accounts of this animal that have hitherto appeared. That they are thus used rests upon the testimony of a multitude of observers, yet some have claimed that this is not one of their functions. Malmgren states most explicitly that these reports are false, and that the tusks are useful only as weapons, and for the far more important service of digging up the mollusks, that almost exclusively constitute their food.* Other writers, however, who appear to have had equally as good opportunity for observation, refer to the tusks as being of considerable service to the animals in climbing. Cranz says: "The use the Sea-cow makes of these tusks seems to be in part to scrape muscles and such kind of shell-fish out of the sand and from the rocks, for these and sea-grass seem to be its only food; and also to grapple and get along by, for he fastens them in the ice or rocks, and thus draws up his unwieldy helpless trunk; and finally 'tis a weapon of defence both against the white bear on the land and ice, and the sword-fish in the sea."†

Most of the other early accounts of the Walrus contain similar statements respecting the use of the tusks as locomotive organs, and many later writers also refer to this use of them. Mr. Brown says: "I have seen it also use them [the tusks] to

---

* Says Malmgren: "In Betreff der eigentlichen Bestimmung der Zähne bin ich im Stande die nöthige Aufklärung zu geben. Es lässt sich nicht bestreiten, dass dieselben als Waffen angewendet werden und als solche auch furchtbar sind; dass sie aber auch als Lokomotionsorgane dienen sollten, ist eine Fabel, und daher der Name *Odontobænus* Steenstr. nicht passend. Gleich den Robben bewegen sich die Walrosse nur mit Hülfe ihrer Füsse, sowohl auf dem Eise als an den sandigen Meeresgestaden, an denen sie bisweilen hinaufsteigen, um zu schlafen, oft zu Hunderten neben einander. Die Bestimmung der Zähne ist eine ganz andere und für die Existenz des Walrosses bei weitem wichtigere, denn nur mit Hülfe derselben kann es zu seiner Nahrung kommen. Ich fand, dass das Walross sich ausschliesslich von zwei Muscheln, *Mya truncata* und *Saxicava rugosa*, nährt, welche in einer Wassertiefe von 10–50 Faden 3–7 Zoll in dem Bodenlehm eingegraben leben. Um an diese zu kommen, muss das Walross sie aus dem Lehm aufgraben."— *Öfversigt Vetensk. Akad. Förhandl. Stockholm*, 1863, p. 131, as translated in Archiv für Naturgesch., 1864, p. 68.

† The History of Greenland, etc., Brethren's Society's English translation, London, 1767, p. 127.

drag its huge body on to the ice. In progressing on shore it
aids its clumsy progression by their means."*

Dr. Kane observes: "Even when not excited, he manages his
tusks bravely. They are so strong that he uses them to grapple
the rocks with, and climbs steeps of ice and land which would be
inaccessible to him without their aid. He ascends in this way
rocky islands that are sixty and a hundred feet above the level
the of sea; and I have myself seen him in these elevated positions basking with his young in the cool sunshine of August and
September." †

ENEMIES.—In respect to the enemies of the Walruses, man is,
of course, their chief foe; but, after man, all writers rank the Polar
Bears as their principal adversaries. In their conflicts with this
formidable antagonist, the Walrus is usually the reputed victor.
Says Mr. Brown: "The Eskimo used to tell many tales of their
battles; and though I have never been fortunate enough to see
any of these scenes, yet I have heard the whalers give most
circumstantial accounts of the Walrus drowning the Bear, etc.
These accounts may be taken merely for what they are worth;
but still this shows that they are not wholly confined to Eskimo
fable, and ought therefore not to be hastily thrown aside. There
is no doubt, however, that the Bear and Walrus are (like all
the Pinnipedia) but indifferent friends." ‡

Captain Hall, however, relates the following story, rife among
the Innuits, of a very ingenious way the Polar Bear has of killing the Walrus. The bear is said to take up his position on a
cliff to which Walruses are accustomed to resort in fine weather
to bask in the sun on the rocks at its base. The Bear, mounted
on the cliff, watches his opportunity, and " throws down upon
the animal's head a large rock, calculating the distance and
the curve with astonishing accuracy, and thus crushing the thick,
bullet-proof skull. If the Walrus is not instantly killed—simply
stunned—the Bear rushes down to the Walrus, seizes the rock,
and hammers away at the head till the skull is broken. A *fat*
feast follows. Unless the Bear is very hungry, it eats only the
blubber of the Walrus, Seal and Whale." Captain Hall accompanies his account with a picture of a Bear in the act of hurling a
stone upon the head of a Walrus !§ The story, doubtless without

* Proc. Zoöl. Soc. Lond., 1868, p. 430.   ‡ Proc. Zoöl. Soc. Lond., 1868, p. 430.
† Arctic Exploration, vol. i, p. 415.   § Arctic Researches, etc., p. 581.

basis in fact, is of interest in its bearing upon the mythical history of the Walrus. In fact, Dr. Kane, on the other hand, says: "The generally-received idea of the Polar Bear battling with the Walrus meets little favor among the Esquimaux of Smith's Straits. My own experience is directly adverse to the truth of the story. The Walrus is never out of reach of water, and, in his peculiar element, is without a rival. I have seen the Bear follow the Ussuk [Bearded Seal, *Erignathus barbatus*] by diving; but the tough hide and great power of the Walrus forbid such an attack." *

The Walrus is also greatly persecuted with parasites. These, according to Brown, are two species of *Hæmatopinus*, one of which invariably infests the base of the mystacial bristles, and the other its body. "I have seen," says this writer, "the Walrus *awuking* loudly on the ice, tumbling about and rushing back from the water to the ice, and from the ice to the water, and then swimming off to another piece, and repeating the same operation, as if in pain. A few hours afterwards I saw a flock of *Saxicola œnanthe* (it was on a land floe, close to the Fru Islands) alight on the spot. On going over, I found the ice speckled with one of these species of *Hæmatopinus*, on which the birds had been feeding; and the unfortunate Walrus seems to have been in the throes of clearing itself of these troublesome friends, after the approved fashion. Subsequently I have seen these and other small birds alight on the back of the Walrus to peck at these insects, just as crows may be seen sitting on the backs of cattle in our fields."† It seems also to be infested with intestinal parasites. Dr. Murie,‡ in his report upon the causes of the death of the specimen in the Zoölogical Society's Gardens, found it infested by a species of *Ascaris* (*A. bicolor*, Baird) to such an extent that it was probably the cause of its death. He states that he removed from its stomach about "half a pailful" of small round worms, two and a half to three inches in length. Their presence had evidently induced chronic gastritis, death resulting from ulceration. Circumstances seemed to indicate that they had not been introduced with its food since its capture, but that it was infested with them before its capture and confinement.

* Arctic Exploration, vol. i, p. 263.
† Proc. Zoöl. Soc. Lond., 1868, p. 430.
‡ Ibid., pp. 67–71.

DOMESTICATION.—The Walrus possesses a high degree of cerebral development, and seems to be easily susceptible of domestication. It appears, however, to be difficult to keep alive in confinement, especially when taken far south of its natural home. Doubtless the long period occupied in its transportation from the Arctic regions to the zoölogical gardens of European cities, during which time it is necessarily subject to very unnatural conditions and unsuitable food, does much toward reducing it to a greatly enfeebled state before it reaches European ports. It appears, however, that three specimens have at different times reached England, while two at least have been taken to Holland and one to St. Petersburg. In each case they were quite young animals, probably less than a year old. The first specimen seen alive in England reached London August 20, 1608. The account of the capture of this specimen and of its arrival in London is thus detailed by Purchas. It was brought in the ship "God-speed," commanded by Thomas Welden, on its return from a voyage to Cherie (now Bear) Island. The account says: "On the twelfth [of July, 1608,] we took into our ship two young Morses, male and female, alive: the female died before we came into *England*: the male lived about ten weeks. When wee had watered, we set sayle for *England* about foure of the clock in the morning. . . . The twentieth of August, wee arrived at *London*, and having dispatched some private business, we brought our living Morse to the Court, where the king and many honourable personages beheld it with admiration for the strangenesse of the same, the like whereof had never before been seene alive in *England*. Not long after it fell sicke and died. As the beast in shape is very strange, so is it of strange docilitie and apt to be taught, as by good experience we often proved."* It hence appears that this specimen lived for only about three weeks after its arrival in London.

Another is reported to have been exhibited alive in Holland in 1612. This specimen was secured with its mother, which died on the voyage to Holland, but its skin was preserved and stuffed, the two forming the originals of Gerard's famous drawing already noticed. Von Baer,† however, raises the question whether the London and Holland specimens were not really the

---

* Purchas his Pilgrimes, etc., 1624, vol. iii, p. 560.
† Loc. cit., p. 131.

same individual exhibited at different times in the two countries, and devises an ingenious explanation for the origin of the
supposed discrepancy of dates. He seems to be led into these
doubts by the similarity of some of the circumstances attending the capture and exhibition of these animals, and the close
agreement of the dates. Master Welden's account of the capture and transportation of his specimen to London, and of its
early death there, seems, however, too explicit to be overthrown
by mere conjecture. There is apparently no reason for supposing that the London specimen was ever seen alive in Holland.

From a statement in Camper's writings, it would appear that
a living specimen reached Amsterdam about or before 1786, as
he refers to having seen the living Walrus in that city.* But
of this specimen there appears to be no further record. The
specimen taken to St. Petersburg from Archangel, and described
by von Baer, lived only a week after its arrival in St. Petersburg.

In 1853, a second living specimen reached London, and was
placed in the Gardens of the Zoölogical Society, where, however, it survived only a few days, dying apparently of improper
and insufficient food. A third specimen, captured in Davis's
Straits, August 28, 1867, reached the Zoölogical Society's Gardens in London about October 28 of the same year, where it
lived till December 19, or for nearly five weeks, when it died of
chronic gastritis induced by the immense number of intestinal
worms (*Ascaris*), by which it was unfortunately infested.† The
first London and Holland specimens were quite young animals,
as were also probably all the others. The second London specimen (1853) was a "very young" female, but I have seen no further statement respecting its probable age or its size. The
third London specimen (1867), a male, was judged to be less
than a year old, but measured 8 feet in length and weighed
about 250 pounds. No other specimen has thus far, so far as I
can learn, been taken alive to any point south of the Scandinavian ports, to which, according to Brown, they have of late
been frequently carried. ‡

That the Walrus, when young, possesses, like the common
Seals, a high degree of docility and intelligence, is amply evi-

* Camper says : " . . . . et que j'en avois vu plusieurs même un
vivant à Amsterdam."—*Œuvres*, tome ii, 1803, p. 481.
† See Proc. Zoöl. Soc. Lond., 1867, p. 818, and 1868, p. 67.
‡ Proc. Zoöl. Soc., 1868, p. 431.

dent from observations made upon it in captivity. In further illustration of this point I quote the following from Mr. Brown's important paper on the Seals of Greenland and Spitzbergen, from which I have already quoted so largely. Mr. Brown says, in referring to the subject of its naturalization in zoölogical gardens: "I cannot better conclude these notes on the habits of the Walrus than by describing a young one I saw on board a ship in Davis's Straits, in 1861, and which, had it survived, was intended for the Zoölogical Society. It was caught near the Duck-Islands off the coast of North Greenland, and at the same time its mother was killed; it was then sucking, and too young to take the water, so that it fell an easy prey to its captors. It could only have been pupped a very few hours. It was then three feet in length, but already the canine tusks were beginning to cut the gums. When I first saw it, it was grunting about the deck, sucking a piece of its mother's blubber, or sucking the skin which lay on deck, at the place where the teats were. It was subsequently fed on oatmeal and water and pea-soup, and seemed to thrive upon this *outré* nourishment. No fish could be got for it; and the only animal food which it obtained was a little freshened beef or pork, or Bear's flesh, which it readily ate. It had its likes and dislikes, and its favorites on board, whom it instantly recognized. It became exceedingly irritated if a newspaper was shaken in its face, when it would run open-mouthed all over the deck after the perpetrator of this literary outrage. When a 'fall'* was called it would immediately run at a clumsy rate (about one and a half or two miles an hour), first into the surgeon's cabin, then into the captain's (being on a level with the quarterdeck), apparently to see if they were up, and then out again, grunting all about the deck in a most excited manner '*awuk! awuk!*' When the men were 'rallying'† it would imitate the operation, though clumsily, rarely managing to get more than its own length before it required to turn again. It lay during the day basking in the sun, lazily tossing its flippers in the air, and appeared perfectly at home and not at all inclined to change its condition. One day the captain tried it in the water for the

---

* "When a boat gets 'fast' to a whale, all the rest of the crew run shouting about the decks, as they get the other boats out, 'A fall! a fall!' It is apparently derived from the Dutch word 'Val', a whale."

† "When a ship gets impeded by loose ice gathering around it, the crew rush in a body from side to side so as to loosen it, by swaying the ship from beam to beam. This is called 'rallying the ship'."

first time; but it was quite awkward and got under the floe, whence it was unable to extricate itself, until, guided by its piteous '*awuking*', its master went out on the ice and called it by name, when it immediately came out from under the ice and was assisted on board again, apparently heartily sick of its mother element. After surviving for more than three months, it died, just before the vessel left for England. As I was not near at the time, I was unable to make a dissection in order to learn the cause of its death."*

Mr. Lamont thus describes a young Walrus he saw on board the Norwegian brig "Nordby," in the possession of Captain Ericson: "Before parting company, we went on board the 'Nordby' to see a young live Walrus ('a leetle boy-Walrus', as Ericson in his broken English called it), which they had on board as a pet. This interesting little animal was about the size of a sheep, and was the most comical fac-simile imaginable of the old Walrus. He had been taken alive after the harpooning of his mother a few weeks ago, and now seemed perfectly healthy, and tame and playful as a kitten. It was, of course, a great pet with all on board, and seemed much more intelligent than I believed; the only thing which seemed to destroy its equanimity was pulling its whiskers, or pretending to use a 'rope's end' to it, when it would sneak off, looking over its shoulder, just like a dog when chastised. They said it would eat salt fish, salt-beef, blubber, or anything offered it; but I strongly advised Ericson to give it, if possible, a mixture of vegetables or sea-weed along with such strong diet. I assured him that, if he succeeded in taking it alive to the Regent's Park or the Jardin des Plantes, he would get a large price for it; but before I left Spitzbergen in September, I heard with regret that the curious little beast had died."†

Mr. Lamont, on one of his later Arctic expeditions, captured several young Walruses, and seems to have had three alive at one time on board the "Diane." The first was captured on May 27, and safely landed on board, "uttering the most discordant cries which ever assailed the ears of man." "A harsh note—or, more properly speaking, noise, something between a grunt and a bark—henceforth, till we were hardened to the annoyance, broke our slumbers at night and destroyed the peace and quiet of the day. Though particularly anxious to secure and carry

* Proc. Zoöl. Soc. Lond., 1868, pp. 431, 432.
† Seasons with the Sea-horses, pp. 39, 40.

home a young specimen of this interesting animal, we soon found
the company of so noisy a shipmate, with the anxiety connected
with its weaning, was not an unmixed blessing." Again he
says: " . . . . we found amusement in attempts to wean
the Walrus-cub, who still proved obstreperous when attempts
were made to inject preserved milk into his guzzle by means of
a special piece of apparatus borrowed from the doctor's case.
In all other respects he comported himself with the 'strange
docilitie' noted by Master Thomas Welden of the God Speed
in 1608. He became a great pet with the men: a dear, loving
little creature, combining the affection of a spaniel with the proportions of a prize pig. What struck us in watching its singular dexterity was that there could be any difference of opinion
as to the hind-flippers of the Walrus being used in conjunction
with the forepaws after the ordinary method of quadrupeds for
walking on land or ice. 'Tommy' also exhibited a marvellous
knack in climbing, or rather wriggling, his supple carcase up on
to casks and packages in the hold." Later two others were captured, and the three were kept in a pen together. The unlucky
fate that finally befell "Tommy" is thus related: "'Tommy',
the first young Walrus picked up at Novaya Zemlya, a month
ago, to the great grief of every one except 'Sailor' and the cook,
was found dead, with his face immersed in a pail of gruel and
one of the others lying on top of him—clearly suffocated. They
were confined in a pen forward well out of the way; for they
lately had become a great nuisance, crawling about the deck,
always in someone's way, and had taken to roaring like bears
down the companion at night. A few nights before his death
this little beast had fallen down the hatchway; this might have
had something to do with his untimely end. Nothing was found
on examination but a total absence of fat, the rest of the dissection was reserved for the anatomical rooms of the University
of Edinburgh, our late companion and playmate being duly
salted and packed in an old pork-barrel." * Of the fate of the
others I find no record, but they evidently did not live to reach
England. "Taking into consideration," says Mr. Lamont, on
an early page of his work last cited, "the facility with which a
Walrus cub may be captured, it seems strange that they are
not more often met with in the zoological gardens of Europe."
After alluding to previous attempts to take them to European
cities, he says: "Until some special vessel, with cows on board,

* Yachting in the Arctic Seas, pp. 47, 48, 62, 218.

or plenty of Swiss preserved milk, visits the Walrus haunts and thus solves the difficulty of weaning, it will not be easy to import a young Walrus in good condition, and many of the interesting habits and traits of this animal will remain unknown. Although the calf of the previous season frequently accompanies the dam with her more recent offspring, at that age the 'half-Walrus' is too unwieldy a beast to be captured alive; if this were practicable, there can be no doubt its nutrition would be a simple matter."*

From the foregoing accounts of the survival for a considerable period in captivity, and from the hardships we are told the third London (1867) specimen † survived during its long voyage to London, it is evident that with a sufficient supply of proper food, and due arrangement for the comfort of the captives during transportation, coupled with a speedy voyage, as by steamship, young Walruses might easily be taken in numbers and brought safely to southern ports. Whether, however, they could long endure the great change of climate they would be thus forced to experience is a matter of more uncertainty, yet they in all probability would not suffer more than the Polar Bear, or the Sea Lions and Sea Bears, which have of late been frequently seen in different zoölogical gardens. A Sea Lion, as is well known, not only survived a voyage from Buenos Ayres,

---

* Yachting in the Arctic Seas, p. 82.

† This specimen was captured in Davis Strait, August 25, "by a noose swung over his head and one fore limb from the ship and hauled on board. For some days the captive was kept tied to a ring-bolt on deck, and refused food altogether. Subsequently he was induced to swallow thin strips of boiled pork, and was thus fed until the vessel reached the Shetlands, when a supply of fresh mussels was provided for its use. A large box with openings at the sides was fabricated; and the animal, secured therein, was brought safely to Dundee on the 26th ult. [October]. From that port to London the Walrus had been conveyed in the steamer 'Anglia' under the care of the society's superintendent."—*Proc. Zoöl. Soc. Lond.*, 1867, p. 819. Mr. Bartlett further says, in referring to the specimen: "As regards the present animal, I may state that on my arrival at Dundee, on the 29th of October, I found the young Walrus in a very restless state, and, as I thought, hungry; it was being fed upon large mussels; about twenty of these were opened at a meal, and the poor beast was fed about three times a day. [!] I immediately told the owners that I thought the animal was being starved. Stevens at once agreed and a codfish was procured from the neighborhood, and by me cut into long thin strips. On offering these pieces of cod to the animal, he greedily devoured them. Since that time I have fed the Walrus upon *fish, mussels, whelks, clams,* and the *stomachs* and *intestines* and other soft parts of fishes cut small; for I find that it cannot swallow anything larger than a walnut."—*Ibid.*, pp. 819, 820.

Misc. Pub. No. 12——10

across the tropics, to London, but lived there for more than a year,[*] and finally died " from natural causes."

Since writing the above I have met with the following from the pen of Mr. Alfred Newton, respecting the feasibility of obtaining living specimens of the Walrus for the Gardens of the London Zoölogical Society. Referring to the specimen taken to London in 1608, Mr. Newton says: "Now surely what a rude skipper, in the days of James I, could without any preparation accomplish, this Society ought to have no difficulty in effecting; and I trust that the example may not be lost upon those who control our operations. From inquiries I have made, I find it is quite the exception for any year to pass without an opportunity of capturing alive one or more young examples of *Triche-chus rosmarus* occurring to the twenty or thirty ships which annually sail from the northern ports of Norway, to pursue this animal in the Spitsbergen seas. It has several times happened that young Walruses thus taken are brought to Hammerfest; but, the voyage ended, they are sold to the first purchaser, generally for a very trifling sum, and, their food and accommodation not being duly considered, they of course soon die. Lord Dufferin brought one which had been taken to Bergen, and succeeded in bringing it alive to Ullapool;[†] and Mr. Lamont mentions another which he saw in the possession of Captain Erichson.[‡] In making an attempt to place a live Walrus in our Gardens, I do not think we ought to be discouraged by the bad luck which has attended our efforts in the case of the larger marine Mammalia. Every person I have spoken with on the subject corroborates the account given by honest Master Welden of the 'strange docilitie' of this beast; and that in a mere financial point of view the attempt would be worth undertaking is, I think, manifest. To the general public perhaps the most permanently attractive animals exhibited in our Gardens are the Hippopotamuses and the Seals. What then would be the case of a species like the Walrus, wherein the active intelligence of the latter is added to the powerful bulk of the former?"[§]

Since Mr. Newton wrote the above, another specimen has reached London, as already detailed, but this was ten years

---

[*] See Murie, Trans. Zoöl. Soc. Lond., vol. vii, 1872, p. 528.

[†] "Letters from High Latitudes, pp. 387–389."

[‡] "Seasons with the Sea-horses, pp. 26, 27."

[§] Proc. Zoöl. Soc. London, 1864, p. 500.

ago. What efforts have been made, if any, since that date, I know not, but the skill, energy, and money which, some fifteen years ago, placed a White Whale (*Beluga catodon*) in the Aquarial Gardens of Boston, and has recently safely brought another to New York, and has taken others alive into the interior nearly to Cincinnati (the latter dying, however, before quite reaching that city), ought certainly, if directed toward securing living specimens of the Walrus for public exhibition, meet with easy success. As to the influence of a change from an Arctic climate to mild temperate latitudes, it may be well to recall the fact that not many centuries since the natural habitat of the Walrus extended to the southern shores of Nova Scotia and Cape Breton.

ODOBÆNUS OBESUS, (*Ill.*) *Allen.*

Pacific Walrus.

"*Wallross*, STELLER, Beschreib. von dem Lande Kamtsch., 1774, 106."

Sea Horse, COOK's Third Voyage, ii, 1784, 456, pl. lii; ibid., abridged ed., iii, 40.

Trichechus rosmarus, SHAW, Gen. Zoöl., i, 1800, 234 (in part), fig. 68* (from Cook).—VON SCHRENCK, Reisen im Amur-Lande, i, 1859, 179, (in part).—LEIDY, Trans. Amer. Phil. Soc., xi, 1860 (in part).—VON MIDDENDORFF, Sibirische Reise, iv, 1867, 934 (in part). Also in part of most recent authors.

Trichechus obesus, ILLIGER, Abhandl. d. Berlin. Akad. (1804–11), 1815, 64, 70, 75 (distribution).

Trichechus divergens, ILLIGER, Abhandl. d. Berlin. Akad. (1804–11), 1815, 68 (based on Cook's description and figure of the Pacific Walrus).

Rosmarus obesus, GILL, Proc. Essex Inst., v, 1866, 13 (in part only).—DALL, Alaska and its Resources, 1870, 503, 577.—SCAMMON, Marine Mam., 1874, 176 (figure of animal).

Trichechus cookii, FREMERY, Bijdrag. tot de naturkuund. Wetensch., vi, 1831, 385.

Rosmarus cooki, GILL, Intern. Exh., 1876, Anim. Resources U. S., No. 2, 1876, 4 ("Pacific Walrus"; no description); Johnson's New Univ. Cycl., iii, 1877, 1725 (no description).

Rosmarus arcticus, PALLAS, Zoöl. Rosso-Asiat., 1831, 269, "pls. xxviii, xxix."—ELLIOTT, Cond. of Affairs in Alaska, 1875, 121, 160 (Prybilov Islands).

Rosmarus trichechus, GILL, Johnson's New Univ. Cyclop., iii, 1877, 633 (in part only).

EXTERNAL CHARACTERS AND SKELETON.—Similar in size (or possibly rather larger) and probably in general contour (though commonly depicted and described as more robust or thicker at the shoulders) to the *Odobænus rosmarus*, but quite different in its facial outline. The tusks are longer and thinner, generally more convergent, with much greater inward curvature; the mystacial bristles shorter and smaller, and the muzzle relatively deeper and broader, in correlation with the greater breadth and depth of the skull anteriorly. The Pacific Walrus has been supposed to further differ from the Atlantic species by the more naked condition of the skin; but this seems to be merely a feature of age, baldness being more or less common in old age to both species. The color of the hair is nearly the same in both. A large old male in the Museum of Comparative Zoölogy, Cambridge, collected at the Prybilov Islands by Capt. Charles Bryant, is entirely destitute of hair, except around the edge of numerous old scars, and on the breast and ventral surface where here and there are patches very thinly clothed with very short hair, hardly sufficient in amount to remove the general impression of almost complete baldness. The longest mystacial bristles are scarcely more than an inch in length, while the greater part barely project beyond the skin. There is another similar specimen in the collection of the National Museum. A much younger specimen (a female) in the collection of Prof. H. A. Ward, of Rochester, is as well clothed with hair as is the Atlantic species at the same age, from which the color of the hair does not appreciably differ. The mystacial bristles are somewhat longer than in the above-described very old specimens, but are rather shorter than in the Atlantic species at the same age. Probably in young individuals the bristles are much longer than in the adult, as is the case in the Atlantic species. The chief external difference between the two species appears to consist in the shape of the muzzle and the size and form of the bristly nose-pad, which has a vertical breadth at least one-fourth greater than in the Atlantic species. Very important differences between the two species are exhibited in the skull, as will be presently described.

The old male Alaskan Walrus in the Museum of Comparative Zoölogy has a length as mounted of 3350 mm. (about 10½ feet), and a circumference at the shoulders (axillæ) of 3050 mm. The skeleton, as measured while the bones were still connected by cartilage, gave a total length of 9½ feet (2646 mm.), of which the skull measured 15½ inches (354 mm.); the cervical vertebræ 13 (330 mm.); the dorsal vertebræ 45 (1130 mm.); the lumbar 15 (370 mm.); and the caudal 23 (580 mm.). The fore limb, from the proximal end of the humerus to the end of the first or longest digit has a length of 40 inches (1010 mm.), and the hind limb, from the proximal end of the femur to the end of the lon-

gest digit, a length of 54 inches (1040 mm.). The scapula has a
length of 16½ inches (420 mm.), and the innominate bones a
length of 13 inches (330 mm.). The measurements more in detail of the principal bones, taken from the skeleton as mounted,
are as follows:

Measurements of an adult male skeleton of Odobœnus obesus.

	mm.
Total length of skeleton	2890
Total length of skull	390
Extreme breadth of skull	305
Length of canines (from plane of molars)	559
Length of lower jaw	290
Breadth at condyles	238
Length of cervical series of vertebræ	400
Length of dorsal series of vertebræ	1170
Length of lumbar series of vertebræ	380
Length of the sacral and caudal series of vertebræ	550
Length of first rib, osseous portion	150
Length of first rib, cartilaginous portion	95
Length of first rib, total	245
Length of second rib, osseous portion	240
Length of second rib, cartilaginous portion	160
Length of second rib, total	400
Length of third rib, osseous portion	310
Length of third rib, cartilaginous portion	180
Length of third rib, total	590
Length of fourth rib, osseous portion	440
Length of fourth rib, cartilaginous portion	190
Length of fourth rib, total	630
Length of fifth rib, osseous portion	480
Length of fifth rib, cartilaginous portion	220
Length of fifth rib, total	700
Length of sixth rib, osseous portion	565
Length of sixth rib, cartilaginous portion	255
Length of sixth rib, total	820
Length of seventh rib, osseous portion	575
Length of seventh rib, cartilaginous portion	285
Length of seventh rib, total	860
Length of eighth rib, osseous portion	580
Length of eighth rib, cartilaginous portion	275
Length of eighth rib, total	855
Length of ninth rib, osseous portion	570
Length of ninth rib, cartilaginous portion	345
Length of ninth rib, total	915
Length of tenth rib, osseous portion	560
Length of tenth rib, cartilaginous portion	400
Length of tenth rib, total	960
Length of eleventh rib, osseous portion	525
Length of eleventh rib, cartilaginous portion	380
Length of eleventh rib, total	905

mm

Length of twelfth rib, osseous portion	500
Length of twelfth rib, cartilaginous portion	320
Length of twelfth rib, total	820
Length of thirteenth rib, osseous portion	450
Length of thirteenth rib, cartilaginous portion	210
Length of thirteenth rib, total	660
Length of fourteenth rib, osseous portion	365
Length of fourteenth rib, cartilaginous portion	120
Length of fourteenth rib, total	485
Length of fifteenth rib, osseous portion	70
Length of fifteenth rib, cartilaginous portion	00
Length of fifteenth rib, total	70
Length of sternum, osseous portion	540
Length of sternum, total	650
Length of scapula	420
Breadth of scapula	245
Greatest height of its spine (at base of acromion)	53
Length of the humerus	390
Transverse diameter of its head	110
Antero-posterior diameter of its head	132
Transverse diameter of distal end	138
Length of radius	273
Length of ulna	362
Longest diameter of proximal end of ulna	130
Length of carpus	48
Length of first digit	124
Length of metacarpal of second digit	87
Length of third digit	68
Length of fourth digit	68
Length of fifth digit	75
Length of femur	250
Circumference of neck of femur	135
Least transverse diameter of shaft	55
Transverse diameter of shaft at end	118
Length of tibia	380
Length of fibula	375
Length of tarsus	172
Length of metatarsal of first digit	142
Length of second digit	126
Length of third digit	123
Length of fourth digit	132
Length of fifth digit	158
Length of innominate bone	430
Greatest width of pelvis anteriorly	320
Length of ilium	475
Length of ischio-pubic bones	245
Length of thyroid foramen	153
Length of os penis	710
Width of manus at base of metacarpus	140
Width of pes at base of metatarsus	130

Respecting the size and external dimensions, Mr. Elliott says, "the adult male is about 12 feet in length from nostrils to tip of tail [probably in a curved line over the inequalities of the surface] and has 10 or 12 feet of girth, and an old bull, shot by the natives on Walrus Island, July 5, 1872, was nearly 13 feet long, with the enormous girth of 14 feet. The immense mass of blubber on the shoulders and around the neck makes the head and posteriors look small in proportion and attenuated."* He estimates the gross weight of a well-conditioned old bull at "two thousand pounds," the skin alone weighing from "two hundred and fifty to four hundred pounds," and the head "from sixty to eighty." The head, he adds, will measure eighteen inches in length from between the nostrils to the occiput.†

Captain Cook says the weight of one, " which was none of the largest," was eleven hundred pounds without the entrails, the head weighing forty-two and the skin two hundred and five. Of this specimen he gives the following measurements:

		Ft.	In.
Length from the snout to the tail		9	4
Length of the neck from the snout to the shoulder-bone		2	6
Height of the shoulder		5	0
Length of the fins {	fore	2	4
	hind	2	6
Breadth of the fins {	fore	1	2½
	hind	2	0
Snout {	breadth	0	5½
	depth	1	3
Circumference of the neck close to the ears		2	7
Circumference of the body at the shoulder		7	10
Circumference near the hind fins		5	6
From the snout to the eyes		0	7‡

This was evidently either a female or not fully grown. The circumference, as here given, is somewhat less than the length.

Respecting the external appearance of the old males as observed in life by Mr. Elliott on Walrus Island, Mr. Elliott says: " I was surprised to observe the raw, naked appearance of the hide, a skin covered with a multitude of pustular-looking warts and pimples, without hair or fur, deeply wrinkled, with dark red venous lines, showing out in bold contrast through the thick yellowish-brown cuticle, which seemed to be scaling off in places as if with leprosy. They struck my eye at first in a

---

* This is well shown in Mr. Elliott's figures.
† Condition of Affairs in Alaska, pp. 161, 162.
‡ Voyage to the Pacific Ocean, etc., vol. ii, p. 459.

most unpleasant manner, for they looked like bloated, mortify-
ing, shapeless masses of flesh; the clusters of swollen, warty
pimples, of a yellow, parboiled flesh-color, over the shoulders
and around the neck, suggested unwholesomeness forcibly." *
The old male, in the Museum of Comparative Zoölogy, of which
measurements are given above, is almost wholly naked, except
about the numerous old healed gashes and scars, which are
generally bordered with very short, stiff, brownish hair. Captain Scammon, however, who has also observed them in their
native waters, states that the hair that covers " most individuals is short and of a dark brown; yet there is no lack of examples where it is of a much lighter shade, or of a light dingy
gray. . . . The young, however, before its cumbrous canines
protrude . . . is of a black color." †

The mystacial bristles appear to vary in length in different
individuals. Pallas's figure of a rather young animal represents
them as thick and long. In the old specimen in the Museum of
Comparative Zoölogy they are very short, and do not form a very
prominent feature of the physiognomy. On the upper part of
the muzzle they are merely short, small-pointed spines, one-
fourth to one-half or three-quarters of an inch in length; they
increase somewhat in length toward the edge of the lip, where
the longest obtain a length of about two inches. They are quite
slender, the coarsest having a diameter of not more than eight
one-hundredths of an inch.

Captain Scammon states that " The cheeks are studded with
four or five hundred spines or whiskers, some of which are rudimentary, while others grow to the length of three or four inches.
They are transparent, curved, abruptly pointed, and about the
size of a straw, but not twisted, as has been stated by some
writers." ‡ Mr. Elliott describes them as being "short, stubbed,
gray-white bristles, from one-half to three inches long." § The
descriptions of the bristles of the Atlantic Walrus, as given by
numerous writers, agree in representing them as much longer
and thicker than in the Pacific species, the dimensions usually
assigned being a length of four or five inches, or even, in some
cases, six, and about one-twelfth of an inch thick. The figures
and descriptions commonly represent them as forming, by their

---

* Condition of Affairs in Alaska, p. 160.
† Marine Mammalia, p. 177.
‡ Marine Mammalia, p. 176.
§ Condition of Affairs in Alaska, p. 161.

size and length, next to the long tusks, one of the most striking features of the physiognomy. In Cook's and Elliott's figures of the Pacific species, how-ever, they are by no means a prominent feature, and there are no such allusions to the formidable aspect they give to the facial ex-pression as are commonly met with in the accounts of the Atlantic species. A direct comparison of speci-mens of corresponding ages shows them to be much shorter than in the Atlan-tic Walrus.

The eyes of the Atlantic Walrus are described as fiery red, one writer com-paring them to glowing coals. Mr. Elliott refers to those of the Pacific species as having the sclerotic coat "of a dirty, mottled coffee-yellow and brown, with an occasional admixture of white; the iris light-brown, with dark-brown rays and spots"; and in no case have I seen any reference to their being "red." While most writers who have de-scribed the Atlantic Wal-rus from life refer to the redness of the eyes as a remarkable and striking feature, Cook, Scammon, and others (Mr. Elliott ex-

FIG. 13.—*Odobænus obesus.*

cepted) make no reference to the color of the eyes, which would hardly have escaped them had they possessed the redness char-acteristic of the Atlantic species.

Mr. Elliott further describes the eyes as small, but prominent, "protruding from their sockets like those of a lobster," and states that the animals have the power of rolling them about in every direction, so that when aroused they seldom move the head more than to elevate it, the position of the eyes near the top of the head giving them the needed range of vision.

The nostrils, as in the Atlantic species, are at the top of the muzzle; they are "oval, and about an inch in their greatest diameter." The auricular opening is placed nearly in a line with the nostrils and eye, and hence near the top of the head in a fold of the skin. The animal is said to have a keen sense of smell and an acute perception of sound, but a limited power of vision. *

An idea of the uncouth and peculiar facial aspect of the Pacific Walrus may be derived from the above-given figures (Fig. 13) drawn by Mr. Elliott, to whose kindness I am indebted for their presentation in the present connection.

I append herewith measurements of a considerable series of skulls, of different ages, one only of which is marked as that of a female, they being mostly skulls of middle-aged or very old males.

---

* See Elliott, l. c., pp. 161, 162.

Measurements of fifteen skulls of ODOBÆNUS OBESUS.

Cat. No.	Locality.	Sex.	Length.	Greatest width at zygoma.	Width at mastoid processes.	Least distances between orbits.	Nasal bones, length.	Nasal bones, width posteriorly.	Nasal bones, width anteriorly.	Anterior border of intermaxillaries to hinder edge of palatines.	Canines, length, measured from plane of crown of the molars.	Canines, circumference at base.	Canines, distance between external edges at base.	Canines, distance apart at tips.	Length of the molariform series of teeth.	Lower jaw, length.	Lower jaw, height at coronoid process.	Age.
*1725	Walrus Island, Alaska	♂	386	261	322	82				231	800	230	222	348	102	294	106	Very old.
*1722	Walrus Island, Alaska	♂	392	247	298	84				223	635	210	198	96	95	294	98	Very old.
*130	Walrus Island, Alaska	♂	375	250	300	78	79	65	59	222	576	203	203	78	97	292	97	Very old.
*	Walrus Island, Alaska	♂	363	235	284	76				218	522	190	195	196	85			Middle-aged.
†	Walrus Island, Alaska	♂	400	256	318	70				228	585	215	227	44	102	252	86	Middle-aged.
†	Alaska	♂	356	208	255	72				218	370	153	167	76	82	230	80	Rather young.
†	Alaska	♂	318	150	241	58				185	472	148	207	162	67	223	81	Rather young.
†	Alaska	♂	318	153	230	63				178	445	140	131	148	77			Rather young.
†7889	Alaska	♂	400	235	298	70				227	415	210	210	438	102	286	104	Old.
†9475	Alaska	♂	380	244	304	80				229	395	212	216	273	100	286	108	Old.
	Alaska	♂		257		74				229								Old.
†11746	Walrus Island, Alaska	♂		257		90	79	64	57	229								Very old.
†14397	Walrus Island, Alaska	♂	368	223	263	65				208	415	223	225	0	95	278	105	Quite young.
†14396	Walrus Island, Alaska	♂	350	244	303	74				233	313	168	168	237	86	274	87	Middle-aged.
†14395	Walrus Island, Alaska																	
‡6780	North Pacific	♀	324	185	246	71				183	356	132	133	115	103	283	102	Old.

* In Museum of Comparative Zoölogy, Cambridge, Mass.   † In Museum of Boston Society Natural History, Boston, Mass.   ‡ In National Museum, Washington, D. C.

DIFFERENTIAL CHARACTERS.—As already stated, the Pacific Walrus differs from the Atlantic Walrus very little in external characters, except in facial outline and in the size and

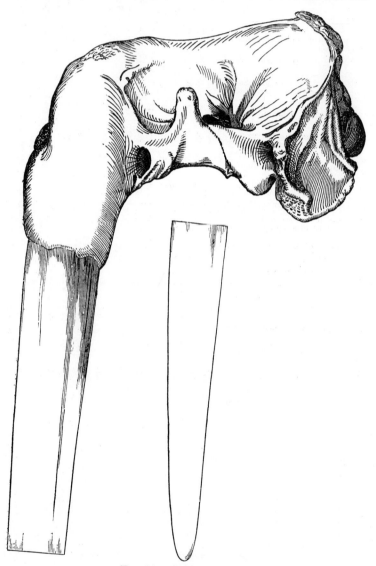

FIG. 14.—*Odobænus obesus.*

"set" or curvature of the tusks. The skulls, however, afford many important differential characters, and on these differences I venture to predicate the existence of two species, using the term "species" in its commonly accepted sense. To show more readily what these differences are, I present herewith a series of figures of skulls of both old and young of the two forms. The skulls selected for this purpose are average examples of a considerable series, the adult skulls being those of males of strictly comparable ages.

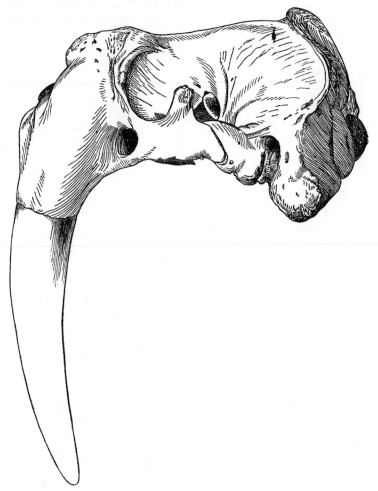

FIG. 15.—*Odobœnus rosmarus.*

The skulls of the two species seen in profile (Figs. 14 and 15) exhibit the following differences: The first and most obvious is perhaps that presented by the tusks, which, in the Pacific species (Fig. 14), are much longer and thicker than in the other

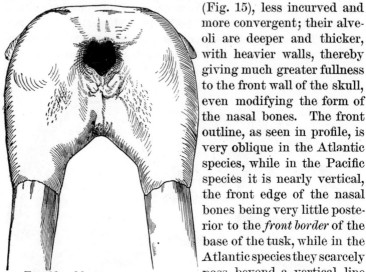

(Fig. 15), less incurved and more convergent; their alveoli are deeper and thicker, with heavier walls, thereby giving much greater fullness to the front wall of the skull, even modifying the form of the nasal bones. The front outline, as seen in profile, is very oblique in the Atlantic species, while in the Pacific species it is nearly vertical, the front edge of the nasal bones being very little posterior to the *front border* of the base of the tusk, while in the Atlantic species they scarcely pass beyond a vertical line

FIG. 16.— *Odobænus rosmarus.*

drawn from the *hinder border* of the tusk. The orbits in the Pacific species are placed more anteriorly than in the other.

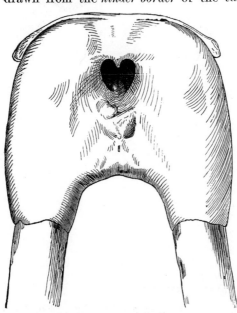

In the front view of the skulls, the muzzle is seen to be much smaller in the Atlantic species (Fig. 16) than in the Pacific (Fig. 17), with, however, not very marked differences in outlines and proportions. The receding upper border in the latter is a marked feature. The difference in size here shown is an important one,

FIG. 17.—*Odobænus obesus.*

since the two skulls compared differ very little in general size, they giving very nearly the same measurements in respect to extreme dimensions of length and breadth. The difference is

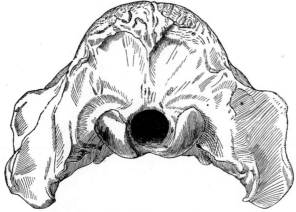

FIG. 18.—*Odobænus rosmarus.*

hence one of proportion, resulting from the far greater development in both breadth and depth of the anterior portion of the skull in the Pacific species. But while the skull of the

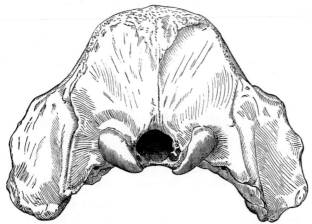

FIG. 19.—*Odobænus obesus.*

Atlantic species is smaller anteriorly than the other, it has the occipital region (Fig. 18) more heavily developed than is the case in the Pacific species (Fig. 19). The difference in development of the mastoid processes is strongly apparent, not only as respects massiveness, but in the general outline. In the Pacific species there is a thinness and anterior deflection not seen in the other.   The sculpturing of the occipital plane (after allowing for a considerable range of individual variation in this respect) is quite different, as well as the relative degree of verticality.   The occipital breadth of the skull, as compared with the total length of the skull, is not greatly different in the two forms.   In the Pacific species, the occipital condyles are narrower than in the other, and are placed at a somewhat different angle, both laterally and vertically.

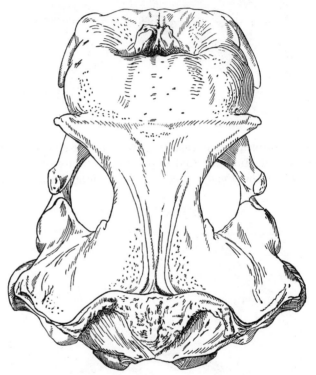

FIG. 20.—*Odobænus rosmarus.*

The difference in relative development of the anterior and posterior portions of the skull in the two species is best seen from above (Figs. 20 and 21).   In this view, the narrow facial breadth in the Atlantic species (Fig. 20) is in striking contrast with its great occipital breadth, whereas in the Pacific species (Fig. 21) the two regions are more equally developed.   Another difference brought out in this view is the greater interorbital constriction in the Pacific species, which is not only relatively but actually much narrower than in the other, while the point of greatest constriction is considerably posterior to the same point in the Atlantic species.

There are also important differences in the form of the different bones of the skull, as shown in the young. In the Pacific species (Fig. 22), the nasals are nearly one-third longer and narrower than in the Atlantic species (Fig. 23), and the frontals have a quite different posterior outline, they being abruptly

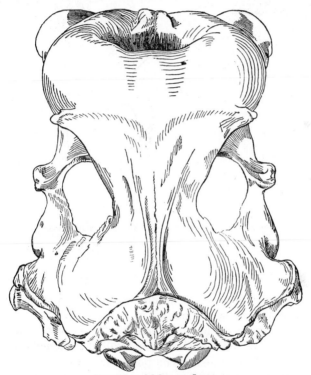

FIG. 21.—*Odobœnus obesus.*

narrowed just behind the orbital fossæ to less than half the breadth they present in the Atlantic Walrus, and extend further posteriorly in a narrow point instead of being rather abruptly truncated. In the Atlantic species, the lateral anterior angle of the frontals is in a line with the most laterally projecting portion of the maxillaries, while in the Pacific species the breadth at this point is considerably greater than at the anterior border of the frontals. While the frontals present in each species a considerable range of variation in respect to their posterior outlines, the average difference is very nearly as here rep-

resented.   The young skulls here compared are of nearly the
same age; but unfortunately the absence of the occipital portion of the skull in the only Pacific specimen of this age (Fig.
22) I am able to figure renders it impossible to compare by figures the occipital region in young specimens.   Other specimens

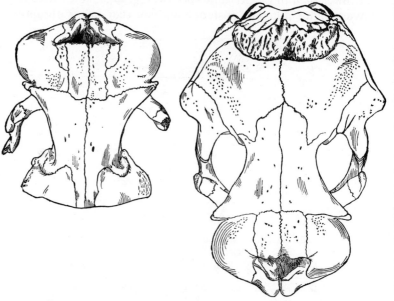

FIG. 22.—*Odobænus obesus.*   FIG. 23.—*Odobænus rosmarus.*

young enough to have the sutures still open show the differences seen in the occipital region of older skulls.

Another difference, but one apparently less constant than the
others, is the presence in the young skull of the Pacific Walrus
(Figs. 22 and 24) of an extension posteriorly of the.intermaxilla-
ries for two-thirds of the length of the nasals.   In the Atlantic
skull (Figs. 23 and 25), the intermaxillaries do not enter into the
dorsal outline of the skull, but terminate at the anterior border of the nasals.   This difference is open to exceptions, and is
not offered as a character of importance, since the same modification or backward prolongation of the intermaxillaries occurs
occasionally in the Atlantic species, and is sometimes absent in
the Pacific species, while in some examples the intermaxillaries
reach the dorsal surface only as isolated ossicles between the
nasals and maxillaries.   As a rule, however, the conditions in
this respect shown in the young skulls here figured appear to
be diagnostic of the two species.

A comparison of the skulls as seen from below (Figs. 26 and 27) shows not only the considerably greater contraction of the skull anteriorly, and the greater massiveness and different form of the mastoid processes in the Atlantic Walrus, but other weighty differences. These are especially seen in the size and form of the auditory bullæ, and, to a less extent, in the form of the occipital condyles, the form of the glenoid cavity, the orbital fossæ, etc. In the Atlantic Walrus (Fig. 26), the auditory

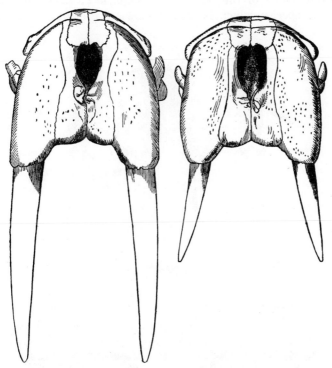

FIG. 24.—*Odobænus obesus.*   FIG. 25.—*Odobænus rosmarus.*

bullæ are relatively larger than in the other (Fig. 27), more quadrilateral in outline, and rather more swollen. The differences in size and outline are very considerable, the auditory bullæ in the Pacific species being, as respects outline, nearly triangular. The inner anterior angle is also strongly developed, being by far the most inwardly salient portion of the bullæ, while in the Atlantic skull it is greatly suppressed. As regards the occipital condyles, they are broader and shorter in the Atlantic species, and less produced anteriorly. The space between them is also

considerably broader than in the other, and the plane of artic-
ulation is more nearly vertical.   This seems correlate with the
greater incurvation of the tusks; these, being almost vertical
in the Pacific species, allow a greater declination of the head.

Another difference apparent in this aspect of the skull is the
relative posterior extension of the condylar portion, which, in the
Pacific species, extends much further beyond the posterior border of the mastoids than in the other.   This is obviously due to
greater length of the basioccipital segment of the skull in the
Pacific species, which is clearly shown in the annexed figures

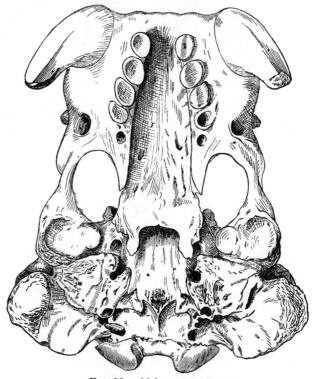

Fig. 26.—*Odobænus rosmarus.*

(Figs. 26 and 27).   The position of the foramina of the basal
portion of the skull is also quite different in the two, as is
especially seen in respect to the condylar foramina, which are
situated more posteriorly in the Atlantic species than in the
other, due, perhaps, to the shortness in this form of the basioccipital region.

Another difference not yet noted consists in the greater length and massiveness of the zygomata in the Pacific species, in which they are fully one-third heavier than in the Atlantic species; they being in the former both deeper and thicker. (This is well shown in the above given figures of the skulls as seen in profile and from above and below, but especially as as seen from below.) The orbital fossæ are also quite different, they being relatively long and narrow in the Pacific, and shorter and broader in the Atlantic Walrus.

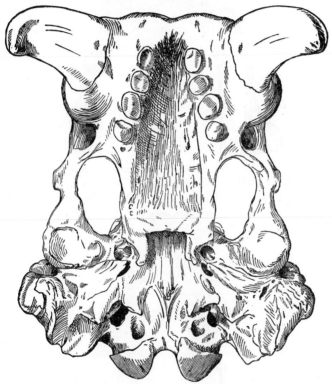

FIG. 27.—*Odobœnus obesus.*

To sum up in a word the above-detailed cranial differences between the two species of Walruses, the skull of the Pacific animal is heavily developed *anteriorly* and relatively much less so posteriorly, while in the Atlantic Walrus just the reverse of this obtains, the skull in the latter being heavily developed *posteriorly* and relatively less so anteriorly. The axis of variation being at the posterior border of the orbital fossæ, the zygomata share the general character of the anterior half of the skull.

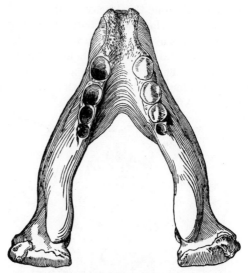

FIG. 28.—*Odobænus rosmarus*. Adult.

But equally striking differences are seen in a comparison of the lower jaws of the two species. These differences correlate

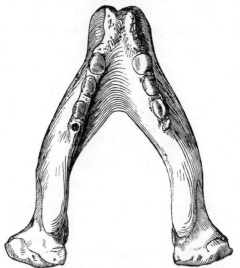

FIG. 29.—*Odobænus obesus*. Adult.

in a most interesting manner with those that characterize the cranium. Thus, in the Atlantic species (Figs. 28 and 30), the mandible is far less massive anteriorly than in the Atlantic Walrus (Figs. 29 and 31), while it is much more massive posteriorly. There is also considerable difference in the mandibles of the two in other respects. Thus, not only is the mandible of the Pacific Walrus much thicker, both laterally and vertically, at the symphysis, but the border of the ramus is widely unlike in the two forms. In the Atlantic Walrus (Fig. 30), the

FIG. 30.—*Odobœnus rosmarus.* Adult.

inferior border of the ramus, from the posterior end of the symphysis to the front of the jaw, rises by a gradual and nearly uniform curve; in the Pacific Walrus (Fig. 31), the inferior border scarcely rises at all, the jaw in front being simply bluntly rounded. In respect to the posterior portion of the ramus, the differences consist in the greater breadth of the condylar portion in the Atlantic species, and the greater thickness of the

FIG. 31.—*Odobœnus obesus.* Adult.

coronoid process. These differences are all strongly pronounced in even quite young skulls, this being especially the case with respect to the inferior border of the symphysial portion of the jaw (Figs. 32 and 34). Another difference consists in the position of the coronoid process, which in the Pacific Walrus, especially in the young, is central to the axis of the ramus, while in the Atlantic species it rises more from the inner edge, and the process itself has an inward curvature not seen in the other (Figs. 33 and 35).

The cranial differences here detailed as obtaining between the Atlantic and Pacific Walruses are borne out by a large series of the skulls of the two species, numbering not less than twelve to fifteen of each. There is in each species a considerable range of individual variation; but the differences presented by the skulls here figured fairly represent average conditions. The only exception to be made is in respect to the tusks of the Pacific specimen figured, which are perhaps above the average in size, while they are remarkably divergent, more so than in any other specimen of this species that I have seen. Ordinarily, *or as a rule*, they are more or less *con*vergent, and sometimes even meet or overlap, while in the Atlantic species they are, as a rule, *di*vergent. While in the Pacific species the tusks descend almost vertically, in the Atlantic species they are quite uniformly strongly incurved.

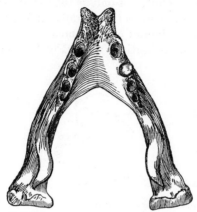

FIG. 32.—*Odobænus rosmarus.* Young.

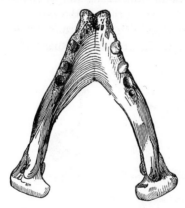

FIG. 33.—*Odobænus obesus.* Young.

In view of the differences in the skulls here described, together with the correlating differences of facial expression, notwithstanding the absence of other very strongly marked external differences, I have little hesitation in according to these two forms specific rank. Added to these differences is the fact of their unquestionably long geographic separation. Whether an individual of one species may not occasionally find its way to the habitat of the other is a question for future consideration. That such an occurrence is not impossible seems evident from the fact of the existence, during portions of the year at least, of areas of open water along those portions of the Arctic coast supposed to separate the habitats of the two species. Further than this, I have seen a skull (now in the Museum of the Boston Society of Natural History) which Capt. Charles Bryant (certainly a trustworthy authority) assures me was taken by his assistant, on Walrus Island, in the summer of 1871 or 1872, that agrees in every particular with the skulls of the Atlantic species. This skull having been somewhat fantastically painted (the lower surface deep red and the upper yellowish-white), led me at first to doubt the correctness of the alleged local-

FIG. 34.—*Odobœnus rosmarus*. Young.

ity, supposing that if really obtained at the Prybilov Islands it might have been brought there from some distant point. This quaint ornamentation proves, however, an aid in fixing the locality of its capture as Walrus Island. It differing so widely from the form usually occurring in those waters, it at once attracted attention, and was mounted on a bracket and preserved as a curiosity, the paint being applied, as Captain Bryant informs me, to facilitate its being kept free of dust! Captain Bryant states (in a letter to the writer) that he has himself "seen two specimens like it," but adds that he "did not succeed in killing them." Hence, of course, their resemblance to the one now in question is only *presumed*, the animals being only *seen alive*. He writes, further, that this "head" was recognized as "different from any before seen there." I will merely add that this skull is indistin-

FIG. 35.—*Odobœnus obesus*. Young.

guishable in any essential detail from skulls of corresponding age from the Atlantic waters, and points to the occasional occurrence of *Odobœnus rosmarus* within the habitat of *Odobœnus obesus*. As von Middendorff has shown (see *anteà*, p. 78), the Walrus (presumably the Atlantic species) has occurred much further to the eastward than the limits assigned it by von Baer, he having traced it, satisfactorily to himself, apparently, to

within thirty degrees of the western limit of the range of the
Pacific animal. In view of these facts, the question arises as
to whether the Atlantic species may not occasionally pass along
the northern coast of Asia so far as to sometimes reach the
habitat of the Pacific species.

NOMENCLATURE.—The first specific name applied to the Pacific Walrus is *obesus*, given by Illiger in 1815, in his "Ueberblick der Säugethiere nach ihrer Vertheilung über die Welttheile."* In this paper this name is three times used as a distinctive appellation for the Pacific Walrus, namely, (1) in his
list of the species of Northern Asia, in which "*Trichechus rosmarus*" and "*Trichechus obesus*" are both given; (2) in his list
of the species of North America; and (3) in his remarks respecting the first-named list. In these remarks (l. c., p. 75) he says,
"Die beiden Arten des Wallrosses, *Trichechus obesus* und [*T.*]
Rosmarus, sind schon bei Nord-Asien vorgekommen." For Europe he gives only *T. rosmarus* (l. c., p. 56), respecting the distribution of which he says, "Der *Trichechus Rosmarus*, das Wallross, lebt an den eisigen Küsten von Nord-Europa, Nord-Asien,
und des ostlichen Nord-America" (l. c., p. 61). It is thus not
quite clear whether he considered his *T. rosmarus* to have a
complete circumpolar range, with *T. obesus* as a second species
occurring only on the northeastern shores of Asia and the northwestern shores of North America, or whether, as is more probable,
he merely meant that *T. rosmarus* ranged eastward along the Arctic coast of the Old World to the northern shore of Western Asia
(as is the fact), and was replaced on the Pacific shores of Asia and
America by *T. obesus*. In either case he recognized as a distinct
species, under the name *T. obesus*, the Walrus of the North Pacific and adjacent portions of the Arctic Ocean. In the same
paper is also a reference to a *Trichechus* "*divergens*," respecting
which he thus observes: "Auser dem schon bei Europa erwähn-
ten Wallross, *Trichechus Rosmarus*, findet sich an der westlichen
Nord-Amerkanischen und nahen Ost-Asiatischen Küste, und
dem Eise dieser Meere, vielleicht aber auch an der ganzen Küste
des Eismeers das von Cook beschriebene und abgebildete Wallross, das ich wegen mehrerer Verschiedenheiten, besonders der
Hauzähne, als eigne Art unter dem Namen *divergens* aufge-
führt habe" (l. c., p. 68). He thus, in the same paper, appears to
recognize two species of Pacific Walruses. The name *divergens*,

---

* Abhand. der Akad. der Wissensch. zu Berlin, 1804–1811, (1815), pp. 64,
70, 75.

however, does not again occur, so far as I can find, either in this paper or in any of the writings of this author. The name *obesus* has several pages priority over *divergens*, and must hence be adopted for the Pacific Walrus.

The next names applied to the Walruses are those used by Fremery, who, in 1831, recognized three species, namely, *Triche-chus rosmarus*, *T. longidens*, and *T. cooki*. The first is the common Walrus of the North Atlantic. The second was founded on a skull with long, slender, and somewhat converging tusks, the locality of which is not stated, but the species is usually considered as based on the skull of a female Atlantic Walrus. The third is obviously the Walrus described and figured by Captain Cook. The latter is hence synonymous with *obesus* (and *divergens*) of Illiger. The second (*longidens*) has generally been, as just stated, considered as based on a female skull of the common Atlantic Walrus.

In 1842, Stannius, while referring all the previously given names to one species, characterized what he believed to be a second species, under the name *dubius*, based on a large skull presenting unusual features of individual variation. I do not find that the locality of this specimen is distinctly given, but von Middendorff appears to consider Stannius's *T. dubius* to have reference to the Pacific Walrus.*

In 1866, Gill, in adopting *Rosmarus* as the generic name of the Walruses, took Illiger's name *obesus* for the specific name of the single species he (Gill) at that time recognized. Later (as already noticed, see *antèa*, p. 22), in naming the two presumed species of Walruses, Gill chose *obesus* as the name of the Atlantic species, and took *cookii* of Fremery for the Pacific species, overlooking the fact that *obesus* was originally applied to the Pacific species, in obvious allusion to its supposed more robust or thicker form as compared with the Atlantic Walrus.

HISTORY.—The Pacific Walrus appears to have received its first introduction into literature through the early exploration of the

---

* Von Middendorff says: "Ersterer verglich [he refers at this point in a footnote to Stannius's paper] Schädel und Gebisse der Walrosse unter einander und fand die Hauer bei den Walrossen der Beringsstrasse etwas länger, dünner und gelinde spiralig gegen einander gekrümmt, im Vergleiche mit denen des atlantischen Eismeeres. Seine eigenen schliesslichen Zweifel spricht aber der vorgeschlagene Name, *Trichechus dubius*, deutlich, genug aus."—*Sibirische Reise*, Bd. iv, p. 792. I do not understand, however, that Stannius's *T. dubius* had any reference to either these characters or to the Pacific Walruses. (Compare Stannius's paper in Müller's Arch. für Anat., 1842, pp. 392, 405-407).

eastern portion of the Arctic coast of Asia, about the middle of the seventeenth century, by the Cossack adventurer Staduchin, who found (about 1645 to 1648) its tusks on the Tschuktschi coast, near the mouth of the Kolyma River. A century later Deschnew found also large quantities of Walrus teeth on the sand-bars at the mouth of the Anadyr. These explorations, so interesting geographically, appear not to have been known in Moscow till Müller, in 1735, discovered the reports of them at Jakutsk which he published in his "Sammlung russischer Geschichte." * Hence not until the last half of the eighteenth century did the Pacific Walrus become fairly known, mainly through the explorations of Steller, Kraschininnikoff, Cook, Kotzebue, Lütke, Billings, Pallas, and others, each of whom referred to or gave more or less full accounts of it. The Pacific Walrus was first figured in Cook's "Last Voyage," and subsequently by Pallas. Later it was noticed by Wrangell on the Tschutkchi coast and by Beechey in Behring's Straits and the neighboring waters. More recently we have notices of it by Dall, Scammon, and Elliott, the two last-named authors giving us by far the most detailed account of these animals which has, to my knowledge, thus far appeared, and from whose writings I have freely borrowed in the preparation of the following pages.

FIGURES.—The first figures of the Pacific Walrus appear to be those published in Cook's "Last Voyage,"† in 1784, when a group of Walruses is represented as resting on the ice. The more prominent of these figures was copied by Shaw‡ in 1800, and later by Godman§ and others. It was also reproduced by Gray in 1853,‖ and is here republished (Fig. 36).

According to von Baer, Pallas, in his "Icones,"¶ gave two illustrations of the Walrus. The one, he says, shows the animal from the side, the other as lying on its back. Von Baer describes these as being far better than any figures of the Walrus that had preceded them, with the exception of Gerard's (1612), already described. The structure of the hind feet, he says, is well represented, except that the nails on all the feet are too long.

* For this history in greater detail, see von Baer, l. c., pp. 175–177.
† Voyage to the Pacific Ocean, etc., under the direction of Captains Cook, Clerke, and Gore, in the years 1766–1780, vol. ii, pl. lii.
‡ General Zoöl., vol. i, 1800, pl. lxviii, facing p. 234. Also Nat. Miscel., pl. lxxyi.
§ Amer. Nat. Hist., vol. i, 1826, pl.
‖ Proc. Zoöl. Soc. Lond., 1853, p. 116.
¶ "Icones ad Pallasii Zoographiam, fasc. ii."

The fore foot, however, he says, is wrongly represented. Von Baer criticises the form of the nose, the front part of which he says is too prominent, and has the angles or wings (Nasenflügel) too distinct, and adds that the coloring is also faulty. But von Baer's comparison is made with the young specimen of the At-

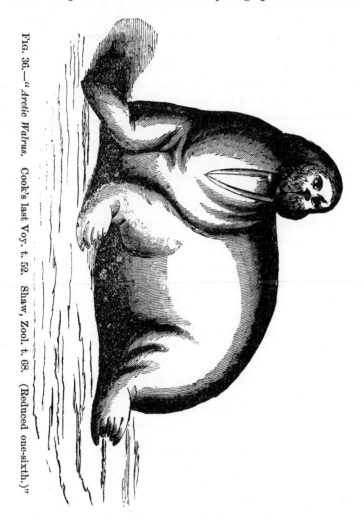

FIG. 36.—"*Arctic Walrus.* Cook's last Voy. t. 52. Shaw, Zool. t. 68. (Reduced one-sixth.)"

lantic Walrus he observed in St. Petersburg, and perhaps indicates the differences between the two species, rather than any incorrectness in Pallas's drawings. Von Baer also refers to the figures in Cook's "Last Voyage" as being somewhat exaggerated in regard to the plump or robust form of the animal, unless, as he says, the Eastern (or Pacific) Walruses are fatter than the Western ones.   Pallas, in his "Zoographia Rosso-Asiatica," cites "tab. xxviii. et xxix." of his accompanying "Icones," but the only copy of the "Icones" I have seen contains only one plate, marked as referring to page 269 of his "Zoographia" (the plates are *not numbered*), where the Walrus is described. This is a most indifferent and badly colored figure of an apparently not half-grown animal, in which the tusks are quite short, the mystacial bristles long and thick, the hind feet extended backward, the tail distinct and prominent, as well as the thighs and shoulders, and all the toes of both the fore and hind limbs are provided with long, conspicuous nails.

The next illustration of the Pacific Walrus appears to have been published by Mr. H. W. Elliott[*] in 1873.   This is the result of a careful study of the animals from life[†] (on Walrus Island, Alaska, in July, 1872), by an artist not only qualified to do justice to the subject from an artistic point of view, but who brings to his work the trained eye of a naturalist.   This illustration represents a group of some ten or more old males quietly reposing on the rocks in a variety of postures.   The figures in the foreground are expressive and detailed, and afford by far the best representations of an adult Walrus yet extant.   The edition of the work embraced only one hundred and twenty-five copies, and can hence, unfortunately, have but a very limited circulation.   Two of the figures seen in the foreground, however, have been reproduced by Scammon[‡] from Mr. Elliott's drawings, and give a good idea of the form of these unwieldy creatures.

I can refer with certainty to no heretofore-published figures of the skull or general anatomy, but some of the representations of the skull already mentioned in the account of the figures of the Atlantic species may possibly represent this species.

GEOGRAPHICAL DISTRIBUTION.—The habitat of the Pacific Walrus embraces a much smaller extent of coast and a much narrower breadth of both latitude and longitude than the Atlantic species.   It is confined on the one hand to a comparatively small stretch of the northern and eastern coasts of Asia, and to

---

[*] Report on the Prybilov Group, or Seal Islands, of Alaska (plates not numbered and text unpaged).   Washington, 1873.

[†] See beyond, p. 179.

[‡] Marine Mammals of the Northwest Coast of North America, 1874, p. 177.

a still smaller portion of the opposite American coast. To the westward the Walrus appears not to have been traced beyond Cape Schelatskoi (157° 30′ east longitude), and to have occurred in large herds only as far west as Koljutschin Island (185° east longitude). These herds are reported as composed almost solely of males, the females rarely passing beyond the mouth of the Kolyma River.* Wrangell, who passed two winters at the mouth of the Kolyma River, asserts confidently that the Walrus of Behring's Straits were abundant at Cape Jakan (176° 30′ east longitude), but only once reached Cape Schelatskoi, while he found them numerous at Koljutschin Island. Thence eastward they form the chief subsistence of the Tschutschi.†

On the eastern coast of Asia, Steller (according to von Baer) reports that as early as 1742 none were killed by the Russians south of Karaginskoi Island in latitude 60°. He reports, however, finding one on the southern point of Kamtschatka, but von Baer questions whether in this isolated instance of its supposed occurrence so far south there may not be some mistake, and that the animal was really a large Seal or a Sea-cow (*Rhytina*).‡ Krashinninikow states that in his day they were confined to the northern seas. He says, "On voit peu de chevaux marins dans les environs de *Kamtschatka*, ou si l'on en trouve, ce n'est que dans les mers qui sont au nord. On en prend beaucoup plus près du cap Tchukotskoi, où ils y sont plus gros & plus nombreux que par-tout ailleurs".§ Lütke found a dead one as far south as Karaginskoi Ostrow (latitude 58°).‖ Higher up the coast from Cape Thaddeus northward and westward, they were met with in great numbers by the early Russian explorers. In the Arctic Sea north of Behring's Straits they have been met with abundantly as far north as ships have penetrated, their northward range being only limited by the unbroken ice sheet.

On the American coast they have been traced eastward only as far as Point Barrow, where they were observed by Beechey

* See von Middendorff, Sibirische Reise, Bd. iv, p. 936, footnote.

† "Auf der Insel Koliutschin werden manchmal eine grosse Menge Wallrosse erlegt, indem die Eingebornen sie, wenn sie aus dem Meere auf das Ufer steigen, plötzlich überfallen, ihnen den Rückweg ins Wasser abschneiden und mit Peitschen und Stöcken weiter hinauftreiben, wo sie sie dann mit leichter Mühe erlegen.—Das Wallross ist dem sitzenden Tschuktschen, wenn auch nicht so unmittelbar, doch fast eben so allgemein nützlich, als dem Nomaden das Rennthier."—*Nordküste von Sibirien*, vol. ii, 1839, pp. 224, 225.

‡ See von Baer, l. c., p. 183.

§ Hist. de Kamtsch., etc., as translated by "M. E. . . ." (Eidous), tom. 1, 1767, p. 283.

‖ Voyage autour du Monde, tom. ii, p. 178.

in 1823. Cook, in 1799, found them numerous in the neighborhood of Icy Cape. They were also met with by Beechey on Diomede and Saint Lawrence Islands, and on other islands more to the southward.* Lütke found great herds at Saint Mathew's Island, in latitude 60°,† where their teeth were seen later by Billings.‡ They formerly resorted in summer in large numbers to Saint Paul's and Saint George's Islands, where, according to Sarytschew, 28,000 pounds of their teeth were obtained in a single year. They still resort, in small numbers, to a neighboring islet (Walrus Island), and even to the easternmost of the Aleutian chain, as will be presently more fully noted. Formerly they were also abundant on Nunivak Island, situated to the eastward of Mathew's Island, and not far from the Alaskan coast.

On the coast of the mainland they have been met with in great herds at different times in Kotzebue and Norton Sounds and in Bristol Bay. Captain Cook appears not to have observed them south of latitude 58° 42', at which point he found them in Bristol Bay, as well as more to the northward.§ There appears to be no certain proof that they were in early times ever met with on the outermost of the Aleutian Islands,|| and no early reference to their occurrence anywhere south of Bristol Bay and the Prybilov Islands. Brown, however, as late as 1868, says: "On the northwest coast of America I have known it to come as far south as 50° north latitude."¶ Of this I can find only a partial confirmation, and think that possibly there is a mistake in respect to the latitude here given.** Elliott says,

* Narrative of a Voyage to the Pacific and Behring's Straits, vol. ii, p. 271.
† Voyage autour du Monde, tom. ii, p. 176.
‡ Sauer's Account of Billings' Exped. to the North Parts of Russia, p. 235.
§ Voyage to the Pacific Ocean, etc., vol. i, pp. 433, 455, 457; vol. ii, pp. 245, 248, 249, 259.
|| On this point, see von Baer, l. c., p. 182.
¶ Proc. Zoöl. Soc. Lond., 1868, p. 432.
** Mr. Brown further states in the same connection that "It [the Walrus] is found all along the circumpolar shores of Asia, America and Europe," and that "It is not unlikely that it may even be found in the Antarctic regions"! L. c., p. 432. This idea I have not seen elsewhere revived since the early part of the present century. (On this point see von Baer, l. c., p. 173, and footnote.) Dr. Gray refers to the reported occurrence by Bonelli of "Sea Horses" on the Island of Saint Lorenzo, Callao. As this author describes "the two great white tusks projecting from the mouth on either side," and further says that "the tusks are of great value and form an important article of commerce," Dr. Gray concludes these remarks "cannot apply to the tusks of the Sea Bear"; but he adds that he had "never heard of the genus Triche-

writing in 1874, that "not more than thirty or thirty-five years ago small numbers of these animals were killed now and then on islands between Kodiak and Oonemak Pass" (lat. 55° to 57°). He adds none "are now found south of the Aleutian Islands."*

Respecting their present distribution, Captain Scammon, writing in 1874, from personal observation, says: "Great numbers of Walruses are found where the waters of the Arctic Sea unite with those of Behring Straits, and also in Behring Sea, and that innumerable herds still resort in the summer months to different points on the southern or central coasts of Alaska, particularly at Amak Island and Point Moller, on the northern shore of the Alaskan peninsula. Within the last ten years many of these animals have been destroyed by the whalers, both in the Arctic and Behring Seas."†

According to Mr. Elliott, the Walruses are now to be seen in the Prybilov Islands only on Walrus Island,‡ they being so shy and timid that they deserted the other islands as they became populated by man. In early days, or when the Russians first took possession, a great many Walruses were found at Northeast Point, and along the south shore of Saint Paul's Island, but with the landing of the traders and seal-hunters the Walruses abruptly took their departure, and Walrus Island alone is now frequented by them, being isolated and seldom visited during the year by the natives. He adds that they are now most numerous, outside of the Arctic circle, in Bristol Bay, where "great numbers congregate on the sandy bars and flats, and where they are hunted to a considerable extent for their ivory."§

They are now far less numerous than formerly, having greatly decreased in numbers within the last fifty years. So numerous were they in Behring's Straits about 1821, that a Russian writer

---

chus living out of the Arctic Ocean, and should have believed that he [Bonelli] had mistaken the Sea Bear (*Otaria leonina*) for the Sea Horse," if he had not so particularly described the tusks.—*Cat. Seals and Whales*, p. 37. The reference by Bonelli to the great white tusks of the " Sea Horses" relates, in all probability, to the large canines of the Sea Elephant, which were formerly employed for a variety of uses.

* Condition of Affairs in Alaska, p. 164, footnote.

† Marine Mammalia, p. 180.

‡ A low rocky island, about half a mile long by one-eighth of a mile in breadth, situated a few miles to the southeastward of the eastern end of Saint Paul's Island.

§ Condition of Affairs in Alaska, pp. 161, 164.

reports meeting with herds there embracing thousands, and even hundreds of thousands, of individuals.*

During recent years, in addition to the number killed by the natives, the whalers are said to have destroyed as many as 12,000 annually, so reducing their numbers that the natives have become anxious lest they shall soon lose this source of subsistence, upon which they are so dependent.

HABITS, FOOD, COMMERCIAL PRODUCTS, AND THE CHASE.— The Pacific Walrus appears to agree quite nearly in habits with its closely allied congener of the Atlantic waters. It has the same gregarious propensity, the same intense affection for its young, the same strong sympathy for a distressed comrade, lives upon similar food, and is limited in its distribution by about the same isotherms. Its leading characteristics were concisely stated nearly a century since by Captain Cook in the following words :

"They lie, in herds of many hundreds, upon the ice; huddling one over the other like swine; and roar or bray very loud; so that, in the night, or in foggy weather, they gave us notice of the vicinity of the ice, before we could see it. We never found the whole herd asleep; some being always on the watch. These, on the approach of the boat, would wake those next to them ; and the alarm being thus gradually communicated, the whole herd would be awake presently. But they were seldom in a hurry to get away, till after they had been once fired at. Then they would tumble one over the other, into the sea, in the utmost confusion. And, if we did not, at the first discharge, kill those we fired at, we generally lost them, though mortally wounded. They did not appear to be that dangerous animal some authors have described; not even when attacked. They are rather more so, to appearance, than in reality. Vast numbers of them would follow, and come close up to the boats. But the flash of a musquet in the pan, or even the bare pointing of one at them, would send them down in an instant. The female will defend the young one to the very last, and at the expense of her own life, whether in the water or upon the ice. Nor will the young one

---

* Von Middendorff says, "Tausende ja Hunderttausende im lebensfrischeren Berings-Eismeere," and cites as authority a Russian writer named Hülsen. Von Middendorff continues, "Im Jahre 1821 über sah er [Hülsen] dort im Dezember Tausende, zu Ende des Juni Hunderttausende von Walrossen zugleich, welche die Luft mit ihrem Stöhnen erfüllten und von denen einige, fruchtlos kratzend, sich bemühten an den Schiffswandungen emporzuklimmen."—*Sibirische Reise*, Bd. iv, p. 913, and footnote.

quit the dam, though she be dead; so that, if you kill one, you are sure of the other. The dam, when in the water, holds the young one between her fore fins."*

In Captain King's continuation of the narrative of Cook's last voyage, reference is made to a "Sea Horse" hunt. "Our people," says the account, "were more successful than they had been before, returning with three large ones, and a young one, besides killing and wounding several others. The gentlemen who went on this party were witnesses of several remarkable instances of parental affection in those animals. On the approach of our boats toward the ice, they all took their cubs under their fins and endeavored to escape with them into the sea. Several, whose young were killed or wounded and left floating on the surface, rose again and carried them down, sometimes just as our people were going to take them up into the boat, and might be traced bearing them to a great distance through the water, which was colored with their blood. We afterward observed them bringing them, at times, above the surface, as if for air, and again diving under it with a dreadful bellowing. The female, in particular, whose young had been destroyed and taken into the boat, became so enraged that she attacked the cutter and struck her two tusks through the bottom of it."†

The accounts given by subsequent observers confirm the general truthfulness of this brief but comprehensive sketch, and supply some further details respecting its interesting history. Mr. H. W. Elliott, recently an agent in the employ of the Treasury Department of the United States Government, stationed at the Prybilov Islands, has made these animals a special study, under opportunities unusually favorable for observation. On Walrus Island, well known as being still a favorite resort for a large herd of old males, he was able to approach within a few yards of a herd of several hundred old bulls, which lay closely packed upon a series of low basaltic tables, elevated but little above the wash of the surf. Here he studied and painted them from life,‡ seated upon a rocky ledge a few feet distant from and above them. He describes these scarred, wrinkled, and almost naked old veterans as of by no means prepossessing appearance. He says they are sluggish

---

* Cook's Voyage to the Pacific Ocean, etc., vol. ii, p. 458.
† Ibid., vol. iii, p. 248.
‡ See *anteà*, p. 174.

and clumsy in the water and almost helpless on land, their immense bulk and weight, in comparison with the size and strength of their limbs, rendering them quite impotent for terrestrial movement. "Like the seal, it swims entirely under water when traveling, not rising, however, quite so frequently to breathe; then it 'blows' not unlike a whale. On a cool, quiet May morning, I watched a herd off the east coast of the island, tracing its progress by the tiny jets of vapor thrown off as the animals rose to respire.

"In landing and climbing over the low rocky shelves," he continues, "this animal is almost as clumsy and indolent as the sloth; they crowd up from the water, one after the other, in the most ungainly manner, accompanying their movements with low grunts and bellowings; the first one up from the sea no sooner gets composed upon the rocks for sleep than the second one comes prodding and poking with its blunt tusks, demanding room also, and causing the first to change its position to another still farther off from the water; and the second is in turn treated in the same way by the third, and so on, until hundreds will be packed together on the shore as thickly as they can lie, frequently pillowing their heads or posteriors upon the bodies of one another, and not at all quarrelsome; as they pass all the time when on land in sluggish basking or deep sleep, they seem to resort to a very irregular method of keeping guard, if I may so term it, for in this herd of three or four hundred bulls under my eye, though all were sleeping, yet the movement of one would disturb the other, which would raise its head in a stupid manner, grunt once or twice, and before lying down to sleep again, in a few moments, it would strike the slumbering form of its nearest companion with its tusks, causing that animal to rouse up for a few minutes also, grunt and pass the blow on to the next in the same manner, and so on, through the whole herd; this disturbance among themselves always kept some one or two aroused, and consequently more alert than the rest.

"In moving on land they have no power in the hind limbs, which are dragged and twitched up behind; progression is slowly and tediously made by a succession of short steps forward on the fore feet. How long they remain out from the water at any one time I am unable to say. Unlike the seals, they breathe heavily and snore.

"The natives told me the walrus of Bering Sea is monogamous, and that the difference between the sexes in size, color, and shape is inconsiderable; that the female brings forth her young, a single calf, in June, usually on the ice-floes in the Arctic Ocean, above Bering Straits; that the calf closely resembles the parent in general proportions and color, but that the tusks which give it its most distinguishing expression are not visible until the end of the second year of its life; that the walrus mother is strongly attached to her offspring, and nurses it later in the season in the sea; that the walrus sleeps profoundly in the water, floating almost vertically, with barely more than the nostrils above water, and can be easily approached, if care be taken, to within easy spearing distance; that the bulls do not fight as savagely as the fur-seal or sea-lion, the blunted tusks of the combatants seldom penetrate the thick hide;* that they can remain under water nearly an hour, or twice as long as the seals, and that they sink like so many stones immediately after being shot."

Mr. Elliott adds: "As the females never come down to the Prybilov Islands, I have never had an opportunity of observing them. . . . The reason why this band of males, many of them old ones, should be here by themselves all through the year is not plain to me; the natives assure me that the females, or their young, never have been seen around the shores of these islands. Over in Bristol Bay great numbers of walrus congregate on the sandy bars and flats, where they are hunted to a considerable extent for their ivory." On Walrus Island, however, they are said to be comparatively unmolested, the natives here " not making any use of their flesh, fat, or hides." They are hence shot here only by the natives of Saint Paul's Island, who visit Walrus Island for the purpose of getting eggs, in June and July, when they often shoot the Walruses wantonly.† Their comparative immunity here from persecution is hence apparently the reason why they select this island as one of their favorite reposing grounds.

Their food is described by Mr. Elliott as consisting exclusively of shell-fish (principally clams), " and the bulbous roots of certain marine grasses and plants, which grow in great abundance in the broad, shallow lagoons and bays of the mainland

* That their blows are at times not lacking in force is sufficiently proven by the too well-known fact of their striking them through the planking of a ship's boat.

† Condition of Affairs in Alaska, pp. 160–164.

coast. I have taken from the paunch of a walrus," he adds, "over a bushel of crushed clams, shells and all, which the animal had but recently swallowed, since digestion had scarcely commenced. Many of the clams in the stomach were not even broken;* and it is in digging these shell-fish that the service rendered by the enormous tusks becomes evident."† Mr. W. H. Dall also says, "They feed principally upon shell-fish which they swallow whole, and the shells, which remain after they have digested the contents, are found in large numbers about the localities they frequent."‡

Among the enemies of the Pacific Walrus are not only to be reckoned man, both savage and civilized, but also the Polar Bear and the Orca or "Killer," while, like the Atlantic species, it is said to be greatly infested with parasites. According to Captain Scammon, the Polar Bear, when meeting with a herd in its prowlings, "selects and seizes one of the smallest individuals with his capacious jaws, and the resisting struggles of the poor victim to free itself are quickly suppressed by repeated blows with Bruin's paws, which cause almost instant death. The murderous beast then quickly tears the skin from the body by means of his long, sharp claws, when the remains are devoured." That carnivorous Cetacean, the Orca, he continues, "also watches for the young cubs of the Walrus, and if there is floating ice at hand, the mother with her charge clambers upon it to avoid the pursuer; if this fails, however, the cub will mount the mother's back as the only place of refuge. But the Killer is rarely baffled in obtaining the object it seeks by this mode of the mother's protection, for the pursuing animal dives deeply, and then comes head up under the old Walrus, with such force as to throw the cub from the dam's back into the water, when it is instantly seized and swallowed by its adversary. Instances have been known, however, when the Orca has paid dearly for its murderous temerity, as the enraged Walrus, when bereft of her young, will sometimes strike her tusks into her foe with such effect as to cause a mortal wound or instant death."§

Captain Scammon says the period of gestation is "about nine

* Compare on this point Malmgren's statement that the Atlantic Walrus rejects the shells, swallowing only the soft parts. See *anteà*, p. 136.
† Condition of Affairs in Alaska, p. 162.
‡ Alaska and its Resources, 1870, p. 504.
§ Marine Mammalia, pp. 180, 181.

months," and that both sexes and the young are often found in company. He adds that the paring season occurs during the "last of the spring months or the first of summer." His general account of their habits is quite in harmony with the early account given by Cook. "The mother and her offspring," he says, "manifest a stronger mutual affection than we have observed in any other of the marine mammals; the cub seeks her protection, clinging to her back whenever there is cause for alarm, and she will at all times place herself between the foe and her helpless charge; frequently has she been known to clasp to her breast the terrified little one, embracing it with her fore flippers, while receiving mortal wounds from the whaleman's lance." Captain Scammon further states, in respect to the affection of the young for its mother, on the authority of Capt. T. W. Williams, an experienced and observing whaling master, that "a female was captured two miles from the ship and the young cub kept close to the boats that were towing its dead mother to the vessel; and when arrived, made every effort to follow her as she was being hoisted on board. A rope with a bowline was easily thrown over it, and the bereaved creature taken on deck, when it instantly mounted its mother's back and there clung with mournful solicitude, until forced by the sailors to again return to the sea; but even then it remained in the vicinity of the ship, bemoaning the loss of its parent by uttering distressful cries." "A male, and a female with her cub," continues our author, "are often seen together; yet herds of old and young, of both sexes, are met with, both in the water and upon the ice. When undisturbed they are quite inoffensive, but if hotly pursued they make a fierce resistance; their mode of attack is by hooking their tusks over the gunwales of the boat, which may overturn them, or they strike a blow through the planking, which has repeatedly been the means of staving and sinking them."*

The commercial products of the Pacific Walrus are, as in the case of the other species, its tusks, oil, and hide. They are, furthermore, to the Tschuktschi what the Greenland Walrus is to the Esquimaux, their most important source of food, utensils, and means of commercial interchange. Cook, Wrangell, and numerous other explorers of the Arctic waters beyond Behring's Straits, unite in the testimony that they form the chief means of support of the coast tribes. To quote the words of a recent

---

* Marine Mammalia, p. 178.

writer, their "flesh supplies them with food; the ivory tusks are made into implements used in the chase, and for other domestic purposes, as well as affording a valuable article of barter; and the skin furnishes the material for covering their summer habitations, planking for their *baidarras*, harness for their dog-teams, and lines for their fishing-gear."*

According to Wrangell, "the Walrus is almost as useful to the settled as the Reindeer is to the nomad Tschuktschi. The flesh and the blubber are both used as food, and the latter for their lamps; the skin is made into durable thongs for harness and other purposes, and into strong soles for boots; the intestines furnish a material for light water-proof upper garments for summer use; a very durable thread is prepared from the sinews; and, lastly, the tusks, which are of the finest ivory, are sometimes formed into long narrow drinking vessels, such as takes a long time to hollow out, but are more frequently sold to the Reindeer Tschuktschi, who dispose of them to the Russians."†

As already incidentally noted in the foregoing pages, their tusks have been an important article of traffic from the earliest times to which the history of this region extends, and the source of this valuable commodity was the "Eldorado" of the Russian adventurers of the middle of the seventeenth century who first explored the Arctic coast of Eastern Asia. Now, as then, the tusks have the highest commercial value of any of the products of the Walrus, and thousands of these animals have annually been sacrificed, for perhaps the greater part of the last two centuries, in order to meet the demand for them. Mr. Dall, writing in 1870 of the Alaskan Walrus, states that "the quantity of Walrus tusks annually obtained will average 100,000 pounds."‡ Allowing the average weight of a pair of Walrus tusks to be 15 to 20 pounds (I have found the weight of large tusks to vary from 6 to 8 pounds each, the very largest 1 have seen weighing less than 9 pounds)—a very high estimate—this enormous quantity implies the destruction of more than six thousand Walruses annually in the waters bordering Behring's Straits.

According to Captain Scammon, the whalers have of late been largely instrumental in the destruction of the Alaskan Walrus, they having, owing to the scarcity of Whales, become more or

* Scammon, Marine Mammalia, p. 180.
† Wrangell's Polar Expl., Harper's Amer. ed., p. 282.
‡ Alaska and its Resources, p. 504.

less interested in Walrus-hunting. According to a quotation given by Captain Scammon from *The Friend** of March 1, 1872, "the whalers first began to turn their attention to Walrus-catching about the year 1868, and the work has continued up to the present time [1874]. Usually, during the first part of every season, there has been but little opportunity to capture whales, they being within the limits of the icy barrier. Hence, much of the whalers' time during the months of July and August has been devoted to capturing the Walrus; and it is estimated that at least 60,000 of these animals have been destroyed by the whale-fishers in the Arctic Ocean and Behring Sea during the last five years, which produced about 50,000 barrels of oil, with a proportionate amount of ivory."† This would make an average annual destruction of 12,000, in addition to the large number habitually destroyed by the natives.

In the "Annual Review" of the products of the North Pacific Whaling Fleet‡ for 1877, it is stated that the whalers arriving at the port of San Francisco during 1877 reported 74,753 pounds of Walrus teeth and 2,178 barrels of Walrus oil. The amount of Walrus ivory "received in the customs district of San Francisco" for 1876 is given as 33,934 pounds. The same authority gives the following statistics for previous years, beginning with 1873:

Year.	Number of vessels.	Pounds of ivory.
1873...............	16	12,142
1874...............	12	7,600
1875...............	11	25,400
1876...............	7	7,000§
1877...............	16	74,000

Total for the last five years, 153,076 pounds, with an estimated value of about $55,000. This amount implies an annual destruction of at least ten to twelve thousand Walruses. It thus appears that for the last ten years the number of Walruses taken

---

* A newspaper published in Honolulu.

† Marine Mammalia, p. 181.

‡ "Commercial Herald and Market Review," vol. xii, No. 531, San Francisco, Cal., Jan. 17, 1878.

§ There is an unexplained discrepancy here, for another statement in the same connection gives the quantity of "Walrus teeth" for 1876 as 33,934 pounds.

by the whalers alone cannot fall far short of one hundred and twenty thousand. It is hence little wonder that these animals are rapidly declining in number, and that the natives manifest alarm at the disappearance of their main reliance for support.

The destruction of the Alaskan Walrus is now largely effected by the use of firearms, even the natives shooting them on shore with rifles and heavy muskets, although they still also practice their former method of pursuing them in the water and there dispatching them with spears and lances.

FAMILY OTARIIDÆ.

Eared Seals.

Phoques à oreilles, BUFFON, Hist. Nat. Suppl., vi, 1782, 305.
Phocacea auriculata, PÉRON, Voy. Terr. Austr., ii, 1816, 37.
Otaria, PÉRON, Voy. Terr. Austr., ii, 1816, 37 (genus).
Otariina, GRAY, Ann. of Phil., 1825, 340 (subfamily).
Otariadæ, "BROOKES, Cat. Anat. and Zoöl. Mus., 1828, 36."—GRAY, Ann. and
    Mag. Nat., Hist., 3d ser., xviii, 1866, 228.—ALLEN, Bull. Mus. Comp.
    Zoöl., ii, 1870, 19.
Otariidæ, GILL, Proc. Essex Inst., v, 1866, 7.
Arctocephalina, GRAY, Charlesworth's Mag. Nat. Hist., i, 1837, 583.
Otaridés, GERVAIS, Hist. Nat. des Mammifères, ii, 1855, 305.

Fore limbs placed far back, and, like the hind limbs, comparatively free and serviceable for terrestrial locomotion; hind feet
susceptible of being turned forward. The digits of the manus
successively decrease very much in size and length from the
first to the fifth, without well-developed nails, and with the
manus bordered with a naked cartiliginous extension. Of
the pes the three middle digits are shorter and weaker than
the others, with well-developed nails; the others strong and
thick, the first rather stouter than the fifth, both with only rudimentary nails; all terminate in hairless, long cartilaginous flaps,
which vary in length in the different genera. Soles and palms
and most of the upper digital surface hairless. Scapula large,
the blade very broad, the crests high, and the acromion greatly
developed. Femur with a trochanter minor, which in adult
males is strongly developed. Pubic bones unanchylosed, and
in the females considerably separated. Ilia long and slender,
not abruptly turned outward posteriorly. Acetabula opposite
the posterior end of the second sacral vertebra. Skull with well-
developed orbital processes, and an alisphenoid canal; mastoid
process strong and salient, distinct from the auditory bullæ, which
are small and but slightly inflated. Incisors always $\frac{3-3}{2-2}$, the two
middle pairs of upper with the crown deeply grooved transversely, the outer caniniform. Dental formulæ: Milk dentition,
I. $\frac{3-3}{2-2}$, C. $\frac{1-1}{1-1}$, M. $\frac{3-3}{3-3}$; permanent dentition, I. $\frac{3-3}{2-2}$, C. $\frac{1-1}{1-1}$, M. $\frac{5-5}{5-5}$,
or $\frac{6-6}{5-5}$, = 34 or 36. Ears with a subcylindrical external conch.
Testes scrotal.

TECHNICAL HISTORY.

HIGHER GROUPS.—The Eared Seals were referred by the older writers to the Linnæan genus *Phoca*. Buffon, in 1782, recognized the Seals as consisting of two groups, characterized by the presence or absence of external ears. Péron, in 1816, first divided the Seals into two genera, he separating the Eared Seals from the earless ones under the name *Otaria*. Later, Brookes, in 1828, raised the group of Eared Seals to the rank of a family, under the name of *Otariadæ*. This classification was not, however, generally adopted till 1866, when it was revived by Gill, and immediately adopted by Gray, and it has been accepted by most subsequent writers. Gray, Turner, and others, had previously considered the Eared Seals as forming a subfamily of the *Phocidæ*, for which Gray, at different times, used the names *Otariina* and *Arctocephalina*, which latter was also adopted for the name of the group by Turner in 1848. In 1870 I divided the Eared Seals into two groups, which I provisionally adopted as subfamilies, with the names *Trichiphocinæ* and *Ouliphocinæ*, in allusion to the nature of the pelage. The characters assigned, while perhaps of small importance, relating mainly to size, character of the pelage, and size and shape of the ear, and insufficient to characterize divisions of this grade, serve to mark two natural groups, the so-called Sea Lions, or Hair Seals, forming the one, and the Sea Bears, or the Fur Seals of commerce, the other.

Dr. Gray, in 1869,* divided the family into five "tribes," which he termed, respectively, *Otariina, Callorhinina, Arctocephalina, Zalophina,* and *Eumetopiina*, mainly with reference to the number of the grinders and the position of the hinder pair. These "tribes" he at the same time combined into two "sections," the one embracing the *Otariina* (consisting of his genus *Otaria*), and the other all the others, this division being based on the posterior extension of the bony palate. To his first primary division ("Section I"), consisting, as just stated, of the single genus *Otaria* as limited by Gray, and, as seems to me, embracing only the single species *O. jubata* of recent authors, he restricted the name "Sea Lions," applying to the other, embracing all the other Eared Seals, the name "Sea Bears." This latter group, however, embraces not only the animals commonly called Sea Bears by other authors, as well as by travelers and

---

*Ann. and Mag. Nat. Hist., 4th ser., vol. iv, pp. 264–270.

sealers (*i. e.*, the "Fur Seals" of commerce), but also the two
Sea Lions (commonly so called) of the northern hemisphere,
and all the Eared Hair Seals of the South, except *Otaria jubata*.
This classification, with scarcely any modification, he followed
also in his papers treating of this group in 1871;[*] but in 1872[†]
he proposed a new arrangement of the "Sea Bears." The subdivision of this group into "tribes" is not here clearly indicated, although he arranges the genera in four unnamed sections. In 1873[‡] he proposed another arrangement of the "Sea
Bears," in which they were placed in two primary divisions, in
accordance with whether the number of molars is $\frac{6-6}{5-5}$ or $\frac{5-5}{5-5}$. His
later modifications were more formally presented in his last gen-
eral account of the group published in 1874,[§] in which the classification then presented differed very much from that adopted
by him in 1868 and 1871. Although a new "tribe" ("Tribe 2,
Gypsophocina") was instituted, his former "tribes," *Callorhinina*, *Arctocephalina*, and *Eumetopiina*, were united into one,
under the name *Arctocephalina*, thus reducing the whole number of "tribes" to four, as follows: 1. *Otariina;* 2. *Gypsophocina;* 3. *Arctocephalina;* 4. *Zalophina*. As before, he recognized
two primary "sections," by means of which *Otaria* is opposed
to all the other genera as a group co-ordinate in rank with all
the rest. Also the "sections," or primary divisions, are still
based on the posterior prolongation of the bony palate, and the
"tribes," or secondary divisions, on the number of the molars
and the position of the hinder pair relative to the "front edge
of the zygomatic arch." It is needless to add that a more purely
artificial and valueless basis could scarcely be devised. In his
later schemes, *Eumetopias* is placed under the division characterized as having the molars $\frac{6-6}{5-5}$, on the wholly theoretical
ground that "the fifth upper molar on each side [is] wanting,"
leaving "the sixth separated from the fourth by a wide space."
On similar grounds his *Phocarctos elongatus*,—based, as I shall
later give reasons for believing, in part on an adult female
Eumetopias stelleri and in part on the young of the Japan species
of *Zalophus*,—is considered as lacking the "fifth grinder" when
adult, though possessing it when young. As late as 1873, *Eumetopias* is placed in a group explicitly characterized as having

---

*Suppl. Cat. Seals and Whales, 1871, p. 11.
†Proc. Zoöl. Soc. Lond., 1872, p. 655.
‡Proc. Zoöl. Soc. Lond., 1873, p. 779.
§Hand-List of Seals.

"thick under-fur"! In his latest notice of the species (in 1874) his synonymy of the species shows that he still believed the skin of a young *Callorhinus ursinus*, referred in 1866 to his *Arctocephalus monteriensis*, belonged to this species, although in 1871* he properly assigned it to *Callorhinus ursinus*, which I had shown in 1870 was its proper allocation.

Dr. Gill, in 1871,† made two primary divisions of the family, the genus *Zalophus* alone constituting one division, which was thus contrasted with all the others. The characters cited as the basis of this division are the rostral profile (whether "more or less decurved," or "straight or incurved") and the sagittal crest. The last distinction was based wholly on a misapprehension of the facts in the case,‡ and the first proves to be open to very obvious exceptions. Although Dr. Gill, in his later papers on this group, retains these divisions as originally proposed by him, he has adduced no additional characters in support of them.

GENERA.—The first generic division of the Eared Seals was made by F. Cuvier in 1824,§ who separated them as "Arctocéphales" (*Arctocephalus*) and "Platyrhinques" (*Platyrhinchus*), with "*Phoca ursina*" (= *Arctocephalus delalandi*, F. Cuvier; *A. antarcticus*, Gray) as the type of the former and "*Phoca leonina*" (= *Otaria jubata* of recent authors) as the type of the latter. Succeeding writers very generally adopted the name *Arctocephalus* for the greater part of the species, while *Platyrhinchus* was considered as equivalent to *Otaria* of Péron, of prior date. *Otaria* has, by some writers, even down to the present time, been used in a generic sense for all the species of the family, sometimes with and sometimes without subgeneric divisions. In 1859, Gray separated generically the Northern Fur Seal from *Arctocephalus* under the name *Callorhinus*, and the group has been since very generally recognized as of generic or subgeneric value. Prior to this date the only commonly recognized genera were *Otaria* and *Arctocephalus*. The next generic subdivisions of the Otaries were instituted by Gill in 1866,‖ namely, *Eumetopias* and *Zalophus*, the former having for its type and only species the Northern Sea Lion, or *Leo marinus* of Steller, while the latter

---

* Suppl. Cat. Seals and Whales, p. 15.
† Amer. Nat., vol. iv, Jan., 1871, p. 681.
‡ See Am. Nat., vol. v, March, 1871, p. 41.
§ Mém. du Mus. d'Hist. Nat., vol. xi, 1824, 205.
‖ Proc. Essex Institute, vol. v, pp. 1–13, March, 1866.

was founded on the *Otaria gillespi* of M'Bain. The genera recognized were five in number, namely: 1. *Otaria* ("Péron, 1816, type *Phoca jubata* Schreber"); 2. *Arctocephalus* ("F. Cuvier, 1824, ... type *Phoca ursina* Linnæus," hence = *Callorhinus*, Gray, 1859, and not *Arctocephalus*, F. Cuvier); 3. *Eumetopias*, Gill (nov. gen., "type *Otaria californiana* Lesson, = *Arctocephalus monteriensis* Gray," the intended type being *Otaria stelleri* of Müller); 4. *Zalophus*, Gill (nov. gen., "type *Otaria Gilliespii*, Macbain"); 5. *Halarctus*, Gill ("type *Arctocephalus Delalandii*, Gray," hence = *Arctocephalus*, F. Cuvier, 1824). Although three new names were proposed, only two new genera were added, *Halarctus* being synonymous with *Arctocephalus* of F. Cuvier, and *Arctocephalus*, as here defined, with *Callorhinus*, Gray, as speedily and almost simultaneously pointed out by Gray* and Peters,† and as has been since freely conceded by Gill. A few months later Professor Peters† adopted, in a subgeneric sense, the genera previously recognized by Gray and Gill, and added two other subgenera, namely, *Phocarctos* and *Arctophoca*. The type of *Phocarctos* was Gray's *Arctocephalus hookeri* (then known to Peters apparently only through Gray's description and figures), with which, however, was associated the *Otaria ulloæ* of von Tschudi, which latter appears to be merely *Otaria jubata*, fem. The type of *Arctophoca* was originally *Otaria philippii*, Peters, sp. nov., probably = *Arctocephalus falklandicus*, fem.; at all events, a *Fur Seal* from the Island of Juan Fernandez. These groups were first established in May, 1866, but the following November, *Phocarctos ulloæ* was removed by Dr. Peters to his section or subgenus *Otaria*, and *Otaria falklandica*, Shaw (= " *O. nigrescens*, Gray"), was taken as the type of *Arctophoca*, to which *O. philippii* was now apparently referred as a subspecies or a doubtful form. Thus *Arctocephalus falklandicus* is here removed from *Arctocephalus*, where he formerly placed it, to become a new type of *Arctophoca!*

In September, 1866, ‡ Gray adopted the above-named generic and subgeneric divisions, to which he added *Neophoca* as a " new genus," based on his *Arctocephalus lobatus*, referred previously by Peters to *Zalophus*, and *Euotaria* and *Gypsophoca* as subgenera of *Arctocephalus*. *Euotaria* was based on his *Arctocephalus nigrescens*, and *Gypsophoca* on his *Arctocephalus cinereus*. In

---

* Ann. and Mag. Nat. Hist., 3d ser., vol. xvii, pp. 444–447, June, 1866.

† Monatsb. d. k. P. Akad. zu Berlin, 1866, pp. 269, 276, 670–672.

‡ Ann. and Mag. Nat. Hist., 3d ser., vol. xviii, pp. 228–237.

1868 * he raised *Euotaria* and *Gypsophoca* to the rank of genera, *ten genera* of Eared Seals being now recognized by this author. In his formal synopsis of the family presented in 1869,† these ten genera were all retained, and are the following:

1. Otaria.
2. Callorhinus.
3. Phocarctos.
4. Arctocephalus.
5. Euotaria.

6. Gypsophoca.
7. Zalophus.
8. Neophoca.
9. Eumetopias.
10. Arctophoca.

In 1871 he again treated two of them (*Euotaria* and *Gypsophoca*) as subgenera of *Arctocephalus*, thereby reducing the number of genera to eight. In 1873 ‡ eight genera of " Sea Bears" alone (*i. e.*, Eared Seals exclusive of *Otaria*) are enumerated, *Euotaria* being omitted. In 1874, § however, both *Euotaria* and *Gypsophoca* are given full generic rank, but no reference is made to *Arctophoca*, the species (*Arctophoca philippii*) formerly referred to it being neither recognized nor accounted for. The number of genera is thus reduced to nine. Dr. Gill, in 1872 || and in 1876, ¶ retained the five generic groups first recognized by him in 1866, with, however, the corrections in nomenclature introduced by Gray and Peters later in the same year. These five genera, namely, *Otaria, Eumetopias, Zalophus, Callorhinus,* and *Arctocephalus*, were adopted by myself in 1870, in my paper on the Eared Seals of the North Pacific.**

Dr. Peters, in 1871,†† referred all the South American Fur Seals (of which he then recognized four, namely, *A. falklandicus, A. nigrescens, A. argentata, A. philippii*) to his subgenus ("Untergattung") *Arctophoca*. Dr. Peters's later views respecting the genera of the *Otariidæ* are given in his paper on the Eared Seals published in August, 1877,‡‡ in which he reduces the genera to three, namely, *Otaria, Eumetopias,* and *Arctocephalus*. The Fur Seals are all united under *Arctocephalus; Otaria* includes only *O. jubata* (to which his *O. leonina* and *O. ulloæ* are referred as " Localrassen "), *Eumetopias* being made to include all the other

---

* Ann. and Mag. Nat. Hist., 4th ser., vol. i, pp. 99–110, Feb., 1868.
† Ann. and Mag. Nat. Hist., 4th ser., vol. iv, pp. 264–270.
‡ Proc. Zoöl. Soc. Lond., 1873, p. 779.
§ Hand-List of Seals.
|| Arrangement of Families of Mammals, p. 69.
¶ Johnson's Cyclopedia, vol. iii, p. 1018.
** Bull. Mus. Comp. Zoöl., vol. ii, No. 1, August, 1870.
†† Monatsb. d. k. P. Akad. d. Wissensch. zu Berlin, 1871, p. 564.
‡‡ Monatsb. d. k. P. Akad. d. Wissensch. zu Berlin, 1877, pp. 505–507.

Hair Seals (= the subgenera *Eumetopias, Zalophus*, and *Phocarctos* of Peters's earlier papers).

SPECIES.—Prior to about the beginning of the present century, the Eared Seals then known were commonly referred to two species, one of which was termed, in common parlance, the Sea Bear, Ours marin, Meerbär, etc., and the other Sea Lion, Lion marin, Meerlöwe, etc. They were hardly more definitely known in technical terminology, the "Sea Bear" being *Phoca ursina*, and the "Sea Lion" the *Phoca jubata*. The first of these names originated with Linné in 1758,* and the other with Forster in 1775.† *Phoca ursina* was based originally on Steller's *Ursus marinus*, and *Phoca jubata* on the Southern Sea Lion, or "Lion marin," of Pernetty, to which species these specific names have of late been properly restricted. Zimmermann, in [1782,‡ named the Southern Sea Bear *Phoca australis* (="Falkland Seal, Pennant II, p. 521," the Sea Bear of Forster), which Shaw, in 1800, renamed *Phoca falklandica*. Both names were based on the "Falkland Isle Seal" of Pennant, but Zimmermann's seems to have been entirely overlooked by subsequent writers. As it has eighteen years' priority, it must be adopted in place of *falklandica*.

During the last half of the last century and the early part of the present, the early voyagers to the southern seas (as Anson, Pernetty, Forster, Weddel, Péron and Lesueur, Quoy and Gaimard, Lesson and Garnot, and Byron, among others) met with different species of Sea Lions and Sea Bears. They described these animals very imperfectly, their accounts relating mainly to their habits and localities of occurrence, and they brought with them to Europe very few specimens.§ Desmarest in 1817, and Lesson in 1828, gave names to the species thus obscurely indicated, the latter renaming several that had already received names. To these authors, and to the often-quoted remark of Péron that he believed there were not less than twenty species of Otaries, we are indebted for much of the confusion and obscurity that must ever be inseparable from the early history of this group. Desmarest alone, in his article on the Otaries in the "Dictionnaire d'Histoire naturelle" (vol. xxv, 1817, pp. 590–603), recognized

---

* Syst. Nat. i, 1758, 37.
† Descrip. Anim., pp. 66, 317.
‡ Geograph. Geschichte, Theil iii, 1782, p. 276.
§ G. Cuvier, according to Gray (Catalogue of Seals, 1850, p. 2), had skulls of only two species of Eared Seals when he wrote the "Ossemens Fossiles."

nine species, only two of which have any tangible basis, or can be determined except conjecturally, and mainly on the basis of their habitat. In fact, it is almost impossible to say whether they are "hair" Seals or "fur" Seals; the descriptions show merely that they were some kind of Eared Seal. Desmarest's species are the following: 1. *Otaria leonina* (= *Otaria jubata*+ *Eumetopias stelleri*); 2. *Otaria ursina* (= *Callorhinus ursinus*); 3. *Otaria peroni* (n. sp., based on a vague account by M. Bailly* of an Eared Seal seen in great numbers on Rottnest Island, west coast of Australia. Desmarest doubtfully refers to it two mounted skins in the Paris Museum, both of very young animals, the larger only about two feet and a half long, brought from "Terres Australes"); 4. *Otaria cinerea* (Péron et Lesueur, Voy. au Terr. Austr., ii, 77; habitat, "Ile Decrès," coast of Australia; an Eared Seal, with rough hair, described only in general terms, and undeterminable; probably = *Zalophus lobatus*); 5. *Otaria albicollis* (Péron et Lesueur, l. c., 118; habitat, "Ile Eugène," coast of Australia; an Eared Seal, eight or nine feet long, characterized by a white spot on the middle and upper part of the neck; perhaps the same as the last, but not certainly determinable); 6. *Otaria flavescens* (Shaw, Mus. Lev.; Gen Zoöl., i, 260, pl. lxxiii; habitat, Straits of Magellan; a "Yellowish Seal, with pointed ears"; not determinable, but probably = *O. jubata*); 7. *Otaria falklandica* (= *Phoca falklandica* Shaw = *Phoca australis*, Zimm.; "Cinereous Seal, with small pointed ears, and the cutting-teeth marked with furrows"; presumably the common Fur Seal of the Falkland Islands); 8. *Otaria porcina* (= *Phoca porcina*, Molina; habitat, coast of Chili; wholly undeterminable); 9. *Otaria pusilla* (= "*Phoca pusilla*, Linn."; a wholly mythical "Otary" as described by Desmarest, *supposed to inhabit the Mediterranean Sea!* † Of these nine species, only one (*Otaria ursina*),

---

* Péron et Lesueur's Voy. Terr. Austr., vol. i, p. 189.

† In view of recent attempts to revive the name *pusilla* as a tenable designation for some species of Eared Seal, it seems desirable to state fully the original basis and early history of this name. It was given originally by Schreber, in 1776, to "Le Petit Phoque" of Buffon, Schreber even copying Buffon's figure (Hist. Nat., xiii, 1765, pl. liii). Buffon introduces his notice of this species as follows: "Le second [espèce] (*planche* LIII) qui est le phoque de la Méditerranée & des mers du Midi, & que nous présumons être le *phoca* des Anciens, paroît être d'une autre espèce, car il diffère des autres par la qualité & la couleur du poil qui est ondoyant & presque noir, tandis que le poil des premiers est gris & rude, il en diffère encore par la forme des dents & par celle des oreilles; car il a une espèce d'oreille externe tres-petite à la vérité . . . ." Then follows a good description of a young Fur Seal; but in

or possibly a second (*O. falklandica*), is positively referable to any particular species as now known.

Three years later (in 1820) Desmarest again, in his "Mammalogie" (Encyclopédie Méthodique, vol. clxxxii, pp. 248–252), redescribed the Otaries, reducing the number of species to eight by uniting his *Otaria pusilla* to his *Otaria peroni* under the latter name, which now relates not only to the Fur Seals of the western coast of Australia, but also to those of the Cape of Good Hope.

Lesson, eight years later, in his article on the Otaries (Dictionnaire classique d'Histoire Naturelle, vol. xiii, 1828, pp. 419–426), raised the number to *fifteen*. One is purely mythical; five or six can be determined as equivalent to species now commonly recognized, but the greater part are not satisfactorily identifiable. His species are the following: 1. *Otaria fabricii* (="*Phoca ursina* Fabricius"; habitat, *Greenland*; wholly undeterminable; certainly not an Eared Seal, and probably wholly mythical); 2. *Otaria stelleri* (=*Leo marinus*, Steller, =*Eumetopias stelleri*, which here receives its first distinctive name); 3. *Otaria californiana* (="jeune Lion marin de la Californie," of Choris, and hence = *Zalophus gillespii* of recent authors, which here received its first specific name*); 4. *Otaria kraschenninikowii* (= *Ursus marinus*,

---

a long footnote to this description he gives quotations from Olaus Magnus, Zorgdrager, Charlevoix, and from collections of voyages, which relate to the Seals of both the Arctic and Antarctic regions, none of which probably refer to any species of Eared Seal. On the following page he says: "C'est par une convenance qui d'abord paroît assez légère, & par quelques rapports fugitifs que nous avons jugé que ce second phoque (*pl.* LIII) étoit le *phoca* des anciens; on nous a assuré que l'individu que nous avons vu venoit des Indes, & il est au moins très-probable qu'il venoit des mers du Levant; . . ." —*Hist. Nat.*, xiii, 1765, pp. 340, 341. Though assumed to be a Mediterranean species, the origin of the specimen here described and figured as "Le Petit Phoque" is avowedly unknown, and a certainly erroneous habitat is assigned to it. This is the sole basis, however, for the *Phoca pusilla* of all the earlier systematists, and of some modern ones. As already stated, Desmarest's *Otaria pusilla* is purely mythical; for while he describes an Eared Seal, he claims for it a Mediterranean habitat, and deems it to be the species described by Aristotle, Pliny, and Ælian, and figured by Belon, and even goes so far as to say, "Buffon et Erxleben paroissent avoir confondu, avec ce phoque, de jeunes individus d'autres espèces particulières aux Terres Australes, et particulièrement à l'ours-marin de l'île de Juan-Fernandez. Quant à lui, il semble propre à la Méditerranée." The *Phoca pusilla* of Erxleben and Gmelin is a heterogenous compound of Eared and Earless Seals from both hemispheres.

* See further remarks, *posteà*, under *Eumetopias stelleri* and *Zalophus californianus*.

Steller=*Callorhinus ursinus*); 5. *Otaria pernettyi* (=*Otaria juba-ta*); 6. *Otaria forsteri* (embraces all the Fur Seals of the Southern Hemisphere); 7. *Otaria mollissina* ("Lesson et Garnot, Zoologie de la Coquille, pl. iii, p. 140"; habitat, "Iles Malouines"; the long• description contains nothing in itself distinctive of any species, but it has been determined, by Nilsson and Gray, from the skull and skin in the Paris Museum, to be a young *Otaria jubata*); 8. *Otaria peroni* (="*Otaria peroni*, Desm., sp. 382"; embraces "*Phoca pusilla* Linn.", "Petit Phoque, Buffon," "Otarie de La-lande, F. Cuvier," and "Loup marin, Pagès"; habitat, Cape of Good Hope; formerly referred by Gray to his *Arctocephalus delalandi*, to which it is mainly referable); 9. *Otaria coronata* ("Desm., spec. 383; *Phoca coronata*, Blainv."; undeterminable, and habitat unknown); 10. *Otaria cinerea* ("Péron et Lesueur," as above); 11. *Otaria albicollis* ("Péron et Lesueur," as above); 12. *Otaria flavescens* (= "*Phoca flavescens*, Shaw," as above; not determinable); 13. *Otaria shawi* (= *Phoca falklandicus*, Shaw, therefore = *Arctocephalus falklandicus*, auct.); 14. *Otaria hau-villii* ("G. Cuvier, Oss. Foss., t. v, p. 220"; = *Arctocephalus falk-landicus*, auct.; habitat, "Iles Malouines"); 15. "*Otaria moli-naii*" (="*Phoca porcina*, Molina"; no tangible description, and wholly undeterminable).

Fischer, in 1829–30,* appears to have recognized fifteen (only twelve have numerals prefixed) species of Eared Seals, which are the same as those described by Lesson in 1828, with the exception that Lesson's *Otaria fabricii* is not admitted, and Gray's *Arctocephalus lobatus* is added.

Hamilton, in 1839,† recognized twelve species, as follows: 1. "Sea-Lion of Steller" (= *Eumetopias stelleri*); 2. "Sea-Lion of Forster" (= *Otaria jubata*); 3. "Sea-Lion of Pernetty" (= *Ota-ria jubata*, mainly); 4. "Pusilla, or Cape Otary" ("*Otaria pusilla*, Desm.," but really based on a skull from the Cape of Good Hope); 5. "Ursine Seal, or Sea Bear of Steller (= *Cal-lorhinus ursinus*); 6. "Ursine Seal, or Sea Bear of Forster" (= *Arctocephalus falklandicus*, auct.); 7. "Sea Bear, from speci-men in the British Museum" (= ? *Otaria jubata*, according to Gray); 8. "Lesson's Otary, *O. mollissina*, Lesson" (=*Arctocepha-lus falklandicus*); 9. and 10. "Ash-coloured and white-necked Otaries" (= *Otaria cinerea* and *O. albicollis*, Péron); 12. "Com-mon Fur-Seal of Commerce" (= *Arctocephalus falklandicus*).

---

* Synopsis Mammalium, pp. 230–234, 374 (*i. e.* 574).

† Amphibious Carnivora, etc. (Jardine's Nat. Library, Mam., vol. viii).

He very judiciously refers to *Otaria porcina, O. coronata, O. delalandi,* and *O. hauvillii* as species so slightly indicated "as still to remain doubtful."

Nilsson, in his celebrated paper on the Seals, published in 1837,* reduced the species to three, reuniting all the Sea Lions (except one) under the name *Otaria jubata,* and all the Sea Bears under the name *Otaria ursina.* His third species is the *Otaria australis* of Quoy and Gaimard, from Australia (= *Arctocephalus lobatus,* Gray, Spicel. Zool., i, 1828). Müller, in his appendix to Nilsson's paper,† recognized five species, as represented in the Berlin Museum, namely: 1. *Otaria stelleri;* 2. *Otaria ursina;* 3. *Otaria platyrhinchus* (= *O. jubata,* auct.); 4. *Otaria chilensis* (described as new from a skull received from Chili, but really = *O. jubata*); 5. *Otaria lamari* (= *Arctocephalus lobatus* Gray, as above). He recognized as "eine sechste Art" the *Otaria australis,* Quoy and Gaimard, and Nilsson.

The next general review of the group is contained in Gray's Catalogue of the Seals of the British Museum, published in 1850, in which eight species are formally recognized. These are: 1. *Arctocephalus ursinus;* 2. *A. falklandicus;* 3. *A. cinereus* (= "? *Otaria cinerea,* Péron," as noticed above); 4. *A. lobatus;* 5. *A. australis* ("Quoy and Gaimard" = *A. lobatus,* Gray); 6. *A. hookeri;* 7. *Otaria stelleri;* 8. *Otaria leonina.* All but *A. australis* probably represent good species. In 1866, in his "Catalogue of Seals and Whales," he raised the number to twelve by adding, 1. *Arctocephalus monteriensis* (first described by him in 1859 = *Eumetopias stelleri, plus* a skin referable to *Callorhinus ursinus*); 2. *A. californianus,* n. sp. (= *A. monteriensis,* Gray, 1859, in part, really = *Eumetopias stelleri*); 3. *A. nigrescens,* first named in Zoöl. Erebus and Terror ; not mentioned in Cat. Seals of 1850, but revived in 1859, when it was really first published (= *A. falklandicus*); 4. *A. delalandi* (= *Petit Phoque,* Buffon, hence *Phoca pusilla,* Schreber, plus *Otaria delalandi,* F. Cuvier, 1828,—the Fur Seal of the Cape of Good Hope); 5. *A. "Gilliespii"* (= *Otaria gillespi,* M'Bain, 1858, = *Otaria californiana,* Lesson, 1828). Of these five, two (*A. monteriensis* and *A. californianus*) are strictly nominal, as is probably a third (*A. nigrescens*); two valid species (*A. "delalandi"*

---

* K. Vet. Akad. Handl. Stockholm, 1837, pp. 235–245. Translated by Peters in Wiegmann's Archiv für Naturgesch., 1841, pp. 301–333, with notes and an appendix by J. Müller.

† Wiegmann's Archiv, 1841, pp. 333, 334.

and *A.* "*gilliespii*") are added to those recognized by this author in 1850.

The same year (1866), Peters* recognized fourteen species (three of them were treated as doubtful), as follows: 1. *Otaria jubata;* "? 2. *Otaria leonina*" (= *O. jubata*); 3. *Otaria godefroyi* (n. sp. = *O. jubata*); 4. "? *Otaria byronia*" (=*Phoca byronia,* Blainville, = *O. jubata*); 5. *Otaria hookeri;* 6. *Otaria ulloæ* (= *O. ulloæ,* von Tschudi, = *O. jubata,* fem.); 7. *Otaria pusilla* (=Petit Phoque, Buffon, *Phoca pusilla,* Schreber, *Otaria delalandi,* F. Cuvier, etc.); 8. *Otaria cinerea* (="*O. cinerea,* Péron and Lesueur, Quoy and Gaimard"; " *O. stelleri,* Schlegel," in part, etc.); ? 9. *Otaria falklandica* (= *Arctocephalus falklandicus,* auct.); 10. *Otaria ursina* (=*Callorhinus ursinus*); 11. *Otaria stelleri* (*Eumetopias stelleri*); 12. *Otaria gillespi* (=*Zalophus californianus*); 13. *Otaria lobata* (= *Arctocephalus lobatus,* Gray, 1828, *Otaria australis,* Quoy and Gaimard, 1830, *O. stelleri,* "Schlegel," in part, = *Zalophus lobatus*); 14. *Otaria philippii* (n. sp. = *Arctocephalus falklandicus,* auct.). Six months later, on again reviewing the group,† the same writer reduced the number of species to ten. In this paper he referred the *O. byronia, O. leonina,* and *O. godeffroyi* of his former paper to *O. jubata,* and his *O. philippii* to *O. falklandica.* *O. ulloæ* is still retained as a valid species, and "*Otaria stelleri,* Schlegel," is determined to be the *O. gillespi,* M'Bain.

In 1868‡ Dr. Gray described as a new species *Arctocephalus nivosus* (= *A. antarcticus,* s. *pusillus*) from the Cape of Good Hope, and Professor Turner added,§ as a new species, *Arctocephalus schisthyperoës* (later corrected to *schistuperus* by Günther), from Desolation Island, considered later by Gray,‖ after an examination of the type, to be referable to his *A. delalandi* (therefore = *A. antarcticus*). M'Bain, the same year,¶ described an imperfect skull of what he called " *O. ulloæ?*" (= *Otaria jubata,* fem.), adding that in case it proved to be a new species it might be called " *O. graii.*"

In 1870** I was able to recognize only six species as well established, but gave two more as probably valid, the latter

---

* Monatsb. d. k. P. Akad. Wissensch. zu Berlin, 1866, pp. 261–281.
† Ibid., 1866, pp. 665–672.
‡ Ann. and Mag. Nat. Hist., 4th ser., vol. i, p. 219.
§ Journ. Anat. and Phys., vol. iii, pp. 113–117.
‖ Ann. and Mag. Nat. Hist., 4th ser., vol. iv, p. 264.
¶ Ibid., pp. 109–112.
** Bull. Mus. Comp. Zoöl., vol. ii, pp. 44, 45.

being *Arctocephalus cinereus*, Gray (Australia and New Zealand), and *A. antarcticus*, Gray (Cape of Good Hope). The others are: 1. *Otaria jubata* (under which was wrongly included *O. hookeri*, Gray); 2. *Eumetopias stelleri*; 3. *Zalophus gillespi*; 4. *Zalophus lobatus*; 5. *Callorhinus ursinus*; 6. *Arctocephalus falklandicus*.

In 1871, Philippi and Peters* added *Arctocephalus* (*Arctophoca*) *argentata*, a Fur Seal from the island of Juan Fernandez (= *Arctocephalus australis*, fem.). The latter here divided the Fur Seals of South America into four species, two of which (*A. falklandica* and *A. nigrescens*) are from the Atlantic Ocean and two (*A. argentata* and *A. philippii*) from the Pacific.

The same year (1871) Gray† recognized thirteen species of Eared Seals, as follows: 1. *Otaria jubata* (embracing *O. leonina* of Gray and Peters, and *O. godeffroyi*, *O. bironia*, *O. ulloæ* of Peters). 2. *Callorhinus ursinus*. 3. *Phocarctos hookeri*. 4. *Arctocephalus antarcticus* (Cape of Good Hope = *Phoca antarctica*, Thunberg, 1811, and *Phoca* [s. *Otaria*] *pusilla* and *delalandi*, auct.). 5. *Arctocephalus nigrescens* (= *A. australis*). 6. *Arctocephalus cinereus*. 7. *Arctocephalus forsteri* ("New Zealand," = "*Phoca ursina*, Forster," = *Otaria forsteri*, Lesson, formerly referred by him to his *A. falklandicus!*). 8. *Arctocephalus falklandicus*. 9. "*Arctocephalus? nivosus*" (= *A. antarcticus*). 10. *Zalophus gillespi*. 11. *Neophoca lobata*. 12. *Eumetopias stelleri* (embracing his *Arctocephalus monteriensis* and *A. californianus*). 13. *Arctophoca philippii* (= *Arctocephalus australis*). *A. forsteri* is the only species added, while no less than six species, recognized by either himself or Peters in 1866, are reduced to synonyms.

Gray, in 1872,‡ added *Gypsophoca tropicalis*, based on a young skull from Auckland Island, to which specimens from North Australia are also referred. This Clark§ believes to be in part based on the young of *Otaria hookeri*, and in part referable to *Arctocephalus cinereus*.

Scott, in 1873, in his account of the *Otariidæ*,|| described (p. 19) what he regarded as two new species of *Arctocephalus*, namely, "*Arctocephalus Grayii*" and "*Arctocephalus eulophus*." The first is equivalent to Gray's *A. falklandicus* of his Cata-

---

* Monatsb. d. k. P. Akad.Wissensch. zu Berlin, 1871, pp. 558–566.
† Suppl. Cat. Seals and Whales.
‡ Proc. Zoöl. Soc. Lond., 1872, pp. 659, 743.
§ Proc. Zoöl. Soc. Lond., 1873, p. 759.
|| Mammalia, Recent and Extinct, an elementary treatise for the use of the public schools of New South Wales. By A.W. Scott, M. A. Sidney, 1873. Section B, Pinnata, Seals, Dugongs, Whales, &c. &c. &c.—Otariidæ, pp. 7–25.

logue of Seals and Whales (1866, p. 55), and the "Supplement"
to the same (1871, p. 25), which Mr. Scott gives as a synonym.
After quoting Gray's description of *A. falklandicus*, he says:
"This is clearly a species distinct from the common Southern
Fur Seal. . . . The specific name *Falklandicus* having
been appropriated almost by general consent for another animal, I beg to substitute that of *Grayii*." The *Arctocephalus
eulophus* is based on verbal information from Mr. Morris, an
experienced sealer, who informed him " that during his sealing
voyages he occasionally met with a fur-seal, which he and
those connected with him in the trade readily recognized as a
distinct kind—by the diminutive size of the adult animal; by
a top-knot of hair on the crown of the head; and by the soft,
beautiful under-fur, unlike in colour to, and much more valuable for articles of ladies' wear than that of any other fur-seal
they were in the habit of capturing." "This seal," continues
Mr. Scott, "appears to be rare, only a few specimens having
been taken; some were seen on the south-east coast of New
Zealand, evidently stragglers driven far away from home. Mr.
Morris has been told that they were formerly common on the
shores of Patagonia and the Island of Juan Fernandez." With
all due deference to the opinions of Mr. Morris and Mr. Scott,
this information hardly forms a satisfactory basis for the erection of a new species in this obscure group, where external
characters, when well known, are of slight distinctive value.
The *Arctocephalus eulophus* can only be assigned to the category
of vaguely described and indeterminable species, of which the
writings of Péron, Desmarest, and Lesson were so prolific half
a century ago. Only six other species were recognized by Mr.
Scott, namely: 1. *Arctocephalus ursinus* (= *Callorhinus ursinus*). 2. *Arctocephalus falklandicus* (embracing all the Southern
Fur Seals, with the exception of his two " new species," already
noticed). 3. *Zalophus gillespi*. 4. *Z. lobatus*. 5. *Otaria stelleri*.
6. *O. jubata* (= *O. jubata* and *Phocarctos hookeri* Gray).

In 1873, Dr. Gray described* a *Eumetopias elongatus*, based
in part on a skull from Japan he had the previous year† referred
to *E. stelleri*, and in part on a young skull, also from Japan,
which, doubtless, is the same as the *Otaria stelleri* of Temminck
(Fauna Japonica).

---

* Proc. Zoöl. Soc., 1873, p. 776.   † Ibid., 1872, p. 738.

In 1874, the same author* added two more "new species" of
Otaria, this time wholly from old material, from unknown localities,which he had had before him in the British Museum for nearly
twenty years, and which he had hitherto uniformly referred to
Otaria jubata! Having, however, found that the lower jaws
differed from those of the other specimens in being "straight,
not bowed on the side, and elongate," and that "the scar of the
temporal muscle is elongate, narrow in front," instead of being
"broad, rounded in front." One of the species, based on the
"skull of an adult male 11½ inches long, and 6½ wide at the
condyles," etc., he calls "*Otaria minor*, the Smaller Sea Lion."
The other, based on "the skull of an adult (female) 9¼ inches long,
and 5¼ broad at the condyles," he calls "*Otaria pygmœa*, the
Pigmy Sea Lion." The last-mentioned skull is "partly broken
behind, and wants all the grinders and the greater part of the
cutting teeth." They are unquestionably referable to the restricted genus *Otaria*, and there is nothing in the descriptions
indicating that 'Dr. Gray's reference of them for twenty years
to *O. jubata* was erroneous. The skull of *Otaria minor* is later
figured in the "Hand-List of Seals" (pl. xvi), and is evidently
that of a young male *Otaria jubata*.

In this year (1874) also appeared the last† of Gray's long
series of publications relating to the Eared Seals, in which we
have his latest views respecting the species of this group. In
this work two other "new species" are added, making in all
eighteen species of *Otariidœ* now recognized by Dr. Gray!
These are: 1. *Otaria jubata*. 2. *Otaria minor* (see above, last
paragraph). 3. *Otaria ulloœ* (= *O. ulloœ*, von Tschudi and
Peters, and *O. pygmœa*, Gray, both formerly, and, I believe, correctly, referred by him to *O. jubata*). 4. *Gypsophoca tropicalis*
(= *Arctocephalus cinereus*). 5. *Phocarctos hookeri*. 6. *Phocarctos elongatus* (= *Eumetopias stelleri*, in part, and *Otaria stelleri*,
Temminck, in part). 7. *Callorhinus ursinus*. 8. *Arctocephalus
antarcticus*. 9. *Euotaria cinerea* (includes *Arctocephalus forsteri*
of Gray's Suppl. Cat. Seals and Whales [see above, p. 199]).
10. *Euotaria nigrescens* (= *Arctocephalus australis*). 11. *Euotaria latirostris* (n. sp., based on a skull supposed to have come
from the Falkland Islands, formerly referred to his *A. nigrescens*. He now says, "The skull may belong to the *Arctocephalus falklandicus*, of which [*i. e.*, his *A. falklandicus*] the skull is

---

*Ann. and Mag. Nat. Hist., 4th ser., vol. xiii, p. 324.   † Hand-List of Seals, etc.

not known, *or it may be a distinct species*"). 12. *Euotaria com-
pressus* (n. sp.; hab. "South Africa ? *Warwick*"; formerly referred by him to *Arctocephalus hookeri* as " ♀ skull, South
Sea, Mr. Warwick's collection "*). 13. *Euotaria schisthyperoës*
(= *Arctocephalus schisthyperoës*, Turner, formerly referred, without reservation, by Gray himself to his *Arctocephalus antarcticus*).
14. *Eumetopias stelleri*. 15. *Zalophus gillespi*. 16. *Neophoca
lobata*. Two other species are also given, as follows : 17. "*Arctocephalus ? nivosus*" (= *A. antarcticus*); 18. "*Arctocephalus ? falklandicus* " (= *A. australis*). These are Fur Seals, referred doubtfully to *Arctocephalus* from lack of knowledge of the skulls.
The first, he says, "may be the skin of *Euotaria compressa* or
schisthyporoës"; to the latter he refers the "*Arctocephalus grayii*"
and "*eulophus*" of Scott (see above, p. 200), the latter, however,
doubtfully.

In 1875 Dr. Peters described† still another species, based on
two specimens, an old male and a young female, brought home
by the German Transit-of-Venus Expedition (supposed by him
to have both come from Kerguelen Island), to which he gave
the name *Arctophoca gazella*. Externally *A. gazella* appears to
differ little from the other Southern Sea Bears, the distinctive
characters resting in the form of the hinder border of the bony
palate, which has a triangular projection at the middle, in the
very small size of the tympanic bones, and in other details of
the skull-structure.‡ Later he found that only one of the specimens on which *A. gazella* was based came from Kerguelen
Island, the other having been brought either from "der Insel
St. Paul oder Amsterdam." In 1876,§ therefore, in referring
again to these specimens, after the discovery of the error in
locality respecting one of the specimens, he renamed the Saint
Paul or Amsterdam Island skin *Otaria (Arctophoca) elegans*.

In 1877, Dr. Peters again reviewed|| the whole group of
Eared Seal, of which he at this time recognized three genera
and thirteen species. He refers to having had access to much
new material, and it is greatly to be regretted that he has not

---

* Cat. Seals, Brit. Mus., 1850, p. 46; Cat. Seals and Whales, 1866, p. 54.

†Monatsb. d. k. P. Akad. Wissensch. zů Berlin, 1875, pp. 393–399.

‡In this paper he refers incidentally to the South American Fur Seals,
stating that in consequence of the reception of more material since the
publication of his last paper respecting them, he is led to unite the *Arctocephalus argentata* with *A. philippii*, and the *A. nigrescens* with *A. falklandica*
(l. c., p. 395).

§ Ibid., 1876, pp. 315, 316.

|| Ibid., 1877, pp. 505–507.

stated of what it consisted, and especially what types it embraced, and that he has not presented the results of his investigations in detail, with more explicit expression of his later views respecting the numerous synonyms of the group, very few of the many nominal species being here definitely allocated. He having here made radical changes of nomenclature, not only from that of his former papers of 1866, but from that of all previous authors, without giving his reasons for such a procedure, such information would have in this connection especial value. Of the restricted genus *Otaria* he recognizes only the single species *O. jubata*. He gives its habitat as extending from the Rio de la Plata and Callao and the Chincha Islands southward. He refers *O. leonina*, F. Cuvier, and *O. ulloœ*, von Tschudi, to this species as "local races," and leaves it to be inferred that his *O. godeffroyi* and Gray's *O. minor* and *O. pygmœa* are regarded by him as purely synonyms. Gray's *Phocarctos elongatus*, he says, belongs, without doubt, to *Eumetopias gillespi*, and gives Japan as falling within its range. Gray's *Zalophus lobatus* he refers to *Otaria cinerea*, Péron, to which he also assigns *O. albicollis* of the same author and *O. australis* of Quoy and Gaimard. He adopts Péron's apparently wholly indeterminable name *cinerea** for this species, without giving his reasons or stating whether he has obtained new light on this intricate matter since 1866, when he referred it to a group having thick under-fur, and associated with it the *Otaria cinerea* of Quoy and Gaimard, and the *Otaria forsteri* of Lesson, both of which he now treats as distinct species belonging to another genus. No reference being made to Turner's *Arctocephalus schisthyperoës*, nor to Gray's *A. nivosus* and *Euotaria compressa*, nor to the *O. peroni*, *O. hauvilli*, etc., of the French writers, it is to be inferred that they are regarded as synonyms, but of what species we are left in doubt. He adopts *Arctocephalus pusillus* (from Schreber) as the name of the South African Fur Seal, on the supposition that Buffon's "Petit

---

* It has been supposed by Gray and others that Péron took with him to France no specimens of his *Otaria cinerea*, but G. Cuvier (Oss. Foss., v, 3d ed., p. 221) refers to a specimen of Otary "vient de Péron (c'est la seule qu'il ait rapportée), elle n'a que deux pieds neuf pouces de long, et est un peu plus blanchâtre que celle du Cap." He adds in a footnote, "C'est probablement celle dont il parle sous le nom d'*otarie cendrée* de l'ile Decrès; *Voy. aux Terres Australes*, t. ii, p. 54." The Otary of the Cape here referred to is the one brought by M. Delalande, which is the Fur Seal of the Cape of Good Hope.

Phoque," on which the name *pusilla* rests,* must have come from the Cape of Good Hope.† The Fur Seals of South America are recognized as belonging to two species, those of the east coast, the Falkland Islands, the southern extremity of the continent, and the west coast northward to Chili being referred to *Arctocephalus falklandicus*, while those from Juan Fernandez and Masafuera Islands are assigned to *A philippi*. We are therefore left to suppose that his and Gray's *A. nigrescens*, his *A. argentata,*‡ Gray's *Euotaria latirostris*, and Scott's *A. grayi* and *A. eulophus*, are regarded by him as synonyms of these species. The Fur Seal of Australia he calls *Arctocephalus brevipes*, citing " *Otaria cinerea* Quoy et Gaimard, Voy. Astrolabe, Zoolog. i, p. 89 (non Péron)." He also recognized *A. elegans* from Saint Paul and Amsterdam Islands (to which he doubtfully referred *A. tropicalis*, Gray); *A. gazella*, from Kerguelen Island; and the *A. forsteri*, Lesson, from New Zealand and the Antarctic Seas to the southward of New Zealand. Four of his species, namely, *Arctocephalus elegans, A. forsteri, A. gazella*, and *A. philippii*, appear to me to be invalid, while under his *Eumetopias gillespi*, I believe he has confounded two quite distinct species, namely, *Zalophus gillespi* and *Z. lobatus*. Peters's thirteen species are the following:

1. Otaria jubata (Forster).	8. Arctocephalus brevipes, Peters.
2. Eumetopias stelleri (Lesson).	9. Arctocephalus elegans, Peters.
3. Eumetopias gillespi (M'Bain).	10. Arctocephalus forsteri, Lesson.
4. Eumetopias cinerea (Péron).	11. Arctocephalus gazella, Peters.
5. Eumetopias hookeri (Gray).	12. Arctocephalus philippi, Peters.
6. Arctocephalus pusillus (Schreber).	13. Arctocephalus ursinus (Linné).
7. Arctocephalus falklandicus, Shaw.	

Five are Hair Seals and eight are Fur Seals. Three only are given as found in the northern seas, while ten are recognized as occurring in the southern.

From the foregoing it will be seen how widely opinions have differed respecting the number of species and their generic affinities among recent writers on this group, and how unstable have been the views of the two leading authorities in this field

---

* See *anteà*, p. 194, second footnote.

† G. Cuvier supposed it to have come from the Cape, because Pagès (see Buffon's Hist. Nat., Suppl., vi, 357) had reported the young Otaries of the Cape as of a black color (Oss. Foss., 3d ed., v, 220); but it is now well known that all Fur Seals are black when young. On the other hand, Daubenton insisted that Buffon's "Petit Phoque" (see Desmarest, Mam., p. 251) came from "l'Inde."

‡ *Anteà*, p. 202, footnote.

during the last ten or twelve years. Peters and Gray have both repeatedly during this time radically modified their views respecting both the number of genera and species; greatly, in the case of Gray at least, out of proportion to the new material they have examined. This fluctuation of opinion shows, in a most emphatic manner, how imperfect our knowledge still is respecting the Otaries of the Southern Hemisphere. Those of the Northern are much better known, the only doubts still existing having relation to those of Japan. Respecting all the others, there has been for the last eight years an almost perfect unanimity of opinion, so far as the question of species is concerned.

In 1870 I could find no satisfactory basis for the discrimination of more than a single species of Fur Seal in the Southern Hemisphere, and to my mind the case is now scarcely better, since I have as yet had opportunity of examining only specimens from South American localities, with the exception of a skin and skull of a very young individual from Australia. I now add one species of Hair Seal to the number I then recognized. These, which will be discussed more fully later, are the following:

Hair Seals or Sea-Lions.	Fur Seals or Sea-Bears.
1. Otaria jubata.	6. Callorhinus ursinus.
2. Eumetopias stelleri.	7. Arctocephalus falklandicus.
3. Zalophus californianus.	?8. Arctocephalus antarcticus.
4. Zalophus lobatus.	?9. Arctocephalus forsteri.
5. Phocarctos hookeri.	

Although taken severely to task by Gray and others for my "conservatism," especially respecting *Otaria hookeri*, auct. (the justness of which in this instance I now concede), but also as regards the Southern Fur Seals, I must still confess my inability to satisfactorily distinguish them by the published figures and descriptions. I find only such differences indicated as a large series of specimens, embracing both skulls and skins, of two allied species (namely, *Callorhinus ursinus* and *Arctocephalus falklandicus*, auct., *australis*, Zimm.) show to have no importance as specific characters. Indeed, I find Gray himself, in his latest reference to two of these species, writing as follows: "The New-Zealand skull ["*Euotaria cinerea*"] is very like the skull of the Southern Fur-Seal (*Arctocephalus nigrescens*) from the Falkland Islands and the south-west coast of Patagonia. It differs in the position and form of the grinders, and in the form of the palate, and its contracted sides and truncated hinder part; it differs considerably from it in the outline and prominence of the temporal bullæ and the os petrosum. The upper surfaces are very much alike, and the orbits are very large and of the same size. The lower jaws are very similar; but the callosity of the Falkland Island specimen is rather longer, and the crown of the teeth is longer and rather more slender—the crown of the New-Zealand specimen being as long as broad, that of the Falkland Island specimen being one-third longer than broad."* I cite the differences here noted by Gray to show how trivial are the grounds of separation. A skull of each of the supposed species only is here compared. The differences are just such as occur between undoubted specimens of *Callorhinus ursinus*, no two of which, even of the same age and sex, can be compared without observing differences, while there is no difficulty in selecting specimens that are very unlike in characters that have been taken, in discussing other species of this group, as having great significance. Again Dr. Gray, in comparing his *Gypsophoca tropicalis* from North Australia with Peters's *Arctophoca argentata* and *A. philippii* from Juan Fernandez and Masafuera, says: "These three skulls have nearly the same teeth, and appear to me to belong to one group; but whether they are three distinct species (two from the west coast of South America and one from North Australia) I will not attempt to determine, as I have only seen the skins and skulls of the one from the latter region; but they are all Fur-Seals and may be distinct."† Dr. Gray says his genus *Gypsophoca* "is most like *Arctophoca* in the position of the teeth; *but the palate is much narrower, the face shorter, and the hinder part of the skull much larger and more ventricose*";‡ but, as Clark has shown,§ and as is evident from Gray's figures, *Gypsophoca* was based on a young skull, and young skulls of Otaries differ from adult ones of the same species in just these characters. It may here be noted that in several instances the so-called "species" of Fur Seals differ from others recognized by the same authors only through differences that can be demonstrated to be, in other well-known allied species, simply sexual. Hence, until writers on this group have learned to discriminate the sexes, and to make due allowance for the great changes in contour and details of structure that result from age in the skulls of Otaries, we can hardly hope to have the subject of species placed on a proper basis.

---

* Hand-List, p. 36.

† Hand-List, p. 28; first printed in Proc. Zoöl. Soc. Lond., 1872, p. 661.

‡ Proc. Zoöl. Soc. Lond., 1872, p. 659.

§ Ibid., 1873, p. 759.

The distribution of the Fur Seals in the Southern Seas presents no obstacle to the supposition of their conspecific relationship. They occur not only on both the Atlantic and Pacific coasts of the South American continent, about its southern extremity, and on all the outlying islands, including not only the Falklands, the South Shetland, and South Georgian, but at other small islands more to the eastward, at Prince Edward's, the Crozets, Kerguelen, Saint Paul, and Amsterdam, the southern and western shores of Australia, Tasmania, New Zealand, and at the numerous smaller islands south of the two last named. They have been found, in fact, at all the islands making up the chain of pelagic islets stretching somewhat interruptedly from Cape Horn and the Falkland Islands eastward to Australia and New Zealand, including among others those south of the Cape of Good Hope, so famous in the annals of the seal-fishery. It has been stated by Gray and others that the Cape of Good Hope Fur Seals (really those of the Crozets and neighboring islands) are far inferior in commercial value to those of other regions; but in tracing the history of the sealing business I have failed to notice any reference to the inferior quality of those from the last-named locality, or that there has been any difference in the commercial value of the Fur Seal skins obtained at different localities in the Southern Seas. The quality differs at the same locality, wherever the Fur Seals are found, with the season of the year and age of the animals, so that skins may come not only from the Cape of Good Hope, but from any other of the sealing-places, that one " might feel convinced could not be dressed as furs," being "without very thick under fur."

In this connection I may add that Gray's figure (Hand-List of Seals, pl. xxiii) of an old male skull of *Arctocephalus antarcticus* so closely resembles an aged male skull (No. 1125, M. C. Z. Coll.) of *Arctocephalus australis* (= *falklandicus*, auct.), that the latter might have served as the original of the figures! while other skulls of the last-named species bear a striking resemblance to Gray's figures of his *Euotaria cinerea* (Hand-List, pl. xxvi) and his *E. latirostris* (ib., pl. xxvii). In fact, the series of skulls of *Arctocephalus australis* in the Museum of Comparative Zoölogy, from the Straits of Magellan and the west coast of South America, presents variations that seem to cover all of Gray's species of *Arctocephalus* and *Euotaria* as figured by him in the Hand-List of Seals.

Synopsis of the Genera and Species.

A. Pelage harsh, without under-fur. Ears short. Molars $\frac{6-6}{5-5} = \frac{12}{10}$, or $\frac{5-5}{5-5} =$ $\frac{10}{10}$. Species generally of large size. Color yellowish-brown; reddish-brown when young. ....TRICHOPHOCACÆ.

I. Genus OTARIA, *Gill ex Péron.*

Otaria, PÉRON, Voy. aux Terr. Austr., ii, 1816, 37, footnote (in part).
[*Platyrhinchus*] *Platyrhinque*, F. CUVIER, Mém. du Mus., xi, 1824, 208, pl. iv, fig. 2.
Platyrhincus,* F. CUVIER, Dict. des Sc. Nat., xxxix, 1827, 555.—LESSON, Man. de Mam., 1827, 203 (in part).
Otaria, GILL, Proc. Essex Inst., v, 1866, 7.

    CHAR. GEN.—Palatine bones extending nearly to the pterygoid processes, deeply concave, truncate behind. Molars $\frac{6-6}{5-5} = \frac{12}{10}$.

1. **Otaria jubata** ("Forster") Blainville.

Phoca jubata, "FORSTER, 1775"; SCHREBER, ERXLEBEN, GMELIN, and other early writers.
Phoca jubata, FORSTER, Descrip. Anim. ad Licht., 1844, 317 ("Terra Statuum; Insula Novi-anni").
Otaria jubata, DESMAREST, Mam., 1820, 248 (in part), and of most recent writers.
? *Phoca flavescens*, SHAW, Gen. Zoöl., i, 1800, 260 (young).
Otaria leonina, PÉRON, Voy. aux Terr. Austr., ii, 1816, 40. Also of DESMAREST, GRAY, PETERS, and some others.
Platyrhincus leoninus, F. CUVIER, LESSON.
Phoca byroni, BLAINVILLE, Journ. de Phys., xci, 1820, 287.—DESMAREST, Mam., 1820, 240 (*fide* Gray, Suppl. Cat. Seals and Whales, 1871, 13).
Otaria mollossina, LESSON *et* GARNOT, Voy. Coq., Zool., i, 1826, 140, pl. iii ("Iles Malouines").
Platyrhyncus mollossinus et *uraniæ*, LESSON, Man. de Mam., 1827, 204.
Otaria pernettyi, LESSON, Dict. Class. d'Hist. Nat., xiii, 1828, 420.
Otaria platyrhinchus et *chilensis*, MÜLLER, Wiegmann's Archiv für Naturgesch., 1841, 333.
Otaria leonina, godeffroyi, byronia, et *ulloæ*, PETERS, Monatsb. Akad. Berlin, 1866, 264, 266, 269, 270, 670, 671.
Otaria ulloæ, VON TSCHUDI, Fauna Peruana, 1842–44, 135, 136, pl. vi.
Arctocephalus falklandicus, BURMEISTER, Ann. and Mag. Nat. Hist., 3d ser., xviii, 1866, pl. ix, figs. 1–4 (at least in part).
Otaria minor, et *pygmœa*, GRAY, Ann. and Mag. Nat. Hist., 3d ser., viii, 326.
Otaria hookeri, SCLATER, Proc. Zoöl. Soc. Lond., 1866, 80.

    HABITAT. — Galapagos Islands (Coll. Mus. Comp. Zoöl., from Hassler Expedition), and coasts of South America from Peru and Chili on the Pacific side, and Rio de la Plata on the Atlantic side southward.

II. *Genus* PHOCARCTOS, *Peters.*

Arctocepnalus, in part, of GRAY, prior to 1866.
Phocarctos (subgenus), PETERS, Monatsb. Akad. Berlin, 1866, 269.

CHAR. GEN.—Palatine bones ending considerably in front of the pterygoid processes, deeply concave in front, narrowed and emarginate behind. Molars $\frac{6-6}{5-5} = \frac{12}{10}$.

2. Phocarctos hookeri (Gray) Peters.

Arctocephalus hookeri, GRAY, "Zoöl. Voy. Erebus and Terror, pll. xiv, xv";
    Cat. Seals and Whales, 1866, 53, fig. 15.
Phocarctos hookeri, GRAY, Suppl. Cat. Seals and Whales, 1871, 15; Hand-
    List Seals, 1874, 29, pl. xx.
Otaria jubata, ALLEN, Bull. Mus. Comp. Zoöl., ii, 1870, 45 (in part).
Otaria hookeri, CLARK, Proc. Zoöl. Soc. Lond., 1873, 754, figs.

HABITAT.—Auckland Islands. (Originally described from specimens *supposed* to have come from the "Falkland Islands and Cape Horn." See Clark, as above cited, and Gray, Ann. and Mag. Nat. Hist., 4th ser., vol. xiv, 1874, pp. 26–30.)

III. *Genus* EUMETOPIAS, *Gill.*

Otaria, in part, of earlier authors.
Eumetopias, GILL, Proc. Essex Inst., v, 1866, 7, 11.

CHAR. GEN.—Palatine bones ending very far in front of pterygoid processes, flat, or nearly so; hinder border hollowed or emarginate. Molars $\frac{5-5}{5-5} = \frac{10}{10}$; fifth pair separated by a considerable space from the fourth pair.

3. Eumetopias stelleri Peters.*

HABITAT.—Pacific coast of North America from California to Alaska; Pacific coast of Asia from Japan northward.

IV. *Genus* ZALOPHUS, *Gill.*

Arctocephalus, in part, of GRAY, prior to 1866.
Zalophus, GILL, Proc. Essex Inst., v, 1866, 7, 11.
Neophoca, GRAY, Ann. and Mag. Nat. Hist., 3d ser., xviii, 1866, 231; Suppl.
    Cat. Seals and Whales, 1871, 28.
Eumetopias, PETERS, Monatsb. Akad. Berlin, 1877, 506 (in part).

CHAR. GEN.—Palatine much as in *Eumetopias.* Sagittal crest very high. Interorbital region greatly constricted. Molars $\frac{5-5}{5-5} = \frac{10}{10}$, in a continuous series.

4. Zalophus californianus (Lesson) Allen.†

HABITAT.—Coast of California.

5. Zalophus lobatus (Gray) Gill.

?? *Otaria albicollis,* PÉRON, Voy. Terr. Austr., ii, 1816, 118.
Otaria cinerea, GRAY, King's Narr. Austral., ii, 413.
Arctocephalus lobatus, GRAY, "Spic. Zoolog., i, 1828, pl. —"; Cat. Seals, 1850,
    44; Cat. Seals and Whales, 1866, 50.

---

* For synonymy, see *infra,* under the general history of *Eumetopias stelleri.*
† For synonymy, see *infra,* under the general history of the species.

Neophoca lobata, GRAY, Ann. and Mag. Nat. Hist., 3d ser., xviii, 1866, 231;
    Suppl. Cat. Seals and Whales, 1871, 28; Hand-List Seals, 1874,
    43, pl. xxx.
Otaria australis, QUOY & GAIMARD, Zool. Voy. Astrolabe, i, 1830, 95; 1833,
    pl. xiv (animal), xv, figg. 3, 4 (skull), "Nouvelle Hollande."
Arctocephalus australis, GRAY, Cat. Seals and Whales, 1866, 57 (not *Phoca
    australis* of Zimmermann and Kerr).
Otaria stelleri, TEMMINCK, Fauna Japon. (at least in part).
Phocarctos elongatus, GRAY, Hand-List of Seals, 1874, 30, pll. xxi, xxii.
Eumetopias cinerea, PETERS (ex Péron), Monatsb. Akad. Berlin, 1877, 506.

HABITAT.—Australian Seas.   Japan??

B.—Pelage soft, with abundant under-fur. Ears longer. Molars $\frac{6-6}{5-5} = \frac{12}{10}$. Size
    smaller.   Color gray; black when young .... OULIPHOCACÆ.

V. *Genus* CALLORHINUS, *Gray.*

Callorhinus, GRAY, Proc. Zoöl. Soc. Lond., 1859, 357.
Arctocephalus, GILL, Proc. Essex Inst., v, 1866, 7, 11 (not of F. Cuvier).

CHAR. GEN.—Facial portion of the skull short, convex.   Molars $\frac{6-6}{5-5} = \frac{12}{10}$.

6. Callorhinus ursinus Gray.[*]

HABITAT.—Shores of the North Pacific, from California and Japan (*Peters*)
northward.

VI. *Genus* ARCTOCEPHALUS, *F. Cuvier.*

[*Arctocephalus*] *Arctocéphales,* F. CUVIER, Mém. du Mus., xi, 1824, 205, pl.
    iv, fig. 1.
Arctocephalus, F. CUVIER, Dict. des Sci. Nat., xxxix, 1827, 554.
Halarctus, GILL, Proc. Essex Inst., v, 1866, 7, 11.
Arctophoca, PETERS, Monatsb. Akad. Berlin, 1866, 276.—GRAY, Suppl. Cat.
    Seals and Whales, 1871, 31.
Euotaria, GRAY, Ann. and Mag. Nat. Hist., 4th ser., iv, 1869, 269; Hand-
    List Seals, 1874, 34.
Gypsophoca, GRAY, Ann. and Mag. Nat. Hist., 4th ser., iv, 1869, 269; Hand-
    List Seals, 1874, 27.

CHAR. GEN.—Facial portion of skull slender, elongated, pointed, gently
declined.   Molars $\frac{6-6}{5-5} = \frac{12}{10}$, much larger than in *Callorhinus.*

7. Arctocephalus australis (Zimmermann) Allen.

Phoca ursina, in part, of various early writers.
Phoca australis, ZIMMERMANN, Geograph. Geschichte, iii, 1782, 276 (= "Falk-
    land Seal, Pennant, ii, 521").—KERR, Anim. King., 1792, 127
    (= "Falkland Seal, Penn., Hist. Quad. N., 378").
Phoca falklandica, SHAW, Gen. Zoöl., i, 1800, 256 (= "Falkland Isle Seal" of
    Pennant—the Fur Seal of the Falkland Islands).
Otaria falklandica, DESMAREST, Dict. d'Hist. Nat., xxv, 1817, 601, and of
    many subsequent writers.
Otaria s. *Arctocephalus falklandicus,* GRAY, PETERS, and others.

[*] For synonymy see *infra*, under the general history of the species.

Otaria shawi et *hauvillei,* LESSON, Dict. Class. d'Hist. Nat., xiii, 1828, 425.
Arctocephalus nigrescens, GRAY, PETERS.
Otaria (Arctophoca) philippii, PETERS, Monatsb. Akad. Berlin, 1866, 276, pl. ii.
Otaria (Arctophoca) argentata, PHILIPPI & PETERS, Monatsb. Akad. Berlin, 1871, 560, pll. i, ii.
Arctocephalus grayi, SCOTT, Mam. Recent and Extinct, 1873, 19.
Euotaria latirostris, GRAY, Hand-List Seals, 1874, 37, pl. xxvii.

HABITAT.—Galapagos Islands (specimens in Mus. Comp. Zoöl., Hassler Exp.*) and shores and islands of South America, from Chili and the Rio de la Plata southward.

* Specimens of both *Otaria jubata* and *Arctocephalus australis* were collected by members of the Hassler Expedition at the Galapagos Islands, showing that they both range much farther northward than has hitherto been generally supposed. For the following observations respecting their numbers and habits I am indebted to my friend Mr. J. H. Blake, artist of the Expedition, who has kindly transcribed them from his note-book:

"*Charles Island, Galapagos Group, June* 10, [1872].—On an island at the eastern side of Post-Office Bay is a Sea Lion rookery, where at almost any time can be seen hundreds of Sea Lions lying at a little distance from the water. Two of our company, in a little boat about ten feet long called the 'Dingy', went near the shore where they were, when the Seals immediately ran into the water and surrounded the boat. The Seals came close to and under the boat, so that there was danger of their capsizing it, some of them being as large as the boat, and some were even larger; hence it was deemed prudent to leave them. Toward evening the Captain, with others, took a larger boat and landed on the shore below the Seals, and while they were running toward the water one measuring six or seven feet in length was shot. Many of them were of enormous size, and great numbers could easily have been killed. They made a noise when rushing to the water louder than the waves on the shore. We saved one skeleton, and next day two half-grown Seals were brought on shipboard and also saved.

"*Jarvis Island, June* 16, 1872.—At this island we saw many Seals, and some were killed, one small one being preserved in alcohol. I went on shore in the second boat, and as our boat landed we were surrounded with Seals of different sizes, which came near the boat. Near where we landed was a mother Seal and her two young ones lying together in a shallow excavation they had made in the sand. They lay very quietly and appeared to be not much disturbed by our presence as we gathered about them, except when we offered to touch the young. . . . . The mother was about six feet long, and of a light grayish color, with the head small and shaped like that of a dog. The young resembled their mother but had shorter noses and were about three or four feet long.

"In walking along the beach I came to another small rookery where there were family groups similar to that above described, lying about in all kinds of positions, and so comfortably situated I did not disturb them. One Seal, about six or seven feet long, which I met with at some distance from the water, I drove some distance to study its movements in walking and running. It would nearly raise itself from the ground and walk like

8. **Arctocephalus antarcticus** (Thunberg) Allen.

Phoca ursina, FORSTER, and in part of many early writers.

? *Phoca pusilla*, SCHREBER, Säuget., iii, [1776?], 314 (=Le Petit Phoque, Buffon, based on a young Fur Seal, from an unknown locality, but supposed to have come from India or the Levant,* but as no Seals exist there, and as many animals which, in former years, purported to have been brought from India were found to be really African, some late writers have assumed that Buffon's "Petit Phoque" must have been also African, but the pertinence of the name *pusilla* to the African Fur Seal is not beyond reasonable doubt).† Also of ERXLEBEN, GMELIN, and others.

? *Otaria pusilla*, DESMAREST, Nouv. Dict. d'Hist. Nat., xxv, 1827, 602 (based on the same).

Otaria pusilla, PETERS, Monatsb. Akad. Berlin, 1866, 271, 671 (name adopted from Schreber).

Arctocephalus pusillus, PETERS, Monatsb. Akad. Berlin, 1877, 506.

? "*Phoca parva*, BODDAERT, Elenchus Anim., pl. lxxvii" (= Buffon's Petit Phoque, as above).

Phoca antarctica, THUNBERG, Mém. Acad. St.-Pétersb., iii, 1811, 222.

Arctocephalus antarcticus, ALLEN, Bull. Mus. Comp. Zoöl., ii, 1870, 45.— GRAY, Suppl. Cat. Seals and Whales, 1871, 17.

? *Otaria peroni*, DESMAREST, Mam., 1820, 250 (= *Otaria pusilla*, Desmarest, as above).

Otaria peroni, "SMITH, South African Quart. Journ. ii, 62."

---

any four-footed animal by bending the fore-flippers and turning the hind-flippers forward as here represented [in some sketches accompanying these notes, but not here reproduced]. They galloped along the sandy shore at quite good speed. In going over the rocks they tumbled about in every way but would still manage to get along with surprising rapidity. I saw many lying on the shore asleep, and there were hundreds more in the water near the shore. On approaching within a few feet of them they would come towards me as if they had been tamed. From a projecting rock I watched their movements in the water—a beautiful sight. They would roll over under water, turning complete somersaults, swim on their backs or sides, and in almost every position would glide about in the most graceful manner around the rock on which I was sitting, looking up at me. They often put their noses together in the most affectionate way. . . . . When annoyed by flies alighting on their noses they would open their mouths widely and snap at them as dogs do.

"Just back of the beach, and separated from the ocean by a row of mangrove trees, was a lagoon of brackish water in which were a number of Seals, while lying about on the border of the lagoon were many skeletons of those that had died."

* Buffon says: " . . . . on nous a assuré que l'individu que nous avons vu venoit des Indes, & il est au moins très-probable qu'il venoit des mers du Levant."—*Hist. Nat.*, tome xiii, p. 341.

† Gray says: "It is as likely to have come from the Falkland Islands as from the Cape, as the French had traffic with Les Iles Malouines, as they call them."—*Suppl. Cat. Seals and Whales*, 1871, p. 19.

Otarie de Delalande, F. CUVIER, Dict. Sci. Nat., xxxix, 1826, 558.*
Arctocephalus delalandi, GRAY, Proc. Zoöl. Soc. Lond., 1859, 107, 369, pl. lxix;
    Cat. Seals and Whales, 1866, 52; Ann. and Mag. Nat. Hist., 3d
    ser., xviii, 1866, 235.
Arctocephalus falklandicus (in part), GRAY, Cat. Seals, 1850, 42.
Arctocephalus schisthyperoës, TURNER, Journ. Anat. and Phys., iii, 1868,
    113, fig.
Arctocephalus nivosus, GRAY, Suppl. Cat. Seals and Whales, 1871, 27.
? Euotaria compressa, GRAY, Hand-List, 1874, 38, pl. xxiv ("South Africa?").
HABITAT.—Cape of Good Hope.

?9. Arctocephalus forsteri (Lesson) Gray.

Phoca ursina, FORSTER, Descrip. Anim. (ad Lichtenstein), 1844, 64 (New
    Zealand)=Sea Bear, Forster, Cook's Second Voyage, 1777,
    = Ours Marin, Buffon, Hist. Nat., Suppl., vi, 1782, 336, pl. xlvii,
    so far as it relates to Forster's figure and notes).
Otaria forsteri, LESSON, Dict. Class. d'Hist. Nat., xiii, 1828, 421 (= Sea Bear
    of Forster, which became, later, *Phoca ursina*, Forster, exclusive
    of references to Steller's *Ursus marinus*).
Arctocephalus forsteri, GRAY, Suppl. Cat. Seals and Whales, 1871, 25.
Otaria cinerea, QUOY & GAIMARD, Zool. Voy. Astrolabe, i, 1830, 89; Atlas,
    1833, pll. xii (animal), xiii (animal), xv, figg. 1, 2, skull ("Nouvelle
    Hollande"; probably not *Otaria cinerea*, Péron, Voy. Terr. Austr.,
    ii, 1866, 54, 77, which, however, is indeterminable).—PETERS,
    Monatsb. Akad. Berlin, 1866, 272, 671 (exclusive of some syno-
    nyms).
Arctocephalus cinereus, GRAY, Cat. Seals, 1850, 43; Cat. Seals and Whales,
    1866, 56; Suppl. Cat. Seals and Whales, 1871, 24, etc.
Euotaria cinerea, GRAY, Hand-List Seals, 1874, 34, pl. xxvi.
? Otaria lamarii, MÜLLER, Wiegmann's Arch. f. Naturges., 1841, 334 (in part
    at least,—*fide* Peters, Monatsb. Akad. Berlin, 1866, 271, 272).
Gypsophoca tropicalis, GRAY, Proc. Zoöl. Soc. Lond., ·1872, 659, figg. 5, 6;
    Hand-List Seals, 1874, 28, pl. xviii.
? Arctophoca gazella, PETERS, Monatsb. Akad. Berlin, 1875, 396 (Kerguelen
    Island).
? Otaria (Arctophoca) elegans, PETERS, Monatsb. Akad. Berlin, 1876, 316 (St.
    Paul and Amsterdam Islands).
Arctocephalus brevipes, forsteri, ? elegans, et *? gazella*, PETERS, Monatsb. Akad.
Berlin, 1877, 507.
    HABITAT.—Australia, New Zealand, Auckland Island; ? Kerguelen Island;
? Saint Paul and Amsterdam Islands.

---

*"*Otaria delalandii*, F. CUVIER, Dict. Sci. Nat., xxxix, 423," cited by
Fischer (Syn. Mam., 232), and repeatedly by Gray and by Peters, is evidently
erroneous, as the article "Phoque" begins on p. 540, and no species of Seal or
Otary is mentioned on p. 423. The correct citation is not "*Otaria delalandii*," but "*Otarie de Delalande*," as given above. G. Cuvier refers (Oss.
Fos., 3d ed., v, 1825, 220, pl. xviii, fig. 5, skull) to it as "Otarie du Cap" "reçu
par M. Delalande."

MYTHICAL AND UNDETERMINABLE SPECIES.

In the preceding pages reference has been made to various species described too imperfectly to admit of recognition. Some of these I have doubtfully allocated as above; others I have made no attempt to determine. Among these are the following:

1. *Phoca pusilla*, SCHREBER, Säuget., iii, [1776?], 314, based, as already stated (see above, p. 194), on Buffon's "Petit Phoque," a young Fur Seal from an unknown locality. Buffon speaks of it as reported to have been brought from the Indies and the Levant (Hist. Nat., xiii, 1765, 341), and later (ib., 345) calls it "le petit phoque noir des Indes & du Levant."

2. *Phoca longicollis*, SHAW, General Zoöl., i, 1800, 256, based on the Long-necked Seal of Grew (Museum, 1686, 95) and Parsons (Phil. Trans., xlvii, 1751–52, pl. vi). Though said by Shaw to be "earless," Gray* contributes the following history: "There formerly existed in the Museum of the Royal Society an Eared Seal without any habitat; it is called the Long-necked Seal in Grew's 'Rarities', p. 95, described and figured under that name by Parsons in the Phil. Trans. xlvii, t. 6, and noticed in Pennant's 'Quadrupeds', ii, p. 274. Dr. Shaw, in his 'Zoology', i, p. 256, translated the name into *Phoca longicollis*, and copied Parsons's figures. The name and the form of the front feet are enough to show that it is an Eared Seal; for the neck of these animals is always long compared with the neck of the Earless Seals or *Phocidæ*. Fischer, in his 'Synopsis', p. 240, overlooking this character and the description of the front feet, considers it as the same as the Sea-Leopard of Weddell (*Phoca Weddellii*) from the Antarctic Ocean, an Earless Seal. Though the habitat is not given, there can be no doubt, when we consider the geographical distribution of the Eared Seal, that it must have been received either from the southern part of South America or from the Cape of Good Hope, as the animals of the North Pacific and of Australia were not known or brought to England in 1686. As no account of the color of the fur is given, it is impossible to determine to which species inhabiting these countries it should be referred. It is most probably the Sea Lion (*Otaria leonina*), as that is the animal which is most generally distributed and commonly brought to England. The sailors sometimes call it the 'Long-necked

---

*Ann. and Mag. Nat. Hist., 4th ser., i, 1868, pp. 217, 218.

Seal'." Gray, however, had formerly referred it doubtfully (Cat. Seals, 1850, 43; Cat. Seals and Whales, 1866, 56) to *Arctocephalus falklandicus.*

3. *Phoca flavescens,* SHAW, Gen. Zoöl., i, 1800, 260, a small, "yellowish" Eared Seal, described from a specimen in the Leverian Museum brought from the Straits of Magellan. It is the "Eared Seal" of Pennant (Quad.; ii, 278), and the *Otaria flavescens* of Desmarest (Mam., 1820, 252). From its size (about two feet long), color, and habitat, it is presumably referable to *Otaria jubata,* but has been referred by Gray to his *Phocarctos hookeri.*

4. *Otaria cinerea,* PÉRON, Voy. Terr. Austr., ii, 1816, 54, 77, is merely referred to in such general terms that it is wholly indeterminable. The name, however, has been commonly referred to the Hair Seal of Australia, for which species the name has been adopted by Peters (see above, p. 203).

5. *Otaria albicollis,* PÉRON, Voy. aux Terr. Austr., ii, 1816, 118. An Eared Seal, eight to nine feet long, distinguished by a large white spot on the middle and upper part of the neck. Observed in great numbers on the islands near Bass Straits. No tangible characters given, and wholly unrecognizable. Referred, however, by Peters, in 1877, to his "*Eumetopias cinerea* (Péron)," the *Zalophus lobatus* of Gray.

6. *Otaria coronata,* DESMAREST, Mam., 1820, 251. Says Desmarest: "*Phoca coronata,* Blainv. Espèce nouvelle observée dans le Muséum de Bullock, à Londres." Locality unknown. Though said to be an Eared Seal, one foot and a half long, black, sparsely and irregularly spotted with yellow, the fore feet are said to have five toes, nearly equal, and armed with very strong, curved, sharp nails, while the hind feet have five nails, "*mais dépassés par des pointes membraneuses*"—a combination of characters unknown in nature.

7. *Otaria porcina,* F. CUVIER, Dict. des Sci. Nat., xxxix, 1826, 559. Based on the *Phoca porcina,* Molina, Hist. Nat. du Chile, 260, recognizable merely as an Eared Seal, which Gray and Peters have thought possibly referable to the *Arctocephalus falklandicus.*

8. *Otaria peroni,* DESMAREST, Mam., 1820, 250. The same as *Phoca pusilla,* Schreber, and the Petit Phoque of Buffon, already noticed.

9. *Otaria fabricii,* LESSON, Dict. Class. d'Hist. Nat., xiii, 1828, 419,= *Phoca ursina,* Fabricius, Faun. Grœnl., 6. Based on a

supposed species of Eared Seal erroneously believed by Fabricius to exist in the Greenland seas, but who never saw the animal, and described it mainly from what were doubtless fabulous reports rife among the Greenlanders. The supposed species is entirely a myth, at least so far as having any relation to an Otary. (See, further, Brown, Proc. Zoöl. Soc. Lond., 1868, pp. 357, 358.)

10. *"Otaria aurita,* HUMBOLDT." This is unknown to me. Peters, in 1877, referred it doubtfully to *Arctocephalus falklandicus.*

11. *Arctocephalus eulophus,* SCOTT, Mam. Recent and Extinct, 1873, 19. Based wholly on the testimony of an "experienced sealer." Not determinable. *Habitat.*— "New Zealand," "Patagonia," "Juan Fernandez"! (See above, p. 199.)

Other species, composite in character, are determinable only by reference to the types, among which are *Otaria stelleri,* Temminck, *Otaria lamari,* Müller, etc., noticed elsewhere.

GEOGRAPHICAL DISTRIBUTION.

The most striking fact in respect to the distribution of the *Otariidæ* is their entire absence from the waters of the North Atlantic.

As already noticed, the Eared Seals are obviously divisible by the character of the pelage, into two groups, which are commercially distinguished as the "Hair Seals" and the "Fur Seals," which are likewise respectively known as the "Sea Lions" and the "Sea Bears." The two groups have nearly the same geographical distribution, and are commonly found frequenting the same shores, but generally living apart. Usually only one species of each is met with at the same localities, and it is worthy of note, that, with the exception of the coast of California, no naturalist has ever reported the occurrence together of two species of Hair Seals or two species of Fur Seals, although doubtless two species of Hair Seals exist on the islands and shores of Tasmania and Australia, as well as on the Californian coast.

The Hair and Fur Seals are about equally and similarly represented on both sides of the equator, but they are confined almost wholly to the temperate and colder latitudes. Of the nine species provisionally above recognized, two of the five Hair Seals are northern and three southern; of the four Fur

Seals, three are southern and one only is northern, but the three southern are closely related (perhaps doubtfully distinct, at least two of them), and are evidently recent and but slightly differentiated forms of a common ancestral stock. Of the two Eared Seals of largest size (*Eumetopias stelleri* and *Otaria jubata*), one is northern and the other southern, and, though differing generically in the structure of the skull, are very similar in external characters, and geographically are strictly representative. *Zalophus* is the only genus occurring on both sides of the equator, but the species are different in the two hemispheres.* The Fur Seals of the north are the strict geographical representatives of those of the south. *Phocarctos hookeri* is Australasian, and has no corresponding form in the Northern Hemisphere. No species of Eared Seal is known from the North Atlantic. Several of the southern species range northward into the equatorial regions, reaching the Galapagos Islands and the northern shores of Australia.

FOSSIL OTARIES.

The only fossil remains unquestionably referable to the Otaries are those found by Dr. Haast† in the Moa Caves of New Zealand. These have been referred by Dr. Haast to the species of Eared Seals still inhabiting the New Zealand coast.‡ Hence no fossil remains have thus far been discovered outside of the present habitat of the group, their supposed occurrence in the Tertiary formations of Europe requiring confirmation. The absence of the *Otariidæ* from the North Atlantic renders any

---

* This statement is made with some reservation, owing to the fact that it is not quite clear what the species are that are found in the Japan Seas. Both *Eumetopias stelleri* and *Callorhinus ursinus* extend southward, apparently in small numbers, along the east coast of Asia to Japan. *Zalophus lobatus* has been accredited to Japan, but apparently on the basis of Temminck's *Otaria stelleri*, which is evidently a composite species, which has been referred, at different times, in part to *Z. lobatus* and *Z. californianus* (=*gillespii*, auct.). The latter has as yet been certainly found nowhere except on the Pacific coast of the United States, and *Z. lobatus* has not been positively identified from any point north of Australia. Temminck's figures 1–4, pl. xxii, of the Fauna Japonica, seem unquestionably to represent skulls of *Zalophus*, but whether the Australian or the Californian species, or a third, as yet unnamed, is apparently by no means settled. If it proves to be the *Z. lobatus*, it forms an exceptional case of the same species occurring on both sides of the equator.

† *Nature*, vol. xiv, pp. 517, 518, Oct. 26, 1876.

‡ Dr. Haast identifies them as "*Arctocephalus lobatus* (?) and *A. cinereus*" and "*Gypsophoca tropicalis*."

indications of their former presence in Europe of special interest, and calls for a critical examination of the supposed evidence of their former existence there.

Gervais, many years since,[*] described and figured a tooth which he referred, with doubt, to *Otaria* ("*Otaria ? prisca*"), but Van Beneden has since determined it to be referable to *Squalodon*. M. E. Delfortrie,[†] in 1872, described two fossil teeth from the bone breccia of Saint-Médard-en-Jalle, near Bordeaux, which he considered as representing two species of Otary, which he named, respectively, *Otaria oudriana* and *Otaria leclercii*. The first is based upon a last upper molar having some resemblance to the last superior molar of *Eumetopias stelleri*; the other is an "incisive inférieure externe," not much unlike the corresponding tooth of some of the Otaries. M. Delfortrie observes that these teeth have a striking analogy to those of *Otaria jubata* figured by Blainville, and to those of *Eumetopias stelleri* and *Callorhinus ursinus* figured by myself. "Cette analogie," says M. Delfortrie, "disons-nous, nous permet d'attribuer sans hésitation à des Otaridés, les deux dents de Saint-Médard, en en faisant toutefois deux espèces distinctes, en raison des caractères bien tranchés qu'elles présentent." Respecting these teeth, Professor Van Beneden remarks: "Ces dents de l'*Otaria Oudriana* me semblant bien se rapprocher de celles de *Pelagius monachus*."[‡]

In another connection, the same writer adds: "Sans avoir vu les originaux nous ne pouvons toutefois nous défendre de l'idée que ces molaires et ces incisives pourraient bien appartenir à un animal fossile voisin du *Pelagius monachus* de la Méditerranée. Nous espérons que l'on pourra bientôt comparer avec le soin nécessaire ces dents intéressantes avec les espèces voisines vivantes et fossiles et nous ne serions pas surpris de voir rencontrer certaines affinités qui échappent jusqu'à présent. Le genre *Palæophoca* que nous décrivons plus loin n'est pas bien éloigné des *Pelagius* de la Méditerranée, et la dent qui a servi de type à l'*Otaria Oudriana* n'est peut être qu'une prémolaire de notre *Palæophoca*; celle sur laquelle est établie l'*Otaria Leclercii*, est peut être une incisive supérieure du même animal."[§] I agree entirely with M. Van Beneden that these teeth cannot be

---

[*] Zoologie et la Paléontologie françaises, 1850–55, p. 276, pl. viii, fig. 8.
[†] Actes de la Société Linnéenne de Bordeaux, xxviii, 4e livr., 1872.
[‡] Ann. du Mus. Roy. d'Hist. Nat. de Belgique, i, prem. part., 1877, p. 25.
[§] Ibid., p. 57.

accepted as satisfactory proof of the presence of Otaries in the Tertiary fauna of Europe.

Van Beneden also refers to a humerus of an Otary in the Museum of the Geological Institute of Vienna, supposed to have been taken from the bed of the Danube, and adds that it bears a close resemblance to the same part in *Otaria jubata*, if indeed it is not referable to that species, but adds: "Cet os, en tout cas, n'est pas fossile." He also refers to a skull found by Valenciennes on the shore in the department of Lande, mentioned by Gervais,[*] and says it is still unknown how it came to be found on the coast.

Van Beneden, however, believes that he has proof of the existence of fossil Otaries in Europe in his *Mesotaria ambigua*,[†] a species presenting many remarkable characters, which ally it, he believes, in some points, to the Otaries. This species is represented by the greater part of the bones of the skeleton and numerous teeth, but the skull is not known.[‡] The teeth, he says, are unlike those of any other genus, while the bones indicate a special mode of life, and a size about equal to or rather larger than that of *Phoca grœnlandica*.[§] The ilium is described as resembling more the same part in the Otaries than the Seals, and as indicating a mode of life more terrestrial than aquatic. The humerus, on the other hand, is stated to more resemble that of the Seals than that of the Otaries.

Of the femur he says: "Nous avons trois fémurs assez complets qui indiquent que cet os s'éloigne par sa conformation des autres Amphitériens. La *tête*, ainsi que le *col*, tiennent de l'Otarie, comme les condyles, et le *grand trochanter*, peu large, ne s'élève pas au-dessus de la *tête* de l'os. La *tête* est comparativement petite. La cavité trochantérique est profonde et étroite vers le milieu de l'os et tout contre le col. Le caractère se rapporte à la position du membre postérieur qui rapproche ainsi des Otaries l'animal qui nous occupe. Les Mésotaries étaient moins aquatiques que les Phoques actuels."

Upon careful comparison of his excellent figures (pl. ix) of the femur, humerus, scapula, and fragment of pelvis, with the

---

[*] Zoologie et la Paléontologie françaises, p. 276.

[†] Ann. du Mus. Roy. d' Hist. Nat. de Belgique, i, 1877, p. 56, pl. i.

[‡] Van Beneden reports having two canines, three molars, seven cervical vertebræ and an axis, six dorsal and seven lumbar vertebræ, a right ilium and a left ischium, the distal end of a scapula, four right and five left humeri, a left and a right femur, six tibiæ, and four metatarsal bones.

[§] The parts of the skeleton figured by Van Beneden correspond very nearly in size with the corresponding parts of *Cystophora cristata*.

corresponding parts of the skeleton in five species of Otaries, representing all the genera of that group, and with the principal types of the Phocids, I fail to appreciate any important approach toward the former, or any marked departure from the latter, especially the subfamily *Cystophorinæ*. In the femur, for example, there is in *Mesotaria* no trace of a *trochanter minor*, which is always strongly developed in the Otaries, as well as in the Walruses, but absent in the Phocids, this feature alone serving to at once distinguish the Gressigrade from the Reptigrade Pinnipeds. The thick short form of the femur in *Mesotaria*, with its greatly enlarged distal extremity, and the great transverse breadth and thickness of the whole bone in proportion to its length, gives it a very close resemblance in its general form and proportions to the same part in *Cystophora* and *Macrorhinus* (*Morunga* of many authors), while it places it in strong contrast with the same bone in any of the Otariids. The scapula is also a very characteristic bone among the Pinnipeds, and even the small portion (the lower extremity) shown in Van Beneden's figure (pl. ix, fig. 7) serves to emphasize and confirm the relationship of *Mesotaria* with the Phocids, and the wide divergence of this type from the Otariids, as shown especially in the obliquity of the articular surface of the glenoid cavity. The portion of the pelvis figured (pl. ix, fig. 8) is decidedly Phocine in its proportions, and in the divergence of the iliac crest, while it is very unlike the same part in the Otariids. Finally, it may be noted that the *tout ensemble* of all the bones of *Mesotaria* represented in Van Beneden's plate is strikingly that met with in the heavier types of Phocids, especially the genera *Cystophora* and *Macrorhinus*, and very unlike that of the Otaries. In all the latter, the bones are relatively small, dense, and slender, and especially is this the case with the bones of the limbs, none of them approaching the thick stout form characterizing these parts in *Mesotaria*. The proportions, to say nothing of details of structure in the principal bones in the Otaries, are so widely different from what is met with in the Phocids, that general contour alone serves at once as a basis for their discrimination.

In view of the foregoing, it seems to me evident that if the distribution of the *Otariidæ* formerly embraced the shores of Europe, we have still to wait for evidence of such a former distribution; and that in Europe, as on the Atlantic seaboard of North America, the only fossil remains of Pinnipeds thus far found are referable to the Phocids on the one hand, and to the Walruses on the other, indicating for the Otariids the same curiously limited habitat as now.

MILK DENTITION.

The milk dentition in the Pinnipeds rarely persists much beyond fœtal life, and is never to any great degree functional, and the dental formula of the temporary teeth is substantially the same in all. In the Walruses, however, two of the posterior upper milk molars and the last lower one often remain till a comparatively late period of life, but all traces of the others disappear soon after birth. The two middle pairs of incisors probably never pierce the gums, and the others scarcely persist beyond the fœtal period. The formula for the temporary dentition of this group is usually recognized as I. $\frac{3-3}{3-3}$, C. $\frac{1-1}{1-1}$, M. $\frac{5-5}{5-5}$ (or M. $\frac{4-4}{4-4}$). In the Seals, however, the number of molars appears to never exceed $\frac{3-3}{3-3}$. In the Earless Seals "the milk-teeth are extremely rudimentary in size and form, and perfectly functionless. The majority of them never cut the gums and are absorbed actually before birth, and certainly within a week after birth scarcely a trace of any of them remains."[*] The milk molars are three in number on each side, both above and below, and are replaced respectively by the second, third, and fourth molars of the permanent set. The canines are all represented in the temporary set. The number of temporary incisors varies in the different genera, it corresponding apparently with the number in the permanent set. In *Phoca vitulina, P. grœnlandica,* and *P. fœtida,* they have been found to be $\frac{3-3}{2-2}$, but the two inner ones of the upper jaw are absorbed long before birth. In the Elephant Seal, Professor Flower found, in a specimen eleven inches long, " a complete set of very minute teeth, viz. I. $\frac{2}{1}$, C. $\frac{1}{1}$, M. $\frac{3}{3}$, on each side; all of the simplest character."[†]

In the Eared Seals, the milk molars are of the same number as in the *Phocinæ,* namely $\frac{3-3}{3-3}$, and hold, approximately at least, the same position relatively to the molars of the permanent set. They are separated by wide diastema, and the middle

[*] FLOWER, " Remarks on the Homologies and Notation of the Teeth of the Mammalia," Journ. Phys. and Anat., iii, 1868, p. 269.
[†] Ibid., p. 271, fig. 4.

molar is. much smaller than either of the others. The middle
incisors are replaced early in fœtal life by the permanent ones,
which are ready to cut the gum at birth. The outer upper
incisor remains much longer, persisting quite till after birth, as
do also the temporary molars, while the canines are not shed
for some weeks, at least five or six weeks. As Professor Flower
has observed, " It is very interesting to note that in the Eared
Seals (genus *Otaria* [or family *Otariidæ*]), which more nearly
approach the terrestrial Carnivora in many points of structure
as well as habits, the milk-teeth are less rudimentary and
evanescent than in the true Seals, the canines especially being
of moderate size and retained for several weeks."*

The milk dentition of the Eared Seals has already been described in two species of *Arctocephalus*, and I am able to add
some account of it in *Eumetopias* and *Zalophus*.

Van Beneden found, in 1871, in a young skull of " *Otaria pusilla*"† ("= *Otaria delalandi*," Cuv.,= *Arctocephalus antarcticus*,
Allen, ex Thunberg), the Fur Seal of the Cape of Good Hope, the
milk dentition to be I. $\frac{3-3}{1-1}$, C. $\frac{1-1}{1-1}$, M. $\frac{3-3}{3-3}$; but he supposed the
absence of the other lower incisors to be due to their having
already fallen. The two inner superior incisors were much
smaller than the outer one, appearing like little white grains
stuck upon the gum. The outer had a long slender root and a
distinct crown. The canines were comparatively large, with
long roots and a lengthened crown, and both the upper and
lower were of similar form. The superior molars were separated by considerable intervals, the first being over the space
between the first and second permanent molars, the second over
the space between the second and third, and the fourth over
the fourth permanent tooth. The middle milk molar he found
to be much smaller than either the first or third, the two last
named being of nearly equal size, but only the third was double-
rooted. The lower milk molars were smaller than the upper,
all single-rooted, and held the same position relatively to the
permanent teeth as the upper ones. The middle one, as in the
upper series, was much smaller than either the first or third.

Later Malm described the milk dentition of *Arctocephalus
nigrescens*‡ (=*Arctocephalus australis*) as existing in a specimen

* Journ. Phys. and Anat., iii, 1868, p. 269.
† "Sur les dents de lait de l'*Otaria pusilla*," Bull. de la Acad. Roy. de Bel-
gique, t. xxxi, 1871, pp. 61–67 (illustrations).
‡ Œfver. af Kongl. Vetensk.-Akad. Förhandl., 1872, No. 7, p. 63.

measuring 730 mm. from the nose to the end of the tail, the skull having a length of 123 mm., the specimen when killed having been probably a few weeks old. The formula of the milk dentition found by Malm in this species is given as I. $\frac{1-1}{0-0}$, C. $\frac{1-1}{1-1}$, M. $\frac{2-2}{3-3}$. The third or last lower molar he describes as standing over the fourth of the permanent set, and as having two diverging roots. The first of the two upper milk molars stands over the third permanent molar, and the second (also double-rooted) over the fourth, these milk molars being probably in reality the second and third respectively, the first having doubtless already fallen, as had all the incisors except the exterior upper ones, owing to the post-fœtal age of the specimens. The formula given by Malm corresponds nearly with that of young skulls of *Zalophus*, presently to be noticed, taken probably from individuals one or two months old.

In a very young skull of *Eumetopias stelleri* (No. 4703, Nat. Mus., San Francisco, Cal., Dr. W. O. Ayres, labelled by the collector as "three or four days old"), the milk teeth have all fallen (probably by maceration), but the alveoli of all but the middle incisors are still distinct, and indicate the following formula: I. $\frac{1-1}{0-0}$, C. $\frac{1-1}{1-1}$, M. $\frac{3-3}{3-3}$. Thus, of the incisors the presence of only an outer pair is indicated, the middle ones, being rootless and probably implanted only on the gum, would leave no trace of their former presence. The alveoli of the molars show that, both above and below, the middle one was much smaller than the others. These alveoli are exterior to the permanent teeth, which do not vertically replace them, that of the first upper milk molar being opposite the space between the first and second permanent molars; the second opposite the space between the second and third permanent molars, while the third is nearly opposite the fourth tooth of the permanent set. In the lower jaw the alveoli of the milk molars are respectively just behind and exterior respectively to the second, third, and fourth permanent teeth.

In three fœtal skulls of *Zalophus* (No. 6156, Mus. Comp. Zoöl., Nos. 15660, ♂, 15661, ♀, Nat. Mus., all from the Santa Barbara Islands), the milk teeth are all still *in situ*, except the middle incisors, which are replaced by permanent incisors that were apparently about ready to pierce the gum. As in *Eumetopias* and *Arctocephalus*, the middle molar, both above and below, is much the smallest, and is placed very close to the third, leaving a very broad interval between the first and second. The first (in No. 15660) stands above the second permanent tooth, the second is just behind the third, while the fourth is a little anterior to, but nearly over, the fourth. In No. 6156 the milk molars stand directly over the second, third, and fourth permanent ones. In several other young skulls, the third milk molar is still in place, while all the others, except the canines, have disappeared. Some of them were probably several weeks old, showing that at least the canines are persistent for a considerable period after birth. In two young skulls of *Callorhinus*, known to have been killed when between four and five weeks old, the milk canines are still in place, and a trace remains of the alveolus of the third superior milk molar.

In all probability, the dental formula is the same in all the Eared Seals, the incisors, except the exterior upper, disappearing before birth. Of the molars the middle is smaller than the others, while the third is longest persistent. The canines appear to remain for several months.

IRREGULARITIES OF DENTITION.

The Eared Seals seem to rather frequently present cases of supernumerary molars, and more rarely cases of suppression of molars. In respect to supernumerary molars, I am able to record the following instances: In *Callorhinus ursinus* I have noted the following irregularities: Skull No. 2922 (M. C. Z. Coll.) has M. $\frac{7-7}{5-5}$, the normal number being $\frac{6-6}{5-5}$; in another (Nat. Mus. Coll., No. 11701), M. $\frac{7-6}{5-5}$; and in still others (M. C. Z. Coll., No. 1787; N. M. Coll., 12270), M. $\frac{5-5}{5-5}$. In each case, the identity of the species is beyond question. In *Zalophus californianus = gillespii*, auct., I have noted the following: two skulls (Nat. Mus. Coll., No. 15254; M. C. Z. Coll., No. 6162) each with M. $\frac{6-5}{5-5}$, and one (M. C. Z. Coll., No.6163) with M. $\frac{6-6}{5-5}$, the normal number being M. $\frac{5-5}{5-5}$. In nearly every case the supernumerary molars were as perfectly developed as the others. About five per cent. of the skulls of these two species (of which I have examined not less than thirty of each) present one or more supernumerary molars. I have also found suppression of molars in *Arctocephalus australis* (M. C. Z. Coll., No. 1131,— M. $\frac{6-5}{5-5}$).

The supernumerary molars are placed (in all the instances I

have seen) behind the last molar of the normal series. They are usually smaller than the normal molars, sometimes almost rudimentary, usually without accessory cusps, and with a smooth or nearly smooth cingulum. They are hence generally recognizable by their size and form. In cases of suppression it is usually the antepenultimate molar that is missing. This molar also frequently falls late in life, but traces of an alveolus in such cases usually attest its former presence.

POSITION OF THE LAST UPPER PERMANENT MOLAR.

In species with the superior molars 5—5, the last (except in *Eumetopias*) is placed opposite the posterior edge of the zygomatic process of the maxillary, varying slightly in position in different individuals belonging to the same species, mainly, however, in consequence of the thickening of the mastoid process in old age. In species having the superior molars 6—6, the last is usually* entirely behind the zygomatic process of the maxillary, the fifth molar holding the same relative position as in the five-molared species. The exception presented by *Eumetopias* seems at first view to favor the theory that the last molar is homologically the sixth, and that the fifth is suppressed, but in reality its position is posterior to that of the last molar in the six-molared species, while the space between it and the fourth is equal to or greater than that occupied by two molars.

GENERAL OBSERVATIONS.

The largest species of the Otaries (genera *Otaria* and *Eumetopias*) are Hair Seals, while the smallest (genera *Callorhinus* and *Arctocephalus*) are Fur Seals; but the species of *Zalophus*, although Hair Seals, are intermediate in size between the other Hair Seals and the Fur Seals. All the Hair Seals have coarse, hard, stiff hair, varying in length with age and season, and are wholly without soft underfur. All the Fur Seals have an abundant soft, silky underfur, giving to the skins of the females and younger males great value as articles of commerce. The longer, coarser overhair varies in length and abundance with season and age. All the Hair Seals are yellowish- or reddish-brown (in *Zalophus* sometimes brownish-black), generally darkest when young, and becoming lighter with age, and also in the same individuals toward the moulting season. There is also

* In *Phocarctos* both the fifth and sixth are behind the posterior edge of the zygomatic process of the maxillary.

considerable range of individual variation in representatives of the same species, so that coloration alone fails to afford satisfactory diagnostic characters. All the Fur Seals are black when young, but they become lighter with age, through an abundant admixture of grayish hairs which vary from yellowish-gray to whitish-gray. The southern Fur Seals are generally, when adult, much grayer than the northern. There is hence a wide range of color variation with age in the same species, as there is also among conspecific individuals of the same sex and age. While some have the breast and sides pale yellowish-gray, others have these parts strongly rufous, the general tint also showing to some extent these differences.

There is also a wonderful disparity in size between the sexes,* the weight of the adult males being generally three to five

---

*The sexual difference in size varies only slightly in the different genera; it is greatest, apparently, in *Otaria*, and least in *Arctocephalus*. It is very much less in *Arctocephalus australis* than in the northern Fur Seal; while relatively small in *Zalophus*, it is very great in both *Otaria* and *Eumetopias*. In *Otaria jubata*, the average dimensions of eight old male skulls are, length 350 mm., breadth 223 mm.; of four old female skulls, length 261 mm., breadth 143 mm. In *Eumetopias stelleri* the average length of ten old male skulls is 375 mm., breadth 221 mm.; of two old female skulls, length 296 mm., breadth 157 mm. In *Zalophus californianus* very old male skulls obtain a length of 290 mm. to 330 mm., while very old females reach 220 mm. to 237 mm. Five old male skulls average, length 269 mm., breadth 157 mm; five old female skulls, length 219 mm., breadth 103 mm. In *Arctocephalus australis* two old male skulls average, length 260 mm., breadth 145 mm., two old female skulls, length 230 mm., breadth 121 mm. In *Callorhinus ursinus* eight adult male skulls (not generally very old) average, length 243 mm., breadth 123 mm.; four female skulls of nearly corresponding age average, length 188 mm., breadth 96 mm. These data may be tabulated as follows, 100, in the column of "Approximate ratio," representing the male sex:

Species.	Sex.	Number of specimens.	Length.	Breadth.	Approximate ratio.	
Otaria jubata	♂	8	350	223	Length 75–100	=69–100.
Do	♀	4	261	143	Breadth 64–100	
Eumetopias stelleri	♂	10	375	221	Length 79–100	=75–100.
Do	♀	2	296	157	Breadth 76–100	
Zalophus californianus	♂	5	289	157	Length 81–100	=74–100.
Do	♀	5	219	103	Breadth 66–100	
Arctocephalus australis	♂	2	260	145	Length 88–100	=85–100.
Do	♀	2	230	121	Breadth 83–100	
Callorhinus ursinus	♂	8	243	123	Length 77–100	=76–100.
Do	♀	4	188	92	Breadth 75–100	

times that of the adult females of the same species. There
are also very great differences in the form of the skull, especially in respect to the development of crests and protuberances
for muscular attachment, these being only slightly developed
in females and enormously so in the males. With such remarkable variations in color and cranial characters, dependent upon
age and sex, it is not a matter of surprise that many nominal
species have arisen through a misappreciation of the real significance of these differences.*

HABITS.

The Eared Seals show also a remarkable resemblance in their
gregarious and polygamous habits. All the species, wherever
occurring, like the Walruses and Sea Elephants, resort in
great numbers to particular breeding stations, which, in sealers' *parlance*, have acquired the strangely inappropriate name
of "rookeries." The older males arrive first at the breeding
grounds, where they immediately select their stations and await
the arrival of the females. They keep up a perpetual warfare
for their favorite sites, and afterward in defense of their harems.
The number of females acquired by the successful males varies
from a dozen to fifteen or more, which they guard with the utmost
jealousy,—might being with them the law of right. The strongest males are naturally the most successful in gathering about
them large harems. The males, during the breeding season,
remain wholly on land, and they will suffer death rather than
leave their chosen spot. They thus sustain, for a period of several weeks, an uninterrupted fast. They arrive at the breeding
stations fat and vigorous, and leave them weak and emaciated,
having been nourished through their long period of fasting
wholly by the fat of their own bodies. The females remain
uninterruptedly on land for a much shorter period, but for a considerable time after their arrival do not leave the harems. The
detailed account given a century ago by Steller, and recently confirmed by Bryant and Elliott, of the habits of the northern Fur
and Hair Seals during the breeding season, is well known to
apply, in greater or less detail, to nearly all the species of the
family, and presumably to all. As the observations by Messrs.
Elliott and Bryant are presented later in this work at length, it
is unnecessary to give further details in the present connection.

---

*Of about fifty synonyms pertaining to the Eared Seals, probably two-thirds
have been based, directly or indirectly, upon differences dependent on sex and
age, and the rest upon the defective descriptions of these animals by travellers.

PRODUCTS.

The products of the Eared Seals vary in importance with the species, the Hair Seals yielding only oil, their skins being almost valueless except to the natives of the countries these animals frequent. The products of the Eared Hair Seals are, consequently, not different from those of the common Earless Seals, and at present are of far less commercial importance, in consequence of the more limited source of supply. The Fur Seals, on the other hand, are hunted almost exclusively for their fur, which forms the well-known and highly-valued "Seal 'fur" of furriers. The fur differs in quality with season and the sex and age of the animals, the most valuable being obtained from the females and rather young males. In the young of the second year taken "in season," the skin "un-plucked" forms a rich and soft fur, the very thick, silky red-dish-brown underfur being slightly overtopped by short, very soft, fine, gray overhair. Later in the season, and especially in the old animals, the overhair is coarser and longer, and even somewhat harsh, beneath which, however, is still the heavy soft underfur. Dealers sort the skins into grades, in accord-ance with the size of the skins and the quality of the fur, these features depending upon the age and sex of the animal, rather than upon the species. Dr. Gray refers to what he calls *Arcto-cephalus falklandicus* as being "easily known from all other Fur Seals in the British Museum by the evenness, shortness, closeness, and elasticity of the fur, and the length of the under-fur. The fur is soft enough to wear as a rich fur, without the removal of the longer hairs, which are always removed in other Fur Seals."* This, however, is not a peculiarity of the Falk-land Island Fur Seal, the overhair in prime young skins of the Alaskan Fur Seal being equally rich and soft. They are also often made up and worn "without the removal of the longer hairs," and are by some preferred to the prepared or "dressed" furs of the furrier. The Australian Fur Seal appears to differ little in the quality or color of its pelage from the Alaskan and Falkland Island species. The Fur Seal of the Cape of Good Hope, although one of the Fur Seals of commerce, appears to have, according to Gray's account of the few examples he has examined, a shorter coat of underfur. I have, however, met with no statement respecting the Cape Fur Seal peltries that indicates that they are inferior in quality to those of other local-

---

* Ann. and Mag. Nat. Hist., 4th ser., i, 1868, p. 103.

ities. As regards color and the variations of color with age, the Cape of Good Hope species appears not to differ appreciably from the others.*

DESTRUCTION OF THE FUR SEALS FOR THEIR PELTRIES.

The value of the peltries of the Fur Seal has led to wholesale destruction, amounting at some localities almost to extermination. The traffic in their skins appears to have begun toward the end of the last century. Captain Fanning, of the ship "Betsey," of New York, obtained a full cargo of choice Fur Seal skins at the island of Masafuera, on the coast of Chili, in 1798, which he took to the Canton market. Captain Fanning states that on leaving the island, after procuring his cargo, he estimated there were still left on the island between 500,000 and 700,000 Fur Seals, and adds that subsequently little less than a million of Fur Seal skins were taken at the island of Masafuera alone,† a small islet of not over twenty-five miles in circumference, and shipped to Canton.‡ Captain Scammon states that the sealing fleet off the coast of Chili, in 1801, amounted to thirty vessels, many of which were ships of the larger class, and nearly all carried the American flag. Notwithstanding this great slaughter, it appears that Fur Seals continued to exist there as late as 1815, when Captain Fanning again obtained them at this island.§

In the year 1800, the Fur Seal business appears to have been at its height at the Georgian Islands, where, in the single season, 112,000 Fur Seals are reported to have been taken, of which 57,000 were secured by a single American vessel (the "Aspasia," under Captain Fanning). Vancouver, at about this date, reported the existence of large numbers of Fur Seals on the southwest coast of New Holland. Attention was at once turned to this new field, and in 1804 the brig "Union," of New York, Capt. Isaac Pendleton, visited this part of the Australian coast, but not finding these animals there in satisfactory numbers, repaired to Border's Island, where he secured only part of a cargo (14,000 skins), owing to the lateness of the season. Later 60,000 were obtained at Antipodes Island. About 1806, the American ship

---

* See Gray, Ann. and Mag. Nat. Hist., 4th ser., i, 1868, pp. 218, 219; Scott, Mam. Recent and Extinct, 1873, pp. 14, 15 ; also Pagès, in Buffon's Hist. Nat., Suppl., vi, p. 357.

† Fanning's Voyages to the South Sea, etc., pp. 117, 118.

‡ Ib., p. 364.

§ Ib., p. 299.

"Catharine," of New York (Capt. H. Fanning), visited the Crozette Islands, where they landed, and found vast numbers of Fur Seals, but obtained their cargo from Prince Edward's Islands, situated a few hundred miles southeast of the Cape of Good Hope, where other vessels the same year obtained full cargoes.

In 1830, the supply of Fur Seals in the southern seas had so greatly decreased that the vessels engaged in this enterprise "generally made losing voyages, from the fact that those places which were the resort of Seals," says Captain Benjamin Pendleton, "had been abandoned by them, or cut off from them," so that the discovery of new sealing grounds was needed. Undiscovered resorts were believed to exist, from the fact that large numbers of Fur Seals were seen while cruising far out at sea, which must repair once a year to some favorite breeding station.[*]

Captain Weddell states that during the years 1820 and 1821 over 300,000 Fur Seals were taken at the South Shetland Islands alone, and that at the end of the second year the species had there become almost exterminated. In addition to the number killed for their furs, he estimates that not less than 100,000 newly-born young died in consequence of the destruction of their mothers. So indiscriminate was the slaughter, that whenever a Seal reached the beach, of whatever denomination, it was immediately killed. Mr. Scott states, on the authority of Mr. Morris, an experienced sealer, that a like indiscriminate killing was carried on at Antipodes Island, off the coast of New South Wales, from which island alone not less than 400,000 skins were obtained during the years 1814 and 1815. A single ship is said to have taken home 100,000 in bulk, which, through lack of care in curing, spoiled on the way, and on the arrival of the ship in London the skins were dug out of the hold and sold as manure! At about the same time there was a similar wasteful and indiscriminate slaughter of Fur Seals at the Aleutian Islands, where for some years they were killed at the rate of 200,000 a year, glutting the market to such an extent that the skins did not bring enough to defray the expenses of transportation. Later the destruction of Fur Seals at these islands was placed under rigid restrictions (see *infra* the general history of the Northern Fur Seal), in consequence of which undue decrease has been wisely pre-

[*] Fanning's Voyages, p. 487.

vented. But nowhere else has there been systematic protection of the Fur Seals, or any measures taken to prevent wasteful or undue destruction.

GENUS EUMETOPIAS, *Gill.*

Otaria, in part of various authors.
Arctocephalus (in part), GRAY, Cat. Seals and Whales, 1866, 51.
Eumetopias, GILL, Proc. Essex Institute, v. 7, 11, July, 1866. Type *" Otaria californianus*, Lesson, = *Arctocephalus monteriensis*, Gray."

Molars $\frac{5-5}{5-5} = \frac{10}{10}$; the upper hinder pair separated from the others by a considerable interval; the last only double-rooted. Postorbital processes quadrate. Palatine surface of the inter-maxillaries flat, only slightly depressed, and greatly contracted posteriorly; the palatals moderately produced, extending about three-fourths of the distance from the anterior end of the zygo-matic arch to the pterygoid process; their posterior margin straight, or slightly or deeply emarginate; rarely deeply so in old age.

Eumetopias differs from *Otaria*, as restricted by Gill, in hav-ing one pair less of upper molars, a much less posterior exten-sion of the palatine bones, and in having the posterior portion of the palatal surface less than one-third, instead of more than one-half, the width of the anterior portion, and but slightly in-stead of deeply depressed; also in the greater depth of the skull anteriorly, and in the less development of the occipital and sagittal crests. In *Eumetopias* the depth of the skull at the anterior border of the orbits is nearly as great as in the plane of the occiput, while in *Otaria* these proportions are as 13 to 18, there being in the latter a marked declination anteri-orly in the superior outline of the skull. The breadth of the skull at the temporal fossæ is also much greater than in *Otaria;* that is, the skull is much less constricted behind the orbits. The postorbital processes also differ considerably in form in the two genera, while another noteworthy difference is the un-usually great development in *Otaria* of the pterygoid hamuli.*

---

* A comparison of adult male skulls of *Eumetopias* and *Otaria*, of strictly corresponding ages, shows the following differences:

Eumetopias stelleri (No. 1765): height of skull in occipital plane 155 mm.; height of skull at anterior edge of orbits 152 mm.

Otaria jubata (No. 1095): height of skull in occipital plane 180 mm.; height. of skull at anterior edge of orbits 130 mm.

Comparing the same skulls in respect to the development of the pterygoid hamuli it is found that when placed on a plane surface the skull of *E. stel-*

Eumetopias differs from *Zalophus* through the presence of a wide space between the fourth and fifth pairs of upper molars, the less emargination of the posterior border of the palatine bones, the quadrate instead of the triangular and posteriorly pointed form of the postorbital processes, the less relative breadth of the posterior nares, and the larger size of the facial angle; also through its much broader muzzle, the less degree of the postorbital constriction of the skull, and its much less developed sagittal crest.

Eumetopias differs too widely from *Callorhinus* and *Arctocephalus*, in dentition and cranial characters as well as in size and pelage, to render comparison necessary. The genus is at once distinguishable from all the others of the family by the wide space between the fourth and fifth upper molars. In distribution it is restricted to the shores and islands of the North Pacific Ocean, ranging from Southern California northward to Behring's Straits. Its geographical representative is the *Otaria jubata* of the Southern Seas, which ranges from the equatorial regions (Galapagos Islands) southward.

EUMETOPIAS STELLERI, *(Lesson) Peters.*

Steller's Sea Lion.

Leo marinus, STELLER, Nov. Comm. Petrop., xi, 1751, 360.
Phoca jubata, SCHREBER, Säugeth., iii, 1778, 300, pl. lxxxiii B (in part only; not *P. jubata*, Forster, with which, however, it is in part confounded).—GMELIN, Syst. Nat., i, 1788, 63 (in part;=*P. jubata*, Schreber).—PANDER & D'ALTON, Skelete der Robben und Lamantine, 1826, pl. iii, figs. *d, e, f.*—HAMILTON, Marine Amphib., 1839, 232 (in part—not the figure of the skull).
Phoca (Otaria) jubata, RICHARDSON, Zoöl. Beechey's Voy., 1839, 6.
Otaria jubata, PÉRON, Voyage Terr. Austr., ii, 1816, 40.—NILSSON, Arch. f. Naturgesch., 1841, 329 (in part only; includes also the true *Otaria jubata*).—? VEATCH, J. R. Browne's Resources of the Pacific Slope, [app.], 150 (probably only in part, if at all).
Otaria stelleri, LESSON, Dict. Class. Hist. Nat., xiii, 1828, 420.—J. MÜLLER, Archiv f. Naturgesch., 1841, 330, 333.—SCHINZ, Synop. Mam., i, 1844, 473.—GRAY, Cat. Seals in Brit. Mus., 1850, 47; Cat. Seals and Whales in Brit. Mus., 1866, 60.—SCLATER, Proc. Zoöl. Soc. Lond., 1868, 190.—SCOTT, Mam. Recent and Extinct, 1873, 22.
Phoca stelleri, FISCHER, Synop. Mam., 1829, 231.

---

leri rests anteriorly on the mastoid processes and the points of the canines, the points of the pterygoid hamuli being several millimetres above the plane of rest, while in *O. jubata* the skull in the same position rests posteriorly on the pterygoid hamuli, which project 5^{mm} below a plane connecting the mastoid processes and the points of the canines.

Otaria (*Eumetopias*) *stelleri*, PETERS, Monatsb. Akad. Berlin, 1866, 274, 671.
Eumetopias stelleri, GRAY, Ann. and Mag. Nat. Hist., 3d ser., xviii, 1866, 233;
    Suppl. Cat. Seals and Whales in Brit. Mus., 1871, 30; Proc. Zoöl. Soc.
    Lond., 1872, 737 (in part), figs. 4, 5 (the young skull on which
    Arctocephalus·monteriensis, Gray, was in part based); Proc. Zoöl. Soc.
    Lond., 1873, 776 (its occurrence in Japan stated to be doubtful);
    Hand-List of Seals, etc., 1874, 40.—ALLEN, Bull. Mus. Comp. Zoöl., ii,
    1870, 46, pll. i, iii, figg. 9-15, and figg. 1-5 in text.—SCAMMON, Marine
    Mam., 1874, 124, four woodcuts of animal pp. 126, 127.—ELLIOTT,
    Report on the Prybilov or Seal Islands of Alaska, 1873 (text not
    paged, five plates); Condition of Affairs in Alaska, 1875, 152; Scrib-
    ner's Monthly, xvi, Oct., 1878, 879 (popular account, with figures).
Stemmatopias stelleri, VAN BENEDEN, Ann. du Mus. Roy. d'Hist. Nat. du Bel-
    gique, pt. i, 1877, 15 (in text),—lapsus pennæ for *Eumetopias stelleri?*
Phoca leonina, PALLAS, Zoog. Rosso-Asiat., i, 1831, 104 (= *P. jubata*, Gmelin).
Arctocephalus monteriensis, GRAY, Proc. Zoöl. Soc. Lond., 1859, 358, 360, pl.
    lxxii, skull (in part only; the skin referable to *Callorhinus ursinus*);
    Cat. Seals and Whales, 1866, 49.
Arctocephalus californianus, GRAY, Cat. Seals and Whales, 1866, 51 (= *A. monte-
    riensis*, Gray, 1859, in part; not = *Otaria californiana*, Lesson).
Eumetopias californianus, GILL, Proc. Essex Inst., v, 1866, 13 (= *Arctocephalus
    monteriensis*, GRAY, 1859, " and possibly also [identical] with *Otaria
    stelleri*, Müller"; hence not = *O. californiana*, Lesson).
? *Eumetopias elongatus*, GRAY, Proc. Zoöl. Soc. Lond., 1873, 776, figs. 1, 2
    (= *E. stelleri*, Gray, ib., 1872, 737, figs. 1-3, Japan?).
? *Phocarctos elongatus*, GRAY, Hand-List Seals, etc., 1874, 30, pl. xxi, xxii.
Meerlöwen, STELLER, Beschreib. von sonderbarer Meerthiere, 1753, 152.
Le Lion marin, BUFFON, Hist. Nat., Suppl., vi, 1782, 337 (in part only).
Leonine Seal, PENNANT, Arctic Zoöl., i, ——, 200 (in part only).
Lion Marin, CHORIS, Voyage Pittoresque, Iles Aléoutienne, 1822, 12 (not
    = *Lion marin de la Californie*, pl. xi, "Port San-Francisco et ses
    Habitants").
Leo marinus, the Sea King, ELLIOTT, Scribner's Monthly, xvi, 879, Oct., 1878.
"*See-Vitchie*," Russian; *Lion marin*, French; *Seelöwe*, German; *Sea Lion,
    Hair Seal*, English.

HABITAT.—Shores of the North Pacific, from Behring's
Straits southward to California and Japan.

EXTERNAL CHARACTERS.—Length of full-grown male eleven
to twelve and a half or thirteen feet, of which the tail forms
three or four inches; girth about eight to ten feet; weight variously estimated at from one thousand to twelve hundred or
thirteen hundred pounds.* The weight of the full-grown female

---

* A skull of this species in the National Museum (No. 4702), collected at
Fort Point, Bay of San Francisco, July, 1854, bears a label with the following legend: "Length 13 ft. 8 in.; weight, by estimate, one ton."

Captain Bryant, in some MSS. notes on this species recently received,
states that the full-grown male measures 13¼ to 14 feet from the tip of the
nose to the end of the *outstretched hind-feet*, and from 7½ to 9 feet in girth

is said to range from four hundred to five hundred pounds, with a length of eight to nine feet. The color varies with age and season. The young are "of a rich dark chestnut-brown." The adults, on their first arrival at their breeding-grounds in spring, present no sexual dissimilarity of color, which is then light brownish-rufous, darker behind the fore limbs and on the abdomen. Later the color changes to "bright golden-rufous or ocher." The pelage is moulted in August, and the new coat, when fully grown in November, is "light sepia or vandyke-brown, with deeper shades, almost dark upon the belly." At this season the females are somewhat lighter-colored than the males, and occasionally specimens of both sexes are seen with patches of dark brown on a yellowish-rufous ground (*Elliott*). In two adult males in the Museum of Comparative Zoölogy, and another adult male in the National Museum, the general color of the upper side of the body varies from pale yellowish-brown to reddish-brown, becoming much darker toward the tail. The sides below the median line are reddish, shading above into the lighter color of the back, and below passing into the dusky reddish-brown of the lower surface, which latter becomes darker posteriorly. The limbs are dark reddish-brown, approaching black, especially externally. The hairs are individually variable in color, some being entirely pale yellowish, others yellowish only at the tips and dark below, while others are wholly dark reddish-brown or nearly black throughout. The relative proportion of the light and dark hairs determines the general color of the body. The pelage consists of two kinds of hair, the one abundant, straight, stiff, coarse and flattened, and constituting the outer coat; the other very short, exceedingly sparse and finer, and in such small quantity as to be detected only on close inspection. The hair is longest on the anterior upper portion of the body, where on the neck and shoulders it attains a length of 40 mm.; it decreases in length posteriorly, and toward the tail has a length of only 15 mm. It is still shorter on the abdomen, becomes still more reduced on the limbs, and disappears entirely toward the ends of the digits. The end of the nose, the soles and palms, the anal region, and the extra-digital cartilaginous flaps are naked and black (in

around the chest, and that the average weight is over one thousand pounds. He gives the length of the full-grown female as 8½ to 9 feet, and the circumference at the shoulders as 4, the females being relatively much slenderer than the males. The weight of the female he states to be one-third that of the male.

life "dull blue-black"). The whiskers are long, slender, and cylindrical, white or brownish-white, and set in four or five rather indistinct rows. Some of the longest have sometimes a length of 500 mm., or about twenty inches, with a maximum thickness of 2 mm. They are set in several rows, and number between thirty and forty, increasing in length from the inner ones to the outer, which are longest. The ears are short and pointed, broader, but only half the length of those of the Northern Fur Seal (*Callorhinus ursinus*).

The fore feet are large, triangular, situated a little in front of the middle of the body. They terminate in a thick, hard, membranous flap, which is slightly and somewhat irregularly indented on the inner side. The terminations of the digits are indicated by small circular horny disks or rudimentary nails. The hind feet are broad, and gradually widen from the tarsus, reaching their greatest breadth at the end of the toes. Their length is short as compared to their breadth, the distance between the ends of the outer toes when spread exceeding half of the length of the foot, measured from the tarsal joint. The toes terminate in strong cartilaginous flaps, covered with a thick leathery naked membrane, which is deeply indented opposite the intervals between the toes, and serves to connect the diverging digits. The three middle toes are provided with long, well-developed nails; the outer toes are without true nails, but in place of them are thickened, horny disks. The outer toes are slightly longer than the three middle ones, which are subequal. The nails on all the feet are bluish horn-color.

The following table of external measurements of two males, one very aged and the other adult, both from St. Paul's Island, Alaska, indicates the general proportions of the body. A part were taken from a moist flat skin before stuffing, and the others from mounted skins.

Measurements from Two Skins of EUMETOPIAS STELLERI.

	No. 2920, Coll. Mus. Comp. Zoöl., ♂, about 10 years old.		No. 2921, Coll. Mus. Comp. Zoöl., ♂, about 15 years old.
	Unmounted.	Mounted.	Mounted.
Length of body.........................	2,750	2,790	3,010
Length of tail	100	100	110
Extent of outstretched fore limbs......	2,362	....	....
Length of hand.........................	575	560	620
Breadth of hand.......................	337	335	360
Length of foot..........................	559	540	610
Breadth of foot at tarsus...............	216	210	230
Breadth of foot at ends of the toe-flaps.	483	445	440
Length of flap of outer toe.............	200	200	220
Length of flap of second toe...........	179	156	210
Length of flap of third toe.............	152	147	190
Length of flap of fourth toe............	164	150	190
Length of flap of inner toe.............	164	150	165
Distance from end of nose to eye.......	215	190	170
Distance from end of nose to ear.......	368	365	380
Distance between the eyes	190	195	210
Distance between the ears..............	372	370	420
Length of the ear......................	37	35	35
Length of longest barbule...	342	342	....
Dist. between points of longest barbules	800	800	....
Circumference of the body at fore limbs.	....	2,250	2,600
Circumference of the body near the tail.	....	1,000	1,020
Circumference of the head at the ears..	....	1,000	980
Length of body to end of hind limbs ..	....	3,450	3,790

Captain Scammon gives the following external measurements of a full-grown male taken at the Farallone Islands, July 17, 1872.

	ft.	in.	mm.
Length from tip of nose to end of hind flippers.............	12	0	=3,660
Length of hind flippers......................................	2	2	= 660
Breadth of hind flippers (expanded)	0	9	= 220
Circumference of body behind fore limbs	7	0	=2,150
From nose to fore limbs......................................	5	0	=1,526
Length of fore flipper.......................................	2	6	= 753
Breadth of fore flipper	1	4	= 470
Distance between extremities of fore limbs...................	10	0	=3,035
Length of ear...		1½=	47
Length of tail		7 =	177
Length of longest whiskers...................................	1	6 =	457
Length of longest hind claw..................................		1¼=	32

SKULL.—The skull varies greatly in different individuals, even of the same sex, not only in its general form, but in the shape of its different bones. In the males the occipital and medial crests are not much developed before the fifth or sixth year. The bones thicken greatly after the animal attains maturity, and the palate becomes more flattened. In the adult male the brain-box may be described as subquadrate, narrower anteriorly, where the skull is abruptly contracted. The greatest diameter of the skull is at the posterior end of the zygoma, and is equal to three-fifths of its length. The postorbital processes are strongly developed and quadrate; the forehead is flat, and the facial profile is either abruptly or gradually declined; the muzzle is broad, its breadth at the canines being rather more than one-fourth the total length of the skull. The palatal surface of the intermaxillaries is flat, or slightly depressed anteriorly, and very slightly contracted posteriorly. Laterally the intermaxillaries reach nearly to the end of the palatals. The latter are much contracted posteriorly, and terminate quite far in front of the hamuli pterygoidei. Both the anterior and posterior nares are a little narrower than high. The nasals are widest anteriorly. The last (fifth) pair of upper molars is placed far behind the fourth pair, the space between them being about equal to that occupied by two molars. The males in old age have exceedingly high occipital and sagittal crests, most developed posteriorly; anteriorly they diverge and terminate in the hinder edge of the postorbital processes.

The lower jaw is massive and strong. Its coronoid processes are greatly developed, as are the tuberosities at the angle of the rami, and a second tuberosity on the lower inner edge of each ramus.

The skull in the female is not only much smaller than in the male, but lacks entirely the high crests seen in the male, and all the processes are much less developed. The teeth, especially the canines, are much smaller, and the bones are all thinner and weaker, the weight of the adult female skull being only about one-third of that of the male of corresponding age. The skull of a full-grown female of this species attains only about the linear dimensions of an adult male skull of *Callorhinus ursinus*.

Measurements of Nine Male Skulls and Two Female Skulls of EUMETOPIAS STELLERI.

Catalogue number.	Locality.	Sex.	Length.	Breadth at mastoid processes.	Greatest breadth at zygomatic arches.	Distance from front edge of intermaxillaries to hamular process of pterygoid.	Distance from front edge of intermaxillaries to hinder border of last molar.	Distance from front border of intermaxillaries to postglenoid process.	Distance from palato-maxillary suture to end of hamular process.	Length of alveolar border of maxillary.	Breadth of palatine surface at posterior end of maxillae.	Length of nasal bones.	Breadth of nasals at front edge.	Breadth of nasals at posterior border.	Breadth of nasals at middle.	Breadth of skull at canines.	Least breadth of skull interorbitally.	Breadth of occipital condyles.	Length of lower jaw.	Depth of lower jaw at coronoid process.	Depth of lower jaw at last molar.	Height of occipital crest.	Breadth of postorbital processes.
*4702	Farallone Islands	♂	393	230	260	253	170	285	110	173	53	61	53	38	33	104	61	87	310	103	66	40	
*3631	Monterey, Cal	♂	395	230	249	286	165	300	150	173	46	66	48	35	32	102	57	87	280	100	61	37	
*6906	....do	♂	383	233	240	255	160	260	125	170	50	63	47	32	26	100	54	83	282	88	60	29	
*4701	Farallone Islands	♂	394	225	235	255	165	286	112	172	42	60	45	35	31	102	57	84	286	104	67	34	137
*13217	....do	♂	400	235	245	264	180	295		178	44	62	52	40	32	110	63	85	300	101	67	28	131
†1767	....do	♂	375	213	232	252	166	300		135	55	55	44	30	32	103	53	85	277	93	55	27	120
†15359	Unalashka	♂	378	225	237		155	280		182	45					102	56	86				25	130
*11675	Saint Paul's Island, Alaska	♂	380	205	220	250	167	275	117	174	38	51	48	31	30	90	57	89	275	63	56	28	122
†2921	....do	♂	385	235	246	247	160	290				64	44	38		110			280	95			130
†2920	....do	♂	374	200	220	243	160	280				60	45	32	20	95			270	85	43		120
*8163	Saint George's Island, Alaska	♀	298	153	173	195	152	217	63	127	37	51	30	22		55	48	70	220	63			
*8162	....do	♀	294	160	145	160	121	218	93	125	39	41	34	33	33	59	47	68					85

* National Museum, Washington, D. C.   † Museum of Comparative Zoölogy, Cambridge, Mass.

TEETH.—The last upper molar is double rooted, and its crown directed backward. All the other molars are single-rooted, with a slight median longitudinal groove on the outside. Their crowns are irregularly conical, pointed, and jut out over their contracted necks; inner side of the crowns hollowed. Surface of the crowns roughened with minute longitudinal grooves and ridges. The upper molars have no trace of the supplemental points to the crowns seen in many species of this family. The lower molars, particularly the third and fourth, have very slight accessory cusps. Necks of the molars uniform in size with the upper part of the fangs. Fangs of the molars gradually tapering, those of the first and second upper much curved inward; that of the third less so; that of the fourth straight; the two fangs of the fifth are directed abruptly forward, the posterior one much the smaller. Canines of both jaws very large; the upper, however, much the larger; the lower more curved. Of the six incisors of the upper jaw, those of the outer pair are much larger than the middle ones, two-thirds as long as the canines, and much like them in form. The middle ones have their antero-posterior diameter nearly twice their lateral diameter, and their crowns are divided transversely. The fangs of the inner pair are slightly bifid. Of the four lower incisors, the outer are much the longer.*

Measurements of the Teeth.†

A.—TEETH OF THE UPPER JAW.

	Molars.					Canines.	Incisors.		
	5th.	4th.	3d.	2d.	1st.		Outer.	Middle.	Inner.
Total length	27	33	36	37	40	84	63	29	25
Length of the crown	9	13	13	13	11	34	23	5	4
" " neck‡	6	6	6	6	6	6	7	7	7
" " root§	12	14	15	18	23	....	....	....	....
Antero-posterior diameter‖	11.5	13	13	13	11.5	24	15	7	6
Lateral diameter‖	6.5	9	10	10	8.5	20	12	5	4

* For figures of the teeth, see Bull. Mus. Comp. Zoöl., vol. ii, pl. i, figg. 5-5ᵉ (one-half natural size).

† These measurements are taken from a middle-aged specimen, in which the dentition is perfect and normal. In old age many of the teeth are usually broken, and a portion of them often entirely wanting, through loss from accident. As the lower canines could not be removed without removing a portion of the jaw, they have not been fully measured.

‡ The distance from the crown to the alveolus.

§ The portion of the tooth inserted in the jaw.

‖ At the base of the crown.

B.—TEETH OF THE LOWER JAW.

	Molars.					Canines.	Incisors.	
	5th.	4th.	3d.	2d.	1st.		Outer.	Inner
Total length	28	42	42	39	30	..	31	25
Length of the crown	10	12	14	12	10	35	8	5
" " neck*	5	5	5	5	5	7	4	4
" " root†	13	25	23	22	15	...	19	16
Antero-posterior diameter‡.	9	13	15	12.5	10.5	26	7	6
Lateral diameter‡	6	9	10	9	8.5	17	9	5

SKELETON.—Vertebral formula: Cervical vertebræ, 7: dorsal,. 15; lumbar, 5; caudal (including the four sacral), variable; probable average, 16.

Ten of the fifteen ribs articulate with the sternum; their sternal portions are entirely cartilaginous. Their osseous portions evidently increase much in length after middle age. The apophyses of the vertebræ are well developed. Of the neural spines of the dorsal vertebræ, the first, second, and third are sub-equal, 130 mm. long; they gradually shorten posteriorly, the last having a length of only 75 mm.

The sternum is normally composed of nine thick and broad osseous segments, the first and last very long, the eighth shortest. Between the eighth and ninth a shorter cartilaginous one is sometimes intercalated (as in specimen No. 2920).

The pelvis is well developed. The ilia are very long and narrow antero-posteriorly. The pubic bones are unanchylosed, they being merely approximate at their posterior extremities. Probably in the females (as in *Callorhinus ursinus*), they are widely separated, and the whole pelvis is much smaller than in the males and differently shaped.

The humerus, as in the other Pinnipeds, is short and thick, with the greater tuberosity enormously developed. The bones of the forearm are also very large and strong, with all their processes greatly developed; in length they but slightly exceed the humerus. The length of neither of the segments of the arm quite equals the length of the bones of the first digit (including its metacarpal bone) of the hand. The first digit of the hand is the longest, twice as long as the fifth, and very thick and strong.

---

* The distance from the crown to the alveolus.
† The portion inserted in the jaw.
‡ At the base of the crown.

The bones of the hinder limbs are also short and thick, especially the femur, which is scarcely more than one-third as long as the tibia. The latter in length about equals the foot. The relative length of the digits is as follows, the longest being mentioned first: 5th, 1st, 2d, 3d, and 4th. The third and fourth are of equal length, and but little shorter than the second. In respect to size, the metatarsal and phalangeal bones of the fifth digit are nearly twice as large as those of the first, while those of the first are about twice the size of those of either of the other three. As previously noticed, the three middle digits of the foot are supplied with long narrow nails; the first and fifth with rudimentary ones.

Measurements of the Bones of the Hand (Metacarpal and Phalangeal).

	Middle-aged specimen.					Very old specimen.				
	1st digit.	2d digit.	3d digit.	4th digit.	5th digit.	1st digit.	2d digit.	3d digit.	4th digit.	5th digit.
Length of metacarpal and phalanges...............	352	310	240	200	177	357	320	250	205	185
Length of metacarpal bone.......	152	110	85	80	80	160	110	90	80	85
Length of 1st phalanx	140	95	70	55	65	140	95	70	60	65
Length of 2d phalanx	60	80	60	45	20	57	80	65	45	18
Length of 3d phalanx	.....	25	25	20	12	.....	35	25	20	17

Measurements of the Bones of the Foot (Metatarsal and Phalangeal).

	Middle-aged specimen.					Very old specimen.				
	1st digit.	2d digit.	3d digit.	4th digit.	5th digit.	1st digit.	2d digit.	3d digit.	4th digit.	5th digit.
Length of metatarsal and phalanges.....................	310	290	290	305	328	320	317	327	350	350
Length of metatarsal bone	120	95	95	110	130	145	110	110	120	130
Length of 1st phalanx.............	140	90	90	90	93	130	100	105	105	110
Length of 2d phalanx	50	75	75	80	70	45	80	85	95	75
Length of 3d phalanx	.....	30	30	25	35	.....	27	27	30	35
Length of nail	.....	40	40	37	.....	.....	50	55	50	.....

The hyoid bone is greatly developed. Each ramus consists of five segments, its two rami being connected together by a transverse segment articulating with the juncture of the fourth and fifth segments. All the parts of the hyoid bone are very thick, especially the transverse and anterior segments; relatively much more so than in *Callorhinus*. In the common *Phoca*,
    Misc. Pub. No. 12——16

the hyoid bone is reduced almost to a bony filament. The length of the hyoid bone in the present species is 270 mm.; of the transverse segment, 65 mm.; circumference of the transverse segment, 45 mm.; of this segment at the thickest part, 95 mm.

The os penis is 170 mm. long, slightly arched, somewhat flattened above, especially posteriorly, sharply convex below, and abruptly expanded and squarely truncate at the end. Its circumference at the base is 72 mm.; just behind the terminal expansion, 32 mm.; of the terminal expansion itself, 65^{mm}.

Measurements of the Skeleton.

	No. 2920, ♂, 10 y'rs old.	No. 2921, ♂, 15 y'rs old.
Whole length of skeleton (including skull)	2,750	2,935
Length of skull	374	385
Length of cervical vertebræ	500	540
Length of dorsal vertebræ	1,050	1,090
Length of lumbar vertebræ	340	400
Length of caudal vertebræ	440	520
Length of first rib	260	240
Length of first rib, osseous portion	130	140
Length of first rib, cartilaginous portion	130	100
Length of second rib	345	305
Length of second rib, osseous portion	175	185
Length of second rib, cartilaginous portion	170	120
Length of third rib	410	410
Length of third rib, osseous portion	230	270
Length of third rib, cartilaginous portion	180	140
Length of fourth rib	470	470
Length of fourth rib, osseous portion	280	330
Length of fourth rib, cartilaginous portion	190	140
Length of fifth rib	535	530
Length of fifth rib, osseous portion	320	370
Length of fifth rib, cartilaginous portion	215	160
Length of sixth rib	580	590
Length of sixth rib, osseous portion	360	420
Length of sixth rib, cartilaginous portion	220	170
Length of seventh rib	640	620
Length of seventh rib, osseous portion	400	440
Length of seventh rib, cartilaginous portion	240	180
Length of eighth rib	670	670
Length of eighth rib, osseous portion	420	480
Length of eighth rib, cartilaginous portion	250	190
Length of ninth rib	710	685
Length of ninth rib, osseous portion	420	485
Length of ninth rib, cartilaginous portion	290	200
Length of tenth rib	750	745
Length of tenth rib, osseous portion	420	485
Length of tenth rib, cartilaginous portion	330	260
Length of eleventh rib, osseous portion only	430	510

Measurements of the Skeleton—Continued.

	No. 2920, ♂, 10 y'rs old.	No. 2921, ♂, 15 y'rs old.
Length of twelfth rib, osseous portion only	490	500
Length of thirteenth rib, osseous portion only	450	470
Length of fourteenth rib, osseous portion only	410	460
Length of fifteenth rib, osseous portion only	340	350
Length of sternum (ossified portion)	700	840
Length of sternum, 1st segment	130	180
Length of sternum, 2d segment	70	90
Length of sternum, 3d segment	70	85
Length of sternum, 4th segment	65	80
Length of sternum, 5th segment	63	85
Length of sternum, 6th segment	60	75
Length of sternum, 7th segment	60	73
Length of sternum, 8th segment	55	65
Length of sternum, 9th segment	70	77
Length of supernumery cartilage (between 8th and 9th)	30	....
Length of scapula	830	370
Breadth of scapula	350	380
Greatest height of its spine	45	52
Length of humerus	300	285
Circumference of its head	300	290
Least circumference of the humerus	170	180
Length of radius	260	260
Length of ulna	310	310
Longest diameter of upper end of ulna	100	130
Length of carpus	80	80
Length of metacarpus and 1st digit	350	360
Length of metacarpus and 2d digit	310	320
Length of metacarpus and 3d digit	240	250
Length of metacarpus and 4th digit	200	205
Length of metacarpus and 5th digit	170	185
Length of femur	170	220
Circumference of neck	125	120
Length of tibia	320	340
Length of fibula	310	330
Length of tarsus	140	160
Length of metatarsus and 1st digit	310	270
Length of metatarsus and 2d digit	290	290
Length of metatarsus and 3d digit	290	270
Length of metatarsus and 4th digit	305	285
Length of metatarsus and 5th digit	227	310
Length of innominate bone	320	360
Greatest width of the pelvis anteriorly	140	160
Length of ilium	140	160
Length of ischio-pubic bones	140	200
Length of thyroid foramen	....	200
Length of os penis	170	170
Width of hand at base of digits	160	....
Width of foot at base of digits	130	140

The above table gives the principal measurements of the bones of the skeleton. Measurements of two specimens are given, as in previous tables, for the purpose of illustrating the variations that occur in the relative size of different parts after maturity is attained, and also for the purpose of illustrating individual variation, which in some particulars these specimens exhibit in a marked degree. The ribs, it will be observed, differ but slightly in total length in the two; not nearly so much as would be expected from the much greater bulk of the body of the older specimen. It will be noticed that the principal differences in the ribs consist in the relative length of the bony to the cartilaginous portions, in the older the ossified portion being much longer and the cartilaginous much shorter than in the younger. An irregularity will be also observed in respect to the sternum, the younger specimen having a supernumerary cartilaginous segment between the eighth and ninth normal ones.

SEXUAL, ADOLESCENT, AND INDIVIDUAL VARIATION.—In respect to external characters, my material, consisting merely of three adult males, does not furnish many facts touching these points. These specimens, however, differ considerably from each other, not only in color, but in size and proportions. Some of these differences are clearly due to age (one of the specimens being much younger than the others), but others equally great cannot be thus explained. The body increases greatly in bulk, and the bones in size and density, after the animal has reached its adult length. The crests of the skull are almost wholly developed after this period, and in great measure also the spines or ridges of the scapula. The tuberosities for the attachment of muscles also increase in size, as do the vertebral or osseous portions of the ribs, as shown by the measurements already given. The teeth also change much in size and form after maturity is attained, and in old age often become much worn and broken by long use. The general form of the skull in the males differs considerably in different individuals of the same age, and also undergoes great modification with age.* As already stated, this consists mainly in the development of the crests and processes for the attachment of muscles, and in the size and form of the teeth.

Mr. Elliott states that the young, when first born, have a weight of about twenty to twenty-five pounds, and a length of

* See Bull. Mus. Comp. Zoöl., vol. ii, pp. 56–60.

about two feet, and describes their color at this age as being "dark chocolate-brown." When they are a year old he says they have the same color as the adults. On their arrival at the Prybilov Islands in spring, Mr. Elliott states that he was unable to discern any marked dissimilarity of coloring between the males and females, and adds that the "young males and yearlings" have the same color as the adults, with here and there an animal marked with irregularly disposed patches of dark brown. After their arrival, the general color gradually becomes somewhat lighter or more golden, and darker again after the moult.

As already noted, the sexual differences in the skull are strongly marked. They are, however, only parallel with those seen in the other species of Otaries. The skeleton of the female is still unknown to me, but may be presumed to differ from that of the male very much, as is found to be the case in the Fur Seal, as described further on.

GEOGRAPHICAL VARIATION.—The material at hand seems to indicate that there is no marked variation in size with locality. A considerable series of skulls from the California coast indicates that the species attains fully as large a size there as at the Prybilov Islands. One of the largest skulls I have seen came from the Farallone Islands, the extreme southern limit represented by the specimens before me.

COMPARISON WITH ALLIED SPECIES.—*Eumetopias stelleri* is the largest of the Eared Seals, very much exceeding in size any of the other species of the family except *Otaria jubata*, which alone it sufficiently resembles in external features to render comparison necessary. While widely distinct from the latter in cranial characters, it seems to quite closely resemble it in external features, so far as may be judged from descriptions. The character of the pelage, the color, and the conformation of the limbs are much the same in both. In neither is there a distinct "mane," so often attributed to them, and especially to the Southern Sea Lion, although the hair on the neck and shoulders is longer than elsewhere, the resemblance to the mane of the Lion being due to the heavy folds of skin over the shoulders when the head is raised, more than to the existence of an abundance of lengthened hair that can in any true sense be considered as forming a mane such as is seen in *Leo*.* The skins

* According to Captain Bryant, "At the fourth year of age the neck and shoulders thicken, from having a thick layer of fat under the skin, the skin

of these two species at my command are in the one case those
of very young animals, and in the other of very old males. A
fine series of the skulls of each enables me, however, to speak
with confidence in respect to the matter of comparative size.
The largest old male skull of *Eumetopias stelleri* has a length of
400 mm., while none fall below 375 mm., the average being about
390 mm. In *Otaria jubata*, the largest old male skull in a series
of a dozen barely reaches 372 mm., and several fall below 340 mm.;
the average being about 355 mm., or about 50 mm. shorter than
the average of a similar series of *Eumetopias stelleri*. Adult
female skulls of the last-named species reach 290 to 300 mm.,
while old female skulls of *Otaria jubata* about 265 mm. Accordingly it seems fair to conclude that the linear measurements of
Otaria jubata are about one-eighth less than those of *Eumetopias
stelleri*, with a corresponding difference in the bulk and weight of
the entire animal in the two species. As very few measurements of the skulls of *Otaria jubata* have been as yet published,
I append the following for comparison with those of *Eumetopias stelleri* already given (*anteà*, p. 238). The wide differences in
dentition and cranial structure have already been sufficiently
indicated.

itself being loose and flabby. When the animal is at rest on a rock with its
hind flippers folded under its body, its head erect and the shoulders thrown
back, the loose skin and fat lies in folds, looking like the mane of a Lion;
hence its name Sea Lion. This thickening of the neck is peculiar to the
adult male."—*MSS. notes.*

Measurements of Twelve Skulls of Otaria jubata.

Catalogue number.	Locality.	Sex.	Total length.	Greatest width at zygomata.	Width at mastoid processes.	Distance between orbits.	Nasal bones, length.	Nasal bones, width posteriorly.	Nasal bones, width anteriorly.	Front border of intermaxillaries to pterygoid hamulus.	Front border of intermaxillaries to glenoid process.	Width at canines.	Height of sagittal crest.	Age.
1095	Paraca Bay, Straits of Magellan	♂	356	227	200	40	55	31	38	253	282	103	30	Old.
1096	Paraca Bay, Straits of Magellan	♂	360	237	175	36	63	36	45	267	277	110	35	Old.
1097	Paraca Bay, Straits of Magellan	♂	350	227	210	33	55	40	47	254	276	110	46	Old.
1112	Straits of Magellan	♂	355	230	215	38	56	30	41	245	280	110	36	Old.
1114	Straits of Magellan	♂	325	213	192	26	65	30	40	223	254	100	20	Middle-aged.
1113	Straits of Magellan	♂	372	227	212	45				260	295	118	30	Old.
1741	Talcahuano, Chili	♂	335	210	195	34	65	27	55	230	260	95	30	Middle-aged.
1743	Talcahuano, Chili	♀	340	213	195					248	265	98	42	Middle-aged.
373	Playa Parda, Patagonia	♀	263	146	125	27	22	30	45	180	197	55	0	Old.
1128	Paraca Bay, Straits of Magellan	♀	252	140	113	26			26	170	185	57	2	Old.
1127	Paraca Bay, Straits of Magellan	♀	277	146	130	15				190	208	53		Old.
1129	Paraca Bay, Straits of Magellan	♀	252	140	112	28				170	190	50	0	Old.

GEOGRAPHICAL DISTRIBUTION.—The known range of this species extends along the west coast of North America from the Farallone Islands, in latitude 37° 40′ N., to the Prybilov Islands. Its northern limit of distribution is not definitely known, but it does not appear to have been met with north of about the latitude of St. Matthew's Island (about latitude 61°). Neither Mr. W. H. Dall nor Mr. H. W. Elliott has met with it above this point, and they have both informed me that they have no reason to suppose it extends any further northward or beyond the southern limit of floating ice. According to Steller, it existed in his time along the whole eastern coast of Kamtchatka and southward to the Kurile Islands. He found it abundant on Behring's and Copper Islands, where it is still well known to exist. If Dr. Gray's *Eumetopias elongatus*, as originally described in 1873 (the same specimen was referred by him in 1872 to *E. stelleri*), be referable, as I believe (see *infra*, p. 252) to the female of *E. stelleri*, the range of this species appears to extend southward on the Asiatic coast as far as Japan.

Although the Sea Lions of the California coast that have of late years attracted so much attention appear to be the smaller species (*Zalophus californianus*), the occurrence of the present species there is also fully established, where it is resident the whole year, and where it brings forth its young, as proven by specimens transmitted some years since by Dr. Ayres to the Smithsonian Institution.

GENERAL HISTORY.—The Northern Sea Lion was first described in 1751 by Steller, who, under the name of *Leo marinus*, gave a somewhat detailed account of its habits and its geographical range, so far as known to him. His description of the animal, however, is quite unsatisfactory. Steller's *Leo marinus*, in size, general form, and color, closely resembles the Southern Sea Lion (*Otaria jubata*), with which Steller's animal was confounded by Pennant, Buffon, and by nearly all subsequent writers for almost a century. Péron, in 1816, first distinctly affirmed the Northern and Southern Sea Lions to be specifically distinct, without, as Temminck says, "avoir vu ni l'une ni l'autre, et sans établir leurs caractères distinctifs."* Lesson, in 1828, gave it the specific name it now bears, in honor of Steller, its first describer. The following year Fischer, on the authority of

---

* Faun. Jap., Mam. Marins, 1842, p. 7.

Lesson, also recognized its distinctness from the southern species. Nilsson, in 1840, in his celebrated monograph of the Seals, reunited them. Müller, however, in an appendix to Dr. W. Peters's translation of Nilsson's essay, published in the Archiv für Naturgeschichte for 1841, separated it again, and pointed out some of the differences in the skulls that serve to distinguish the two species. Gray, in his Catalogue of the Seals, published in 1850, also regarded it as distinct. But one is led to infer that he had not then seen specimens of it, and that he rested his belief in the existence of such a species mainly on Steller's account of it, as he himself expressly states in his later papers. The skull received subsequently at the British Museum from Monterey, California, and figured and described by Gray, in 1859, as a new species, under the name *Arctocephalus monteriensis*, proved, however, to be of this species, as first affirmed by Dr. Gill, and later by Professor Peters and by Gray himself. With the exception of the figures of an imperfect skull of Steller's Sea Lion from Kamtchatka, given by Pander and D'Alton in 1826, Dr. Gray's excellent figure* (a view in profile) is the only one of its skull published prior to 1876. The only specimens of the animal extant, up to about ten years since, in the European museums, seem to have consisted of the two skulls and a stuffed skin in the Berlin Museum mentioned by Peters, and the skull in the British Museum figured and described by Gray.

With the Monterey skull above mentioned, Dr. Gray received another very young skull, and the skin of a Fur Seal, both of which were said to have belonged to one animal, and which he hesitatingly referred to his *Arctocephalus monteriensis*.[†] Later, however, he regarded them as representing a new species,[‡] which he called *Arctocephalus californianus*. Still later he referred his *A. californianus* to *Eumetopias stelleri* § (=*Arctocephalus monteriensis*, Gray, of earlier date), and in 1872 || published figures of this young California skull. Concerning the skin above referred to he remarked at one time as follows: "If the skin sent last year by Mr. Taylor to Mr. Gurney, and by that gentleman presented to the Museum, is the young of this species [*A. mon-*

---

* Proc. Zoöl. Soc. Lond., 1859, pl. lxii.

† Proc. Zoöl. Soc. Lond., 1859, p. 358.

‡ Cat. Seals and Whales, 1866, p. 49.

§ Ann. and Mag. Nat. Hist., 3d series, 1866, vol. xviii, p. 233; Hand-List of Seals, etc., 1874, 40.

|| Proc. Zoöl. Soc. Lond., 1872, pp. 740, 741.

teriensis], the young animal is blackish, silvered by the short white tips to the short black hairs; those on the nape and hinder parts of the body with longer white tips, making those parts whiter and more silvery. The under-fur is very abundant, reaching nearly to the end of the hair. The end of the nose and sides of the face are whitish. The whiskers are elongated, rigid, smooth, and white. The hind feet are elongate, with rather long flaps to the toes. The skull is small for the size of the skin, and I should have doubted its belonging to the skin if it were not accompanied by the following label: 'Skull of the Fur Seal I sent last year. It is very imperfect, from my forgetting where I had put it; but it must do until accident throws another in the way; the other bones were lost.— A. S. T.'"* Dr. Gray, in his "Hand-List," published in 1874, refers the skulls of both *A. monteriensis* and *A. californianus* to *Eumetopias stelleri*, but makes no reference to the skin. As he seems, however, to have become settled in his opinion that this skin is identical with his *A. monteriensis*, this may account for the statement made by him in 1866,† and subsequently reiterated,‡ that the *Eumetopias stelleri* is a species in which "the fur is very dense, standing nearly erect from the skin, forming a very soft, elastic coat, as in *O. falklandica* and *O. stelleri*, which," he erroneously says, "are the only Seals that have a close, soft, elastic fur." §

Lesson gave the name *Otaria californiana* to a supposed species of Eared Seal based solely on a figure entitled "Jeune lion marin de la Californie," published by Choris.‖ The following is the only allusion Choris makes to this animal, in this connection, in his text: "Les rochers dans le voisinage de la baie San-Francisco sont ordinairement couverts de lions marins. Pl. XI." In his chapter on the "Iles Aléoutiennes," in describing the "Lions marins," he says: "Ces animaux sont aussi très-communs au port de San-Francisco, sur la côte de Californie, où

---

* Proc. Zoöl. Soc. Lond., 1850, p. 358.

† Ann. and Mag. Nat. Hist., 4th series, 1866, vol. i, p. 101.

‡ Ibid., p. 215.

§ Dr. Gray's mistake seems to have misled others in respect to the real characters of *Eumetopias stelleri*, which Dr. Veatch, on the authority of Gray, refers to as the "*fur-coated* Eumatopias," which he supposed to be the proper name of the Fur Seal of the North. (See "Report of Dr. John A. Veatch on Cerros or Cedros Island," in J. Ross Browne's "Resources of the Pacific Slope," [appendix], p. 150, 1869.)

‖ Voyage Pittoresque, pl. xi, of the chapter entitled "Port San-Francisco et ses habitants." The date of this work is 1822.

on les voit en nombre prodigieux sur les rochers de la baie. Cette espèce m'a paru se distinguier de ceux qui fréquentent les îles Aléoutiennes; elle a la corps plus fluet et plus allongé, et la tête plus fine: quant à la couleur, elle passe fortement au brun, tandis que ceux des îles Aléoutiennes sont d'une couleur plus grise, ont le corps plus rond, les mouvements plus difficiles, la tête plus grosse et plus épaisse; la couleur du poil des moustaches plus noirâtre que celui des îles Aléoutiennes."*

It thus appears that Choris clearly recognized the larger and the smaller Sea Lions of the west coast of North America, and correctly pointed out their more obvious points of external difference. Hence Lesson's name *Otaria californiana*, founded on Choris's "Lion marin de la Californie," must be considered as applying exclusively to what has till now been commonly known as *Zalophus gillespii*.

Dr. Gill, however, in his "Prodrome," adopted provisionally Lesson's name (*californiana*) for the present species, but at the same time asserted its identity with the *Arctocephalus monteriensis* of Gray (1859), and also suggested its probable identity with the so-called *Otaria stelleri* of Müller. Peters, a few months later, came to the conclusion that Gill's suggestion was correct, since which time the name *stelleri* has been universally accepted for the larger northern Hair Seal. The *Otaria stelleri* of Temminck,† formerly supposed by Gray‡ and also by Peters§ to include both the Australian Eared Seals (viz, *Arctocephalus cinereus* and *Zalophus lobatus*), has finally been referred by the latter, after an examination of the original specimens in the Leyden Museum, to the so-called *Zalophus gillespii*.‖ I believe, however, that the skull of the young female figured in Fauna Japonica (pl. xxii, figg. 5 and 6) belongs to some other species. It certainly differs greatly in proportions, as well as in dentition, from the other skulls figured in that work (same plate), and called *O. stelleri*.

The northern Sea Lion having become generally recognized as specifically distinct from the Sea Lion of the southern seas, Dr. Gill, in 1866, separated the two generically. This had indeed already been done practically by Dr. Gray, inasmuch as

---

* Voy. Pittor. aut. du Monde, Iles Aléoutienues, p. 13.

† Fauna Japonica, Mam. marins, p. 10.

‡ Ann. and Mag. Nat. Hist., 3d series, 1866, vol. xviii, p. 229.

§ Monatsberichte Akad. Berlin, 1866, pp. 272, 276.

‖ Ibid., p. 669. See further on this point *posteà*, under *Zalophus californianus*.

he placed his *A. monteriensis* (= *O. stelleri*, auct.) in the genus *Arctocephalus*, and the southern Sea Lion in *Otaria*, with which he associated the *O. stelleri*. He failed, however, to recognize the identity of his *A. monteriensis* with his *O. stelleri*, and hence the entire generic diversity of the northern and southern Sea Lions seems to have escaped his observation. The latter fact was first pointed out by Dr. Gill in his "Prodrome," as above stated.

Dr. Gray has recently described and figured the skull of what he at first regarded as a second species of *Eumetopias* from Japan, and which he called *Eumetopias elongatus*,* but he subsequently transferred it to his "genus" *Phocarctos*.† In his first mention of it, however, he referred it to *Eumetopias stelleri*.‡ The "*Phocarctos elongatus*" was first described from a "nearly adult" skull (pl. xxi, Hand-List), eleven inches long and seven and a half broad at the condyles, and placed "in the genus *Eumetopias*, because it had a space in the place of the fifth upper grinder." Judging from the figures § and Dr. Gray's description, it seems to differ in no important point from the skull of an adult female, *E. stelleri*. Later he received from Japan a younger skull (pl. xxii, Hand-List), "seven and a half inches long and four and a half inches broad," which agrees in general form with the other, but has a "shorter palate," six upper molars (instead of five), and differs "in the form of the internal nostrils." He considered the two as both belonging to the same species, and, from the presence of six upper molars in the young skull, transferred the species to "*Phocarctos*."

Judging from Dr. Gray's figure of this skull (Hand-List of Seals, pl. xxii), it seems to be referable to *Zalophus* (the Japan species, probably *Z. lobatus*), the last pair of upper molars being in all probability supernumerary, as they are smaller than the others and differ from those preceding them just as do the supernumerary molars in skulls of *Zalophus californianus*.

Dr. Gray seems to have believed that *Eumetopias* has in early life six upper molars on each side, and that the fifth, or last but one, is deciduous, thus leaving a vacuity between the last two molars on either side. Of this I have seen no evidence; on the contrary, I have found in a very young skull the same number of molars as in the adult. Thus skull No. 4703 (National Museum),

---

* Proc. Zoöl. Soc. Lond., 1873, 776, figg. 1, 2.
† Hand-List Seals, etc., 1874, 30, pll. xxi, xxii.
‡ Proc. Zoöl. Soc. Lond., 1872, 737, figg. 1 (head), 2, and 3 (skull).
§ Proc. Zoöl. Soc. Lond., 1872, pp. 738, 739; Hand-List of Seals, pl. xxi.

from San Francisco, Cal., labelled by the collector, Dr. Ayres, as "3 or 4 days old," shows distinctly the alveoli of the milk dentition, with the permanent molars, five in number, just cutting the gum. The last (fifth) upper molar is placed but little further from the fourth than the fourth is from the third. The broader space between the fourth and fifth molars is already indicated, but is, even relatively, much less than in the adults. The last molar stands close to the end of the maxilla, and hence has the same relative position at this early period that it has in old age. As the size of the skull increases, however, the space between the fourth and fifth molars becomes enlarged. Dr. Gray says that in a "fœtal skull" of this species "from California the hind upper grinder is at a considerable distance from the others, as in the very old skull in the [British] Museum and the two adult skulls figured by Mr. Allen; but there is to be observed on each side a concavity in the place of the fifth grinder—on the right side it is a shallow, small cavity which has enclosed a rudimentary tooth;* on the other side the concavity is larger, but not so evidently the cavity for a tooth.†"

As is well known, the *Otaria stelleri* of Temminck's "Fauna Japonica" is a *Zalophus* (at least in part).‡

Since the publication of my paper on the Eared Seals, in 1870, our knowledge of this species has greatly increased, mainly through the published observations of Captain Scammon and Mr. H. W. Elliott. Captain Scammon, however, seems to have not distinguished the two species occurring in California, since he gives no distinct account of the smaller Californian species, although he appears to have given measurements of a female of the latter, and evidently blends, in a general way, the history of the two. Mr. Elliott has not only published a very full account of its habits, as observed by him during several years' residence at Saint Paul's Island, but also a most admirable series of sketches of the animals, drawn from life. In the fol-

---

*As already stated, Dr. Gray appears to have believed that the last or fifth grinder is homologically the sixth, because it has two roots, and that the fifth is deciduous, a theory I believe unsupported. Was not the small cavity he here refers to as having enclosed a tooth merely the alveolus of the last milk molar, which I have found to occupy just this position? Dr. Gray himself, in previously referring to the same skull, alludes to "a small pit" "at the back edge of the fourth grinder," "from which no doubt a small rudimentary tooth has fallen out."—*Suppl. Cat. Seals and Whales*, pp. 29, 30.

†Hand-List of Seals, p. 41.

‡For further remarks on "*Otaria stelleri*" Temminck see *infra*, under *Zalophus californianus*.

lowing pages I shall borrow largely from his excellent account of its habits.*

HABITS.—Aside from Steller's early account of the northern Sea Lion, little had been published relating to the habits of this species prior to 1870. Now, however, with possibly one exception, none of our Pinnipeds is better known.

Steller gave a very full description of the habits of the Sea Bear (*Callorhinus ursinus*), and remarked that, with some few exceptions (which he specifies), those of the Sea Lion closely resemble those of that animal. Choris states: "On y [l'île Saint-Georges] tue une grand quantité de Lions Marins; mais seulement des mâles, à cause de leur grandeur; on se sert de leur peau pour recouvrir les canots, et des intestins pour faire le *kamleyki*, espèces de blouses que l'on endosse par-dessus les autres vêtements lorsqu'il pleut pour ne pas se mouiller. La chair, que l'on fait sécher, est dure; c'est une bonne nourriture pour l'hiver. . . . . Les jeunes sont très-tendres et ont le goût de poisson."

"Le rivage était couvert de troupes innombrables de lions marins. L'odeur qu'ils répandent est insupportable. Ces animaux étaient alors dans le temps du rut. L'on voyait de tous

---

* Mr. Elliott's account was first printed in his "Report on the Prybilov Group, or Seal Islands, of Alaska," in 1873. The work is an oblong quarto of about 130 pages, interleaved with about 40 photographic plates. The text, however, is unpaged, and the plates are not numbered, so that it is almost impossible to cite it definitely. As the edition was limited to one hundred and twenty-five copies, and was privately distributed, it is almost inaccessible, and can hardly be said to have been published. [*] The text, however, was reprinted, in substance, in 1875, in octavo form, as one of the Reports of the Treasury Department (of which Mr. Elliott was Assistant Special Agent at the Fur Seal Islands), under the title "A Report upon the Condition of Affairs in the Territory of Alaska." This edition is the one quoted in the present work. The quarto report contains five plates devoted to the Sea Lion. The first gives a nearly front view of an adult male. The second shows several natives creeping along the shore in order to get between a herd of Sea Lions and the water to intercept their retreat. A third is entitled "Capturing the Sea Lion—Springing the Alarm," and indicates the stage of the hunt when the hunters expose themselves to view and rush upon the herd to drive them inland. A part are retreating land-ward, while others are plunging precipitately into the sea. The fourth, "Shooting Sea Lion Bulls," represents the killing of the old males with firearms. The fifth and last depicts the slaughter of the females, and is entitled, "Spearing Sea Lion Cows, 'The Death Whirl.'"

[*It is well known that opinions of "what constitutes publication?" differ. I have the author's permission to record here my own view, which is, that a printed work is "published" if a single copy is placed in a public library.—ELLIOTT COUES.]

côtés les mâles se battre entre eux pour s'enlever les uns aux autres les femelles. Chaque mâle en rassemble de dix à vingt, se montre jaloux, ne souffre aucun autre mâle, et attaque ceux qui tentent de s'approcher; il les tue par ses morsures ou s'en fait tuer. Dans le premier cas, il s'empare des femelles du vaincu. Nous avons trouvé plusieurs mâles étendus morts sur la plage, des seules blessures qu'ils avaient reçues dans les combats. Quelques femelles avaient déjà des petits. Les Aléoutes en prirent plusieurs douzaines pour nous. L'animal n'est pas dangereux; il fuit à l'approche de l'homme, excepté depuis la mi-mai jusqu'à la mi-juin, qui est le plus fort temps du rut, et où les femelles mettent bas leurs petits; alors il ne se laisse pas approcher et il attaque même."* Choris's plates (Nos. XIV and XV of the chapter on the Aleutian Islands, the work is not regularly paged) doubtless give a very good idea of the appearance of these animals and the Sea Bears when assembled on the land. Plate XIV, entitled "Lions Marins dans l'île de St.-Georges," gives a view of a large assemblage of these animals, in which the various attitudes are duly represented, the animals in the foreground being depicted with considerable accuracy of detail.

In 1870 I was able to add the following remarks by Captain Bryant: "The Sea Lion visits St. Paul's Island in considerable numbers, to rear its young. It is one of the largest of the Seal family, the male frequently measuring thirteen feet in length, and weighing from fifteen to eighteen hundred pounds.[†] Its habits are the same as those of the Fur Seal. When roused to anger it has a very marked resemblance, through the form of its head and neck, to the animal from which it is named, and its voice, when roaring, can be heard to a great distance. Its body is thickly covered with fine, short, dark [?] brown hair, without any fur. Its skin is of considerable value as an article of commerce in the Territory, it being used in making all kinds of boats, from a one-man canoe to a lighter of twenty tons' burden. The natives of all the Aleutian Islands and of the coast as far east as Sitka, besides those of many ports on the mainland to the north, rely on this island for a supply of the skins of this animal. The rookery is on the northeast end of the island, and the animals have to be driven ten or eleven miles to the village

---

* Voy. Pittoresque autour du Monde, Iles Aléoutiennes, pp. 12, 13.
† See *anteà*, p. 233, second paragraph of footnote, for Captain Bryant's later statements respecting size and weight.

to bring their skins to the drying-frames. It sometimes requires five days to make the journey, as at frequent intervals they have to be allowed to rest. It is a somewhat dangerous animal, and the men frequently get seriously hurt by it in driving and killing it. They are driven together in the same manner as the Fur Seals are; and while confining each other by treading upon each others' flippers the small ones are killed with lances, but the larger ones have to be shot.

"This animal is the most completely consumed of any on the island. Their flesh is preferred to that of the Seal for drying for winter use. After the skins are taken off (two thousand of which are required annually to supply the trading-posts of the Territory), they are spread in piles of twenty-five each, with the flesh side down, and left to heat until the hair is loosened; it is then scraped off, and the skins are stretched on frames to dry. The blubber is removed from the carcass for fuel or oil, and the flesh is cut in strips and dried for winter use. The linings of their throats are saved and tanned for making the legs of boots and shoes, and the skin of the flippers is used for the soles. Their stomachs are turned, cleaned, and dried, and are used to put the oil in when boiled out. The intestines are dressed and sewed together into water-proof frocks, which are worn while hunting and fishing in the boats. The sinews of the back are dried and stripped to make the thread with which to sew together the intestines, and to fasten the skins to the canoe-frames. The natives receive thirty-five cents apiece for the skins when ready for shipment. But these skins are not so much valued by the trader for the profit he makes on their sale, as for the advantage it gives him in bargaining with the hunters, since by buying these they are able to secure a right to the purchase of the hunter's furs on his return, the natives always considering such contracts binding."*

The following careful description of their movements on land was also communicated to me by Mr. Theodore Lyman in 1870, who had recently observed the Sea Lions on the "Seal Rocks" near San Francisco. His remarks may, however, relate in part to the smaller species.

"These rocks," he says, "are beset with hundreds of these animals,—some still, some moving, some on the land, and some in the water. As they approach to effect a landing, the head only appears decidedly above water. This is their familiar

*Bull. Mus. Comp. Zoöl., ii, pp. 64, 65.

element, and they swim with great speed and ease, quite unmindful of the heavy surf and of the breakers on the ledges. In landing, they are apt to take advantage of a heavy wave, which helps them to get the forward flippers on *terra firma*. As the wave retreats, they begin to struggle up the steep rocks, twisting the body from side to side, with a clumsy worm-like motion, and thus alternately work their flippers into positions where they may force the body a little onward. At such times they have a general appearance of *sprawling* over the ground. It is quite astonishing to see how they will go up surfaces having even a greater inclination than 45°, and where a man would have to creep with much exertion. When the surface is nearly horizontal, they go faster, and often proceed by gathering their hind-quarters under them, raising themselves on the edges of their fore-limbs and then giving a push, whereby they make a sort of tumble forwards. In their onward path they are accompanied by the loud barking of all the Seals they pass; and these cries may be heard a great distance. Having arrived at a good basking-place, they stretch themselves out in various attitudes,—often on the side, sometimes nearly on the back, but commonly on the belly, with the flippers somewhat extended. They seem much oppressed with *their own weight* (which is usually supported by the water), and it seemed an exertion for them even to raise the head, though it is often kept up for a long time. They play among themselves continually by rolling on each other and feigning to bite. Often, too, they will amuse themselves by pushing off those that are trying to land. All this is done in a very cumbrous manner, and is accompanied by incessant barking. As they issue from the water, their fur is dark and shining; but, as it dries, it becomes of a yellowish brown. Then they appear to feel either too dry or too hot, for they move to the nearest point from which they may tumble into the sea. I saw many roll off a ledge at least twenty feet high, and fall, like so many huge brown sacks, into the water, dashing up showers of spray."*

From the accounts given by various observers, the Sea Lions evidently move with much less facility on land than do the Fur Seals, doubtless mainly from their much greater size. The young and the females of several of the different species of these animals are described as walking with much greater ease and rapidity than the half-grown and the more unwieldy old males.

---

* Bull. Mus. Comp. Zoöl., ii, pp. 66, 67.

Captain Bryant states that the Fur Seal may be driven at the rate of a mile and a half per hour, while, according to the same authority, the Sea Lions can be driven with safety but about two miles a day.

Captain Scammon, in 1874, published a very interesting account of the Sea Lions of the Aleutian Islands, particularly as respects the methods employed in their capture, portions of which will be quoted later. His account is devoted largely, however, to the Sea Lions of the California coast, and certainly includes the history of the smaller species, if in fact this part does not relate mainly to the latter. At about the same time appeared Mr. H. W. Elliott's more detailed history of the northern species, which is so full and explicit that I transcribe it almost entire.

The Sea Lion, he says, " has a really leonine appearance and bearing, greatly enhanced by the rich, golden-rufous of its coat, ferocity of expression, and bull-dog-like muzzle and cast of eye, not round and full, but showing the white, or sclerotic coat, with a light, bright-brown iris.

" Although provided with flippers to all external view as the fur-seal, he cannot, however, make use of them in the same free manner. While the fur-seal can be driven five or six miles in twenty-four hours, the sea-lion can barely go two, the conditions of weather and roadway being the same. The sea-lions balance and swing their long, heavy necks to and fro, with every hitch up behind of their posteriors, which they seldom raise from the ground, drawing them up after the fore feet with a slide over the grass or sand, rocks, &c., as the case may be, and pausing frequently to take a sullen and ferocious survey of the field and the drivers."

"The sea-lion is polygamous, but does not maintain any such regular system and method in preparing for and attention to its harem like that so finely illustrated on the breeding-grounds of the fur-seal. It is not numerous, comparatively speaking, and does not 'haul' more than a few rods back from the sea. It cannot be visited and inspected by man, being so shy and wary that on the slightest approach a stampede into the water is the certain result. The males come out and locate on the narrow belts of rookery-ground, preferred and selected by them; the cows make their appearance three or four weeks after them, (1st to 6th June,) and are not subjected to that intense jealous supervision so characteristic of the fur-seal

FIG. 37.—*Eumetopias stelleri.* Adult male, females, and young.

harem.   The bulls fight savagely among themselves, and turn off from the breeding-ground all the younger and weak males.

"The cow sea-lion is not quite half the size of the male, and will measure from 8 to 9 feet in length, with a weight of four and five hundred pounds. She has the same general cast of countenance and build of the bull, but as she does not sustain any fasting period of over a week or ten days, she never comes out so grossly fat as the male or 'see-catch.'

"The sea-lion rookery will be found to consist of about ten to fifteen cows to the bull. The cow seems at all times to have the utmost freedom in moving from place to place, and to start with its young, picked up sometimes by the nape, into the water, and play together for spells in the surf-wash, a movement on the part of the mother never made by the fur-seal, and showing, in this respect, much more attention to its offspring.

"They are divided up into classes, which sustain, in a general manner, but very imperfectly, nearly the same relation one to the other as do those of the fur-seal, of which I have already spoken at length and in detail; but they cannot be approached, inspected, and managed like the other, by reason of their wild and timid nature. They visit the islands in numbers comparatively small, (I can only estimate,) not over twenty or twenty-five thousand on Saint Paul's and contiguous islets, and not more than seven or eight thousand at Saint George. On Saint Paul's Island they occupy a small portion of the breeding-ground at Northeast Point, in common with the *Callorhinus*, always close to the water, and taking to it at the slightest disturbance or alarm.

"The sea-lion rookery on Saint George's Island is the best place upon the Seal Islands for close observation of these animals, and the following note was made upon the occasion of one of my visits, (June 15, 1873 :)

"'At the base of cliffs, over 400 feet in height, on the east shore of the island, on a beach 50 or 60 feet in width at low water, and not over 30 or 40 at flood-tide, lies the only sea-lion rookery on Saint George's Island—some three or four thousand cows and bulls. The entire circuit of this rookery-belt was passed over by us, the big, timorous bulls rushing off into the water as quickly as the cows, all leaving their young. Many of the females, perhaps half of them, had only just given birth to their young. These pups will weigh at least twenty to twenty-five pounds on an average when born, are of a dark, chocolate-brown, with the eye as large as the adult, only being a suffused, watery, gray-blue, where the sclerotic coat is well and sharply defined in its maturity. They are about 2 feet in length, some longer and some smaller. As all the pups seen to-day were very young, some at this instant only born, they were dull and apathetic, not seeming to notice us much. There are, I should say, about one-sixth of the sea-lions in number on this island, when compared with Saint Paul's. As these animals lie here under the cliffs, they cannot be approached and driven; but should they haul a few hundred rods up to the south, then they can be easily captured. They have hauled in this manner always until disturbed in 1868, and will undoubtedly do so again if not molested.

" 'These sea-lions, when they took to the water, swam out to a distance of fifty yards or so, and huddled all up together in two or three packs or squads of about five hundred each, holding their heads and necks up high out of water, all roaring in concert and incessantly, making such a deafening noise that we could scarcely hear ourselves in conversation at a distance from them of over a hundred yards. This roaring of sea-lions, thus disturbed, can only be compared to the hoarse sound of a tempest as it howls through the rigging of a ship, or the playing of a living gale upon the bare branches, limbs, and trunks of a forest grove.' They commenced to return as soon as we left the ground.

"The voice of the sea-lion is a deep, grand roar, and does not have the flexibility of the *Callorhinns*, being confined to a low, muttering growl or this bass roar. The pups are very playful, but are almost always silent. When they do utter sound, it is a sharp, short, querulous growling.

"The natives have a very high appreciation of the sea-lion, or *see-vitchie*, as they call it, and base this regard upon the superior quality of the flesh, fat, and hide, (for making covers for their skin boats, *bidarkies* and *bidarrahs*,) sinews, intestines, &c.

"As I have before said, the sea-lion seldom hauls back far from the water, generally very close to the surf-margin, and in this position it becomes quite a difficult task for the natives to approach and get in between it and the sea unobserved, for, unless this silent approach is made, the beast will at once take the alarm and bolt into the water.

"By reference to my map of Saint Paul's [not here repro-

duced] a small point, near the head of the northeast neck of
the island, will be seen, upon which quite a large number of
sea-lions are always to be found, as it is never disturbed except
on the occasion of this annual driving.   The natives step down
on to the beach, in the little bight just above it, and begin to
crawl on all fours flat on the sand down to the end of the neck
and in between the dozing sea-lion herd and the water, always
selecting a semi-bright moonlight night.   If the wind is favorable, and none of the men meet with an accident, the natives
will almost always succeed in reaching the point unobserved,
when, at a given signal, they all jump up on their feet at once,
yell, brandish their arms, and give a sudden start, or alarm, to
the herd above them, for, just as the sea-lions move, upon the
first impulse of surprise, so they keep on.   For instance, if the
animals on starting up are sleeping with their heads pointed in
the direction of the water, they keep straight on toward it; but
if they jump up looking over the land, they follow that course
just as desperately, and nothing turns them, *at first*, either one
way or the other.   Those that go for the water are, of course,
lost, but the natives follow the land-leaders and keep urging
them on, and soon have them in their control, driving them
back into a small pen, which they extemporize by means of
little stakes, with flags, set around a circuit of a few hundred
square feet, and where they keep them until three or four hundred, at least, are captured, before they commence their drive
of ten miles overland down south to the village.

"The natives, latterly, in getting this annual herd of sea-
lions, have postponed it until late in the fall, and when the animals are scant in number and the old bulls poor.   This they
were obliged to do, on account of the pressure of their sealing-
business in the spring, and the warmth of the season in August
and September, which makes the driving very tedious.   In this
way I have not been permitted to behold the best-conditioned
drives, *i. e.*, those in which a majority of the herd is made up
of fine, enormously fat, and heavy bulls, some four or five hundred in number.

"The natives are compelled to go to the northeast point of
the island for these animals, inasmuch as it is the only place
with natural advantages where they can be approached for the
purpose of capturing alive.   Here they congregate in greatest
number, although they can be found, two or three thousand of
them, on the southwest point, and as many more on 'See-
vitchie Cammin' and Otter Island.

" Capturing the sea-lion drive is really the only serious business these people on the islands have, and when they set out for the task the picked men only leave the village. At Northeast Point they have a barrabkie, in which they sleep and eat while gathering the drove, the time of getting which depends upon the weather, wind, &c. As the squads are captured, night after night, they are driven up close by the barrabkie, where the natives mount constant guard over them, until several hundred animals shall have been secured, and all is ready for the drive down overland to the village.

" The drove is started and conducted in the same general manner as that which I have detailed in speaking of the fur-seal, only the sea-lion soon becomes very sullen and unwilling to move, requiring spells of frequent rest. It cannot pick itself up from the ground and shamble off on a loping gallop for a few hundred yards, like the *Callorhinus*, and is not near so free and agile in its movements on land, or in the water for that matter, for I have never seen the *Eumetopias* leap from the water like a dolphin, or indulge in the thousand and one submarine acrobatic displays made constantly by the fur-seal.

" This ground, over which the sea-lions are driven, is mostly a rolling level, thickly grassed and mossed over, with here and there a fresh-water pond into which the animals plunge with great apparent satisfaction, seeming to cool themselves, and out of which the natives have no trouble in driving them. The distance between the sea-lion pen at Northeast Point and the village is about ten miles, as the sea-lions are driven, and occupies over five or six days under the most favorable circumstances, such as wet, cold weather; and when a little warmer, or as in July or August, a few seasons ago, they were some three weeks coming down with a drove, and even then left a hundred or so along on the road.

" After the drove has been brought into the village on the killing-grounds, the natives shoot down the bulls and then surround and huddle up the cows, spearing them just behind the fore-flippers. The killing of the sea-lions is quite an exciting spectacle, a strange and unparalleled exhibition of its kind. . . . . The bodies are at once stripped of their hides and much of the flesh, sinews, intestines, (with which the native water-proof coats, &c., are made,) in conjunction with the throat-linings, (*œsophagus*,) and the skin of the flippers, which is exceedingly tough and elastic, and used for soles to their boots or ' *tarbosars*.'

"As the sea-lion is without fur, the skin has little or no commercial value; the hair is short, and longest over the nape of the neck, straight, and somewhat coarse, varying in color greatly as the seasons come and go. For instance, when the *Eumetopias* makes his first appearance in the spring, and dries out upon the land, he has a light-brownish, rufous tint, darker shades back and under the fore flippers and on the abdomen; by the expiration of a month or six weeks, 15th June, he will be a bright golden-rufous or ocher, and this is just before shedding, which sets in by the middle of August, or a little earlier. After the new coat has fairly grown, and just before he leaves the island for the season, in November, it will be a light sepia, or vandyke-brown, with deeper shades, almost dark upon the belly; the cows, after shedding, do not color up so dark as the bulls, but when they come back to the land next year they are identically the same in color, so that the eye in glancing over a sea-lion rookery in June and July cannot discern any noted dissimilarity of coloring between the bulls and the cows; and also the young males and yearlings appear in the same golden-brown and ocher, with here and there an animal spotted somewhat like a leopard, the yellow, rufous ground predominating, with patches of dark-brown irregularly interspersed. I have never seen any of the old bulls or cows thus mottled, and think very likely it is due to some irregularity in the younger animals during the season of shedding, for I have not noticed it early in the season, and failed to observe it at the close. Many of the old bulls have a grizzled or slightly brindled look during the shedding-period, or, that is, from the 10th August up to the 10th or 20th of November; the pups, when born, are of a rich, dark chestnut-brown; this coat they shed in October, and take one much lighter, but still darker than their parents', but not a great deal.

"Although, as I have already indicated, the sea-lion, in its habit and disposition, approximates the fur-seal, yet in no respect does it maintain and enforce the system and regularity found on the breeding-grounds of the *Callorhinus*. The time of arrival at, stay on, and departure from the island is about the same; but if the winter is an open, mild one, the sea-lion will be seen frequently all through it, and the natives occasionally shoot them around the island long after the fur-seals have entirely disappeared for the year. It also does not confine its landing to these Prybilov Islands alone, as the fur-seal unquestionably does, with reference to our continent; for it has been and is often shot upon the Aleutian Islands and many rocky islets of the northwest coast.

"The sea-lion in no respect whatever manifests the intelligence and sagacity exhibited by the fur-seal, and must be rated far below, although next, in natural order. I have no hesitation in putting this *Eumetopias* of the Prybilov Islands, apart from the sea-lion common at San Francisco and Santa Barbara, as a distinct animal; and I call attention to the excellent description of the California sea-lion, made public in the April number for 1872 of the Overland Monthly, by Capt. C. M. Scammon, in which the distinguishing characters, externally, of this animal are well defined, and by which the difference between the *Eumetopias* of Bering Sea and that of the coast of California can at once be seen; and also I notice one more point in which the dissimilarity is marked—the northern sea-lion never barks or howls like the animal at the Farralones [*sic*] or Santa Barbara. Young and old, both sexes, from one year and upward, have *only* a deep *bass growl*, and *prolonged, steady roar;* while at San Francisco sea-lions break out incessantly with a 'honking' bark or howl, and *never roar*.

"I am not to be understood as saying that *all* the sea-lions met with on the Californian coast are different from *E. stelleri* of Bering Sea. I am well satisfied that stragglers from the north are down on the Farralones, but they are not migrating back and forth every season; and I am furthermore certain that not a single animal of the species most common at San Francisco was present among those breeding on the Prybilov Islands in 1872–'73.

"According to the natives of Saint George, some fifty or sixty years ago the *Eumetopias* held almost exclusive possession of the island, being there in great numbers, some two or three hundred thousand; and that, as the fur-seals were barely permitted to land by these animals, and in no great number, the Russians directed them (the natives) to hunt and worry the sea-lions off from the island, and the result was that as the sea-lions left, the fur-seals came, so that to-day they occupy nearly the same ground covered by the *Eumetopias alone* sixty years ago. This statement is, or seems to be, corroborated by Choris, in his description of the Iles S.-George's et S.-Paul's [*sic*], visited by him fifty years ago; * but the account given by Bishop Ven-

"*Voyage Pittoresque autour du Monde."

iaminov,* . . . . . differs entirely from the above, for by it
almost as many fur-seals were taken on Saint George, during
the first years of occupation, as on Saint Paul, and never have
been less than one-sixth of the number on the larger island.
. . . . I am strongly inclined to believe that the island of Saint
George never was resorted to in any great numbers by the fur-
seal, and that the sea-lion was the dominant animal there until
disturbed and driven from its breeding-grounds by the people,
who sought to encourage the coming of its more valuable relative by so doing, and making room in this way for it.

"The sea-lion has but little value save to the natives, and is
more prized on account of its flesh and skin, by the people living upon the islands and similar positions, than it would be
elsewhere.   The matter of its preservation and perpetuation
should be left entirely to them, and it will be well looked after.
It is singular that the fat of the sea-lion should be so different
in characters of taste and smell from that of the fur-seal, being
free from any taint of disagreeable flavor or odor, while the
blubber of the latter, although so closely related, is most repug-
nant.   The flesh of the sea-lion cub is tender, juicy, light-colored, and slightly like veal; in my opinion, quite good.   As the
animal grows older, the meat is dry, tough, and without flavor."†

Captain Scammon gives a few particulars respecting the
"drive," not especially referred to by Mr. Elliott.   "This
'drive,'" he says, "to the good-natured Aleuts, is what the
buffalo hunt has been to the red-skins on the plains of the
Platte, or *matanza*-time with the old Californians, for the party
starts out as on a sporting foray, and at night they stealthily
get between the herd of Sea Lions and the water; then, with
professional strategy, they manage to 'cut out' six or eight of
the largest at a time, and drive them a short distance inland,
where they are guarded until a band of two or three hundred
are assembled.   Formerly the implement used in driving was a
pole with a small flag at the end; but, since our adopted coun-
try-folk have become Americanized, that Yankee production, a
cotton umbrella, has been substituted, and it is said that any
refractory *siutch* in the 'drive' is instantly subdued by the sudden expansion and contraction of an umbrella in the hands of
a pursuing native.

"To collect the desired number for the yearly supply involves

"* Zapeeskie ob Ostrovah Oonahlashkenskaho Otdayla, St. Petersburg,
1840."

† Report upon Condition of Affairs in Alaska, pp. 152–159.

several days; therefore a throng of villagers, it is said, sets out prepared with everything needful for the campaign. As the work of driving goes on only at night, the day is passed in sleeping and cooking their food by smoldering fires of drift-wood and seal-fat, sheltered by their umbrellas, or a sort of tent contrived by spreading blankets and garments over whales' ribs in lieu of tent-poles—never forgetting in their repast the fragrant *chi*, which is quaffed in numberless cups from the steaming *sam-o-var*. At length, the whole troop of animals being assembled, a flash of umbrellas here and there, with the call of the herdsmen, brings all into a moving phalanx. But the time for driving must be either at night, after the dew is fallen, or upon a dark, misty, or rainy day; as the thick mat of grass that covers the land must be wet, in order that the animals may easily slip along in their vaulting gait over the green road to their place of execution. Under the most favorable circumstances, the march does not exceed six miles in the twenty-four hours; and it being a distance of four leagues or more to the village, three days and nights, or more, are spent before they arrive at the slaughtering-place. There they are allowed to remain quiet for a day or two, to cool their blood, which becomes much heated by the tedious journey; after which they are killed by shooting. The dead animals are then skinned, and their hides packed in tiers until fermented suf-ficiently to start the hair, when they are stretched on frames to dry, and eventually become the covering or planking for the Aleutian *baidarkas* and *baidarras*. The fat is taken off and used for fuel, or the oil is rendered to burn in their lamps. The flesh is cut in thin pieces from the carcass, laid in the open air to dry, and becomes a choice article of food. The sinews are extracted, and afterward twisted into thread. The lining of the animal's throat is put through a course of tanning, and then made into boots, the soles of which are the under covering of the Sea Lion's fin-like feet. The intestines are carefully taken out, cleaned, blown up, stretched to dry, then tanned, and worked into water-proof clothing. The stomach is emptied of its contents, turned inside out, then inflated and dried for oil-bottles, or is used as a receptacle for the preserved meat; and what remains of the once formidable and curious animal is only a mutilated skeleton."

Captain Scammon adds the following respecting their cap-ture on the Asiatic coast: "Crossing Behring and the Okhotsk

Seas, to the coast of Siberia, including the peninsula of Kamschatka and the island of Saghalien, the mode of capture by the
natives changes from that of the eastern continental shores.*
The inlets and rivers of these Asiatic regions swarm with salmon from June to September, and at this season the Seals follow and prey upon them as they ascend the streams. The
natives then select such places as will be left nearly bare at
low tide, and then set their nets—which are made of seal-
thongs—to strong stakes, so placed as to form a curve open to
the confluence of the stream. These nets are similar to gill-
nets, the meshes being of a size to admit a Seal's head,—which
gives free passage to the shoals of fish—and the pursuing animal, as soon as entangled in the net, struggles forward in his
efforts to escape, but is held firmly in the meshes, where it remains till low water, when the natives, in their flat-bottomed
skin-boats, approach and dispatch the victim with their rude
bone implements. As the season becomes warm, the animals
of both sexes congregate in their favorite rookeries, and the
females climb to the most inaccessible places among the rocks
and crags to bring forth and nurture their offspring. But here
they are hunted by the natives accustomed to the use of firearms, who shoot them for the skins of the young ones, which
are used for clothing.

"In this region also, during the spring and fall, after the
'net-sealing' is over, great numbers of Sea Lions are captured
upon the floating ice, with gun or spear; and during the rigorous months, the seal-hunters cut through the congealed mass
what they term 'breathing-holes'. Through these the Seals
emerge, to the frosted surface, and, if the sun peers through the
wintry clouds, the creatures, warmed into new life, may stroll
hundreds of yards away; the watchful hunter, secreted behind
a cake of ice or a bank of snow, rushes out from his covert, and
places a covering over the hole, effectually preventing the animal's escape, and then dispatches it with knife and spear. Its
skin is stripped off, scraped clean, closely rolled, and laid away

---

* Although Captain Scammon purports to be speaking of "Sea Lions," I
have recently become convinced (since the copy of this article was sent to
the printer) that very little, if anything, in this paragraph and the next
relates to any species of Eared Seal. In the first place, the locality is one
not known to be frequented, except casually, by Otaries, while the account
of the capture in nets and in the ice, and especially the reference to
"breathing holes," renders it almost certain that the animals referred to
are Phocids.

until the hair starts—this process is called 'scouring'; then the hair is scoured off and the bare hide is stretched to season—a process usually requiring about ten days—when it is taken down and rubbed between the hands to make it pliable; this completes the whole course of dressing it. The prepared skins are then converted into harness for the sledge-dogs and reindeer, and water-proof bags; if wanted for the soles of moccasins, or to cover their skin-boats, they are dried with the hair on, and become nearly as stiff as plates of iron. The blubber of the animals, if killed in the fall or winter, is preserved by freezing, and is used for food, fuel, and lights, as desired; while the same part of those taken in the spring and summer is put in the skins of young Seals, and placed in earthen vaults, where it keeps fresh until required for consumption. The residue of the animal is tumbled into a reservoir, sunk below the surface of the ground, where it is kept for the winter's supply of food for the dogs, which live upon the frozen flesh and entrails of the Seals, whose skin furnishes the tackle by which they transport the primitive sledge over the snow-clad wastes of Siberia and Kamschatka."[*]

Since the foregoing was transmitted for publication I have received from Captain Charles Bryant a very full account of this species, based on his many years' observations as United States Treasury Agent at the Fur Seal Islands, and kindly prepared by my request for use in the present connection. Although so much space has already been devoted to the history of this species, it seems desirable to give Captain Bryant's report nearly in full, although repeating in substance some of the details which have already been presented, since it contains some new points, and is at least based on long experience. Some portions, relating especially to the products of the Sea Lion and their uses, are omitted, since they are fully anticipated by what has already been given.

"From fifteen to twenty thousand Sea Lions," says Captain Bryant, "breed annually on the Prybilov or Fur Seal Islands. They do not leave the islands in winter, as do the Fur Seals, to return in spring, but remain during the whole year. They bring forth their young a month earlier than the Fur Seals, landing during the months of May and June. They advance but little above high tide-mark, and those of all ages land together. The strongest males drive out the weaker and mono-

---

[*] Marine Mammalia, pp. 136–138.

polize the females and continue with them till September. They
go with them into the water whenever they are disturbed, and
also watch over the young. When in the water they swim
about the young and keep them together until they have an
opportunity to land again. The females also keep near, rushing
hither and thither, appearing first on one side and then on the
other of the groups of young, constantly uttering a deep hoarse
growl at the intruder whenever they come · to the surface.
When left undisturbed they all soon land again, preferring to
spend the greater portion of their time at this season on the
shore. During the breeding season they visit the same parts
of the shore as the Fur Seals, but the Sea Lions, by their superior size and strength, crowd out the Seals, the latter passively
yielding their places without presuming to offer battle to their
formidable visitors. After having been disturbed the Sea Lions
continue for some time in a state of unrest, occasionally uttering
a low moaning sound, as though greatly distressed. Even after
the breeding season they keep close to the shore near the breeding station until the severe weather of January. After this
time they are seen only in small groups till the shores are free
from snow and ice in the spring.

"The capture of these animals is laborious and hazardous,
and must be managed by the most skilful and experienced of
the natives. They are so sensitive to danger and so keenly on
the alert that even the screaming of a startled bird will cause
the whole herd to take to the water.

"The only place frequented by the Sea Lions that, by the
nature of the ground, is practicable for their capture, is ten or
twelve miles from the village where all the natives reside.
They keep so near the shore that the favorable time to get
between them and the water is when the tide is lowest; and
they are so quick of scent that the wind must blow from them
toward the sea, so they may not smell the hunters as they attempt to approach them. The chiefs select a party of fifteen or
twenty of the best men, who leave the village prepared for an
absence of a week or ten days, for the place selected for the
hunt. Near this they have a lodging-house, where they wait
for favoring conditions of wind and tide. Under cover of the
darkness of night, the chief takes the lead and the men follow,
keeping a little distance apart, creeping noiselessly along the
shore at the edge of the receding tide until they get between
the Sea Lions and the water. At a given signal the men start

up suddenly, fire pistols, and make all the noise possible. The animals thus suddenly alarmed immediately start in whatever direction they chance to be headed; those facing the water rush precipitately into it. These the hunters avoid, letting them pass them, and start at once after those heading inland, shouting at them to keep them moving until some distance from the shore. The Sea Lions, when once fairly in motion, are easily controlled and made to move in the desired direction till they reach some convenient hollow, where they are guarded by one or two men stationed to watch their movements and prevent their escape until enough have been obtained to make a herd for driving, numbering usually two or three hundred individuals. They sometimes capture in this way forty or fifty in a single night, but oftener ten or twenty, and many times none at all. As at this season Sea Lions of all ages and sizes congregate together, it often happens that females are caught while their cubs escape, or the reverse, but as the capture is continued for several successive nights at the same place, and the new captives are driven to the herd already caught, the mothers and their young are again brought together. They recognize each other by their cries long before they meet, and it makes lively work for the herders to prevent the herd from rushing to meet the new comers. When the recruits join the herd the mothers and cubs rush together with evident pleasure, the mothers fondling their young, and the latter, hungered by separation, struggle to nurse them. After a sufficient number have been thus obtained they are driven to the village for slaughter, in order that all parts of the animal may be utilized.

"The distance to the village is, as already stated, about ten or twelve miles, and the route lies near the shore. Along the way are several small ponds through which they pass and which serve to refresh them on their slow toilsome journey. The journey is necessarily slow and tedious, for the Sea Lions are less well fitted for traveling on land than the Fur Seals, which are able to raise their bodies from the ground and gallop off like a land animal. The Sea Lions travel by bending the posterior part of the body to the right or left, extending their long flexible necks in an opposite direction to balance themselves, and then slowly raising their bodies by their fore limbs and plunging forward, by which movement they thus gain an advance of only half a length at a time. When they arrive in sight of the ponds they make a hurried scramble for them, and, rushing in pell-mell, roll and tumble in the water as though it afforded immense relief to their heated and wearied bodies. When it is convenient to do so they are allowed to rest over night in the water, by which they acquire fresh vigor for the completion of the journey. This severe and unnatural exertion overheats and exhausts these poor beasts and necessitates long halts to enable them to rest and cool. It usually requires five days to make the journey, averaging two miles per day. Three men conduct the herd, and camp at night with their charge. On starting they kill a young cub for their subsistence, using the flesh for food and the blubber for fuel in cooking it and making their tea.

"After two days' travel the animals become very tired, and as soon as they are permitted to halt they drop at full length with their limbs extended. But their rest is not peaceful, for some restless one soon starts up and flounders over the others as if seeking a better place. This disturbs the whole herd, which constantly keeps up a low moaning apparently expressive of sore distress. A most apt description of such a scene was once given by a military officer who was seated with me on the edge of a sand-dune watching a herd resting in this condition. After a long silence he observed, 'This is the first thing I have ever seen or heard that realizes my youthful conception of the torment of the condemned in purgatory.'"

"When the herd is once fairly halted and at rest it requires from half an hour to an hour to get it moving again in marching order. The process is quite novel and worth describing. The Sea Lions have now become so accustomed to their captors that they will sooner fight than run from them, and they are too much deafened by their own noise to hear or fear any other sound. As they lie on the ground in a compact mass, one of the men takes an umbrella (before the introduction of umbrellas a flag was used) and goes twenty to thirty yards to the rear of the herd and approaching stealthily until he is quite near suddenly expands the umbrella and runs with it along the edge of the herd; then closing it he retires to repeat the maneuver. This has the effect to rouse the rear rank, which thus suddenly alarmed plunges forward and arouses those in front, which immediately begin struggling and biting. The return of the man with the umbrella communicates another shock and adds another wave to the sluggish mass. This is repeated at intervals of four or five minutes till the successive shocks have aroused the whole herd, when, with much roaring and bellowing, the whole mass begins to move, gradually extending itself in a long irregular line in open order, each animal lumbering along as best it can. By shouting and waving flags at the rear and on the flanks of the herd, they are kept moving until it is necessary to halt them again for rest. Seen when thus moving in a long irregular line, the slow heaving motion of their bodies and the swaying of their long flexible necks give a grotesque appearance to the scene and suggest anything but a herd of Lions. The island, being composed of volcanic rock, is full of subterranean fissures covered thinly with soil and vegetation, and the earth so resounds with the noise of the tread of the Sea Lions that the sound may be heard to the distance of two miles. The approach of a herd to the village is always an occasion of interest and excitement to all of the inhabitants, who go out *en masse* to meet them and escort them to the slaughtering ground, where they are allowed to rest and cool before they are killed.

"The Sea Lions are too formidable to be killed with clubs like the Fur Seals. When all is ready for the slaughter the herd is started up a sloping hillside; the hunters follow, armed with rifles, and shoot the full-grown males from behind, the back of the skull being the only part a ball can penetrate. After all of these have been killed, the head of the column is checked and turned back so that the animals become massed together, and piled on each other five or six deep. In this way those below are held by those above while the hunters, armed with short lances, watch their opportunity to rush up to the struggling mass and thrust their lances into some vital part of the doomed beasts. This is attended with some danger to the hunters, who sometimes receive serious wounds from being hit with the lances that the Sea Lions, in their death agonies, seize in their mouths and wrench from the hands of the hunters.

"Nearly every part of these animals is valuable to the natives, but they have no commercial value outside of Alaska. Their skins are indispensable to the Sea Otter hunters of the Aleutian Islands, for the covering of their canoes in which they hunt these animals. The natives also use them for covering their large boats used in loading and unloading vessels. . . . Its flesh is preferred for food to that of the Fur Seal, that of the full-grown animal being finer in texture, lighter in color, and of a sweeter

Misc. Pub. No. 12——18

flavor, and it dries more readily in preserving it for winter use; the flesh of the young at the age of four months is esteemed a great luxury by the natives and is not easily distinguished from veal by educated palates. . . . Only the skeleton is left to waste.

"The stomachs of the full-grown Sea Lions are found to always contain from six to ten pounds of stones, varying in size from that of a hen's egg to a large apple. These stones are the same as those found on the beaches, worn round and smooth by the surf. The natives say they take these stones into the stomach for ballast when they leave the breeding-grounds, and cast them out again when they land in the spring. I have, however, had no means of verifying this, as the only season when they are taken is during the winter.

"As soon as the animals have all been killed the men proceed to remove the skins and blubber, and the other useful parts, which the chiefs divide and distribute among the several families. . . . Only a few of the skins are required for use on the island, the remainder being shipped to Ounalashka and other points where they are sold to the Sea Otter hunters. The value of the skins at the island is sixty cents each. About eight hundred are annually taken at St. Paul's Island, without apparently any decrease in the stock.

"There are many other places in the Territory where these animals bring forth their young, but as they resort mostly to outlying rocks and ledges they cannot be captured in any considerable numbers.

"The Sea Lion of Alaska, so far as my opportunities of observation have enabled me to judge, is a much larger species than that of California, the largest males I have ever seen at San Francisco and vicinity being not much larger than the full-grown female at the Fur Seal Islands, while I have seen at San Francisco females with young that were not much larger than a yearling of the species found at St. Paul's."

The food of the Sea Lion is well known to consist, like that of the other species of Eared Seals, of fish, mollusks, and crustaceans, and occasionally birds. As shown by animals kept in confinement, they require an enormous quantity. Captain Scammon states that the daily allowance of a pair kept in Woodward's Gardens, San Francisco, amounted to forty or fifty pounds of fresh fish.

Genus ZALOPHUS, *Gill.*

Arctocephalus (in part), GRAY, Cat. Seals and Whales, 1866, 55.
Zalophus, GILL, Proc. Essex Institute, July, 1866, v, 7, 11. Type *Otaria gillespii,* McBain.
Neophoca, GRAY, Ann. and Mag. Nat. Hist., 3d series, 1866, xviii, 231 Type *Arctocephalus lobatus,* Gray.

Molars $\frac{5-5}{5-5}$, large, closely approximated, the last under the hinder edge of the zygomatic process of the maxillary. Muzzle narrow. Superior profile, from the postorbital process anteriorly, gently declined. Bony palate moderately contracted posteriorly, and but slightly depressed. Hinder edge of the palatals deeply concave. Pterygoid hooks slender. Posterior nares broader than high; anterior higher than broad. Postorbital cylinder narrow and elongate. The postorbital constriction of the skull is deep and abrupt, giving a quadrate or subquadrate form to the brain-box, which varies to triangular through the varying degree of prominence of its latero-anterior angles. The postorbital processes are triangular, developed latero-posteriorly into a rather slender point. The sagittal crest, in very old males, forms a remarkably high, thin, bony plate, unparalleled in its great development in any other genus of the family. The general form of the skull is rather narrow, much more so than in *Eumetopias,* and nearly as much so as in *Arctocephalus,* the breadth to length being as 60 to 100.

Zalophus, so far as the skull is concerned, is the most distinct generic form of the family, it being thoroughly unlike all the others. In general form, as in size, it more nearly resembles *Arctocephalus* than any other genus, but differs from it in the dental formula, as well as in its enormously produced crests. It differs from *Otaria* in having one pair less of upper molars, in the slight depression of the bony palate, the less extension posteriorly of the palatines, the much narrower muzzle, the much less abrupt declination of the facial profile, its much higher sagittal and occipital crest, and in its narrower and more elongated form.

It differs from *Eumetopias,* as already pointed out, in having all the upper molars closely approximated, in the greatly concave outline of the posterior border of the palatines, and otherwise much as it differs from *Otaria.*

Zalophus differs from *Callorhinus* in its smaller number of upper molars, its high crests, narrower and more elongated muzzle, and in the more declined profile of the face. In the nature of its pelage, and in other external features, it is radically distinct from the whole group of Fur Seals, as it is also in its high sagittal crest. In size it is nearest *Arctocephalus*. The body is rather slender, and the head is narrow, long, and pointed, and with this slenderness of form is coördinated a corresponding litheness of movement.

The genus is restricted, so far as known, to the shores of the North Pacific and the Australian Seas, and is apparently represented by two species, the one confined mainly, if not wholly, to the western coast of the United States, and the other to temperate (and tropical?) portions of the eastern coast of Asia, from Japan southward, and the northern shores of Australia. The genus is thus southern or subtropical in its distribution, occurring on both sides of the equator, but not in the colder waters of either hemisphere.

ZALOPHUS CALIFORNIANUS (*Lesson*) *Allen*.

Californian Sea Lion.

Otaria californiana, LESSON, Dict. Class. d'Hist. Nat., xiii, 1828, 420 (based on "Le jeune Lion marin de la Californie," Choris, pl. xi).—SCHINZ, Synop. Mam., i, 1844, 473 (from Lesson).

Phoca californiana, FISCHER, Syn. Mam., 1829, 231 (= *Otaria californiana*, Lesson).

Otaria gillespii, M'BAIN, Proc. Edinb. Roy. Soc., i, 1858, 422.

Arctocephalus gilliespii,[*] GRAY, Proc. Zoöl. Soc. Lond., 1859, 110, 360, pl. lxx (from cast of the skull described by M'Bain); Cat. Seals and Whales, 1866, 55.

Zalophus gillespii, GILL, Proc. Essex Inst., v, 1866, 13.—GRAY, Ann. and Mag. Nat. Hist., 3d ser., xviii, 1866, 231; Suppl. Cat. Seals and Whales, 1871, 28; Hand-List Seals, etc., 1874, 41.—SCOTT, Mam. Rec. and Extinct, 1873, 20.—THOMPSON, Forest and Stream (newspaper), xii, 1879, 66 (habits and breeding in confinement).

Otaria (Zalophus) gillespii, PETERS, Monatsb. Akad. Berlin, 1866, 275, 671.

? *Otaria jubata*, VEATCH, J. R. Browne's Resources of the Pacific Slope, [app.], 1869, p. 150 (mainly, if not wholly; Cerros Island, L. Cal.).

Lion Marin de la Californie, CHORIS, Voy. Pittoresque.

Sea Lion [*of California*], SCAMMON, Overland Monthly, viii, 1872, 266 (in part).—GURNEY, Zoölogist, 1871, 2762 (Southern California).—STEARNS, Amer. Nat., x, 1876, 177 (in part).

Lobo marino, Spanish ; *Sea Lion* and *California Sea Lion*, English.

HABITAT.—Coast of California.

EXTERNAL CHARACTERS.—Color dark chestnut-brown, darker (blackish-brown) on the limbs, ventral surface, and the

---

[*] Spelled "*gilliespii*" by Gray and most other writers, but "*gillespii*" by M'Bain, who named it for his friend Dr. Gillespie.

extreme posterior part of the body, but varying greatly in different individuals and at different seasons. Whiskers whitish or yellowish-white, a few of them usually dusky at base. Length of adult male 7 to 8 feet; of adult female about 5.75 feet. Pelage short, harsh, and stiff.

A series of a dozen specimens varies greatly in color—from yellow through various shades of brownish-yellow to dark reddish-brown and even blackish-brown. At the season of moult they change from reddish-brown to yellowish or golden-brown. An adult female (M. C. Z. Coll., No. 5787) taken about September 1, 1877, is golden brownish-yellow, passing into dark brown on the limbs and ventral surface. Top of the nose, between and around the eyes, anterior edge of hand, and outer edge of foot, pale yellowish-white. Said by the collector (Mr. Paul Schumacher) to have just shed its coat. A nearly adult male (M. C. Z. Coll., No. 5785), taken at the same date, is dull dark yellowish-brown, passing into blackish-brown on the limbs and ventral surface; around base of hind limbs and tail and behind the axillæ nearly black. A third (M. C. Z. Coll., No. 5788) is dingy yellowish-brown, lighter on top of head, hind neck, and over the shoulders, and darker posteriorly, beneath, and on the limbs, where the color becomes very dark chestnut-brown, and blackish around the eyes and nostrils. A fourth, a very old male (M. C. Z. Coll., No. 5786), is dark yellowish-brown above, varied with dusky, and with small dots and narrow streaks of white, the white streaks and spots indicating the position of wounds received in fighting. A large whitish spot on the back of the neck.* Lower surface pale yellowish posteriorly, passing into darker anteriorly. A sixth (M. C. Z. Coll., No. 5677), an adult female, has the body everywhere dark yellowish-brown, passing into darker on the limbs and ventral surface. Still another (M. C. Z. Coll., No. 5785), a nearly full-grown male, differs from the foregoing in being light yellowish on the chin and about the mouth, very dark or blackish on the throat and sides of neck; breast yellowish-white; sides of body and ventral surface very pale yellowish-white, as is also the central portion of upper surface of both fore and hind limbs; top of head and greater part of dorsal surface very dark brown or blackish, slightly varied with white; shoulders and breast washed with gray; edges of the flippers very dark brown.

---

* In this specimen the atlas is firmly anchylosed with the skull, the result, doubtless, of injury in early life, to which, perhaps, this whitish spot is due.

Three specimens observed alive in Central Park, New York City, in April, 1878, differed very much in color. One (a male) was quite pure gray along the back, rather darker on the sides, and yellowish-gray on sides of belly; throat and breast pale yellowish-brown; ventral surface and limbs dark brown; sides of nose pale yellowish-white. Another (male) was dark brown varied with black. The third (female) was deep brownish-yellow on the throat and breast, blackish over the ventral surface and limbs; general color above, deep brownish-black.

Captain Scammon says:* "The color of the adult male is much diversified; individuals of the same rookery being quite black, with scattering hairs tipped with dull white, while others are of a reddish-brown, dull gray, or of light gray above, darker below. The adult female is not half the bulk of the male, and its color is light brown." He refers particularly to one specimen as being "black above, a little lighter below, with scattering hairs of light brown or dull white."

YOUNG.—Captain Scammon says: "The young pups, or whelps, are of a slate or black color, and the yearlings of a chestnut-brown." In the Museum of Comparative Zoölogy are several young specimens taken at the Santa Barbara Islands by Mr. Paul Schumacher (M. C. Z. Coll., Nos. 5678 and 5679) that are everywhere nearly uniform dark reddish-brown. The skulls show that they were quite young, the milk canines and last milk superior molars on each side being still in place; they were probably not more than two or three months old. The Museum also has a foetal specimen, received from the same locality and collector. In this (M. C. Z. Coll., No. 5839) the body is nearly uniform dark gray, with the top and sides of the head and the nape darker. Nose and face, to and around the eyes, black. Limbs brownish-black. The whiskers are mostly grayish-white, dusky at the base; some of the shorter ones entirely blackish.

PELAGE.—In the adult animal the pelage is short, stiff, and harsh, especially the new hair about the time of the moult.

---

* Under the head of "*Eumetopias stelleri*" (Marine Mam., p. 128), but, judging from the context, I think his remarks are based on the Sea Lions of the Santa Barbara Islands, and really refer to the present species. The specimens sent by him to the National Museum under this name from these islands are really *Zalophus californianus*. He spent much time at these islands, and his only detailed reference to the animals as seen by him in life relates to these islands and unquestionably to this species.

In the fœtal specimen the pelage is longer, and very soft to the touch, feeling like fur, but is simply soft straight hair, not at all like the underfur of the Fur Seals, or even the first coat of *Callorhinus ursinus*, under the long soft hair of which is an abundance of soft silky underfur. In the fœtal *Zalophus* a very slight admixture of fine curly underfur can be detected on close inspection; but in no sense is the first coat in this species comparable, in respect to underfur, with the first coat in the Fur Seals. In the older specimens of *Zalophus*, above described, which have already acquired their second coat, the pelage is still longer and softer than in the adults.

SIZE.—I am unable to give the dimensions of very old males. A male (M. C. Z. Coll., No. 5786), in which the crests of the skull are well developed, and the teeth slightly worn, but which is evidently only middle-aged, gives the following measurements: Total length from nose to end of tail* 2160 mm.; to end of outstretched hind-flippers, 2542 mm. (collector's measurement from fresh specimen, "8 ft. 4 in."); hind foot (from body), 380 mm.; fore foot (from axilla), 360 mm.; tail, 110 mm; ear, 35 mm.; longest whisker, 225 mm. The collector gives the girth behind the axillæ as 1337 mm. ("4 ft. 5 in."). Another specimen (M. C. Z. Coll., No. 5789, young male), with the crests of the skull wholly undeveloped, gives a length from nose to end of tail of 2140 mm.; to end of outstretched hind limbs, 2480 mm. (collector's measurement from fresh specimen, "8 ft. 2 in."); hind-flipper, 340 mm.; fore-flipper (from axilla), 370 mm.; tail, 80 mm.; ear, 33 mm.; longest whisker, 190 mm. The collector gives the girth behind the axillæ as 1220 mm. ("4 ft.").

A fully-adult female (M. C. Z. Coll., No. 5787) gives a length (from tip of nose to end of tail) of 1800 mm.; to end of outstretched hind-flippers, 2054 mm. (collector's measurement from fresh specimen, "6 ft. 9 in."); girth behind axillæ, 1247 mm. (collector's measurement, "3 ft. 9 in."); hind-flipper, 270 mm.; fore-flipper, 310 mm.; tail, 70 mm.; ear, 30 mm.; longest whisker, 110 mm.

Another adult female (M. C. Z. Coll., No. 5788) gives the following: Nose to end of tail, 1570 mm.; nose to end of outstretched

---

* This measurement is by estimate based on the collector's measurement of the total length to end of outstretched hind-flipper, taken from the fresh specimen, the calculation being based on a study of the skeleton. The total length of head and body, as taken from the mounted specimen, is obviously much too short.

hind-flippers, 1996 mm. (collector's measurement from fresh specimen, "6 ft. 7 in."); girth behind axillæ, 1068 mm. (collector's
measurement, "3 ft. 6 in."); tail 80 mm.; ear, 34 mm.; longest
whisker, 100 mm. The skeleton of an adult female (M. C. Z.
Coll., No. 6) has a total length to end of tail of 1706 mm.; to end
of phalanges of hind flipper, 1908 mm.

The collector's measurements of the young (two or three
months' old) specimens are: Male, nose to end of outstretched
hind-flippers, "4 ft." (1220 mm.); girth, "2 ft." (610 mm.).
Female, nose to end of outstretched hind-flippers, "3 ft. 8 in."
(1120 mm.); girth, "1 ft. 11 in." (583 mm.). The fœtal specimen
(M. C. Z. Coll., No. 5839, stuffed), already described, measures
from nose to end of tail 850 mm.; from nose to end of outstretched hind-flippers, 970 mm.; hind-flippers (from heel),
115 mm.; fore-flipper (from axilla), 200 mm.; tail, 45 mm.;
ear, 25 mm.; longest whisker, 55 mm.

Captain Scammon gives the following measurements of "Sea
Lions," taken at Santa Barbara Island in April and May,
1871–73. They include an "adult female" (column 1); a male
(column 2), "about ten months old," taken April 4, 1872; a
female (column 3), "supposed to be a yearling"; and "a new-
born female Sea Lion pup" (column 4), taken May 3, 1873.

	1*		2†		3‡		4§	
	ft.	in.	ft.	in.	ft.	in.	ft.	in.
Length from tip of nose to end of hind-flippers	6	4	4	10	4	10	2	4
Length from tip of nose to base of tail	......		3	10½	3	10½	1	11
Length of hind-flippers	1	1	0	11½	0	11	0	5½
Length of fore-flippers	1	4	1	3	1	2¼	0	7
Girth behind axillæ	3	3	2	8	2	7	1	3
Girth at base of hind-flippers	1	6	0	11½	1	1	0	6¾
From tip of nose to eye	0	3¾	0	3₁⁵₆	0	3	0	1¾
From tip of nose to ear	0	8	0	7	0	6	0	4
Length of ear	0	1¼	0	1	0	1¼	......	
Length of tail	0	2	0	2½	0	2½	0	1½
Length of longest whiskers	0	6	0	5½	......		......	
From base of tail to posterior teats	1	2	......		......		......	
From base of tail to anterior teats	1	10	......		......		......	
Distance between posterior teats	0	5	......		......		......	
Distance between anterior teats	0	8	......		......		......	
Thickness of blubber	0	0¾	0	0½	0	0¾	0	0¼

* Adult female.
† Male, ten months old.
‡ Female, about one year old.
§ Female, newly born.

*Measurements of the Skeleton of an Adult Female.**

	mm.
Whole length of skeleton (including skull)	1706
Length of skull	236
Length of cervical vertebræ	320
Length of dorsal vertebræ	640
Length of lumbar vertebræ	230
Length of caudal vertebræ (+ sacral)	280
Length of first rib, total	140
Length of first rib, osseous portion	75
Length of first rib, cartilaginous portion	65
Length of second rib, total	173
Length of second rib, osseous portion	100
Length of second rib, cartilaginous portion	73
Length of third rib, total	240
Length of third rib, osseous portion	158
Length of third rib, cartilaginous portion	82
Length of fourth rib, total	280
Length of fourth rib, osseous portion	185
Length of fourth rib, cartilaginous portion	95
Length of fifth rib, total	335
Length of fifth rib, osseous portion	220
Length of fifth rib, cartilaginous portion	115
Length of sixth rib, total	370
Length of sixth rib, osseous portion	250
Length of sixth rib, cartilaginous portion	120
Length of seventh rib, total	395
Length of seventh rib, osseous portion	270
Length of seventh rib, cartilaginous portion	125
Length of eighth rib, total	445
Length of eighth rib, osseous portion	295
Length of eighth rib, cartilaginous portion	150
Length of ninth rib, total	445
Length of ninth rib, osseous portion	290
Length of ninth rib, cartilaginous portion	155
Length of tenth rib, total	430
Length of tenth rib, osseous portion	280
Length of tenth rib, cartilaginous portion	150
Length of eleventh rib, total	413
Length of eleventh rib, osseous portion	280
Length of eleventh rib, cartilaginous portion	133
Length of twelfth rib, total	395
Length of twelfth rib, osseous portion	260
Length of twelfth rib, cartilaginous portion	135
Length of thirteenth rib, total	362
Length of thirteenth rib, osseous portion	247
Length of thirteenth rib, cartilaginous portion	115
Length of fourteenth rib, total	310
Length of fourteenth rib, osseous portion	215

* No. 6159, Collection of Museum of Comparative Zoölogy.

	mm.
Length of fourteenth rib, cartilaginous portion	95
Length of fifteenth rib, total	220
Length of fifteenth rib, osseous portion	180
Length of fifteenth rib, cartilaginous portion	40
Length of sternum, ossified portion	550
Length of sternum, 1st segment	110
Length of sternum, 2d segment	50
Length of sternum, 3d segment	53
Length of sternum, 4th segment	50
Length of sternum, 5th segment	48
Length of sternum, 6th segment	47
Length of sternum, 7th segment	46
Length of sternum, 8th segment	38
Length of sternum, 9th segment	55
Length of scapula	180
Greatest breadth of scapula	250
Greatest height of its spine	18
Length of humerus	155
Antero-posterior diameter of proximal end of humerus	63
Transverse diameter of proximal end of humerus	65
Transverse diameter of distal end of humerus	57
Length of radius	155
Length of ulna	194
Longest diameter of upper end of ulna	67
Length of carpus	40
Length of 1st metacarpus and its digit	218
Length of 2d metacarpus and its digit	188
Length of 3d metacarpus and its digit	150
Length of 4th metacarpus and its digit	120
Length of 5th metacarpus and its digit	90
Width of manus at base of metacarpals	80
Total length of fore limb (excluding scapula)	568
Length of femur	90
Longest diameter of proximal end of femur	46
Longest diameter of distal end of femur	48
Least antero-posterior diameter of shaft of femur	13
Length of tibia	185
Length of tarsus	40
Length of 1st metatarsus and its digit	222
Length of 2d metatarsus and its digit	187
Length of 3d metatarsus and its digit	180
Length of 4th metatarsus and its digit	180
Length of 5th metatarsus and its digit	183
Width of pes at base of metatarsals	57
Total length of hind limb	537
Length of innominate bone	176
Greatest width of pelvis anteriorly	100
Length of ilium	70
Length of ischio-pubic bones	100

*Measurements of the Metacarpal and Phalangeal Bones of an Adult Female.**

	1st digit.	2d digit.	3d digit.	4th digit.	5th digit.
Length of manus to end of	258	226	187	147	112
Length of metacarpal of.............	90	58	55	48	42
Length of 1st phalanx of	78	66	55	35	31
Length of 2d phalanx of.............	50	47	36	23	15
Length of 3d phalanx of.............	....	20	15	12	10

*Measurements of the Metatarsal and Phalangeal Bones of an Adult Female.**

	1st digit.	2d digit.	3d digit.	4th digit.	5th digit.
Length of pes (posterior end of os calcis) to end of	280	260	255	250	250
Length of metacarpal of	79	61	60	60	65
Length of 1st phalanx of	75	60	57	54	58
Length of 2d phalanx of	40	46	48	48	37
Length of 3d phalanx of	....	16	16	16	22
Length of nail of	....	16	18	15	....

SKULL.—The skull in *Zalophus californianus*, as compared with the skull in allied genera, is remarkable for the narrowness and great elongation of the facial portion, which is even much more elongated ·and slenderer than in *Arctocephalus*. In its general configuration (excepting, of course, the great development of the sagittal and occipital crests in the very old males) it more resembles the Arctocephaline type than any other. The maximum breadth (*i. e.*, at the zygomata) in the females barely equals or falls a little short of half the length, while in the old males it rather exceeds this proportion. In very old males the crests of the skull are enormously developed, and, contrary to what usually obtains in the other genera of this family, are considerably developed in very old females. The superior outline (in old males) slopes rapidly from the high sagittal crest to the end of the nasals. The postorbital processes are long and rather narrow, and are directed backward in old age; the nasals are long and narrow, decreasing in width posteriorly. The superior edge of the intermaxillæ is very narrow, and is prolonged backward nearly to the middle of the nasals. The postorbital cylinder is long and narrow, and often abruptly contracted posteriorly. The bony palate is nearly flat, but little depressed, and is rather deeply emarginate posteriorly. The

*Specimen No. 6159, Coll. Mus. Comp. Zoölogy.

palato-maxillary suture is about opposite the hinder edge of the last molar. The pterygoid hamuli are small. The posterior narial opening is wider than deep; the anterior has these two dimensions about equal.

In *Zalophus* the superior aspect of the skull, before the development of the crest, is strikingly like that of *Arctocephalus*, as indeed is also the inferior aspect, aside from the dental formula. In *Zalophus* the auditory bullæ are rather less swollen than in *Arctocephalus*, but in all other respects there is a striking resemblance. The anteorbital portion of the skull, however, is more attenuated, and relatively much longer. With this exception there is little difference in the general conformation of the skull in middle-aged females of these two genera, while both differ widely from *Otaria*, *Eumetopias*, and *Callorhinus*. The great development of the crests of the skull late in life in *Zalophus* gives it at that time a highly peculiar conformation.

Measurements of Ten Skulls of ZALOPHUS CALIFORNIANUS.

Catalogue number.	*261	*15254	*15255	†—	‡6146	‡14506	*14507	†6150	†6151	†6152
Locality.	California	San Nicolas Island	...do	California	Santa Barbara Island	...do	...do	...do	...do	...do
Sex.	♂	♂	♂	♂	♂	♀	♀	♀	♀	♀
Length.	290	270	?300	330	255	212	195	237	233	220
Greatest breadth at zygomatic arch.	170	140	180	180	145	82	78	120	128	110
Breadth at mastoid processes.	163	132	170	165	127	99	94	100	110	95
Distance from anterior edge of intermaxillaries to end of hamular process of pterygoid.	180	170	183	190	165	125	....	148	142	134
Distance from anterior edge of intermaxillaries to last molar.	97	104	105	100	95	82	72	85	83	82
Distance from anterior edge of intermaxillaries to postglenoid process.	....	188	205	....	185	140	131	168	167	156
Distance from palato-maxillary suture to end of hamular process.	....	76	76	....	72	50	....	65	58	65
Length of maxilla along alveolar border.	....	112	117	....	90	97	85	87	90	86
Breadth of palatine surface at posterior end of maxillæ.	....	40	45	....	43	27	28	30	30	27
Length of nasal bones.	56	61	59	61	50	34	32	41	40	41
Breadth of nasal bones anteriorly.	27	22	28	30	27	17	19	19	20	19
Breadth of nasal bones at fronto-maxillary suture.	20	15	27	28	16	11	6	10	13	12
Breadth of skull at canines.	60	58	61	70	47	35	33	39	36	36
Least breadth of skull at temporal fossæ.	31	31	....	....	34	35	31	22	23	27
Breadth of occipital condyles.	....	63	....	....	....	59	50	52	57	55
Length of upper molar series.	....	....	58	....	53	52	46	52	50	51
Length of lower molar series.	....	....	58	....	50	48	44	48	47	47
Length of lower jaw to condyles.	200	184	210	240	187	130	121	160	150	140
Height of occipital crest.	29	23	....	38	18	0	0	6	3	1
Age.	Very old	Middle-aged	Do.	Very old	Do.	Middle-aged	Do.	Very old	Do.	Middle-aged

* National Museum, Washington, D. C.
† Chicago Academy of Sciences.
‡ Museum of Comparative Zoölogy, Cambridge, Mass.

DENTITION.—The teeth in *Zalophus* are all strongly developed and very firmly implanted. All are single-rooted except the last molar, which is imperfectly double-rooted. The molars all have a distinctly beaded cingulum on the inner side. The lower molars and the fifth and sometimes the fourth upper molars have a small but distinct anterior cusp. The canines and the incisors present the usual form seen among the Otaries.

The teeth of a middle-aged male skull present the following measurements :

Measurements of the Teeth.

A.—TEETH OF THE UPPER JAW.

	Molars.					Canines.	Incisors.		
	5th.	4th.	3d.	2d.	1st.		Outer.	Middle.	Inner.
Antero-posterior diameter.	9	9	10	10	7	17	6	5	5
Transverse diameter	7	7	9	8	5	13	4	3	2.5
Height of crown	8	8.5	10	9.5	9	26	15	5	4

B.—TEETH OF THE LOWER JAW.

	Molars.					Canines.	Incisors.	
	5th.	4th.	3d.	2d.	1st.		Outer.	Inner.
Antero-posterior diameter ...	9	12	11	10	8	18	5	4
Transverse diameter	6.5	8	9	7	7	11	6	4
Height of crown	8	9	9	7.5	6	25	6	5

The molars are usually closely approximated, but sometimes there is a small space between the two hindermost of the series, and occasionally they are all slightly and evenly spaced. The hinder edge of the last upper molar is generally anterior to or about even with the posterior border of the zygomatic process of the maxillary. When more than five upper molars are present, the sixth or supernumerary is posterior to the fifth, and is usually smaller than the fifth (sometimes almost rudimentary) and lacks the accessory cusps seen in the fifth.

The milk dentition (fully represented in three skulls before me and partly so in five others) does not differ from that of the other species of the family, and has been already fully described (see *anteà*, p. 223).

SEXUAL DIFFERENCES.—From the testimony of Captain Scammon, and from the material I have been able to examine, the female differs from the male in color in being rather lighter, or of a more yellowish-brown. The most notable difference is in size, the female being very much the smaller, but not quite so great a sexual disparity in size obtains in this species as in *Eumetopias stelleri* and *Callorhinus ursinus*.[*] Unfortunately the material at my command will not enable me to give full statistics on this point. Most of the male specimens in a large series sent to the Museum of Comparative Zoölogy, by Mr. Schumacher, from Santa Barbara Islands, are young or middle-aged, only one having the teeth perceptibly worn or the crests of the skull very highly developed. A comparison of very old skulls of both sexes shows that nearly the usual amount of sexual variation in size common to the Otaries obtains in the present species. The table of measurements (on page 285) of ten skulls—five male and five female—all fully adult and most of them very old, gives all the information I am able to offer respecting sexual variation in size.

As usual in this group, the dental armature (especially the canines and caniniform incisors) is much weaker in the female than in the male, by means of which the skulls of females can be readily distinguished from those of males of about the same size. The whole skull is slighter and weaker, and all the processes and ridges for the attachment of muscles much less developed. There are, however, in very old female skulls, distinct, but comparatively low, and thin sagittal and occipital crests, which attain the height of 3 to 6 mm., while in the males they sometimes rise to 35 or 40 mm. The limbs are also much weaker and slenderer, as of course are all the bones of the skeleton.

VARIATION WITH AGE.—As already noticed, the color of the young at birth is dark gray or slaty, and the pelage has at this time a delicate softness, due to the silky texture of the hair. The pelage is wholly devoid of a second coat of true under-fur, like that of the Fur Seals, but from its softness might readily be mistaken on casual observation for true fur. This is very soon replaced by a coarser and harsher, but still quite soft pelage, in comparison with that of adults, of nearly uniform chestnut or dark reddish-brown color. This is succeeded by the harsh, stiff pelage of the middle-aged and adult animals.

---

* See *anteà*, p. 226.

The skull changes greatly in its proportions with age. The capacity of the brain-case does not greatly increase after birth, it enlarging mainly by the thickening of its walls. The anterior half of the skull develops rapidly and alters very much in form. At birth the inter- and anteorbital portions of the skull are very short, they together forming rather less than half the length of the skull, while in full-grown skulls they comprise about two-thirds of its length. In a young skull (Nat. Mus., No. 15660, taken from an animal killed a few days after birth), which has a total length of 146 mm., the brain-case alone has a length of 78 mm., the interorbital region* a length of 31 mm., and the anteorbital a length of 37 mm., giving a total length from the anterior wall of the brain-case to the front border of the intermaxillæ of 68 mm. In this skull (as in several others before me of about the same age) the occipital condyles are wholly anterior to the plane of the occiput. In a very old female skull (M. C. Z. Col., No. 6150), with a total length of 233 mm., the occipital condyles project 15 mm. behind the occipital plane. Of the remaining 218 mm. of the length, the brain-case occupies 83 mm., the interorbital region 65 mm., and the anteorbital 71 mm., and the two regions together 136 mm. In the first the ratio of the length of the brain-case to that of the rest of the skull is as 78 to 68; in the last as 83 to 136. In a middle-aged male skull, the total length is 282 mm., of which the condylar extension is 22 mm. Of the remaining 260 mm., the brain-case occupies 95 mm., the interorbital region 78 mm., and the anteorbital 87 mm., making the proportionate length of the brain-case to the rest of the skull as 95 to 165. The ratios between the different regions in these three skulls are as follows:

	Young.	Female.	Male.
Ratio of brain-case to whole skull	53.5—100	35.6—100	33.7 —100
Ratio of interorbital region to whole skull	21.2—100	28 —100	30.85—100
Ratio of anteorbital region to whole skull	25.3—100	30.5—100	27.66—100
Condylar extension to whole skull.................	....	6 —100	7.7 —100

The width of the brain-case in these skulls is respectively 90 mm., 97 mm., and 107 mm.

In adult skulls the breadth of the interorbital region is relatively, and generally absolutely, much less in adult skulls than at birth, and the point of greatest constriction is placed much

---

* That is, the narrow portion of the skull bounded laterally by the temporal fossæ and orbits.

more posteriorly, being in the adult at the posterior end of the temporal fossæ, and in the young at the orbits. The breadth of the skull just in front of the brain-box in very young skulls (those taken a few days after birth) is 40 to 42 mm., in those three or four months old, 38 mm.; in adult females, usually 22 to 30 mm.; in adult males, about 30 to 35 mm. The amount of constriction varies somewhat in adult skulls of the same sex, the constriction increasing with the advance of age. There is a corresponding contraction posteriorly of the palatal region. In very young skulls, the palate is widest at the pterygoid hamuli; in those a few months old it is nearly straight, but later in life becomes narrowed posteriorly, the contraction being greatest in aged specimens, in which the width at the pterygoid hamuli is a third less than it is at the last molar.

The crests of the skull do not begin to develop until the animal reaches adult size, and attain their highest development in very old specimens. In a series of thirty skulls, only two have the crests remarkably developed, these being the two old male skulls described by me in 1870.* In only one of the skulls of the series, aside from the two above mentioned, are the teeth much worn. The two very old skulls show, by their large size and rugose character, that the deposition of bony matter is continued to a very late period in life.

COMPARISON WITH ALLIED SPECIES.—*Zalophus californianus* is too distinct in cranial characters and dentition to require comparison with any of the Hair Seals of other genera, while its pelage and color afford obvious points of difference from the Fur Seals. As respects the conformation of the skull, it finds its nearest allies in *Arctocephalus*, from which, however, it is readily distinguished by its more elongated muzzle and dental formula. It appears to closely resemble its congener, *Z. lobatus*, both in size and color. Having no specimens of that species at command, I am unable to state the points of difference between the two. The descriptions and figures of *Z. lobatus* indicate their close alliance.

GEOGRAPHICAL DISTRIBUTION.—The exact boundaries of the habitat of *Zalophus californianus* cannot at present be given. The only specimens I have seen are from the coast of California and its islands, from San Diego and San Nicolas Island northward to the Bay of San Francisco. Captain Scammon (see

---

* Bull. Mus. Comp. Zoöl., vol. ii, p. 69; see measurements at p. 70.

infra, pp. 301, 302) twice alludes incidentally to its presence "along the Mexican and Californian coasts," and Dr. Veatch states that "Sea Lions" (which he calls "*Otaria jubata*, but which are, almost beyond doubt, the present species) had populous breeding stations twenty years ago, and doubtless have still, on Cerros or Cedros Island, in about the latitude of 28½°, off the Lower California coast. Whether they occur southward of this point at the present time I am unable to state, but should infer that such was the case from Scammon's allusion to their capture along the "Mexican" coast. In any case, it appears probable that in Dampier's time they ranged as far south as the Chametly and Tres Marias Islands, respectively in latitudes about 23° and 21°, at which points he saw "Seals" in the year 1686. In describing the Chametly Islands (the most northerly of the two groups mentioned by him under this name), situated off the West coast of Mexico in latitude 23° 11', he says: "The Bays about the Islands are sometimes visited with Seals; and this was the first place where I had seen any of these Animals, on the North side of the Equator, in these Seas. For the Fish on this sandy Coast lye most in the Lagunes or Salt-Lakes, and Mouths of Rivers; For this being no rocky Coast, where Fish resort most, there seems to be but little Food for the Seals, unless they will venture upon Cat-Fish."*

He also met with Seals at the Tres Marias Islands (in latitude "21° 5'"), and consequently two degrees south of the Chametly Islands, in describing one of which islands, named by him St. George's Island, he says: "The Sea is also pretty well stored with Fish, and Turtle or Tortoise, and Seal. This is the second place on this Coast where I did see any Seal: and this place helps to confirm what I have observed, that they are seldom seen but where there is plenty of Fish."†

It is of course not certain that the Seals here alluded to are *Zalophus californianus*, since the Sea Elephant of the California coast also occurs at Cedros Island, and probably still further south, the two species having apparently about the same range. If they had been the latter, Dampier would probably have made some allusion to their large size.

The species of *Zalophus* occurring in Japan has been by some writers considered to be the same as the Californian one; but, though doubtless closely allied, its affinities, as will be noticed

---

* A New Voyage round the World, 5th ed., vol. i, 1703, pp. 263, 264.
† Ibid., p. 276.

later (see *infra*, p. 293), appear to be not as yet satisfactorily determined. As *Zalophus californianus* has not yet been detected on the American coast north of California, its occurrence on the Asiatic coast seems hardly to be expected.

GENERAL HISTORY.—This species has hitherto been believed to be free from any serious complications of synonymy, and to have been first brought to the notice of the scientific world by M'Bain in 1858. The only synonym hitherto quoted has been *Otaria stelleri*, "Schlegel" (*i. e.*, Temminck), which Dr. Peters* stated, after an examination of the original specimens preserved in the Leyden museum, to be identical with the *O. gillespii* of M'Bain. A re-examination of the subject, in the light of much new information and material, shows that the first notice of the species was published by Choris in 1822, under the name of "Lion marin de la Californie," who gave a rather poor figure of it in plate XI of his chapter entitled "Port San-Francisco et ses Habitants." As already stated under the head of *Eumetopias stelleri*, his only reference to it in the text of this chapter is as follows: "Les rochers, dans le voisinage de la baie San-Francisco sont ordinairement couverts de lions marins, pl. XI." In his account of the Aleutian Islands, however, he again refers to it, and clearly indicates its characteristic external features. He says: "Ces animaux [Lions marins] sont aussi très-communs au port de San-Francisco, sur la côte de Californie, ou on les voit en nombre prodigieux sur les rochers de la baie. Cette espèce m'a paru se distinguer de ceux qui fréquentent les îles Aléoutiennes; elle a le corps plus fluet et plus allongé, et la tête plus fine: quant à le couleur, elle passe fortement au brun, tandis que ceux des îles Aléoutiennes sont d'une couleur plus grise, ont le corps plus rond, les mouvements plus difficiles, la tête plus grosse et plus épaisse; la couleur du poil des moustaches plus noirâtre que celui des îles Aléoutiennes."†

The importance of this reference turns upon its being an explicit indication of the character of his "Lion marin de la Californie," the subject of "Pl. XI"; this being, as is well known, the basis of Lesson's *Otaria californiana*, which has hitherto been referred to *Eumetopias stelleri*, but which is really the same as the so-called *Zalophus gillespii*. Lesson says: "Cette espèce, d'après la figure de Choris, a le pelage ras, uniformément fauve-brunâtre, les moustaches peu fournies; le museau assez pointu;

---

* Monatsb. Akad. Berlin, 1866, p. 669.

† Voy. pittoresque, Iles Aléoutiennes, p. 15.

les membres antérieurs sont réguliers, plus grands que les postérieurs. Cinq rudimens d'ongles occupent l'extrémité des
phalanges, et sont débordés par une large bande de la membrane. Les pieds postérieurs sont minces, ayant trois ongles au
milieu et deux rudimens d'ongles internes et externes. Cinq
festons lancéolés et étroits dépassent de cinq à sex pouces les
ongles. La queue est trés-courte. Des côtes de la Californie."*
His sole reference is "jeune Lion marin de la Californie, Choris,
Voy. pittoresq., pl. 11," and his description seems to be based
wholly upon this figure. Immediately preceding this is his
description of the "Otarie de Steller, *Otaria Stellerii,* N.; Lion
Marin, *Leo marinus,* Steller, *de Bestiis Marinis,*" etc., which
closes with "Peut-être l'Otarie de Steller est-il identique avec
l'Otarie suivant?" While it may be urged that the *Eumetopias
stelleri* also occurs in San Francisco Bay, Choris does not seem
to have recognized it there, while he did observe a species that
seemed to him to be different from the Sea Lions and Sea Bears
of the Aleutian Islands, and in describing these differences he
has indicated most clearly the distinctive points of difference, as
seen in the living animals, between these species. Furthermore,
it turns out that the *Zalophus gillespii,* auct., is still the common
species of that locality and of the California coast generally.
On this point Mr. Elliott, who has had ample opportunity of
observing both species in life,† says: "I have no hesitation
in putting this *Eumetopias* of the Prybilov Islands apart from
the Sea Lion common at San Francisco and Santa Barbara, as a
distinct animal," but adds, "I am not to be understood as saying
that *all* the Sea Lions met with on the Californian coast are different from *E. stelleri* of Bering Sea. I am well satisfied that
stragglers from the north are down on the Farallones, but they
are not migrating back and forth every season; and I am furthermore certain that not a single animal of the species most
common at San Francisco was present among those breeding on
the Prybilov Islands in 1872–'73."‡

If I am right in considering the *Zalophus gillespii,* auct., as
identical with *Otaria californiana* of Lesson, of which I think
there is no reasonable doubt, the synonymy of this species has
narrowly escaped further complications, Dr. Gill, in his first
mention of *Eumetopias,* saying: "Type, *Otaria californiana*

---

*Dict. class. d'Hist. Nat., xiii, 1828, 420.
† I have in hand colored drawings of both species, made by him from life,
which he has kindly placed at my disposal.
‡ Cond. of Affairs in Alaska, p. 158.

Lesson = *Arctocephalus monteriensis* Gray." But he cites as type of *Zalophus*, in the same connection, "*Otaria Gillespii* Macbain," and subsequently, in the same paper, so character·izes his genera *Eumetopias* and *Zalophus* as to leave no doubt that *Eumetopias* relates to the *Otaria stelleri* of Müller, and *Zalophus* to the *Otaria gillespii* of M'Bain. He further says that his *Eumetopias californianus* "is identical with the *Otaria monteriensis* of Gray, and possibly also with *Otaria Stelleri* Müller."*

The *Otaria stelleri* of the Fauna Japonica is unquestionably a *Zalophus* and not a *Eumetopias*, but is probably not identical with the *Zalophus* of the California coast, although, as already stated, so considered by Peters.† Not only do the skulls figured by Temminck show that the species is not *Eumetopias stelleri*, but his comparative remarks respecting its relationship to *O. jubata* indicate unmistakably the same thing. Although I at one time accepted Peters's determination of Temminck's *Otaria stelleri*, a subsequent examination, in the light of much new material and information, has led me to doubt its correctness. The range of *Zalophus californianus* (= *gillespii*) has not been reported as extending northward on the American coast beyond California, and no specimens of this species (except one cited by Gray, the identification of which seems open to question) have been thus far recognized from Japan or any portion of the Asiatic coast. Temminck, with good series of the Japan species and of the *Zalophus lobatus* before him (he seems not to have had the true *E. stelleri*), was unable to recognize any appreciable differences between them. In comparing his *Otaria stelleri* with the *Otaria australis* of Quoy and Gaimard, he says: "Un crâne absolument semblable à celui figuré par les voyageurs dont nous venons de parler [Quoy et Gaimard] a été décrit sous le nom d'Arctocephalus lobatus, par Gray, Spic. Zool., I, p. 1, pl. 4, fig. 2 et 2 *a*; ce crâne provenant de la collection de feu Brookes fait maintenant partie du Musée des Pays-Bas; il ne se distingue en effet par aucun caractère essentiel de celui de

---

* Proc. Essex Inst., v, 1866, pp. 7, 11, and 13 (footnote).

† Peters says: "Uebrigens zweifle ich jetzt auch gar nicht mehr daran, dass *O. Gilliespii* Macbain und *O. japonica* Schlegel [Ms. = *O. stelleri*, Fauna Japonica] zu derselben Art gehören, da die Schädel beider nicht allein in der Form, sondern auch in der Grösse miteinander übereinstimmen. Denn der alte Schädel von *O. Gilliespii* ist 0.m 295 lang, während alte Schädel des Leidener Museums von *O. japonica* 0.m 270 bis 0.m 310 lang sind."—*Monatsb. Akad. Berlin*, 1866, p. 669.

l'Otaria australis et de ceux de l'Otarie de Steller, tirés de nos individus du Japon. Le Musée des Pays-Bas enfin vient de recevoir, comme nous l'avons constaté plus haut, un très-jeune individu d'une Otarie, prise sur les îles Houtman près de la côte occidentale de la Nouvelle Hollande, et qui ne paraît différer ni de l'Otarie australe de Quoy et Gaimard, ni du Lion marin de Steller. Il paraît résulter de ces données que l'Otarie de Steller n'habite pas seulement le nord de l'océan pacifique, mais qu'elle se trouve aussi dans les parties australes de cette mer."* It appears to me probable that if we change the phrase "l'Otarie de Steller" in the last sentences above quoted to read *Zalophus lobatus,* we have the case correctly stated.† Indeed, Gray, in his earlier papers (down to 1866), positively referred the *Otaria stelleri* of Temminck to his *Arctocephalus lobatus.* Later‡ he says it "includes both the Australian Eared Seals, viz, *Arctocephalus cinereus* and *Neophoca lobata,*" but finally § doubt-

---

* Faun. Jap., Mam. Marins, p. 8.

† Just what Temminck's young skulls referred to *Otaria stelleri* are seems not so clear, they having six superior molars on each side. As elsewhere stated, I have found supernumerary molars in about one skull in ten in adult specimens of *Zalophus californianus,* and occasionally in other species of Eared Seals, but Temminck describes all his four young skulls as having each six superior molars on each side, or alveoli indicating their recent presence, but the probabilities are entirely against the sixth being supernumerary. In referring to his "*Otaria stelleri,*" he says: "la sixième molaire de la mâchoire supérieure est sujette à tomber à l'époque de l'apparition des dents permanents," and gives this as one of the characters which distinguish it from *O. jubata.* What he had before him is hard to recognize, for the skulls he described had long passed the age when all traces of the temporary dentition are lost. It is only supposable that the young skulls belonged to some six-molared species; for no species of Otary is known to lose at any stage the hinder pair of upper permanent molars, and thus undergo a change in the dental formula from M. $\frac{6-6}{5-5}$ to M. $\frac{5-5}{5-5}$. At one time (Bull. Mus. Comp. Zoöl., vol. ii, p. 62) I thought it probable that the young skull here figured (as well as the other young skulls Temminck describes) might have been that of *Callorhinus ursinus,* but the form of the nasals and the frontal extension of the intermaxillaries in the one figured show that such could not have been the case. Dr. Gray at one time referred it without doubt to *Arctocephalus cinereus,* which is probably its correct allocation, although later he doubtfully assigned it to his *Phocarctos elongatus* (Hand-List, 1874, p. 31), but a little further on in the same work (p. 42) he says, "figures 5 and 6 [of Temminck's plate xxii] are evidently *Gypsophoca,*" but thinks they may belong to an undescribed species.

‡ Suppl. Cat. Seals and Whales, 1871, p. 24.

§ Hand-List of Seals, 1876, p. 42.

fully accepted Peters's reference of it to the *Zalophus gillespii.*
Peters himself first (in 1867) referred the species (except the
figure of the young skulls) to *Zalophus lobatus*, which were assigned to *Arctocephalus cinereus*, but later, as above stated, he
identified it with *Zalophus gillespii.**
The references to this species are still very few. Aside from
Choris's account, and Lesson's (based on Choris's) and Fischer's
(based on Lesson's) and Temminck's, the first of importance is
M'Bain's description of a skull from California in 1858, which
specimen was redescribed and figured from a cast by Gray, in
1859. Dr. Gray, as late as 1871,† appears to have seen only this
specimen, but in 1874‡ cites (without full description) a skull
from Japan.§ Aside from this its Japan record still rests
wholly on Dr. Peters's determination of "Schlegel's" (*i. e.*,
Temminck's‖) specimens in the Leyden Museum. Dr. Gill, in
1866, had examined a skull from California in the museum of
the Smithsonian Institution, which led him to separate the
species generically from the other Eared Seals. This skull,
and another (belonging to the museum of the Chicago Academy of Sciences), also from California, I was able to describe in
detail in 1870.¶ These Californian skulls are the only ones thus
far described,** but Scammon, in his "Marine Mammalia," under
the name "*Eumetopias stelleri*," has given detailed measure-

---

* Monatsb. der Akad. der Wissensch. zu Berlin, 1866, (1867), pp. 272, 276,
668.

† Suppl. Cat. Seals and Whales, p. 28.

‡ Hand-List of Seals, p. 41.

§ "1589b. Skull, 12¼ inches long, with canines very large; no other teeth;
no lower jaw; frontal crest very high. Japan, 73. 3. 12. 1."

‖ Dr. Peters cites Schlegel as the author of that part of the "Fauna Japonica" relating to the Mammals, although published as "par C. J. Temminck." Misled by Peters, I made the same error in my paper on the
Eared Seals, published in 1870.

¶ They were not figured in the regular edition of my paper on the Eared
Seals (Bull. Mus. Comp. Zoöl., vol. ii, No. 1), but two photographic plates,
representing both specimens, were added to a few of the author's copies (about
twenty-five), which were sent to some of the more prominent workers in this
field. These interpolated plates have been referred to by Dr. Gray (Hand-
List of Seals, p. 42) as though they formed a part of the original work.

** During the last year, I may here add, as an indication of the amount of
material relating to this species now accessible, that I have examined not
less than a dozen skins, representing adults of both sexes, and young of
various ages from a fœtal specimen upward, and more than twenty skulls,
likewise embracing young, even with the milk dentition, and both sexes of
various ages, and two complete skeletons.

ments of what I take to be examples of this species from the Farallone and Santa Barbara Islands. *

HABITS.—Several more or less full accounts of the habits of the Californian Sea Lions have been given by different writers, who have, however, failed to distinguish the two species occurring along the Californian coast, and consequently their descriptions are not wholly satisfactory. The large Northern species certainly occurs, and rears its young, as far south as the Farallones, but probably exists there only in small numbers, while I have seen no evidence of its presence at Santa Barbara Island. Even Captain Scammon, in his account of the Sea Lions of California, has not distintly recognized the two species occur-

---

* Captain Scammon published his first account of the Sea Lions in the "Overland Monthly" magazine (vol. viii, pp. 266–272, March, 1872), in an article entitled "About Sea Lions," which is substantially the same as that in the "Marine Mammalia," with the omission of figures and about two pages of tabulated measurements and other details, based on specimens subsequently obtained at the Farallone and Santa Barbara Islands. In a footnote in the "Marine Mammalia" (p. 125), he refers to his former article as follows: "Since the publication of the article 'About Sea Lions,' in the 'Overland Monthly' of September, 1871 [*lege* March, 1872!], we have had opportunity of making additional observations upon these animals at the Farallone Islands, where we saw the largest females we have ever met with on the California coast. Hence, what we have formerly taken to be the *Eumetopias Stelleri* may prove to be the *Zalophus Gillespii?*; but if such be the fact, both species inhabit the coast of California, at least as far south as the Farallones. Moreover, both species, if we may be allowed the expression, herd together in the same rookeries. On making a series of observations upon the outward forms of Sea Lions, it will be found that a confusing variety exists in the figures of these very interesting animals, especially in the shape of the head—some having a short muzzle with a full forehead [*Eumetopias stelleri*]; others with forehead and nose somewhat elongated [*Zalophus californianus* = *gillespi*, auct.]; and still others of a modified shape, between the two extremes [*E. stelleri*, female ?]." In this connection it may be noted that four of the five specimens of which Captain Scammon gives measurements in the "Marine Mammalia," were taken *after* the publication of the article in the "Overland Monthly," namely, No. 1, "full-grown male," Farallones, July 17, 1872; No. 3, male "about ten months old," Santa Barbara Island, April 4, 1872; No. 1 *bis*, female, supposed to be a yearling, and No. 2 *bis*, female, new-born pup, same locality, May 3, 1873. The other, No. 2 (referred to in the "Overland Monthly" paper), adult female, Santa Barbara Island, April 12, 1871. The first (No. 1, full-grown male) I refer with little hesitation to *E. stelleri*, and the second (No. 2, adult female), to *Z. californianus*, especially as I find skulls in the National Museum, received from Captain Scammon, agreeing respectively with these in locality, sex, and age.

ring there, and his description doubtless refers in part to both species, but unquestionably relates mainly to the present one.* His "Sketch of a sealing season upon Santa Barbara Island," in 1852, presumably relates exclusively to *Zalophus califor-nianus*, but in addition to this I quote a few paragraphs from his general account of "the Sea Lion," since it is the testimony of a trustworthy eye-witness. "On approaching an island, or point, occupied by a numerous herd," he observes, "one first hears their long, plaintive howlings, as if in distress; but when near them, the sounds become more varied and deafening. The old males roar so loudly as to drown the noise of the heaviest surf among the rocks and caverns, and the younger of both sexes, together with the 'clapmatches,' croak hoarsely, or send forth sounds like the bleating of sheep or the barking of dogs; in fact, their tumultuous utterances are beyond description. A rookery of matured animals presents a ferocious and defiant appearance; but usually at the approach of man they become alarmed, and, if not opposed in their escape, roll, tumble, and sometimes make fearful leaps from high precipitous rocks to hasten their flight. Like all the others of the Seal tribe, they are gregarious, and gather in the largest numbers during the 'pupping season,' which varies in different latitudes. On the California coast it is from May to August, inclusive, and upon the shores of Alaska it is said to be from June to October, dur-ing which period the females bring forth their young, nurse them, associate with the valiant males, and both unite in the care of the little ones, keeping a wary guard, and teaching them, by their own parental actions, how to move over the broken, slimy, rock-bound shore, or upon the sandy, pebbly beaches, and to dive and gambol amid the surf and rolling ground-swells. At first the pups manifest great aversion to the water, but soon, instinctively, become active and playful in the element; so by the time the season is over, the juvenile crea-tures disappear with the greater portion of the old ones, only a few of the vast herd remaining at the favorite resorts through-out the year. During the pupping season, both males and fe-males, so far as we could ascertain, take but little if any food, particularly the males, though the females have been observed

---

* That Captain Scammon confounded the two species of Northern Sea Lions is evident not only from his published writings, but from his having transmitted to the National Museum specimens of *Zalophus* from Santa Bar-bara Island, labelled by him "*Eumetopias stelleri*."

to leave their charges and go off, apparently in search of subsistence, but they do not venture far from their young ones.
That the Sea Lion can go without food for a long time is unquestionable. One of the superintendents of Woodward's Gardens informed me that in numerous instances they had received
Sea Lions into the aquarium which did not eat a morsel of nourishment during a whole month, and appeared to suffer but little
inconvenience from their long fast.

"As the time approaches for their annual assemblage, those
returning or coming from abroad are seen near the shores, appearing wild and shy. Soon after, however, the females gather
upon the beaches, cliffs, or rocks, when the battles among the
old males begin for the supreme control of the harems; these
struggles often lasting for days, the fight being kept up until
one or both become exhausted, but is renewed again when sufficiently recuperated for another attack; and, really, the attitudes assumed and the passes made at each other, equal the
amplification of a professional fencer. The combat lasts until
both become disabled or one is driven from the ground, or perhaps both become so reduced that a third party, fresh from his
winter migration, drives them from the coveted charge. The
vanquished animals then slink off to some retired spot as if dis-
graced. Nevertheless, at times, two or more will have charge
of the same rookery; but in such instances frequent defiant
growlings and petty battles occur. So far as we have observed
upon the Sea Lions of the California coast, there is but little attachment manifested between the sexes; indeed, much of the
Turkish nature is apparent, but the females show some affection for their offspring, yet, if alarmed when upon the land, they
will instantly desert them and take to the water. The young
cubs, on the other hand, are the most fractious and savage little
creatures imaginable, especially if awakened from their nearly
continuous sleeping; and frequently, when a mother reclines to
nurse her single whelp, a swarm of others will perhaps contend
for the same favor.

"To give a more detailed and extended account of the Sea
Lions we will relate a brief sketch of a sealing season on
Santa Barbara Island. It was near the end of May, 1852,
when we arrived, and soon after the rookeries of 'clapmatches,'
which were scattered around the island, began to augment,
and large numbers of huge males made their appearance,
belching forth sharp, ugly howls, and leaping out of or dart-

ing through the water with surprising velocity, frequently
diving outside the rollers, the next moment emerging from
the crest of the foaming breakers, and waddling up the
beach with head erect, or, with seeming effort, climbing some
kelp-fringed rock, to doze in the scorching sunbeams, while
others would lie sleeping or playing among the beds of sea-
weed, with their heads and outstretched limbs above the surface. But a few days elapsed before a general contention with
the adult males began for the mastery of the different rookeries, and the victims of the bloody encounter were to be seen on
all sides of the island, with torn lips or mutilated limbs and
gashed sides, while now and then an unfortunate creature
would be met with minus an eye or with the orb forced from its
socket, and, together with other wounds, presenting a ghastly
appearance. As the time for 'hauling-up' drew near, the island
became one mass of animation; every beach, rock, and cliff,
where a seal could find foothold, became its resting-place, while
a countless herd of old males capped the summit, and the
united clamorings of the vast assemblage could be heard, on a
calm day, for miles at sea. The south side of the island is high
and precipitous, with a projecting ledge hardly perceptible
from the beach below, upon which one immense Sea Lion managed to climb, and there remained for several weeks—until the
season was over. How he ascended, or in what manner he retired to the water, was a mystery to our numerous ship's crew,
as he came and went in the night; for 'Old Gray,' as named
by the sailors, was closely watched in his elevated position during the time the men were engaged at their work on shore.*

"None but the adult males were captured, which was usually
done by shooting them in the ear or near it; for a ball in any
other part of the body had no more effect than it would in a
Grizzly Bear. Occasionally, however, they are taken with the
club and lance, only shooting a few of the masters of the herd.

"* Relative to the Sea Lions leaping from giddy heights, an incident occurred at Santa Barbara Island, the last of the season of 1852, which we
will here mention. A rookery of about twenty individuals was collected
on the brink of a precipitous cliff, at a height at least of sixty feet above
the rocks which shelved from the beach below; and our party were sure
in their own minds, that, by surprising the animals, we could drive them
over the cliff. This was easily accomplished; but, to our chagrin, when we
arrived at the point below, where we expected to find the huge beasts helplessly mutilated, or killed outright, the last animal of the whole rookery
was seen plunging into the sea."

This is easily accomplished with an experienced crew, if there is sufficient ground back from the beach for the animals to retreat. During our stay, an instance occurred, which not only displayed the sagacity of the animals, but also their yielding disposition, when hard pressed in certain situations, as if naturally designed to be slain in numbers equal to the demands of their human pursuers. On the south of Santa Barbara Island was a plateau, elevated less than a hundred feet above the sea, stretching to the brink of a cliff that overhung the shore, and a narrow gorge leading up from the beach, through which the animals crowded to their favorite resting-place. As the sun dipped behind the hills, fifty to a hundred males would congregate upon the spot, and there remain until the boats were lowered in the morning, when immediately the whole herd would quietly slip off into the sea and gambol about during the day, returning as they saw the boats again leave the island for the ship. Several unsuccessful attempts had been made to take them; but at last a fresh breeze commenced blowing directly from the shore, and prevented their scenting the hunters, who landed some distance from the rookery, then cautiously advanced, and suddenly yelling, and flourishing muskets, clubs, and lances, rushed up within a few yards of them, while the pleading creatures, with lolling tongues and glaring eyes, were quite overcome with dismay, and remained nearly motionless. At last, two overgrown males broke through the line formed by the men, but they paid the penalty with their lives before reaching the water. A few moments passed, when all hands moved slowly toward the rookery, which as slowly retreated. This maneuvre is termed 'turning them,' and, when once accomplished, the disheartened creatures appear to abandon all hope of escape, and resign themselves to their fate. The herd at this time numbered seventy-five, which were soon dispatched, by shooting the largest ones, and clubbing and lancing the others, save one young Sea Lion, which was spared to see whether he would make any resistance by being driven over the hills beyond. The poor creature only moved along through the prickly pears that covered the ground, when compelled by his cruel pursuers; and, at last, with an imploring look and writhing in pain, it held out its fin-like arms, which were pierced with thorns, in such a manner as to touch the sympathy of the barbarous sealers, who instantly put the sufferer out of its misery by a stroke of a heavy club. As soon as the animal is

killed, the longest spires of its whiskers are pulled out, then it
is skinned, and its coating of fat cut in sections from its body
and transported to the vessel, where, after being 'minced,' the
oil is extracted by boiling. The testes are taken out, and, with
the selected spires of whiskers, find a market in China—the
former being used medicinally, and the latter for personal orna-
ments.

"At the close of the season—which lasts about three months,
on the California coast—a large majority of the great herds,
both males and females, return to the sea, and roam in all
directions in quest of food, as but few of them could find sustenance about the waters contiguous to the islands, or points
on the mainland, which are their annual resorting-places. They
live upon fish, mollusks, crustaceans, and sea-fowls; always
with the addition of a few pebbles or smooth stones, some of
which are a pound in weight.* Their principal feathery food,
however, is the penguin in the southern hemisphere, and the
gulls in the northern; while the manner in which they decoy
and catch the *gaviota* of the Mexican and California coasts displays no little degree of cunning. When in pursuit the animal
dives deeply under water and swims some distance from where
it disappeared; then, rising cautiously, it exposes the tip of its
nose above the surface, at the same time giving it a rotary motion, like that of a water-bug at play. The unwary bird on the

* "The enormous quantity of food which would be required to maintain
the herd of many thousands, which, in former years, annually assembled
at the small island of Santa Barbara, would seem incredible, if they daily
obtained the allowance given to a male and female Sea Lion on exhibition at Woodward's Gardens, San Francisco, California, where the keeper
informed me that he fed them regularly, every day, forty pounds of fresh
fish. . . . ."

[That the destruction of fish by the Sea Lions on the coast of California
is very great is indicated by the following item, which recently went the
rounds of the newspapers: "In a recent meeting at San Francisco of the
Senate Committee on Fisheries, the State Fish Commissioners, and a committee representing the fishermen of the coast, the question as to the destructive performances of the sea-lions in the harbor was actively discussed.
One of the fishermen's representatives said that it was estimated that there
were 25,000 sea-lions within a radius of a few miles, consuming from ten to
forty pounds each of fish per day; the sea-lions were protected while the
fishermen were harassed by the game laws. Another witness declared that
salmon captured in the Sacramento river often bore the marks of injury
from sea-lions, having barely escaped with life; but it was supposed that
the salmon less frequently fell victims to the amphibian than did other
fishes that cannot swim as fast."—*Country*, January 26, 1878.]

wing, seeing the object near by, alights to catch it, while the
Sea Lion at the same moment settles beneath the waves, and
at one bound, with extended jaws, seizes its screaming prey,
and instantly devours it.[*]

"A few years ago great numbers of Sea Lions were taken
along the coast of Upper and Lower California, and thousands
of barrels of oil obtained. The number of Seals slain exclusively for their oil would appear fabulous, when we realize the
fact that it requires on an average, throughout the season, the
blubber of three or four Sea Lions to produce a barrel of oil.
Their thick, coarse-grained skins were not considered worth
preparing for market, in a country where manual labor was so
highly valued. At the present time, however, they are valued
for glue-stock, and the seal-hunters now realize more comparative profit from the hides than from the oil. But while the
civilized sealers, plying their vocation along the seaboard of
California and Mexico, destroy the *Lobo marino*, for the product
of its oil, skin, testes, and whiskers, the simple Aleutians of
the Alaska region derive from these animals many of their indispensable articles of domestic use. . . . ."†

To Captain Scammon's graphic account I add a few lines from
the pen of a non-scientific writer respecting the Sea Lions of
the Farallones: "The Sea Lions, which congregate by thousands upon the cliffs, and bark and howl and shriek and roar
in the caves and upon the steep sunny slopes, are but little disturbed, and one can easily approach them within twenty or
thirty yards. It is an extraordinarily interesting sight to see
these marine monsters, many of them bigger than an ox, at
play in the surf, and to watch the superb skill with which they
know how to control their own motions when a huge wave
seizes them, and seems likely to dash them to pieces against
the rocks. They love to lie in the sun upon the bare and warm
rocks; and here they sleep, crowded together, and lying upon
each other in inextricable confusion. The bigger the animal
the greater his ambition appears to be to climb to the highest
summit; and when a huge, slimy beast has with infinite squirm-
ing attained a solitary peak, he does not tire of raising his

---

[* This account appeared originally in Captain Scammon's account of the
"Islands off the West coast of Lower California," in J. Ross Browne's "Resources of the Pacific Slope," second part, p. 130 (1869), and has been quoted
by Mr. Gurney in the "Zoölogist" for 1871, p. 2762.]

†Marine Mammalia, pp. 130–135.

sharp-pointed, maggot-like head, and complacently looking
about him. They are a rough set of brutes,—rank bullies, I
should say; for I have watched them repeatedly, as a big fellow shouldered his way among his fellows, reared his huge front
to intimidate some lesser seal which had secured a favorite spot,
and first with howls, and if this did not suffice, with teeth and
main force, expelled the weaker from his lodgment. The smaller
Sea Lions, at least those which have left their mothers, appear
to have no rights which any one is bound to respect. They get
out of the way with abject promptness, which proves that they
live in terror of the stronger members of the community; but
they do not give up their places without harsh complaint and
piteous groans."*

Dr. John A. Veatch, in his account of the Cerros or Cedros
Island, situated off the coast of Lower California (between the
parallels of 28° and 29°), doubtless refers to this species under
the name of *Otaria jubata.* He says: " He [the Sea Lion] is more
prolific [than the Sea Elephant], and there are fewer induce-
ments for his destruction. He is, however, by no means beyond
danger from the oil-man. At certain seasons, when the Lion
chances to have a little fat on his bones, he is slaughtered
most mercilessly. Fortunately for him his skin is nearly worthless, or there would be double inducement for his destruction.
Toward the north end of the island there is a great breeding-
place for these animals. It is a small bay, two or three miles
in length, and perhaps three-fourths of a mile in breadth, surrounded on the land by a perpendicular cliff, and on the ocean-
side by a belt of kelp. It is thus protected both from winds
and waves. It is bordered with a sandy beach, some 200 paces
in breadth. The access by land is exceedingly difficult, and can
only be gained by careful clambering down where breaks and
fissures offer hand and foot-hold. This sequestered and quiet
place is the comfortable and appropriate resort of the lionesses
to bring forth and rear their young. It is, indeed, a great seal-
nursery. My first visit to this interesting locality was in the
latter part of the month of July [1859]. Seals, in countless
numbers, literally covered the beach. They were of every con-
ceivable size, from the young ones, seemingly a few days old,
up to the full-grown animal. So unconscious of danger were
the little ones, that they scarce made an effort to get out of the

* Charles Nordhoff, " The Farallon Islands," in Harper's Magazine, vol.
xlviii, p. 620, April, 1874.

way. I picked up many of them in my hands; after a brief struggle, the little captive would yield, and seemed to fear no further harm. Hundreds slept so soundly that I rolled them over before they could be induced to open their great baby eyes. While thousands slept and basked on the shore, an equal number floated lazily in the water, or dipped and dived about in sport.

"The mother-seals were more timid than their young, but seemed less alarmed than surprised at my approach. The look of startled inquiry was so human and feminine—nay, lady-like, that I felt like an intruder on the privacy of the nursery.

"I could not discover any individual claim set up by the mother for any particular little lion, but, like a great socialistic community, maternal love seemed to be joint-stock property, and each infant communist had a mother in every adult female.

"The *fathers* of the great family appeared, in point of numbers, to be largely in the minority; counting, as I judged, not the hundredth part of the adult animals. A few bearded, growling old fellows tumbled about in the water, yelling and howling in a most threatening manner at me, and approaching within a few feet of where I stood. A pebble tossed at one of them, however, would be answered by a plunge beneath the surface and reappearance at a safer distance.

"I witnessed an unexpected act of tenderness on the part of one of the hugest and most boisterous old threateners for a little one that seemed to claim him for papa. He was blowing and screaming at me fearfully, when a young one at my feet hustled into the water, glided off to the old one, and, childlike, placed its mouth up to his. The old savage ceased his noise, returning the caress, and seemed, for several seconds, to forget his wrath at the unwelcome intruder. This show of affection saved his life. I was, at the moment, rifle in hand, waiting a chance to dislocate his neck. I wanted the skull of an otaria for my collection, and his huge size suggested him as an appropriate victim. I at once lost all murderous desire, and left him to the further enjoyment of paternal felicity.

"The noise and uproar of the locality can scarcely be imagined. A hundred thousand seals grunting, coughing, and shrieking at the same instant, made a phocine pandemonium I shall never forget. I will observe here that the male was four times as large as the female."*

* J. Ross Browne's Resources of the Pacific Slope. Sketch of the Settlement and Exploration of Lower California, 1869, p. 150.

Mr. Elliott, in referring to the differences between the Californian and Alaskan Sea Lions, calls attention to the dissimilarity of their voices. The Northern Sea Lion, he says, "never barks or howls like the animal at the Farallones or Santa Barbara. Young and old, both sexes, from one year and upward, have *only* a deep *bass growl*, and *prolonged, steady roar;* while at San Francisco Sea Lions break out incessantly with a 'honking' bark or howl, and *never roar."* *

The Californian Sea Lion is now a somewhat well-known animal with the public, various individuals having been at different times on exhibition at the Central Park Menagerie in New York City, and at the Zoölogical Gardens at Philadelphia and Cincinnati, as well as Woodward's Gardens in San Francisco. They have also formed part of the exhibition of different travelling shows, especially that of P. T. Barnum. They have also been carried to Europe, where examples have lived for several years at the Zoölogical Gardens of London, Paris, and elsewhere. Their peculiar "honking" bark, referred to by Mr. Elliott, is hence not unfamiliar to many who have never met with the animal in a state of nature. Their various attitudes and mode of life on the Farallones have also been made familiar to many by the extensive sale of stereoscopic views of the animals and their surroundings. The Sea Lions that have been exhibited in this country all, or nearly all, belong to the present species, although often wrongly labelled "*Eumetopias stelleri.*" The true *E. stelleri* has, however, at least in one instance, been exhibited in Eastern cities.

The Californian Sea Lion seems generally not to suffer greatly in health by confinement, if properly cared for, although deaths from tuberculosis have repeatedly occurred. They are always objects of great attraction to visitors, and various accounts of their habits in confinement have been published. Mr. Henry Lee, in referring to two that had been for a short time at the Brighton (England) Aquarium, says: "They have grown so much, and are so plump and sleek, that a visitor seeing them now [February, 1876] for the first time since the day of their arrival, would hardly recognize in them the pair of lean, ill-conditioned animals, with ribs as visible as those of an old cab-horse, which waddled out of their travelling crates to follow Lecomte and a herring on the 13th of October last. What their rations had

* Condition of Affairs in Alaska, p. 158.

been since they left their home I am unable to say, but I am inclined to suspect it was often like a midshipman's half-pay, 'nothing a-day,' and as they had no means of 'finding themselves,' they probably had many a 'banyan day' whilst on their way to Europe. Fortunately their capability of fasting is very great. Mr. Woodward, the proprietor of 'Woodward's Gardens,' San Francisco, with whom I have recently had the pleasure of becoming acquainted, tells me that in numerous instances he has received sea-lions which have not eaten a morsel during a whole month, and appeared to suffer little inconvenience in consequence. Fearing, however, that it would tell injuriously upon the health of one which persisted longer than usual in total abstinence, he had the beast lassoed and held fast whilst food was forced into its stomach down an india-rubber hose-pipe. As the males are believed to take no sustenance for three or four months together during the breeding-season, this was probably unnecessary. We had no trouble of this kind with ours. They ate with appetite immediately, and although when they arrived they looked like the omnibus horses in *Punch*, which, as their driver informed an outside passenger, had been fed on butter-tubs, and showed the hoops, '*nous avons changé tout cela.*' Nearly half-a-hundredweight of fish a day for the last sixteen weeks has been gradually converted into sea-lion flesh and blubber, and the result is apparent in the greatly increased size and weight of these valuable animals. Herrings and sprats are the food which they like best, and which we prefer to give them, both because they are very nutritious, and because, as they are netted fish, there is no fear of their containing hooks. When herrings cannot be obtained, whiting are generally substituted; but these have to be opened, one by one, and carefully searched for fish-hooks which may have been left in them; for it may be remembered that the first Otaria possessed by the Zoölogical Society died in great pain, in 1867, from having swallowed a hook which had escaped discovery among its food. As these animals do not masticate, they are, of course, unable to detect and reject from the mouth any foreign substances concealed within the body of a fish. . . . . When one of the sea-lions takes a fish from his keeper, the head is no sooner inside its mouth than the tail disappears after it, before one can say the proverbial 'Jack Robinson'. There is not a moment's pause for deglutition; one after another the fish, whole and unbitten, disappear from sight as instantaneously as so many letters slipped into a pillar-box. It is therefore easy to administer physic whenever medical treatment may be thought desirable; and the necessity for it has occasionally occurred. Soon after her arrival the female exhibited symptoms of distemper, a disease to which these animals, like dogs, are liable. . . . ."*

In captivity these animals appear to become strongly attached to each other, so much so that in case one of a pair dies, the other is very apt to die soon after, of grief. They have also been known to propagate in confinement, an instance of which is related by Mr. F. J. Thompson in his interesting paper recently published in "Forest and Stream," on "The Habits and Breeding of the Sea Lions in Captivity,"† based on observations made at the Zoölogical Garden in Cincinnati, Ohio. As his paper contains, besides a general account of the habits of these interesting animals, several novel points, I give it place in full:

"In the early part of June, 1877, I went, sent by the Zoölogical Society of Cincinnati, to Chicago to receive some black sea-lions (*Zalophus gillespiei*) which had arrived there from the southern coast of California. On my arrival I found that the female had calved on the previous night, therefore thought it best to lie over for a day in order that the young might acquire a little extra strength to bear the fatigue of the railway journey to Cincinnati. They all arrived in the garden in fine condition, but had to be kept in their shipping crates for the first few days, until an old beaver pond could be arranged as temporary quarters for them, while the large basin intended for their permanent home could be built. During this time, on account of a heavy freshet in the Ohio River, the water in the pond became quite muddy, which affected them so much that they were unable to retain their food, invariably vomiting up their fish some one or two hours after feeding. By giving small doses of Rochelle salts for a few days, all recovered, but the calf died from a violent attack of *cholera infantum*, caused no doubt by its mother's milk being affected by the muddy water.

"A short time before the calf was taken sick my attention was attracted to the peculiar appearance of the mother on emerging from the water after taking her customary bath. She was completely covered with a whitish oleaginous substance, about the consistency of semi-fluid lard, which seemed

---

* Land and Water, Feb. 5, 1876, p. 104.
† Forest and Stream (newspaper), vol. xii, p. 66, Feb. 23, 1879.

to ooze out all over her. As soon as she got into the crate with the young one, she commenced rolling, so that in a short time the young one and the inside of the crate were completely covered with it. The calf seemed to enjoy it hugely, and rolled about until his coat glistened as if he had just left the hands of a first-class tonsorial artist. It instantly struck me that his mother had been preparing him for the water, and I immediately tested the matter by taking him out and placing him on the edge of the pond, when in a few moments he began to paddle about in the water, something he had never before attempted although he had been almost daily placed in the same position.

"As soon as the large basin was completed, and they were transferred to it, I had a fine opportunity of observing the tyrannical attentions of the male toward the female during rutting season. He constantly swam back and forth along the partition, which separated him from another male, frequently endeavoring to get through, splintering and tearing the rails with his powerful canine teeth. If the female attempted to approach the division she was immediately forcibly driven back, when he would redouble his efforts to get through, barking and roaring as if beside himself with rage. This would be kept up until late at night, when the female was allowed to go into the house situated in the centre of the basin, when he would follow and place himself immediately in the doorway so as to prevent her egress. He never seemed to sleep soundly, as he invariably kept up a series of grunts and muffled roars, as if he were fighting his battles over again in his sleep. I would frequently annoy him by stealing up softly and then suddenly scraping the gravel with my foot, when he would instantly start up, plunge into the basin, swim rapidly back and forth, barking with all his might, until he was satisfied there was no interloper about, when he would sullenly return to his post and gradually drop off again into his troubled sleep. Frequently at night the two males would climb to the roof of the house, and in their efforts to get at each other through the partition, would raise such a din that persons living at quite a distance from the garden would frequently ask me the cause of the uproar.

"At the end of some two months there was a change, when the female commenced playing and coquetting with the male, frequently pinching him so sharply as to make him snarl with pain, and if he seemed to be much out of humor she would soothe him by swimming up and giving him a good old-fashioned conjugal kiss. Finally they quieted down to the humdrum of regular wedded life, and early in October I noticed that the female was suffering from a violent catarrh, which gradually disappeared, followed by a dry cough, particularly at night. It was in March when I first thought she showed signs of pregnancy, and in May, from her appearance when out of the water, I became convinced of it. On June 25 the young one was born, making the period of gestation as nearly as I could judge about ten months, and it was some days before the mother would allow me to handle it, and when I did succeed in so doing it was always at the risk of getting a nip, as he was certainly the most ill-tempered, snarling little brute with which a dry nurse could be vexed. I soon found out that there was but one way of handling him with impunity, and that was by suddenly catching him just back of the flippers and quickly lifting him clear of the floor, when he would snap and struggle for a few moments and then quietly give up. I frequently took him out of the house for the purpose of showing him to friends, and for the first three or four weeks he never made the slightest attempt to get into the water, although I invariably placed him on the lip outside of the door and loosed my hold in order that he could be fully seen. During this period the mother was let out for a bath twice daily, and after she had played about as long as she wished she would swim up to the closed door, rear up on the sill and bellow until she was allowed to get in to her calf. Invariably in the morning, so soon as I would start across the bridge in order to turn her out, the male would swim up to the door and await her appearance, always exacting his morning kiss before he would allow her to plunge into the water. After playing with her for a few minutes he would commence sentry duty, back and forth along the partition, occasionally making fierce rushes if the other approached too near to it.

"In the meanwhile, as the young one never showed the slightest inclination to go into the water, in spite of frequent opportunities to do so, I began to watch for a second appearance of the oleaginous matter. During the fifth week after birth, on going into the house one morning, I found marks of grease in every direction, and the youngster shone as if he had just emerged from an oil tank. Taking a bucket, I filled it with water, placed it in his way, and he immediately stuck his head to the bottom of it. Fearing an accident, as the water in the basin only reached within about a foot of the top of the lip surrounding the house, I had the carpenter construct a small, shallow, wooden tank inside the larger one, with a sloping platform leading into it. So soon as the door was opened connecting with it he followed his mother, and in a short time was having high jinks swimming and diving to his full bent. When he tired he would quietly rest in the water with his head lying across his mother's neck, or he would scramble up on the platform, stretch himself, have a short nap, and then commence his play again. So soon as I thought he had gained sufficient strength the small tank was removed, and he was allowed the run of the larger one, when his wonderful swimming powers came into full play. I have frequently seen him dash off with such velocity that the water would part and fly from each side of his neck with a fairly hissing sound. Again he would dive, and then suddenly make a succession of salmon-like leaps with such rapidity that I could easily imagine with what little difficulty he would be able to capture the swiftest of fish. One of the favorite ways of amusing himself was by taking a chip— several of which were always kept in the basin—out on the lip, lying on his back, and playing with it with front flippers and mouth, almost precisely as an infant would act with a common rattle. At first he was rather shy of the old male, but gradually took the greatest delight in swimming about with, and trying to induce him to join in a game of romps; but the old fellow was proof against all his wiles, and always good-naturedly endeavored to get rid of him.

"I noticed that the female's cough disappeared immediately after the birth of the young one; but about the middle of August both her appetite and actions became variable, some days feeding and seeming lively as usual, on others she would either take but little or entirely refuse her food. She gradually grew worse until September 8, when, on going to the basin in the morning, I missed her, and found the male busily engaged in diving just at one particular spot. He finally succeeded in bringing the body to the surface, and when the keepers attempted to remove it he repeatedly charged, and it was only by great care and watchfulness that they avoided being bitten. On dissection it was found that *tuberculosis,* that scourge of all zoölogical collections, was the cause of her death.

"The young one did not seem to notice the loss of his mother until about twenty-four hours after her death, when he commenced to sulk, and obstinately refused to eat, in spite of every effort and strategem to induce him to do so. He gradually wasted away, and finally died of starvation on October 16, having viciously attempted to bite me a few hours before his death.

"The old male grieved so over the loss of his mate that for some time I was afraid we would lose him also, and at the end of about six weeks he became so thin that I thought it best to remove him to a small tank in-doors. Since, he has been improving slowly up to within ten days, since when he shows a marked improvement."

The Otaries, wherever occurring, appear to closely agree in their habits, especially during the breeding season. As an interesting supplement to the history of the two Northern species of Sea Lions already given I transcribe the following concise account of the great Southern Sea Lion (*Otaria jubata*), based on recent observations made at the Falkland Islands, without, however, endorsing the author's "ballasting" theory:

"The Sea Lion attains its full growth at nine years, and annually comes back to the place it was born to breed and shed its hair. The former operation occurs between the 25th of December and the 15th of January, the latter in April and May. The Lions commence to arrive at their 'rookery' in November to wait for the females, who do not haul up until within two or three days of pupping. They are fatter at this time than at any other, and have to take in a quantity of ballast to keep them down, without which they could not dive to catch fish. I have opened them at this time, and found, in a pouch they have inside, upwards of twenty-five pounds of stones, some as large as a goose-egg. As they get thin they have the power of throwing these stones up, retaining only a sufficient quantity to keep them from coming up too freely to the surface.

"They are very savage in the breeding-season, and are continually fighting, biting large pieces out of each other's hide, and sometimes killing the females. At this time they become an easy prey to man, as they will stand and be killed without trying to get away.

"The Lioness has her first pup at three years of age, never more than one at a time, and comes up to have intercourse with

the Lion at two, and as soon as the pup is born. . . . . They
suckle their young five months before they are taken to the
sea, by which time the pup has shed its first hair. Before the
mother takes her pup to fish she has to ballast it, and I have
seen a Lioness trying for hours to make her pup swallow small
stones at the water's edge.

"The female keeps her pup with her until two or three weeks
before the next breeding-season, when she drives it from her.
About this time the yearlings will be found some few miles
from the old rookery. . . .

"The Lions stay as long as two months on shore, during the
breeding-season, without going into the water. During that
time their fat gives them sufficient nourishment. After the
season is over some of them are so thin and weak that they
are but just able to crawl into the water. I have killed them in
this state, and not one particle of stone have I found in them."[*]

Genus CALLORHINUS, Gray.

Callorhinus, GRAY, Proc. Zoöl. Soc. Lond., 1859, 359. Type "*Arctocephalus
     ursinus*, Gray," = *Phoca ursina*, Linné.
Arctocephalus, GILL, Proc. Essex Institute, v, 1866, 7, 11. Type "*Phoca
     ursina*, Linnæus." Not *Arctocephalus*, F. Cuvier, 1824.

Molars $\frac{6-6}{5-5}$, small. Facial portion of skull short, broad, convex, and but slightly depressed; nasals short, rapidly narrowing posteriorly. Palatal surface short, narrowed behind, with
the hinder border rather deeply concave. Toe-flaps very long,—
nearly as long as the rest of the foot.

Callorhinus, in coloration, character of the pelage, size, general form, and dental formula, is rather closely allied to *Arctocephalus*, from which, however, it is readily distinguished by the
form of the facial portion of the skull, which in *Arctocephalus* is
narrower, longer, and much less convex, with much longer nasals. From the other genera of the Otaries it is distinguishable not only by coloration and the character of the pelage, but
by its weaker dentition, and the strongly marked cranial differences, which are too numerous and obvious to require detailed
enumeration. It is the only North American genus which has
the upper molars 6—6.

Very young skulls and skulls of females of the different species of Otaries differ from each other very little in general form,
and in some cases are not readily distinguishable, especially in

---

[*] Letter from Captain Henry Pain, of the S. S. "Scanderia" to Mr. F. Cole-
man of the Falkland Islands Company, Proc. Zoöl. Soc. Lond., 1872, pp.
681, 682.

figures. In all, the interorbital region is relatively broad and short, and becomes relatively more and more narrowed and lengthened in the adult. In young specimens, and always in the females, except in *Arctocephalus*, the mastoid processes remain almost wholly undeveloped. Among the North American species, *Eumetopias* is easily recognized at any age by its dental formula and large size. The young and females of *Callorhinus* and *Zalophus* are easily separated, aside from differences of dentition, by the shape of the muzzle, but especially by the ascending limb of the intermaxillary. In *Zalophus* it gradually narrows posteriorly and ends in a slender point near the middle of the nasals. In *Callorhinus* it widens posteriorly and ends abruptly quite near the anterior border of the nasals.

The distribution of the genus is almost exactly the same as that of *Eumetopias*,—namely, the shores of the North Pacific,— and, like that genus, is represented by only a single species, the well-known Alaskan " Fur Seal."

CALLORHINUS URSINUS, (*Linné*) *Gray*.

Northern Fur Seal; Sea-Bear.

Ursus marinus, STELLER, Nov. Comm. Acad. Petrop., ii, 1751, 331, pl. xv.

Phoca ursina, LINNÉ, Syst. Nat., i, 1758, 37 (from Steller).—SCHREBER, Säugth., iii, 1758, 289.—SHAW, Gen. Zoöl., i, 1800, 265, pl. lxii.—GODMAN, Amer. Nat. Hist., i, 1826, 346 (in part).—FISCHER, Synop. Mam., 1829, 231.—PALLAS, Zoog. Rosso-Asiat., i, 1831, 102.

Otaria ursina, PÉRON, Voy. Terr. Austr., ii, 1816, 39, 41.—DESMAREST, Nouv. Dict. Hist. Nat., xxv, 1817, 595 ; Mam., i, 1820, 249.—HARLAN, Faun. Amer., 1825, 112.—GRAY, Griffith's Cuvier's An. Kingd., v, 1829, 182.—HAMILTON, Marine Amphib., 1839, 253, pl. xxi.—NILSSON, Arch. f. Naturg., 1841, 331 (in part only).—MÜLLER, Arch. f. Naturg., 1841, 333.—WAGNER, Schreber's Säugt., vii, 1846, 65 (in part only) ; Arch. f. Naturg., 1849, 39.—VON SCHRENCK, Amur-Lande, i, 1859, 189.

Phoca (Otaria) ursina, RICHARDSON, Zoöl. Beechey's Voy., 1839, 6.

Otaria (Callorhinus) ursinus, PETERS, Monatsb. Akad. Berlin, 1866, 373, 672.

Arctocephalus ursinus, LESSON, Man. de Mam., 1827, 203.—GRAY, Cat. Seals, 1850, 41 (not of F. Cuvier, or only in part) ; Proc. Zoöl. Soc. Lond., 1859, 103, 107, pl. lxxiii (skull).—GILL, Proc. Essex Institute, v, 1866, 13.—SCOTT, Mam. Recent and Extinct, 1873, 8.—CLARK, Proc. Zoöl. Soc. Lond., 1878, 271, pl. xx (colored figures of male, female, and young).

Callorhinus ursinus, GRAY, Proc. Zoöl. Soc. Lond., 1859, 359, pl. lviii (skull) ; Cat. Seals and Whales, 1866, 44, fig. 16 (skull) ; Ann. and Mag. Nat. Hist., 3d ser., xviii, 1866, 234 ; Suppl. Cat. Seals and Whales, 1871, 15 ; Hand-List Seals, etc., 1874, 32, pl. xix (skull).—ALLEN, Bull. Mus. Comp. Zoöl., ii, 73, pll. ii, iii (skull, etc.).—SCAMMON, Marine Mamm., 1874, 141, pl. xxi, figg. 1, 2, and figg. 1-6 in text (animal).— ELLIOTT, Cond. Affairs in Alaska, 1874, 123.

Phoca nigra, PALLAS, Zoog. Rosso-Asiat., i, 1831, 107 (young).
Otaria krachenninikowi, LESSON, Dict. Class. d'Hist. Nat., xiii, 1828, 420 (= *Ursus marinus,* Steller.
? *Otaria fabricii,* LESSON, Dict. Class. d'Hist. Nat., xiii, 1828, 419 (= *Phoca ursina,* Fabricius, F. Grœnl. 6—"Greenland").
Arctocephalus monteriensis, GRAY, Proc. Zoöl. Soc. Lond., 1857, 360 (in **part** only).
Arctocephalus californianus, GRAY, Cat. Seals and Whales, 1866, 51 (in **part** only = *A. monteriensis*).
Meerbär, STELLER, Beschreib. von sonderbaren Meerthieren, 1753, 107.
Le Chat marin, KRASCHENNINIKOW, Hist. Kamtsch., i, 1764, 316.
Ours marin, BUFFON, Hist. Nat., Suppl., vi, 1782, 336, pl. xlvii (in part).
Ursine Seal, PENNANT, Synop. Quad., 1771, 344 (based mainly on Steller); Hist. Quad., ii, 1793, 281 (in part).
Fur Seal, SCAMMON, Overland Monthly, iii, 1869, 393 (habits).

DESCRIPTION.*

COLOR.—(*Male.*)—The general color above, except over the shoulders, is nearly black, varying in different individuals of equal age from nearly pure black to rufo-grayish black. Over the shoulders the color is quite gray. The sides of the nose and the lips are brownish, as is a considerable space behind the angle of the mouth, and a small spot behind the ear. The neck in front is more or less gray. The breast and the axillæ are brownish-orange. The limbs are reddish-brown, especially near their junction with the body, as is also the abdomen. The hairs individually vary considerably in color, some being entirely black nearly to their base, and others entirely light yellowish-brown; others are dark in the middle and lighter at each end. The naked skin of the hind limbs, the nose, and the anal region is black.

(*Female.*)—The general color of the female is much lighter than that of the male. Above it is nearly uniformly gray, varying to darker or lighter in different individuals and with age. The color about the mouth is brownish, varying to rufous, of which color are the axillæ, the breast, and the abdomen. The sides are brownish-gray. At the base all the hairs are usually brownish, like the under-fur, with a broad subterminal bar of black, and tipped for a greater or less distance with gray. The variation in different individuals in the general color results from the varying extent of the gray at the ends of the hairs.

* The technical matter here following includes that previously given in my former paper on the *Otariidæ,* with here and there slight verbal changes, and the addition at a few points of considerable new matter, especially in the tables of measurements, which are based almost wholly on an examination of new material. The remarks on individual variation might be amplified by reference to many other specimens, but this has not been thought necessary.

(*Young.*)—The general color of the upper surface of the body in the young, previous to the first moult, is uniformly glossy black. The region around the mouth is yellowish-brown. The neck in front is grayish-black. The axillæ are pale yellowish-brown; a somewhat darker shade of the same color extends posteriorly and inward toward the median line of the belly, uniting on the anterior portion of the abdomen. The greater part of the lower surface, however, is dusky brownish-gray, the rest being black, but less intensely so than the back. Specimens of equal age vary much in color, some specimens corresponding nearly with the above description, while others are much darker. On the head and sides of the neck a portion of the hairs are found on close inspection to be obscurely tipped with gray. After the first moult the pelage becomes gradually lighter, through the extension of the gray at the tips of the hairs, especially in the females, the two sexes being at first alike. Contrary to what has been asserted, the young are provided from birth with a long coat of silky under-fur, of a lighter color and sparser than the under-fur of the adults.

PELAGE.—The pelage in this species consists of an outer covering of long, flattened, moderately coarse hair, beneath which is a dense coat of long fine silky fur, which reaches on most parts of the body nearly to the ends of the hairs. The hairs are thicker toward the end than at the base, but their clavate form is most distinctly seen in the first pelage of the young. In length the hair varies greatly on the different parts of the body. It is longest on the top of the head, especially in the males, which have a well-marked crest. The hair is much longer on the anterior half of the body than on the posterior half, it being longest on the hinder part of the neck, where in the males it is very coarse. On the crown the hair has a length of 42 mm.; on the hinder part of the neck it reaches a length of 50 to 60 mm. From this point posteriorly it gradually shortens, and near the tail has a length of only 20 mm. It is still shorter on the limbs, the upper side of the digits of the hind limbs being but slightly covered, while the anterior limbs are quite naked as far as the carpus. The males have much longer hair than the females, in which it is much longer than in *Eumetopias stelleri.**

* From the accounts given by most writers it would seem that *Otaria jubata* is provided with a conspicuous mane, but in the few accurate descriptions in which the length of the longest hairs is stated, the so-called "flowing mane"—which refers only to the greater length of the hair on the neck and shoulders as compared with the other regions of the body—does not appear

The whiskers are cylindrical, long, slender, and tapering, and vary with age in length and color. In the young they are black; later they are light colored at the base, and dusky at the ends. In mature specimens they are either entirely white, or white at the base and brownish-white toward the tips.

SIZE.—Mr. Elliott has given a table showing the weight, size, and rate of growth of the Fur Seal, from the age of one week to six years, based on actual weight and measurement, with an estimate of the size and weight of specimens from eight to twenty years of age. From this table it appears that the pups when a week old have a length of from twelve to fourteen inches; and a weight of six to seven and a half pounds. At six months old the length is two feet and the weight about thirty pounds. At one year the average length of six examples was found to be thirty-eight inches, and the weight thirty-nine pounds, the males and females at this time being alike in size. The average weight of thirty males at the age of two years is given as fifty-eight pounds, and the length as forty-five inches. Thirty-two males at the age of three years were found to give an average weight of eighty-seven pounds, and an average length of fifty-two inches. Ten males at the age of four averaged one hundred and thirty-five pounds in weight, and fifty-eight inches in length. A mean of five examples five years old is: weight, two hundred pounds; length, sixty-five inches. Three males at six years gave an average weight of two hundred and eighty pounds, and a length of six feet. The estimated average weight of males from eight years and upward, when fat, is given as four hundred to five hundred pounds, and the average length as six feet three inches to six feet eight inches. Mr. Elliott further adds that the average weight of the females is from eighty to eighty-

---

to be any more truly a mane than in *Eumetopias stelleri, Callorhinus ursinus, Zalophus californianus, Arctocephalus "falklandicus"*, or in any of the *Arctocephali*. All the Sea Bears and Sea Lions, according to authors, have the hair much longer on the anterior than on the posterior half of the body; and in the Hair Seals it is not longer than in the Fur Seals. The resemblance to the mane of the Lion, with which in several species this long hair has been compared, is doubtless partly imaginary and partly due to the loose skin on the neck and shoulders being thrown into thick folds when these animals erect the head. I have not, however, seen the distinct crest formed by the long hairs on the crown of the males of *C. ursinus* mentioned as occurring in the other species, unless it is alluded to in the specific name *coronata*, given by Blainville to a South American specimen of Fur Seal, and in the name *eulophus* of Scott. It is certainly not possessed by the *E. stelleri*.

Fig. 38.—*Callorhinus ursinus.* Adult males, female, and young.

five pounds, but that they range in weight from seventy-five to
one hundred and twenty pounds, and that the five and six year
old males, on their first appearance in May and June, when fat
and fresh, may weigh a third more than in July, or at the time
those mentioned in the table were weighed, which would thus
indicate an average maximum weight of about 375 pounds for
the six-year-old males. According, however, to my own measurements of old males, from mounted and unmounted specimens,
the length is between seven and eight feet; and of a full-grown
female, about four feet. Captain Bryant states* that the males
attain mature size at about the sixth year, when their total
length is from seven to eight feet, their girth six to seven feet,
and their weight, when in full flesh, from five to seven hundred
pounds. The females, he says, are full grown at four years old,
when they measure four feet in length, two and a half in girth,
and weigh eighty to one hundred pounds. The yearlings, he
says, weigh from thirty to forty pounds. The relative size of
the adults of both sexes and the young is well shown in the
accompanying cut (see p. 317), drawn by Mr. Elliott.

The subjoined table of external measurements may be taken
as indicating the general size of the adult males and females,
and the young at thirty-five days old. In some respects the
dimensions are only approximately correct, being taken from
mounted specimens; in the main, however, they are sufficiently
accurate. A few measurements taken from the soft skin are
also given; I accidentally omitted to make a complete series of
measurements of the skins before they were mounted. In addition to the six specimens of Captain Bryant's collection, I am
indebted to Mr. W. H. Dall for measurements of a male and a
female, taken by him† from the animals immediately after they
were killed. The female (said by Mr. Dall to be six years old)
is evidently adult, but the male, being but little larger, seems
not to have been fully grown. In the last column of the table
a few measurements are given of a male specimen of the *Arctocephalus "falklandicus,"* taken by Dr. G. A. Maack from a fresh
specimen collected by him at Cabo Corrientes, Buenos Ayres.
This specimen appears also to have not been fully grown.

---

* Bull. Mus. Comp. Zoöl., vol. ii, p. 95.
† At Saint George's Island, Alaska, August, 1868.

External Measurements.

Measurement	No. 2922, Adult ♂ (Unmounted)	No. 2922, Adult ♂ (Mounted)	No. 2923, Adult ♂ (Mounted)	No. 2924, Adult ♀ (Mounted)	No. 2925, Adult ♀ (Unmounted)	No. 2925, Adult ♀ (Mounted)	No. 2926, Young ♀ (35 days old) (Mounted)	No. 2927, Young ♀ (35 days old) (Mounted)	Young ♂ (Animal)	Adult ♀ ("6 years old") (Animal)	Young ♂ Arctocephalus falklandicus (Animal)
Length of body	2,311	2,390	2,470	1,350	1,118	1,160	840	860	1,270	1,092	1,180
Length of tail		53	47	54		50	15	18	26	50	
Length from nose to end of outstretched hind limbs	472	2,740	2,860	1,790		1,750	1,015	1,020			
Length of fore limb	452	470	460	320	317	315	170	190	357	357	550
Breadth of hand	229	225	220	140		130	75	85	135	101	
Length of hind limb		515	500	400	432	390	175	200	406	357	450
Breadth of foot at tarsus		145	135	75		80	55	57	75	76	
Breadth of foot at ends of toes		250	210	150		130	110	120	127	101	
Length of toe-flaps of hind feet (average)		225	200	190		160	80	75			
Distance from end of nose to eye	96	95	85	75		70	50	52	76	62	
Distance from end of nose to ear	254	255	260	180		190	120	130	178	152	
Distance between the eyes	127	137	105	78		75	53	55			
Distance between the ears	240	360	315	225		205	150	156	152	?152	
Length of the ear	50	44	50	35		33	34	33	38	38	
Length of the longest barbule	180	180	185	175		140	65	65			
Distance between the ends of outstretched fore limbs	2,083				1,321						
Circumference of body in front of fore limbs		1,720	1,650	930		900	555	500	914	711	970
Circumference of body in front of hind limbs		680	670	410		460	260	260	357	266	
Circumference of head at the ears		770	820	490		550	315	330			

EARS.—The ears are long, narrow, and pointed, being absolutely longer than those of the *E. stelleri*, though the latter
animal is two or three times the larger.

FORE LIMBS.—The hands are very long and narrow, with a
broad cartilaginous flap extending beyond the digits, which
has a nearly even border. Both surfaces are naked the whole
length; not covered above with short hair, as in *Eumetopias* and
Otaria. The nails are rudimentary, their position being indicated by small circular horny disks, as in all the other Eared
Seals.

HIND LIMBS.—The feet are very long, nearly half their length
being formed by the cartilaginous flaps that project beyond the
ends of the toes. They widen much less from the tarsus to the
ends of the toes than these parts do in *E. stelleri*, and the length
of the toe-flaps is relatively many times greater than in the
latter species. The toes of the posterior extremities are of
nearly equal length. The outer are slightly shorter than the
three middle ones. The nails of the outer toes are rudimentary and scarcely visible; those of the middle toes are strong
and well developed.

SKULL.—In adult specimens the breadth of the skull is a little
more than half its length, the point of greatest breadth being at
the posterior end of the zygomatic arch. The muzzle or facial portion is broad and high, or greatly produced, much more so even
than in *Eumetopias*.* The postorbital processes vary from subquadrate to sub-triangular, sometimes produced posteriorly into
a latero-posteriorly diverging point, as in *Zalophus*. The postorbital cylinder is broad and moderately elongated. The postorbital constriction is well marked, giving a prominently quadrate form to the brain-case, the latero-anterior angles of which
vary somewhat in their sharpness in different specimens. The
sagittal and occipital crests are well developed in the old males,
nearly as much as in *Eumetopias*, as are also the mastoid processes. The palatine bones terminate midway between the last
molar teeth and the pterygoid hamuli; their posterior outline
is either slightly concave, or deeply and abruptly so. The palatal surface is flat, but slightly depressed posteriorly, and but
moderately so anteriorly. The zygomatic foramens are broad,
nearly triangular, and truncate posteriorly. The posterior and
anterior nares are of nearly equal size in the males, with their

* See figs. 39–41, female, rather young, about ¾ natural size. Specimen
No. 6537, National Museum.

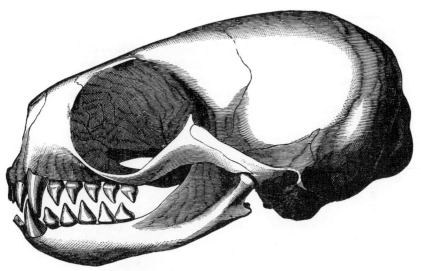

FIG. 39.—*Callorhinus ursinus.* Female.

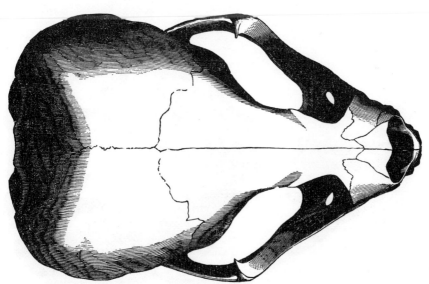

FIG. 40.—*Callorhinus ursinus.* Female.

transverse and vertical diameters equal; in the females the posterior nares are depressed, their transverse diameter being greater than the vertical. The nasal bones are much broader in front than behind.

The lower jaw is strongly developed, but relatively less massive than in *Eumetopias*. The coronoid processes are high and pointed, but much more developed in the males than in the females. The ramial tuberosities are greatly produced, especially the hinder one.

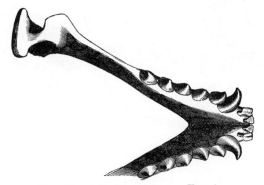

FIG. 41.—*Callorhinus ursinus.* Female.

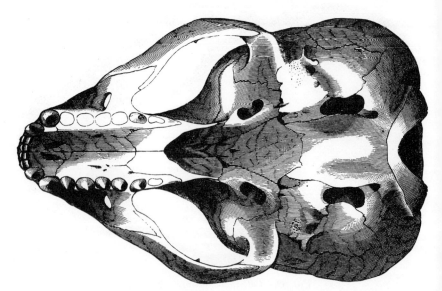

FIG. 42.—*Callorhinus ursinus.* Female.

Measurements of Twelve Skulls of CALLORHINUS URSINUS.

Catalogue number.	Locality.	Sex.	Length.	Greatest breadth at zygomatic arch.	Breadth at mastoid processes.	Distance from anterior edge of inter-maxillaries to end of hamular process of pterygoid.	Distance from anterior edge of inter-maxillaries to hinder edge of last molar.	Distance from anterior edge of inter-maxillaries to postglenoid process.	Distance from palato-maxillary suture to end of hamular process.	Length of maxillary along alveolar border.	Breadth of palatine surface at posterior end of maxillæ.	Length of nasal bones.	Breadth of nasal bones anteriorly.	Breadth of nasal bones at fronto-maxillary suture.	Breadth of skull at canines.	Least breadth of skull at orbits.	Breadth of occipital condyles.	Length of upper molar series.	Length of lower molar series.	Length of lower jaw to condyles.	Length of lower jaw to last molar.	Depth of lower jaw at coronoid process.	Height of occipital crest.	Age.
*11736	Pribilov Isds., Alaska.	♂	245	137	148	148	83	165	88	91	38	49	34	18	57	38	56			174	60	33	12	Old.
*11695	.....do........	♂	237	130	145	144	83	163	72	85	32	42	34	17	52	38	54	55	55	163	59	31	6	Middle-aged.
*11089	.....do........	♂	228	122	132	142	83	165	74	97	29	42	30	18	49	33	58	49	43	149	54	30	6	Do.
*11715	.....do........	♂	240	115	137	151	87	171	77	92	32	40	33	14	51	34	58	52	48	160	58	32		Do.
*11733	.....do........	♂	240	118	134	140	82	163	71	92	33	39	30	15	49	39	55	52	42	158	54	31	2.5	Rather young.
*11701	.....do........	♂	235	118	137	138	86	160	67	90	32	39	31	16	47	32	52	55	43	157	56	33	7.5	Middle-aged.
†2922	.....do........	♂	245		145	140	88	153	68		30	40	32	20	51					160				Old.
†2923	.....do........	♂	275		155	165	97	165	75		26	46	40	22	56					176				Do.
*9079	Strait of Juan de Fuca.	♀	171	89	102	100	61	119	50	71	23	28	20	12	30	37	47	40	31	108	32	17	0	Adult.
*9080	.....do........	♀	186	98	105	108	68		56	78	24				31	36	46	46	35	122	40	19	0	Old.
†2924	Pribilov Isds., Alaska.	♀	185		115	120	63	120	58		20	33		12	34					120			0	Do.
†2925	.....do........	♀	200		117	124	75	135	58		20		22		33					126			0	Do.

* National Museum, Washington, D. C. † Museum of Comparative Zoölogy, Cambridge, Mass.

TEETH.—The dentition is relatively much weaker than in either *Eumetopias* or *Zalophus*, or even in *Arctocephalus*. As usual in the Otaries the outer pair of upper incisors is much larger than the others and caniniform; the two central pairs are flattened antero-posteriorly, and in youth and middle age their crowns are deeply divided by a transverse groove. The lower incisors are smaller than the upper and are hollowed on their inner face but are not grooved. The canines are large and sharply pointed, the lower somewhat curved. The molars are small and closely approximated, with sharply conical crowns and all single-rooted. They have no accessory cusp, or only very minute ones in early life. The roots are usually grooved both externally and internally, sometimes slightly so, but sometimes so deeply that the fang seems to consist of two connate roots. The distinctness of these grooves varies not only in different individuals, but in the corresponding teeth of the two sides of the mouth in the same skull,* so that it is not improbable that teeth may be found in which the grooves of the fangs may be entirely obsolete, or so deep as to nearly or quite divide the fang into two distinct roots. The roots of the molars are very short, and but partly fill their alveoli; hence when the periosteum is removed they fit so loosely that they require to be cemented in to prevent their constantly falling out whenever the skull is handled. The canines and the incisors have much longer roots, which more nearly fill their sockets.

SKELETON.—Vertebral formula: Cervical vertebræ, 7; dorsal, 15; lumbar, 5; sacral, 3; caudal, 8 to 10.

The skeleton in its general features resembles that of *Eumetopias stelleri*, already described. The bones of *C. ursinus* are, however, all slenderer, or smaller in proportion to their length, than in that species, the general form of the body being more elongated. The scapulæ are shorter and broader than in *E. stelleri*, the proportion of breadth to length being in the one as 11 to 10 and in the other as 13 to 10. The pelvis is more contracted opposite the acetabula in *C. ursinus* than in *E. stelleri*, and the last segment of the sternum is also longer and narrower. The differences in the skull of the two forms have already been pointed out in the generic comparisons. In proportions, the principal difference, aside from that already mentioned as existing in the form of the scapula, consists in the

* See Bull. Mus. Comp. Zoöl., vol. ii, pl. ii, figs. 6 *b* and 7 *e*.

longer neck and longer hind feet in *C. ursinus;* the ratio of the length of the cervical vertebræ to the whole length of the skeleton being as 15 to 100 in *E. stelleri,* and as 23 to 100 in *C. ursinus;* and the ratio of the length of the foot to the tibia being in the former as 13 to 10, and in the other as 16 to 10. The following measurements of two adult males and two adult females indicate the length of the principal bones, and of the different vertebral regions.

Measurements of the Skeleton.

	Adult ♂ No. 2922.	Adult ♂ No. 2923.	Adult ♀ No. 2925.	Adult ♀ No. 2924.
Whole length of skeleton (including skull)	2,040	1,840	1,370	1,215
Length of skull.................................	275	245	200	185
Length of cervical vertebræ	430	360	200	172
Length of dorsal vertebræ......................	770	680	520	470
Length of lumbar vertebræ.....................	270	245	185	173
Length of sacral vertebræ	160	145	105	95
Length of caudal vertebræ.....................	140	145	160	120
Length of first rib	212	178	120	110
Length of first rib, osseous portion	112	105	55	55
Length of first rib, cartilaginous portion.......	100	73	65	55
Length of third rib	395	370	205	175
Length of third rib, osseous portion...........	265	210	140	115
Length of third rib, cartilaginous portion......	130	90	65	60
Length of sixth rib...........................	465	400	323	265
Length of sixth rib, osseous portion..........	350	295	230	190
Length of sixth rib, cartilaginous portion	115	105	93	75
Length of tenth rib...........................	590	....	405	335
Length of tenth rib, osseous portion	360	340	265	215
Length of tenth rib, cartilaginous portion	230	....	140	120
Length of twelfth rib, osseous portion only....	345	320	210	200
Length of fifteenth rib, osseous portion only...	210	205	150	130
Length of sternum............................	640	590	385	370
Length of sternum, 1st segment................	135	127	76	73
Length of sternum, 2d segment................	68	54	37	34
Length of sternum, 3d segment................	65	57	39	36
Length of sternum, 4th segment	65	55	40	36
Length of sternum, 5th segment	60	57	40	37
Length of sternum, 6th segment	58	55	40	36
Length of sternum, 7th segment	63	57	43	409
Length of sternum, 8th segment	115	110	70	70
Length of scapula.............................	250	217	140	120
Breadth of scapula............................	295	285	170	160
Greatest height of its spine...................	35	27	14	12
Length of humerus............................	220	220	130	130
Length of radius..............................	205	195	128	128
Length of ulna...............................	243	223	160	157
Length of carpus	55	55	35	35

Measurements of the Skeleton—Continued.

	Adult ♂ No. 2922.	Adult ♂ No. 2923.	Adult ♀ No. 2925.	Adult ♀ No. 2924.
Breadth of carpus	100	80	60	55
Length of 1st digit* and its metacarpal bone	250	250	180	177
Length of 2d digit and its metacarpal bone	245	235	178	....
Length of 3d digit and its metacarpal bone	215	195	155	....
Length of 4th digit and its metacarpal bone	170	150	125	....
Length of 5th digit and its metacarpal bone	127	115	100	....
Length of femur	150	135	82	85
Length of tibia	250	225	167	157
Length of fibula	230	210	145	150
Length of tarsus	87	84	57	60
Breadth of tarsus	67	65	40	37
Length of 1st digit† and its metatarsal bone	270	260	200	....
Length of 2d digit and its metatarsal bone	265	260	....	....
Length of 3d digit and its metatarsal bone	265	260	....	....
Length of 4th digit and its metatarsal bone	264	255	....	....
Length of 5th digit and its metatarsal bone	290	280	....	....
Length of innominate bone	234	210	145	140
Greatest (external) width of pelvis anteriorly	115	110	70	75
Width of posterior end of pubic bones	17	14	30	25
Length of ilium	100	95	60	60
Length of ischia-pubic bones	134	110	75	73
Length of thyroid foramen	67	63	45	45
Breadth of thyroid foramen	34	25	20	20

* Fore limb.   † Hind limb.

SEXUAL DIFFERENCES.—The sexes differ in color, as already stated, in the females being much lighter than the males, or grayer. In respect to the skeleton they differ extraordinarily in the form of the pelvis, as already described, all the parts of which in the female are greatly reduced in size, and instead of the pubic bones meeting each other posteriorly, as they do in the males, they are widely separated. The innominate bones are also much further apart in the females, and the bones forming the front edge of the pelvis are less developed, so that the pelvis in the female is entirely open in front. In consequence of the remarkable narrowness of the pelvis in the male, the form of this portion of the skeleton is necessarily varied in the female, to permit of the passage of the fœtus in parturition. As already remarked, no such sexual differences are seen in the *Phocidæ*.

In respect to other parts of the skeleton, the absence of the great development of the sagittal and occipital crests seen in the males has already been noticed. The bones of all parts of the skull are much smaller and weaker, especially the lower jaw and the teeth. The attachments for the muscles are correspondingly less developed throughout the skeleton. The most striking sexual difference, however, is that of size; the weight of the full-grown females, according to Captain Bryant, being *less than* ONE-SIXTH that of the full-grown males. This estimate Mr. Elliott has since found to be correct by actual weight of large series of specimens.

DIFFERENCES RESULTING FROM AGE.—The differences in color between the young and the adult consist, as already stated, in the young of both sexes during the first three or four months of their lives being glossy black, and gradually afterwards acquiring the color characteristic respectively of the adult males and females. In respect to the differences in the skeleton that distinguish the young, I can speak only of the skull, in which the relative development of its different regions differs widely from what is seen in the adult of either sex. The two young skulls before me, said to be from specimens thirty-five days old, are both females, but at this age the sexes probably do not differ in osteological features, especially in those of the skull. In these young specimens the anterior or facial portion of the skull is but little developed in comparison with the size of the brain-case. The muzzle is not only excessively short, but the orbital space is small, and the postorbital cylinder is correlatively reduced almost to zero, the postorbital processes being close to the brain-case. The zygomatic arch is hence very short; the zygomatic foramen is as broad as long, instead of being nearly twice as long as broad, as in the adult. On the other hand, the brain-case is exceedingly large, the greatest breadth of the skull being at the middle of the brain-case instead of at the posterior end of the zygomatic arch. As will be seen by the table of measurements of the skull already given, *the brain-case is nearly as large as in the adults,* and the bones being thinner, it must have a capacity as great as that of the skulls of the adult males and females, there being, in respect to this point, *but slight difference in the sexes.* As the young advance in age, the anterior portion of the skull, or that part in advance of the brain-case, greatly elongates, especially the postorbital cylinder, and increases also in breadth, the skull in a great measure losing the triangular form and the narrow pointed muzzle characteristic of the young. The postorbital processes also greatly change their form as they further develop.

The limbs are also relatively much larger than in the adult, as mentioned by Quoy and Gaimard in respect to the *Arctocephalus cinereus* of Australia,* which enables them to move on land with greater facility than the adult, as the above-mentioned authors have stated to be the case in the Australian species.

It is not true, however, that the young of *C. ursinus* are devoid of under-fur, as some writers have affirmed.†

INDIVIDUAL VARIATION.—The two males were both not only full-grown, but quite advanced in age, though in all probability the crests of even the older skull (No. 2922) would have been still further developed. The other male (No. 2923) was somewhat younger, but already had the sagittal crest considerably pro- duced; the teeth, however, were but moderately worn, the mid- dle upper incisors still retaining the groove dividing the surface of the crowns. In the younger male skull the posterior outline of the palatines is but slightly concave, whereas in the other it is deeply and abruptly emarginate in the middle,—as deeply so as in the young (one month old) skulls;—showing that differ- ences in this respect do not necessarily depend upon differences in age. They also differ in the form of the postorbital pro- cesses, in the younger they having nearly the same form as in *Eumetopias*, whereas in the older nearly that seen in *Zalophus*. The postorbital cylinder is also much shorter in the younger, though these two skulls do not present nearly the great differ- ence in this respect exhibited by two very old male skulls of *Zalophus* already described. Another difference is seen in the parieto-maxillary suture. In the younger specimen it is nearly straight and directed forward, the nasals extending consider- ably beyond it. In the other it curves at first moderately back- ward, and then abruptly in the same direction; the maxillaries extending in this case slightly beyond the nasals, instead of end- ing considerably in front of the end of the latter. The nasals

---

* Voyage de l'Astrolabe, Zoologie, tome i, p. 89.

† It may be added that the young specimens above described had not fully shed their milk teeth. The incisors appear to have been renewed, but both the first and second sets of canines were still present (see Bull. Mus. Comp. Zoöl., vol. ii, pl. iii, fig. 5), the permanent ones being in front of the others. The three molars of the first set have been replaced by the perma- nent ones, the first and second of which are already quite large. The hinder molars are in one of the specimens but just in sight, and doubtless had not cut through the gum. In the other they are a little more advanced. The middle one is quite prominent; the first is much smaller, while the last or sixth molar is far less advanced than either of the others.

themselves are much narrower in the younger specimen, especially anteriorly, and hence have a very different form in the two examples.

In respect to the teeth, it may be added that the older skull has *seven* upper molars on one side and *six* on the other, the normal number being six on each side. I have before me two other skulls in which the molars are $\frac{7}{6}-\frac{6}{6}$, and two others in which they are $\frac{5}{6}-\frac{5}{5}$! The form of the molar teeth, especially of the fangs, differs markedly in the two skulls; those of the younger having the longitudinal grooves of the fangs of nearly all the teeth almost wholly obsolete, while in the other specimen the roots of nearly all the molars are more or less strongly grooved.

Of the two female skulls one is very aged,* as shown by the closed sutures and the greatly worn and defective teeth. The younger, however, is also quite advanced in years. Differences of a similar character to those seen in the males also occur between these, but they are less marked.

There are also considerable variations in color. Not only is one of the young females much darker below and about the face than the other, but similar variations are seen in the old females and in the males.

COMPARISON WITH ALLIED SPECIES.—In cranial characters *Callorhinus ursinus*, as already noticed, is widely different from any of the species of *Arctocephalus*, to which, however, it is most closely allied. While in the latter the facial portion of the skull is narrow and depressed, in *C. ursinus* it is broad, high, and short. The ascending portion of the intermaxillæ is also much broader and shorter, and the whole dental armature is much weaker. In the character of the pelage the Northern and Southern Fur Seals present no marked differences, but in respect to color the latter are much grayer than the former. Another obvious difference consists in the great elongation of

---

* Respecting the age of these specimens of Fur Seals, Captain Bryant has responded to my inquiries as follows: "The grown females (the mothers of the pups) were average specimens. The only means I had of determining their age was by the evidences afforded by dissection. These were that the older female had given birth to seven young and the other five, which would make their ages respectively ten and eight years. The two grown males were also selected as average specimens in size and color. Judging from their general appearance and color, I estimated them to be ten years old. The two pups were thirty-five days old, and in that time had doubled their size from birth. They were both females."

the toe-flaps of the posterior limbs, which in *C. ursinus* are greatly developed, their extension beyond the digits being nearly equal to the length of the rest of the foot—in adult females 140 mm. beyond the toes. In *Arctocephalus australis* (= *falklandicus*, auct.), on the other hand, they extend, as in most of the other Otaries, only a short distance beyond the toes, the indentations between them being but little posterior to the ends of the toes.

The extremes of sexual variation seem to be presented by *Callorhinus ursinus* and *Arctocephalus australis ;* the former presenting the greatest and the latter the least amount of sexual variation in size of any of the five American species of Otaries, to which my comparisons are limited by lack of material. The adult males of these two species differ little in size, while the female of *Arctocephalus australis* is very much larger than the female of *C. ursinus.* The average length of the skull in the adult males of both species is about 245 mm.; of the females of *A. australis* about 233 mm., against 185 mm. in *C. ursinus.*

For purposes of comparison I submit the following measurements of skulls of *A. australis* for corresponding measurements of *C. ursinus* (see *anteà*, p. 323):

Measurements of Six Skulls of ARCTOCEPHALUS AUSTRALIS.

Catalogue number.	Locality.	Sex.	Total length.	Greatest width at zygomata.	Width at mastoid processes.	Between orbits.	Nasal bones, length.	Nasal bones, width behind.	Nasal bones, width before.	From front edge of intermaxillaries to pterygoid hamulus.	From front edge of intermaxillaries to glenoid process.	Breadth of canines.	Height of crest.	Remarks.
1126	Straits of Magellan	♂	233	140	126	26	37	16	28	150	170	48	1	Rather young.
1125	Straits of Magellan	♂	235	145	133	34	35	18	30	150	166	52	7	Middle-aged.
1108	Paraca Bay, Straits of Magellan	♂	250	150	145	30	...	...	...	150	190	57	17	Old.
1131	Galapagos Islands	♂	265	140	188	33	...	...	...	162	186	50	5	Middle-aged.
1130	....do	♀	237	119	95	23	...	...	...	143	148	38	2	Old.
1133	Patagonia	♀	225	122	97	29	30	...	19	131	155	38	2	Do.

GEOGRAPHICAL DISTRIBUTION AND MIGRATION.—The Fur
Seal is well known to have been formerly abundant on the
western coast of North America, as far south as California, but
the exact southern limit of their range I have been unable to
determine. Captain Scammon speaks of having seen them " on
one of the San Benito Islands, on the coast of Lower California," and again says, "On the coast of California many beaches
were found fronting gullies, where [Fur] Seals in large numbers
formerly gathered; and, as they there had plenty of ground
to retreat upon, the sealers sometimes drove them far enough
back to make sure of the whole herd, or that portion of them
the skins of which were desirable."* He also states that the
"Fur Seal and Sea Elephant once made the shores [of Guadalupe
Island] a favorite resorting-place," and refers to their former
occurrence on Cedros Island, in latitude 28°.† Although at
one time abundant on the California coast, they are by no means
numerous there now, having been nearly exterminated by un-
restricted destruction by the sealers. The writer above cited
refers also to their capture by the Indians at the mouth of the
Strait of Juan de Fuca. The Seals appear here and on the
neighboring coast, he adds, "some years as early as the first of
March, and more or less remain till July or August; but they
are most plentiful in April and May. During these two months
the Indians devote nearly all their time to sealing when the
weather will permit." He reports their increase there in later
years, and that while only a few dozens were annually taken
there from 1843 to 1864, fully five thousand were taken in 1869.‡
Captain Bryant has given a similar report, referring especially
to their abundance along the coasts of Oregon, Washington
Territory, and British Columbia in 1869, as compared with former years. He says those taken "were mostly very young
Seals, none appearing to be over a year old. Formerly in
March and April the natives of Puget Sound took large numbers of pregnant females,§ but no places where they have
resorted to breed seem to be known off this coast." He thinks
it probable, however, that they may occupy rocky ledges off

---

* Marine Mamm., pp. 152, 154.
† J. Ross Browne's Resources of the Pacific Slope, second part, p. 128.
‡ Marine Mamm., p. 154.
§ There are six skulls in the National Museum from Puget Sound and the
neighboring coast (collected at several different points by Messrs. Scammon
an Swan), *all of which are females.*

shore which are rarely visited by boats.* In his MS. report just received he states that a half-breed hunter told him that he found in summer " on Queen Charlotte's Island, groups of these animals consisting of two or more beach-masters with a dozen or more females and pups, but no half-grown males."

As is well known, the Prybilov or so-called " Fur Seal Islands," off the coast of Alaska, form the great breeding-ground of the Fur Seals, to which hundreds of thousands annually resort to bring forth their young. The Prybilov Group consists of four small islands, known respectively as Saint Paul's, Saint George's, Otter, and Walrus Islands. The two last named are of small size, and are not used as breeding-grounds by the Seals, although Otter Island is visited by a large number of " non-breeding Seals." Saint Paul's Island is the largest, containing an area of about 33 square miles, and having a coast-line of about forty-two miles, nearly one-half of which is sand-beach. Of this, sixteen and a half miles, according to Mr. Elliott, are occupied in the breeding-season by the Fur Seals. Saint George's Island is somewhat smaller, with only twenty-nine miles of shore-line. It presents a bold coast, a grand wall of basalt extending continuously for ten miles, with no passage-way from the sea. It has, in all, less than a mile of sand-beach, and only two and a quarter miles of eligible landing grounds for the Seals.

A few old male Fur Seals are said to make their appearance at the rookeries on these islands between the 1st and 15th of May, they acting, as it were, the part of pioneers, since their number is not much increased before the first of June. At about this date, and with the setting in of the humid, foggy weather of summer, the male Seals begin to land by " hundreds and thousands," to await the arrival of the females, which do not appear before about July first. The young are born soon after, and toward the last of this month the rookeries begin to lose their compactness and definite boundaries, but they are not fully broken up till about the middle of September. The Seals begin to leave the islands about the end of October, the greater proportion departing in November, while some remain till the end of the following month, and even later.

The number of Fur Seals present on Saint Paul's Island in July, 1872, was estimated by Mr. Elliott to exceed *three million*, and on Saint George's Island in July, 1873, at about one hun-

---

* Bull. Mus. Comp. Zoöl., ii, p. 88.

dred and sixty-three thousand.*   Although these islands form
by far their most populous resorts, they are said to occur in
considerable numbers on some of the islands to the northward,
but I am unable to find definite statements as to their numbers
or favorite stations.   Mr. Elliott, after examining Saint Mathew's
and Saint Lawrence Islands, became convinced that they were
not only not resorted to as breeding stations by the Fur Seals,
but that these islands, by their constitution and climatic conditions, were unsuitable for this purpose, and adds, " it may be
safely said that no land of ours in the north is adapted to the
wants of that animal, except that of Saint Paul and Saint
George."†   Mr. W. H. Dall states that " They have never been
found in Bering Strait, or within three hundred miles of it."‡   In
early times these animals are well known to have been abundant
on Behring's and Copper Islands.   According to Krascheninikow, they were so numerous upon Behring's Island about the
middle of the last century as to cover the whole southern shore
of the island.   Their range on the Asiatic coast is given by Steller and others as extending southward along the Kamtschatkan
coast to the Kurile Islands.   Krascheninikow states that they

---

* As of interest in the present connection, I quote the following from Dampier respecting the abundance of the Southern Fur Seal at the Island of
Juan Fernandez, two hundred years ago, or about a century before the
beginning of the Seal slaughter there, which in less than a generation nearly
exterminated the species at that locality.   Dampier and his party spent
fifteen days on this island in the year 1683.   He says: " Seals swarm as
thick about this Island [" of John Fernando," as he terms it,] as if they had
no other place in the World to live in ; for there is not a Bay nor Rock that
one can get ashoar on, but is full of them.   . . .   These at John Fernando's have fine thick short Furr ; the like I have not taken notice of any where
but in these Seas.   Here are always thousands, I might say possibly millions
of them, either sitting on the Bays, or going and coming in the Sea round the
Island, which is covered with them (as they lie at the top of the Water playing and sunning themselves) for a mile or two from the shore.   When they
come out of the Sea they bleat like Sheep for their young ; and though they
pass through hundreds of others young ones before they come to their own,
yet they will not suffer any of them to suck.   The young ones are like
Puppies, and lie much ashore, but when beaten by any of us, they, as well
as the old ones, will make towards the Sea, and swim very swift and
nimble ; tho' on shore they lie very sluggishly, and will not go out of our
way unless we beat them, but snap at us.   A blow on the Nose soon kills
them.   Large Ships might here load themselves with Seals Skins and Trane-
oyl; for they are extraordinary fat."—A New Voyage Round the World, " fifth
edition, corrected," 1703, vol. i, pp. 88, 90.

† Cond. of Aff. in Alaska, pp. 217, 224.

‡ Alaska and its Resources, p. 493.

appeared there, however, only in spring and in September, none being seen there from the beginning of June till the end of August, at which time he says they *return from the south* with their young. Von Schrenck speaks of their occurrence in the Ochotsk Sea and the Tartarian Gulf as far south as the 46th degree of latitude, or to the southern point of Saghalien Island. The natives reported to him the occurrence of great numbers of the animals on the eastern coast of that island. Captain Scammon also refers to their abundance twenty years since on the eastern side of Saghalien.*

Except during the season of reproduction, these animals appear to lead a wandering life, but the extent and direction of their migrations are not yet well known. Steller spoke of their migrations as being as regular as those of the various kinds of sea-fowl, and they are recorded as arriving with great regularity at the Prybilov Islands, but where they pass the season of winter is still a matter of conjecture.

GENERAL HISTORY.—The Northern Fur Seal was first made known to science by Steller in 1751, under the name *Ursus marinus*. During his visit to Kamtschatka and its neighboring islands, in 1742, he met with these animals in great numbers

---

* Captain Scammon relates in an off-hand way, merely as an interesting incident of sealing life, the following: "In the midst of the Crimean War, an enterprising firm in New London, Connecticut, fitted out a clipper bark, which was officered and manned expressly for a sealing voyage in the Okhotsk Sea. The captain was a veteran in the business, and many thought him too old to command, but the result of the voyage proved him equal to the task. The vessel proceeded to Robin Island—a mere volcanic rock, situated on the eastern side of the large island of Saghalien. Many outlying rocks and reefs are about it, making it dangerous to approach, and affording but slight shelter for an anchorage. Here the vessel (of about three hundred tons) lay, with ground-tackle of the weight for a craft of twice her size. Much of the time fresh winds prevailed, accompanied by the usual ugly ground-swell; and, in consequence of her being long, low, and sharp, the deck was at such times frequently flooded; nevertheless, she 'rode-out the whole season, though wet as a half-tide rock,' and a valuable cargo of skins was procured, which brought an unusually high price in the European market, on account of the regular Russian supply being cut off in consequence of the war. This is only given as one of the many that may be related of sealing life."—*Marine Mammalia*, pp. 150-152. In this connection I can hardly help adding that it is to be regretted that Captain Scammon has not favored us with more of these "incidents," from the important bearing they have upon the former distribution and abundance of this and other species of Seal, and that he has not given more explicit references to the localities at which the Fur Seals were formerly hunted on the southern portion of the North American coast and elsewhere.

at Behring's Island, where he spent some time among them, and carefully studied their habits and anatomy, a detailed account of which appeared in his celebrated memoir entitled "De Bestiis Marinis," in the Transactions of the St. Petersburg Academy for the year 1749.* This important essay was the source of nearly all of the accounts of this animal that appeared prior to the beginning of the present decade. The twenty-eight quarto pages of Steller's memoir devoted to this species, gave not only a detailed account of its anatomy, with an extensive table of measurements, but also of its remarkable habits, and figures of the animals themselves. A little later Krascheninikow, in his History of Kamtschatka,† under the name "Sea Cat," gave also a long account of its habits, apparently based mainly on Steller's notes,‡ but it embraces a few particulars not given in "De Bestiis Marinis." Steller's description of the habits of this animal has been largely quoted by Buffon, Pennant, Schreber, Hamilton, and other general writers.

Buffon, Pennant, Schreber, Gmelin, and nearly all writers on the Pinnipeds, down to about 1820, confounded the Northern Fur Seal with the Fur Seals of the Southern hemisphere, blending their history as that of a single species. Péron, in 1816, first recognized it as distinct from its southern allies, and it was so treated somewhat later by Demarest, Lesson, Fischer, Gray, and other systematic writers, § but its distinctive characters were not

---

* Nov. Comm. Acad. Petrop., ii, pp. 331–359, pl. xv, 1751. This, as is well known, is a posthumous paper, published six years after Steller's death, Steller dying of fever November 12, 1745, while on his way from Siberia to St. Petersburg. The description of the Sea Bear was written at Behring's Island in May, 1742.

† Hist. Kamtchatka (English edition), translated from the Russian by James Grieve, pp. 123–130, 1764.

‡ Krascheninikow, it is stated, "received all of Mr. Steller's papers," to aid him in the preparation of his "History of Kamtschatka."

§ Nilsson and Müller in 1841, and Wagner in 1846 and 1849, on the other hand, still considered all the Sea Bears as belonging to a single species. Wagner, in 1849 (Arch. für Naturg., 1849, pp. 37–49), described the osteological characters of the Northern species from three skeletons in the Munich Museum received from Behring's Sea. One of these was apparently that of a full-grown female ; a second was believed to be that of a half-grown male, while the third belonged to a very young animal, in which the permanent teeth were still not wholly developed. Wagner compares the species with Steller's Sea Lion, and with the figures of the skulls of the Southern Sea Bears given by F. Cuvier, Blainville, and Quoy and Gaimard, and notes various differences in the form of the teeth and skull, but believes that these differences must be regarded as merely variations dependent upon age.

clearly set forth till 1859, when Dr. J. E. Gray described and figured its skull, and showed that the Northern species was not even congeneric with the Sea Bears of the South. Very few specimens of either the Northern or Southern Sea Bears appear to have reached European museums prior to about that date, so that naturalists had not previously been able to make a direct comparison of this species with any of its Southern affines. Dr. Gray, in referring to this point in 1859, wrote as follows: "I had not been able to see a specimen of this species in any of the museums which I examined on the Continent or in England, or to find a skull of the genus [*Arctocephalus*] from the North Pacific Ocean, yet I felt so assured, from Steller's description and the geographical position, that it must be distinct from the Eared Fur Seals from the Antarctic Ocean and Australia, with which it had usually been confounded, that in my 'Catalogue of Seals in the Collection of the British Museum' [1850] I regarded it as a distinct species, under the name of *Arctocephalus ursinus*, giving an abridgment of Steller's description as its specific character." "The British Museum," he adds, "has just received, under the name *Otaria leonina*, from Amsterdam, a specimen [skull and skin] of the Sea Bear from Behring's Straits, which was obtained from St. Petersburg";* which is the specimen already spoken of as figured by Dr. Gray. From the great differences existing between this skull and those of the southern Sea Bears, Dr. Gray, a few weeks later, separated the northern species from the genus *Arctocephalus*, under the name *Callorhinus*.†

It seems, however, that there were two skulls of Steller's Sea Bear in the Berlin Museum as early as 1841,‡ and three skeletons of the same species in the Museum of Munich in 1849,§ yet Dr. Gray appears to have been the first to compare this animal with its southern relatives, and to positively decide its affinities.

Misled, however, by erroneous information respecting specimens of Eared Seals received at the British Museum from California, a skin of the *Callorhinus ursinus* was doubtfully described by this author, in the paper in which the name *Callorhinus* was proposed, as that of his *Arctocephalus monteriensis*, which is a

---

* Proc. Zoöl. Soc. Lond., 1859, p. 102.
† Ibid., 1859, 359.
‡ See Archiv für Naturgesch., 1841, p. 334.
§ Ibid., 1849, 39.

Hair Seal. This skin was accompanied by a young skull, purporting, by the label it bore, to belong to it, but Dr. Gray
observes that otherwise he should have thought it too small to
have belonged to the same animal. Seven years later,* he described the skull as that of a new species (*Arctocephalus californianus*), still associating with it, however, the skin of the
Callorhinus ursinus. The skull he subsequently considered as
that of a young *A. monteriensis* (= *Eumetopias stelleri*); and referring his *A. californianus* to that species, he was consequently
led into the double error of regarding the *Eumetopias stelleri* as
a Fur Seal (as already explained under that species and elsewhere in the present paper), and of excluding the *Callorhinus
ursinus* from the list of Fur Seals. To this I called attention
in 1870, and in 1871 Dr. Gray correctly referred his *A. monteriensis* and *A. californianus* in part (the "skin only") to *Callorhinus ursinus*.†

What may be termed the second or modern epoch in the
general history of this species began in 1869, when Captain C.
M. Scammon published a highly important contribution to its
biology,‡ he describing at considerable length, from personal
observation, its habits, distribution, and products, as well as
the various methods employed for its capture. The following
year Mr. W. H. Dall devoted a few pages§ to its history, in
which he made many important suggestions relative to the
sealing business. During the same year I was able to add not
only something to its technical history, || but also to make pub-
lic an important communication on its habits kindly placed at
my disposal by Captain Charles Bryant,¶ Government agent in
charge of the Fur Seal Islands of Alaska. In 1874, Captain
Scammon republished his above-mentioned paper,** adding
thereto a transcript of Captain Bryant's observations already
noted. Almost simultaneously with this appeared Mr. H. W.
Elliott's exhaustive Report on the Seal Islands of Alaska,†† in

---

* Cat. Seals and Whales, 1866, p. 51.
† Suppl. Cat. Seals and Whales, p. 15; Hand-List of Seals, 32.
‡ Overland Monthly, vol. iii, Nov. 1869, pp. 393–399.
§ Alaska and its Resources, 1870, pp. 492–498.
|| Bull. Mus. Comp. Zoöl., ii, pp. 73–89.
¶ Ibid., pp. 89–108.
** Marine Mammalia, 1874, pp. 141–163.
†† Report on the Prybilov Group, or Seal Islands of Alaska, 4to, unpaged,
1873 [1874].

which the present species properly comes in for a large share
of the author's attention. The work is richly illustrated with
photographic plates, taken from Mr. Elliott's sketches, about
twenty-five of which are devoted to the Fur Seal. The text of
this rare and privately distributed work has been since reprinted,* with some changes and additions, and has been
widely circulated. It contains very little relating to the Fur
Seal that is strictly technical, but the general history of its life
at the Prybilov Islands is very fully told, while the commercial
or economic phase of the subject is treated at length. A few
minor notices of this species have since appeared (mostly popular articles in illustrated magazines, chiefly from the pen of Mr.
Elliott), but nothing relating to its general history requiring
special notice in the present connection.

FIGURES.—The first figures of the Northern Sea Bear were
given by Steller, in his paper already cited. They represent an
adult male, in a quite natural attitude, and a female reclining
on her back. In respect to details, these early figures were
naturally more or less rude and inaccurate. They were copied,
however, by Buffon, Schreber, Pennant, and other early writers,
and are the only representations of this species known to me
that were made prior to about the year 1839, except Choris's
plate of a group of these animals entitled "Ours marins dans
l'île de St. Paul",† published in 1822. This represents three
old males, surrounded by their harems, and indicates very faithfully the mode of grouping and the variety of attitudes assumed by these animals when assembled on the rookeries.
Hamilton, in 1839, gave a figure of the "Sea Bear of Steller
(*Otaria ursina*)"‡ which he tells us is "from the engraving of
the distinguished Naturalist of the Rurick",§ the original of
which I have not seen. This represents a male and female, the
latter reclining on its side, with a pup resting on its right
flipper.

The first figure of the skull is that published by Gray in
1859, ||—a view in profile of the skull of an adult male. A
wood-cut of the same was given in 1866,¶ and a fine lithographic

* Condition of Affairs in Alaska, 1875, pp. 107–151.
† Voy. pittor. autour du Monde, Iles Aléoutiennes, pl. xv.
‡ Marine Amphibiæ, pl. xxi.
§ Ibid., p. 266.
|| Proc. Zoöl. Soc. Lond., 1859, pl. lxviii.
¶ Cat. Seals and Whales, p. 45, fig. 16.

plate in 1874,* representing the skull in profile, from above and from below.†

In 1870 I gave figures of two adult male skulls (two views of each), of an adult female skull (three views), of a very young skull (three views), and of the scapula, dentition, etc. These, so far as known to me, are the only figures of the skull or other details of structure thus far published.

In 1874 Captain Scammon gave figures of the animal, ‡ a zincograph of an old male, § from a sketch by Mr. Elliott, a wood-cut of the head of a female seen from below (drawn by Elliott),‖ two outline figures representing the female as seen from below and in profile, and two others in outline illustrating "attitudes of the Fur Seals."¶ Mr. Elliott, in his first Report on the Seal Islands, ** in a series of over two dozen large photographic plates (from India ink sketches from nature) has given an exhaustive presentation of the phases of Fur Seal life so faithfully studied by him at St. Paul's Island. Among these may be mentioned especially those entitled "The East Landing and Black Buttes—The beach covered with young Fur Seals"; "The North Shore of St. Paul's Island" (giving an extensive view of the rookeries); "Lukannon Beach" (Fur Seals playing in the surf, and rookeries in the distance); "Old male Fur Seal, or 'Seecatch'" (as he appears at the end of the season after three months of fasting); "Fur Seal Harem" (showing the relative size of males, females, and young, various attitudes, positions, etc.); 'Fur Seal males, waiting for their 'harems'" (the females beginning to arrive); Fur Seal "Rookery" (breeding-grounds at Polavina Point); "Fur Seal Harem" (Reef Rookery, foreground showing relative size of males and females); "Fur Seal Pups at Sleep and Play"; "Hauling Grounds" (several views at different points); "Capturing Fur Seals"; "Driving Fur Seals"; "Killing Fur Seals—Sealing gang at work," etc.

The only other pictorial contributions to the history of the

---

* Hand-List of Seals, pl. xix.

† I infer this to be the same specimen in each case, not only from the resemblance the figures bear to each other, but from Dr. Gray, so far as I can discover, referring to only the single skull from Behring's Strait, received in 1859.

‡ Marine Mammalia, pl. xxi, two figures.

§ Ibid., p. 143.

‖ Ibid., p. 145.

¶ Ibid., p. 149.

** Report on the Prybilov Group, or Fur Seal Islands, of Alaska, unpaged, and plates not numbered.

Fur Seal of noteworthy importance is Mr. Clark's colored plate,* on which are represented a nearly full-grown male, a female and a pup, prepared from skins sent to the British Museum by the Alaska Commercial Company. In these the attitudes are excellent and the coloring fair.

HABITS.—The habits of the Fur Seal of the north seem to have been well known to Steller and his companions a century and a quarter ago, and their seemingly marvellous accounts of them prove to have been only to a slight degree erroneous. As a matter of historic interest, and for comparison with our present knowledge of the subject, as well as in some respects supplementary to it, I herewith subjoin a few extracts from the account left us by Krascheninikow, based partly, apparently, on his own observations, but largely on those of his fellow-traveller, Steller. "The Sea Cat", says Krascheninikow,† "is about half the size of the Sea Lion; in form resembling the Seal, but thicker about the breast, and thinner towards the tail. They have the snout longer than the Sea Lions, and larger teeth; with eyes like cow's eyes, short ears, naked and black paws, and black hair mixed with gray, which is short and brittle. Their young are of a bluish black color.

"The Sea Cats are caught in the spring and in the month of *September*, about the river *Sheepanova;* at which time they go from the *Kurilskoy* Island to the *American* coast; but the most are catched about the cape of *Kronotzkoy,* as between this and the cape *Shupinskoy* the sea is generally calm and affords them properer places to retire to. Almost all the females that are caught in the spring are pregnant; and such as are near their time of bringing forth their young are immediately opened and the young taken out and skinned. None of them are to be seen from the beginning of *June* to the end of *August,* when they return from the south with their young. The natives were formerly at a loss to conceive where such great herds of pregnant fat animals retired in spring, and why they returned so weak and lean was owing to their fatigue.

. . . . "The male and female differ so much in the form and strength of their bodies, that one who does not carefully examine them would take them for different species of animals; besides the females are wild and fearful. The male has from

---

* Proc. Zoöl. Soc. London, 1878, 271, pl. xx.

† I use here Grieve's English translation from the Russian, published in 1764.

eight to fifteen, and even sometimes fifty females, whom he
guards with such jealousy that he does not allow any other to
come near his mistresses: and though many thousands of them
lie upon the same shore, yet every family keeps apart; that is,
the male with his wives, young ones, and those of a year old,
which have not yet attached themselves to any male; so that
sometimes the family consists of 120. They likewise swim at
sea in such droves. Such as are old, or have no mistresses, live
apart; and the first that our people found upon *Bering's* Island
were such old ones, and all males, extremely fat and stinking.
These sometimes lie asleep a whole month without nourishment,
and are the fiercest of all, attacking all that pass them, and their
pride or obstinacy is such that they will rather die than quit
their place. When they see a man coming near them some of
them rush upon him and others lie ready to sustain the battle.
They bite the stones that are thrown at them, and rush the
most violently upon him who throws them; so that though you
strike out their teeth with stones, or put out their eyes, yet,
even blind, they will not quit their place: nay, they dare not
leave it, for every step that any one moves off he makes a new
enemy, so that though he could save himself from the attacks
of men, his own brethren would destroy him; and if it happens that any one seems to retire the least, then others draw
near no [to] prevent his running away; and if any one seems
to suspect the courage of another, or his design to run away, he
falls upon him. This suspicion of one another is sometimes carried so far, that for a whole verst one sees nothing but these
bloody duels; and at such time one may pass between them
without any manner of danger. If two fall upon one, then some
others come to support the weakest; for they do not allow of
unequal combat. During these battles the others that are swimming in the sea raise their heads, and look at the success of
the combatants; at length becoming likewise fiercer, they come
out and increase the number.

. . . . "When two of them only fight the battle lasts frequently for an hour: sometimes they rest awhile, lying by one
another; then both rise at once and renew the engagement.
They fight with their heads erect, and turn them aside from
one another's stroke. So long as their strength is equal they
fight with their fore paws; but when one of them becomes weak
the other seizes him with his teeth, and throws him upon the
ground. When the lookers on see this they come to the assist-

ance of the vanquished. The wounds they make with their teeth are as deep as those made with a sabre; and in the month of *July* you will hardly see one of them that has not some wound upon him. After the end of the battle they throw themselves into the water to wash their bodies. The occasions of their quarrels are these:—The first and most bloody is about their females, when one endeavors to carry off the mistress of another, or the young ones that are females; the females that are present follow the conqueror. The second is about their places, when one comes too near that of another, which they don't allow, either for want of room, or because they are jealous of their coming too near their mistresses. The third is owing to their endeavouring to do justice, and end the quarrels of others.

. . . . "Another reason of the Sea Cats going in the spring to the eastwards to the Desert Islands must be, that resting and sleeping without nourishment for three months, they free themselves from the fat which was troublesome to them, in the same manner as the bears who live the whole winter without nourishment; for in the months of *June, July,* and *August,* the old ones do nothing but sleep upon the shore, lying in one place like a stone, now and then looking at one another, and yawning and stretching, without meat or drink; but the young ones begin to walk in the beginning of *July.* When this animal lies upon the shore and diverts himself, his lowing is like that of a cow; when he fights he growls like a bear; when he has conquered his enemy, he chirps like a cricket; but being vanquished or wounded, he groans or mews like a cat; coming out of the water, he commonly shakes, strokes his breast with his hinder paws, and smooths the hair upon it. The male lays his snout to that of the females, as if he was kissing her. When they sleep in the sun they hold up their paws, wagging them as the dogs do their tails. They lie sometimes upon their backs, at other times like a dog upon their bellies; sometimes contracting, at other times extending themselves. Their sleep is never so sound but that they awake at the approach of any person, how softly soever he goes, and are presently upon their guard; besides their smell and hearing are surprisingly acute.

" They swim so fast that they can easily make ten versts in an hour; and when they happen to be wounded at sea they seize the boats of the fishers with their teeth, and drag them along with such swiftness that they appear to fly and not to swim upon the water. By this means the boat is frequently over-turned and the people drowned, unless he who steers it be very skillful, and observes the course of the animal. . . . . They fasten their fore paws in the rocks, and thus draw up their body, which they can move but slowly in such places, but upon a plain, one is in danger of being overtaken by them. Upon *Bering's* Island there are such numbers of them that they cover the whole shore; so that travellers are frequently obliged for safety to leave the sands and level country and go over the hills and rocky places. It is remarkable that in this island the Sea Cats are found only upon the south coast which looks towards *Kamtschatka*. The reason for this may be, that this is the first land they meet with going east from the *Kronotzkoy* pass."*

Steller and Krascheninikow both evidently considered the "Sea Cats" dangerous to man, both on land and in the sea. They also attributed to them a degree of magnanimity and intelligence in relation to their contests with each other uncon-firmed by modern observers. In several respects the accounts of these authors—in the main virtually identical—border upon the mythical, but, generally speaking, are remarkably free from exaggeration, considering the times at which they were written.

As already stated, they formed the source of all our knowl-edge of these strange beasts prior to the beginning of the pres-ent decade. Choris makes only very brief mention of them and says very little about their habits.† Veniaminov, in his "Zapieska" published at Saint Petersburg, in Russian, in 1840, and known to me only as quoted and translated by Mr. H. W. Elliott,‡ has given valuable statistical information respecting the sealing business as prosecuted by the Russians at the Pry-bilov Islands, but seems to have given no detailed account of their habits. Our first important recent information respecting the economy of these animals is that given by Captain Charles

---

*Krascheninikow's Hist. Kamtschatka, Grieve's English translation, pp. 123, 131.

†His account in full is as follows: "L'ours marins, en russe *sivoutch*, cou-vre par milliérs les rivages des îles Kotoviya [Islands of Saint Paul and Saint George], où sont jetées abondamment des plantes marines (*fucus*). On entend de très-loin le cri de ces animaux, lorsqu'on est en mer. Les femelles sont beaucoup plus petites que les mâles; elles ont le corps plus fluet et de couleur jaunâtre. Les mâles ont jusqu'à six pied de haut lors-qu'ils lèvent la tête; les jeunes sont ordinairement d'un brun noir; il parait que les femelles ne font jamais plus d'un petit."—From the description of "Iles S. Georges et S. Paul," in "Voy. pittoresq. autour du Monde."

*Condition of Affairs in Alaska, pp. 241-242.

Bryant, in 1870.* Mr. Elliott's account, published three or four years later, is far more detailed, and respecting most points may be considered as fairly exhaustive of the subject, more than thirty pages of his report being devoted exclusively to the habits of the species. Captain Bryant has now kindly placed at my disposal a communication embodying the results of his eight years' observations on these animals, prepared by my request expressly for the present work. While replete with new information, it does not, to any great extent, duplicate the account of the habits of the species published by Mr. Elliott, being devoted mainly to a detailed history of the changes in the relative preponderance of the different classes of Seals resulting from the different systems of selecting the animals to be killed for their furs, and to other features of the general subject not hitherto fully presented. Its importance as a contribution to the economic phases of the subject can scarcely be overrated, while at the same time it forms a most valuable contribution to the biology of the species. Believing it desirable to present in the present connection a full and connected history of the species, I offer no apology for the copious extracts from Mr. Elliott's graphic account of the habits of the Fur Seal which here follow:

"The fur seal (*Callorhinus ursinus*), which repairs to these islands to breed in numbers that seem almost fabulous, is by far the highest organized of all the Pinnipedia, and, indeed, for that matter, when land and water are fully taken into account, there is no other animal superior to it from a purely physical point of view; and few creatures that can be said to exhibit a higher order of instinct, approaching even intelligence, belonging to the animal kingdom. . . . .

"Observe it as it comes leisurely swimming on toward the land; how high above the water it carries its head, and how deliberately it surveys the beach, after having *stepped* upon it; it may be truly said to step with its fore flippers, for they regularly alternate as it moves up, carrying the head well above them, at least three feet from the ground, with a perfectly erect neck.

. . . . "We observe as the seal moves along that, though it handles its fore limbs in a most creditable manner, it brings up its rear in quite a different style; for after every second step ahead with the fore feet it arches its spine, and with it drags

* Bull. Mus. Comp. Zoöl. ii, pp. 89–108.

and lifts together the hinder limbs to a fit position under its body for another movement forward, by which the spine is again straightened out so as to take a fresh hitch up on the posteriors. This is the leisurely and natural movement on land when not disturbed, the body being carried clear of the ground.

"The radical difference in the form and action of the hinder feet cannot fail to strike the eye at once. They are one-seventh longer and very much lighter and more slender; they, too, are merged in the body like those anterior; nothing can be seen of the legs above the tarsal joint. . . . .

"Now, as we look at this fur seal's progression, that which seems most odd is the gingerly manner (if I may be allowed to use the expression) in which it carries these hind flippers. They are held out at right angles from the body directly opposite the pelvis, the toe-ends and flaps slightly waving and curling or drooping over, supported daintily, as it were, above the earth, only suffering its weight behind to fall upon the heels, which are opposed to each other scarcely five inches apart.

" We shall, as we see him again later in the season, have to notice a different mode of progression, both when lording it over his harem or when he grows shy and restless at the end of the breeding-season, and now proceed to notice him in the order of his arrival and that of his family, his behavior during the long period of fasting and unceasing activity and vigilance and other cares which devolve upon him, as the most eminent of all polygamists in the brute world; and to fully comprehend this exceedingly interesting animal, it will be necessary to refer to my drawings and paintings made from it and its haunts. [*]

"The adult males are first to arrive in the spring on the ground deserted by all classes the preceding year.

"Between the 1st and 5th of May, usually, a few bulls will be found scattered over the rookeries pretty close to the water. They are at this time quite shy and sensitive, not yet being satisfied with the land, and a great many spend day after day before coming ashore idly swimming out among the breakers a little distance from the land, to which they seem somewhat reluctant at first to repair. The first arrivals are not always the oldest bulls, but may be said to be the finest and most ambitious of their class. They are full grown and able to hold their stations on the rocks, which they immediately take up after coming ashore.

[* See Mr. Elliott's "Report on the Prybilov Group, or Fur Seal Islands, of Alaska," especially the plates already mentioned at p. 340.]

"I am not able to say authoritatively that these animals come back and take up the same position on the breeding-grounds occupied by them during the preceding season. From my knowledge of their action and habit, and from what I have learned of the natives, I should say that very few, if any of them, make such a selection and keep these places year after year. One old bull was pointed out to me on the Reef Gar-butch Rookery as being known to the natives as a regular visitor at, close by, or on the same rock every season during the past three years, but he failed to re-appear on the fourth; but if these animals came each to a certain place and occupied it reg-ularly, season after season, I think the natives here would know it definitely; as it is, they do not. I think very likely, how-ever, that the older bulls come back to the same rookery-ground where they spent the previous season, but take up their posi-tions on it just as the circumstances attending their arrival will permit, such as fighting other seals which have arrived before them, &c.

"With the object of testing this matter, the Russians, during the early part of their possession, cut off the ears from a given number of young male seals driven up for that purpose from one of the rookeries, and the result was that cropped seals were found on nearly all the different rookeries or 'hauling-grounds' on the islands after. The same experiment was made by agents two years ago, who had the left ears taken off from a hundred young males which were found on Lukannon Rookery, Saint Paul's Island; of these the natives last year found two on No-vastosh-nah Rookery, ten miles north of Lukannon, and two or three from English Bay and Tolstoi Rookery, six miles west by water; one or two were taken on Saint George's Island, thirty-six miles to the southeast, and not one from Lukannon was found among those that were driven from there; and, prob-ably, had all the young males on the two islands been driven up and examined, the rest would have been found distributed quite equally all around, although the natives say that they think the cutting off of the animal's ear gives the water such access to its head as to cause its death; this, however, I think requires confirmation. These experiments would tend to prove that when the seals approach the islands in the spring they have nothing but a general instinctive appreciation of the fitness of the land as a *whole*, and no especial fondness for any *partic-ular spot.*

"The landing of the seals upon the respective rookeries is influenced greatly by the direction of the wind at the time of approach to the islands. The prevailing winds, coming from the northeast, north, and northwest, carry far out to sea the odor or scent of the pioneer bulls, which have located themselves on different breeding-grounds three or four weeks usually in advance of the masses; and hence it will be seen that the rookeries on the south and southeastern shores of Saint Paul's Island receive nearly all the seal-life, although there are miles of eligible ground on the north shore.

"To settle this question, however, is an exceedingly difficult matter; for the identification of individuals, from one season to another, among the hundreds of thousands, and even millions, that come under the eye on a single one of these great rookeries, is really impossible.

"From the time of the first arrivals in May up to the 1st of June, or as late as the middle of this month, if the weather be clear, is an interval in which everything seems quiet; very few seals are added to the pioneers. By the 1st of June, however, or thereabouts, the foggy, humid weather of summer sets in, and with it the bull-seals come up by hundreds and thousands, and locate themselves in advantageous positions for the reception of the females, which are generally three weeks or a month later, as a rule.

"The labor of locating and maintaining a position in the rookery is really a serious business for those bulls which come in last, and for those that occupy the water-line, frequently resulting in death from severe wounds in combat sustained.

"It appears to be a well-understood principle among the able-bodied bulls that each one shall remain undisturbed on his ground, which is usually about ten feet square, provided he is strong enough to hold it against all comers; for the crowding in of fresh bulls often causes the removal of many of those who, though equally able-bodied at first, have exhausted themselves by fighting earlier, and are driven by the fresher animals back farther and higher up on the rookery.

"Some of these bulls show wonderful strength and courage. I have marked one veteran, who was among the first to take up his position, and that one on the water-line, where at least fifty or sixty desperate battles were fought victoriously by him with nearly as many different seals, who coveted his position, and when the fighting season was over, (after the cows have mostly all hauled up,) I saw him, covered with scars and gashes raw and bloody, an eye gouged out, but lording it bravely over his harem of fifteen or twenty cows, all huddled together on the same spot he had first chosen.

"The fighting is mostly or entirely done with the mouth, the opponents seizing each other with the teeth and clenching the jaws; nothing but sheer strength can shake them loose, and that effort almost always leaves an ugly wound, the sharp canines tearing out deep gutters in the skin and blubber or shredding the flippers into ribbon-strips.

"They usually approach each other with averted heads and a great many false passes before either one or the other takes the initiative by griping; the heads are darted out and back as quick as flash, their hoarse roaring and shrill, piping whistle never ceases, while their fat bodies writhe and swell with exertion and rage, fur flying in air and blood streaming down—all combined make a picture fierce and savage enough, and from its great novelty, exceedingly strange at first sight.

"In these battles the parties are always distinct, the offensive and the defensive; if the latter proves the weaker he withdraws from the position occupied, and is never followed by his conqueror, who complacently throws up one of his hind flippers, fans himself, as it were, to cool himself from the heat of the conflict, utters a peculiar chuckle of satisfaction and contempt, with a sharp eye open for the next covetous bull or 'see-catch.'*

"The period occupied by the males in taking and holding their positions on the rookery offers a favorable opportunity in which to study them in the thousand and one different attitudes and postures assumed between the two extremes of desperate conflict and deep sleep—sleep so sound that one can, by keeping to the leeward, approach close enough, stepping softly, to pull the whiskers of any one taking a nap on a clear place; but after the first touch to these whiskers the trifler must jump back with great celerity, if he has any regard for the sharp teeth and tremendous shaking which will surely overtake him if he does not.

"The neck, chest, and shoulders of a fur-seal bull comprise more than two-thirds of his whole weight, and in this long thick neck and fore limbs is embodied the larger portion of his strength; when on land, with the fore feet he does all climbing

---

* "'See-catch,' native name for the bulls on the rookeries, especially those which are able to maintain their position."

over rocks, over the grassy hummocks back of the rookery, the hind flippers being gathered up after every second step forward, as described in the manner of walking; these fore feet are the propelling power when in water, almost exclusively, the hinder ones being used as rudders chiefly.

"The covering to the body is composed of two coats, one being of short, crisp, glistening over-hair, and the other a close, soft, elastic pelage, or fur, which gives distinctive value to the pelt.

"At this season of first 'hauling up' in the spring, the prevailing color of the bulls, after they dry off and have been exposed to the weather, is a dark, dull brown, with a sprinkling of lighter brown-black, and a number of hoary or frosted-gray coats; on the shoulders the over-hair is either a gray or rufous-ocher, called the 'wig;' these colors are most intense upon the back of the head, neck, and spine, being lighter underneath. The skin of the muzzle and flippers, a dark bluish black, fading to a reddish and purplish tint in some. The ears and tail are also similar in tint to the body, being in the case of the former a trifle lighter; the ears on a bull fur-seal are from an inch to an inch and a half in length; the *pavilions* tightly rolled up on themselves so that they are similar in shape and size to the little finger on the human hand, cut off at the second (phalangeal) joint, a shade more cone-shaped, for they are greater in diameter at the base than at the tip.

"I think it probable that the animal has and exerts the power of compressing or dilating this scroll-like *pavilion* to its ear, accordingly as it dives deep or rises in the water; and also, I am quite sure that the hair-seal has this control over the *meatus externus*, from what I have seen of it; but I have not been able to verify it in either case by observation; but such opportunity as I have had, gives me undoubted proof of the greatest keenness in hearing; for it is impossible to approach one, even when sound asleep; if you make any noise, frequently no matter how slight, the alarm will be given instantly by the insignificant-looking auditors, and the animal, rising up with a single motion erect, gives you a stare of astonishment, and at this season of defiance, together with incessant surly roaring, growling, and 'spitting.'

"This spitting, as I call it, is by no means a fair or full expression of the most characteristic sound and action, peculiar, so far as I have observed, to the fur-seals, the bulls in particular.

It is the usual prelude to their combats, and follows somewhat in this way: when the two disputants are nearly within reaching or striking distance, they make a number of feints or false passes at one another, with the mouth wide open and lifting the lips or snarling, so as to exhibit the glistening teeth, and with each pass they expel the air so violently through the larynx as to make a rapid choo-choo-choo sound, like the steam-puffs in the smoke-stack of a locomotive when it starts a heavy train, and especially when the driving-wheels slip on the rail.

"All the bulls now have the power and frequent inclination to utter four entirely distinct calls or notes—a hoarse, resonant roar, loud and long; a low gurgling growl; a chuckling, sibilant, piping whistle, of which it is impossible to convey an adequate idea, for it must be heard to be understood; and this spitting, just described. The cows* have but one note—a hollow, prolonged, *bla-a-ting* call, addressed only to their pups; on all other occasions they are usually silent. It is something like the cry of a calf or sheep. They also make a spitting sound, and snort, when suddenly disturbed. The pups '*bla-at*' also, with little or no variation, the sound being somewhat weaker and hoarser than that of their mothers for the first two or three weeks after birth; they, too, spit and cough when aroused suddenly from a nap or driven into a corner. A number of pups crying at a short distance off bring to mind very strongly the idea of a flock of sheep '*baa-aa-ing.*'

" Indeed, so similar is the sound that a number of sheep brought up from San Francisco to Saint George's Island during the summer of 1873 were constantly attracted to the rookeries, running in among the seals, and had to be driven away to a good feeding-ground by a small boy detailed for the purpose.

---

* "Without explanation I may be considered as making use of misapplied terms in describing these animals, for the inconsistency of coupling 'pups' with 'cows' and 'bulls,' and 'rookeries' with the breeding-grounds of the same, cannot fail to be noticed; but this nomenclature has been given and used by the English and American whalemen and sealing-parties for many years, and the characteristic features of the seals suit the odd naming exactly, so much so that I have felt satisfied to retain the style throughout as rendering my description more intelligible, especially so to those who are engaged in the business or may be hereafter. The Russians are more consistent, but not so 'pat.' The bull is called 'see-catch,' a term implying strength, vigor, &c.; the cow, 'matkah,' or mother; the pups, 'kotickie,' or little seals; the non-breeding males, under six and seven years, 'holluschickie,' or bachelors. The name applied collectively to the fur-seal by them is 'morskie-kot,' or sea-cat."

" The sound arising from these great breeding-grounds of the fur-seal, where thousands upon thousands of angry, vigilant bulls are roaring, chuckling, piping, and multitudes of seal-mothers are calling in hollow, bla-ating tones to their young, which in turn respond incessantly, is simply indescribable. It is, at a slight distance, softened into a deep booming, as of a cataract, and can be heard a long distance off at sea, under favorable circumstances as far as five or six miles, and frequently warns vessels that may be approaching the islands in thick, foggy weather, of the positive, though unseen, proximity of land. Night and day, throughout the season, the din of the rookeries is steady and constant.

" The seals seem to suffer great inconvenience from a comparatively low degree of heat; for, with a temperature of 46° and 48° on land, during the summer, they show signs of distress from heat whenever they make any exertion, pant, raise their hind flippers, and use them incessantly as fans. With the thermometer at 55°–60°, they seem to suffer even when at rest, and at such times the eye is struck by the kaleidoscopic appearance of a rookery, on which a million seals are spread out in every imaginable position their bodies can assume, all industriously fanning themselves, using sometimes the fore flippers as ventilators, as it were, by holding them aloft motionless, at the same moment fanning briskly with the hind flipper, or flippers, according as they sit or lie. This wavy motion of flapping and fanning gives a peculiar shade of hazy indistinctness to the whole scene, which is difficult to express in language; but one of the most prominent characteristics of the fur-seal is this fanning manner in which they use their flippers, when seen on the breeding-grounds in season. They also, when idling, as it were, off shore at sea, lie on their sides, with only a partial exposure of the body, the head submerged, and hoist up a fore or hind flipper clear of the water, while scratching themselves or enjoying a nap; but in this position there is no fanning. I say ' scratching,' because the seal, in common with all animals, is preyed upon by vermin, a species of louse and a tick, peculiar to itself.

"All the bulls, from the very first, that have been able to hold their positions, have not left them for an instant, night or day, nor do they do so until the end of the rutting-season, which subsides entirely between the 1st and 10th of August, beginning shortly after the coming of the cows in June. Of necessity, therefore, this causes them to fast, to abstain entirely from food of any kind, or water, for three months at least, and a few of them stay four months before going into the water for the first time after hauling up in May.

"This alone is remarkable enough, but it is simply wonderful when we come to associate the condition with the unceasing activity, restlessness, and duty devolved upon the bulls as heads and fathers of large families. They do not stagnate, like bears in caves; it is evidently accomplished or due to the absorption of their own fat, with which they are so liberally supplied when they take their positions on the breeding-ground, and which gradually diminishes while they remain on it. But still some most remarkable provision must be made for the entire torpidity of the stomach and bowels, consequent upon their being empty and unsupplied during this long period, which, however, in spite of the violation of a supposed physiological law, does not seem to affect them, for they come back just as sleek, fat, and ambitious as ever in the following season.

"I have examined the stomachs of a number which were driven up and killed immediately after their arrival in spring, and natives here have seen hundreds, even thousands, of them during the killing-season in June and July, but in no case has anything been found other than the bile and ordinary secretions of healthy organs of this class, with the exception only of finding in *every* one a snarl or cluster of worms (*Nematoda*), from the size of a walnut to that of one's fist, the fast apparently having no effect on them, for when three or four hundred old bulls were slaughtered late in the fall, to supply the natives with 'bidarkee' or canoe skins, I found these worms in a lively condition in every paunch cut open, and their presence, I think, gives some reason for the habit which these old bulls have of swallowing small bowlders, the stones in some of the stomachs weighing half a pound or so, and in one paunch I found about five pounds in the aggregate of larger pebbles, which in grinding against one another must destroy, in a great measure, these intestinal pests. The sea-lion is also troubled in the same way by a similar species of worm, and I have preserved a stomach of one of these animals in which are more than ten pounds of bowlders, some of them alone quite large. The greater size of this animal enables it to swallow stones which weigh two and three pounds. I can ascribe no other cause for this habit among these animals than that given, as

Misc. Pub. No. 12——23

they are of the highest type of the carnivora, eating fish as a regular means of subsistence; [*] varying the monotony of this diet with occasional juicy fronds of sea-weed, or kelp, and perhaps a crab, or such, once in a while, provided it is small and tender, or soft-shelled.

" Between the 12th and 14th of June the first of the cow-seals come up from the sea, and the bulls signalize it by a universal, spasmodic, desperate fighting among themselves.

" The strong contrast between the males and females in size and shape is heightened by the air of exceeding peace and amiability which the latter class exhibit.

" The cows are from 4 to 4½ feet in length from head to tail, and much more shapely in their proportions than the bulls, the neck and shoulder being not near so fat and heavy in proportion to the posteriors.

" When they come up, wet and dripping, they are of a dull, dirty-gray color, darker on the back and upper parts, but in a few hours the transformation made by drying is wonderful; you would hardly believe they could be the same animals, for they now fairly glisten with a rich steel and maltese-gray luster on the back of the head, neck, and spine, which blends into an almost pure white on the chest and abdomen. But this beautiful coloring in turn is altered by exposure to the weather, for in two or three days it will gradually change to a dull, rufous ocher below, and a cinereous-brown and gray-mixed above; this color they retain throughout the breeding-season up to the time of shedding the coat in August.

" The head and eye of the female are really attractive; the expression is exceedingly gentle and intelligent; the large, lustrous eyes, in the small, well-formed head, apparently gleam with benignity and satisfaction when she is perched up on some convenient rock and has an opportunity to quietly fan herself.

" The cows appear to be driven on to the rookeries by an accurate instinctive appreciation of the time in which their period of gestation ends; for in all cases marked by myself, the pups are born soon after landing, some in a few hours after, but most usually a day or two elapses before delivery.

[* The habit of swallowing stones is one apparently common to all of the Pinnipeds. The common belief among sealers and others is that they take in these stones as ballast. Compare on this point a quotation already given respecting the Southern Sea Lion (*Otaria jubata*), *anteà*, p. 311. Mr. Elliott's explanation appears to be more reasonable than most that have been proposed.]

" They are noticed and received by the bulls on the water-line station with much attention; they are alternately coaxed and urged up on to the rocks, and are immediately under the most jealous supervision; but owing to the covetous and ambitious nature of the bulls which occupy the stations reaching way back from the water-line, the little cows have a rough-and-tumble time of it when they begin to arrive in small numbers at first; for no sooner is the pretty animal fairly established on the station of bull number one, who has installed her there, he perhaps sees another one of her style down in the water from which she has just come, and in obedience to his polygamous feeling, he devotes himself anew to coaxing the later arrival in the same winning manner so successful in her case, when bull number two, seeing bull number one off his guard, reaches out with his long strong neck and picks the unhappy but passive creature up by the scruff of hers, just as a cat does a kitten, and deposits her on his seraglio-ground; then bulls number three, four, and so on, in the vicinity, seeing this high-handed operation, all assail one another, and especially bull number two, and have a tremendous fight, perhaps for half a minute or so, and during this commotion the cow generally is moved or moves farther back from the water, two or three stations more, where, when all gets quiet, she usually remains in peace. Her last lord and master, not having the exposure to such diverting temptation as had her first, he gives her such care that she not only is unable to leave did she wish, but no other bull can seize upon her. This is only one instance of the many different trials and tribulations which both parties on the rookery subject themselves to before the harems are filled. Far back, fifteen or twenty stations deep from the water-line sometimes, but generally not more on an average than ten or fifteen, the cows crowd in at the close of the season for arriving, July 10 to 14, and then they are able to go about pretty much as they please, for the bulls have become greatly enfeebled by this constant fighting and excitement during the past two months, and are quite content with even only one or two partners.

" The cows seem to haul in compact bodies from the water up to the rear of the rookeries, never scattering about over the ground; and they will not lie quiet in any position outside of the great mass of their kind. This is due to their intensely gregarious nature, and for the sake of protection. They also select land with special reference to the drainage, having a great dislike to water-puddled ground. This is well shown on Saint Paul.

"I have found it difficult to ascertain the average number of cows to one bull on the rookery, but I think it will be nearly correct to assign to each male from twelve to fifteen females, occupying the stations nearest the water, and those back in the rear from five to nine. I have counted forty-five cows all under the charge of one bull, which had them penned up on a flat table-rock, near Keetavie Point; the bull was enabled to do this quite easily, as there was but one way to go to or come from this seraglio, and on this path the old Turk took his stand and guarded it well.

"At the rear of all these rookeries there is always a large number of able-bodied bulls, who wait patiently, but in vain, for families, most of them having had to fight as desperately for the privilege of being there as any of their more fortunately-located neighbors, who are nearer the water than themselves; but the cows do not like to be in any outside position, where they are not in close company, lying most quiet and content in the largest harems, and these large families pack the surface of the ground so thickly that there is hardly moving or turning room until the females cease to come up from the sea; but the inaction on the part of the bulls in the rear during the rutting-season only serves to qualify them to move into the places vacated by those males who are obliged to leave from exhaustion, and to take the positions of jealous and fearless protectors for the young pups in the fall.

"The courage with which the fur-seal holds his position, as the head and guardian of a family, is of the very highest order, compared with that of other animals. I have repeatedly tried to drive them when they have fairly established themselves, and have almost always failed, using every stone at my command, making all the noise I could, and, finally, to put their courage to the full test, I walked up to within 20 feet of a bull at the rear and extreme end of Tolstoi Rookery, who had four cows in charge, and commenced with my double-barreled breech-loading shot-gun to pepper him all over with mustard-seed or dust shot. His bearing, in spite of the noise, smell of powder, and pain, did not change in the least from the usual attitude of determined defense which nearly all the bulls assume when attacked with showers of stones and noise; he would dart out right and left and catch the cows, which tim-

idly attempted to run after each report, and fling and drag
them back to their places; then, stretching up to his full height,
look me directly and defiantly in the face, roaring and spitting
most vehemently. The cows, however, soon got away from
him; but he still stood his ground, making little charges on me
of 10 or 15 feet in a succession of gallops or lunges, spitting
furiously, and then retreating to the old position, back of which
he would not go, fully resolved to hold his own or die in the
attempt.

"This courage is all the more noteworthy from the fact that,
in regard to man, it is invariably of a defensive character. The
seal, if it makes you turn when you attack it, never follows
you much farther than the boundary of its station, and no ag-
gravation will compel it to become offensive, as far as I have
been able to observe.

"The cows, during the whole season, do great credit to their
amiable expression by their manner and behavior on the rook
ery; never fight or quarrel one with another, and never or sel
dom utter a cry of pain or rage when they are roughly handled
by the bulls, who frequently get a cow between them and tear
the skin from her back, cutting deep gashes into it, as they
snatch her from mouth to mouth. These wounds, however,
heal rapidly, and exhibit no traces the next year.

"The cows, like the bulls, vary much in weight. Two were
taken from the rookery nearest Saint Paul's Village, after they
had been delivered of their young, and the respective weights
were 56 and 101 pounds, the former being about three or four
years old, and the latter over six. They both were fat and in
excellent condition.

"It is quite out of the question to give a fair idea of the positions in which the seals rest when on land. They may be said
to assume every possible attitude which a flexible body can be
put into. One favorite position, especially with the cows, is to
perch upon a point or top of some rock and throw their heads
back upon their shoulders, with the nose held aloft, then, closing their eyes, take short naps without changing, now and then
gently fanning with one or the other of the long, slender hind
flippers; another, and the most common, is to curl themselves
up, just as a dog does on a hearth-rug, bringing the tail and
the nose close together. They also stretch out, laying the head
straight with the body, and sleep for an hour or two without
moving, holding one of the hinder flippers up all the time, now
and then gently waving it, the eyes being tightly closed.

"The sleep of the fur-seal, from the old bull to the young pup, is always accompanied by a nervous, muscular twitching and slight shifting of the flippers; quivering and uneasy rolling of the body, accompanied by a quick folding anew of the fore flippers, which are signs, as it were, of their having nightmares, or sporting, perhaps, in a visionary way far off in some dreamland sea; or disturbed, perhaps more probably, by their intestinal parasites. I have studied hundreds of all classes, stealing softly up so closely that I could lay my hand on them, and have always found the sleep to be of this nervous description. The respiration is short and rapid, but with no breathing (unless your ear is brought very close) or snoring sound; the heaving of the flanks only indicates the action. I have frequently thought that I had succeeded in finding a snoring seal, especially among the pups, but a close examination always gave some abnormal reason for it, generally a slight distemper, by which the nostrils were stopped up to a greater or less degree.

"As I have said before, the cows, soon after landing, are delivered of their young.

"Immediately after the birth of the pup, (twins are rare, if ever [occurring],) it finds its voice, a weak, husky *blaat*, and begins to paddle about, with eyes wide open, in a confused sort of way for a few minutes until the mother is ready to give it attention, and, still later, suckle it; and for this purpose she is provided with four small, brown nipples, placed about eight inches apart, lengthwise with the body, on the abdomen, between the fore and hinder flippers, with some four inches of space between them transversely. The nipples are not usually visible; only seen through the hair and fur. The milk is abundant, rich, and creamy. The pups nurse very heartily, gorging themselves.

"The pup at birth, and for the next three months, is of a jet-black color, hair, eyes, and flippers, save a tiny white patch just back of each fore foot, and weighs from 3 to 4 pounds, and 12 to 14 inches long; it does not seem to nurse more than once every two or three days, but in this I am most likely mistaken, for they may have received attention from the mother in the night or other times in the day when I was unable to watch them.

"The apathy with which the young are treated by the old on the breeding-grounds is somewhat strange. I have never seen a cow caress or fondle her offspring, and should it stray but a short distance from the harem, it can be picked up and killed before the mother's eyes, without causing her to show the slightest concern. The same indifference is exhibited by the bull to all that takes place outside of the boundary of his seraglio. While the pups are, however, within the limits of his harem-ground, he is a jealous and fearless protector; but if the little animals pass beyond this boundary, then they may be carried off without the slightest attention in their behalf from their guardian.

"It is surprising to me how few of the pups get crushed to death while the ponderous bulls are floundering over them when engaged in fighting. I have seen two bulls dash at each other with all the energy of furious rage, meeting right in the midst of a small 'pod' of forty or fifty pups, trampling over them with their crushing weights, and bowling them out right and left in every direction, without injuring a single one. I do not think more than 1 per cent. of the pups born each season are lost in this manner on the rookeries.

"To test the vitality of these little animals, I kept one in the house to ascertain how long it could live without nursing, having taken it immediately after birth and before it could get any taste of its mother's milk; it lived nine days, and in the whole time half of every day was spent in floundering about over the floor, accompanying the movement with a persistent hoarse blaating. This experiment certainly shows wonderful vitality, and is worthy of an animal that can live four months without food or water and preserve enough of its latent strength and vigor at the end of that time to go far off to sea, and return as fat and hearty as ever during the next season.

"In the pup, the head is the only disproportionate feature when it is compared with the proportion of the adult form, the neck being also relatively shorter and thicker. I shall have to speak again of it, as it grows and changes, when I finish with the breeding-season now under consideration.

"The cows appear to go to and come from the water quite frequently, and usually return to the spot, or its neighborhood, where they leave their pups, crying out for them, and recognizing the individual replies, though ten thousand around, all together, should blaat at once. They quickly single out their own and attend them. It would be a very unfortunate matter if the mothers could not identify their young by sound, since their pups get together like a great swarm of bees, spread out upon the ground in 'pods' or groups, while they are young, and not very large, but by the middle and end of September, until they leave in November, they cluster together, sleeping and frolicking by tens of thousands. A mother comes up from the water, where she has been to wash, and perhaps to feed, for the last day or two, to about where she thinks her pup should be, but misses it, and finds instead a swarm of pups in which it has been incorporated, owing to its great fondness for society. The mother, without at first entering into the crowd of thousands, calls out, just as a sheep does for her lambs, listens, and out of all the din she—if not at first, at the end of a few trials—recognizes the voice of her offspring, and then advances, striking out right and left, and over the crowd, toward the position from which it replies; but if the pup at this time happens to be asleep she hears nothing from it, even though it were close by, and in this case the cow, after calling for a time without being answered, curls herself up and takes a nap, or lazily basks, and is most likely more successful when she calls again.

"The pups themselves do not know their mothers, but they are so constituted that they incessantly cry out at short intervals during the whole time they are awake, and in this way a mother can pick, out of the monotonous blaating of thousands of pups, her own, and she will not permit any other to suckle.

"Between the end of July and the 5th or 8th of August the rookeries are completely changed in appearance; the systematic and regular disposition of the families, or harems, over the whole extent of ground has disappeared; all order heretofore existing seems to be broken up. The rutting-season over, those bulls which held positions now leave, most of them very thin in flesh and weak, and I think a large proportion of them do not come out again on the land during the season; and such as do come, appear, not fat, but in good flesh, and in a new coat of rich dark and gray-brown hair and fur, with gray and grayish-ocher 'wigs' or over-hair on the shoulders, forming a strong contrast to the dull, rusty-brown and umber dress in which they appeared during the summer, and which they had begun to shed about the 15th of August, in common with the cows and bachelor seals. After these bulls leave, at the close of their season's work, those of them that do return to the land do not come back until the end of September, and do not haul up on the rookery-grounds as a rule, preferring to herd together, as do the young males, on the sand-beaches and other rocky points close to the water. The cows, pups, and those bulls which have been in retirement, now take possession, in a very disorderly manner, of the rookeries; also, come a large number of young, three, four, and five year old males, who have not been permitted to land among the cows, during the rutting-season, by the older, stronger bulls, who have savagely fought them off whenever they made (as they constantly do) an attempt to land.

" Three-fourths, at least, of the cows are now off in the water, only coming ashore to nurse and look after their pups a short time. They lie idly out in the rollers, ever and anon turning over and over, scratching their backs and sides with their fore and hind flippers. Nothing is more suggestive of immense comfort and enjoyment than is this action of these animals. They appear to get very lousy on the breeding-ground, and the frequent winds and showers drive and spatter sand into their fur and eyes, making the latter quite sore in many cases. They also pack the soil under foot so hard and solid that it holds water in the surface depressions, just like so many rock basins, on the rookery; out and into these puddles they flounder and patter incessantly, until evaporation slowly abates the nuisance.

" The pups sometimes get so thoroughly plastered in these muddy, slimy puddles, that their hair falls off in patches, giving them the appearance of being troubled with scrofula or some other plague, at first sight, but they are not, from my observation, permanently injured.

" Early in August (8th) the pups that are nearest the water on the rookeries essay swimming, but make slow and clumsy progress, floundering about, when over head in depth, in the most awkward manner, thrashing the water with their fore flippers, not using the hinder ones. In a few seconds, or a minute at the most, the youngest is so weary that he crawls out upon the rocks or beach, and immediately takes a recuperative nap, repeating the lesson as quick as he awakes and is rested. They soon get familiar with the water, and delight in it, swimming in endless evolutions, twisting, turning, diving, and when exhausted, they draw up on the beach again, shake themselves as young dogs do, either going to sleep on the spot, or having a lazy frolic among themselves.

" In this matter of learning to swim, I have not seen any

'driving' of the young pups into the water by the old in order to teach them this process, as has been affirmed by writers on the subject of seal life.

"The pups are constantly shifting, at the close of the rutting-season, back and forth over the rookery in large squads, some-times numbering thousands. In the course of these changes of position they all come sooner or later in contact with the sea; the pup blunders into the water for the first time in a most awkward manner, and gets out again as quick as it can, but so far from showing any fear or dislike of this, its most natural element, as soon as it rests from its exertion, is immediately ready for a new trial, and keeps at it, if the sea is not too stormy or rough at the time, until it becomes quite familiar with the water, and during all this period of self-tuition it seems to thoroughly enjoy the exercise.

"By the 15th of September all the pups have become familiar with the water, have nearly all deserted the background of the rookeries and are down by the water's edge, and skirt the rocks and beaches for long distances on ground previously un-occupied by seals of any class.

"They are now about five or six times their original weight, and are beginning to shed their black hair and take on their second coat, which does not vary at this age between the sexes. They do this very slowly, and cannot be called out of molting or shedding until the middle of October, as a rule.

"The pup's second coat, or sea-going jacket, is a uniform, dense, light pelage, or under-fur, grayish in some, light-brown in others, the fine, close, soft, and elastic hairs which compose it being about one-half of an inch in length, and over-hair, two-thirds of an inch long, quite coarse, giving the color by which you recognize the condition. This over-hair, on the back, neck, and head, is a dark chinchilla-gray, blending into a white, just tinged with a grayish tone on the abdomen and chest. The upper lip, where the whiskers or mustache takes root, is of a lighter-gray tone than that which surrounds. This mustache consists of fifteen or twenty longer or shorter whitish-gray bristles (one-half to three inches) on each side and back of the nostrils, which are, as I have before said, similar to that of a dog.

"The most attractive feature about the fur-seal pup, and upward as it grows, is the eye, which is exceedingly large, dark, and liquid, with which, for beauty and amiability, together with intelligence of expression, those of no other animal can be compared. The lids are well supplied with eyelashes.

"I do not think that their range of vision on land, or out of the water, is very great. I have had them (the adults) catch sight of my person, so as to distinguish it as a foreign character, three and four hundred paces off, with the wind blowing strongly from them toward myself, but generally they will allow you to approach very close indeed, before recognizing your strangeness, and the pups will scarcely notice the form of a human being until it is fairly on them, whereupon they make a lively noise, a medley of coughing, spitting, snorting, blaating, and get away from its immediate vicinity, but instantly resume, however, their previous occupation of either sleeping or playing, as though nothing had happened.

"But the power of scent is (together with their hearing, before mentioned) exceedingly keen, for I have found that I would most invariably awake them from soundest sleep if I got to the windward, even when standing a considerable distance off.

"To recapitulate and sum up the system of reproduction on the rookeries as the seals seem to have arranged it, I would say, that—

"First. The earliest bulls appear to land in a negligent, indolent way, shortly after the rocks at the water's edge are free from ice, frozen snow, &c. This is generally about the 1st to the 5th of May. They land first and last in perfect confidence and without fear, very fat, and of an average weight of five hundred pounds; some staying at the water's edge, some going away back, in fact all over the rookery.

"Second. That by the 10th or 12th of June, all the stations on the rookeries have been mapped out, fought for, and held in waiting for the cows by the strongest and most enduring bulls, who are, as a rule, never under six years of age, and sometimes three, and even occasionally four times as old.

"Third. That the cows make their first appearance, as a class, by the 12th or 15th of June, in rather small numbers, but by the 23d and 25th of this month they begin to flock up so as to fill the harems very perceptibly, and by the 8th or 10th of July they have most all come, stragglers excepted; average weight eighty pounds.

"Fourth. That the rutting-season is at its height from the 10th to the 15th of July, and that it subsides entirely at the end of this month and early in August, and that it is confined entirely to the land.

"Fifth. That the cows bear their first young when three years of age.

"Sixth. That the cows are limited to a single pup each, as a rule, in bearing, and this is born soon after landing; no exception has thus far been witnessed.

"Seventh. That the bulls who have held the harems leave for the water in a straggling manner at the close of the rutting-season, greatly emaciated, not returning, if at all, until six or seven weeks have elapsed, and that the regular systematic distribution of families over the rookeries is at an end for the season, a general medley of young bulls now free to come up from the water, old males who have not been on seraglio duty, cows, and an immense majority of pups, since only about 25 per cent. of their mothers are out of the water at a time.

"The rookeries lose their compactness and definite boundaries by the 25th to 28th July, when the pups begin to haul back and to the right and left in small squads at first, but as the season goes on, by the 18th August, they swarm over three and four times the area occupied by them when born on the rookeries. The system of family arrangement and definite compactness of the breeding-classes begins at this date to break up.

"Eighth. That by the 8th or 10th of August the pups born nearest the water begin to learn to swim, and by the 15th or 20th of September they are all familiar more or less with it.

"Ninth. That by the middle of September the rookeries are entirely broken up; only confused, straggling bands of cows, young bachelors, pups, and small squads of old bulls, crossing and recrossing the ground in an aimless, listless manner; the season is over, but many of these seals do not leave these grounds until driven off by snow and ice, as late as the end of December and 12th of January.

"[*Hauling-grounds.*"—] This recapitulation is the sum and substance of my observations on the rookeries, and I will now turn to the consideration of the hauling-grounds, upon which the yearlings and almost all the males under six years come out from the sea in squads from a hundred to a thousand, and, later in the season, by hundreds of thousands, to sleep and frolic,.going from a quarter to half a mile back from the sea, as at English Bay.

"This class of seals are termed 'holluschukie' (or 'bachelor seals') by the natives. It is with the seals of this division that these people are most familiar, since they are, together with a

few thousand pups and some old bulls, the only ones driven up
to the killing-grounds for their skins, for reasons which are excellent, and which shall be given further on.

"Since the 'holluschukie' are not permitted by their own
kind to land on the rookeries and rest there, they have the
choice of two methods of landing and locating.

"One of these opportunities, and least used, is to pass up
from and down to the water, through a rookery on a pathway
left by common consent between the harems. On these lines
of passage they are unmolested by the old and jealous bulls,
who guard the seraglios on either side as they go and come;
generally there is a continual file of them on the way, traveling up or down.

"As the two and three year old holluschukie come up in small
squads with the first bulls in the spring, or a few days later,
these common highways between the rear of the rookery-ground
and the sea get well defined and traveled over before the arrival
of the cows; for just as the bulls crowd up for their stations, so
do the bachelors, young and old, increase. These roadways
may be termed the lines of least resistance in a big rookery;
they are not constant; they are splendidly shown on the large
rookeries of Saint Paul's, one of them (Tolstoi) exhibiting this
feature finely, for the hauling-ground lies up back of the rookery, on a flat and rolling summit, 100 to 120 feet above the sea-
level. The young males and yearlings of both sexes come
through the rookery on these narrow pathways, and, before
reaching the resting-ground above, are obliged to climb up an
almost abrupt bluff, by following and struggling in the little
water-runs and washes which are worn in its face. As this
is a large hauling-ground, on which fifteen or twenty thousand
commonly lie every day during the season, the sight always, at
all times, to be seen, in the way of seal climbing and crawling,
was exceedingly novel and interesting. They climb over and
up to places here where a clumsy man might at first sight
say he would be unable to ascend.

"The other method by which the 'holluschukie' enjoy themselves on land is the one most followed and favored. They, in
this case, repair to the beaches unoccupied between the rookeries, and there extend themselves out all the way back from
the water as far, in some cases, as a quarter of a mile, and even
farther. I have had under my eye, in one straightforward
sweep, from Zapad-nie to Tolstoi, (three miles,) a million and a

half of seals, at least, (about the middle of July.) Of these I estimated fully one-half were pups, yearlings, and 'holluschu-kie.' The great majority of the two latter classes were hauled out and packed thickly over the two miles of sand-beach and flat which lay between the rookeries; many large herds were back as far from the water as a quarter of a mile.

"A small flock of the younger ones, from one to three years old, will frequently stray away back from the hauling-ground lines, out and up onto the fresh moss and grass, and there sport and play, one with another, just as puppy-dogs do; and when weary of this gamboling, a general disposition to sleep is suddenly manifested, and they stretch themselves out and curl up in all the positions and all the postures that their flexible spines and ball-and-socket joints will permit. One will lie upon his back, holding up his hind flippers, lazily waving them in the air, while he scratches or rather rubs his ribs with the fore hands alternately, the eyes being tightly closed; and the breath, indicated by the heaving of his flanks, drawn quickly but regularly, as though in heavy sleep; another will be flat upon his stomach, his hind flippers drawn under and concealed, while he tightly folds his fore feet back against his sides, just as a fish will sometimes hold its pectoral fins; and so on, without end of variety, according to the ground and disposition of the animals.

"While the young seals undoubtedly have the power of going without food, they certainly do not sustain any long fasting periods on land, for their coming and going is frequent and irregular; for instance, three or four thick, foggy days will sometimes call them out by hundreds of thousands, a million or two, on the different hauling-grounds, where, in some cases, they lie so closely together that scarcely a foot of ground, over acres in extent, is bare; then a clearer and warmer day will ensue, and the ground, before so thickly packed with animal-life, will be almost deserted, comparatively, to be filled again immediately on the recurrence of favorable weather. They are in just as good condition of flesh at the end of the season as at the first of it.

"These bachelor seals are, I am sure, without exception, the most restless animals in the whole brute creation; they frolic and lope about over the grounds for hours, without a moment's cessation, and their sleep after this is short, and is accompanied with nervous twitchings and uneasy movements; they seem to be fairly brimful and overrunning with warm life. I have never observed anything like ill-humor grow out of their playing together; invariably well pleased one with another in all their frolicsome struggles.

"The pups and yearlings have an especial fondness for sporting on the rocks which are just at the water's level, so as to be alternately covered and uncovered by the sea-rollers. On the bare summit of these water-worn spots they struggle and clamber, a dozen or two at a time, occasionally, for a single rock; the strongest or luckiest one pushing the others all off, which, however, simply redouble their efforts and try to dislodge him, who thus has, for a few moments only, the advantage; for with the next roller and the other pressure, he generally is ousted, and the game is repeated. Sometimes, as well as I could see, the same squad of 'holluschukie' played around a rock thus situated, off 'Nah Speel' rookery, during the whole of one day; but, of course, they cannot be told apart.

"The 'holluschukie,' too, are the champion swimmers; at least they do about all the fancy tumbling and turning that is done by the fur-seals when in the water around the islands. The grave old bulls and their matronly companions seldom indulge in any extravagant display, such as jumping out of the water like so many dolphins, describing, as these youngsters do, beautiful elliptic curves, rising three and even four feet from the sea, with the back slightly arched, the fore flippers folded back against the sides, and the hinder ones extended and pressed together straight out behind, plumping in head first, re-appearing in the same manner after an interval of a few seconds.

"All classes will invariably make these dolphin-jumps [*] when they are suddenly surprised or are driven into the water, turning their heads, while sailing in the air, between the 'rises' and 'plumps,' to take a look at the cause of their disturbance. They all swim with great rapidity, and may be fairly said to dart with the velocity of a bird on the wing along under the water; and in all their swimming I have not been able yet to satisfy myself how they used their long, flexible, hind feet, other than as steering mediums. The propelling motion, if they have any, is so rapid, that my eye is not quick enough to catch it; the fore feet, however, can be very distinctly seen to work,

[* Mr. J. H. Blake, who accompanied Professor Agassiz on the Hassler Expedition to South America in 1871, as artist of the expedition, observed the Southern Sea Lions (*Otaria jubata*) performing similar evolutions.]

feathering forward and sweeping back flatly, opposed to the water, with great rapidity and energy, and are evidently the sole propulsive power.

"All their movements in the water, when in traveling or sport, are quick and joyous, and nothing is more suggestive of intense satisfaction and great comfort than is the spectacle of a few thousand old bulls and cows, off and from a rookery in August, idly rolling over, side by side, rubbing and scratching with the fore and hind flippers, which are here and there stuck up out of the water like lateen-sails, or 'cat-o'-nine tails,' in either case, as it may be.

"When the 'holluschukie' are up on land they can be readily separated into two classes by the color of their coats and size, viz, the yearlings, and the two, three, four, and five year old bulls.

" The first class is dressed just as they were after they shed their pup-coats and took on the second the previous year, in September and October, and now, as they come out in the spring and summer, the males and females cannot be distinguished apart, either by color or size; both yearling sexes having the same gray backs and white bellies, and are the same in behavior, action, weight, and shape.

"About the 15th and 20th of August they begin to grow 'stagey,' or shed, in common with all the other classes, the pups excepted. The over-hair requires about six weeks from the commencement of the dropping or falling out of the old to its full renewal.

" The pelage, or fur, which is concealed externally by the hair, is also shed and renewed slowly in the same manner; but, being so much finer than the hair, it is not so apparent. It was to me a great surprise to 'learn,' from a man who has been heading a seal-killing party on these islands during the past three years, and the Government agent in charge of these interests, that the seal never shed its fur; that the over-hair only was cast off and replaced. To prove that it does, however, is a very simple matter, and does not require the aid of a microscope. For example, take up a prime spring or fall skin, after every single over-hair on it has been plucked out, and you will have difficulty, either to so blow upon the thick, fine fur, or to part it with the fingers, as to show the hide from which it has grown; then take a 'stagey' skin, by the end of August and early in September, when *all the over-hair is present, about one-third to one-half grown*, and the first puff you expend upon it easily shows the hide below, sometimes quite a broad welt. This under-fur, or pelage, is so fine and delicate, and so much concealed and shaded by the coarse over-hair, that a careless eye may be pardoned for any such blunder, but only a very casual observer could make it.

"The yearling cows retain the colors of the old coat in the new, and from this time on shed, year after year, just so, for the young and the old cows look alike, as far as color goes, when they haul up on the rookeries in the summer.

"The yearling males, however, make a radical change, coming out from their 'staginess' in a uniform dark-gray and gray-black mixed and lighter, and dark ocher, on the under and upper parts, respectively. This coat, next year, when they come up on the hauling-grounds, is very dark, and is so for the third, fourth, and fifth years, when, after this, they begin to grow more gray and brown, year by year, with rufous-ocher and whitish-gray tipped over-hair on the shoulders. Some of the very old bulls become changed to uniform dull grayish-ocher all over.

"The female does not get her full growth and weight until the end of her fourth year, so far as I have observed, but does the most of her growing in the first two.

"The male does not get his full growth and weight until the close of his seventh year, but realizes most of it by the end of the fifth, osteologically, and from this it may be, perhaps, truly inferred that the bulls live to an average age of eighteen or twenty years, if undisturbed in a normal condition, and that the cows attain ten or twelve under the same circumstances. Their respective weight, when fully mature and fat in the spring, will, I think, strike an average of four to five hundred pounds for the male and from seventy to eighty for the female.

" From the fact that all the young seals do not change much in weight, from the time of their first coming out in the spring till that of their leaving in the fall and early winter, I feel safe in saying, since they, too, are constantly changing from land to water and from water to land, that they feed at irregular but not long intervals during the time they are here under observation. I do not think the young males fast longer than a week or ten days at a time as a class.

"They leave evidences of their being on these great reproductive fields, chiefly on the rookeries, such as hundreds of the dead carcasses of those of them that have been infirm, sick, killed, or which have crawled off to die from death-wounds received in some struggle for a harem ; and over these decaying, putrid bodies, the living, old and young, clamber and patter, and by this constant stirring up of putrescent matter give rise to an exceedingly disagreeable and far-reaching 'funk,' which has been, by all the writers who have spoken on the subject, referred to as the smell which these animals have in rutting. If these creatures have any such odor peculiar to them when in this condition, I will frankly confess that I am unable to distinguish it from the fumes which are constantly being stirred up and rising out from these decaying carcasses of old seals and the many pups which have been killed accidentally by the old bulls while fighting with and charging back and forth against one another.

"They, however, have a peculiar smell when they are driven and get heated; their steaming breath exhalations possess a disagreeable, faint, sickly tone, but it can by no means be confounded with what is universally understood to be the rutting-odor among animals. The finger rubbed on a little fur-seal blubber will smell very much like that which is appreciated in their breath coming from them when driven, only stronger. Both the young and old fur-seals have this same breath-smell at all seasons.

"By the end of October and the 10th of November the great mass of the 'holluschukie' have taken their departure; the few that remain from now until as late as the snow and ice will permit them to do, in and after December, are all down by the water's edge, and hauled up almost entirely on the rocky beaches only, deserting the sand. The first snow falling makes them uneasy, as also does rain-fall. I have seen a large hauling-ground entirely deserted after a rainy day and night by its hundreds of thousands of occupants. The falling drops spatter and beat the sand into their eyes, fur, &c., I presume, and in this way make it uncomfortable for them.

"The weather in which the fur-seal delights is cool, moist, foggy, and thick enough to keep the sun always obscured so as to cast no shadows. Such weather, continued for a few weeks in June and July, brings them up from the sea by millions; but, as I have before said, a little sunlight and the temperature as high as 50° to 55°, will send them back from the hauling-grounds almost as quickly as they came. These sunny, warm days are, however, on Saint Paul's Island, very rare indeed, and so the seals can have but little ground of complaint, if we may presume that they have any at all."*

THE CHASE.—The manner of capturing the Fur Seals has greatly varied at different times and at different localities. Krascheninikow states that on Behring's Island, a century and a quarter ago, the common way of killing them was to first strike out their eyes with stones, and then beat out their brains with clubs. This he says was a work of so much labor that "three men were hardly able to kill one with 300 strokes." In consequence of their seldom landing on the Kamtschatka coast, the same writer states that the natives were accustomed to pursue them in boats, "and throw darts or harpoons at them." He says they had to be particularly cautious not to let the wounded Seal "fasten upon the side of the boat and overturn it," to prevent which he says some of the fishermen stood ready "with axes to cut off his paws."†

Captain Scammon thus describes the pursuit of the Fur Seal by the Indians of Vancouver Island: "When going in pursuit of seals, three or four natives embark in a canoe at an early hour in the morning, and usually return the following evening. The fishing-gear consists of two spears, which are fitted to a pronged pole fifteen feet in length; to the spears a line is attached, which is fastened to the spear-pole close to, or is held in the hand of, the spearman, when he darts the weapon. A seal-club is also provided, as well as two seal-skin buoys—the latter being taken in the canoe to be used in rough weather; or if a seal, after being speared, can not be managed with the line in hand, a buoy is 'bent on', and the animal is allowed to take its own course for a time. Its efforts to escape, by diving repeatedly, and plunging about near the surface of the water, soon exhaust the animal somewhat; and when a favorable time is presented, the spearman seizes the buoy, hauls in the line until within reach of the seal, and it is captured by being clubbed. But generally the line is held in the hand when the spear is thrust into the seal; then the pole is instantly withdrawn, and the canoe is hauled at once to the floundering creature, which is dispatched as before described. Indians from the Vancouver shore frequently start in the night, so as to be on the best seal-

---

* Condition of Affairs in Alaska, pp. 123–150.
† Hist. Kamtsch. (English ed.), p. 130.

ing ground in the morning. This locality is said to be south-west of Cape Classet, five to fifteen miles distant."*

In hunting Seals for their commercial products the common method of killing them appears to have generally been by club-bing them, as is at present practiced on the Seal Islands of Alaska, one or two heavy blows upon the head being sufficient to dispatch them. The method of attack is very much like that practiced in destroying herds of Walruses, already described. A large party cautiously land, when possible, to the leeward of a rookery, and then, at a given signal, rush upon the Seals, with loud shouting, and with their clubs soon destroy large numbers. It has generally been practiced without system or restraint, resulting in the speedy destruction of large rookeries. As is well known, the Southern Sea Bears or Fur Seals (*Arctocephalus* "*falklandicus*," *A. forsteri, A.* "*cinereus*," etc.) were long since practically exterminated at many localities where they were formerly very abundant, as has been the case with the Northern Fur Seal on our own Californian coast.   At one time the same destructive and ruinous policy was pursued by the Russians at the Prybilov Islands, but the folly of such a practice was soon perceived, and through government interference their extermi-nation there has been happily prevented.   Their destruction is at present regulated by the United States Government, the whole matter being judiciously and systematically managed. The manner of taking and killing the Seals, and the method adopted to prevent their decrease, has been described in detail by Mr. Elliott, and is here appended.

"TAKING THE SEALS.—By reference to the habits of the fur-seal, it is plain that two thirds of all the males that are born (and they are equal in number to the females born) are never per-mitted by the remaining' third, strongest by natural selection, to land upon the same ground with the females, which always herd together *en masse*. Therefore, this great band of bachelor seals, or 'holluschuckie,' is compelled, when it visits land, to live apart entirely, miles away frequently, from the breeding-grounds, and in this admirably perfect manner of nature are those seals which can be properly killed without injury to the rookeries selected and held aside, so that the natives can visit and take them as they would so many hogs, without disturbing in the slightest degree the peace and quiet of the breeding-grounds where the stock is perpetuated.

* Marine Mammalia, pp. 154, 155.

"The manner in which the natives capture and drive the holluschuckie up from the hauling-grounds to the slaughtering-fields near the villages and elsewhere, cannot be improved upon, and is most satisfactory.

"In the early part of the season large bodies of the young bachelor seals do not haul up on land very far from the water, a few rods at the most, and the men are obliged to approach slyly and run quickly between the dozing seals and the surf, before they take alarm and bolt into the sea, and in this way a dozen Aleuts, running down the long sand-beach of English Bay, some driving-morning early in June, will turn back from the water thousands of seals, just as the mold-board of a plow lays over and back a furrow of earth. As the sleeping seals are first startled they arise, and seeing men between them and the water, immediately turn, lope and scramble rapidly back over the land; the natives then leisurely walk on the flanks and in the rear of the drove thus secured, and direct and drive them over to the killing-grounds.

"A drove of seals on hard or firm grassy ground, in cool and moist weather, may with safety be driven at the rate of half a mile an hour; they can be urged along with the expenditure of a great many lives in the drove, at the speed of a mile or a mile and a quarter even per hour, but this is highly injudicious and is seldom ever done. A bull seal, fat and unwieldy, cannot travel with the younger ones, but it can lope or gallop as it were over the ground as fast as an ordinary man can run for a hundred yards, but then it falls to the earth supine, utterly exhausted, hot and gasping for breath.

"The seals, when driven thus to the killing-grounds, require but little urging; they are permitted to frequently halt and cool off, as heating them injures their fur; they never show fight any more than a flock of sheep would do, unless a few old seals are mixed in, which usually get so weary that they prefer to come to a stand-still and fight rather than to move; this action on their part is of great advantage to all parties concerned, and the old fellows are always permitted to drop behind and remain, for the fur on them is of little or no value, the pelage very much shorter, coarser, and more scant than in the younger, especially so on the parts posteriorly. This change in the condition of the fur seems to set in at the time of their shedding, in the fifth year as a rule.

"As the drove progresses the seals all move in about the same way, a kind of a walking-step and a sliding, shambling gallop, and the progression of the whole body is a succession of starts, made every few minutes, spasmodic and irregular. Every now and then a seal will get weak in the lumbar region, and drag his posterior after it for a short distance, but finally drops breathless and exhausted, not to revive for hours, days perhaps, and often never. Quite a large number of the weaker ones, on the driest driving-days, are thus laid out and left on the road; if one is not too much heated at the time, the native driver usually taps the beast over the head and removes its skin. This will happen, no matter how carefully they are driven, and the death-loss is quite large, as much as 3 or 4 per cent. on the longer drives, such as three and four miles, from Zapadnie or Polavina to the village on Saint Paul's, and I feel satisfied that a considerable number of those rejected from the drove and permitted to return to the water die subsequently from internal injuries sustained on the drive from overexertion. I therefore think it improper to extend drives of seals over any distance exceeding a mile or a mile and a half. It is better for all parties concerned to erect salt-houses and establish killing-grounds adjacent to all of the great hauling-grounds on Saint Paul's Island should the business ever be developed above the present limit. As matters now are, the ninety thousand seals belonging to the quota of Saint Paul last summer were taken and skinned in less than forty days within one mile from either the village, or salt-house on Northeast Point.

"KILLING THE SEALS.—The seals when brought up to the killing-grounds are herded there until cool and rested; then squads or 'pods' of fifty to two hundred are driven out from the body of the drove, surrounded and huddled up one against and over the other, by the natives, who carry each a long, heavy club of hard wood, with which they strike the seals down by blows upon the head; a single stroke of a heavy oak bludgeon, well and fairly delivered, will crush in at once the slight, thin bones of a seal's skull, laying the creature out lifeless; these strokes are usually repeated several times with each animal, but are very quickly done.

"The killing-gang, consisting usually of fifteen or twenty men at a time, are under the supervision of a chief of their own selection, and have, before going into action, a common understanding as to what grades to kill, sparing the others which are

unfit, under age, &c., permitting them to escape and return to
the water as soon as the marked ones are knocked down; the
natives then drag the slain out from the heap in which they
have fallen, and spread the bodies out over the ground just free
from touching one another so that they will not be hastened in
'heating' or blasting, finishing the work of death by thrusting
into the chest of each stunned and senseless seal a long, sharp
knife, which touches the vitals and bleeds it thoroughly; and
if a cool day, another 'pod' is started out and disposed of in
the same way, and so on until a thousand or two are laid out, or
the drove is finished; then they turn to and skin; but if it is a
warm day, every 'pod' is skinned as soon as it is knocked down.

"This work of killing as well as skinning is performed very
rapidly; for example, forty-five men or natives on Saint Paul's
during June and July, 1872, in less than four working-weeks
drove, killed, skinned, and salted the pelts of 72,000 seals.

"The labor of skinning is exceedingly severe, and is trying
to an expert, requiring long practice before the muscles of the
back and thighs are so developed as to permit a man to bend
down to and finish well a fair day's work.

"The body of the seal, preparatory to skinning, is rolled over
or put upon its back, and the native makes a single swift cut
through the skin down along the neck, chest, and belly, from
the lower jaw to the root of the tail, using for this purpose a
large, sharp knife. The fore and hind flippers are then successively lifted, and a sweeping circular incision is made through
the skin on them just at the point where the body-fur ends;
then, seizing a flap of the hide on either one side or the other of
the abdomen, the man proceeds to rapidly cut the skin clean
and free from the body and blubber, which he rolls over and
out from the skin by hauling up on it as he advances with his
work, standing all the time stooping over the carcass so that
his hands are but slightly above it or the ground. This operation of skinning a fair-sized seal takes the best men only a minute and a half, but the average time on the ground is about
four minutes.

"Nothing is left of the skin upon the carcass save a small patch
of each upper lip, on which the coarse mustache grows, the skin
on the tip of the lower jaw, the insignificant tail, together with
the bare hide of the flippers.

"The blubber of the fur-seal is of a faint yellowish white, and
lies entirely between the skin and the flesh, none being depos-

ited in between the muscles. Around the small and large intestines a moderate quantity of hard, firm fat is found. The blubber possesses an extremely offensive, sickening odor, difficult to wash from the hands. It makes, however, a very fair oil for lubricating, burning, &c.

"The flesh of the fur-seal, when carefully cleaned from fat or blubber, can be cooked, and by most people eaten, who, did they not know what it was, might consider it some poor, tough, dry beef, rather dark in color and overdone. That of the pup, however, while on the land and milk-fed, is tender and juicy but insipid.

"The skins are taken from the field to the salt-house, where they are laid out open, one upon another, 'hair to fat,' like so many sheets of paper, with salt profusely spread upon the fleshy sides, in 'kenches' or bins. After lying a week or two salted in this style they are ready for bundling and shipping, two skins to the bundle, the fur outside, tightly rolled up and strongly corded, having an average weight of twelve, fifteen, and twenty-two pounds when made up of two, three, and four year old skins respectively.

"The company leasing the islands are permitted by law to take one hundred thousand, and no more, annually; this they do in June and July; after that season the skins rapidly grow worthless by shedding, and do not pay for transportation and tax. The natives are paid forty cents a skin for the catch, and keep a close account of the progress of the work every day, as it is all done by them, and they know within fifty skins, one way or the other, when the whole number have been secured each season. This is the only occupation of some three hundred and fifty people here, and they naturally look well after it. The interest and close attention paid by these Aleuts on both islands to this business was both gratifying and instructive to me while stationed there."

In regard to the preparation and value of the skin Mr. Elliott states as follows:

"The common or popular notion regarding seal-skins is that they are worn by those animals just as they appear when offered for sale. This is a very great mistake; few skins are less attractive than the seal-skin as it is taken from the creature. The fur is not visible, concealed entirely by a coat of stiff over-hair, dull gray, brown, and grizzled. The best of these raw skins are worth only $5 to $10, but after dressing they bring from $25 to $40; and it takes three of them to make a lady's sack and boa."*

As an interesting supplement to this portion of the subject, I transcribe a letter from George C. Treadwell & Co., leading furriers, and long familiar with the manner of preparing the skins, addressed to Mr. Elliott (dated Albany, October 22, 1874), in which the process of dressing the skins for market is very clearly set forth. The letter (extracted from Mr. Elliott's Report) is as follows:

"The Alaska Commercial Company sold in London, December, 1873, about sixty thousand skins taken from the islands leased by our Government of the catch of 1873. The remainder of the catch, about forty thousand, were sold in March. This company have made the collection of seal from these islands much more valuable than they were before their lease, by the care used by them in curing the skins, and taking them only when in season. We have worked this class of seal for several years—when they were owned by the Russian American Fur Company, and during the first year they were owned by our Government.

"When the skins are received by us in the salt, we wash off the salt, placing them upon a beam somewhat like a tanner's beam, removing the fat from the flesh-side with a beaming-knife, care being required that no cuts or uneven places are made in the pelt. The skins are next washed in water and placed upon the beam with the fur up, and the grease and water removed by the knife. The skins are then dried by moderate heat, being tacked out on frames to keep them smooth. After being fully dried they are soaked in water and thoroughly cleansed with soap and water. In some cases they can be unhaired without this drying process, and cleansed before drying. After the cleansing process they pass to the picker, who dries the fur by stove-heat, the pelt being kept moist. When the fur is dry he places the skin on a beam, and while it is warm he removes the main coat of hair with a dull shoe-knife, grasping the hair with his thumb and knife, the thumb being protected by a rubber cob. The hair must be pulled out, not broken. After a portion is removed the skin must again be warmed at the stove, the pelt being kept moist. When the outer hairs have been mostly removed, he uses a beaming-knife to work out the finer hairs, (which are shorter,) and the remaining

* Condition of Affairs in Alaska, pp. 80–85.

coarser hairs. It will be seen that great care must be used, as the skin is in that soft state that too much pressure of the knife would take the fur also; indeed, bare spots are made; carelessly-cured skins are sometimes worthless on this account. The skins are next dried, afterward dampened on the pelt side, and shaved to a fine, even surface. They are then stretched, worked, and dried; afterward softened in a fulling-mill, or by treading them with the bare feet in a hogshead, one head being removed and the cask placed nearly upright, into which the workman gets with a few skins and some fine hard-wood saw-dust, to absorb the grease while he dances upon them to break them into leather. If the skins have been shaved thin, as required when finished, any defective spots or holes must now be mended, the skin smoothed and pasted with paper on the pelt side, or two pasted together to protect the pelt in dyeing. The usual process in the United States is to leave the pelt sufficiently thick to protect them without pasting.

"In dyeing, the liquid dye is put on with a brush, carefully covering the points of the standing fur. After lying folded, with the points touching each other, for some little time, the skins are hung up and dried. The dry dye is then removed, another coat applied, dried, and removed, and so on until the required shade is obtained. One or two of these coats of dye are put on much heavier and pressed down to the roots of the fur, making what is called the ground. From eight to twelve coats are required to produce a good color. The skins are then washed clean, the fur dried, the pelt moist. They are shaved down to the required thickness, dried, working them some while drying, then softened in a hogshead, and sometimes run in a revolving cylinder with fine sawdust to clean them. The English process does not have the washing after dyeing.

"I should, perhaps, say that, with all the care used, many skins are greatly injured in the working. Quite a quantity of English-dyed seal were sold last season for $17, damaged in the dye.

"The above is a general process, but we are obliged to vary for different skins; those from various parts of the world require different treatment, and there is quite a difference in the skins from the Seal Islands of our country—I sometimes think about as much as in the human race." *

HISTORY AND PROSPECTS OF THE FUR SEAL BUSINESS AT

---

* Condition of Affairs in Alaska, pp. 85, 86.

THE PRYBILOV ISLANDS.—From the speedy extermination of the Fur Seals of the Southern hemisphere at many points where they existed a century ago in apparently inexhaustible numbers,[*] the preservation of the Northern Fur Seals at the two small islands that now, so far as known, form their principal breeding-stations, becomes a matter of much zoölogical interest as well as of practical importance. The islands of Saint George and Saint Paul were discovered, respectively, in 1786 and 1787, and immediately after, it is stated, as many as six companies established themselves at these islands, all vieing with each other in the destruction of the Seals in consequence of the great commercial value of the skins. No record appears to have been kept of the number annually killed between 1787 and 1805, at which time the number of Seals frequenting the islands had greatly decreased. Then follows for two years a cessation of the slaughter, which was resumed in 1808. Up to 1822 the destruction of Seal life was indiscriminate and wholly without restriction from government or other sources. In this year it was ordered that young Seals should be spared each year for the purpose of keeping up the stock. This order was so honestly enforced that in four years the number of Seals on Saint Paul's Island increased tenfold. The number annually taken these years was only 8,000 to 10,000, instead of 40,000 to 50,000, the number formerly killed yearly. Subsequently the killing was allowed to greatly increase, which prevented any augmentation in the number of Seals. In 1834 the number allowed to be killed on Saint Paul's Island was reduced from 12,000 to 6,000. After this date the conditions of increase were more carefully studied and more carefully regarded, so that there was a gradual numerical increase from 1835 to 1857, when the rookeries are said to have become very nearly as large as now, the natives believing, however, that there has been, since the last-named date, a very gradual but steady increase. The great diminution seems to have set in about 1817, and to have continued till 1834, when, as Mr. Elliott expresses it, "hardly a tithe of the former numbers appeared on the ground." From 1835 to 1857 there was a steady increase, when the maximum then reached appears to have been maintained.

In regard to the number now present on these islands, Mr. Elliott estimated, from a careful survey of the breeding-grounds,

---

[*] See *anteà*, p. 334, footnote, *e. g.*, respecting their former abundance and early almost total extirpation at the Island of Juan Fernandez.

that in 1873 there were on the Prybilov Islands "*over four million seven hundred thousand*" Fur Seals, and that one million are born there annually, divided about equally between males and females. So many of these are destroyed by their natural enemies during the following six months that only about one-half return the succeeding spring. During the next winter about one-tenth of the remainder are also destroyed at sea, after which very few appear to die from natural causes. Only one-fifteenth of the annual increase of males can, in consequence of the peculiar habits of the animals, share in the office of reproduction. Assuming the above statement to be a fair estimate of the number of Seals annually born on the islands, Mr. Elliott states it as his belief that, after making due allowance for the number that perish at sea during early life, and for the perpetuation of the stock, 180,000 young male Seals may be annually taken for their skins.

"With regard to the *increase* of the seal-life," says Mr. Elliott, "I do not think it within the power of human management to promote this end to the slightest appreciable degree beyond its present extent and condition in a state of nature; for it cannot fail to be evident, from my detailed description of the habits and life of the fur-seal on these islands during a great part of the year, that could man have the same supervision and control over this animal during the *whole* season which he has at his command while they visit the land, he might cause them to multiply and increase, as he would so many cattle, to an indefinite number, only limited by time and means; but the case in question, unfortunately, takes the fur-seal six months out of every year far beyond the reach, or even cognizance, of any one, where it is exposed to known powerful and destructive natural enemies, and many others probably unknown, which prey upon it, and, in accordance with a well-recognized law of nature, keep it at about a certain number which has been for ages, and will be for the future, as affairs now are, *its maximum limit of increase*. This law holds good everywhere throughout the animal kingdom, regulating and preserving the equilibrium of life in a state of nature. Did it not hold good, these Seal Islands and all Bering Sea would have been literally covered, and have swarmed with them long before the Russians discovered them; but there were no more seals when first seen here by human eyes in 1786-'87 than there are now, in 1874, as far as all evidence goes."

"What can be done to promote their increase ? We cannot cause a greater number of females to be born every year; we do not touch or disturb these females as they grow up and live, and we save more than enough males to serve them. Nothing more can be done, for it is impossible to protect them from deadly enemies in their wanderings for food."

"In view, therefore, of all these facts," continues Mr. Elliott, "I have no hesitation in saying quite confidently that, under the present rules and regulations governing the sealing interests on these islands, the increase or diminution of the life will amount to nothing; that the seals will continue for all time in about the same number and condition." *

ENEMIES OF THE FUR SEALS.—Man, of course, stands first in importance as an enemy of the Fur Seals, but under the restrictions respecting the killing these animals now enforced at the Prybilov Islands, does not appear to have a very marked influence in effecting their decrease. That they suffer greatly from other animals is evident from the fact that only about one-half of the Seals annually born at the Seal Islands ever return there again. What these enemies are is not as yet well known, since it is only within a few years that the matter has been so closely studied as to render it apparent that there is this very large decrease of young Seals during their absence from the islands. It has been known, however, for many years that Killer Whales (different species of *Orca*) prey habitually upon the young, from these having been found in their stomachs. Michael Carroll, Esq., in his "Seal and Herring Fisheries of Newfoundland" and in the reports on the Canadian Fisheries, alludes to the great destruction of young Seals on the Atlantic coast by this animal and by sharks and sword-fishes, and also by their being crushed in the ice. The *Orca* and the sharks are alluded to by Mr. Elliott as preying extensively upon the young Seals, and it may be that many others are destroyed by enemies not at present well known.

Since the foregoing was prepared for publication, I have received from Captain Bryant the subjoined account, based on long personal experience at Saint Paul's Island. Although in some points anticipated by Mr. Elliott's published Report, and covering to a great extent the same phases of the subject, it contains so much additional matter that at the expense of some

* Condition of Affairs in Alaska, pp. 88, 89.

reiteration I have deemed it best to introduce it entire. The report is addressed as a personal communication to me in response to my earnest solicitation for the final results of his many years of observation upon the Alaskan Fur Seal. By way of explanation of the character of his report he observes:

"The object I wish to attain in writing these notes is to put on record the result of my observations on the Fur Seals of Saint Paul's Island during eight years' residence as Treasury agent in charge of the interest of the United States Treasury Department. In order to do this some account of their habits and the condition of affairs on my first arrival there seems necessary as a starting point, in order that the changes that have since occurred may be more clearly understood. As you have had the result of my first season's observations there, [*] I need not be so diffuse in my descriptions as would be otherwise necessary, and you will understand that where any of my former statements are omitted or changed it is due to correction made necessary by my longer experience. I shall endeavor to make this report as brief as is consistent with the successful attainment of the objects before stated."

"HISTORY OF THE FUR-SEAL FISHERY AT THE PRYBILOV ISLANDS, ALASKA, FROM 1869 TO 1877.—PRELIMINARY AND GENERAL OBSERVATIONS.—The island of Saint Paul is of purely volcanic origin, consisting of a collection of elevated cones and elongated ridges, connected by low valleys composed of beds of marine sand that has gradually been thrown on the shores by the action of the waves. This sand is of so light a character that when dry it readily drifts over the hills, thereby covering the lava surface. It also washes into the coves formed by the projecting points of land, where it constitutes broad, low beaches. The shores of the points and ridges which extend out into the sea are mostly composed of irregular masses of broken rock, washed by the surf and rains, so that no sand accumulates on them except in an occasional crack or gully. These rocky slopes are selected by the breeding Seals as the places for bringing forth their young, they having a repugnance to occupying the sandy spaces.

"The male Fur Seal attains its full growth and strength at the age of six or seven years, when it weighs, at the time of land-

---

[* See Bull. Mus. Comp. Zoöl., vol. ii, 1870, pp. 89–108.]

ing, from three hundred and fifty pounds to four hundred; in exceptional cases a weight of four hundred and fifty pounds is attained. The males acquire the power of procreation in the fourth year, and at five years share largely in the duty of reproduction.

"The females bring forth young in their fourth year, and then weigh from fifty-five to sixty-five pounds. They continue to increase in size until the sixth year, often attain a weight of ninety pounds, and, in exceptional cases, even one hundred and eight, the general average being eighty pounds. It will be thus seen that the greater strength and weight of the males enable them to control the females, which they do absolutely when on the breeding-places. The young Seals at birth weigh six pounds, and the young males, when they leave the island at the age of four and a half months, weigh thirty-eight pounds, but a large portion of this weight consists of excessive fat, so that when they return to the island the following year, although they have grown longer, they have lost their superabundant fat, and weigh only forty-two pounds. At the age of two years their average weight is sixty-one pounds; at three one hundred and seven. After this they increase in weight much more rapidly, attaining their full size at six. Subsequently, their increase in weight is due to excessive fatness, rather than to continued growth.

"In spring a careful watch is kept for the arrival of the first Seals, which come with great uniformity, the record showing only four days' variation in the last seven years in the time of their being first seen in the water near the island. The time of landing, however, varies with the condition of the shores, some seasons the beaches being obstructed by snow and ice. As a rule, a few effect a landing within five or six days after their first appearance. The males invariably come first and entirely by themselves. The first arrivals are of old Seals, which coast along the shore for two or three days, and are at first exceedingly sensitive to disturbing influences, but soon after landing become torpid and indifferent to objects approaching to within eight or ten rods. They continue in this state until they become so numerous as to begin to crowd on each others' premises. After the first fortnight they arrive quite rapidly. The groups are then composed of Seals of all ages, from two years upward, with a few yearlings, but those of full size predominate. Most of the yearlings arrive with the females in July.

"As before stated, the Seals select and occupy for their breed-ing-stations the rocky slopes of the projecting head-lands. [*] On their arrival at the island the full-grown Seals separate from the younger, the former hauling up on the shore singly or in groups of two or three, separated by quite wide intervals. The young gather in a single body where the shore is smooth and spend their time in play, pushing and tumbling over each other, or gathering in groups of from three to ten around some rock near the shore, passing hours in apparently trying to crowd each other off the rock, of which each seems to be striving to gain possession, to the exclusion of the others. Later, as the number of Seals on the beaches increases, the young ones are crowded back to the upland, and find access to the water by passing along the sandy belts which extend down to the sea. As the shore line becomes completely occupied those which are old and strong enough fight their way to a place on the breed-ing-grounds, while the younger and weaker seek the sandy openings and crawl up to join their own class. Here they spend the time alternately in playing and sleeping, usually going into the water for an hour or two every day. It is only the 'beachmasters,' or breeding bulls, on the rookery that remain continuously in their places, for if they were to leave them they would be immediately occupied by some other beach-master, and they could regain possession only by a victory over the trespasser. The struggle among the old bulls goes on until the breeding-grounds are fully occupied, averaging one old male to each square rod of space, while the younger, meantime, find their way to the upland. During the latter portion of the land-ing time there is a large excess of old males that cannot find room on the breeding places; these pass up with the younger Seals and congregate along the upper edge of the rookery, and watch for a chance to charge down and fill any vacancies that may occur. These, to distinguish them from the beachmasters, are called the 'reserves,' while those younger than five years are denominated by the natives 'holluschucke,' a term denot-ing bachelors or unmarried Seals. It is from these latter that the Seals are selected to kill for their skins.

"By the middle of June all the males, except the great body of the yearlings, have arrived; the rookery is filled with the

[* This statement, as well as the following account of the habits of the Fur Seal, relates to the state of the rookeries as observed in 1869, as is stated later (posteà, p. 388) in the present report.]

beachmasters; the 'reserves' all occupy the most advantageous position for seizing upon any vacancies, and the bachelors spread over the adjoining uplands. At this time the first females make their appearance. They are not observed in the water in any numbers until they appear on the shore. Immediately on landing they are taken possession of by the nearest males, who compel them to lie down in the spaces they have reserved for their families. For a few days the females arrive slowly, but by the 25th of the month thousands land daily. As soon as the males in the line nearest to the shore get each seven or eight females in their possession, those higher up watch their opportunity and steal them from them. This they accomplish by seizing the females by the neck as a cat takes her kitten. Those still higher up pursue the same method until the entire breeding space is filled. In the average there are about fifteen females to one beachmaster. Soon after the females have landed each gives birth to a single young one. During parturition the female lies extended on the rocks, and keeps up a fanning motion with her hind flippers. They appear to suffer little in labour. The young Seal remains in the placenta until liberated by the mother, who rends the envelope with her teeth, which she sometimes does before parturition is completed. Once freed from the sac, the little fellow is very active and soon learns to nurse. The mother suckles her young while lying on her side; the teats being situated on the belly. Two days after the birth of the young the female is in heat and receives the male. During copulation the female extends herself on the rocks in the same manner as when giving birth to her young. The act of coition continues for from seven to ten minutes, during which, at intervals of two or three minutes, occur rapid vibrations of the body of the male, accompanied by a fanning movement of the hind flippers by the female, who is otherwise quiescent. Ordinarily the operation is similar to that of the cat, but in some instances, when a male and female are by themselves, without danger of interruption, I have seen the male deliberately turn the female on her back and copulate in that manner. This, however, happens more frequently in the water than on the land. It is often observable that while the females are landing in great numbers they come in heat faster than the males on the rookeries can cover them. In such cases some of the females break away and escape into the water to meet fresher and more vigorous mates. It is in this way that the class of young males of four and five years of age perform a most important service. While sufficiently developed to be fully able to serve the females, they lack the physical strength to successfully contend for a place on the rookery. They haul up with the bachelors at night, but during the day are in the water swimming along the shore of the rookery, always on the alert for the females that seek the water as above stated. On meeting them they immediately accompany them to a little distance from the shore and then perform the act of coition. The females, after remaining for a short time in the water, again return to the shore to their former places. The old males finding they have been served express their disgust in a most evident manner. The jealous watchfulness of the male over the female ceases with her impregnation, after which she is allowed to go at will about the rookery. From that time she lies either sleeping near her young or spends her time floating or playing in the water near the shore, returning occasionally to suckle her pup. The male, meanwhile, watches over the young, and makes additions to his harem as long as the landing season continues. The females, after giving birth to their young, temporarily repair again to the water, and are thus never all on shore at once, so that by the end of the season there will be twice as many young Seals on shore as there are females. As the season advances, or by the 15th of July, the earliest-born young Seals gather in large groups of from three hundred to five hundred in number on the upper edge of the breeding-places, thus separating themselves in a measure from the beachmasters. They spend their time in play until tired, when they fall asleep, often sleeping so soundly that one can almost lift them from the ground by the flipper without awaking them.

"By the 25th of July the females have all arrived and given birth to their young. At this time the beachmasters, after having been confined to the same rock for an average period of ninety days, without eating or drinking, fighting and struggling with each other for their places, have become so lean and exhausted as to present a remarkable contrast to the fat and sleek condition in which they arrived at the island. They are now mere skeletons, almost too weak to drag themselves into the water; they now crawl away, and are seen only in small numbers hanging about the shores away from the breeding-places. As these leave, the reserves and younger Seals come in to take their places, covering any straggling female that may have arrived late or missed impregnation earlier. The withdrawal of the beach-masters leaves the breeding-grounds in possession of the younger males, with the pups gathered in masses on the upper side.

"As already stated, the females now mostly spend their time in the water, returning on shore only to suckle their young as they require food. On landing, the mother calls out to her young with a plaintive bleat like that of a sheep calling to her lamb. As she approaches the mass several of the young ones answer and start to meet her, responding to her call as a young lamb answers its parent. As she meets them she looks at them, touches them with her nose as if smelling them, and passes hurriedly on until she meets her own, which she at once recognizes. After caressing him she lies down and allows him to suck, and often falls into a sound sleep very quickly after.

"By the 20th of August the young, then forty or forty-five days old, move down to the edge of the water, where they begin to learn to swim. The greater part of the young seem to resort to the water from a natural instinct, but some require to be urged in by the older ones, and I have in a few instances observed the parents take them by the neck and carry them into the water, and when they have become tired return with them to the shore again. When once in the water the young Seals soon appear to delight in it, spending most of their time there in play, tumbling over each other like shoals of fish. It seems strange that an animal like this, born to live in the water for the greater portion of its life, should be at first helpless in what seems to be its natural element; yet these young Seals, if put into it before they are five or six weeks old, will drown as quickly as a young chicken. They are somewhat slow, too, in learning to swim, using at first only the fore flippers, carrying the hind ones rigidly extended and partially above water. As soon as they are well able to swim (usually about the last week of August) they move from the breeding-places on the exposed points and headlands to the coves and bays, where they are sheltered from the heavy surf, and where there are low sand-beaches. Here they occupy a belt of shore near the water entirely separated from their parents, where they play until weary, and then haul up on to the beach to rest and sleep, often covering an area of several acres in extent in one compact body. The mothers lie apart (when not in the water) at a convenient distance, for the young to find them to nurse. Thus they remain until October, when the oldest and strongest begin to leave for the winter, and others soon follow. By the middle of December all, both young and old, are gone, and are seen no more until the next season, when they return to repeat the cycle above described.

"Having now carried the breeding Seals through their annual round, we will return to the young males, or holluschucke that were left in June spread out in the rear of the breeding Seals. This class is made up of a very small number of yearlings (the greater part of these coming later, as before stated), and those of all ages between two and six years old, with a few superannuated males, which, being unable to hold their place on the rookeries, retire here with the younger Seals for quiet and rest. All of the Seals between four and six years of age pass a large portion of their time during the day in the water, returning to the shore at night. While in the water they swarm along the shore of the breeding-places, watching for opportunities of mating with any females that may chance to be in the water. To this class I shall have to return later, when I come to refer to the changes in the movements of the Seals growing out of the effect of the present mode of taking them for their skins.

"It is from the holluschucke class that the animals are selected and killed for their skins. As the process of driving has been so often described in detail, I shall refer to it only so far as is necessary to explain its effects under the present management. In the foregoing description I have followed the observations made during the first year of my residence on the island (1869), as the normal conditions then existed in a greater degree than afterwards, when other influences came into operation.

"RECENT CHANGES IN THE HABITS AND RELATIVE NUMBERS OF THE DIFFERENT CLASSES OF SEALS.—In order to be able to understand fully what the changes are that have occurred, it will be necessary to go back to a date still earlier. According to information derived from the natives, and hence somewhat meagre and vague, it appears that in the year 1842 large quantities of ice and snow accumulated on the island and remained on the breeding-places when the Seals arrived. They landed and brought forth their young on it, but a large portion of them were lost by the breaking up of the ice, the young being drowned, while thousands of females were crushed by the sliding of the masses of snow from the higher grounds. The number of Seals became thus so reduced that the natives for two years were not allowed to kill them for food. From that time up to the transfer of the islands to the United States great care was given to their increase, at which time were established the methods in practice when I arrived on the island in 1869, and which still continue with little modification. The islands were then in charge of Kazean Shisenekoff, a creole born on the island and educated in the school at Sitka. He appears to have been a man of great natural ability. He left a family of sons, part of whom inherit their father's talent, the oldest one being pontenori or arch-priest for the diocese of the Territory. This Kazean governed the islands twenty-seven years, and his memory is revered by the people like that of a saint. He kept a record in manuscript of his observations and left it on the island at his death, but before my arrival there it had been used to paste over the cracks in the ceiling of the hut of one of the natives and so was lost. During the administration of this able governor these nurseries of the Seals had been developed from almost nothing to the condition in which they were at the transfer of the islands to the United States. For many years they were able to kill only a small number, but the Seals gradually increased so that they killed as many as 40,000 in one year. The result of this judicious system was seen in the condition of affairs in the spring of 1867, when, knowing the islands were to be surrendered to the United States, the Russians took all the Seals they could, amounting to 75,000. During the season of 1868, when there was no legal protection for the Seals, 250,000 were taken.

"This brings us to the year of 1869, the date of my first visit; and on that year's observation is based the foregoing description of the habits of the Seals. One of the first objects to be attained was an approximate determination, at least, of the number of Seals frequenting the islands; but to count them was impossible. After the rookeries were filled I discovered that on the breeding-grounds there were no open spaces; that, as a rule, they began to fill at the water-line and extended no further back than they could occupy in a compact body. Making as careful a calculation as possible of the space occupied, and ascertaining the average number to the square rod, I found that this gave the astonishing number of 1,130,000 for the breeding Seals alone. The other or non-breeding Seals—that is, the males not on the breeding-grounds—were at that time occupying the upland in the rear of the females in groups of from five or six hundred to as many thousands. These being more restless in their habits, it was not so easy to calculate their numbers; but after comparing these groups with the masses of breeding Seals in their vicinity, and estimating their proportional numbers, I found that they were nearly as numerous as the breeding Seals, numbering at least one million. Adding to these the young of the year, nearly equal in number to the females, it became evident that there were on the island at that time not less than 3,230,000 Seals.

"Under the Russian *régime* the work was all done by the hand labor of the natives, the Seals being not only driven in, killed, and skinned by them, but the skins were carried on their backs to the salt-houses. The work of salting and preparing for shipment was necessarily slow, tedious, and exhausting, and as skins of young animals were smaller to take off and lighter to carry, and the choice of animals being left to the natives, they seldom killed any over three years of age, and only a small portion of this age. As a natural consequence, the killing falling on this younger and more numerous class, a larger number of males than were really necessary for breeding purposes escaped to grow up, so that at this date more than 30 per cent. of the male non-breeding Seals were of procreative age. Owing to the large number of young males constantly in the water about the rookeries, in addition to the beachmasters, all the females were impregnated before the 10th of August.

"The number of full-grown males at this date may be considered as three times greater than the number required, or equal to one full-grown male to every three or four females. In consequence of this large excess of males, and their strong desire to possess the females, they crowded the rookeries to the extent of leaving only fighting room, and kept up a continuous struggle for the mastery, regardless of both mother and young, and often destroying each other. There being always a large reserve on the alert, the contending forces were recruited as fast as the combatants became crippled or exhausted, so that there was no cessation in the strife, day nor night, while the noise of the mingled voices could be heard at the distance of five miles from the rookeries.

"The Russians contracted with dealers in Europe for a given number of skins at a fixed rate per skin, and then ordered them taken at the islands. The killing being left to the parties there, they, for their own convenience (as before stated), killed mostly from the younger class. The killing commenced on the 1st of

June, O. S., or the 12th of our style, and continued through the entire season, or until the number ordered was obtained. During June and July, the breeding season, the greater part of the four-, five-, and six-year-old Seals being in the water, the killing naturally fell heaviest on the two- and three-year-olds. After the arrival of the yearlings, they being a more numerous class, the killing fell largely on them for the remainder of the season. This system prevailed not only during 1868 and 1869, when the natives were allowed to kill for food and to sell for supplies, but the same practice was followed during the season of 1870. Although the lease bears this date, it was not put in practical operation until 1871, when all this became changed.

"Until this year (1871) the Fur Seal skins that had been sold in the market of London had varied greatly in price, ranging from one dollar to sixteen dollars per skin, but only a very small percentage brought the latter price, the average price being about four and a half dollars each.

"Having now stated the condition and numerical proportions of the different classes of Seals on the islands at the time the United States Government leased the right to take one hundred thousand skins per annum to the Alaska Commercial Company, a brief statement of the effects of this provision will throw further light on the habits of the Seals. Owing to the erroneous information prevailing at the time the lease was made, respecting the proportionate number of Seals at that time visiting the islands of Saint Paul and Saint George, 75,000 of the annual quota were assigned to Saint Paul's and 25,000 to Saint George's.

"The parties having the lease paying a tax of a certain sum per skin, and as it cost as much to get a poor skin to market as a good one, pains were taken to determine at what age the skin was of most value. It proved that Seals of the ages of three, four, and five years were the most desirable, and the lessees having the right to select their skins, took only Seals of those ages.

"This matter, however, was not fully understood until the season of 1873, when it was found that the skins of highest value were those taken from animals three years old, those older yielding skins of less value, while those older than five years were not worth taking. From this date only the three-year-old Seals have been taken. The selection of this class instead of the younger animals was a great change, the effect of which soon became manifest, as I shall presently show.

".When the agent and employés of the company came to the islands in 1871, they had no knowledge of the business, and had to learn it of the natives, so that they naturally at first followed the old routine, with only the difference that instead of confining the killing to the younger classes, as before, a larger percentage of 'half-bulls,' or four- and five-year-olds, were killed. The 75,000 Seals killed by the Russians in 1867, the 250,000 killed by various parties in 1868, and the 85,000 taken by the natives in 1869, being mostly young animals, the markets had become so overstocked with small skins as to render them unsalable, and the manufacturers in London notified the agent of the Alaska Commercial Company that only large skins were desirable; hence the agent selected for killing all the larger Seals available.   Seventy thousand of the quota of seventy-five thousand were taken during the months of June and July, the remainder being left to be supplied by the skins of animals required for food by the natives during the remainder of the season.   During this year (1871) no material changes were observed in the movements of the Seals as compared with former years.

" This brings the history of the subject to the year 1872.   The product of 1871 had been sold in Europe, and the demand for larger skins had become more imperative than before, and it being in the interest of the lessees to suit their customers, they instructed their agent residing at the islands, whose duty it was to select the animals for killing, to take *only large skins*.   Under these instructions their agent, as far as possible, confined the killing for skins to Seals of from four to six years old, and often a seven-year-old got killed by straggling into the younger groups to rest.   The effect of killing the class that formed so important an element in the reproduction of the species showed itself in the diminished number doing service in the water along the shore.   The reserves also showed quite a perceptible decrease in number, in comparison with their number in 1869. The female breeding Seals showed, through the increased space occupied by them, an increase in numbers equivalent to 15 per cent. over their number in 1869, or an increase of 5 per cent. a year, while the selection of the four-, five-, and six-year-olds, instead of the younger as formerly, had spared so large a number under four years of age that when the yearlings came on shore the two classes united seemed to flood the island with their living masses, thronging the beaches and spreading up the hillsides, their moving troops looking like armies. This year a part of the reserves located on new places, and by gathering a few females around them appeared to be forming new rookeries. At the breaking up of the rookeries, during the last days of July and the early part of August, all the females with their young did not go to the coves as before, but a considerable number remained, herding with the young bulls, while the pups learned to swim on the shore of the breeding-ground. The weather proved exceptionally fine in November, December, and January, and a part of the females remained with their young a month later than usual, and groups of two-, three-, and four-year-olds were seen in the water near the shore as late as February.

"During the latter part of the winter and spring of the following year (1873), great masses of ice from the north passed the island, coming from the northwest and drifting toward the southeast, keeping the island nearly enclosed until the 23d of May, and remaining in scattered belts for seven or eight days later. The earliest arrival of the Seals that landed was May 15, and all that arrived in May showed by their exhausted condition that they had encountered obstructions in coming. At the usual time, however, June 15, the rookeries were occupied by the beachmasters, but there were a smaller number to a given area than formerly, the great body of the reserves of 1869 having become reduced one-half. The females showed the same average increase of about 5 per cent. over the previous year, but none of the attempts to form new rookeries were continued. The increased number of females found room by filling up the spaces between the old rookeries through which the young Seals had been in the habit of passing to the uplands to the rear of the reserves, and where such spaces were not to be found the females crowded over the ridge into the inner slopes, in some places actually locating on clear sand-beaches. The closing of the passes by the breeding Seals had the effect of forcing the young Seals to coast along the shore entirely past the rookery, where they found resting places in the beaches by themselves, thus rendering their separation from the breeding Seals more complete than before, so that when wanted for driving they were found in large bodies instead of small groups as formerly, when they remained in the rear of the rookeries, and when each group had to be driven separately before they could be massed for the general drive.

This change has ever since remained permanent, and greatly facilitates the gathering of the droves for slaughter. It was also apparent that the killing of so many half-bulls the two previous years had reduced to a minimum the number that hovered in the vicinity of the breeding-rookeries, keeping the beachmasters in continual alarm. The effect of this change could also be perceived in the lessening of the noise resulting from the fighting on the breeding-grounds.

" When the season for the breaking up of the rookeries arrived, only a small part of the females moved to the coves with their young, the remainder lingering on the breeding-places with their pups, and gathering around the half-bulls and remnant of the reserves that had not left the shore, as if their first covering had failed of impregnation, and they had again become in heat and were seeking the males.

" The introduction of mules and carts for the purpose of hauling the seal-skins had greatly lessened the amount of physical labor for the natives, and the full quota of 75,000 seal-skins were ready for shipment by the 1st of August. The killing then ceased, except for fresh food for the natives, amounitng to about two hundred and fifty Seals a week. For this purpose care was taken to kill, as far as possible, only animals whose skins would be accepted by the company as a part of the quota for the next season, it being for the interest of all concerned to obtain the necessary quota of 75,000 skins and feed the natives with as little waste of Seal life as possible. In September and October a few females were seen to land and bear their young—females which had been covered out of season the previous year. This was the first time this had been observed, and was spoken of by the natives, who are thoroughly familiar with every detail, as exceptional.

" The dissatisfaction of the London manufacturers with the quality of the skins sent to market still continuing, the company, during the winter of 1872 and 1873, sent their agent to London to find out under what conditions the skins were of greatest value. This investigation established the fact previously noted that the best skins were obtained from three-year olds. At four the value has already depreciated, while skins of large-sized two-year-olds and five-year-olds are of still poorer quality than those of the four-year-olds. On this basis was established the rule by which the killing has since been regulated.

" During the fall of 1873 the weather was again very mild, and thus continued into the winter. The females, consequently, lingered in small numbers, with their young, till into January, while some of the young bulls, in groups of ten or twenty, were seen as late as February 10.

"The rapid decrease of the reserves, with the attendant changes in the movements of the Seals, caused considerable anxiety. The wise ones among the natives shook their heads ominously, and said they had predicted this from slaughtering so many half-bulls during the previous three years. I felt this, but could not order differently, the company having the right to select their own animals; but at the same time I thought that this might not be the whole cause. I watched, as did all on the island, the coming of the Seals in 1874, with intense anxiety. In the spring of that year the shores became, at the usual time, fairly clear of ice and other obstructions, and on April 13, the usual time for their arrival, the chief reported that Seals had been seen in the water. Soon after two or three beachmasters landed, and these were followed, on succeeding days, by scattered groups of three to five at a time. By the 23d of May enough young bulls had landed on the point to make a drive for the purpose of obtaining fresh food for the people.

"The changes that had been observed in the movements of the Seals during the year 1873 were noticeable in a more marked degree. The beachmasters took their positions on the breeding-grounds farther apart than formerly, and there being less cause for fighting there was less noise and tumult. The reserves appeared in about the same numbers as in 1873, but there was an increase in the proportion of the younger over the older animals, as if a larger number of the former were coming forward to take the place of the old stock of the period before the leasing of the island. There was, on the whole, an evident gain over the previous year, which gave us hope that the crisis of depletion had passed. When the females came it was found that their numbers had not materially changed. When the time arrived for the breaking up of the rookeries they all remained, only moving up farther from the water, where the reserves and half-bulls met them, forming families in the same manner as on their first landing earlier in the season; and they remained here with their young until the time of leaving the island for the winter, going from here instead of from the bays, as formerly. This has now become their fixed habit, they remaining on, and going from, the breeding-places direct.

"At this point it is necessary to take up another thread of this subject. I have already stated that, under erroneous information as to the relative proportion of the numbers of Seals breeding on Saint George's and Saint Paul's Islands, the quota had been fixed at 75,000 for the latter and 25,000 for the former island. Samuel Falconer, my assistant in charge of Saint George's Island, had reported a rapid decrease in the number of Seals of the quality desired for their skins. Assistant Agent H. W. Elliott, who had resided with me on Saint Paul's during the season of 1872, I now assigned to assist Mr. Falconer on Saint George during the season of 1873. He, by his residence on Saint Paul's, was able to give the relative difference in the proportionate number of Seals on the two islands. After his examination it was found necessary to change the original division, and assign to Saint Paul's 90,000 and to Saint George's only 10,000. This, however, disturbed the relative compensation allowed the natives for the support of their families, and twelve of the sealers were removed from Saint George's to Saint Paul's to assist in taking the skins on that island.

" In 1875 the sealing began, as usual, June 1, and with this additional assistance and the improved facilities for doing the work, 85,000 skins were taken by July 24, leaving the balance to be supplied from Seals killed for food. The agent of the company, whose duty it was to select the skins, having become convinced that it was detrimental to the future increase of the Seals to kill the half-bulls, confined the killing to those less than five years old. This left a larger number to mature as breeding males. This year more two-year-olds were taken than previously since 1870. This proved to be an important change, resulting in the sparing of a much larger percentage to mature. The movements of the females were the same as in 1873; that is, they occupied the breeding-grounds with their young until the time of leaving the island, and when departing left directly from the breeding-ground. But it was observable that there were many young Seals born in August, or later in the season than formerly, showing that a portion of the females had been covered out of season. The weather proved favorable, continuing warm till into January, thus affording the Seals born in August time to learn to swim and get strong enough to insure their safety at sea. Many of the bachelor class also remained around the island until February, when the ice coming down drove them away

" The season of 1876 was marked by no special change in the movements of the Seals from that of the preceding year. The Seals came at their usual time, beginning to arrive the 12th of April, and the same conditions of location obtained as in 1874 and 1875. The beachmasters, by occupying the entire length of the old breeding-ground, compelled the younger Seals to pass completely beyond to the bays and sand-beaches, while the increased number of females, through lack of space on the old grounds, began to occupy the sand-beaches nearest the rookeries.*

" The average time of landing of the females was a little later, or, rather, a portion landed after the 20th of July to have their young, showing that they were not covered in their first heat the previous season.

" The beachmasters and reserves showed an increase in number over the previous year, due to young bulls just matured. The old stock of the year prior to the lease had apparently nearly died out, leaving a new and more vigorous stock to supply their places.

" In the autumn the weather, which for three years had been so mild, proved unusually rough and severe. October 30th there was a severe gale, accompanied with snow, which covered the breeding-grounds to a depth of ten inches, and drove all the Seals, both young and old, into the water, and only a comparatively small number returned again to the shore. Among these were large numbers of females which had lost their young, and for several days they went about the breeding-grounds plaintively calling for their pups. In November, when the time had arrived for driving the young Seals to kill for the supply of winter food for the natives, it was found that only half the number (five thousand) requisite for that purpose could be obtained. Undoubtedly great numbers of the young Seals which were driven to the water by the storm must have become separated from their parents and lost. As I was relieved before the time for their return the next season, there was no one on the island experienced enough to perceive to what extent the

---

" * I may here state that the repugnance the females have to occupying the sandy beaches for a breeding-ground appears to arise from the evident discomfort they experience from the sand getting into their hair and fur while obliged to remain there, especially when rains occur. The young Seals also appear to suffer in health if rains occur before they are old enough to take to the water, they becoming scurfy precisely as young pigs do when compelled to live in muddy places."

injury prevailed, and as the product of 1877 will not arrive at
the proper age for killing until 1880, it must be left to the
future to determine the extent of the loss.

"CAUSES OF THE CHANGES IN THE HABITS OF THE SEALS,
ETC.—It will now be well to try and trace some of the causes that
operated to produce changes in what had been the usual habits
of these animals. At the date of the transfer of these islands to
the United States the non-breeding Seals (and by this class I
mean the males of all ages not in active service on the breeding-places) were, as nearly as can be ascertained, equal to
the whole number of both beachmasters and females. Thirty
per cent. of this non-breeding class were capable of procreation. During the years 1867, 1868, and 1869 there were
taken 410,000 Seals, mostly of the product of 1866, 1867, and
1868. This large number killed in so short a time, left only a
small portion of the product of those years to mature, to furnish
the half-bulls in 1871, 1872, and 1873. During these years (the
first years under the lease), the demand, for reasons before
stated, was for large skins, and it had to be met by killing four-,
five-, and six-year-olds. This destruction of the remnant that
escaped from the excessive killing in the years of 1868 and 1869
had the effect to exterminate the product of those years and
create a chasm that had to be bridged over by the products
of years prior to 1865, which had to supply the males necessary for breeding, until the products of later years could mature. Again, during the season of 1870 the natives, to purchase supplies and for their own food, killed 85,000, mostly one-
and two-year-old Seals. This operated in the same direction,
reducing to a minimum the products of 1868 and 1869, and
rendering the breach still wider. There was consequently only
a limited number to fill this gap until those spared by the company in 1871, 1872, and 1873, when only half-bulls were taken,
had matured. Before these had time to attain maturity the
large surplus of reserves of the year 1869 became so reduced
in numbers by natural causes as to induce the changes we have
noted in the movements of the female breeding Seals. The old
males, having become weakened and exhausted, failed to im-
pregnate the females in their first heat and forced them to seek
the younger males in their second heat instead of going to the
beaches with their young as formerly. This caused many females to bear their young later in the season, and consequently,

during this and the following years, resulted in considerably delaying the time of their impregnation. Owing to the mild weather late in autumn, the mothers of these late pups were able to stay until their young were old enough and strong enough to insure their safety.

"The decrease in the number of breeding males may be considered as having reached its minimum in 1876. In 1877, the last season I spent at the islands, there was an evident increase in the number of this class. A review of the different classes will now assist us in drawing our conclusions. The reserves and beachmasters belong to a single class, the only difference being that those which get on shore first and hold a place for a family are denominated beachmasters, while those of the same class that arrive too late for this purpose are termed reserves. In 1869 the beachmasters were numerous enough to occupy the breeding-ground in the proportion of one to the square rod, leaving a surplus or reserve of double this number, or three times as many as could find space on the breeding-ground. There being so large an excess of males of breeding age, they crowded each other to the extent of leaving only fighting-room, averaging one beachmaster to seven females. The beachmasters were continually fighting for the possession of the females, often killing each other in their struggles, while many more became so crippled as to have to retire from the breeding-grounds, so that during the season the injured and exhausted amounted to fully 30 per cent. This condition continued until the effect of the excessive killing in 1865 became apparent, resulting in the reduction of the reserves from natural causes. Those already old died out, and the products of 1866 and 1867 being reduced by overkilling in 1868, fewer were left to mature to make up for the natural loss. Consequently, in 1870, or before the Alaska Commercial Company began to take the animals for their skins, this class had perceptibly decreased. During the succeeding three years, nearly all of the half-bulls being killed, there was no new stock to replace the natural decrease from old age and exhaustion, nor many half-bulls to assist in the duties of reproduction. This rendered the season of service for the old Seals more protracted. In three years, or by 1873, the old reserves had become so reduced in number that when the rookeries were fully occupied there were only half as many beachmasters there as formerly, or only one to two square rods of area, while each beachmaster had on the average about fifteen females. The reserves were now only about half as numerous as the beachmasters. As a result of these conditions, the females were imperfectly covered, and instead of going in August to the beaches remained on the breeding-grounds until their second heat, herding with the younger or less matured males, while the young Seals learned to swim from the shores of the rookeries. Another marked result was that on the rookeries the beachmasters were so far separated and had each so many more females that there was less occasion for fighting, and consequently less uproar and destruction from wounds. At this date occurred the change in the system of killing, the younger males being taken instead of the half-bulls. Since this time the relative number of breeding males has been steadily increasing. It is still, however, below the proper proportion, and under the present system will require three years at least to supply the deficiency. The period of gestation being nearly one year, it is necessary that the females should be all impregnated in their first heat, lest, as was the case in 1876, early storms occur and force the late-born young to enter the sea before they have acquired sufficient strength and endurance to insure their safety. Similar changes in their movements and relative numbers were noticed in the holluschucke or bachelor class. This class includes all under six years old, with a very small number of superannuated males, and also the half-bulls. The custom of killing the younger Seals for their skins, by the Russians, had allowed so large a number of half-bulls to mature every year that in 1869 the proportion of half-bulls to the whole number was fully 20 per cent., but in 1871, 1872, and 1873, when this class were taken for their skins, it decreased to less than 5 per cent., and did not show any perceptible increase until 1876.

"In 1869, the rookeries, where they extended along the shore, were not continuous, but broken into sections by the small gullies formed by the streams from the melting snows in spring. These open spaces appeared to be regarded by all classes as neutral ground, and all the Seals not old enough to maintain a position on the breeding-places passed through them to the uplands in the rear of the rookery, going and returning at will. When on the upland these younger Seals occupied places near the reserves. When they were wanted for their skins the men passed rapidly between them and the reserves, cutting them off from these open passages and turning them inland. Here the small squads collected from the different divisions were gathered into herds of from six hundred to eight hundred each; that number being as many as can be driven to advantage in one flock.

"The decrease of the reserves from 1870 to 1872 gave ample room on the old breeding-places without forming new rookeries. Later the increase of the females made it necessary for them to occupy the open places which had before afforded passages for the young males from the water to the uplands, so that the young males on their arrival, after trying in vain to find landing-places as before, passed the rookeries and occupied the beaches of the coves and bays beyond. This began in 1872, and in 1875 had become general, and may now be considered as a fixed habit. As it saves gathering them in small groups, it greatly facilitates the process of obtaining the drives without detrimental effect to the rookeries.

" When I made my estimate of the number of the Seals in 1869, the proportionate number these groups bore to the breeding-rookery near which they were located suggested the inquiry whether the young returned to the exact point where they were born. I found on questioning the natives that they believed they did thus return. To test this matter I had, in November, 1870, fifty young males selected from one rookery, and marked on the right ear, and fifty more selected from another rookery, two miles distant from the first, were marked on the left ear. The result was that in 1873, when they were of the proper age to be taken for their skins, four of them were killed on Saint Paul's Island, at points more or less distant from the place where they were marked, and two were found on the island of Saint George.

" Passing now to the consideration of the females, we meet with greater difficulties and arrive at less satisfactory results, owing to the fact that our knowledge of them during the first three years is less definite. During the first four months after birth the sexes do not appreciably differ. When the Seals are driven to the uplands in November for the purpose of selecting young males for the winter supply of food for the natives, the sexes, as nearly as can be judged, are equal in numbers; but at this time the females average at least one-tenth smaller than the males.

"At this stage they leave the island for the winter, and very few appear to return to the island until they are three years old,

Misc. Pub. No. 12——26

at which age they seek the males for sexual intercourse. On the other hand, the males return the following year with the mature females. On their arrival they land on the island and pass some distance inland, where they repose in large herds during the breeding-season, and linger about the island until after the females and their young are gone. They continue to return each year, arriving earlier as they approach to maturity, until old enough to become beachmasters. But the young females, as already stated, are not seen in numbers until they are three years old, when they arrive about the height of the breeding-season. Considerable numbers land on the breeding-places, but a larger portion are covered by the males before they have time to land. In the females there are no definite external indications of age as there are in the males, and in assigning the age of three years I have accepted the judgment of the natives, who are familiar with every phase of Seal life, and are governed mainly in their opinion by the appearance of the teeth. In the few I have had killed from time to time for examination the differences have been pointed out to me, but I do not consider myself competent to judge, and in the absence of more definite evidence I accept the statements of the natives. At this age they weigh about forty-five pounds, are of a steel-gray color on the back, with pearly white breast and belly, the gray of the back gradually shading into the white on the sides, and their coat has a very soft, velvety appearance. When they return the following year to give birth to their first young they average sixty pounds, and the color of the back is of a deeper shade and extends lower down on the sides. They still continue to grow for two or three years, and attain an average weight of eighty pounds. At six or seven years old the color of the back has become brown and extends to the belly, which is then only a few shades lighter than the back. When the young females first land their color is bright and soft, but in two or three days the white tint gradually becomes of a rusty shade, so that when visiting the rookeries daily it is easy to distinguish the Seals that have just landed. After they have been on shore ten days they become all of the same shade, and are individually undistinguishable. As the females are never killed, but are left to die from old age or natural causes, there are no means of ascertaining their length of life. In exceptional cases they become barren and haul up with the young males.

"ALBINOS AND SEXUALLY ABNORMAL INDIVIDUALS.—Three
or four albinos are found every year, which are of a pinkish
white color, sometimes mottled with liver-colored spots, and the
eyes and the skin covering the flippers are also pinkish. On
one or two occasions I have seen the young pup black, except
a narrow light stripe extending from the corners of the mouth
along the sides to the posterior extremities. These conditions
are very unusual.

"Among the males we find sometimes an imperfect development of the organs of generation. Individuals thus defective
are not distinguishable until the fourth year, when instead of
the neck thickening and the hair on it growing curly and longer
as in the perfect male, they retain the slim form of the neck,
characteristic of the female, non-development of the testes having the same effect upon their development that castration has
upon the domestic bull.

"Occasional instances of hermaphroditism also occur, in which
the same individual has a nearly perfect development of the
organs of both sexes. These herd with the males, but are
readily distinguishable from them by their having the posterior
part of the body fuller and thicker as in the full-grown female.

"DESCRIPTION OF THE YOUNG; VARIATION IN COLOR WITH
AGE, ETC.—The young Seals are all born with a coat of short,
stiff black hair covering the whole body. When sixty days old
this is replaced by a very soft and silky covering three-fourths of
an inch in length. This is a fine steel-gray on the back and
white on the throat, breast, and abdomen. This coat of overhair
is shed annually in August and September, becoming coarser
and darker with age each year. At seven years the back has
attained a dark brown, shading gradually to two or three shades
lighter on the belly. At the fifth year the hair on the neck and
shoulders grows coarser, curling at the ends, and on many of
these the curly tips are white, giving a grizzled or silver-gray
appearance. The fur commences to grow with the first coat of
overhair, as a soft light down, the overhair entirely covering
and concealing it. On the male this continues to grow to three-
eighths of an inch in length, increasing evenly in thickness and
fineness all over the body until the third year, when it is in its
greatest perfection. After this, as the male develops the characteristics of his sex, as the thickening of the neck and shoulders, the fur also becomes longer and thicker, while, as the

animal grows older, the fur on the posterior portion of the back gets thinner. The skin thus deteriorates in value; the fur being unequally developed on different parts of the body, prevents the use of the whole skin in the same garment. The color of the fur is not indicated by the overhair, and as a rule shows greater variation in shade than the latter, varying from a smutty white to a rich maroon, the latter shade being the most rare.

"MOULTING.—A diversity of opinion exists on the island as to whether or not the fur is shed with the overhair. I have given close attention to the subject and find that all the evidence is against the opinion that the fur is shed. The great quantity of overhair annually shed by this immense number of animals covers the ground like dead leaves in a forest. It is blown by the winds around the rocks, and becomes trodden into the soil, so that when the earth is dry if a piece be taken and broken the whole mass is found to be permeated with it like the hair in dried plaster. The difference between the fibers of the overhair and the fur is plainly apparent to the eye. I have, however, gathered parcels of it at all times during the shedding season and subjected it to microscopic examination, but have always failed to detect the presence of fur in sufficient quantity to warrant the belief that any of it is shed naturally. The shedding of the overhair begins about the middle of August, and the Seals are not fully clothed with the new coat until the end of September, and it does not attain its full length before the end of October. The first indications of shedding are noticed around the eyes and fore flippers and in the wrinkles or folds of the skins. The new overhair appears in the fur as short black points, and as it grows out the old coat is gradually cast. The whole process covers a period of forty days, during which time the skins are in a condition denominated by fur-dealers as 'stagey,' and are of inferior value. This, however, is not due to any defect in the fur, but to the condition of the overhair, which is so short as to render the process of plucking too slow and laborious by the usual methods to be remunerative. In the first shipments of skins under the lease complaints were made of the number of 'stagey' skins that were sent. As this was a term that conveyed no explanation of the defect, it was necessary to send to London for a package of stagey skins. This was done during the third year, and on their reception at the island the cause was at once understood and no

more stagey skins were shipped. Two years after, inquiries
were made by parties in London for stagey skins, or rather why
there were no more to be had in the market. It was ascertained that parties there had been making it a special business to
manufacture stagey skins. The low price at which they were
sold in the raw state enabled them to bestow the extra labor
necessary to pluck them and realize a large profit thereby, the
skins after plucking being of prime value. This gives further
proof that the animals were not shedding their fur. Whenever Seals are wounded before the shedding season the wound
heals very quickly and the scar is covered with a coat of fur
immediately after, but no overhair grows on the wound until
the shedding season arrives, when nature wholly repairs the injury. I have had such animals killed in the shedding season and
found the new overhair showing in the fur of the wound just
as on the rest of the body. In the spring of 1873 a fine three-
year-old landed with a wound on its body as large as two
hands, apparently caused by the animal getting pinched in the
ice. The wound, though fresh, soon healed and became covered with fur. This Seal was several times driven to the kill-
ing-ground and allowed to go back, on account of the blemish
on its skin. In August, when taking Seals for food, this Seal
was killed and unmistakable evidences of the new overhair
covering the wound were found.

"SEXUAL ORGANS AND COPULATION.—As before stated, the
male is born with the testes enclosed in the body. These descend
in the second year but do not become fully developed until the
fourth. In the fifth year the scrotum becomes distended and
the testes show like those of the dog. The vaginal orifice of
the female being within the anus there is but one external opening; hence the difficulty of distinguishing the sexes at birth. The
female has four teats, two on each side of the middle line of the
belly, equidistant from the fore and hind flippers. During
lactation they are half an inch in length, but do not protrude
beyond the overhair. The mode of copulation on land has
already been described. When there was a full supply of
breeding males copulation occurred mainly on the breeding-
grounds, the half-bulls participating to only a limited extent,
and was rarely seen to occur in the water. Since 1874, owing
to the decrease in the number of breeding males, a much
larger proportion of the females receive the males in the water,
so that on any still day after the 20th of July, by taking a canoe

and going a little off-shore, considerable numbers may be seen pairing, and readily approached so near as to be fully observed. They are then found in single pairs, swimming in circles, sometimes the one sometimes the other leading. They come together in approaching the surface from below, the male shooting onto the back of the female and firmly clasping her between his fore flippers. The time of contact is shorter than on land, not exceeding five minutes, but the operation is repeated two or three times, at intervals of fifteen or twenty minutes. They then separate, each going in a different direction.

"POWER OF SUSPENDING RESPIRATION.—As these creatures spend so great a portion of their life at sea, it is interesting to know how long they are capable of remaining below the surface. When full-grown males, sleeping on the edge of the beach, are frightened into the water so suddenly that they do not recognize the nature of the disturbance, they invariably plunge and swim beneath the surface till obliged to rise to breathe. In such cases they remain from two to two and a half minutes under the surface and come up from one hundred and fifty to two hundred yards distant. If a boat passes among them they will follow it at a distance of ten or fifteen yards. On coming to the surface they will stand erect with the whole of the body anterior to the fore limbs above the water, and in this position remain perfectly still for several seconds, then with a summersault and a splash, disappear for a minute or so to reappear again in some other direction, apparently enjoying the fun; in no case have I timed them when they remained over two and a half minutes under water. I do not think their power of remaining below the surface equal to that of an experienced and well-trained pearl-oyster diver. This seems to indicate that they must feed on fishes living near the surface; at least not on bottom-fish in deep waters.

"NATURAL ENEMIES.—From the birth of the young Seals until they leave the island at four and a half months old, the loss of life from natural causes is very slight, not exceeding one-half of 1 per cent. At the time of their departure they are excessively fat and clumsy, and easily fall a prey to the small Whale known as the Killer, their only positively-known enemy. These grow to a length of fifteen to eighteen feet, and go in schools of from five to a dozen or more, frequently attacking and killing full-grown Right Whales by eating out their tongues; hence the name Killer, applied to them by whalemen. When the season arrives for the young Seals to enter the water these animals are seen near the island, creating great consternation among the Seals both young and old. They rarely venture near shore, but in three cases where they have been caught young Seals have been found in their stomachs, leaving no doubt of their object in approaching the island. I have also been informed by the natives of Bristol Bay that these same animals are formidable enemies to the young Walruses and Hair Seals. The Killers doubtless follow the Seals to their winter feeding-grounds and prey upon them. During the time the young Seals are absent from the island fully 60 per cent. of their number are destroyed by their enemies before they arrive at the age of one year, and during the second year about 15 per cent. more are lost. Later they appear to be better able to protect themselves, but before they arrive at maturity at least 10 per cent. more are destroyed; so that if left entirely to themselves only 10 or 15 per cent. of the annual product would mature or reach the age of seven years. To what age the males attain, there is no means of definitely ascertaining. In the records of Shisenekoff, to which I have before alluded, it is stated that he observed one male occupy the same rock for fourteen successive years. Only in five instances have I been able to identify the same Seal as occupying the same place. Four of these returned four years in succession, and the other, five years. They were probably eight years old when first observed, so that they attained at least to twelve years, which I think may be considered as their average length of life. As I have before stated, the large surplus of full-grown males existing in 1869 nearly all disappeared in about six years; and when we consider the fact of their severe labors during the breeding season, when they pass from ninety to one hundred and twenty days without food, engaged in a constant struggle for their positions, and performing the most exhaustive function of physical life, six or seven years would seem to be the limit of the active period of their lives.

"EFFECT OF CLIMATIC INFLUENCES.—It remains now to notice the effect of the climate on these animals. The climate is very uniform at the islands during the period the Seals remain here. The months of April, May, June, and July are the most important portions of the year, as in August the Seals are all in a condition to go in the water and avoid the influences most

injurious to them, namely, sunshine and rain.   I here furnish
an abstract from the meteorological tables of C. P. Fish, signal-
officer resident on the island in 1874, this being the warmest
year and the one most unfavorable to the Seals during my
eight years' residence at the islands.

Mean average for the month.	April.	May.	June.	July.	August.	September.	October.	November.
Thermometer..........	34°	38°	44°	49°	50°	47°	40°	37°
Relative humidity	80.4	77.4	80.6	87.4	83.8	79.2	76.2	80.8
Proportion of cloudiness	83.6	76.1	88.0	97.0	82.2	75.5	78.0	73.2
Amount of rainfall.....	0.84	0.58	2.91	3.81	2.62	3.01	4.82	9.28
Number of days on which precipitation occurred..............	29	21	25	27	29	25	31	28
Proportion of foggy days	2.8	0.12	0.05	61.7	22.2	2.6	1.0	1.3

"During the months of May, June, and July the sun's rays are
generally obscured, the sky has a leaden appearance, but there
is very little fog until about the end of July, at which time also
there are usually two or three days of heavy rain.   In June and
July there are occasionally days when the sun shines clearly for
two or three hours, but rarely longer.   These days are dreaded by
the sealers.   At a temperature of 40°, with obscuration of the
sun, the Seals lie quietly; at 42° they manifest signs of discomfort from heat, and lie on their sides, fanning themselves with
their hind flippers, occasionally changing sides.   At 45° they
are decidedly uncomfortable, and all that can go into the water
to bathe and cool themselves, remaining there an hour or two.
"The beachmasters, and the little Seals that have not yet
learned to swim, remain on the land.   When the sun shines for
two or three hours, and the rocks become heated, there are occasional deaths among the beachmasters and very young pups
from sunstroke, the symptoms being a nervous jerking of the
limbs, followed by convulsions and death.   Fortunately these
occurrences are rare, and it was only in 1874 that any appreciable number were lost from this cause.   That year many young
Seals died about the first of August.   With light rain or thick
fog they endure a temperature of 50° without inconvenience.
The same fatal results occur from overheating when driving,
in which case if the animals are not skinned immediately the

fur is loosened and the skin becomes valueless. It occasionally happens that when a herd has been driven in during the night from a distance of five or six miles, they do not get sufficiently cool before seven o'clock. If the sun shines, in half an hour the whole herd show symptoms of discomfort, and soon become entirely unmanageable, breaking away from the watchers and rushing in all directions, heedless of obstacles, running into the village, entering open doors, and attempting to climb up the sides of the houses, piles of lumber, or any other object in their way, keeping on until convulsions and death result. At such times every available hand is required with club and knife to follow, knock down, and skin before the pelt is damaged by the heat. The only remedy is to get them into the nearest water as fast as possible, as all thus treated are saved. After a bath of half an hour they can be driven again, and if allowed to lie quietly no further trouble is experienced.

"They are also greatly disturbed by rain. They pay little attention to a slight drizzle, but when copious showers occur they all resort to the water until the rain is over, preferring this to the shore.

"It thus appears that for the successful breeding of these animals certain meteorological conditions are indispensable, and that they require a place where they can land and remain undisturbed for half the year. The geographical situation of these islands affords not only the isolation they require, but the requisite humidity and exemption from extremes of heat and cold. If the islands had been especially created for the Seals they seemingly could not have been better adapted to their requirements. Hence, if not rudely disturbed by man, there need be no fear of their changing their place of breeding.

"NUMBER OF SEALS REQUIRED FOR THE SUBSISTENCE OF THE NATIVES.—The principal subsistence of the native population is Seal-flesh. When the Seals arrive in spring, if the winter supply is not exhausted, it has become so stale as to render it necessary to commence killing fresh food as soon as enough young males for the purpose have landed, which usually occurs by the 20th of May. The skins of these animals are salted, and on the 1st of June (the legal time for taking Seals for their skins) all the skins of animals that have been taken for food are turned over by the Treasury agent to the agent of the company as a part of the quota for that year. While the company are taking these animals for their skins, the natives subsist on the flesh of the animals so killed. When the company have obtained their full quota their control of the killing ceases, the Treasury agent directing the killing for food. From about the end of July until the Seals leave the island, there are required for the subsistence of the natives three hundred animals per week. Care is taken in killing for this purpose to take such animals as will yield skins acceptable by the company as a part of the next year's quota, but during the shedding season the skins are valueless. As this period lasts from the middle of August to October, the number of skins so lost is about two thousand. Besides this loss, in November, before the young Seals leave the islands for the winter, about 8,000 four-and-a-half-months-old young males are taken for a supply of blubber and Seal-flesh for use in winter while the Seals are absent from the islands. This is necessary, because at this season the older Seals that have been so long ashore as to have become quite thin and poor, yield little blubber, while their flesh is tough and stringy. The blubber of the nursing Seals is quite different from that of the older Seals, being finer in texture and firmer, with less proportion of oil, and is far preferable for food-purposes. The carcasses of these young Seals are dressed and suspended on poles in the open air, and are kept fresh nearly all winter by being frozen. It will be seen by this that the total number of animals killed annually for all purposes is 110,000. This, in the six years that have elapsed since the beginning of the lease, amounts to a total of 660,000 male Seals. Allowing the sexes to be produced in equal numbers (and so far as can be judged this appears to be the fact), there have been added to the original stock of breeding females 660,000 over the number existing at the beginning of the lease; and this agrees very nearly with the increased area now occupied by them, which shows a total of not much less than 1,800,000 breeding females.

"WINTER RESORTS AND HABITS WHILE ABSENT FROM THE ISLANDS.—Of the life of these animals while absent from the islands but little is known, nor is it known where their principal feeding-grounds are. We know that the greater part pass through Ounimak and Aukootan Straits, going east in autumn and west in spring; and that in December, about six weeks after they leave the islands, fishing parties of Indians at Sitka

1,500 miles east of the Seal Islands, occasionally shoot and bring in the young Seal pups (they probably shoot the mothers, but they having too little blubber to float them they lose them). During the winter months considerable numbers of Seal-skins are taken by the natives of British Columbia; some years as many as two thousand. These find their way to the San Francisco market. On examining parcels of them I have found them to be mostly very young Seals, with no male skins among them old enough to show the sex. When at Victoria I made special inquiries about the sealing, and found that most of the skins obtained there were taken in the water, but a half-breed hunter told me he had found in summer, on Queen Charlotte Island, groups of these animals consisting of two or more beachmasters, with a dozen or more females and pups, but no half-grown males.

"Nor is it known whether the different sexes associate during the period of their absence from the islands. The males invariably come to the island first, take up their positions and wait the arrival of the females, which come after the males have all arrived. They not only come by themselves, but they all remain till after the males have gone. I have made constant inquiries of all masters of vessels cruising for trade or whales in Behring's Sea with reference to the occurrence of these animals in those waters, but in only one instance can I learn that they have been observed. In 1870 a vessel, becalmed for nearly a week one hundred miles north of the islands, in the month of August, reported seeing many Seals, nearly all old bulls. As at that time this class was largely in excess, it is possible that these males were off to feed. The Alaska Commercial Company have a general depot of supplies at Onalaska, whence the merchandise for their trade is distributed in schooners to the different points on the main coast and the islands. The masters and officers of these schooners, who are familiar with the Seals, say they see small groups of small (apparently one- and two-year-old) Seals at all times during July and August. These, I think, may be young females, which, as already stated, do not visit the island till they are three years old."

FAMILY PHOCIDÆ.

EARLESS SEALS.

Les Phoques sans oreilles ou Phoques proprement dits, BUFFON, Hist. Nat. Suppl., vi, 1782, 306.

Phocacea inauriculata, PÉRON, Voy. aux Terr. Austr., ii, 1816, 37, foot-note.

Phocidæ, GRAY, Ann. of Phil., xxvi, 1825, 340, in part, and also (in part only) of numerous writers prior to about 1870.

Phocidæ, "BROOKES, Cat. Mus., 1828, 36"; GILL, Proc. Essex Inst., v, 1866, 5.—GRAY, Ann. and Mag. Nat. Hist., 4th ser., iv, 1869, 268, 342, 344.—ALLEN, Bull. Mus. Comp. Zoöl., ii, 1870.—Also of most authors since 1870.

Fore limbs placed well forward; neck rather short; hind limbs not susceptible of being turned forward, and not capable of use in terrestrial locomotion. Manus and pes entirely hairy; nails of all the digits usually well developed (rudimentary in *Stenorhynchinæ*). Digits of the manus subequal, usually decreasing slightly in size from the first to the fifth; of the pes the first and fifth stouter than the three middle ones. Scapula small, the superior posterior angle rounded, the crests small, and the acromion process slightly developed. Femur with the trochanter minor undeveloped. Pubic bones approximated in the females, and in the males appressed posteriorly for about one-third of their length. Ilia short and broad, abruptly turned outward and recurved anteriorly. Acetabula opposite first sacral vertebra. Skull with the postorbital process generally wholly undeveloped or rudimentary; mastoid process swollen, continuous with the auditory bullæ; no alisphenoid canal. Auditory bullæ greatly inflated. Incisors conical, variable in number ($\frac{3-3}{2-2}$, $\frac{2-2}{2-2}$, or $\frac{2-2}{1-1}$). Dental formula: Milk dentition, I. (variable, as in the adult, and probably of the same number), C. $\frac{1-1}{1-1}$, M. $\frac{3-3}{3-3}$; Permanent dentition, I. $\frac{3-3}{2-2}$, $\frac{2-2}{2-2}$, or $\frac{2-2}{1-1}$, C. $\frac{1-1}{1-1}$, M. $\frac{5-5}{5-5}$. No external ear. Testes enclosed within the body.

TECHNICAL HISTORY.

HIGHER GROUPS.—As noticed in the history of the preceding families, the group formerly termed *Phocidæ* was coextensive with the suborder *Pinnipedia*. Although Péron in 1816 divided the Pinnipeds into *Phocacea auriculatæ* and *Phocacea inauriculatæ*, and although F. Cuvier in 1824 separated them

into two primary unnamed groups in accordance with whether
the molar teeth were single-rooted or had several roots, and
later into three, based on the same characters coupled with the
number of incisors, in his later classification his first division
is equivalent to the *Phocinæ* and *Stenorhynchinæ* of recent
authors, his second, to the *Cystophorinæ* of late authors, and
his third comprises the Otaries. Nilsson, in 1837, made two
primary divisions of the *Pinnipedia*, the first comprising the
genera *Stenorhynchus*, *Pelagius*, and *Phoca*, and the other the
genera *Halichœrus*, *Trichechus*, *Cystophora*, and *Otaria*, or the
Gray Seal, the Walruses, and the Otaries. Brookes, in 1828,
was the first to accord to the Earless Seals the rank of a family. Gill, in 1866, however, was the first to effectually make
clear their true position and relations to the other Pinnipeds,
since which date their family rank has been very generally conceded, and the term *Phocidæ* has been restricted to the Earless
Seals. Turner, in 1848, presented the same scheme of primary
grouping of the Pinnipeds as that adopted by Gill, and pointed
out at the same time the leading distinctions of the three
groups, but he allowed to them merely the rank of subfamilies,
the term *Phocidæ* still covering the whole of the Pinnipeds.

The early classifications of Pinnipeds having been already presented somewhat in detail (see *anteà*, pp. 9–12), it is unnecessary
to repeat them at length in the present connection. It may
suffice to state that the classifications of the *Phocidæ*, with the
exceptions already named, prior to 1866, generally embraced
both Eared Seals and Earless Seals in the same primary division. This was the case in Gray's schemes of 1821 and 1825, in
F. Cuvier's in 1824, in Nilsson's in 1837, in Wagner's in 1846,
and in Giebel's in 1855, the two authors last named separating
the Walruses as one family, and combining all the rest of the
Pinnipeds as another, called by them *Phocina*, with no subdivisions higher than genera. Gray, in 1825, divided the *Phocidæ* (=*Pinnipedia*, exclusive of the Walruses and with the addition of *Enhydris*) into two primary groups and five secondary
groups, of which latter the Earless Seals formed four and the
Otaries a fifth. In 1837 he replaced the Walruses in *Phocidæ*
(but excluded *Enhydris*), and adopted the classification employed by him till 1866. This includes two primary divisions
termed "Sections," and five secondary divisions termed "Subfamilies," as follows: Section I. Subfamily 1. *Stenorhynchina*,
with, originally, the genera *Leptonyx*, *Pelagius*, and *Stenorhynchus*, to which were added later *Lobodon* and *Ommatophoca*.

Subfamily 2. *Phocina*, with, originally, the genera *Phoca* and *Callocephalus*, to which were added later *Pagomys*, *Pagophilus*, and *Halicyon*. Subfamily 3. *Trichisina* (or *Trichechina*, as spelled later), with the genera *Halichœrus* and *Trichecus* (or *Trichechus*, as spelled later). Subfamily 4. *Cystophorina*, with the genera *Cystophora* and *Morunga*. Subfamily 5. *Arctocephalina* (including all the Eared Seals).

In 1866 Gill restricted the family *Phocidœ* to the Earless Seals—a group equivalent to Turner's subfamily *Phocina*—and divided it into three subfamilies, as follows : 1. *Phocinœ*, including the genera *Phoca*, *Pagomys*, *Pagophilus*, *Erignathus*, *Halichœrus*, and *Monachus* ; 2. *Cystophorinœ*, including the genera *Cystophora* and *Macrorhinus* ; 3. *Stenorhynchinœ*, with the genera *Lobodon*, *Stenorhynchus*, *Leptonyx*, and *Ommatophoca*. Gill's *Phocinœ* is the equivalent of Gray's *Phocina* and *Trichechina*, with the addition to the former of *Monachus* and the exclusion from the latter of *Trichechus*, while Gill's *Cystophorinœ* and *Stenorhynchinœ* are the exact equivalents of Gray's groups of similar name, except that *Monachus* is excluded from the latter.

In 1869 Gray separated from the *Phocidœ* the Walruses and the Otaries as distinct families, thereby restricting the *Phocidœ* to the Earless Seals, as Gill had previously done, but divided the *Phocidœ* into five "tribes." His classification of the group as presented by him in 1871 is as follows: Family *Phocidœ*. Tribe I. *Phocina*, with the genera *Callocephalus*, *Pagomys*, *Pagophilus*, *Halicyon*, *Phoca*. Tribe II. *Halichœrina*, = genus *Halichœrus*. Tribe III. *Monachina*, = genus *Monachus*. Tribe IV. *Stenorhynchina*, with the genera *Stenorhynchus*, *Lobodon*, *Leptonyx*, *Ommatophoca*. Tribe V. *Cystophorina*, with the genera *Morunga* and *Cystophora*. The difference between the two schemes consists (1) in the equivalency of Gill's *Phocinœ* with Gray's first three "tribes," and (2) in the designations "subfamilies" and "tribes." Gill's scheme of division of the family into three subfamilies has been adopted by most subsequent writers, even Gray himself adopting it in 1874. The three "subfamilies" now so currently accepted seem to be well-marked natural groups, but whether entitled to the rank thus accorded may perhaps be open to question.

GENERA.—The first dismemberment of the Linnæan genus *Phoca*, after the removal of the Eared Seals by Péron in 1816,[*]

---

[*] For a discussion of *Pusa*, Scopoli, 1777, see *posteà* under the genus *Halichœrus*. The term was generically applied to what seems to have been *Phoca fœtida*, but fortunately slumbered for a century, when it was unhappily revived.

was made by Nilsson in 1820,* who separated from the other
Earless Seals the Gray Seal and the Crested Seal, which formed
the types and sole representatives, respectively, of his genera
Halichœrus (type, *Halichœrus griseus*, Nils., = *Phoca grypus*,
Fabr.), and *Cystophora* (type, *Cystophora borealis*, Nils. = *Phoca
cristata*, Erxl.). These genera became soon widely recognized,
and have been (with the exceptions soon to be noticed) since
currently adopted. The next dismemberment was virtually
effected by Fleming† in 1822 by suggesting that the Monk or
Mediterranean Seal might require, with possibly others, to be
placed in a genus *Monachus*, with the *Phoca monachus* of Hermann as the type. ‡

F. Cuvier§ in 1824, either ignorant of, or ignoring, the generic
separations made by Nilsson and Fleming (as he makes no reference to them), divided the genus *Phoca* of preceding authors
into five genera, namely: 1. *Callocephalus* (type, *Phoca vitulina*,
Linn.); 2. *Stenorhynchus* (type, *Phoca leptonyx*, Blainv.); 3. *Pelagius* (type, *Phoca monachus*, Hermann); 4. *Stemmatopus* (type,
Phoca cristata, Erxleben); 5. *Macrorhinus‖* (type, *Phoca pro-

---

* "Skand. Fauna, i, 1820, pp. 376, 382."

†Philosophy of Zoölogy, vol. ii, 1822, p. 187, foot-note.

‡ Fleming *suggested* rather than *constituted* the genus *Monachus*, for he
simply says of it in a foot-note under the genus *Phoca :* "Some seals, as Ph.
monachus, are said to have four incisors in each jaw. Such will be probably constituted into a new genus, under the title Monachus." This is the
whole basis of Fleming's genus *Monachus*, which is allowed precedence over
F. Cuvier's genus *Pelagius*, based on a detailed account of the distinctive
cranial characters of *Phoca monachus*, together with a figure of the skull.
Other Seals than *Phoca monachus* are not only "said to have," but are well
known to have "four incisors in each jaw", and mention of *Phoca monachus*
is all that saves Fleming's genus, for it cannot be said to be characterized,
and ought not to be recognized to the prejudice of *Pelagius*. Upon Dr. Gray
(Cat. Seals, Brit. Mus., 1850) appears to fall the responsibility of reviving
the generic name *Monachus*, and renaming the species *Monachus albiventer*
(ex Boddaert), although Nilsson appears to have previously employed it in
1837, but speedily abandoned it for *Pelagius*, F. Cuvier.   Says Nilsson, "Ich
hatte in der Vet. Academiens Handl. för 1837, p. 235, die hierher gehörige
Art *Monachus mediterraneus* genannt, aber seitdem ich erfahren, dass Fr.
Cuvier dieselbe schon im Dict. d'hist. nat. unter dem Namem *Pelagius monachus* beschrieben, scheint mir dieser Name wegen seiner Priorität beibehalten
werden zu müssen."—*Wiegmann's Arch. für Naturg.*, 1841, i, p. 308, foot-note.

§ Ann. du Mus. d'Hist. Nat., vol. xi, 1824, pp. 174–200, pll. xii–xiv.

‖ F. Cuvier's names appeared here only under the gallicized forms respectively of *Callocéphale*, *Sténorhinque*, *Pelage*, *Stemmatope*, and *Macrorhine*. They
were first Latinized as above by the same author in 1826 (Dict. des Scien.
Nat., vol. xxix, 1826, pp. 544–552), but naturalists generally concur in assigning 1824 as the date of the introduction of these genera into systematic
literature.

boscidea, Péron = *Phoca leonina*, Linné, 1758). Of these genera
two are synonyms, *Stemmatopus* being antedated four years
by *Cystophora* of Nilsson, and *Pelagius** two years by *Monachus* of Fleming, in each case the two later names being the
exact equivalents of the earlier ones. Yet *Stemmatopus* and
Pelagius had for a time considerable currency, particularly with
French and English writers. The name *Stenorhynchus* was
doubly preoccupied for genera of Articulates, and was thus untenable in its present connection. *Callocephalus*, as originally
used by its author, and later for some years by him and others,
embraced not only *Phoca vitulina*, but also *fœtida*, *grœnlandica*,
and *barbata*, as well as numerous nominal species referable to
these. In this connection it should be noted that nothing was
now left to represent the old Linnæan genus *Phoca*, which thus
became wholly set aside. In 1827 Gray† proposed the genus
Mirounga (later changed by him to *Morunga*‡), including under
it *Phoca cristata*, Erxl., and *Phoca proboscidea*, Péron, besides
three nominal species referable in part or wholly to the latter.
In 1830 Wagler§ renamed this genus *Rhinophora* (type, "*Phoca
proboscidea*, Péron"). Dr. Gray, in 1837, ‖ proposed the genus
Leptonyx for Lesson's "*Otaria weddelli*", which has been since
generally current for that species, but with which, however, some
authors (as Wagner, 1846, Giebel, 1855) have associated the
Monk Seal of the Mediterranean and all the Antarctic Phocids,
except the Sea-Elephant. The name, however, is antedated by
Leptonyx, Swainson, 1832, applied to a genus of birds, and is
hence untenable as used by Gray and others for a genus of Seals.

In 1844 Dr. Gray introduced four additional genera among
the *Phocidæ*, which have since been more or less commonly
adopted, either in a generic or subgeneric sense. These are
Pagophilus (introduced originally as "a subgenus of *Callocephalus*") for the Harp Seal (*Phoca grœnlandica*); *Lobodon* (type,
Phoca carcinophaga, Homb. and Jacq.) for the Antarctic Crab-
eating Seal, and *Ommatophoca* (type, *O. rossi*, n. sp.), based on
a species from the Antarctic seas here first described. He at
the same time revived the Linnæan name *Phoca*, separating the

---

* This name is variously written by different authors, as *Pelagias*, *Pelagios*,
and *Pelagius*.

† Griffith's Anim. Kingd., vol. v, p. 179.

‡ List Osteolog. Spec. in Brit. Mus., 1847, p. 33 (species, "*Morunga elephantina*").

§ Natur. Syst. Amph., 1830, p. 27.

‖ Charlesworth's Mag. Nat. Hist., vol. i, 1837, p. 582.

Bearded Seal from *Callocephalus* to bear this name. Hence Gray's genus *Phoca* (type and sole species *Phoca barbata*) had nothing in common with the Linnæan genus of that name, being assigned to a species unknown to Linné, so that *Phoca,* Gray, as here defined, was virtually a new genus.

In 1854 the same author* proposed the genus *Heliophoca* for a nominal species which he himself ten years later referred to *Monachus albiventer.* In 1864† he added *Halicyon* and *Pagomys,* the first based on what he described as a new species from the west coast of North America, but which is merely the *Phoca vitulina* from the Pacific, and the other on the Ringed Seal (*Phoca fœtida*), which is here made the type of a new genus. *Pagomys,* as will be shown later, is antedated by *Pusa* of Scopoli.

In 1866 Dr. Gill, in his "Prodrome of a Monograph of the Pinnipedes,"‡ instituted the genus *Erignathus* for the *Phoca barbata* of authors (= *Phoca,* Gray), and insisted on the restoration of *Phoca* for the group represented by *Phoca vitulina,* and in defense of his position offered the following: "In the Syst. Nat., 10th ed., 1758, the first in which the binomial system was introduced, four species were included by Linnæus in the genus *Phoca*: 1. *P. ursinus,* = *Arctocephalus ursinus;* 2. *P. leonina,* = *Macrorhinus leoninus;* 3. *P. rosmarus,* = *Rosmarus obesus;* 4. *P. vitulina.* The name *Phoca* must be retained for one of these, and as the third, second, and first species were successively elevated to the rank of generic types, and the genus was thus by elimination restricted to the fourth species, for that and its allies the generic name must necessarily be reserved."

In regard to the above it may be added that Linné, in 1766, (Syst. Nat., 12th ed.) removed his *Phoca rosmarus* to *Trichechus,* while Péron, as early as 1816, and Desmarest in 1817 as well as in 1820§ referred the *Phoca ursina* to *Otaria,* thus leaving under *Phoca,* as early as 1816, only *Phoca leonina* and *Phoca vitulina.* These, F. Cuvier, in 1824, made respectively the types of his genera "Macrorhine" (*Macrorhinus*), and "Callocéphale" (*Callocephalus*), both in his paper entitled "De quelques espèces de phoques et des groupes generiques entre lesquels elles se partagent,"‖ but *Callocephalus* is the first genus mentioned, and *Phoca*

* Proc. Zoöl. Soc. Lond., 1854, p. 43.   † Ibid., 1864, pp. 28, *et seq.*
‡ Proc. Essex Inst., vol. v, 1866, pp. 4–13.
§ Péron and Leseur's Voy. au Terres Aust., ii, 1816, p. 41; Nouv. Dict. d'Hist. Nat., vol. xxx, 1817, p. 595; Mamm., 1820, p. 249.
‖ Mem. du Mus. d'Hist. Nat. xi, 1824, pp. 174–214.

vitulina is especially designated as the type. As both genera were published simultaneously, preference should be claimed for *Callocephalus*, as it occurs eighteen pages earlier in the paper than *Macrorhinus*. Thus the process of elimination necessitates, on the principle above implied, and in accordance with a commonly recognized rule in such cases, the restriction of *Phoca* to Linné's *Phoca leonina*. This, however, seems so contrary to the traditions of *Phoca*, which from 1735 to the present day has been generally associated by the majority of writers with *vitulina* and its nearest allies, that it seems an act of violence to transfer it to what is logically its legitimate connection with *leonina*, thereby making *Macrorhinus* a synonym of the restricted genus *Phoca*, notwithstanding that it has been universally accepted for half a century for the Sea-Elephants, while *Phoca* has not for an equal length of time been looked upon as having any intimate relationship with that group. In view of the tradition and usage of the case it seems best to waive the technicality here involved and suffer *Phoca* to retain its time-honored associations.*

As regards further subdivision of the *Phocidæ*, Dr. Gray, in 1866,† proposed *Haliphilus* as a generic name for Peale's *Halichœrus antarcticus* (= *Phoca pealei*, Gill), while Dr. Gill, in 1872,‡ substituted *Leptonychotes* for *Leptonyx*, Gray, (1836, nec

---

* In this connection reference may be very properly made to Prof. Alfred Newton's Paper "On the Assignation of a Type to Linnæan Genera, with especial reference to the Genus *Strix*" (Ibis, 3d ser., vi, 1876, pp. 94–105), in which he very reasonably maintains that, as Linné had no notion of a type species as commonly used by modern systematists, we should make him "the interpreter of his own intentions" by imagining him "put in our place and called on to show which he would consider his type species according to modern ideas." This he claims can be accomplished by giving some degree of attention to the works of Linné's predecessors, which will enable one to hunt down almost every name used by him, since by far the greater part of Linné's generic names were adopted by him from preceding authors, "by whom the majority were used absolutely and in a specific sense. When this was the case," continues Professor Newton, "there can scarcely be a reasonable doubt that Linnæus, had he known our modern practice, would have designated as the type of his genus that species to which the name he adopted as generic had formerly been specifically applied." As regards the present case, there can be no doubt that under this rule the proper type of the Linnæan genus *Phoca* is the common small Seal of the European shores, the *Phoca vitulina* of Linné, and that *Callocephalus* is strictly a synonym of *Phoca*.

†Ann. and Mag. Nat. Hist., vol. xvii, 1866, p. 446.

‡Families of Mammals, p. 70.

Swainson, 1821), and in 1873* instituted the genus *Histriophoca* for the *Phoca fasciata* of Shaw (= *P. equestris*, Pallas).

Peters, in 1875,† finding that the old generic name *Stenorhynchus* of F. Cuvier, which had been currently received since 1824 as the generic designation of the Sea-Leopard of the Antarctic Seas, was preoccupied for a genus of Crustaceans to which it was applied by Lamark in 1819, as well as for a genus of insects in 1823, proposed in its place the name *Ogmorhinus*. Finally Gill has recently revived the name *Pusa* (Scopoli, 1777) for the Gray Seal in place of *Halichœrus*, Nilsson, 1820, but, as will be shown later, *Pusa* is not tenable in this connection.

At this point the following *résumé* of the subject may be presented, synonyms and untenable names being designated by thick type:

1735—*Phoca*, Linné, covering four species, belonging to three families and four distinct genera.

1777—*Pusa*, Scopoli, = *Phoca fœtida*, Fabricius.

1820 { *Halichœrus*, Nilsson, = *Phoca grypus*, Fabricius.
{ *Cystophora*, Nilsson, = *Phoca cristata*, Erxleben.

1822—**Monachus**, Fleming, = *Phoca monachus*, Hermann.

1824 { *Callocephalus*, F. Cuvier, type, *Phoca vitulina*, Linné.
{ **Stenorhynchus**, F. Cuvier, = *Phoca leptonyx*, Blainville. **Preoccupied in carcinology and entomology.**
{ **Pelagius**, F. Cuvier, = *Phoca monachus*, Hermann.
{ **Stemmatopus**, F. Cuvier, = *Phoca cristata*, Erxleben.
{ *Macrorhinus*, F. Cuvier, = *Phoca leonina*, Linné.

1827—**Mirounga**, Gray, = *Cystophora*, Nilsson, and *Macrorhinus*, F. Cuvier.

1830—**Rhinophora**, Wagler, = *Phoca leonina*, Linné.

1836—**Leptonyx**, Gray, = *Stenorhynchus weddelli*, F. Cuvier. **Preoccupied in ornithology.**

1844 { *Pagophilus*, Gray, = *Phoca grœnlandica*, Fabricius.
{ *Lobodon*, Gray, = *Phoca carcinophaga*, Homb. and Jacq
{ *Ommatophoca*, Gray, = *O. rossi*, n. sp.,, Gray.
{ **Phoca**, Gray, nec Linné, = *Phoca barbata*, Fabricius.

1854—**Heliophoca**, Gray, = *H. atlantica* n. sp., Gray, = *Phoca monachus*, Hermann.

1864 { **Halicyon**, Gray, = *H. richardsi* n. sp., Gray = *Phoca vitulina*, Linné.
{ **Pagomys**, Gray, = *Phoca fœtida*, Fabricius.

1866 { *Erignathus*, Gill, = *Phoca barbata*, Fabricius.
{ **Haliphilus**, Gray, = *Halichœrus antarcticus*, Peale, = *Phoca vitulina*, Linné.

1872—*Leptonychotes*, Gill, = *Leptonyx*, Gray, nec Swainson.

1873—*Histriophoca*, Gill, = *Phoca fasciata*, Shaw.

1875—*Ogmorhinus*, Peters, = *Stenorhynchus*, F. Cuvier, nec Lamark.

Of the above-given list of twenty-five generic names, two

---

* Amer. Nat., vol. vii, 1873, p. 179.

† Monatsb. Köngl. Preuss. Akad. d Wissens. zu Berlin, 1875, p. 393.

(*Sten or hynchus* and *Leptonyx*) were preoccupied in other departments of zoölogy, and consequently untenable; ten were exclusively based on four species, so that six of these ten may be set down as pure synonyms. The remaining thirteen have at times been more or less current, either in a generic or subgeneric sense. But leading writers have employed several of them with widely differing scopes. *Callocephalus*, as at first used, was synonymous with *Phoca*, as commonly interpreted, *Callocephalus* being employed for all of the smaller species of the family (*vitulina*, *fœtida*, *grœnlandica*, *fasciata*, *caspica*, &c, with all their numerous synonyms, as well as for *barbata*). In the subjoined history of the species of the family, the varying significance attached by different writers to the more prominent generic names will become sufficiently apparent, but in the present connection a few examples may be cited.

In 1820, Desmarest placed all the Pinnipeds, except the Walruses, under *Phoca*, which he divided into two subgenera ("sous-genre")—*Phoca* (including all the *Phocidæ* of recent authors) and *Otaria* (= *Otariidæ* of late writers), in this following Péron's classification of 1816. F. Cuvier, in 1824, divided the Earless Seals into five genera, which classification was followed by the same writer in 1826, and by Lesson in 1827, when he adopted strictly the generic nomenclature of F. Cuvier, but abandoned it in 1828. Gray, at about the same date, employed only two genera, *Phoca* and *Mirounga*, while Fischer, in 1829, placed all the species once more in *Phoca*.

Nilsson, in 1837, recognized five genera,—*Stenorhynchus*, *Pelagius*, *Phoca*, *Halichœrus*, and *Cystophora*. Gray, in 1844 and in 1850, adopted ten,—*Lobodon*, *Leptonyx*, *Ommatophoca*, *Stenorhynchus*, *Pelagius*, *Callocephalus*, *Pagophilus*, *Phoca* (with one species), *Halichœrus*, *Morunga*, and *Cystophora*. Wagner, in 1846, adopted four,—*Halichœrus*, *Phoca* (with six species), *Leptonyx* (with five species), and *Cystophora*. Turner, in 1849, admitted eight,—*Morunga*, *Cystophora*, *Halichœrus*, *Ommatophora* (typographical error for *Ommatophoca*), *Lobodon*, *Leptonyx*, *Stenorhynchus*, and *Phoca*. Giebel, in 1855, adopted the same genera as Wagner in 1846. Gray, in 1866 and in 1871, admitted thirteen,—*Lobodon*, *Leptonyx*, *Ommatophoca*, *Stenorhynchus*, *Monachus*, *Callocephalus*, *Pagomys*, *Pagophilus*, *Halicyon*, *Phoca*, *Halichœrus*, *Morunga*, and *Cystophora;* Gill, in 1866 and 1872, twelve,—*Phoca*, *Pagomys*, *Pagophilus*, *Erignathus*, *Halichœrus*, *Monachus*, *Cystophora*, *Macrorhinus*, *Lobodon*, *Stenorhynchus*, *Leptonyx*, and *Ommatophoca;* in 1877, twelve,—*Phoca*, *Pagophilus*, *Erignathus*,

Histriophoca, Pusa, Monachus, Cystophora, Macrorhinus, Lobodon, Ogmorhinus, Leptonychotes, and *Ommatophoca.* While Dr. Gill recognized the same *number* of genera in 1877 as in 1866, the nomenclature is quite different; but this is due mainly to simply changes of names, as the substitution of *Pusa* for *Halichœrus,* of *Ogmorhinus* for *Stenorhynchus,* and of *Leptonychotes* for *Leptonyx;* but in the later enumeration *Pagomys* is omitted and *Histriophoca* is added.

So far as the number of genera is concerned, the greatest difference of opinion has always obtained in respect to the *Phocinœ,* all the members of which group are confined to the Northern Hemisphere. Gray, after 1864 (1864–1874), uniformly recognized seven; Gill, 1866–1877, six, only two of which (*Halichœrus* and *Monachus,* about which authors generally have for many years been in unison) were the exact equivalents of Gray's genera; but the chief disagreement consisted in Gill's use of *Phoca* for what Gray termed *Callocephalus,* and of *Erignathus* for what Gray termed *Phoca.* Lilljeborg*, in 1874, referred all of the species of the *Phocinœ,* except *Halichœrus grypus* (and *Monachus albiventer,* which latter is not there treated), to the genus *Phoca,* and von Heuglin the same year did the same, except that *Pagomys, Pagophilus,* and *Callocephalus* (the latter being applied to *C. barbata*) were recognized as subgenera under *Phoca.* The classification and nomenclature of Giebel (1855), Blasius (1857), Malmgren (1863), and Holmgren (1865) are, generically, the same as Lilljeborg's in 1874. The tendency has, in short, been to refer all the species of *Phocinœ,* with the two exceptions already specified, to the Linnæan genus *Phoca.*

SPECIES.—Although Seals have figured in works on natural history since the time of Rondelet, Olaus Magnus, and Gesner (1554–1555), it is unnecessary in the present connection to refer in detail to those earlier works, since down to the time of Steller (1751), all the Phocids or Earless Seals known to systematic writers were referred to the common Seal (*Phoca vitulina,* auct.) of the shores of Middle and Northern Europe. This indeed was the only species recognized by Linné, from the Northern Hemisphere, even in the last (1766) edition of his "Systema Naturæ." But other species had been incidentally and vaguely described by the early Greenland missionaries, and by explorers and travellers in both the Arctic and Antarctic regions, to which refer-

---

* Fauna Sveriges och Norges Rygggradsdjur, i, Däggdjuren, pp. 667–729.

ence is necessary, since these descriptions became later the basis. in part or wholly, of various systematic names.

As early as 1741 the Harp Seal and the Crested Seal were figured (or caricatured) by Egede* under the names respectively of *Svartsiide* and *Klapmüts*. He says in the accompanying text that Seals are of different sorts and sizes, but have all the same shape, except the *Klapmüts*, which is the only species he expressly distinguishes in the text. Ellis,† in 1748, again rudely figured these two species under the names "Blackside Seal" and "Seal with a Cawl". Although he gives of them no descriptions, subsequent systematic writers have seen fit to cite the names and figures given by both these writers, but their interest is purely historic.

The same year (1748) was also published in Anson's "Voyage"‡ the first specially important account of the Southern Sea-Elephant ("Sea Lion" of Anson), since it became later the basis of Linné's *Phoca leonina*, and, besides, is one of the fullest and most explicit descriptions of the habits of the species extant.

Steller, in his memoir entitled "De Bestiis Marinis", published in 1751,§ distinguished three species of Seals as follows: "Distinguo autem phocas ratione magnitudinis in tres species, in maximam, quae magnitudine Taurum superat, ac solummodo in oceano Orientali a gradu latitudinis 56. ad 59. occurrit, ac in colis Kamtschaticis Lachtak vocatur. Mediae magnitudinis, quae omnes Tigridum instar, multis exiguis maculis variae sunt, 3. infimae magnitudinis, ut Oceanica, quae tam in mari Balthico, quam circa portum Sti Archangeli, in Suecia, Norwegia, America et Kamtschatka capitur, et lacustris dulcium aquarum monochroa seu unicolor, ut Baikali ea coloris argentei." The first two of the species here thus briefly mentioned, have been quoted by Schreber, Erxleben, Gmelin, and by some later writers, as respectively, "*Phoca maxima*, Steller" (also "*Lachtak*, Steller"), and "*Phoca oceanica*, Steller," the first being referred to *Phoca barbata*, and the other sometimes to *Phoca vitulina* and sometimes to *Phoca grœnlandica*.

In 1744 Parsons published a paper entitled "Some Account

---

* Det gamle Grønlands nye Perlustration, eller Natural-Historie, og Beskrivelse over det gamle Grønlands Situation, Luft, Temperament og Beskaffenhed, etc. 1741. Plate facing p. 46. He also figures the common Seal under the name *Spraglet*.

† A Voyage to Hudson's Bay, etc., 1748, plate facing p. 134.

‡ A Voyage around the World in 1740–1744, p. 172.

§ Nov. Comm. Acad. Petrop., tom. ii, p. 290.

of the *Phoca*, Vitulus marinus, or Sea-Calf, shewed at Charing-Cross, in February, 1742–'43",* containing a plate illustrative of the animal. Figure one, it is said by the author, "Represents the Phoca lying upon the right side, that the belly and Parts of Generation may be the better observed." He says the animal was very young, "though Seven Feet and half in Length, having scarce any Teeth, and having Four Holes regularly placed about the naval, as appears by the Figure, which in time become *Papillæ*." This account, as will be noticed later, has figured very prominently in relation to the history of the Bearded Seal (*Phoca barbata*, auct.), especially in reference to its right to a place in the British Fauna.

In 1753 the same author, in a paper entitled "A Dissertation upon the Class of the Phocae Marinae,"† formally described five "species" under the term *Phoca*, of which three were caricatured in figures. In the way of criticism of his paper, it is perhaps enough to say that one of his species is a compound of the Manatee and the Sea-Elephant (*i. e.*, "Manati, De Laet" and "Sea-lion, Lord Anson"). Another, based on Grew's "Long-necked Seal", from an unknown locality, is evidently some kind of Otary. His "Common Seal", his "Tortoise-headed Seal", and his "Long-bodied Seal", are evidently Phocids, but the short diagnoses give no distinctive characters, and the species, as here described, are consequently unrecognizable. The last named, which was originally described, as already noticed, in 1744, has been usually referred, but generally with doubt, to the *Phoca barbata*, mainly on account of its large size, but his figure gives other characters that render it pretty certain that this is a correct allocation. His "Common Seal" has been presumed to be the *Phoca vitulina*. Two of these were introduced into technical nomenclature by Kerr in 1792, on which Shaw imposed additional names in 1800.

In 1758 Linné, in the tenth edition of the "Systema Naturæ", gave four species under the genus *Phoca*, namely; 1, *Phoca ursina* (based exclusively on Steller's "*Ursus marinus*"); 2, *Phoca leonina* (based exclusively on Anson's "Sea Lion"); 3, *Phoca rosmarus* (the Walrus); and 4, *Phoca vitulina*, with its habitat defined as "Mari Europæo". In the twelfth edition of the same work (1766) the third species above named was removed to the genus *Trichechus*, and Ellis's figure of the "Seal with a

* Phil. Trans., vol. xlii, for the years 1742 and 1743 (1744), p. 383, pl. i.
† Ibid., vol. xlvii, 1751–1752 (1753), pp. 109–122, pl. vi.

Cawl" is cited under *Phoca leonina*, thereby incepting the confusion of the Crested Seal of the Arctic Seas with the Sea-Elephant of the Southern Hemisphere which prevailed more or less
generally for the next quarter of a century. *Phoca vitulina* was
thus the only northern Phocid here distinctively recognized.

A second notice of Seals on the basis of Steller's observations,
and one that has figured prominently in the history of the subject, is contained in Kraschinenikow's "History of Kamtschatka", published in 1764, *—a work avowedly based largely
on Steller's MSS. Grieve's translation of the passage relating
to the Seals is as follows: " There are reconed to be four sorts
of this animal; the very largest of which is catched from 56° to
64° of north latitude. This sort differs from the others in its
bulk, which exceeds that of a large ox. The second species is
about the size of a yearling bullock. Their skin is of different
colours, something like the skin of a tyger, having several spots
of equal largeness on the back, with a white and yellowish belly.
Their young ones are as white as snow. The third is yet less
than the former. Its skin is yellowish, with large cherry-col-
oured circles, which take up near the half of its surface. The
fourth kind is seen in the large lakes of *Baikaal* and *Oronne*. Its
size is like those that are found near *Archangel;* and their colour
is whitish." These indications, though so vague, have served,
either in part or solely, as the basis of several of the species of
the later systematic writers, they being referred to numerically
as the " First sort of Seal", the " Fourth sort of Seal", etc.

The first really important account of the Seals of the Northern Seas is that given by Cranz in 1765, in his "Historie von
Grönland," in which he enumerates and briefly characterizes
all of the five species of Seals hunted or commonly met with in
Greenland. Although his descriptions are in most cases meagre,
and relate more to the habits of the species and to their
useful products than to their external characters, his species
are, from one circumstance or another, so easily recognized
that there has never been much uncertainty in regard to them.

---

* I cite Grieve's (English) translation (1 vol., 4to) from the original Russian, published in 1764, wherein the matter relating to the Seals appears at
page 116. There is also a French translation (2 vols. 12mo) published in
1767, which is often quoted by French authors. The work quoted by German writers as Steller's "Beschreibung von den Lande Kamtschatka"
(1 vol. 8vo, 1774—which I have not been able to see), seems to be, so far as
the matter relating to the Seals is concerned, merely a German version of
the same work.

They are, 1. The *Kassigiak* (= *Phoca vitulina*); 2. *Attarsoak* (= *Phoca grœnlandica*); 3. *Neitsek* (= *Phoca fœtida*); 4. *Neiter-soak*, called also *Clapmutz* (= *Cystophora cristata*); and 5. *Utsuk* (= *Erignathus barbatus*). The Neitsek or Ringed Seal (*Phoca fœtida*) appears to be here for the first time indicated.

Pennant, in 1771, formally introduced three species into his "Synopsis of Quadrupeds" under English names, the Neitsek appearing under the name "Rough Seal." His description of this species is based wholly on Cranz, and those of the "Harp" and "Hooded" Seals on Egede and Cranz. In 1776 these species all received systematic names at the hands of Fabricius, in an inedited MS. in Müller's "Zoölogiæ Danicæ Prodromus" (p. viii of the Introduction, received after the main body of the work was printed), except the long previously named Kassigiak (*Phoca vitulina*). Fabricius's names, however, were unaccompanied by descriptions, but carried with them the common Icelandic and Greenlandic names of the species indicated, by means of which they are susceptible of strict identification, aside from their being identified later by Fabricius's own descriptions and references to them. The following is a literal transcript of Fabricius's inedited list:

"PHOCA *leonina* capite antice cristato., I. *Blandruselr.* Gr. *Neitsersoak.*

"Ph. *fœtida*, I. *Utselr.* Gr. *Neitsek, Neitsilek.*

"Ph. *grœnlandica*, I. *Vadeselr.* Gr. *Atak.*

"Ph. *barbata*, I. *Gramselr.* Gr. *Urksuk.*" *

Here is the origin of the names still in current use for three of the four species here named by Fabricius. †

Simultaneously with the publication of Müller's "Prodromus" must have appeared the first fasciculus of the third part of Schreber's "Säugthiere" (as appears by contemporaneous evidence, although the completed part bears date 1778), in which all these and two other species of Seals are described, in addition to the common *Phoca vitulina*. In the text they are mentioned only under vernacular names, but the plate of the

---

* It is worthy of note in this connection that Müller himself, on page 1 of the "Prodromus," under *Phoca vitulina*, cites the names of "Klapmüts" and "Svartside." He then gives a list of Icelandic, Greenlandic, and other vernacular names of Seals, respecting which he says information is desirable, and adds "varietates an species?" Yet Müller is quite commonly quoted as the authority for these Fabrician names.

† The first species of the list bears the name previously given by Linné to the Sea-Lion of the Antarctic Seas.

"Neitsek" bears the name *Phoca hispida*, between which and
Fabricius's *Phoca fœtida* there is consequently a troublesome
question of priority.* In Schreber's work are first formally
introduced into a general systematic treatise, the Siberian Seal
and the Caspian Seal (based wholly respectively on previous
descriptions of the same by Steller and Gmelin †), and "Der
graue Seehund," commonly referred to *Halichœrus grypus*.
Schreber also described, under the head of *Phoca*, two species
of Otary. His species are the following: 1. Der Seebär, *Phoca
ursina* (= *Ursus marinus*); 2. Der glatte Seelöwe, *Phoca leonina*
(= primarily Anson's Sea Lion); 3. Der zottige Seelöwe, *Phoca
jubata* (= primarily Steller's *Leo marinus*, but including also the
Southern Sea-Lion); 4. *Phoca vitulina*; 5. Der graue Seehund
(= *Erignathus barbatus*); 6. Der sibirische Seehund (= *Phoca
sibirica*); 7. Der caspische Seehund (= *Phoca caspica*); 8. Der
Schwarzside (= *Phoca grœnlandica*); 9. Der rauhe Seehund
(named *Phoca hispida* on the plate); 10. Der Klappmüze
(= *Cystophora cristata*); 11. Der grosse Seehund (= *Phoca
barbata* + *H. grypus*); 12. Der kleine geöhrte Seehund, *Phoca
pusilla*. These twelve species, excepting the last, all represent
valid species, nine of which belong to the present family.
He mentions, however, Olafsen's Gramm-Selur as still another
"grosse Seehundsart," but does not formally notice it as a species. His Der grosse Seehund, it should be further noticed, is
based on the Utsuk of Cranz and the Ut-Selur of Olafsen, combining an account of the habits of the latter with a description
of the external characters of the other.

Olavsen (or Olafsen, as more commonly written), in his account
of his travels in Iceland, published in 1772, ‡ repeatedly alludes
to the various species of Seals met with in Iceland. As already
noticed, Olafsen is quoted by Schreber, and quite frequently by
later writers. While he describes quite fully their habits, distribution, and products, he has very little to say of their external

---

* For a discussion of this point see *posteà* under *Phoca fœtida*.

† Schreber's "Der caspische Seehund" is based on the account of the
Caspian Seal given by Gmelin in 1770, in the third volume (p. 246) of his
Reise durch Russland zur Untersuchung der drey Naturreiche.

‡ I cite here the German edition entitled "Des Bice-Lavmands Eggert
Olafsens und des Landphysici Biarne Povelsens Reise durch Island, veran-
staltet von der Königlichen Societät der Wissenschaften in Kopenhagen
und beschrieben von bemeldtem Eggert Olafsen. Aus dem Dänischen übersetzt. Mit 25 Kupfertafeln und einer neuen Charte über Island versehen.
Kopenhagen und Leipzig bey Heinecke und Faber, 1774." Zwei Theilen, 4°.

characters. Yet in the light of present knowledge it is not difficult to determine with considerable certainty the species he mentioned.

His "Land-Selur," called also "Wor-Selur" or Spring Seal, because it brings forth its young in the spring (l. c. §§ 83, 329, 524, 651–655,) is evidently the *Phoca vitulina*, at least in greater part, but may be a general term for the smaller Seals found in Iceland, and hence refer in part to *Phoca fœtida.* His "Ut-Selur," called also "Wetrar-Selur," or Winter Seal, because it brings forth its young at the beginning of winter (l. c. §§ 329, 651–655) is without doubt the *Halichœrus grypus.* He says it is much larger than the Land-Selur, but resembles it in appearance, and brings forth its young on the island at the time of the withering of the grass in the month of November (l. c. § 329). He also mentions not only the Walrus as of rare occurrence in Iceland, but enumerates three other species of Seals, two of which are identifiable. One of these is the "Vade-Säl" or "Hav-Säl," said to be as large as the Ut-Selur, or four elles long, but thicker and fatter, with a very strong skin. It is described as being black in color, with large round spots, which are smaller on the back than on the sides. It swims in a straight line, in great troops, and close together, in a certain order, whence it derives its name "Vada," signifying a swimming herd. One of them, commonly the largest, takes the lead, and is called "Säle Konge" (King of the Seals). This species is never seen on the land, but only on the drift-ice, where it is hunted with harpoons, particularly on the Northern Coast. It has its young in April, on the remote outlying rocks and islands, for it goes away in March, and when it comes back in May it has its young with it. This account, in almost every particular, points to the Bearded Seal (*Phoca barbata*, auct.) as the species indicated, with which the size, coloration, and habits sufficiently agree.

Another is the "Blaudru-Selur" or "Blase-Seehund," which is here rarely met with and killed. It has a protuberance resembling a bladder on the head over the nose, where the skin is loose, so that the animal can suddenly draw it with the fat down upon the nose. He says it is uncertain whether this is the *Phoca leonina* of Linné, for the character *capite antice cristato* does not agree. He also raises the question whether it can be the Sea Bear, and decides it in the negative, and gently criticises the above-named author for referring all the Seals to one species, under the name *Phoca vitulina*. In the characters here given we have certainly indicated the Hooded or Crested Seal (*Cystophora cristata*).

The third additional species he gives is called "Gramm-Selur," which he says is also known in Iceland, and refers to its being mentioned in the "Speculo Regali, p. 177," and in "Olaf Tryggesen's Saga, p. 263." He says it is called commonly "Grâm-Selur," and is counted as large as some kinds of whales. "Gram" signifies in the old poets a king; the Gram-Selur may be twelve or fifteen Icelandic elles long, and is rare in Iceland; still they sometimes find it in Westland where examples have been killed on the outermost rocks of Breedefiord. They find it also thrown up on the shore dead, but then no further information or description of it can be obtained than is found in Olaf Tryggesen's Saga, namely, that it has long hair on the head, particularly around the mouth, therefore it is perhaps a Sea Lion, or the great species which lives in the American Antilles (see, he says, "Joh. Sam. Hallens Natur-Geschichte der Thiere, p. 593 und 581"), which is also credible. It thus appears that Olafsen, like the other early writers who refer to the "Gram-Selur," had no personal knowledge of it and spoke of it only from report. It may doubtless therefore be safely treated as a myth.

It will be noticed that two species known to frequent Iceland are not here mentioned, namely, the Harp Seal and the Rough Seal. The latter may have been confounded with the Land Seal, but the former can hardly be thus accounted for, especially as Olafsen on one occasion distinctly refers to its occurrence in Greenland. Only four species are thus apparently referred to by Olafsen as inhabitants of Iceland. He alludes further to the manner of capturing the different kinds of seals in Iceland and the value and uses of their products, and recounts the fables current among the Icelanders respecting the Seal-tribe, and the estimation in which the animals are held among them.

In Erxleben's admirable compilation, entitled "Systema Regni Animalis," published in 1777, the Earless Seals recognized and briefly diagnosed, are the following: 1. *Phoca vitulina*, with three unnamed varieties (of which latter more will be said later); 2. *Phoca. grœnlandica*; 3. *Phoca hispida*; 4. *Phoca cristata* (the Klapmütz of Egede and the Neitsersoak of Cranz = *Phoca leonina*, capite antice cristato, Fabricius, 1780, but not of Linné, 1758); 5. *Phoca barbata*. The Hooded Seal (*Cystophora cristata*) here receives its first tenable specific name.

In 1778 Lepechin* described two species of Seals under the names *Phoca oceanica* and *Phoca leporina*. The first is unquestionably the *Phoca grœnlandica* of Fabricius and Erxleben, while the other is usually regarded as having been based on the young of the *Phoca barbata* of the same authors, although in both cases the incisors are described as four in each jaw.

The following year (1779) appeared Hermann's elaborate memoir † (of fifty pages and two plates) on the Monk Seal ("Münchs-Robbe,—*Phoca monachus*") of the Mediterranean Sea—the first explicit account of the species, and a very admirable monograph for this early date.

The next year (1780) Fabricius published his "Fauna Grœnlandica," in which all the Seals named in Müller's "Prodromus" are quite satisfactorily described, under the names there first proposed. He, however, erroneously includes among the Seals of Greenland Steller's Sea Bear, under the name *Phoca ursina*, and concludes his account of the Greenland Seals by mentioning four other marine animals he had heard of from the Greenlanders, but of which he had never seen either skins or skulls, and of which he knew nothing with certainty, namely *Singuktop, Imab-ukallia, Atarpiak*, and *Kongeseteriak.* ‡

In 1780, in his "Synopsis der Quadrupeden" (Geographische Geschichte, etc., Theil ii, pp. 419–423), Zimmermann gave the same species that Erxleben did in 1777, and under the same names. In 1782, however, in an appendix to the "Synopsis" (ibid., Theil iii, 1782, pp. 276–278), he added three species under "*Phoca*," two of them based on Pennant's "History of Quadrupeds," published in 1781, as follows: 1. "*Phoca australis*. Falkland Seal Pennant ii, 521"; 2. "*Phoca fasciata*. Rubbon Seal Pennant ii, p. 523"; 3. *Phoca leporina*, Lepech. The first two are for the first time named; the first, however, is an Otary. Zimmermann also says, "Le phoque à Ventre blanc, Buffon Suppl. vi, pl. 44, p. 310, ist wohl *Phoca monachus*"; yet subsequent writers of less discrimination held it for a distinct species.

Buffon, in 1782, in the sixth volume of the "Supplement" of his "Histoire naturelle," recognized eight species of "Les Phoques sans oreilles ou Phoques proprement dits" (some of which,

---

*Act. Acad. Sci. Imp. Pétrop., i, 1777 (1778), pp. 259,264, pll. vi–ix.

† Berschaft. d. Berlinische Gesselschaft naturf. Freunde, Band iv, 1779, pp. 456–509, pll. xii, xiii (external characters).

‡ Respecting these see *posteà*, pp. 432,433.

however, really covered several distinct species), as follows: 1.
Le grand Phoque à museau ridé (=*Macrorhinus leoninus*); 2.
Le Phoque à ventre blanc (=primarily *Monachus albiventer*,
with an original description and a good figure, from a specimen
taken October 28, 1777, in the Adriatic Sea, but to which he erroneously referred Parsons's Long-bodied Seal, he giving a translation of Parsons's description and a copy of his figure, and also
the Utsuk of Cranz, and a large Seal mentioned by Charlevoix
as found on "les côtes de l'Acadie"); 3. Le Phoque à capuchon (=*Cystophora cristata*); 4. Le Phoque à croissant (=*Phoca
grœnlandica*, at least mainly); 5. Le Phoque Neit-soak (*Phoca
fœtida*); 6. Le Phoque Laktak de Kamtschatka (=*Erignathus
barbatus*); 7. Le Phoque Gassigiak (= the Seal "appelée *kassigiak* par les Groënlandois"; consequently *Phoca vitulina*); 8.
Le Phoque commun (= primarily *Phoca vitulina*, but with allusions to other species). Of these eight species two are composite,
and one is purely nominal.*

In the same year (1782) also appeared Molina's work on the
natural history of Chili,† in which, under the head of *Phoca*, are
described four species, all claimed by the author to be new.
These are: 1. "L'Urique, *Phoca lupina*" (a Fur-Seal, or at least
an Otary); 2. "Il Porco marina, *Phoca porcina*" (probably the
young of the next); 3. "Il Lame, *Phoca elephantina*" (=*Phoca
leonina*, Linné, 1758 and 1766); 4. "Il Leon marin, *Phoca leonina*" (=*Otaria jubata*, auct.).

In 1784 Boddaert appears to have added (I have not the work‡
at hand) four synonyms, as follows: 1. *Phoca albiventer* (=*P.
monachus*, Herm.); 2. *Phoca semilunaris* (=*P. grœnlandica*); 3.

*This enumeration, however, is a great improvement upon that given by
the same author in 1765, in the thirteenth volume of his "Histoire naturelle",
where all the Seals then known are referred to four species. " . . . . le
premier (*pl. xlv*) est le phoque de notre océan, dont il y a plusieurs varié-
tés"; called also "le Veau marin ou Phoque de nos mers". The second, supposed to be "le *phoca* des anciens", and which is figured in pl. liii, is a young
Eared Seal, the *Phoca pusilla* of later writers, of which he says, "on nous a
assuré que l'individu que nous vu venoit des Indes", etc. Later it is
called "le petit phoque noir des Indes & du Levant". (See further, *anteà*,
p. 194.) The third is the Seal described by Parsons in 1743—the Long-bodied
Seal of this and many subsequent authors—here called "le grand phoque
des mers du Nord". The fourth is Anson's "Sea Lion", but which here
covers also "les grands phoques des mers du Canada, dont parle Denis, sous
le nom de loups marins", to which he is also inclined to refer the larger Seal
described by Parsons!

†Saggio Sulla Storia Naturale del Chili, pp. 275–290, 341.
‡Elenchus Animalium, vol. i, 1784, pp. 170, 171.

Phoca cucullata (=*P. cristata*, Erxl.); 4. *Phoca maculata* (? = *P. vitulina*).

The next work of importance in this connection is Gmelin's 'Systema Naturæ", which appeared in 1788. Here Erxleben's five species appear without change of name, and in addition to them the Monk Seal (*Phoca monachus*) of the Mediterranean.* Under *Phoca vitulina* are three named varieties, to wit, *botnica*, *sibirica* ("colore argenteo. *Habitat in lacubus* Baikal *et* Orom") and *caspica* ("colore vario"), which are respectively Erxleben's varieties *a*, *β*, and *γ*, and Schreber's "graue Seehund", "sibirische Seehunde", and "caspische Seehund".

In 1790 and 1791 Fabricius published his celebrated memoir on the Seals of Greenland,† in which all the Greenland species are described in great detail, and the skulls of *Phoca grœnlandica*, *Cystophora cristata*, *Erignathus barbatus*, and *Halichœrus grypus* are for the first time figured, while the last-mentioned species is for the first time named. In this series of papers the general subject is exhaustively treated in all its bearings, nearly eighty pages being devoted to the "Svartside" (Harp Seal, *Phoca grœnlandica*) alone; twenty-four to the Fiordsæl (Ringed Seal, *Phoca fœtida*); twenty-two (including nearly four pages of bibliographical references) to the "Spraglede Sæl" (Harbor Seal, *Phoca vitulina*); about the same number each to the "Klapmydsen" (Hooded Seal, *Cystophora cristata*); and the "Remmelsæl" (Bearded Seal, *Erignathus barbatus*). In the bibliography of these species are, however, given various references that are not pertinent, particularly under *Phoca cristata*, under which name are confounded the Sea-Elephant of the Southern Hemisphere with the Crested Seal of the Northern. The *Halichœrus grypus* is mentioned (but not fully described), and the skull figured, under the names "Krumsnudede Sæl (*Phoca grypus*)". As regards changes in nomenclature, he abandons the names *fœtida* and *leonina* respectively for *hispida* and *cristata*. His memoir is greatly marred by the introduction into its closing portion of various species (already referred to in his "Fauna Grœnlandica") that are either mythical or have no relation to the Greenland fauna, as the Sea-Bear ("Søebiørne, *Phoca ursina*)", the "Sviinsæl (*Phoca porcina*)", the "Søehare (*Phoca leporina*)",

---

*Ex "Hermann, Act. Nat. Scrutat. Berol. iv, p. 246, t. xii, xiii."

†"Udførlig Beskrivelse over de Grønlandske Sæle", Skriv. af Naturh. Selsk., 1ste Bind, 1ste Hefte, 1790, pp. 79–157, 2det Hefte, 1791, pp. 73–170, Tab. xii, xiii.

and the "Atârpiak", making nine species formally introduced
(besides "*Phoca grypus*", which is treated incidentally at the
close of the account of the "Sviinsæl"). The *Phoca leporina* of
Lepechin is a synonym of his *Phoca barbata*, but the *Phoca leporina* of Fabricius has doubtless no foundation except in the
imagination of the Greenlanders. These doubtful or mythical
species have been especially investigated by Mr. Robert Brown.
The *Phoca ursina* was based on a part of a cranium which was
"full of holes", but so much uncertainty prevailed in Fabricius's
mind respecting the nature of the creature it represented that
he makes the *same fragment* the basis for the introduction into
the Greenland fauna ("Fauna Grœnlandica", p. 6) of two species,—Steller's Sea-Cow (" *Trichechus manatus*",=*Rhytina gigas*,
auct.) as well as Steller's Sea-Bear. Says Mr. Brown, "Whatever it is, there can, I think, be scarcely a doubt as to the exclusion of *Trichechus manatus* and *Phoca ursina* from the Greenland fauna; nor can their place as yet be supplied by any other
species. Prof. Steenstrup thinks that it was a portion of the
skull of a Sea-wolf (*Anarrhichas*). The situation of the teeth
and the nature of this fish's cellular skull well agree with his
description of the skull as 'full of holes'. . . . . Hr. Bolbroe,
who understands the Eskimo language intimately, tells me that
the word [*Auvekœjak*] means a 'little walrus', and that in all
probability it was only the skull of a young walrus, an animal
not at all familiar to Fabricius, as they are chiefly confined to
one spot, and the natives fear to go near that locality. Fabricius may have only written the description from recollection;
and memory, assisted by preconceived notions, may have led
him into error in the description of the long teeth, which after
all might, without great trouble, be made to refer to the dentition of the young walrus. . . . . This opinion is strengthened by
a passage in Fabricius's account of the walrus, where he again
is in doubt whether a certain animal is the young of the walrus
or the dugong. . . . . So that, after all, perhaps the *Auvekœjak*
was only the young of the walrus; and this opinion I am on
the whole inclined to acquiesce in".*

The other species of Fabricius's supposed *Phocæ* are thus referred to by Mr. Brown: "Fabricius has notified in his fauna
[and noticed them more at length in his later memoir on the
Greenland Seals already cited] many species of supposed Seals,

---

* Proc. Zoöl. Soc. London, 1868, p. 358; Man. Nat. Hist., etc., Greenland,
Mam., p. 29.

&c., under various Eskimo names, but which he was unable to decipher. Hr. Fleischer, Colonibestyrer of Jacobshavn, has aided me in resolving these:

" 1. *Siguktok,* having a long snout and a body similar to *Phoca grœnlandica,* perhaps *P. ursina.* This is apparently some Eskimo perversion, if interpreted properly; for I am assured that it is only the name of the Eider Duck (*Somateria mollissima*). [In his memoir in the ' Skrivter af Naturhistorie-Selskabet' (vol. i, part ii, p. 163) it is called ' Sviinsælen (*Phoca porcina*)', and he refers to it such diverse creatures as Molina's '*Phoca porcina*' and Pennant's ' Bottle-nosed Seal,' and devotes nearly four pages to its consideration.]

"2. *Imab-ukullia,* a Seal with a snow-white coat, 'the eye presenting a red iris, probably *P. leporina*', is a rare albino of the *Netsik* (*Pagomys fœtidus*). The meaning of the word is the Sea-hare. [In the ' Skrivter' (l. c. p. 168) this is called 'Søeharen (*Phoca leporina*)', of which Lepechin's '*Phoca leporina*' and Schreber's 'sibirische Seehund' are cited as synonyms.]

" 3. *Atarpiak* or *atarpek,* 'the smallest species of Seal, not exceeding the size of the hand, of a whitish color, and a blackish spot of the form of a half-moon on each side of the body.' This description does not correspond to the meaning of the word, which is 'the Brown Seal'. [This in the 'Skrivter' (l. c. p. 169) appears as the ' Niende Art, Atârpiak', without a Latin name or synonyms.] Hr. Fleischer thinks that it is only a myth, as is—

"4. *Kongesteriak* [not mentioned in the 'Skrivter'], which has, ' according to the description given by the natives, some resemblance to the Sea-ape described by Mr. Heller'. This is one of the northern myths."*

In 1792 appeared Kerr's "Animal Kingdom", the title-page of which states it to be "A translation of that part of the Systema Naturæ, as lately published by Professor Gmelin of Gœttingen, together with numerous additions from more recent zoological writers and illustrated with copper plates." In this work (pp. 121–128) nineteen species appear under the generic name *Phoca*, with five additional varieties, among which we find the real origin of quite a number of names currently attributed to much later authors. The species and varieties here enumerated are the following, the new names being distinguished by the use of thick type: 1. *Phoca ursina;* 2. *P. leonina* (= Bot-

---

* Proc. Zoöl. Soc. London, 1868, p. 360; Man. Nat. Hist., etc., Greenland, pp. 31, 32.

tled-nosed Seal of Pennant); 3. *P. jubata;* 4. *P. vitulina* (with vars. *botnica, sibirica, caspica*); 5. *P. monachus;* 6. *P. grœnlandica* (with var. **nigra**); 7. *P. hispida* (with var. **quadrata** = "Square Phipper, Arct. Zool. i, 161"); 8. *P. cristata;* 9. *P. barbata;* 10. *P. pusilla* (an Eared Seal); 11. *P.* **chilensis** (= *P. porcina*, Molina); 12. *P.* **mutica** (= Long-necked Seal of Parsons—an Eared Seal); 13. *P. australis* (= "Falkland Seal, Penn. Hist. Quad., n. 378"); 14. *P.* **testudo** (= Tortoise-head Seal, Pennant); 15. *P. fasciata* ("Harnassed Seal" = Rubbon Seal, Pennant); 16. *P.* **laniger** (= *Phoca leporina*, Lepechin); 17. *P.* **punctata** ("Is speckled all over the body, head, and limbs. Penn. Hist. Quad. p. 523. Inhabits the seas about Kamtschatka and the Kurile Islands"); 18. *P.* **maculata** ("The body is spotted with brown. Penn. Hist. Quad. p. 523. Inhabits the coasts about the Kurile Islands. This species is very scarce"); 19. *P.* **nigra** ("Has a peculiar conformation of the legs. Penn. Hist. Quad. p. 523. Inhabits the coast about the Kurile Islands. This and the two last species are mentioned by Mr. Pennant as being obscurely described in the manuscripts of Steller. What the peculiarity in the conformation of the hind legs, in the Black Seal, consists of, is not said").

In this case the "numerous additions" are all from Pennant, and embrace nine new names, seven of which are specific and two varietal. In each case explicit reference is made to Pennant's species. Kerr's work has been so completely overlooked or ignored by subsequent writers that most of his new names have been attributed to Shaw and other still later sources.

Pennant, in 1793, published the third and last edition of his "History of Quadrupeds." Although he employed only vernacular names, his descriptions are well drawn, and some of them are important from their being the basis, wholly or in part, of several technical names imposed by later writers. Although almost exclusively a compilation, the matter relating to the Seals reflects fairly the then present state of knowledge respecting these animals. As it was, furthermore, the last general account of the Seals published prior to the year 1800, it may well be taken as an exponent of the subject as known a few years prior to that date.

Under the term "Seal" Pennant embraced all of the Pinnipeds then known, except the Walruses, the term being equivalent to "*Phoca*" of more technical writers of the same period. The species he recognized are the following :* 1. "Common Seal,"

---

* L. c., vol. ii, pp. 270–291.

of which he says, "Inhabit most quarters of the globe, but in greatest multitudes towards the North and the South; swarm near the *Arctic* circle, and the lower parts of *South America*, in both oceans; near the southern end of *Terra del Fuego;* and even among the floating ice as low as *south lat.* 60. 21. Found in the *Caspian* Sea, in the lake *Aral*, and lakes *Baikal* and *Oron*, which are fresh waters. They are lesser than those which frequent salt waters; but so fat that they seem almost shapeless. In lake *Baikal* some are covered with silvery hairs; others are yellowish, and have a large dark-colored mark on the hind part of the back, covering almost a third of the body." 2. "Pied Seal" (= Le Phoque à ventre blanc, Buffon, = *Monachus albiventer*); 3. "Mediterranean Seal" (= *Phoca monachus*, Hermann, = *Monachus albiventer*); 4. "Long-necked Seal" (of "Grew's Museum, 95," and of Parsons = some indeterminable species of Otary); 5. "Falkland Isle Seal" (= *Arctocephalus australis*=*A. falklandicus*, auct.); 6. "Tortoise-headed Seal" (= Tortoise-headed Seal, Parsons,—undeterminable); 7. "Rubbon Seal," based on a description (?) and a drawing communicated to him by "Doctor Pallas," of a mutilated skin received by the latter "from one of the remotest *Kuril* islands." The drawing "is engraven on the title of Division III, Pinnated Quadrupeds." This is a species described much later by Pallas as *Phoca equestris;* 8. "Leporine Seal" (= *Phoca leporina*, Lepechin); 9. "Great Seal" (= *Phoca barbata*, at least mainly); 10. "Rough Seal" (= Neitsek, Cranz; *Phoca hispida*, Schreber); 11. Porcine Seal, (*Phoca porcina*, Molina); 12. "Eared Seal" (described from a specimen in Parkinson's Museum, from the Straits of Magellan, and probably a young *Otaria jubata*); 13. "Hooded Seal" (= *Cystophora cristata*, unmixed with any other species); 14. "Harp Seal" (*Phoca grœnlandica*); 15. "Little Seal" (= Le petit Phoque, Buffon; based on a specimen of a young Eared Seal, originally supposed to have come from India—undeterminable); 16. "Ursine Seal" (a compound of all the Sea-Bears or Fur-Seals); 17. "Bottle-nose Seal" (= Sea Elephant of the Southern Seas. A good description is given of the external characters of both sexes, that of the female being based on a "well-preserved specimen in the Museum of the Royal Society"—the specimen previously described by Parsons under the name Manatie!); 18. "Leonine Seal" (= a compound of the Northern and Southern Sea Lions); 19. "Urigene Seal" (= *Phoca lupina*, Molina).

Of these nineteen species six are Otaries; of the remaining thirteen one is described for the first time; three are pure synonyms, and two are not certainly determinable. Three valid species are confounded under the name "Common Seal," which embraces (a) *Phoca vitulina*, (b) *P. caspica*, (c) *P. sibirica*, the two last already thus named as varieties of *Phoca vitulina* by Gmelin. With this exception, all of the species of Earless Seals of the Northern Hemisphere, up to that time indicated, were duly recognized and clearly distinguished by Pennant, as well also as one from the Southern Seas.

In 1798 Thunberg, in a small work on the mammalian fauna of Sweden,* recognized five species of Seals, three of which appeared under new names, but they are so briefly described it is nearly impossible to determine to what his various names relate, especially as he gives no synonyms, and no subsequent author appears to have been able to positively identify them. The names are accompanied by short Latin diagnoses, which I here transcribe: 1. "*P*[*hoca*] *hispida*: corpore pallido fuscomaculato." He adds the Swedish names "*Skål, Gråskål, Hofsskål*." Perhaps the *Phoca* "*hispida*" of authors, but of this there is no certainty. 2. "*P. sericea*: corpore albido immaculato." "*Stȧt-skål*." Undeterminable. 3. "*P. canina*: corpore griseo immaculato." "*Vikare-skål och Grȧ Vikare-skål.*" 4. "*P. vitulina*: corpore fusco." "*Svart Vikare-skål.*" 5. "*P. variegata*: corpore griseo nigro-maculato." "*Morunge.*" This by some authors is judged to be *Phoca vitulina*. Eight of the eleven pages devoted to the Seals in this work are occupied with the account of the present species. Thus Thunberg obtained the distinction of adding five species to the numerous list of Seals too inadequately described for recognition, and of contributing three new names to the synonymy of the subject.

Pausing now for a hasty retrospect, we find that prior to the year 1800 the following species (exclusive of synonyms and unidentifiable " species"), named in the order of their first recognition in technical nomenclature, had already made their appearance in works on systematic zoölogy: 1. *Phoca vitulina;* 2. *Macrorhinus leoninus;* 3. *Cystophora cristata* (1–3 as early as, or prior to, 1758); 4. *Phoca fœtida;* 5. *Phoca grœnlandica;* 6. *Erignathus barbatus* (4–6, 1766); 7. *Monachus albiventer* (1779); 8.

---

* Beskrifning på Svenske Djur. Första Classen, om Mammalia eller Dåggandejuren, af Carl Peter Thunberg, Upsala, 1798. 8 vo. pp. 100. Seals, pp. 85–96.

Phoca fasciata (1783); 9. *Halichœrus grypus* (1791). Two other species had been distinguished as varieties, namely, 10. *Phoca caspica;* 11. *Phoca sibirica* (both 1788). As will be seen later, only two northern species (*Macrorhinus angustirostris* and the West Indian Seal), and three others from the Southern Seas, remained to be added, although the literature of the subject has since been burdened by the addition of not less than sixty synonyms!

In the year 1800 Shaw, in his "General Zoölogy" (Quadrupeds, vol. i, pp. 250–272) redescribed Pennant's nineteen species, under Latin as well as English names, bestowing new Latin names upon five of them, none of which, however, have proved to be valid species. So far as the Seals are concerned, his work is little more than an abridged paraphrase of Pennant, being strictly a compilation (based almost wholly on Pennant), with the most of the bibliographical references omitted (he cites usually only Gmelin and Pennant), with the form of the matter changed by throwing the descriptions of the external characters into brief Latin diagnoses, duplicated in English.* His species are the following, the new names added being here printed in heavy type: 1. *Phoca vitulina;* 2. *Phoca* **bicolor** (=Le Phoque à ventre blanc, Buffon, hence *Monachus albiventer*); 3. *Phoca monachus;* 4. *Phoca* **longicollis** (= Long-necked Seal, Pennant); 5. *Phoca* **falklandica** (= Falkland Isle Seal, Pennant = *Phoca australis,* Zimm.); 6. *Phoca* **testudinea** (Tortoise-head Seal, Pennant, ex Parsons = *Phoca testudo,* Kerr); 7. *Phoca fasciata* (= Rubbon Seal, Pennant); 8. *Phoca leporina* (= Leporine Seal, Pennant; Lepechin not cited); 9. *Phoca barbata;* 10. *Phoca hispida* (= Rough Seal, Pennant); 11. *Phoca porcina* (= Porcine Seal, Pennant; Molina not cited); 12. *Phoca* **flavescens** (= Eared Seal, Pennant); 13. *Phoca cristata;* 14. *Phoca grœnlandica;* 15. *Phoca pusilla;* 16. *Phoca ursina;* 17. *Phoca leonina* (= Anson's Sea-Lion); 18. *Phoca jubata* (= all the Sea-Lions then known); 19. *Phoca lupina* (Urigene Seal, Pennant; Molina not cited).

---

* The author of the work under consideration thus expresses its *raison d'être:* "The general history of quadrupeds has been so often detailed in the various works on Natural History, that a fresh publication on the subject must of necessity labour under peculiar disadvantages. The valuable works of the Count de Buffon and Mr. Pennant have diffused such a degree of information on these subjects, that it does not seem an easy task to improve upon their plan otherwise than by the introduction of the Linnæan method of arrangement, the rectification of errors relative to synonyms, the addition of proper specific characters, and the introduction of new species. . . . ."— *Gen. Zoöl.,* Introd. to vol. i, pp. vii, viii.

Under the generic term *Phoca* are here of course included the Otaries as well as the Phocids. These nineteen species are simply those of Pennant, with the addition of Latin names.

In 1806 Turton brought out his "General System of Nature" (the dedication is dated 1800) in which (vol. i, pp. 38–40) nineteen species of *Phoca* are given, they being the selfsame nineteen enumerated by Kerr in 1792.

Péron,* in 1816, described in great detail the Sea-Elephant of the Southern Seas under the name *Phoca proboscidea*, claiming that the Linnæan name was not strictly tenable. He also named† Buffon's "Phoque à ventre blanc" *Phoca leucogaster* (= *Phoca monachus*, Hermann), and gave the name *Phoca resima*‡ to "le grand Phoque des îles St.-Pierre et St.-Paul d'Amsterdam, dont Macartney, Cox et Mortimer nous ont successivement donné l'intéressante histoire" (= *Macrorhinus leoninus*, fem.), thus introducing three synonyms.

In the same year (1816) appeared the second part of Oken's "Lehrbuch der Naturgeschichte", in which sixteen species are enumerated under *Phoca*. The only noteworthy points are bestowal upon Molina's *Phoca lupina* of the name *tetradactyla*, the omission of all of Kerr's and Shaw's new names, and the inclusion of three of Thunberg's (namely *Phoca sericea*, *P. canina*, and *P. variegata*).

Desmarest gave in 1817§ a very fair monograph of the Seals, especially considering the date of publication. He distinguishes, first, with commendable discrimination, seven species "sans oreilles" which he considers are not sufficiently well known to take a place in a list of the species, or to be referred with certainty to other species. These are very properly: 1. *Phoca longicollis*, Shaw; 2. *P. testudinea*, Shaw; 3. *P. fasciata*, Shaw; 4. *P. punctata*, "Encycl. angl.";‖ 5. *P. maculata*, "Encycl. angl.";

---

* Voy. aux Terr. Austr. vol. ii, 1816, pp. 32–66, pl. xxxii.

† L. c., p. 47, footnote.

‡ L. c., p. 66.

§ Nouv. Dict. d'Hist. Nat., vol. xxv, article "Phoque", 1817, pp. 544–590.

‖ The "Encycl. angl." here quoted by Desmarest, and later by F. Cuvier and Lesson, and by Fischer as "Enc. Brit.," is doubtless Rees's "Cyclopædia; or, Universal Dictionary of Arts, Sciences, and Literature," in which I find under "PHOCA" (vol. xxviii, of the "first American edition," without date, but given in catalogues as published 1806–1824), the names here cited by Desmarest, as well as additional ones cited by Fischer (see beyond, p. 446). The authorship of the article is not given, but the editor of the work states that the zoölogical portions were chiefly prepared by Donovan. The matter relating to the Seals could scarcely be more noxious, the ac-

6. *Phoque tigré*, Krascheninikow; 7. *Le Phoque grumm-selur* des Islandaises, "Ann. d'Olaf Tryggesen et le Speculum Regale". Eleven species are admitted as valid, among which appear two under new names; eight may be considered as representatives of valid species, the other four being synonyms, but not in all

counts of even the more common species being very meagre and erroneous, while many of the less known are introduced with such inadequate descriptions that from these alone they are mostly indeterminable. Although virtually anonymous, as well as worthless, they have been dragged to light by the above-named and other writers, but they always appear in the waste lumber of unidentifiable species. To show the nature of this rubbish, and for the purpose of elucidating the references to it which follow, I quote verbatim and entire the portions in question, and adding thereto the real origin and basis of the names here appearing.

Phoca "*mutica;* Long-necked Seal. Body slender, without claws on the forefeet." [The "Long-necked Seal" of Parsons ($=Phoca\ mutica$, Kerr), which, as already stated, is some undeterminable species of Otary.]

Phoca "*testudo;* Tortoise Seal. Head resembling a tortoise ; neck slender. It is said to inhabit many European shores; the species is, however, but little known." [The Tortoise-headed Seal of Parsons and the *Phoca testudo* of Kerr.]

Phoca "*laniger;* Leporine Seal. It has four fore-teeth in each jaw; the upper lip is thick, with long, thick whiskers; the fur is soft and uneven; the feet have nails, and its length is about six feet and one-half. It inhabits the White Sea, Iceland, and the Frozen Ocean." [The *Phoca leporina* of Lepechin ($=Phoca\ laniger$, Kerr).]

Phoca "*punctata;* Speckled Seal. Body, head, and limbs speckled. It inhabits the seas of Kamtschatka, and the Kurile islands." [The *Phoca punctata* of Kerr.]

Phoca "*maculata;* Spotted Seal. Body spotted with brown. It inhabits the Kurile seas, and is very scarce." [The *Phoca maculata* of Kerr.]

Phoca "*nigra;* Black Seal. Hind legs peculiarly formed. It is found on the coasts of the Kurile seas; but the structure of its legs has not been accurately ascertained." [The *Phoca nigra* of Kerr.]

Kerr's references show that the last three species were unquestionably derived from the following passage in Pennant: "Other obscure species in those [Kurile] seas, which are mentioned in Steller's MSS., are, I. A middle-sized Seal, elegantly speckled in all parts; II. One with brown spots, scarcer than the rest; III. A black species with a peculiar conformation of the hind legs."— PENNANT'S *History of Quadrupeds*, third edition (1793), vol. ii, p. 276.

None of these species make their first appearance here, they all occurring in Kerr. As Kerr (see *antèa*, pp. 433, 434) cites Pennant in each case, and also Parsons and Lepechin respectively in the three instances where Pennant's species are based on these authors, the above-given names are thus strictly identifiable. Whether these names and diagnoses were here copied from Kerr or from Turton it is impossible to say, as Turton also gives them, but entirely without reference to previous authors. It thus happens that neither Turton nor the "Cyclopædia" gives us any direct clue to their origin.

cases referable with certainty to other species. Three nominal
species of previous authors are correctly allocated. All are introduced under the generic name (or rather "sous-genre" as
he terms it) *Phoca*, and all are true Phocids. His accepted
species are the following : 1. *Phoca proboscidea* ; 2. *Phoca coxii*
(sp. nov. = "Sea Lion, John Henry Cox, Description of the Island
called St.-Paulo by the Dutch, and by the English" = *P. resima*,
Péron = *Macrorhinus leoninus*, female) ; 3. *Phoca monachus*
(correctly covers *P. bicolor*, Shaw, and *P. albiventer* Bodd.) ; 4.
Phoca grœnlandica (includes *P. oceanica*, Lepechin); 5. *Phoca
cristata ;* 6. *Phoca leporina* (ex Lepechin); 7. *Phoca vitulina ;* 8.
Phoca maculata (ex Boddacrt); 9. *Phoca hispida* (= *P. fœtida*);
10. *Phoca lakhtak* (sp. nov. = Lachtak, Krascheninikow); 11.
Phoca lupina (ex Molina). Only six of these are valid, to which
may be added one from the list of doubtful species, making
seven in all.

In 1820 Desmarest gave a second account of the Earless
Seals in the "Encyclopédie méthodique" (vol. clxxxii, Mammalogie, Part i, 1820, pp. 237–247, Part ii, 1822, 541), recognizing
thirteen species as valid, all of which are referred, as before, to
Péron's "subgenus" *Phoca*. They are: 1. *Phoca proboscidea*
(= *Macrorhinus leoninus*); 2. *Phoca ansoni* (cites "*Phoca leonina*,
Linn., Gmel., Erxl."), based in part on a skull of *Otaria jubata*,
and in part on Anson's Sea Lion, which is his *Phoca proboscidea ;* 3. *Phoca byroni* (ex Blainville MSS., based, it is stated,
on a skull in the Hunterian collection labelled "Sea Lion from
the Island of Tinian, by Commodore Byron"; = *Macrorhinus
leoninus*); 4. *Phoca cristata ;* 5. *Phoca monachus ;* 6. *Phoca oceanica ;* 7. *Phoca leporina ;* 8. *Phoca vitulina ;* 9. *Phoca grœnlandica;* 10. *Phoca fœtida;* 11. *Phoca barbata;* 12. *Phoca leptonyx*
("Blainv."); 13. *Phoca albicauda* (= *Phoca grœnlandica*). Of these
thirteen species three (*Phoca ansoni, P. byroni,* and *P. albicauda*)
are here first named; all are nominal. Two other nominal
species are Lepechin's *Phoca oceanica* and *P. leporina,* leaving
eight valid species. Desmarest appends a list of eleven species,
briefly characterized, "qui sont bien moins connus" than those
more formally recognized, these being as follows: 1. *Phoca
lupina ;* 2. *Phoca coxii ;* 3. *Phoca longicollis ;* 4. *Phoca testudinea;* 5. *Phoca fasciata ;* 6. *Phoca punctata ;* 7. *Phoca maculata ;*
8. *Phoca nigra ;* 9. *Le Phoque lakhtak* (= *Phoca laktak,* Desm.,
1817); 10. *Le Phoque tigré ;* 11. *Le Phoque grumm-selur.*

The present enumeration differs much from the same author's

former one. It includes as valid one more species, while the list
of doubtful ones contains three more, including the two "new
species" (*P. coxii* and *P. lackhtak*) described by him three years
before. *Phoca oceanica* is raised from a synonym to the rank of
a valid species; *Phoca foetida* appears in place of *P. hispida;*
Phoca lupina is transferred to the doubtful list, and two nominal species are added. Altogether there is an increase of two
valid species (*P. barbata* and *P. leptonyx*), making eight in all
(or nine with *P. fasciata*, given as doubtful).

Blainville, the same year (1820,) himself published descrip-
tions of the species accredited to him by Desmarest, namely,
*Phoca byronia** and *Phoca leptonyx* (= *Stenorhynchus leptonyx*,
F. Cuvier), both based on specimens in the Museum of the Royal
College of Surgeons of London.

Nilsson, in 1820, in his "Skandinavisk Fauna", described
Phoca vitulina under the name *Phoca variegata, Phoca foetida* as
Phoca annellata, Halichœrus grypus as *Halichœrus griseus,* and
Cystophora cristata as *Cystophora borealis.*

In 1822, Choris, in his "Voyage pittoresque autour du
Monde," figured and described a Seal under the name "Chien
de mer du détroit de Behring" (pl. viii of the livraison treating
of Kamtschatka, etc.). The figure is exceedingly inartistic, but
the coloration agrees very well with a common phase of *Phoca
vitulina.* Its only importance turns on the fact that it later
became the *Phoca chorisi* of Lesson. The figure is often referred
to as being unaccompanied by a description, but at p. 12 of
the livraison above cited, occurs the following: "Phoque du
détroit de Behring, blanc, tacheté de petites marques noires; il
diffère cependant de celui des îles Aléoutiennes qui est d'un
blanc sale, et n'a presque point de taches. Dans les îles Kou-
riles on en trouve encore une autre espèce, mais tout-à-fait
noire, marquetée de petites taches blanches en form d'annelets
[*Phoca foetida?*]. Sa grandeur toutefois est généralement de
quatre pieds à quatre pieds et demi."

In 1824, F. Cuvier, in his paper on the classification of the
Seals, already noticed,† in which he divided the Earless Seals
into five generic groups, gave to the young Seal figured and
described in his livraison ix of his "Histoire Naturelle des
Mammifères" as the "Phoque commun", the specific name *discolor* (= *Phoca foetida*), referring it at the same time to his

---

* Journ. de Physique, vol. xci, 1820, pp. 287, 288.

† Ann. du Mus. d'Hist. Nat., tom. xi, pp. 174–200, pll. xii–xix. See *antèa*,
pp. 415, 417.

genus *Callocephalus*. He discusses to some extent the characters of several of the species of Earless Seals (as well as Otaries)
and figures the skulls of five of them.

The same year (1824) Thienemann * published his observations on the Seals collected and observed by him in Iceland,
in which he renamed most of the species, and published a full
description and colored illustrations of the animals, and figured
the skulls of most of them.   He recognized seven species, which
are as follows, with the allocations usually assigned them by
recent authorities : † 1. *Phoca littorea* (= *P. vitulina*); 2. *Phoca
annellata* (= *Phoca fœtida*); 3. *Phoca grœnlandica*; 4. *Phoca
barbata ;* 5. *Phoca halichœrus* (= *Halichœrus grypus*); 6. *Phoca
scopulicola* (= *Halichœrus grypus*, young); 7. *Phoca leucopla*
(= *Cystophora cristata*).   Of these seven species two are purely
nominal, and three others represent species previously described.   Thienemann here adds in all four synonyms and no
new species.‡

G. Cuvier, in his "Ossemens fossiles", § gave a somewhat
extended but informal review of the "Phoques vivantes," in
which are described two new species, namely, *Phoca lagura*
(= *Phoca grœnlandica*, young), from "Terre-Neuve," received
from M. de la Pilaye.   It is evidently based on a quite young
animal, having a length of "trois pièd trois pouces," and covered with "laine blanche".   He also describes a *Phoca mitrata*
(ex "Camper" MSS.)|| based on a skull of *Cystophora cristata*

---

* Naturhistorische Bemerkungen gesammelt auf einer Reise im Norden von
Europa, vorzülich in Island in d. Jahren 1820 bis 1821, I, Säugeth. 1824.
Quoted at second hand, as cited by various authors, the work being inac-
cessible to me.

† Based mainly on the identifications of Gray, Lilljeborg, and von Heuglin.

‡ According to Ferussac's Bull. des Sci. Nat. v, 1825, pp. 260-262, Thienemann gives the following figures : *Phoca barbata*, pl. i, ♀ adult; pl. ii, ♂ 2
years old; pl. iii, ♂ 1 year old ; pl. iv, skull.   *Phoca scopulicola*, pl. v, ♂ ad.
Phoca littorea, pl. vi, ♂; pl. vii, skull; pl. viii, anatomy.   *Phoca annellata*,
pl. ix, ♀ ad.; pl. x, juv.; pl. xi, skull; pl. xii, anatomy.   *Phoca leucopla*,
pl. xiii.   *Phoca grœnlandica*, pl. xiv, ♂ ad.; pl. xv, ♀; pl. xvi, ♂ 2 years
old; pl. xvii, ♂ 1 year old; pl. xviii, young 8 days old; pl. xix, skull; pl.
xx, anatomy; pl. xxi, attitudes in the water.

§ I cite here the third (the author's last) edition, tom. v, 1825, pp. 206
et seq.

|| The name *mitrata*, derived from the same specimen, appears to have
been previously made public by Desmarest in 1820, who says "M. Milbert,
correspondent du Muséum, dans les États-Unis, vient d'envoyer à cet établisse-
ment, sous le nom de *Phoca mitrata*, la tête d'un phoque qui diffère essen-
tiellement de celui-ci [" *Phoca cristata*"] par le marque de crête et par le
nombre de dents. . . ."—*Mammalogie*, p. 241, footnote.

sent by M. Milbert to the Paris Museum from New York. The author's remarks on the species of this group abound with judicious suggestions and form an important contribution to the subject. He discusses at length the *Phoca oceanica* of Lepechin, and refers it unquestionably to *Phoca grœnlandica*.

Harlan, in 1825, in his "Fauna Americana" (pp. 102–112), recognized five species, as follows: 1. *Phoca cristata;* 2. *Phoca vitulina;* 3. *Phoca grœnlandica;* 4. *Phoca fœtida;* 5. *Phoca barbata*. All are valid; all stand under their correct specific names; the few synonyms given are all correctly referred; and only one species (*Halichœrus grypus*) known at that time to inhabit North America is omitted.

Godman, the following year (1826), in his "American Natural History" (vol. i, pp. 310–346,) recognized also the same number of North American species, and under the same names, but gave a much more extended account of them.

In 1826, in the article "Phoque," F. Cuvier gave, in the "Dictionnaire des Sciences Naturelles" (vol. xxxix, pp. 540–553), a systematic revision of the Seals, respecting which he says, "Nous réunirons donc dans cet article, mais d'une manière fort succincte, tout ce qui a rapport aux phoques considérés comme ordre, comme genres et comme espèces." In this revision he adopts the genera proposed by him two years earlier (see *anteà*, p. 415), and recognizes the following species: 1. *Callocephalus vitulinus;* 2. *C. leporinus;* 3. *C. discolor* (= "Phoque commun," Hist. nat. des Mamm., 9e livraison); 4. *C. lagurus;* 5. *C. grœnlandicus;* 6. *C. hispidus;* 7. *C. barbatus;* 8. *Stenorhynchus leptonyx;* 9. *Pelagius monachus;* 10. *Stemmatopus cristatus;* 11. *Macrorhinus proboscideus*.

In addition, under the caption "*Phoques privées d'oreilles externes*," he cites the following as too little known to enable him to recognize their generic characters, explains the basis on which each rests, and gives such brief notices of their characters as he was able to glean: 1. *Phoca coxii*, Desm.; 2. *P. maculata*, Bodd.; 3. *P. lakhtak*, Desm.; 4. *P. lupina*, Molina; 5. *P. byroni*, "Blainv."; 6. *P. ansoni*, "Blainv."; 7. *P. oceanica*, Lepech.; 8. *P. testudinea*, Shaw; 9. *P. longicollis*, Shaw; 10. *P. fasciata*, "Shaw"; 11. *P. punctata*, "Encycl. angl."; 12. *P. maculata*, "Encycl. angl."; 13. *P. nigra*, "Encycl. angl."; 14. *Phoque tigré*, Krasch.

Of the eleven species accepted as valid, two only (*C. leporinus* and *C. discolor*) are nominal; of the fourteen provisionally given, one only (*P. "fasciata"*) is valid, making nine valid species now enumerated, six of which stand under their legitimate specific names.

Lesson,[*] in 1826, named Weddell's Sea Leopard *Otaria weddelli*, supposing it to be an Eared Seal, but the following year (1827), in his "Manuel de Mammalogie", referred it to F. Cuvier's genus *Stenorhynchus*. In this work, "Manuel de Mammalogie" (pp. 196–208), he treats formally nineteen species of Earless Seals (Nos. 529–546, and 551), and under the caption "§ 1, *Point d'oreilles externes,* Phoques des auteurs," gives a list of eleven additional species, the last all under the genus *Phoca,* which he was unable to rigorously determine. The work is admittedly based largely on that of Desmarest[†] ("Encyclopédie méthodique," vol. clxxxii), but is stated to have grown out of his systematic studies of the collections made during the voyage of the corvette *la Coquille* (1822–1825), the results of which were just then published. He adopts the generic divisions previously instituted by F. Cuvier and Nilsson, and on the whole presents a very judicious summary of the subject. The species and genera, concisely and fairly characterized, are as follows: 1. *Calocephalus oceanicus ;* 2. *C. grœnlandicus ;* 3. *C. vitulinus* (includes *P. littorea,* Thienemann); 4. *C. albicauda* (ex Desmarest); 5. *C. leporinus ;* 6. *C. discolor* (ex F. Cuvier); 7. *C. lagurus* (ex G. Cuvier); 8. *C. barbatus ;* 9. *C. scopulicolus* (ex Thienemann); 10. *Stenorhynchus leptonyx ;* 11. *S. weddelli* (= *Otaria weddelli,* Lesson, 1826); 12. *Pelagius monachus* (correctly covering *Phoca bicolor,* Shaw, and *P. leucogaster,* Péron); 13. *Stemmotopus cristatus ;* 14. *Macrorhinus proboscideus ;* 15. *M. ansoni ;* 16. *M. byroni ;* 17. *Halichœrus griseus.* No new names are introduced, but several of the species are placed under new generic relations.

Among the eleven undetermined species referred to under *Phoca* are two (*P. longicollis,* Shaw, *P. nigra,* "Encycl. angl.") Otaries. The other seven are as follows: 1. *P. coxii,* Desm.; 2. *P. lupina,* Molina; 3. *P. maculata,* Bodd.; 4. "*Phoque lakhtak,*" Desm. ex Krasch.; 5. *P. testudinea,* Shaw; 6. *P. fasciata,* "Shaw"; 7. *P. punctata,* "Encycl. angl."; 8. *P. maculata,* "Encycl. angl."; 9. *P. tigrina* (ex Krasch., apparently here first named).

---

[*] Férussac's Bull. des Sci. Nat., vol. vii, 1826, p. 438.
[†] See dedicatory note addressed to M. A. G. Desmarest, l. c., p. vii.

Gray,* the same year (1827), in Griffith's "Cuvier's Animal Kingdom" (vol. v, pp. 175–181), recognized thirteen species, as follows: 1. *Phoca vitulina*, with varieties "a. *bothionica*" ("Lin. Faun. Sue."); "b. *sebrica*" ("*sibirica*, Gmelin, Syst. Nat."); "c. *caspica*" ("Gmelin, Syst. Nat."); "d. *maculata*" ("*maculata*, Bodd."); 2. *Phoca "Leporinus"* (= *leporina*, Lepechin); 3. *Phoca discolor* (= *discolor*, F. Cuv.); 4. *Phoca lagura* (= *lagura*, G. Cuv.); 5. *Phoca grœnlandica* (covers also *oceanica*, Lepech., and *semilunaris*, Bodd.); 6. *Phoca fœtida* (covers *hispida*, Schreber, and *Halichœrus griseus*, Nilss., "His., 1824, 810"!); 7. *Phoca barbata;* 8. *Phoca leptonyx* (covers *leptonyx*, Blainville, *Le Phoque à ventre blanc*, Buffon, and also Hermann's plate of *Phoca monachus!*); 9. *Mirounga* (n. gen.) *cristatus* (= *cristata*, "Gmel.", *leonina*, Fabric., and *mitrata*, "Camper"); 10. *Mirounga proboscidea;* 11. *Mirounga patagonica* (n. sp. = "*Phoque de Patagone*, F. Cuv., Mém. Mus. iv, 203" (= *Macrorhinus leoninus*); 12. *Mirounga ansoni* (= *leonina*, "Gmel.", *ansoni*, Desm.); 13. *Mirounga byroni* (= *Phoca byronia*, Blainv.). Of these thirteen species only six are valid, while two then well-known species (*Halichœrus grypus* and *Monachus albiventer*) are confounded with others.

In 1828 Lesson† added a large number of synonyms by deliberately renaming species previously described, a large part of the species so renamed being also merely nominal. He gives a synopsis of the genera proposed by F. Cuvier in 1824, in his general history of the group, but enumerates the species all under the old generic name of *Phoca*. His review of the group is made with discrimination, but is greatly marred by the free indulgence of his love for coining new names. In the list of his species here following the new names are printed in thick type: 1. *Phoca cristata* (covering *leonina*, Fabr., *cucullata*, Bodd., and *mitrata*, "DeKay"); 2. *Phoca* **mulleri** (covering *grœnlandica*, "Müller", *oceanica*, Lepechin, and *semilunaris*, Boddaert); 3. *Phoca* **schreberi** (covering *hispida*, Schreber, *fœtida*, Müller, and *annellata*, Nilsson); 4. *Phoca* **parsonsi** (= "Phoca major, Parsons," to which is referred *barbata*, "Müller"); 5. *Phoca* **thienemanni** (= *scopulicola*, Thienemann); 6. *Phoca leucopla* (ex Thienemann); 7. *Phoca* **linnæi** (= *vitulina*, Linné); 8. *Phoca littorea* (ex Thienemann); 9. *Phoca* **lepechini** (= *leporina*, Lepechin);

* The authorship is not distinctly stated in the volume, so far as I have been able to find, but is uniformly claimed by Gray in his subsequent works.

† Dict. class. d'Hist. Nat., tome xiii, art. *Phoque*, Janvier, 1828, pp. 400–426.

10. *Phoca* **frederici** (= *discolor*, F. Cuvier); 11. *Phoca* **pylayi** (= *lagura*, G. Cuvier); 12. *Phoca* demaresti (= *albicauda*, Desmarest) ; 13. *Phoca* **hermanni** (= *monachus*, Hermann, to which are referred also *albiventer*, Bodd., and *leucogaster*, Péron); 14. *Phoca* **chorisi** (= Chien de Mer, Choris); 15. *Phoca byroni* (ex Blainville); 16. *Phoca* **homei** (= *leptonyx*, Blainville); 17. *Phoca weddelli* (*Otaria weddelli*, Less.); 18. *Phoca proboscidea* (includes *leonina*, Linn., *elephantina*, Molina, and *ansoni*, Desm., in part).

Of these eighteen " species " nine are purely nominal; eleven are needlessly renamed, and in addition to which Choris's "Chien de Mer" is for the first time introduced into systematic nomenclature, thereby adding in all twelve synonyms in a notice of eighteen supposed species, representing only nine valid ones. The only redeemable feature is the proper allocation of twelve nominal species of preceding authors. He also considers it probable that Péron's *Phoca resima*, Desmarest's *Phoca coxii*, and Molina's *Phoca lupina*, should be referred to his *Phoca proboscidea*. Of the nine valid species only two stand under tenable specific names.

Fischer, in 1829, in his " Synopsis Mammalium," recognized eighteen species of "*Phocæ*" under his division "†† *Auriculis nullis* " (l. c., pp. 234–242, " 375–378," *i. e.*, 575–578), which are as follows: 1. *Phoca leonina*, Linn. (= *P. proboscidea*, Péron); 2. *Phoca ansoni*, Desm.; 3. *Phoca byroni*, Blainv. (the three preceding all referable, either wholly or in part, to *Macrorhinus leoninus*); 4. *Phoca monachus*, Herm.; 5. *Phoca vitulina*, Linn.; 6. *Phoca leporina*, Lepech.; 7. *Phoca discolor* (= *Callocephalus discolor*, F. Cuv.); 8. *Phoca scopulicola*, Thienm.; 9. *Phoca leucopla*, Thienm.; 10. *Phoca lagura*, G. Cuv.; 11. *Phoca grœnlandica*, "Müll."; 12. *Phoca grypus*, Fabr.; 13. *Phoca hispida*, Schreber; 14. *Phoca barbata*, "Müll."; 15. *Phoca leptonyx*, Blainv.; 16. *Phoca weddelli*, "Less."; 17. *Phoca cristata*, Erxl.; 18. *Phoca chorisi*, Less. Thirteen others are given as doubtful or not well determined, one of which is here first named. These are: 1. *Phoca dubia* (n. sp., = *Macrorhinus leoninus*, juv.*); 2. *Phoca oceanica*, Lepech.; *Phoca lupina*, Molina; 4. *Phoca sericea*, Thunb.; 5. *Phoca canina*, Thunb.; 6. *Phoca vitulina*, Thunb.; 7. *Phoca testudinea*, Shaw; 8. *Phoca fasciata*, Shaw; 9. *Phoca punctata*, " Ency. Brit."; 10. *Phoca maculata*, " Ency. Brit."; 11. *Phoca nigra*, " Ency. Brit."; 13. *Phoca antarctica*, Thunb. (= *Arctocephalus antarcticus*). In the " Addenda" (1830), how-

* See Nilsson, Wiegmann's Archiv, 1841, pp. 324, 325.

ever, *Phoca oceanica* is referred to *P. grœnlandica*, and *Phoca fœtida*, "Müll.," is substituted for *P. hispida*. "*Phoca dubia*" is apparently the only new name given. Not only are all these referred to *Phoca*, but this name is made to cover also all of the Eared Seals.

Of the eighteen species here formally recognized ten only are valid, to which one may be added from the list of doubtful species, making eleven in all, nine of which have correct specific designations—a great improvement upon Lesson's work of the previous year.

In 1831* Pallas, in his "Zoographia Rosso-Asiatica" (vol. i, pp. 100–119), described twelve species of marine mammals under the generic name *Phoca*, as follows: 1. *Phoca lutris* (= *Enhydris lutris*); 2. *Phoca ursina* (= *Callorhinus ursinus*); 3. *Phoca leonina* (=*Eumetopias stelleri*); 4. *Phoca nigra* (=*Callorhinus ursinus*, juv.); 5. *Phoca nautica* (= ? *Erignathus barbatus*); 6. *Phoca albigena* (= *Erignathus barbatus*); 7. *Phoca equestris* (= *Histriophoca fasciata*); 8. *Phoca dorsata* (= *Phoca grœnlandica*); 9. *Phoca monacha* (= *Monachus albiventer*); 10. *Phoca largha* (a young Earless Seal, species indeterminable); 11. *Phoca canina* (= *Phoca vitulina, Phoca caspica*, and *Phoca sibirica*); 12. *Phoca ochotensis* (indeterminable; probably = *P. vitulina*). Of these twelve species seven only are Phocids, none of which are for the first time named; two (*P. nautica* and *P. largha*) are not with certainty determinable. The author himself identifies five of his species with species previously described, yet in each case bestows a new name. In short, Pallas's twelve supposed species of "*Phoca*" add seven pure synonyms, three indeterminable species, and not one tenable name to the literature of the subject. His *Phoca ochotensis* (by some later authors, as von Schrenck, recognized as a valid species) presents a combination of characters thus far unknown in nature. His diagnosis begins "P. subauriculata", and in his description he says, "Auriculae externae minutae, nigricantes", on which account it has been sometimes regarded as an Otary, but he describes the molars as "supra infraque utrinque quini, primo minore *subbicuspidato;* reliqui acute *tricuspidati*, medio majore, conico"; and also says, "Palmarum ungues terminales magni, incurvi, robusti," etc., which certainly cannot be said of an Otary. There is nothing in the account

---

* The date on the title page is 1831, but the work seems to have been printed as early as 1811. The first volume, however, is quoted by Fischer in the "addenda" to his "Synopsis Mammalium", dated 1830, and is not quoted in the work itself, dated 1829.

of the pelage or coloration, either of the young or adult, that might not apply, for instance, to *Phoca vitulina*, while the general drift of the description certainly indicates an Earless Seal.

Nilsson, in 1837,* published an important revision of the *Pinnipedia*, which, so far as the Phocids are concerned, is one of the most important contributions to the subject that has yet appeared, the variations dependent upon age and individual peculiarities being discussed at length, while a number of the nominal species of preceding authors (some of them for the first time) take their proper stations. *Phoca caspica* is here first established as a species,—the only new species added. Characters strictly specific are sharply contrasted among allied species not previously well understood. Only a limited amount of synonymy is presented, but that is well considered, and has stood the test of subsequent researches. Only ten species of the family *Phocidæ*, as now restricted, were recognized, as follows: 1. *Stenorhynchus leptonyx*; 2. *Pelagius monachus*; 3. *Phoca vitulina* (to which is referred Thienemann's *P. littorea*); 4. *Phoca annellata* (= *Phoca fœtida*, to which is referred F. Cuvier's *Callocephalus discolor*); 5. *Phoca caspica* (n. sp.); 6. *Phoca grœnlandica* (to which are referred Lepechin's *P. oceanica* and G. Cuvier's *P. lagura*); 7. *Phoca barbata* (to which are referred Lepechin's *P. leporina* and Pallas's *P. nautica* and *P. albigena*); 8. *Halichœrus grypus*; 9. *Cystophora proboscidea* (to which Fischer's *Phoca dubia* is referred; *Phoca ansoni* is again shown to be a compound of this species and *Otaria leonina* [= *O. jubata*, auct.], and *Phoca byroni* is declared to have been based on an old skull without the lower jaw of "*Otaria jubata*"); 10. *Cystophora cristata* (to which are referred *Phoca mitrata* of "Fischer", *P. leucopla* of Thienemann, and *Cystophora borealis* of Nilsson, Skand.-Fauna, 1, 1820, 283). No reference is hence made to several valid species, and a multitude of nominal ones, previously described.

Gray, in 1837,† described some kind of Hair Seal "forty-seven inches" long from the "Cape of Good Hope", under the name *Phoca ? platythrix*. He seemed to be thus in doubt as to whether it was a true *Phoca*, but it was doubtless an Earless Seal, or he would not have at this date referred it in any way to *Phoca*. I find no subsequent reference to it, either by Gray

---

* "Utkast till en systematisk indelning af Phocaceerna. <K. Vet. Akad. Handl. Stockholm, 1837, pp. 235–340". Translated by Dr. W. Peters in Wiegmann's Archiv für Naturgeschichte, 1841, Bd. 1, pp. 401–332. This is the version commonly cited, and the one used in the present work.

† Charlesworth's Mag. Nat. Hist., vol. i, 1837, p. 582.

or any other writer. It may have been a young Sea-Elephant, this being the only Phocid reported from that locality.

In 1839 Hamilton, in his "Natural History of the Amphibious Carnivora"* (pp. 124–227, 279, 280), recognized fifteen species (adopting F. Cuvier's genera) of Earless Seals, as follows: 1. *Calocephalus vitulina;* 2. *C. discolor* (=*fœtida*); 3. *C. barbata;* 4. *C. bicolor* (=*monachus*); 5. *C. grœnlandica;* 6. *C. oceanica* (= *grœnlandica*); 7. *C. hispida* (=*fœtida*); 8. *C. leporina* (=*barbata*); 9. *Halichœrus griseus* (=*grypus*); 10. *Stenorhynchus leptonyx;* 11. *S. leopardina* (=*weddelli*); 12. *Pelagius monachus;* 13. *Stemmatopus cristatus;* 14. *S. mitratus* (=*cristata*); 15. *Macrorhinus proboscideus* (=*leoninus*). Of these fourteen species nine only are valid, and only six of these stand under their correct specific names. He also gives a list of four doubtful ones, only the first of which is described. These are the following: 1. *Phoca fasciata;* 2. *P. coxii;* 3. *P. lupina;* 4. *P. punctata.*

The same year (1839) Kutorga† gave a detailed account of *Phoca fœtida*, under the name *Phoca communis*, characterizing two new varieties, which he called *octonotata* and *undulata.*

Temminck, in 1842, in the "Fauna Japonica" (Mammifères Marins, pp. 1–4), passed in review the Seals of the North Pacific, discussing especially those indicated by Steller and Pallas. The only species particularly described is Pallas's *Phoca largha*, which he renamed *Phoca nummularis.*

DeKay, in the same year (New York Zoölogy, part i, 1842, p. 53), based the name *Phoca concolor* upon New York examples of *Phoca vitulina*, he believing the American animal to be specifically distinct from the European.

In 1843 Lesson‡ described a specimen of *Cystophora cristata* taken on the coast of France, under the name *Phoca isidorei.*

In the same year (1843) Owen§ redescribed *Lobodon carcinophaga* under the name *Stenorhynchus serridens.*

In 1844 Gray, in the "Zoölogy of the Erebus and Terror", described *Ommatophoca rossi*, a valid new species.

Schinz, in the same year (1844), in his "Systematische Verzeichniss aller bis jetzt bekannters Säugethiere; oder Synopsis Mammalium" (i, pp. 429–486), recognized twelve species of *Phoca*, and mentioned three additional doubtful ones, as follows:

---

* Forming vol. viii of the Mammalia of Jardine's "Naturalist's Library".
† Bull. Soc. Imp. des Nat. de Moscou, année 1839, pp. 178–196, pll. xiii–xviii.
‡ Revue Zoolgique, 1843, p. 256; Echo du Monde Savant, 1843, p. 228.
§ Proc. Zoöl. Soc. London, 1843, p. 131.

1. *Phoca proboscidea ;* 2. *P. monachus ;* 3. *P. vitulina ;* 4. *P. caspica;* 5. *P. barbata;* 6. *P. annellata* (=*fœtida*); 7. *P. grœnlandica ;* 8. *P. grypus ;* 9. *P. lagura ;* 10. *P. leptonyx;* 11. *P. weddelli ;* 12. *P. cristata.* Eleven of these represent valid species, one only (*P. lagura*) being nominal. The doubtful ones are, 1. *P. chorisi ;* 2. *P. sericea,* Thunb.; 3. *P. testudinea,* Shaw.

In 1846 Andreas Wagner, in his continuation of Schreber's "Säugthiere " (Theil vii, pp. 5–51) presented an important revision of the Earless Seals. The fourteen species recognized by him he refers to four genera, as follows: 1. *Halichœrus grypus* (to which he refers *Phoca hispida,* Schreber!); 2. *Phoca barbata ;* 3. *Phoca grœnlandica ;* 4. *Phoca nummularis* (ex "Schlegel", *i. c.* Temminck, Fauna Japon. = *P. fœtida*); 5. *Phoca vitulina ;* 6. *Phoca annellata* (ex Nilsson ; *P. fœtida* and *P. hispida,* Fabric. are given as synonyms!); 7. *Phoca caspica ;* 8. *Leptonyx serridens* (= *Stenorhynchus serridens,* Owen, and *Lobodon carcinophaga,* Gray); 9. *Leptonyx leopardina* (ex Jameson MSS. apud Hamilton ; " *Phoca leptonyx,* Blainville" = *Leptonyx weddelli*); 10. *Leptonyx weddelli ;* 11. *Leptonyx rossi ;* 12. *Leptonyx monachus ;* 13. *Cystophora proboscidea ;* 14. *Cystophora cristata.* Of these, two (*Phoca nummularis* and *Leptonyx leopardinus*) are nominal. Although a highly important, and in most respects a judicious review of the subject, it presents several strange allocations of synonymy, as above noted. Under *Leptonyx weddelli,* for example, the only references he cites he had just previously given under *S. leopardinus,* and appears to separate the two species on the basis of erroneous drawings of the hind feet. Neither does he explain why he refers Schreber's *Phoca hispida* to *Halichœrus grypus,* or why he allows *annellata* of Nilsson to take precedence of *fœtida* of Fabricius.

Peale,* in 1848, misled by the transposition of a label, described specimens of *Phoca vitulina* from the Pacific coast of North America, under the name *Halichœrus antarcticus,* supposing the specimens came from the Desolation Islands.

In 1849 Dr. J. E. Gray† "received", he says, "from the West Indies the skin and skull of a Seal which evidently belongs to the same genus as the crested seal of the northern hemisphere", which he described under the name *Cystophora antillarum.* He refers to another "imperfect skin of a seal from Jamaica, which

---

* Rep. U. S. Ex. Exp., vol. viii, 1848, p. 30.

† Proc. Zoöl. Soc. London, 1849, p. 93.

was brought home by Mr. Gosse", which the following year*
became the basis of his *Phoca tropicalis.*

Gray, in 1850 (Cat. of Seals in Brit. Mus.) recognized eighteen
species of Earless Seals, distributed among eleven genera, of which
fourteen of the species are doubtless valid.   The species † recognized are the following: 1. *Lobodon carcinophaga;* 2. *Stenorhynchus léptonyx;* 3. *Leptonyx weddelli;* 4. *Monachus albiventer;* 5.
Ommatophoca rossi; 6. *Callocephalus vitulinus;* 7. *C. hispidus;*
8. *C. fœtida* (7 and 8 are the same); 9. *C. caspicus;* 10. *C. dimidiatus;* 11. *C. largha* (10 and 11 nominal); 12. *Pagophilus grœnlandicus;* 13. *Phoca barbata;* 14. *Phoca tropicalis;* 15. *Halichœrus grypus;* 16. *Morunga elephantina;* 17. *Cystophora cristata;*
18. *C. antillarum.*   The only new names are *Callocephalus dimidiatus* (n. sp., ex Schlegel MS.), and *Phoca tropicalis* (n. sp.), and
the only innovation in nomenclature is *Monachus albiventer (albiventer* ex Bodd.).

The same year (1850) Drs. Hornschuch and Schilling,‡ after
an examination of some sixty skulls of *Halichœrus,* proposed a
division of the genus into three species, namely, *H. grypus*
(Fabr. *= griseus* Nilss.), *H. macrorhynchus,* and *H. pachyrhynchus,*
the last two being added as new species.   Subsequent writers,
however, have not considered them as entitled to specific recognition.

In 1854 Gray§ described a specimen of *Monachus albiventer*
from Madeira under the name *Heliophoca atlantica,* basing on
it a new genus as well as new species.

In 1855 Giebel, in his "Säugethiere" (pp. 129–143), gave a
noteworthy account of the animals here under consideration.
It is concise and discriminative, and though closely following
Wagner, is an admirable exposition of the state of knowledge
respecting this group at the date of its publication, nearly a
quarter of a century ago.   Although dealing to only a small
extent with the bibliography of the subject, the principal synonyms of the species are given in footnotes, with generally a
brief reference to their character.   The species recognized,

---

* Cat. Seals Brit. Mus., 1850, p. 28.

† The synonymy he here gave is substantially the same as that of his later
(1866) "Catalogue of Seals and Whales", for a notice of which see below, p.
453

‡ "Kurze Notizen über die in der Ostsee vorkommenden Arten der Gattung
Halichœrus, Nilss. Greifswald, 1850". Abstract in Wiegmann's Archiv für
Naturgesch., 1851, Bd. 2, p. 22. The original brochure I have not seen.

§ Proc. Zoöl. Soc. London, 1854, p. 43.

thirteen in number, are referred to four genera, as follows: 1. *Halichœrus grypus;* 2. *Phoca barbata* (with *leporina,* Lepech., *albigena* and *nautica,* Pallas, as synonyms); 3. *Phoca grœnlandica* (with *oceanica,* "Steller," and *dorsata,* Pallas, as synonyms, to which also *ochotensis,* Pallas, is doubtfully added); 4. *Phoca nummularis,* "Schlegel" (= *fœtida*); 5. *Phoca vitulina* (with the synonyms *variegata,* Nilss., *littorea* and *scopulicola,* Thienemann, *concolor,* DeKay, etc.); 6. *Phoca annellata* (with *fœtida* and *hispida,* Fabr., *discolor,* F. Cuv., and *octonotata* and *undulata,* Kutorga, as synonyms); 7. *Phoca caspica;* 8. *Leptonyx serridens* (= *carcinophaga*); 9. *Leptonyx leopardinus* (= *leptonyx,* Blainv.); 10. *Leptonyx weddelli;* 11. *Leptonyx monachus;* 12. *Cystophora proboscidea* (with *ansoni,* Desm., and *dubia,* Fisch., as synonyms); 13. *Cystophora cristata* (with *leonina,* Linn., *borealis,* Nilss., *leucopla,* Thienmann, and *mitrata,* "Fisch.," as synonyms). Of the thirteen species one only (*nummularis*) is nominal, and nearly all stand under their proper specific names, while the various synonyms are in every case correctly referred.

Von Schrenck, in 1859, in his "Reisen und Forschungen im Amur-Lande" (vol. i, pp. 180–188, pl. ix) recognized four species as occurring on the Amoor coast of the Ochots Sea, namely: 1. "*Phoca nummularis,* Schleg." (= "*Phoca largha,* Pallas"); 2. *Phoca barbata* (= "*Ph. nautica* und *Ph. albigena,* Pallas"); 3. "*Phoca ochotensis,* Pallas"; 4. "*Phoca equestris,* Pallas" (= *P. fasciata.,* Zimm.). While none are assumed by the author to be new, the last is for the first time adequately described and figured. Although the existence of this remarkable species was indicated by Pennant in 1781, on information and a drawing furnished him by Pallas, it had hitherto been seen by no subsequent author, and had generally figured as a synonym of other species or in the lists of the doubtful or indeterminable ones. Von Schrenck, however, not only gave detailed descriptions of the dentition and external characters, with measurements of the old and young of both sexes, but also colored figures of the adult male and female.

In 1862, Radde, in his "Reisen im Süden von Ost-Sibirien" (Theil i, pp. 296–304, pl. xiii) described at length a skull of a young female of the Lake Baikal Seal under the name *Phoca annellata,* and incidentally in comparison therewith a skull of the Caspian Seal (*Phoca caspica*) and three skulls of *Phoca fœtida* ("*annellata*") from the East Sea, all of which he referred to the *Phoca annellata* of Nilsson. His article is of importance

:as affording the first detailed description of an authentic specimen of the Baikal Seal.

In 1864, Dr. Gray * described as new a species of Seal from the west coast of North America, under the name *Halicyon richardi* (lege *richardsi*) based on a skeleton from Frazer's River and a skull from Vancouver's Island. He was so impressed with its distinctness that he created for it the new genus *Halicyon*. In the same paper *Heliophoca atlantica* is referred to *Monachus albiventer;* and he enumerates fifteen species which he claims are fully established on osteological as well as external characters.

In 1866 Dr. Theodore Gill † first made known the California Sea-Elephant under the name *Macrorhinus angustirostris.*

Gray, in 1866, in his "Catalogue of Seals and Whales in the British Museum" (pp. 8–35, 38–43, 367, 368), enumerated nineteen species, placed in fourteen genera, as follows: 1. *Lobodon carcinophaga* (includes *serridens*, Owen, and *antarctica*, Peale); 2. *Leptonyx weddelli* (*leopardinus*, Jameson); 3. *Ommatophoca rossi;* 4. *Stenorhynchus leptonyx;* 5. *Monachus albiventer* (includes *bicolor*, Shaw, *leucogaster*, Péron, *atlantica*, Gray, etc.); 6. *Monachus tropicalis;* 7. *Callocephalus vitulinus* (includes *variegata*, Nilsson, *littorea*, Thienemann, *canina*, Pallas, *communis*, vars. *octonotata* and *undulata*, Kutorga, etc., the last incorrectly); 8. "*Callocephalus? caspicus*"; 9. "*Callocephalus? dimidiatus*" (ex Schlegel MS.; locality "Norway"—not determinable); 10. *Pagomys fœtidus* (includes *hispida*, *annellata*, *discolor*, etc., and, incorrectly, "*fasciata*, Shaw," ? *concolor*, De Kay, "? *equestris*, Pallas"); 11. "*Pagomys ? largha*" (= *fœtida;* includes *chorisi* and *tigrina*, Lesson, and *nummularis*, Temm.)'; 12. *Pagophilus grœnlandicus* (includes *oceanica*, Lepech., *semilunaris*, Bodd., *dorsata*, Pallas, *annellata*, Gaimard, etc.); 13. *Halicyon richardsi* (*grœnlandica*, Middendorff = *vitulina*); 14. *Phoca barbata* (*leporinus*, Lepech., etc.); 15. *Halichœrus grypus* (includes *halichœrus*, Thienem., *griseus*, Nilsson, etc., and incorrectly *hispida*, Schreber, and *scopulicola*, Thienem.); 16. *Morunga elephantina* (includes *leonina*, Linné, *proboscidea*, Nilsson, *ansoni*, Desm., *patagonica*, Gray, *dubia*, Fisch.); 17. *Cystophora cristata* (includes *cucullata*, Bodd., *mitrata*, Milbert, MS., *borealis*, Nilsson, *leucopla*, Thienem., etc.); 18. *Cystophora antillarum;* 19. *Halicyon californica* (p. 367, n. sp.; nominal).

---

* Proc. Zoöl. Soc. Lond., 1861, p. 28.
†Proc. Chicago Academy of Sciences, vol. i, p. 33.

This is the largest number, both of species and genera, recognized by one author since Lesson, who, in 1827, gave the same number of species but only six genera. Of these nineteen species, doubtless five (*antillarum, dimidiatus, largha, richardsi,* and *californica*) are nominal, leaving fourteen that are valid, or an average of one to each genus! One of the genera, however, (*Halicyon*) is based on a nominal species; the second species (*Cystophora antillarum*) of the only genus which contains two, may be regarded as doubtful (see *posteà*, under the general history of *Monachus ? tropicalis*). Compared with the same author's revision made in 1850 (see *anteà*, p. 451) there is an increase of three genera and one species; only one "new species" (*californica*) being here added. This the author seems to have again recognized only once,[*] remarking that he considers Gill's *Phoca pealei* as identical with his "*Halicyon ? californica*".

In the same year (1866) Dr. Theodore Gill published his "Prodrome of a Monograph of the Pinnipedes",[†] in which he proposed one new generic name (*Erignathus = Phoca,* Gray), and introduced one new specific name by renaming Peale's *Halichœrus antarcticus, Phoca pealei.* As regards this species, he found that the type specimen, through the misplacement of a label, was wrongly assigned a habitat in the Antarctic Seas, whereas it undoubtedly came from the Pacific coast of North America, thereby rendering the name *antarcticus* undesirable as perpetuating a grave error. He furthermore found that the species had nothing to do with the genus *Halichœrus*, but was a true *Phoca.* The paper introduced several important changes from the Grayian nomenclature, particularly in substituting *Phoca* for *Callocephalus, Erignathus* for *Phoca,* and *Macrorhinus* for *Morunga,* although, as above stated, only one new generic name was introduced. Diagnoses are given of the genera, and lists of the species found respectively on the east and west coasts of North America. The species of North American *Phocidæ* given are the following:

"*Eastern North America.*"

1. Phoca vitulina, Linn.
2. Pagomys fœtidus, Gray.
3. Pagophilus grœnlandicus, Gray.
4. Erignathus barbatus, Gill.
5. Halichœrus grypus, Nilss.
6. Cystophora cristata, Nilss.

"*California, Oregon, &c.*"

1. Phoca richardsi, Gill ex Gray.
2. Phoca pealei, Gill.
3. Macrorhinus angustirostris, Gill.

As the author himself now freely admits, in the light of the material he has since had opportunity of examining, *Phoca richardsi* and *Phoca pealei* are both synonyms of *Phoca vitulina.*

Gray, later in the same year, in a short paper* devoted to a notice of Gill's "Prodrome," proposed to refer Peale's *Halichœrus antarcticus* to a new genus which he named *Haliphilus.*

Gray in 1871, in his "Supplement to the Catalogue of Seals and Whales in the British Museum" (pp. 2–5), raised the number (nineteen) recognized by him in 1866 to twenty-two, omitting two and adding five, mainly by separating the species of the North Pacific from their allies of the North Atlantic. The species added are: 1. "*Halicyon* ? *pealei*" (ex Gill, = *Halichœrus antarcticus*, Peale); 2. "*Pagophilus* ? *equestris*" (ex Pallas, covering *fasciata*, "Shaw," and *annellata*, Radde); 3. "*Pagophilus*? *ochotensis*" (ex Pallas); 4. *Phoca* "*naurica*" (*nautica* and *albigena*, Pallas); 5. *Morunga angustirostris* (*Macrorhinus angustirostris*, Gill). Of these, two only ("*equestris*" and *angustirostris*) are valid. He also tabulated the species according to their distribution, as follows:

"*North Atlantic.*	"*North Pacific.*
"Callocephalus vitulinus.	"Halicyon richardsi.
"Callocephalus dimidiatus.	"Halicyon pealei.
"Pagomys fœtidus.	"Pagophilus ? equestris.
"Pagophilus grœnlandicus.	"Pagophilus ? ochotensis.
"Phoca barbata.	"Phoca naurica.
"Halichœrus grypus.	"Morunga angustirostris."
"Cystophora cristata."	

"*Halicyon californica*" is thus omitted, and is nowhere mentioned in the "Supplement," and the same is the case also with "*Pagomys* ? *largha*". Whether the former was accidentally overlooked, or intentionally retracted, does not appear. From Dr. Gray's "Hand-List of Seals, Morses, Sea-Lions, and Sea-Bears in the British Museum," published in 1874, in which "it is proposed to give an account of all the specimens" of these animals in the British Museum, it appears that the only specimens of North Pacific Phocids there represented were the three referred to *Halicyon richardsi.* Five out of six of his North Pacific species were apparently unknown to him except through authors' descriptions, and are, as I hope later to satisfactorily show, merely nominal. The *Phoca annellata* of Radde, referred by him to his "*Pagophilus*? *equestris*", relates not at all to this

---

* Ann. and Mag. Nat. Hist., 1866, 3d ser., vol. xvii, p. 446.

species, Radde's specimen being *Phoca caspica*, while Radde's *annellata*, as he understood it, is the *fœtida* of authors.

In 1873 Dr. Dybowski* gave a detailed account of the Lake Baikal Seal, with figures, under the name *Phoca baicalensis*, for the first time clearly setting forth its distinctive characters, although the species had been vaguely known, chiefly through incidental notices by travellers, for a century, and as early as 1788 had received, at the hands of Gmelin, the varietal name *sibirica*, he referring it, however, as have many subsequent writers, to *Phoca vitulina*.

In 1875 Dr. Peters† proposed the recognition of five species of Sea Elephants, as follows: 1. *Cystophora leonina*, Linn. (=the Sea Lion of Anson); 2. *C. falklandica* (= the Sea Lion of Pernety); 3. *C. proboscidea* (ex Péron); 4. *C. angustirostris* (ex Gill); 5. *C. kerguelensis* (the species occurring at Kerguelen Island). Nos. 1, 2, 3, and 5 are doubtless synonyms of *Macrorhinus leoninus*. Two new names are proposed, namely, *falklandica* and *kerguelensis*.

The foregoing review has been intentionally limited to works or papers that either (1) ostensibly relate to the whole family, or (2) to the species of the North American fauna, or (3) to those which introduce "new species" or new synonyms. Consequently, reference to many important papers or memoirs treating of particular groups, or of special subjects, is wholly omitted; but the greater part of these will be found cited in subsequent pages under the species to which they particularly relate. No special reference has been made, for example, to Bell's "History of British Quadrupeds", to Blasius's "Naturgeschichte der Säugethiere Deutschlands", Lilljeborg's "Fauna öfver Sveriges och Norges Ryggradsdjur", etc., or to the special memoirs on the Seals of the Arctic Seas by Brown, Malmgren, von Heuglin, etc., or the various papers relating to the anatomy, milk-dentition, etc., of the different species.

For convenience of reference, I present the following chronological summary of the foregoing analysis, premising that the names following the sign of equality are those adopted in the present monograph. The names under which valid species are first introduced are designated by the use of thick type, synonyms by italic type, and indeterminable names by plain type. Only the Phocids are here taken into account.

---

*Arch. für Anat. u. Phys., 1873, pp. 109, *et seqq.* pll. ii, iii.
†Monatsb. Akad. d. Wissensch. zu Berlin, 1875, p. 394, footnote.

Synonymatic Résumé.

1758— ⎱
1766— ⎰ LINNÉ. - - - - - - ⎰ **Phoca leonina** = Macrorhinus leoninus.
                   ⎱ **Phoca vitulina** = Phoca vitulina.

1776—FABRICIUS ....
  ⎰ *Phoca leonina* = Cystophora cristata.
    Phoca fœtida = Phoca fœtida.
    Phoca grœnlandica = Phoca grœnlandica.
  ⎱ **Phoca barbata** = Erignathus barbatus.

1776—SCHREBER, *Phoca hispida* = Phoca fœtida.

1777—ERXLEBEN, **Phoca cristata** = Phoca leonina, Fabr., iu part, neo leonina Linné.

1778—LEPECHIN - - - - - ⎰ *Phoca oceanica* = Phoca grœnlandica.
                ⎱ *Phoca leporina* = Erignathus barbatus.

1779—HERMANN, **Phoca monachus** [*] = Monachus albiventer.

1782—MOLINA - - - - - - - ⎰ *Phoca porcina* =? Macrorhinus leoninus, juv.
             ⎱ *Phoca elephantina* = Macrorhinus leoninus.

1783—ZIMMERMANN, **Phoca fasciata** = Histriophoca fasciata.

1784—BODDAERT - - - -
  ⎰ *Phoca albiventer*[*] = Monachus albiventer.
    Phoca semilunaris = Phoca grœnlandica.
    Phoca cucullata = Cystophora cristata.
  ⎱ *Phoca maculata* = ? Phoca vitulina.

1788—GMELIN - - - - - - -
  ⎰ *Phoca vitulina* var. *botnica* = ? Phoca vitulina.
    Phoca vitulina var. **sibirica** = Phoca sibirica.
  ⎱ *Phoca vitulina* var. **caspica** = Phoca caspica.

1791—FABRICIUS, **Phoca grypus** = Halichœrus grypus.

1792—KERR - - - - - - - - -
  ⎰ *Phoca grœnlandica* var. *nigra* = ? P. grœnlandica.
    Phoca hispida var. *quadrata* = ? **Halichœrus grypus.**
    Phoca chilensis = ? Macrorhinus leoninus, juv.
    Phoca mutica = ?
    Phoca testudo = ?
    Phoca laniger = ? Erignathus barbatus, juv.
    Phoca punctata = ?
  ⎱ Phoca maculata = ?

1798—THUNBERG - - - -
  ⎰ Phoca hispida = ?
    Phoca sericea = ?
    Phoca canina = ?
    Phoca vitulina = ?
  ⎱ Phoca variegata = ? ? Phoca vitulina.

1800—SHAW - - - - - - - - - ⎰ *Phoca bicolor* = Monachus albiventer.
            ⎱ Phoca testudinea = ?

1816—PÉRON - - - - - - -
  ⎰ *Phoca proboscidea* = Macrorhinus leoninus.
    Phoca leucogaster = Monachus albiventer.
  ⎱ *Phoca resima* = ? Macrorhinus leoninus.

1817—DESMAREST - - - ⎰ *Phoca coxii* = Macrorhinus leoninus.
             ⎱ *Phoca lakhtak* = Erignathus barbatus.

1820—BLAINVILLE - - - ⎰ *Phoca byroni* = Macrorhinus leoninus.
             ⎱ **Phoca leptonyx** = Ogmorhinus leptonyx.

---

[*] When the name *monachus* was taken for the generic name of the species, it became untenable in a specific sense, and *albiventer*, originally a synonym, was taken for the species.

458 FAMILY PHOCIDÆ.

1820—DESMAREST ...
$\left\{\begin{array}{l}\textit{Phoca ansoni} \text{ (Blainville MS.)} = \text{Macrorhinus leoni-} \\ \text{nus} + \text{Otaria jubata.} \\ \textit{Phoca albicauda} = \text{Phoca grœnlandica.}\end{array}\right.$

1820—NILSSON ....
$\left\{\begin{array}{l}\textit{Phoca variegata} = \text{Phoca vitulina.} \\ \textit{Phoca annellata} = \text{Phoca fœtida.} \\ \textit{Halichœrus griseus} = \text{Halichœrus grypus.} \\ \textit{Cystophora borealis} = \text{Cystophora cristata.}\end{array}\right.$

1824—F. CUVIER, *Callocephalus discolor* = Phoca fœtida.

1824—THIENEMANN..
$\left\{\begin{array}{l}\textit{Phoca littorea} = \text{Phoca vitulina.} \\ \textit{Phoca halichœrus} = \text{Halichœrus grypus.} \\ \textit{Phoca scopulicola} = \text{Halichœrus grypus, juv.} \\ \textit{Phoca leucopla} = \text{Cystophora cristata.}\end{array}\right.$

1825—G. CUVIER ....
$\left\{\begin{array}{l}\textit{Phoca lagura} = \text{Phoca grœnlandica, juv.} \\ \textit{Phoca mitrata} \text{ (Milbert MS.)} = \text{Cystophora cristata.}\end{array}\right.$

1826—LESSON, **Otaria weddelli** = Leptonychotes weddelli.

1827—GRAY ....
$\left\{\begin{array}{l}\text{Phoca vitulina var. bothionica} = \text{?} \\ \textit{Phoca vitulina} \text{ var. } \textit{sebrica} = \text{Phoca sibirica.} \\ \textit{Morounga patagonica} = \text{Macrorhinus leoninus.}\end{array}\right.$

1828—LESSON ....
$\left\{\begin{array}{l}\textit{Phoca mulleri} = \text{Phoca grœnlandica.} \\ \textit{Phoca schreberi} = \text{Phoca fœtida.} \\ \textit{Phoca parsonsi} = \text{? Erignathus barbatus.} \\ \textit{Phoca thienemanni} = \text{Halichœrus grypus, juv.} \\ \textit{Phoca linnaei} = \text{Phoca vitulina.} \\ \textit{Phoca lepechini} = \text{Erignathus barbatus.} \\ \textit{Phoca frederici} = \text{Phoca fœtida.} \\ \textit{Phoca pylayi} = \text{Phoca grœnlandica, juv.} \\ \textit{Phoca desmaresti} = \text{Phoca grœnlandica.} \\ \textit{Phoca hermanni} = \text{Monachus albiventer.} \\ \textit{Phoca chorisi} = \text{? Phoca vitulina.} \\ \textit{Phoca homei} = \text{Ogmorhinus leptonyx.}\end{array}\right.$

1829—FISCHER, *Phoca dubia* = Macrorhinus leoninus, juv.

1831—PALLAS ....
$\left\{\begin{array}{l}\textit{Phoca nautica} = \text{? Erignathus barbatus.} \\ \textit{Phoca albigena} = \text{Erignathus barbatus.} \\ \textit{Phoca equestris} = \text{Histriophoca fasciata.} \\ \textit{Phoca dorsata} = \text{Phoca grœnlandica.} \\ \textit{Phoca largha} = \text{?} \\ \textit{Phoca canina} = \text{Phoca vitulina} +. \\ \textit{Phoca ochotensis} = \text{? Phoca vitulina.}\end{array}\right.$

1837—NILSSON, **Phoca caspica** * = Phoca caspica.

1837—GRAY, Phoca? platythrix = ?

1839—HAMILTON, *Stenorhynchus leopardinus* (Jameson, MS.) = Leptonychotes weddelli.

1839—KUTORGA .....
$\left\{\begin{array}{l}\textit{Phoca communis} \text{ var. } \textit{octonotata} \\ \textit{Phoca communis} \text{ var. } \textit{undulata}\end{array}\right\} = \text{Phoca fœtida.}$

1842—TEMMINCK *Phoca nummularis* = Phoca fœtida.

1842—DEKAY, *Phoca concolor* = Phoca vitulina.

1843—LESSON, *Phoca isidorei* = Cystophora cristata.

1843—OWEN, *Stenorhynchus serridens* = Lobodon carcinophaga.

1844—GRAY, **Ommatophoca rossi** = Ommatophoca rossi.

---

* First recognized as a species; = *Phoca vitulina* var. *caspica*, Gmelin.

1848—PEALE, *Halichœrus antarcticus* = Phoca vitulina.

1849—GRAY, Cystophora antillarum = ?

1850—GRAY .... { **Phoca tropicalis** = Monachus tropicalis.
{ Callocephalus dimidiatus (Schlegel MS.) = ?

1850—HORNSCHUCH & SCHILLING. { *Halichœrus macrorhynchus* = H. grypus.
{ *Halichœrus pachyrhynchus* = H. grypus.

1854—GRAY, *Heliophoca atlantica* = Monachus albiventer.

1864—GRAY, *Halicyon richardsi* = Phoca vitulina.

1866—GILL .... { **Macrorhinus angustirostris** = Macrorhinus angustirostris.
{ *Phoca pealei* = Phoca vitulina.

1866—GRAY, *Halicyon californica* = Phoca vitulina.

1873—DYBOWSKI, *Phoca baicalensis* = Phoca sibirica.

1875—PETERS .... { *Cystophora falklandicus* } = Macrorhinus leoninus.
{ *Cystophora kerguelensis* }

One hundred and three distinct specific and varietal names have thus been bestowed upon sixteen species, leaving eighty-seven of the names as synonyms,—an average of about six to a species. Fourteen names appear to be wholly indeterminable, while fourteen others can be referred only with more or less doubt. Of the fifty-nine remaining synonyms, about the identification of which there can be but little doubt, *Phoca vitulina* and *Phoca fœtida* have each eleven; *Phoca grœnlandica* has eight, and *Macrorhinus leoninus* nine; *Halichœrus grypus* and *Cystophora cristata* have each six; *Monachus albiventer* has five; and *Erignathus barbatus* seven. Five other species have each one, and three (*Phoca caspica*, *Macrorhinus angustirostris*, and *Ommatophoca rossi*) are apparently without synonyms.

The above summary is exclusive of the generic changes that have been rung on these sixteen species. Regarding each different generic combination as a synonym, would raise the total number of distinct names to probably nearly *four hundred*, or an average of at least twenty to each species, with a maximum for some of the species of at least thirty.

It may be further observed that Lesson has the unenviable distinction of having added thirteen (nearly one-fourth) of the fifty-nine identifiable synonyms, and only one valid species and one tenable specific name out of a total of the fourteen specific names for which he is responsible. Pallas comes next with seven specific names, only four of which are identifiable, and none of them tenable. Next follows Gray with ten, covering two and possibly three new species, and three unidentifiable ones, with the result of seven and probably eight synonyms.

In respect to the general subject, it may be noted that there have been four periods of unusual fertility in respect to the literature of the *Phocidæ*. The first covers the time of Egede, Cranz, Anson, Steller, and Parsons (1741–1765), and antedates nearly all of the systematic literature of the subject, but for which it formed the ground-work of the early portion. The second (1776–1792) may be termed the period of Fabricius, Schreber, Erxleben, Molina, Gmelin, and Kerr, or that of the early technical writers. The third may be denominated the Encyclopædic period, covering the work of Desmarest, F. Cuvier, Lesson, Gray (his first general review of the species only), to which may be added (in point of time) Péron, Nilsson, Fischer, and Pallas (1816–1831). During this period originated more than one-half of the synonyms with which the literature of the subject is burdened, out of nearly forty names only two representing valid new species. Within this period were published no less than eight mongraphic revisions of the Pinnipeds, prepared by the leading mammalogists of that time. The fourth period may be regarded as extending from 1837 to 1873, but the different portions of this interval were not equally prolific in important general memoirs. Of special note in the light of a general revision of the subject are those of Nilsson (1837), Gray (1844), Wagner (1846), Gray (1850), Giebel (1855), Gray (1866, 1871, 1873), and Gill (1866).

CLASSIFICATION.

As already noted, three subfamilies of the *Phocidæ* are now commonly recognized, while the number of genera admitted by two leading authorities who have recently revised the group is respectively twelve (Gill) and thirteen (Gray), with, in the majority of instances, only a single valid species to each. Nine of Gill's genera are monotypic, while of the others two have two species each. The generic affinities of one—the little-known West Indian Seal—have yet to be determined. As will be shown later, only sixteen species can be considered as satisfactorily established. Consequently the question naturally arises whether generic division among the Phocids has not been carried to an excessive degree, and if so, whether the groups termed subfamilies are really entitled to that rank. In the *Pinnipedia* differentiation, it is true, has been carried to such a degree that not only are the family types sharply circumscribed, but the species are so far specialized as to form types that at least some naturalists look upon as types of generic, or at least subgeneric, value. Of the six Otarian genera, four are certainly monotypic, and if the other two are not also mono-typic, the species respectively composing them have not as yet reached the point of well-pronounced specific divergence nor of geographic isolation. As regards the Phocids, we have already seen that in nearly every instance each species has been made the type of a distinct genus. Conservative writers, how-ever, agree in referring four species (*vitulina, grœnlandica, fœtida, barbata*) to the genus *Phoca* (*Callocephalus*, F. Cuvier), yet each species differs from the others to such a degree in cranial and other important osteological characters that, if we allow to such differences the value usually accorded them among the terrestrial *Ferœ*, each of these species may be regarded as the type of a distinct subgenus, or even genus. In the present revision I feel constrained to separate the *Phoca barbata* of authors as generically distinct from the other species of the restricted genus *Phoca*, and to associate with the remaining species *Phoca caspica* and *Phoca sibirica*. The *Phoca fasciata* of authors (= *equestris*, Pallas) Gill has made the type of his genus *Histriophoca*. The species is remarkable for its peculiar pattern of coloration, and von Schrenck compares its dentition to that of *Halichœrus*, between which and the ordinary *Phocœ* it holds an intermediate position as regards the structure of the molar teeth. The other genera of the Phocids will be provision-ally received as now commonly accepted. *Monachus* stands widely aloof from the other genera of the *Phocinœ*, with which, however, it seems more closely allied than with any of the gen-era of the *Stenorhynchinœ*, although all of these have been re-ferred by some systematists (as Wagner and Giebel) to a single genus under the name *Leptonyx*. Without feeling sure that the Phocids are susceptible of subdivision into trenchantly-marked subfamilies, or into groups really entitled to such rank, they will in the present connection be provisionally adopted in their current acceptation.

Synopsis of Subfamilies and Genera.

I. Zygomatic process of the maxillary with the posterior border subvertical, not extending far backward beneath the malar; the latter short. Intermaxillaries prolonged upward, meeting the nasals. Nasals long, nearly reaching to the middle of the orbits, greatly narrowed posteriorly, and wedged between the frontals. Supraorbital pro-cesses wholly obsolete or (in *Erignathus*) rudimentary. Interorbital region very narrow. Incisors usually $\frac{3-3}{2-2}$, exceptionally (in *Mo-nachus*) $\frac{2-2}{2-2}$. Nails of all the digits well developed; outer digits of the pes not much prolonged beyond the others.......**PHOCINÆ.**

1. Muzzle narrow, regularly declined. Incisors $\frac{3-3}{2-2}$, simple, conical; molars, except first, 2-rooted, and mostly 3-lobed. Digits of manus
   slightly decreasing in length from first to fifth, or first and second
   subequal. Mammæ 2 ....**Phoca.**
   a. Skull broad, massive; general form thick; the limbs short; nose
      broad. Molar teeth large, crowded, obliquely implanted, especially
      in youth and in the lower jaw. Nasals considerably prolonged pos-
      teriorly. Posterior nares narrow, the septum incompletely ossified.
      Palatines deeply emarginate. Scapula sickle-shaped, the post-scap-
      ular fossa greatly developed.....................(Subgenus) *Phoca.*
   b. Skull thin, light; nose pointed, and general form slender. Teeth
      small, slightly separated. Palatines, posterior nares, and narial
      septum nearly as in the subgenus *Phoca.* Nasals less prolonged
      posteriorly. Digits of pes subequal. Scapula nearly as in *Phoca.*
                                                  (Subgenus) *Pusa.*
   c. Skull, teeth, and general form nearly as in the subgenus *Pusa.* Pos-
      terior nares broad (nearly twice as broad as high), the narial sep-
      tum complete. Palatines truncate, or slightly emarginate; never
      deeply so as in *Phoca* and *Pusa.* Scapula nearly as in the typical
      terrestrial *Feræ*—not sickle-shaped, and with a broad pre-scapular
      fossa. Sexes, when adult, widely different in coloration.
                                              (Subgenus) *Pagophilus.*
2. Muzzle broad; forehead convex. Rudimentary supraorbital processes.
   Dentition weak; the molars much separated, slightly implanted,
   and partly deciduous or abortive in old age. Palatines broad,
   emarginate. Middle digit of manus the longest. Limbs small.
   Scapula with no acromion process. Whiskers smooth, attenuated.
   Mammæ 4...**Erignathus.**
3. Cranial characters unknown. Dental formula as in *Phoca.* Molars,
   except the first, 2-rooted, somewhat separated, with the crowns
   simple and directed backward, as in *Halichœrus.* Sexes, when
   adult, widely different in color ....**Histriophoca.**
4. Muzzle broad; skull much arched, increasing in height anteriorly.
   Molars single-rooted, except the last lower and the last two upper,
   nearly simple or 1-lobed, conical. Whiskers crenulated. Digits of
   manus as in *Phoca* ....**Halichœrus.**
5. Muzzle elongate, depressed; nasals, short; skull somewhat depressed
   posteriorly. Incisors $\frac{2-2}{2-2}$, notched transversely on the inner side of
   the crown. Canines large. Molars thick, strong, obliquely and
   closely implanted, imperfectly lobed, and only the three posterior 2-
   rooted. Whiskers flat, smooth, tapering. Claws small, especially
   those of the pes ....**Monachus.**
II. Zygomatic process of the maxillary and the malar bones nearly as in
    the *Phocinœ.* Intermaxillaries not prolonged to meet the nasals.
    Nasals very small. Supraorbital processes distinct, prominent, but
    small. Incisors $\frac{2-2}{1-1}$. Molars simple or plaited, not lobed, with a
    single club-shaped root.....................**CYSTOPHORINÆ.**
6. Palatines short, slightly emarginate, somewhat arched or vaulted.
   Auditory bullæ square in front. Adult males with an inflatable
   sack extending from the nose to the occiput. All of the digits with
   claws strongly developed ....**Cystophora.**

7. Palatines very short, deeply emarginate, and deeply vaulted. Auditory bullæ concave in front. Adult males with an elongated tubular proboscis. Claws small, those of the pes rudimentary.

Macrorhinus.

III. Zygomatic process of the maxillary prolonged backward beneath the malar, the latter elongate. Intermaxillaries not (usually) reaching the nasals. Supraorbital processes rudimentary. Nasals generally greatly prolonged posteriorly, widely expanded anteriorly, and usually early consolidated by anchyloses. Incisors $\frac{2-2}{2-2}$. Molars lobed (in two genera acutely multi-lobed). Claws of the hind limbs rudimentary, and the outer digits lengthened. **STENORHYNCHINÆ.**

8. Skull elongate, narrow anteriorly. Nasals greatly narrowed posteriorly. Molars 4- or 5-lobed, the principal lobe large, pointed, recurved, with a smaller one in front of it and two (in the first and second molars) or three (in the others) slender, pointed, recurved lobes behind it. Lower jaw abruptly angular behind...**Lobodon.**

9. General form of the skull much as in the last. Nasals greatly prolonged posteriorly. Molars 3-lobed, the central lobe cylindrical, high, pointed, recurved, with a smaller lobe in front and another behind the principal one. Lower jaw gently rounded posteriorly ....**Ogmorhinus.**

10. Skull broad; muzzle short and broad, with very short, small nasals. Intermaxillaries prolonged upward, meeting the nasals. Molars small, separated, with a central prominent point, and a smaller one (in unworn teeth) behind it. Lower jaw slender, with a short symphysis and no prominent posterior angle .....**Leptonychotes.**

11. Skull very broad (in general outline much as in the *Cystophorinæ*), with a broad, short muzzle, and very large orbital fossæ. Nasals very broad in front, greatly prolonged and gradually narrowed posteriorly. Molars small, 3-lobed, the central lobe much the largest and slightly recurved ....**Ommatophoca.**

SYNONYMATIC LIST OF THE SPECIES.*

I. *Genus* PHOCA, *Linné.*

SYN.—*Pusa*, SCOPOLI; *Callocephalus*, F. CUVIER; *Pagophilus, Pagomys, Halicyon, Haliphilus*, GRAY.

1. **Phoca vitulina,** *Linné.*

SYN.—*Phoca variegata*, NILSSON, 1820.
Phoca littorea, THIENEMANN, 1824.
Phoca linnæi, tigrina, chorisi, LESSON, 1828.
Phoca canina, PALLAS, 1831.
Phoca concolor, DEKAY, 1842.
Halichœrus antarcticus, PEALE, 1848.
Lobodon carcinophaga, CASSIN, 1858.

---

*For full citation of the synonymy of the North American Phocids see *posteà*, in the general history of the species.

Halicyon richardsi, GRAY, 1864.
Phoca pealei, GILL, 1866.
Halicyon? californica, GRAY, 1866.

HAB.—North Atlantic, from New Jersey and the Mediterranean northward to the Arctic regions; North Pacific, from Southern California and Kamtschatka northward to Arctic regions.

2. Phoca grœnlandica, *Fabricius.*

SYN.—*Phoca oceanica,* LEPECHIN, 1778.
Phoca semilunaris, BODDAERT, 1785.
Phoca albicauda, DESMAREST, 1822.
Phoca lagura, G. CUVIER, 1825.
Phoca mülleri, desmaresti, pilayi, LESSON, 1828.
Phoca dorsata, PALLAS, 1831.

HAB.—North Atlantic, from Newfoundland and the North Sea northward, and the Arctic Seas; North Pacific.

3. Phoca fœtida, *Fabricius.*

SYN.—*Phoca hispida,* SCHREBER, 1776.
Phoca annellata, NILSSON, 1820.
Phoca discolor, F. CUVIER, 1824.
Phoca frederici, schreberi, LESSON, 1828.
Phoca largha, PALLAS, 1831.
Phoca communis, vars. *octonotata* et *undulata,* KUTORGA, 1839.
Phoca nummularis, TEMMINCK, 1842.
? *Callocephalus dimidiatus,* GRAY, 1850.

HAB.—North Atlantic, North Pacific, and Arctic Seas.

4. Phoca caspica (*Gmelin*), *Nilsson.*

SYN.—*Phoca vitulina* var. *caspica,* GMELIN, 1788.
Phoea canina [var. *caspica*], PALLAS, 1831.

HAB.—Caspian and Aral Seas.

5. Phoca sibirica, *Gmelin.*

SYN.—*Phoca vitulina* var. *sibirica,* GMELIN, 1788.
Phoca annellata, RADDE, 1862 (in part).
Phoca baicalensis, DYBOWSKI, 1873.

HAB.—Lakes Baikal and Oron.

II. *Genus* HISTRIOPHOCA, *Gill.*

6. Histriophoca fasciata (*Zimmermann*), *Gill.*

SYN.—*Phoca equestris,* PALLAS, 1831.
HAB.—North Pacific.

III. *Genus* ERIGNATHUS, *Gill.*

SYN.—*Phoca,* GRAY, 1850.

7. Erignathus barbatus (*Fabricius*), *Gill.*

SYN.—*Phoca leporina*, LEPECHIN, 1778.
Phoca lachtak, DESMAREST, 1817.
Phoca lepechini, parsonsi, LESSON, 1828
Phoca albigena, nautica, PALLAS, 1831.
Phoca naurica, GRAY, 1871.

HAB.—North Atlantic, North Pacific, and Arctic Seas.

IV. *Genus* HALICHŒRUS, *Nilsson.*

8. Halichœrus grypus (*Fabricius*), *Nilsson.*

SYN.—*Halichœrus griseus*, NILSSON, 1820.
Phoca halichœrus, scopulicola, THIENEMANN, 1824.
Phoca thienemanni, LESSON, 1828.
Halichœrus macrorhynchus, pachyrhynchus, HORNSCHUCH and SHIL-
LING, 1850.

HAB.—North Atlantic, from Newfoundland and Western Islands northward.

V. *Genus* MONACHUS, *Fleming.*

SYN.—*Monachus*, FLEMING, Phil. Zoöl., ii, 1822, 187, footnote.—Type, *Phoca
monachus*, Hermann.
Pelagius ("*Pelage*"), F. CUVIER, Mém. du Mus., xi, 1824, 193.—Type,
Phoca monachus, Hermann.
Heliophoca, GRAY, Proc. Zoöl. Soc. Lond., 1854, 43.—Type, *Heliophoca
atlantica*, Gray, = *Monachus albiventer*, juv.

9. Monachus albiventer (Boddaert), Gray.

SYN.—*Phoca monachus*, HERMANN, Beschaft. d. Berlinishche Gesells. Na-
turf. Freunde, iv, 1779, 456, pll. xii, xiii.
Phoque à ventre blanc, BUFFON, Hist. Nat., Suppl., vi, 1782, pl. xliv.
"*Phoca albiventer*, BODDAERT, Elen. Anim., 1785, 170" (from Buffon, as
above).
Phoca bicolor, SHAW, Gen. Zoöl., i, 1800, 254.
Phoca leucogaster, PÉRON, Voy. aux Terr. Austr., ii, 1817, 47.
Pelagius monachus, F. CUVIER, Dict. des Sci. Nat., xxxix, 1826, 550.
Phoca hermanni, LESSON, Dict. Class. d'Hist. Nat., xiii, 1828 (= *Phoca
monachus*, Hermann).
"*Monachus mediterraneus*, NILSSON, Kongl. Vet. Akad. Handl. Stock-
holm, 1837, 235" (see Wiegmann's Arch. f. Naturg., 1841, i, 308,
footnote).
Heliophoca atlantica, GRAY, Proc. Zoöl. Soc. Lond., 1854, 43 (young).

HAB.—Mediterranean, Adriatic, and Black Seas; Madeira and Canary
Islands; east coast of Africa?

VI. *Genus* CYSTOPHORA, *Nilsson.*

SYN.—*Stemmatopus*, F. CUVIER; *Mirounga*, GRAY (in part).

11. Cystophora cristata (*Erxleben*), *Nilsson.*

SYN.—*Phoca leonina*, LINNÉ 1766 (in part; not *Phoca leonina*, Linné, 1758).
Phoca cucullata, BODDAERT, 1785.

466 FAMILY PHOCIDÆ.

Cystophora borealis, NILSSON, 1820.
Phoca mitrata, G. CUVIER (ex Milbert, MSS.), 1823
Phoca leucopla, THIENEMANN, 1824.
Phoca isidorei, LESSON, 1843.

HAB.—North Atlantic and Arctic Seas.

VII. *Genus* MACRORHINUS, *F. Cuvier*

SYN.—*Mirounga,* GRAY (in part); *Rhinophora,* WAGLER; *Morunga,* GRAY.

12. Macrorhinus leoninus (*Linné*).

SYN.—*Phoca leonina,* LINNÉ, Syst. Nat., i, 1758, 38; *ibid.,* i, 1766, 38 (in part).
Phoca elephantina, MOLINA, Sagg. sul. Stor. Nat. del Chili, 1782, 280.
? *Phoca porcina,* MOLINA, ibid., 279 (young).
Phoca proboscidea, PÉRON, Voy. aux Terr. Austr. ii, 1817, 34, pl. xxxii.
Phoca ansoni, DESMAREST, Mam., 1820, 239 (in part).
Phoca byroni, DESMAREST, ibid., 240.
Phoca dubia, FISCHER, Synop. Mam., 1829, 235.
Mirounga patagonica, GRAY, Griffith's An. King., v, 1827, 186.
Cystophora leonina, falklandica, proboscidea, kerguelensis, PETERS, Monatsb. K. P. Akad. Wissensch. zu Berlin, 1875, 394, footnote.

HAB.—Southern portions of the South Pacific and Indian Oceans and the Antarctic Seas.

13. Macrorhinus angustirostris, *Gill.*

HAB.—Coast of Western Mexico and Southern California.

VIII. *Genus* OGMORHINUS, *Peters.*

SYN.—*Stenorhynchus* ("*Sténorhynque*") F. CUVIER, Mém. du Mus., xi, 1824, 190 (preoccupied in Carcinology and Entomology).—Type, *Phoca leptonyx,* Blainville.
Ogmorhinus, PETERS, Monatsb. K. P. Akad. Wissensch. zu Berlin, 1875, 393, footnote.

14. Ogmorhinus leptonyx (*Blainville*), *Peters.*

SYN.—*Phoca leptonyx,* "BLAINVILLE, Journ. de Physique, xci, 1820, 288."
Phoca homei, LESSON, Dict. Class. d'Hist. Nat., xiii, 1828, 417.

HAB.—"New Zealand; Lord Howe's Island," *Gray;* Desolation Islands.

IX. *Genus* LOBODON, *Gray.*

15. Lobodon carcinophaga, *Gray.*

SYN.—*Phoca carcinophaga,* HOMB. and JACQ., d'Urville's Voy. au Pôle sud, Atlas, 1842? (1842–1853) Mam., pl. x (animal), x A (skull), (not described).—JACQ., Zool., iii, 1855, 27.
Stenorhynchus serridens, OWEN, Ann. & Mag. Nat. Hist., xii, 1863, 331.

HAB.—"Antarctic Seas," *Gray.*

X. *Genus* LEPTONYCHOTES, *Gill.*

SYN.—*Leptonyx,* GRAY, Ann. and Mag. Nat. Hist., x, 1836, 582 (preoccupied in Ornithology).—Type, *Leptonyx weddelli.*
Leptonychotes, GILL, Arrang. Fam. Mam., 1872, 70 (= *Leptonyx,* Gray).

16. Leptonychotes weddelli (*Gray*), *Gill.*

SYN.—*Otaria weddelli,* LESSON, Férussac's Bull. des Sci. Nat., vii, 1826, 438.
Stenorhynchus weddelli, LESSON, Man. de Mam., 1827, 200.
Leptonyx weddelli, GRAY, Ann. and Mag. Nat. Hist., x, 1836, 582;
"Zoöl. Erebus and Terror, Mam., 2, pl. v (animal), plate vi (skull)."
Phoca leopardina, JAMESON, Hamilton's Mar. Amphib., 1839, 183.

HAB.—Antarctic Seas.—"East coast of Patagonia", ·Gray.

XI. *Genus* OMMATOPHOCA, *Gray.*

SYN.—*Ommatophoca,* GRAY, "Zoöl. Erebus and Terror, Mam."—Type, *O. rossi.*

17. Ommatophoca rossi, *Gray.*

SYN.—*Ommatophoca rossi,* GRAY, "Zoöl. Erebus and Terror, Mam., 3, pl. vii (animal), pl. viii (skull)"; Cat. Seals Brit. Mus., 1850, 31; Hand-List Seals, 1874, 15, pl. xi.

HAB.—"Antarctic Seas," *Gray.*

GEOGRAPHICAL DISTRIBUTION.

The *Phocidæ* are found along the seashores of all parts of the temperate and colder portions of the globe, but those of the Southern Hemisphere belong (with one exception) to different genera from those whose habitat is in the Northern Hemisphere, and for the most part to a distinct subfamily not elsewhere represented. All the members of the so-called " subfamily" *Stenorhynchinæ* are confined to the south-temperate and Antarctic Seas. The *Phocinæ,* on the other hand—by far the most numerous division of the family—are strictly northern, only two or three of the species reaching the middle-temperate latitudes. Of the *Cystophorinæ,* consisting of two genera, one genus (*Cystophora*) is boreal, and the other (*Macrorhinus*) has one representative on the coast of Lower California, and another on the islands and shores of the southern part of South America, South Africa, and the Crozet and Desolation Islands in the Indian Ocean. Of the *Phocinæ,* one species, the Monk Seal (*Monachus albiventer*), is found on both shores of the Mediterranean Sea, in the Adriatic and Black Seas, and at the Madeira and Canary Islands, and probably on the neighboring Atlantic coast of

Africa. An apparently near relative and geographical representative of this species is found on the shores of Yucatan, Cuba, Jamaica, the Bahamas, and the Florida Keys. None of the remaining members of the *Phocinœ* occur in the North Atlantic, except as stragglers, south of the British Islands and Spain, on the European coast, or of New Jersey on the American, or of Japan and Lower California in the North Pacific. The species having the widest distribution is the common *Phoca vitulina*, which occurs not only in both the North Atlantic and North Pacific Oceans, as far southward as the limits just given, but reaches Greenland, Finmark, and the northern coast of Europe generally, and is also found in Behring's Straits. Other species, as *Erignathus barbatus*, *Phoca fœtida*, and *Phoca grœnlandica*, extend beyond its habitat to the northward, but have a much more limited range to the southward, the British Islands and the coast of the United States being quite beyond their usual southern limit of distribution. Like *Phoca vitulina* these species also occur in the North Pacific. Two other species are restricted to the North Atlantic, namely, *Halichœrus grypus* and *Cystophora cristata*, neither of which ranges so far northward as the others, and the latter only casually wanders to the southward of Newfoundland and the southern coast of Scandinavia, while the former reaches Nova Scotia and Ireland. *Phoca fœtida* and *Erignathus barbatus* are the most northern of all, both being winter residents of the icy shores of Davis's Strait and Jan Mayen Island. It thus appears that of the six species found on the northern shores of Europe, Greenland, and the Atlantic coast of North America, two only are confined to the North Atlantic, the other four being common also to the North Pacific. The *Histriophoca fasciata*, on the other hand, is limited to the North Pacific, and is the only species occurring there that is not also found in the North Atlantic. Consequently about one-half of the commonly recognized species of the *Phocidæ* of the Northern Hemisphere have a circumpolar distribution.

A species (*Phoca caspica*) formerly regarded by writers as identical with *Phoca vitulina*, and by others a nearly allied but distinct species, inhabits the Caspian Sea, and another (*Phoca sibirica*), similarly referred by most writers to *Phoca fœtida*, inhabits Lake Baikal. These great interior and almost isolated seas have been for so long a time separated, the Caspian Sea wholly, and Lake Baikal nearly, from the great oceans or any other large body of water communicating with the sea, that if originally derived from the marine species to which they are allied, it may well be supposed that the peculiar conditions of environment to which they have been for so long a time subjected have not been powerless in effecting slight changes of structure, as they have certainly led to well-marked changes in habits.

As already noted, *Macrorhinus* is the only genus having representatives on both sides of the equator, the two species of which are nevertheless separated by wide areas, the one occurring on the Pacific coast of North America between the parallels of 23° and 35° north latitude, while the other is restricted to the shores and islands of the southern extremity of the South American continent, New Zealand, and a few groups of pelagic islands in the southern parts of the Indian Ocean.

Of the *Stenorhynchinæ* only four species are recognized, all of large size, and all confined to the cold-temperate or subfrigid southern waters.

FOSSIL REMAINS.

NORTH AMERICA.—In North America teeth or other remains attributed to Seals have been reported as occurring at various localities, in Tertiary and Quaternary deposits, from Maine and Canada southward to Virginia and South Carolina. In several instances, merely the finding of such remains has been recorded, the specimens themselves having never been described, or even specifically determined, so that it is impossible to assign them to any particular species, or even to say whether they were correctly identified as the remains of Seals. In other cases, remains described as Phocine are unquestionably referable to Squalodont Cetaceans. In only two or three instances are the supposed remains of Seals obviously Phocine, and in each of these cases they were found in deposits of Post-pliocene age, and referred (usually with some doubt) to existing species. The subject may, therefore, be conveniently treated under the following heads, namely: 1. Remains supposed to be Phocine, but which are not specifically determinable. 2. Squalodont remains described as Phocine. 3. Remains doubtfully referred to existing species. 4. Extinct species.

I. *Remains supposed to be Phocine, but not specifically determinable.*—1. Newbern, North Carolina.—Under this head must be placed the incidental reference by Dr. Harlan * to the remains

---

* Am. Journ. Sci., vol. xliii, 1842, p. 143.

of a "Seal" found associated with those of Mastodon, Elephant, Horse, Deer, Elk, etc., in the Post-pliocene deposits of Newbern, North Carolina, in his description of his "*Sus Americanus*". The specimens here referred to appear to have never been described, and the only information we have respecting the occurrence of Phocine remains at this locality is Dr. Harlan's casual reference to the matter, as above indicated.

2. Martha's Vineyard, Massachusetts.—Sir Charles Lyell, in a paper "On the Tertiary Strata of the Island of Martha's Vineyard in Massachusetts", in enumerating the organic remains collected by him at that locality, mentions,* under the head of Mammalia, "A tooth, identified by Prof. Owen as the canine tooth of a Seal, of which the crown is punctured. It seems nearly allied to the modern *Cystophora proboscidea*". As no description is given, its positive determination is impossible. No other Seal remains, so far as known to me, have been found at that locality.

3. Richmond, Virginia.—As will be presently noticed more fully, some supposed Phocine remains were described by the late Professor Wyman from the Tertiary deposits underlying the city of Richmond, Virginia. They came from two localities, and consisted of quite different materials. The specimens are at present unknown, so that their reëxamination is impossible. A part of these remains were in all probability Squalodont, while others may have been Phocine. A detailed account of these specimens, with the original descriptions in full, is given below, under the heading "*Phoca wymani*".

4. South Berwick, Maine.—Professor Wyman, in 1850,† referred briefly to some Seal bones found at South Berwick, Maine, in "marine mud", at a depth of thirty feet from the surface, in digging a well. They "proved to be an ulna and a radius", but no description of them is given, they being mentioned simply as "bones of a Seal". Professor Leidy‡ has conjecturally referred them to *Phoca grœnlandica*.

II. *Squalodont Remains described as Phocine.*—No less than three species referred originally to "*Phoca*" are in all probability referable, in part or wholly, to *Squalodon*, as is more or less explicitly admitted by their original describer. These are *Phoca wymani*, *P. debilis*, and *P. modesta*, of Leidy. The first

---

* Proc. Geol. Soc. Lond., vol. iv, 1843–1845, p. 32; Amer. Journ. Sci., vol. xlvii, 1844, p. 319; Phil. Mag., vol. xxxiii, 1843, p. 188.

† Amer. Journ. Sci. and Arts, 2d ser., vol. x, 1850, p. 230, footnote.

‡ Extinct Mam. N. Amer., 1869, p. 415.

was based originally on remains from the Tertiary deposit at Richmond, Virginia; the others on teeth from the Ashley River beds of the same age in South Carolina.

1. *"Phoca wymani"*.—The Richmond remains were first described by the late Professor Wyman in 1850,[*] who merely referred them to "an animal belonging to the family of *Phocidæ*". The bones are spoken of as fragile, and as having "evidently been crushed previous to exhumation". "The pieces in my possession", says Professor Wyman, "consist of two temporal bones nearly entire, a fragment including a portion of the parietal and occipital bones, and in addition a part of the base of the skull. The reëntering angle of the occiput, the well-marked depressions corresponding with the cerebral convolutions on the parietal bones, the form of the cranial cavity, the deep fossa above the internal auditory foramen, the vascular canals opening on the occiput, and the inflated tympanic bones, all indicated an affinity to the *Phocidæ*. The size varied but little from that of the common Harp Seal (*Phoca grœnlandica*). The presence of an interparietal crest, indicating a large development of the temporal muscles, offers a diagnostic sign by which it may be distinguished from *P. barbata*, *P. grœnlandica*, *P. hispida*, *P. mitrata*, and *P. vitulina*. From those species of Seals which are provided with a crest the fossil presents a well-marked difference in having the mastoid process much larger, more rounded and prominent, nearly equalling the tympanic bone in size. The entrance to the carotid artery is in full view when the base of the skull is turned upwards. The imperfectly divided canal which lodges the Eustachian tube and the tensor tympani muscle is of remarkable dimensions, especially when compared with that of *P. grœnlandica*. The interparietal crest, extending from the occiput to the anterior edge of the frontals, is most narrow posteriorly where it is but slightly elevated above the surrounding bones".

In the description above given there is nothing to prevent the supposition that these cranial fragments are referable to a small species of Squalodont. If, however, they are really Phocine, they represent a type very unlike anything at present known, either existing or extinct. But other remains are described by Professor Wyman, from the same locality, and in the same paper, which do not seem to admit of such an interpretation. Thus, to continue the quotation: "The fragments

---

[*] Amer. Jour. Sci., 2d ser., vol. x, 1850, p. 229.

of cranium above described were found in the Shockoe Creek ravine near the base of Church Hill. In the ravine at the eastern extremity of the city, and in the neighborhood of the penitentiary, Dr. Burton obtained several other portions of the skeleton of another Seal. These consisted of an imperfect cervical vertebra, a lumbar vertebra nearly entire, a fragment of the sacrum, coccygeal vertebra, fragments of ribs, and the lower extremity of a fibula. Their generic characters have been satisfactorily made out by comparison with recent bones.

"In figure 1, page 232, I have represented the coccygeal vertebra which corresponds in its general characters very accurately with recent bones of *P. grœnlandica* from the same region of the vertebral column. The small size of the vertebral canal and the imperfect transverse process, the wide-spread articulating processes, and the blunted spinous process indicate its affinity to the Seals. The fragment of a left fibula (figs. 2 and 3), presents at its lower extremity (fig. 3) an oblique, regularly concave articulating surface on its inner face, and on its outer (figs. 2 and 3), an elevated ridge or crest, on either side of which is a groove for the passage of a tendon."

The specimens here described do not appear to have been preserved, or to have been seen by subsequent writers, but Professor Wyman was an osteologist of too well-known proficiency to admit of the supposition that these remains did not present well-marked Phocine affinities. Indeed, his description and rude figures of the fibula above mentioned show clearly that its affinities were rightly interpreted. The vertebra is not so evidently Phocine. Three years later the description of these remains became the basis of Dr. Leidy's *"Phoca wymanii"*, who, in proposing the name,* merely cited Wyman's description. In 1856† he referred to it a tooth "apparently an inferior canine from the miocene deposit of Virginia." This tooth he describes as being "14 lines, and about as robust in its proportions as the corresponding tooth of *P. barbata*. The crown is $4\frac{1}{2}$ lines long and $3\frac{1}{2}$ broad at base, and it presents an anterior and a posterior ridge, of which the former is denticulated, and bifurcates half way towards the base. The enamel is rugose, especially towards the base of the crown internally; and at one or two points in front presents a short inconspicuous tubercle."

In 1867 Professor Cope referred *Phoca wymani*, Leidy, to

---

* Ancient Fauna of Nebraska, 1853, p. 8.
† Proc. Acad. Nat. Sci. Phila., 1856, p. 265.

Squalodon*, of which he says: "Of this, the smallest species of the genus. three premolar teeth are in the collection [made by Mr. James T. Thomas, in Charles County, Maryland from beds of the Yorktown epoch], and the type specimen [Dr. Leidy's?] is in the Academy's Museum. The teeth are remarkable for the abrupt posterior direction of their crowns. The roots are curved, one of them abruptly so, and flattened."

The *Squalodon wymani* of Cope thus, inferentially at least, includes the remains described by Wyman, though direct reference seems to be made only to the tooth referred by Leidy to his *Phoca wymani* in 1856, and which is that of a Squalodont. The *Phoca wymani*, if not originally a composite species, as was in all probability the case, certainly became so in 1856. In 1869 Dr. Leidy retained, under the name *Phoca wymani*, the specimens above mentioned as described by Wyman in 1850, separating the tooth referred by him to this species in 1856 under the name *Delphinodon wymani*.†

2. "*Phoca debilis*".—In 1856‡ Dr. Leidy gave a description of his *Phoca debilis*, of which the following is a transcript in full: "A species of Seal is apparently indicated by three specimens of molar teeth obtained by Capt. Bowman, U. S. A., from the sands of the Ashley River, South Carolina. The teeth bear considerable resemblance to the corresponding ones of *Otaria jubata*, having small, compressed conical crowns, tuberculate in front and behind, and single, long, gibbous fangs. The smallest specimen is 5½ lines long, and the largest, when perfect, was about an inch long".

In 1867 this species was referred by Professor Cope to *Squalodon*, who says:—"A species still smaller than *S. wymanii* has been described by Leidy as *Phoca debilis*, from the Pliocene of Ashley River of S. Carolina. It will no doubt be found to be allied to Squalodon".§ It had, in fact, been apparently already referred by Cope in the early part of the same paper to *Squalodon*, where (on page 144) he gives, in his list of species, "*Squalodon debilis* Cope, Pliocene". Dr. Leidy himself, in 1869, admitted that Professor Cope's suspicions of their Squalodont affinities might be correct, but adds that these teeth "may belong to a Dolphin".‖

* "Squalodon wymanii m. Phoca wymanii Leidy. Proceedings Academy N. Sci., 1856, 265."—*Proc. Acad. Nat. Sci. Phila.*, 1867, p. 152.
† Ext. Mam. N. Amer.), p. 426.
‡ Proc. Acad. Nat. Sci. Phila., 1856, p. 265.
§ Ibid., 1867, p. 153.
‖ Ext. Mam. N. Am., p. 475.

3. "*Phoca modesta*".—This species, described by Leidy in 1869,* is based on a small tooth from the Ashley River deposits of South Carolina, and, says this author, "is referred to a Seal, though it is not improbable it may belong to a Squalodont"; as, in fact, I have little doubt is the case.

III. *Remains referred to Existing Species.*—In 1856 Professor Leidy described and figured† some fossil remains of Seals found in the "township of Gloucester, county of Carleton, Canada West, about nine miles east of the city of Ottawa", in a bed of blue clay containing boulders and marine shells and fishes. The shells found embrace, according to Mr. E. Billings, *Tellina grœnlandica, Mytilus edulis, Saxicava rugosa*, and a small species allied to *Leda;* while the fishes are *Mallotus villosus* and *Cyclopterus lumpus*, and the clays containing them are regarded as of Postpliocene age. "The bones," says Dr. Leidy, "proved on examination to be those of the greater portion of the hinder extremities of a young Seal, but whether of a species distinct from those now found in the neighboring seas, is only to be determined by careful comparison with the corresponding parts of the recent animals. The soft distal extremities of the tibia and fibula are crushed together. The bones of the ankle and foot are well preserved, but the epiphyses of the latter are separated and only partially developed. The matrix in the vicinity of the bones is marked by the impression of the hairs and skin which enveloped them."

Dr. Leidy has since‡ referred these remains provisionally to *Phoca grœnlandica.*

Dr. Leidy's account of these remains was also published in the "Canadian Naturalist and Geologist" (i, 1857, pp. 238, 239, pl. iii). Twenty years later some further notice of fossil Seal remains from the same locality was given by Dr. Dawson,§ in which, referring to the former account, he says: "A good figure and description were published in the first volume of the *Naturalist* in 1856. No further information bearing directly on this fossil was secured until the present year, when the bone now exhibited [before the Natural History Society of Montreal, October 29, 1877], was obtained by Dr. Grant, from a boy who had collected it at the same place and in the same bed in which the first-mentioned specimen was found. It is the left ramus of the

---

* Ext. Mam. N. Amer., 1869, p. 415, pl. xxviii, fig. 14.
† Proc. Acad. Nat. Sci. Phila., 1856, pp. 90, 91, pl. iii.
‡ Ext. Mam. N. Amer., 1869, p. 415.
§ Canad. Nat., 2 ser., vol. viii, 1877, pp. 340, 341.

lower jaw of a young Seal, containing a canine and four molar teeth, with an impression of a fifth. It enables us now to affirm that the species is *Phoca Groenlandica*—(*Pagophilus Groenlandicus* of Gray's Catalogue), the common Greenland Seal, and it is of such size that it may have belonged to the same individual which furnished the bones described in 1856, or at least an animal of the same species and of similar age."

IV. *Extinct Species.*—Another reference to fossil remains apparently referable to a Seal is of special interest as indicating, if there is no mistake respecting the origin of the specimen, the former presence on our Atlantic coast of a Phocine type existing at present only in the Antarctic Seas. The species was described by Dr. Leidy in 1853, under the name *Stenorhynchus vetus.* The description is based entirely on *an outline drawing* of a tooth purporting to be from the "*green sand* of the Cretaceous series, near Burlington, New Jersey". The specimen was never seen by the describer of the species, and was long since lost. The tooth is said to have been found by Mr. Samuel A. Wetherill, who gave it to Mr. T. A. Conrad, by whom was made the drawing. "The figure", says Dr. Leidy, "represents a double-fanged tooth, with a crown divided into five prominent lobes. It is, without doubt, the tooth of a mammal, and resembles very much one of the posterior molars of *Stenorhynchus serridens*, Owen, an animal of the Seal tribe. It may have belonged to a Cetacean allied to *Basilosaurus*, but until further evidence is obtained I propose to call the species indicated by the tooth *Stenorhynchus vetus*".* Later the same writer referred the species to *Lobodon*, and adds, "The specimen purports to have been derived from the green sand, but is probably of miocene age and accidental in its position in relation with the preceding formation. The original of the tooth I have not seen, but it was in the possession of Timothy Conrad, the well-known naturalist, who made an outline drawing of it the size of nature, which is represented in a wood-cut, of the same size, on page 377 of the Proceedings of this Academy for 1853. The specimen has been lost. The drawing of it so nearly resembles the representations of the molar teeth of the Crab-eating Seal, *Lobodon carcinophaga* of Gray, or the *Stenorhynchus serridens* of Owen, that it may be regarded as an indication of an extinct species of the same genus".† The close resemblance of the figure to the tooth of

---

* Proc. Acad. Nat. Sci. Phila., 1853, p. 377 (wood-cut).
† Extinct Mam. N. Amer., 1869, p. 416.

Lobodon carcinophaga is certainly unquestionable, but the history of the tooth which served as the original of the drawing, in reference to the locality of its assumed discovery, seems not altogether satisfactory. Dr. Leidy discards its Miocene origin, but seems to have no doubt respecting its discovery at the locality named. Dr. Gray, in his synonymy of *Lobodon carcinophaga*,* says, "See *Stenorhynchus vetus*, Leidy, ... tooth, said to be found in the greensand of New Jersey", seemingly implying not only its close resemblance to *Lobodon carcinophaga*, but doubt as to the correctness of the assumed locality. In view of the possible extralimital origin of the tooth, I hesitate to formally include the species in the list of North American Pinnipeds.†

EUROPE.—While fossil remains of Seals have been found so rarely in North America, not a single extinct species having been certainly determined, the Tertiary deposits of Europe, particularly those of Belgium, have yielded abundant remains of

---

* Cat. Seals and Whales, 1866, p. 10.

† Mr. Andrew Murray, in commenting (Geog. Distr. Mam., p. 124, 1866) upon this species (Leidy's *Stenorhynchus vetus*) observes, as follows: "Sir Charles Lyell tells us [Elements of Geology, sixth ed., London, 1865, p. 336] that that gentleman [Mr. Samuel R. Wetherill] related to him and Mr. Conrad, in 1853, the circumstances under which he met with it, associated with *Ammonites placenta*, *Ammonites Delawarensis*, *Trigonia thoracica*, &c., and he adds that although the tooth had been mislaid, it was not so until it had excited much interest, and been carefully examined by good zoölogists. . . . . There seems to be no reason to doubt that the tooth was found where Mr. Wetherill said it was, nor is there any question here of misplaced labels, but there is certainly room for doubting its determination, because we see where and how an error might easily enough have arisen. In the first place, it is referred to a living genus of mammals, and we know of no genus which has subsisted through so many cycles. The presumption is therefore against it on that score. In the next place, there is a certain resemblance between the teeth of Sharks and some Seals, and it is precisely in the genus *Stenorhynchus* that the resemblance is most marked. . . . . It is possible, therefore, that the supposed Seal's tooth may have been a very much rubbed and worn Shark's tooth ; and although Lyell says it was carefully examined by good zoölogists, the only one of known competence whom he mentions as having had to do with it is Dr. Leidy, who did not see it, but described it from a drawing. The objections to the supposed mesozoic Seal's tooth, therefore, appear to be too well founded to require us to devote much time to a speculation founded upon its authenticity." Mr. Murray gives comparative views of Shark and Seal teeth, to show how close is the resemblance of the teeth of *Stenorhynchus* to those of certain Sharks, but if Mr. Murray had taken the trouble to consult the original figure of the tooth of *S. vetus* he would have seen, first, that it was not a "much worn and rubbed tooth", and, secondly, that it was not a *three*-pointed tooth like those he figures, but a *five*-pointed tooth, representing *Lobodon* and not *Stenorhynchus*.

these animals, Professor J. P. Van Beneden having already indicated thirteen supposed species from the Anvers Basin alone. Quite a number of species have also been described from various localities in France, Germany, Italy, and the borders of the Black Sea. Various remains of Seals have also been obtained from the Quaternary, especially in the British Islands and in Norway, but all such prove to be closely allied to if not identical with species still existing in the neighboring or more northerly seas. No remains of Seals have been reported from beds older than the Upper Miocene, while the greater part have been obtained from deposits referable to the Pliocene. While a detailed account of the extralimital species of extinct Phocids is hardly required in the present connection, a brief *résumé* of the subject may be of interest. This will be based mainly on the elaborate memoirs on this subject recently published by Van Beneden. *

Until quite recently very few extinct species of true Phocids had been described, most of the remains attributed to this group by the earlier palæontologists proving on later examination to be mainly referable to Squalodont, Delphinoid, or Xiphoid Cetaceans. The two fragments considered by Cuvier to be Phocine were found by Blainville to be Sirenian. Of the various suppositive remains of Seals described by Blainville, Van Beneden claims that in one instance only do they belong positively to this group, this being the foot preserved in the Museum of Pesth, described under the name *Phoca halitschensis*, which is said to somewhat resemble the corresponding part of the common *Phoca vitulina*. H. von Meyer's *Phoca rugidens* turns out to be referable to *Squalodon*. The same author's *Phoca ambigua* is allied to *Phoca vitulina*. Pictet's genus *Pachyodon*, Van Beneden says is Squalodont and not Seal, while the bones referred by the same author to *Phoca ambigua*, Van Beneden believes was not a fortunate reference. Staring's *Phoca ambigua*, Van Beneden refers to his own *Palæophoca nysti*. Some of the bones of Seals from various localities in France referred by Gervais to *Pristiophoca occitana* are thought by other authorities to be those of Delphinoid or Xiphoid Whales, while Van Beneden considers the *Phoca pedroni*, Gervais, to be probably also Xiphoid. The *Phoca pontica* of Nordmann is closely related to *P. vitulina*

* See especially this author's magnificently illustrated work on the Fossil Pinnipeds of the Basin of Anvers, forming part one of volume one of the Annales du Musée royal d'Histoire naturelle de Belgique, 1877, where the historical portion of the subject is presented with considerable detail.

(from which it is said to differ in size), while his *P. mœotica* is allied to *Monachus albiventer*, of which latter Guiscardi's *Phoca gaudini* seems to have been the progenitor.

In 1853, M. J. P. Van Beneden described an extinct species of Seal under the name *Palæophoca nysti*, based mainly on specimens from the vicinity of Anvers. In 1876 the same writer, in his memoir on "Les Phoques fossiles du Bassin d'Anvers",[*] added twelve species to those previously indicated, all from the environs of Anvers, making thirteen described by him from that locality. They are based usually on numerous specimens, consisting generally of vertebræ and the bones of the limbs and pelvis. They are generally more or less fragmentary, and the most characteristic parts of the skeleton, as the cranium and dentition, are not represented. These species were redescribed in greater detail the following year, and illustrated with a splendid suite of plates, in which the more important specimens were figured of the size of nature, several views being given of each.[†] Five are from the Upper Miocene, and eight from the Pliocene. None of them depart very widely from existing types, although with one exception all are referred to extinct genera. One (*Mesotaria ambigua*), Professor Van Beneden thinks, presents characters indicative of Otarian affinities, and that this form probably represented the Otaries in the Tertiary seas of Europe, but neither the description nor the figures seem to me to evince such an alliance. On the contrary, *Mesotaria ambigua* appears to be[†] not remotely allied to the *Cystophorinæ* (see *anteà*, pp. 219, 220). All the other species, so far as can be judged by their fragmentary remains, exhibit affinities, more or less remote, with one or another of the species still existing in the European seas. The extinct species of this family considered by Van Beneden as fairly entitled to recognition, are the following:[‡]

1. **Mesotaria ambigua**, Van Beneden. Anvers. Pliocene. Allied to *Cystophora cristata*, or at least referable in all probability to the *Cystophorinæ*.

2. **Palæophoca nysti**, Van Beneden. Elsloo; Boltringen; Anvers. Pliocene. Allied to *Monachus albiventer*.

3. **Pristiophoca occitana**, Gervais. Central France. Allied to *Monachus albiventer*.

---

[*] Bull. de l'Acad. roy. de Belgique 2me, sér. 1, xli, No. 4, April, 1876.

[†] Descriptions des ossements fossiles des Environs d'Anvers, folio, 1877, with an Atlas of eighteen plates.= Ann. du Mus. roy. d'Hist. nat. de Belgique, tome i, prem. part.

[‡] The authority for the localities, geological age, and affinities (except in the case of *Mesotaria*) is M. Van Beneden.

4. **Phoca gaudini,** Guiscardi. Italy, from caverns. Allied to *Monachus albiventer.*

5. **Phoca mœotica,** Eichwald. Basin of the Black Sea. Allied to *Monachus albiventer.*

6. **Phoca ambigua,** H. von Meyer. Osnabruk. Tertiary. Allied to *Phoca vitulina.*

7. **Platyphoca vulgaris,** Van Beneden. Anvers. Pliocene. Allied to *Erignathus barbatus.*

8. **Callophoca obscura,** Van Beneden. Anvers. Pliocene. Allied to *Phoca grœnlandica.*

9. **Gryphoca similis,** Van Beneden. Anvers. Pliocene. Allied to *Halichœrus grypus.*

10. **Phocanella pumila,** Van Beneden. Anvers. Pliocene. Allied to *Phoca fœtida.*

11. **Phocanella minor,** Van Beneden. Anvers. Pliocene. Allied to *Phoca fœtida.*

12. **Phoca vitulinoides,** Van Beneden. Anvers. Pliocene. Allied to *Phoca vitulina.*

13. **Phoca pontica,** Eichwald. Basin of the Black Sea. Allied to *Phoca vitulina.*

14. **Phoca halitschensis,** Blainville. Valley of the Danube. Allied to *Phoca vitulina.*

15. **Monatherium delongii,** Van Beneden. Anvers. Upper Miocene. So far as can be judged the genus *Monatherium* (known only from vertebræ) is allied to *Monachus.*

16. **Monatherium aberratum,** Van Beneden. Anvers. Upper Miocene.

17. **Monatherium affinis,** Van Beneden. Anvers. Upper Miocene.

18. **Prophoca rousseaui,** Van Beneden. Anvers. Upper Miocene.

19. **Prophoca proxima,** Van Beneden. Anvers. Upper Miocene. Although the genus *Prophoca* is a true Phocid, its affinities with any one of the existing types rather than with another are not apparent.

It thus appears that each of the existing species is represented by one or more allied forms among the extinct species, the greater part of the extinct forms, however, clustering about the Monk Seal (*Monachus albiventer*) of the Mediterranean, and the common Vituline or Harbor Seal (*Phoca vitulina*).* As

---

* The nearest living affines of the extinct genera may be thus tabulated:

Extinct.	Living.
Mesotaria, represented by, or allied to	*Cystophora.*
Pristiophoca, *Palœophoca,* } represented by, or allied to	*Monachus.*
Callophoca, represented by, or allied to	*Pagophilus.*
Platyphoca, represented by, or allied to	*Erignathus.*
Gryphoca, represented by, or allied to	*Halichœrus.*
Phocanella, represented by, or allied to	"*Pagomys.*"
Phoca vitulinoides, represented by, or allied to	*Phoca vitulina.*
Monatherium, represented by, or allied to	*Monachus.*
Prophoca, represented by, or allied to	———

Van Beneden has remarked, the extinct species of *Phocidæ* present already the distinctive characters of the group; the species, however, were more numerous, and they were of larger size.* The remains of Seals discovered in deposits of Quaternary age have all been referred to existing species; those from the Tertiary bear a strong resemblance to existing types, the genus *Prophoca*, of the Miocene of Anvers, alone, having no very closely related existing representative. The materials on which are based many of the species above enumerated are so scanty, and in many cases so imperfectly preserved, that doubtless additional specimens may show the necessity of somewhat reducing the number, while, on the other hand, others may be added.

By far the greater part of the remains of Pinnipeds thus far known have been found at the single locality of Anvers, where not only most of the species have been found, but where probably more than nine-tenths of all such remains have thus far been obtained. The Royal Museum of Belgium alone contains upward of five hundred specimens from this locality, which M. Van Beneden has referred to sixteen species and twelve genera. With these remains are associated those of *Halitherium*, and of various types of Cetaceans. The whole series of the beds containing these fossils are regarded by some geologists as Pliocene, but by other good authorities the lower ones of the series are regarded as Upper Miocene. The great Tertiary sea, beneath whose waters these deposits were formed, covered the greater part of Holland, part of Germany, and extended to the counties of Norfolk and Suffolk in England, over all of which region the waters prevailed till the close of the Tertiary epoch.

As has already been shown, in North America few remains of Pinnipeds have been found, and these, with two exceptions, are all from the Quaternary, and are referable to existing species. The exceptions are the so-called *Phoca wymani*, based in part at least upon veritable Phocine remains from the Miocene of Richmond, Virginia, and the enigmatical *Lobodon vetus*, based on a tooth purporting to have been found at Burlington, New

* "Nous finirons par cette observation que si tous ces Thalassothériens présentent déjà les caractères propres de leur groupe, la seule différence de quelque importance se rapporte à leur nombre qui si considérablement réduit et à leur taille qui a notablement diminué. . . . À l'exception des ossements recueillis dans la sable noir [the Miocene genus *Prophoca*], tous les autres se rapportent à des espèces qui rapellent celles qui vivaient encore dans notre hémisphère, depuis la *Floe rat* jusqu'au *grand Phoque*."
—*Descrip. des Ossem. fos. des Environs d'Anvers, pp.* 85, 86.

Jersey, in beds of Cretaceous age. An error of horizon, if not of locality, being admitted in case of the last named, it appears thus far that no traces of Pinnipeds have been met with in beds older than Miocene.

MILK-DENTITION.

In the *Phocidæ*, as in the other *Pinnipedia*, the milk-teeth are very small, are perfectly functionless, and persist for only a short period beyond fœtal life. As in the other Carnivores, the number of milk-teeth in the molar series is three, and their position is the same as that of the deciduous molars of the fissipede *Feræ*, standing respectively over the second, third, and fourth of the permanent set. The incisors and canines are each preceded by deciduous teeth, always minute, and generally absorbed prior to the birth of the animal, as are also, in most cases, the deciduous molars. It is consequently difficult to obtain dry specimens that retain the minute milk-teeth, they being usually partly or wholly (especially the very small incisors) lost in the preparation of the specimen, or by subsequent handling. Alcoholic or fresh specimens alone afford satisfactory material, and these are not often accessible. The milk-dentition of most of the northern genera of Phocids has, however, been described, but I have met with no reference to that of any of the genera of the *Stenorhynchinæ*. As already stated, each incisor of the permanent dentition is preceded by a milk-tooth, and the number of temporary incisors thus varies in the different genera in accordance with the number of permanent ones. The *Stenorhynchinæ* will doubtless be found to afford no exception as regards the relative number and position of the deciduous teeth.

Steenstrup, * in 1861, described and figured the milk-dentition of *Erignathus barbatus* and *Phoca grœnlandica*, and described also that of *Phoca fœtida*. Nordmann, † at about the same date, described that of *Halichœrus*, and in 1865 the milk-dentition of *Cystophora cristata* was described and figured by Rein-

---

* Mælketandsættet hos Remmesælen, Svartsiden og Fjordsælen (*Phoca barbata* O. Fabr., *Ph. grönlandica* O. Fabr., og *Ph. hispida* Schreb.), og i Anledning deraf nogle Bemærkninger om Tandsystemet hos to fossile Slægter (*Hyænodon* og *Pterodon*). Af Professor Japetus Steenstrup. Vid. Medd. fra den naturh. Forening i Kjöbhavn, 1860 (1861), pp. 251–264, pl. v.

†"Palaeontologie Südrusslands, iv Abth. vorgetr. in d. Finnl. Soc. d. Wiss., 1860, pp. 306–308."—The only copy of Nordmann's work accessible to me is imperfect and unfortunately lacks the iv Abth.

hardt,* while Steenstrup† further discussed that of *Erignathus barbatus*. Flower,‡ in 1869, figured and described the milk-teeth of "*Morunga proboscidea*" (= *Macrorhinus leoninus*).

So far as known to me the observations here cited embrace all the original descriptions of the milk-dentition of the *Phocidæ*.

While the number of the milk-molars is in each case three on each side, both above and below, their size, as well as that of the other deciduous teeth, varies in the different genera, being very small in *Macrorhinus* and *Cystophora*, and larger in *Phoca*, *Erignathus*, and *Halichœrus*. Professor Flower, in describing the teeth of a fœtal specimen of *Macrorhinus* says: "The jaws contained a complete set of very minute teeth, viz. $i. \frac{2}{1}, c. \frac{1}{1}, m. \frac{3}{3}$, on each side, all of the simplest character. The incisors and canines were cylindrical, and open at the base. The upper canine, which was the largest tooth, and of which the whole of the crown and greater part of the root were calcified, measured in length 0.1″ and in greatest thickness 0.04″. The second upper incisor was about half this size, and the first still smaller. The molars consisted only of a rounded crown, about the size of a small pin's head, the roots were not calcified. As the crowns of teeth once calcified never enlarge in diameter, we may presume that these rudimentary teeth had attained their full dimensions, except perhaps as to the root of some of them."§

Professor Reinhardt's description and figures of the milk-teeth of *Cystophora cristata* represent them as correspondingly small, except the last molar, which is broader and thicker than the others. With this exception they appear to be equally simple and rudimentary.

The milk-teeth of *Erignathus barbatus*, as figured by Steenstrup, agree in number, relative size, and form very nearly with those of *Phoca grœnlandica*, as described and figured by the same author, with, however, one important discrepancy, namely, a large fourth, probably caducous, upper molar, many times larger and otherwise quite unlike the true milk-teeth. This has the appearance of being an abnormal or supernumerary

---

* Om Klapmydsens ufödte Unge og dens Melketandsæt. Af J. Reinhardt. Vid. Medd. f. d. Naturh. Forening i Kjobhavn, 1864 (1865), pp. 248–264.

† Yderligere Bemærkninger om Mælketandsættet hos Remmesælen (*Phoca barbata*). Ibid., pp. 269–274.

‡ Remarks on the Homologies and Notation of the Teeth of the Mammalia. Journ. Anat. and Phys., vol. iii, 1869, pp. 270, 271, fig. 4.

§ Or, rather, had not the roots been already absorbed? See remarks below under *Phoca fœtida*.

tooth,* intermediate in size between the permanent and decid-
uous molars. The true milk-teeth are I. $\frac{3-3}{2-2}$, C. $\frac{1-1}{1-1}$, M. $\frac{3-3}{3-3}$;
the upper are all much smaller than their representatives of the
lower series, the canine and incisors forming merely minute
cylindrical calcified points. The molars are also very small,
consisting of little rootless dentinal caps, the middle one the
larger, with indications of two roots. The lower molars are not
only several times larger, but have conical, pointed crowns,
and two long, distinct roots, especially the first and second.

Steenstrup gives the milk-dentition of both *Phoca grœnland-
ica* and *P. fœtida* as I. $\frac{3-3}{2-2}$, C. $\frac{1-1}{1-1}$, M. $\frac{3-3}{3-3}$. The lower molars
and the second and third upper are distinctly two-rooted; the
upper, however, are many times smaller than the lower, the
first upper consisting of merely a minute rootless crown.

Four fœtal specimens of *Phoca fœtida,* collected by Mr. Ludwig Kumlien on the late Howgate Polar Expedition, show the
upper, as well as the lower, molars to be all two-rooted, but
through the process of absorption the anterior fang of the upper molars early disappears, in one specimen the upper molars
consisting of minute crowns, with an oblique posterior fang.
The lower molars are several times larger than the corresponding teeth of the upper series, and are all distinctly two-rooted;
the third, or posterior, is much the larger, and is distinctly
tricuspid, there being a well-developed secondary cusp on each
side of the larger principal one. In the specimen in which the
fangs of the upper molars have become partly absorbed, the canines and incisors are wholly wanting. In the others the canines
are present; one has all the lower incisors, but only one of the
upper incisors. Most of the incisors, both above and below,
had, in the other examples, either been wholly absorbed, or were
lost in the preparation of the specimens. The upper canines are
directed horizontally forward, forming a large angle with the
permanent canines beneath them, and consist of small cylinders
two-tenths of an inch long and about two-one-hundredths in
thickness. The lower canines are somewhat larger, and are
directed obliquely upward and forward.

In *Halichœrus,* according to Nordmann, a fœtal specimen gave
the following formula: I. $\frac{3-3}{2-2}$, C. $\frac{1-1}{1-1}$, M. $\frac{3-3}{3-3}$. The milk-molars

---

* This is the view maintained by Reinhardt (Vid. Medd. f. d. Naturh.
Forening f. 1864 (p. 259), and to which Steenstrup (Ibid., pp. 269–274)
seems to substantially accede.

stand respectively over the second, third, and fourth of the permanent set, those of the lower series being also much larger than those of the upper.*

The species of the family *Phocidæ* agree, almost without exception, in possessing strong social instincts and in being almost unsurpassed in their affection for their young. Many of the species are gregarious, at least during the breeding season, while some associate at all seasons in large herds. They are, in general, patient and submissive creatures, and harmless to man, to whose power and love of gain doubtless not less than a million to a million and a half fall victims each year. The Crested Seal of the North Atlantic is one of the few species that will habitually resist an attack, or whose power is in any degree dangerous. As regards their reproduction, the female, as a rule, brings forth but a single young one, and the period of gestation is supposed to range from nine to nearly twelve months. The Sea Elephants are well known to visit, for the purpose of reproduction, particular breeding stations on land, assembling in large numbers at their favorite resorts, and, like the Otaries and Walruses, crawl up some distance on to sandy shores or rocky islands, to remain for weeks without food and without visiting the sea. Others, like the common Seal (*Phoca vitulina*), select outlying rocky islands or rocky points of the mainland for their breeding stations, and never congregate in large numbers. The Greenland Seal (*Phoca grœnlandica*) is at all times gregarious, assembling in immense numbers in particular districts to bring forth its young on the ice-floes. The Caspian Seal (*Phoca caspica*) possesses similar instincts, and is said to be always found in immense herds. While most of the species are confined to the neighborhood of shores or firm ice, others are almost pelagic, though rarely found far from floating ice.

Seals are very fond of basking in the sunshine, and spend a large part of their time on sand-bars, rocks, or on the ice, according to the season, the species, or the locality. They are very voracious, their food consisting chiefly of fishes, but in part of crustaceans and mollusks. Nearly all the species, and in fact all the Pinnipeds, are known to swallow small stones, often in con-

* This reference is based on Hensel's "Bericht über die Leistungen in der Naturgeschichte der Säugethiere während des Jahres 1864," in Arch. für Naturg., 1861, ii, pp. 99, 100.

siderable quantities, the purpose of which habit is still a matter of conjecture. Sailors, and even some intelligent naturalists, believe they serve as ballast, and some affirm that a larger quantity is swallowed when the animals are fat than when they are lean, and that when they are very fat they require them to give their bodies the proper specific gravity to enable them to remain easily under water. Whatever may be the cause, the strange fact rests on abundant and trustworthy evidence. Most species of Seals are strongly attracted by musical sounds, but whether their interest is merely that of curiosity or real fondness for such sounds may be fairly judged to be an open question. That they possess a great deal of curiosity admits of no doubt.

One of their most remarkable traits is the great length of time they are able to remain under water. Mr. R. Brown states that the average time is five to eight minutes, and that he never saw them remain below the surface for more than fifteen minutes, but other observers give from twenty minutes to half an hour. Various theories have been offered in explanation of this remarkable power in a warm-blooded, air-breathing animal, but none seems satisfactory. It has by some been supposed to be due to the large size of the venous system of circulation; by others to venous sinuses in the liver and surrounding parts, which serve as reservoirs for the venous blood; by others to the large size of the foramen ovale; while still others deem it to be wholly physiological and not structural. Some of the Arctic species have the habit of forming breathing-holes through the ice, through which they not only rise to breathe, but ascend to bask on the ice. These are circular, with smooth sides, and are kept open by constant use, and are believed to be made while the ice is forming. Other species keep near natural openings formed by the winds and currents and never construct breathing-holes.

Strange as it may seem, it is a well-established fact that the young Seals take to the water reluctantly and have to be actually taught to swim by their parents. The young of some species remain entirely on the ice for the first two or three weeks of their lives, or until they have shed their first or soft woolly coat of hair. Those that are brought forth on land, as in the case of the Elephant Seals, are, like the Otaries, timid of the water, swim at first awkwardly, and tire easily in their first efforts.

Seals utter a variety of cries, from which they have derived

such various names as Sea-dogs, Sea-calves, Sea-wolves, etc.
Some have a barking note, others a kind of tender bleat, or a
cry more or less resembling that of a child. The cry of the
young is usually more or less pathetic, while that of the adults
is heavier and hoarser. None appear to produce the loud barking or roaring so characteristic of most of the Sea-Lions and
Sea-Bears.

FOOD.

The food of Seals is known to consist largely of fish, but some
of the species are believed to subsist mainly upon mollusks and
crustaceans, particularly the latter. Malmgren states explicitly
that this is the food of the Rough and Bearded Seals, as he has
found by an examination of their stomachs. The Harbor and
Greenland Seals are supposed to subsist almost exclusively upon
fish, of which they destroy enormous quantities. Mr. Carroll
estimates that not less than three millions to four millions of
Seals annually congregate around the island of Newfoundland,
remaining there for a period of not less than one hundred and
twenty days. Allowing that each Seal consumes only one codfish a day, they would each destroy during this interval not less
than a quintal of fish, making in the aggregate some three millions to four millions of quintals of codfish killed by Seals during
about one-third of the year. Startling as this may seem, it is
unquestionably a low estimate. Indeed, the destruction of these
fish by Seals is believed to account, in part at least, for the "short
catch" of codfish at the various fishing stations around the
island. They, however, do not restrict themselves to codfish,
but doubtless vary their fare as circumstances may favor, they
being known to wage a furious warfare upon the white-fish.
As long as white-fish are " in with the land," in passing down
from the Labrador coast, " so sure will Seals of every description be there." Late in autumn the white-fish always pass
through the Straits of Belle Isle, followed by " all kinds of Seals
known to ice-hunters." *

Mr. Robert Brown states that all of the different species of
Seals "live on the same description of food, varying this at
different times of the year and according to the relative abundance or otherwise of that article in different portions of the
Arctic seas. The great staple of food, however, consists of
various species of Crustacea which swarm in the northern seas.

---

* Carroll, Seal and Herring Fisheries of Newfoundland, 1873, p. 18.

During the sealing-season in the Spitzbergen sea I have invariably taken out of their stomachs various species of *Gammarus* (*G. sabini*, Leach, *G. loricatus*, Sab., *G. pinguis*, Kr., *G. dentatus*, Kn., *G. mutatus*, Lilljeb., etc.), collectively known to the whalers under the name of 'Mountebank Shrimps,' deriving the name from their peculiar agility in the water. This 'seals' food' is found more plentiful in some latitudes than in others, but in all parts of the Greenland sea, from Iceland to Spitzbergen; I have seen the sea at some places literally swarming with them. Again, in the summer in Davis's Strait I have found in their stomach remains of whatever species of small Fish happened to be just then abundant on the coast, such as the *Mallotus arcticus*, *Salmo* (various species), etc. I have even known them to draw down small birds swimming on the surface; but their chief food is Crustacea and Fish. They also feed on Medusæ and Cuttlefish (Squids)."* That Seals vary their fare with an occasional gull or duck is attested by numerous observers; but birds form, of course, but an insignificant portion of their diet. Malmgren also refers especially to the occurrence in the stomachs of *Erignathus barbatus* and other Seals of various species of crustaceans and mollusks, and sometimes of fishes.†

ENEMIES.

Man is undoubtedly the Seal's chief enemy, but many fall a prey to the Polar Bear, and doubtless, also,—particularly in the case of the young—to sharks, and to that carnivorous Cetacean, the Orca. They are also greatly subject to the attacks of intestinal parasites. Many are also destroyed by the elements, thousands being sometimes ground to pieces by the ice. They are said to avoid rough water, but when amidst the ice-floes are frequently killed by the jamming together of the ice in a heavy sea. "At times during the spring, if there is a heavy sea, the Seals are sure to mount the ice, and whilst on it, provided it runs together, they are certain to be jammed", at which times many old Seals, as well as young, are destroyed.‡

MIGRATIONS OF SEALS.

The periodical movements of Seals have long been noticed, and it has been found that a proper semi-annual migration is

---

* Proc. Zoöl. Soc. London, 1868, p. 411; Man. Nat. Hist. Greenland, etc., Mammals, 1875, pp. 40, 41.

† Arch. für Naturg., 1864, pp. 75–84.

‡ Carroll, Seal and Herring Fisheries of Newfoundland, pp. 19, 20.

common to several of the species. Others, however, are sedentary. The common Harbor Seal (*Phoca vitulina*) is so strictly nonmigratory that wherever it occurs at all it is reported to be found at all seasons. The Rough or Hispid Seal (*Phoca fœtida*) is to only a small extent, if at all, migratory, and the same is true of the Bearded Seal (*Erignathus barbatus*). On the other hand, the Hooded Seal (*Cystophora cristata*) and the Harp or Greenland Seal (*Phoca grœnlandica*) move southward in winter and northward in summer. Most Arctic explorers have noted these movements, which in point of regularity have been compared to migrations of ducks and geese and other boreal waterfowl. Their passage along the Labrador coast and arrival in the Straits of Belle Isle, and other portions of the Newfoundland coast, have long been a matter of record, as well as the periodical departure and return of certain species on the west coast of Greenland. The Greenland Seal especially makes long journeys in spring to its favorite breeding-grounds, and later disperses to other haunts. Drs. Koldewey and Pansch, in referring to the assembling in spring of this species in the icy seas westward of Jan Mayen and Spitzbergen, observe as follows: "The appearance of these Seals reminded us that we were now in the neighborhood of the Seal-catchers, that is, in that part of the northern icy seas where, from the end of March to the end of April, the Seals come in thousands to the smooth floating ice to cast their young ones. These 'Seal-coasts' change their position somewhat every year, and range between 68° and 74° N. Lat., and from 2° to 16° W. Long. It is a highly interesting sight to see the Seals assembled from all quarters at this time. It is said that they not only come from the coasts of Spitzbergen and Greenland, but even swim in flocks from Nova Zembla." [*]

Mr. H. Y. Hind, in referring to their movements along the eastern coast of North America, says that in autumn, before the ice forms, "they 'hug' the shore, either of Labrador or Newfoundland, penetrating into all the bays and never going far from land. During the colder winter months they strike into the Gulf, looking for ice-floes, on which they give birth to their young in March, and [where they] continue for two or three months. In May and June they congregate near the coasts, and return to the main ocean for the summer." [†]

[*] German Arctic Exped., 1869–70, English ed., 1874, p. 61.
[†] Expl. in the Interior of the Labrador Peninsula, vol. ii, 1863, p. 202.

The general subject of the movement of Seals along our northern Atlantic coast having been presented quite fully by Mr. J. C. Steavenson, I herewith subjoin a transcript of his account, premising, however, that his remarks have a more limited application to Seals in general than the writer appears to have supposed, they doubtless having reference mainly to the two species already mentioned as being preëminently migratory.

"Independently of his constant motion in pursuit of his prey, the migrations of the Seal are most extensive. During the summer and autumn numbers of these creatures are met with, scattered in small parties, in all parts of the Northern Ocean visited by the whalers and other fishermen, where they remain until the severity of the Arctic winter warns them to retreat southward. Mariners who have been beset amongst the ice, or for other reasons have passed the winter in these hyperborean seas, remark that few Seals are met with during the winter, and some of them chronicle the time at which they first appeared on their return. Our information with regard to their general motions is not limited to these somewhat vague records. The habits of the genus (for it consists of many species) are so visible that we must conclude the scattered Seals met with during the dark winter of the Pole are only stragglers [in reality, the non-migratory species] left behind when the main body moved southward. As the severity of the weather increases it is evident that, like swallows, an instinctive move ment must commence, communicated to and understood by the whole family, like a masonic sign, prompting a general assembly of the clans at some long-frequented, well-known spot of their wide domain, where, it is to be supposed, they enjoy their sport until the gathering is completed. At length the frost commences, and the army is set in motion. This proceeding is keenly looked forward to and watched by the inhabitants of the coast, whose interest is much involved in their passing visit, and who fail not to levy tribute in kind. A fisherman, posted as sentinel on some headland commanding an extensive seaview, communicates to the hamlet the first indication of the approaching host, the vanguard of which invariably consists of small detachments of from half a dozen to a score of Seals; such parties continue to pass at intervals, gradually increasing in frequency and numbers during the first two or three days of the exodus, by the end of which time they are seen in companies of one or more hundreds. The main body is now at hand, and during the greater part of the next two days one continuous, uncountable crowd is constantly in sight. The whole procession coasts along at no great distance from the shore, presenting to an eye-witness a most extraordinary scene. In all quarters, as far as the eye can carry, nothing is visible but Seals—the sea seems paved with their heads. Some idea may be formed of the vast multitude when we consider the time occupied in passing, and the rate at which the animals are hurried along by the ceaseless, rapid stream which forms the highway of their long though expeditious voyage. The rear is brought up by small parties, such as formed the leading detachments. In one short week the whole host passes, consisting of many hundreds of thousands. The current of which these sagacious voyagers take advantage is the well-known polar current which proved so inimical to the success of our North-West Passage discoverers, and which sets through Hudson's Bay, and sweeps the coast of Labrador in a south-east direction; running at all seasons at the rate of several knots an hour, hurling with it, during the winter and spring, quantities of ponderous field-ice, together with numerous icebergs of various size, and frequently of most grotesque shapes. By it the Seals continue their passage steadily on in one unbroken course until the island of Belleisle presents an obstacle—situated in the entrance of the Straits of Belleisle, into which a branch of the current sets, carrying with it a portion of the force towards the Gulf of St. Lawrence. The main body continue onward until they reach the Gulf Stream, on the banks of Newfoundland. Here they arrive about the end of December or early in January, and halt for a time in the more still and warmer waters of that locality, resting until the time for bringing forth their young arrives; nor is the rest of long duration. About the end of January it becomes necessary to turn northward. During the southerly migration no ice encumbered the way—all circumstances were favorable; but now the new projected movement is undertaken under many impediments; the animals, heavy with young, must stem the strong current; the bed on which their snow-white cubs are to be laid is solid ice. Onward they struggle until they fall in with the immense continent of this material—one part of which is formed on the shores and a much larger portion hurried forward by the Polar Stream. This now covers the identical sea along which they so recently passed, and is to be their home until the duties of the nursery are performed, and their sleek progeny are strong enough to accompany the herd. The detachments which we left on their way up the Straits of Belleisle [have] met their own difficulties: the fishermen waylay them here most assiduously—net after net awaits the toiling emigrants, which are turned to good purpose. Several thousands are taken at the many stations planted on all parts of the shore from Cape Charles to the Gulf of St. Lawrence. In the Gulf many of them pass the winter and bring forth on the ice formed near the shores of this sea; a few of the young are taken by the inhabitants of the Magdalene and other islands; but a considerable section of the original stock circumnavigate Newfoundland, and join the great body on the banks. Those which winter in the Gulf of St. Lawrence quit their quarters in that sea about the end of June, and on their way down the Straits of Belleisle reward the watchful fishermen with a few additional thousands of their much-prized carcasses. These are now accompanied by their young, all but as round and bulky as their parents. After clearing the Straits little more is seen of them. It is believed that, in order to avoid the adverse current, they make their passage north to their old summer haunts at a much greater distance from the land." *

LOCOMOTION ON LAND.

As first pointed out by Dr. Murie,† the Pinnipeds present three distinct modes of terrestrial locomotion. The common or Earless Seals are usually described as progressing belly-wise, by a wriggling motion of the body, with the hind-limbs directed backward and held in opposition, and the fore-limbs drawn close to the body. This seems to have been hitherto commonly considered as the only mode of progression on land or ice possessed by any member of this group,‡ the Elephant Seal perhaps excepted. The plantigrade walk of the Otaries and the Walruses has of

---

* J. C. Steavenson, in the "Field" of November 28, 1863; quoted in (and here transcribed from) the "Zoölogist," vol. xxii, 1864, pp. 8873, 8874.

† Proc. Zoöl. Soc. London, 1870, p. 605, pl. xxxii.

‡ Bell, writing of the *Phocœ* in 1837 says, "Their movements on land are ludicrously awkward. They make no use of their feet in terrestrial progression, but throw themselves forward by plunges, the anterior part of the body and the posterior being alternately applied to the ground. In this way they make their way at a moderate pace along a tolerably even surface."— *Hist. Brit. Quad.*, p. 257.

late been repeatedly described, while their considerable power of locomotion on land has been known for a century. Dr. Murie, however, in 1870, described a mode of progression among the common Seals intermediate in character between that usually recognized as characteristic of these animals and that of the Otaries and Walruses. From observations made on a living Greenland Seal or "Saddle-back" (*Phoca grœnlandica*) in the Gardens of the London Zoölogical Society, Dr. Murie has described this "third sort of land-movement," and given figures of the various attitudes assumed by this Seal, particularly when moving on land. The Greenland Seal, he says, "very often uses its fore limbs, placing these on the ground in a semigrasping manner, and by an alternate use of them drags its body along. The hind legs meantime are either trailed behind slightly apart, or with opposed plantar surfaces slightly raised and shot stiffly behind. On uneven ground, or in attempting to climb, a peculiar lateral wriggling movement is made; and at such times, besides alternate palmar action, the body and the hind limbs describe a sinuous spiral track." He also states that he has seen the Crested Seal (*Cystophora cristata*) assume similar attitudes, and says that in the Harp Seal (*Phoca grœnlandica*), as well as the Crested Seal, "the fore legs and paws, and, to a very moderate extent, the hind limbs are freely brought into action," while in the common Seal (*Phoca vitulina*) the "limb-appendages on land are of slight subservience to progression, the fore paws only occasionally being used among rocks."[*]

Other observers report that other kinds of Earless Seals possess a considerable power of locomotion. Michael Carroll, in his account of the Harp Seal, says[†] that when Seals get "embayed" and cannot get into the water owing to the ice being jammed, "they begin to travel out in a direct line for the water. . . . Much depends upon the character of the ice they have to travel on, as to their rate of speed; they travel principally by night. I have killed them with the hair and skin worn off the fore flippers and bleeding. Were it not for the fore flippers they could not mount the ice or travel over it. All kinds of Seals known in Newfoundland travel to that degree so as to overheat themselves; then the fur or hair is loosened and the skin becomes almost valueless. In a cool night Seals will average about one mile per hour. Much depends on the character of the ice they have to

---

[*] Proc. Zoöl. Soc. London, 1870, p. 606.
[†] Seal and Herring Fisheries of Newfoundland, 1873, pp. 24, 25.

travel on; they travel by lifting themselves from off the ice on their fore legs or fore flippers and hitching their body after them with a kind of sidelong loping gallop. An old Seal when on level ice will outstrip a smart fellow in a distance of sixty yards, provided the Seal is ten or twelve feet ahead of him." This account, though couched in rather untechnical language, indicates a speed and manner of progression in the common Phocine Pinnipeds not as yet generally recognized, and certainly surprising from its rapidity.

Scoresby also long since observed that although the Seals "cannot be said to walk, as they do not raise their bodies off the ground; yet they shuffle along, especially over the ice, with surprising speed."*

Mr. H. W. Elliott has recorded a similar mode and rate of progression on land in the common Seal of Alaska (*Phoca vitulina*) that has not, to my knowledge, been noted by other observers. Says Mr. Elliott, "I desire also to correct a common error, made in comparing *Phocidæ* with *Otariidæ*, where it is stated that, in consequence of the peculiar structure of their limbs, in the former, their progression on land is '*mainly accomplished* by a wriggling, serpentine motion of the body, slightly assisted by the extremities.' This is not so; for, when excited to run or exert themselves to reach the water suddenly, they strike out quickly with *both fore feet*, simultaneously lift and drag the whole body, without any wriggling whatever, from 6 inches to a foot ahead and slightly from the earth, according to the violence of the effort and the character of the ground; the body then falls flat, and the fore-flippers are free for another similar action, and this is done so earnestly and rapidly that in attempting to head off a young nearhpah from the water I was obliged to leave a brisk walk and take to a dog-trot to do it. The hind feet are not used when exerted in rapid movement at all, and are dragged along in the wake of the body, perfectly limp. They do use their posterior parts, however, when leisurely climbing up and over rocks, or playing one with another, but it is always a weak effort, and clumsy. These remarks of mine, it should be borne in mind, apply only to the *Phoca vitulina*, that is found around these islands at all seasons of the year, but in very small numbers." Mr. Elliott adds that he thinks this mode of locomotion on land will be found to characterize all the species of the genus *Phoca.*†

* Account of the Arctic Regions, vol. i, p. 509.
† Condition of Affairs in Alaska, p. 122.

Mr. Lloyd relates several well-attested instances of Seals making long journeys on land, the most remarkable of which is the following: He says, "'During the winter of 1829,' so we read in *Jägare Förbundets Tidskrift* of 1832—and the truth of the story is certified by the signature of several most respectable individuals—'a young Gray Seal took to the land from the Skärgård, near the village of Grönö, and, striking into the forest in a southerly direction, passed, in its way, the hamlets of Sund and Wahlnäs, the church and iron-forge of Leufsta, and the hamlets of Elinge and Fählandbo. Near the last-named it met with a small river, then hard frozen over. This it followed for a while, but was unable to find an opening in the ice. It then took to the forest in a south-westerly direction to the Flo Lakes, in the parish of Tegelsmora, where it was also unsuccessful in obtaining access to water. Hence it proceeded south-east, crossing in its progress the Lake Wika, in the parish of Film, on the opposite side of which it again entered the forest, and finally entered the hamlet of Andersbo, situated about three (English) miles from Dannemora (the celebrated iron-mines), where it was overtaken by its pursuers and killed. The peregrinations of this Seal are believed to have occupied nearly a week, it being, as is imagined, without nourishment of any kind; and during which period it must have gone over at least thirty (English) miles of country. The ground, it should be remarked, was then covered with a foot and a half, or more, of newly-fallen snow, which, no doubt, very greatly facilitated the animal's movements.'"*

The Sea-Elephants (genus *Macrorhinus*) are well known to resort to the land for reproduction very much in the manner of the Otaries or Eared Seals, but I have met with no very clear statement of their manner of progression. Captain Scammon, in his history of the Sea-Elephant of the California coast (*M. angustirostris*, Gill) gives the following account of their power of movement on land: "When coming up out of the water they were generally first seen near the line of surf; then crawling up by degrees, frequently reclining as if to sleep; again, moving up or along the shore, appearing not content with their last resting-place. In this manner they would ascend the ravines, or 'low-downs,' half a mile or more, congregating by hundreds. They are not so active on land as the seals; but when excited to inordinate exertion, their motions are quick, the whole body

---

* The Game Birds and Wild Fowl of Sweden and Norway, 1867, p. 403.

quivering with their crawling, semivaulting gait, and the animal at such times manifesting great fatigue. Notwithstanding their unwieldiness, we have sometimes found them on broken and elevated ground, fifty or sixty feet above the sea."*

In describing the "Leopard" Seal of the California coast ("*Phoca pealei*? Gill "=*P. vitulina*), the same author says: "Its terrestrial movements, however, are quite different from those of the Sea Lion, having a quick, shuffling, or hobbling gait, only using its pectorals to draw itself along with, while a small portion of the animal's belly alternately rests upon the ground, the posterior part of the body, including the hind flippers, being turned a little upward. The head and neck are slightly elevated, also, when the animal is in its land-traveling attitude, but the creature is not so erect as, nor does it present the imposing appearance of, the Sea Lion, in its habits upon shore." † No direct statement is made as to the extent of its land journeys, but one is led to infer that it is not often seen far from the water.

From the foregoing it appears that the Phocine Seals generally have considerable power of movement upon land, though using only the fore limbs in terrestrial locomotion; and that not only the Sea-Elephants, but the common Seals of the North Atlantic are capable of moving quite freely when out of the water; and that their manner of progression at such times differs mainly from that of the other Pinnipeds in their using only the fore limbs, and in their not being able to raise the body fully from the ground or ice.

The Walruses and the Otaries, as mentioned in the account of these animals, not only use their hind limbs as a means of locomotion on land, turning them freely forward, but move the fore limbs alternately, actually "stepping" with them, as one writer terms it. while the hind limbs are carried forward simultaneously by arching the back and "hitching" them up beneath the body. In the Fur Seals, and in some of the smaller Sea Lions, the walk is not only plantigrade with all the feet, but the body is raised clear of the ground. The same is generally true also of the larger Sea Lions and the Walruses, but according to some writers the latter partly lose this power in old age, either from indolence, obesity, or decrepitude. The larger species appear to be simply less agile, both in the water and out,

---

* Marine Mammalia, p. 117.
† Ibid., p. 166.

than their smaller affines, and do not voluntarily retire so far inland.

SEAL-HUNTING.

The pursuit of Seals for their commercial products, forms, as is well known, a highly important branch of industry, giving employment for a considerable part of each year to hundreds of vessels and thousands of seamen, as well as to many of the inhabitants of the Seal-frequented coasts of Newfoundland, Greenland, and Northern Europe. Although these animals are destitute of the fine soft coat of under-fur that gives to the Fur Seals their great economic importance, their oil and skins render them a valuable booty. Seals have been hunted from time immemorial, but until within the last hundred years their pursuit was limited to the vicinity of such inhabited coasts as they were accustomed to frequent. For nearly a century, however, a greater or less number of vessels have been constantly employed in their capture on the ice-floes of the Arctic seas, or on the uninhabited coasts and islands of the far North. This industry, therefore, plays an important part in the history of the species here under consideration, and is, moreover, of such high commercial importance as to render a somewhat detailed account of the general subject indispensable in the present connection. As all the species hunted in the northern waters belong to the North American fauna, the consideration of the subject involves other hunting-grounds than those geographically connected with the North American continent. Although the principal sealing-grounds are in the North Atlantic and Arctic Oceans, there were formerly other important sealing-stations on the coast of Lower California and in the Antarctic waters. Here, however, the business was mainly limited (aside from the Fur-Seal fishery, already considered) to Sea-Elephant hunting, which has of late greatly declined in importance in consequence of the well-nigh practical extermination of the species hunted, through indiscriminate and injudicious over-hunting. Yet some notice of what Sea-Elephant hunting has been, as well as its present status, may not be out of place in the present general consideration of the subject.

SEAL-HUNTING DISTRICTS IN THE NORTH ATLANTIC AND ARCTIC WATERS.—The principal "Sealing-grounds" in the North Atlantic and Arctic Oceans are: (1) the West Greenland coasts; (2) Newfoundland, the coast of Labrador, and the islands and shores of the Gulf of Saint Lawrence, but especially the ice-floes to the eastward of these coasts; (3) the Spitzbergen and Jan Mayen seas; (4) Nova Zembla and the adjacent waters; (5) the White Sea. In addition to these districts (6) the Caspian Sea affords an important seal-fishery.

1. *West Greenland.*—Along the West Greenland coasts seal-hunting is mainly prosecuted by the natives of the country, and is their chief means of support. Dr. Rink states that the average annual catch amounts to about 89,000 Seals. Of these 2,000 to 3,000, belonging mainly to the larger species, are consumed as food; the remainder consist chiefly of Harp Seals (*Phoca grœnlandica*), but embrace many Ringed Seals (*Phoca fœtida*), and Harbor Seals (*Phoca vitulina*). Rather more than one-half of the skins taken are exported, while the rest are used by the inhabitants of Greenland. The Greenlanders, hunting chiefly with the harpoon and kayak, or the rifle, are confined in their operations to the immediate vicinity of the coast.*

2. *Newfoundland District.*—Many Seals are taken at the Magdalen and other islands at the mouth of the Gulf of Saint Lawrence as well as along the shores of Newfoundland, in nets or with the gun, but by far the greater part are captured on the floating ice to the eastward of Newfoundland, to which several hundred vessels annually repair at the proper season, and where alone the yearly catch aggregates about half a million Seals. This, indeed, is the sealing-ground *par excellence* of the world, twice as many Seals being taken here by the Newfoundland fleet alone as by the combined sealing-fleets of Great Britain, Germany, and Norway in the icy seas about Jan Mayen, or the so-called "Greenland Sea" of the whalemen and sealers.

According to Charlevoix (see beyond, p. 554) thousands of Seals were taken along the shores of the Gulf of Saint Lawrence as early as the beginning of the last century, but a high authority on the subject—Mr. Michael Carroll,† of Bonavista, Newfoundland—states that the seal-fishery was not regularly prosecuted, at least in vessels especially equipped for the purpose, prior to the year 1763. As early as 1787 the business had already begun to assume importance, during which year nearly five thousand Seals were taken. Twenty years later (1807)

---

* Danish Greenland, its People and its Products, pp. 129, 130.
† Seal and Herring Fisheries of Newfoundland, 1873, p. 7.

thirty vessels from Newfoundland alone were engaged in the prosecution of sealing voyages, and subsequently the number became greatly increased. In the year 1834 one hundred and twenty-five vessels, manned by three thousand men, sailed from the single port of St. John's; two hundred and eighteen vessels, with nearly five thousand men, from Conception Bay, and nineteen from Trinity Bay, besides many others from other ports, making in all not less than three hundred and seventy-five, with crews numbering in the aggregate about nine thousand men.* To these are to be added a considerable number from Nova Scotia (chiefly from Halifax), and the Magdalen Islands. In 1857 the Newfoundland sealing-fleet exceeded three hundred and seventy vessels, their "united crews numbering 13,600 men." The total catch of Seals for that year was five hundred thousand, valued at £425,000, provincial currency.† The business at this date seems to have attained its maximum so far as the number of men and vessels are concerned, the number of vessels subsequently employed falling to below two hundred, which has since still further decreased. Yet the number of Seals annually captured has not apparently diminished, the business being prosecuted in larger vessels, which secure larger catches. According to statistics furnished by Governor Hill, C. B., of Newfoundland, to the home government,‡ it appears that in 1871 the whole number of vessels employed in sealing was one hundred and forty-six sailing-vessels and fifteen steamers, manned by 8,850 men. The exports of Seal products for that year from Newfoundland were 6,943 tuns of oil, valued at $972,020, and 486,262 skins, valued at $486,262, the catch for the year being about 500,000 Seals, which were sold for the aggregate sum of $1,458,282. The single steamship "Commodore," of Harbour Grace, brought in 32,000 Seals, valued at £24,000 sterling. While the number of vessels employed in the Newfoundland Seal slaughter had at this time declined more than one-half, and the number of men engaged was one-third less, it appears that the annual catch was equal to that of average seasons twenty years earlier.

Prior to about 1866 the sealing-fleet consisted wholly of sailing-vessels, but since that date a small but steadily increasing

---

* Bonnycastle, Newfoundland in 1842, vol. i, p. 159.

† Carroll, Seal and Herring Fisheries of Newfoundland, p. 7.

‡ Papers relating to Her Majesty's Colonial Possessions. part ii, 1873, pp. 143, 145.

number of steamships have been added. In 1873, of the one hundred and seven sealing-vessels fitted out from the ports of Newfoundland, nearly one-fifth were steamers. Notwithstanding, however, this comparatively small number of vessels, the "catch" for that year is said to have been 526,000.*

The number of vessels sailing from other provincial ports is usually small in comparison with the number from Newfoundland, and they are generally of smaller size.†

3. *Jan Mayen or "Greenland" Seas.*—The icy seas about Jan Mayen are the sealing-grounds *par excellence* of the English and continental seal-hunters. According to Moritz Lindeman, the chief sealing-district is a circular area four hundred miles in diameter, the central point of which is the island of Jan Mayen, but it varies somewhat in different years in consequence of the unstable position of the ice-fields. The point where the greatest slaughter occurs is a small area on the meridian of Greenwich, between 72° and 73° N. lat., about two hundred miles northeast of Jan Mayen.‡ In this limited district are taken annually about 200,000 Seals.

Lindeman, in his history of the Arctic Fisheries,§ has sketched in considerable detail the general history of the whale- and seal-fisheries of the North Atlantic. From this exceedingly interesting and important memoir we learn that sealing was prosecuted as early as 1720 from the ports of the Weser, and that in the year 1760 nineteen sailing-vessels from Hamburg took 44,722 Seals; and that in 1790 the Hamburg fleet took 45,000. In 1850 twelve vessels returned with 48,800. Few statistics, however, can be gathered respecting the early history of the seal-fishery, those given relating generally to the amount and value of the cargo rather than to the number of Seals taken. The vessels were engaged principally in the whale-fishery, pursuing, however, either Seals, Whales, or Walruses, as opportunity favored. Lindeman gives a few detailed statistics relating to the

---

* Carroll, Seal and Herring Fisheries of Newfoundland, p. 7.

† In 1856, about five thousand Seals were taken by vessels from the Magdalen Islands; in 1867, about three thousand two hundred, and the following year only about eight hundred and fifty.—*Ann. Rep. of Depart. of Marine and Fisheries for* 1868.

‡ Peterm. Mittheil., Ergänzungsheft Nr. 26, 1869, Taf. 1, 2.

§ Die Arktische Fischerei der Deutschen Seestädte 1620-1868. In Vergleichender Darstellung von Moritz Lindeman, Ergänzungsheft Nr. 26, zu Petermann's Geographischen Mittheilungen, pp. vi, 118, mit zwei karten von A. Petermann, 1869.

capture of Seals in later years. The results of the seal-fishery for the year 1868, he states as follows:

Taken by 5 German ships.......................... 17, 000
Taken by 5 Danish ships 5, 000
Taken by 15 Norwegian ships 63, 750
Taken by 22 British ships 51, 000

making a total of 136,750, while 100,000 more were taken in Greenland. Great numbers are also killed about Nova Zembla by the Russians, whose sealing-fleet in 1865 is said to have numbered twenty-six vessels. According to Spörer, three hundred Seals were taken at Nova Zembla in three days in three nets.*

Lindeman gives a tabular statement of the number of vessels engaged in seal-hunting from the ports of Southern Norway for the five years ending with 1868, together with their tonnage, size of the crews, value of the ships, number of Seals taken by each, etc., from which I compile the following:

Year.	Number of ships.	Number of men.	Number of Seals taken.		Value in thalers.
			Young.	Old.	
1864........................	16	714	23, 364	24, 723	152, 000
1865........................	16	714	41, 758	18, 724	172, 000
1866	16	728	39, 576	8, 106	144, 000
1867........................	15	688	50, 931	23, 292	247, 000
1868........................	15	684	49, 533	14, 224	184, 284

From the foregoing it appears that the number of Seals taken by the same fleet in different years is exceedingly variable, ranging from about 48,000 in 1864 to upward of 83,000 in 1867, and that while in 1864 more old Seals were taken than young ones, there were taken in 1866 five times as many young ones as old ones. In 1870, according to Captain Jakob Melsom,† of Tönsberg, eighteen vessels (three of them screw-steamships) from Southern Norway captured 55,375 young Seals and 30,-390 old ones, or 85,765 in all. The greatest number brought in by a single vessel was 9,400. This year and the year 1867 are said by Captain Melsom to be the best years the Norwegian Seal-hunters had experienced up to that date.

The British sealing-vessels are mainly from Dundee. According to statistics given by Mr. Robert Brown the number from

---

* Petermann's Mittheil., Nr. 21, 1867.
† Petermann's Geogr. Mittheil. 17 Band, 1871, p. 340.

this port varies greatly in different years, being only four in
1865, and twelve in 1868, but notwithstanding the fact that the
number of vessels was three times greater in 1868 than in 1865,
the catch was less than one-third as large as in 1865.   Mr.
Brown gives* the following statistics relating to the Dundee
sealers :

Year.	Number of vessels.	Number of Seals taken.
1865...................	4	63,000
1866...................	7	58,000
1867...................	11	56,000
1868...................	12	16,670
1869...................	11	45,600
1870...................	9	90,450
1871†..................	6	62,000

In 1874, the Dundee sealers are said to have taken 46,252
Seals;‡ in 1875, 45,295 Seals, valued at £27,026; and in 1876,
53,776 Seals, valued at £34,332.§
The Spitzbergen sealing-fleet from British ports, says Mr.
Brown, "meets about the end of February in Bressa Sound off
Lerwick, in Zetland; it leaves for the north about the first week
in March, and generally arrives at the ice in the early part of
that month.   The vessels then begin to make observations for
the purpose of finding the locus of the Seals, and this they do
by crawling along the edge of the ice, and occasionally penetrating as far as possible between 70° and 73° N. lat.; then
continue sailing about until they find them, which they generally
do about the first week of April.   If they do not get access to
them, they remain until early in May, when, if they intend to
pursue the whaling in the Spitzbergen sea that summer, they
go north to about 74° N. lat. to the 'old sealing,' or further still
(even to 81° N.) to the whaling.   Most of them, however, if not
successful by the middle of April, leave for home to complete
their supplies in order to be off by the first of May, to the
Davis's Strait Whale fishery.   .   .   .
"The number of Seals taken yearly by the British and Continental ships (principally Norse, Dutch, and German) in the

---

* Man. Nat. Hist., Geol., etc., of Greenland, 1875, Mam., p. 68, footnote.
† "Up to the 11th of April," and hence including only part of the catch of
that year.
‡ Geograph. Mag., vol. i, 1874, p. 386.
§ Baird's Ann. Rec. Sci. and Indust., 1876, p. 389.

Greenland sea, when they get among them, will average upwards of 200,000, the great bulk of which are young 'Saddle-
backs', or, in the language of the sealer, 'white-coats'".*

According to Lindeman (l. c., p. 81) the seal-hunters leave
the ports of the Weser and the Elbe about the end of February,
or, at latest, by the beginning of March, and reach the hunting-
grounds about the third week of March, sailing to the northwestward between the Shetland Islands and Norway, and
thence through the "funnel" ("Trechter") of the "Spanish Sea",
and eastward to Jan Mayen, varying their course according to
the winds and the ice, reaching this rocky island in from eight
days to four weeks, according to the favorableness of the season. The steamships delay their voyages so as to reach the sealing-grounds at the same time as the sailing-vessels.

About the 18th of March, as a rule, the "bay-ice" ("pan-ice"
of the English) begins to form, at about which time the ships
reach this latitude. The bay-ice first takes the form of small
round flakes, of the size of a tea-plate, varying in thickness
from an inch to a foot. In it the ships find protection from
storms, it considerably lessening the force of the sea, and serving to keep down the waves. Where the bay-ice has formed the
surface of the sea looks as if oil had been poured over it. The
flakes increase in size, and if sharp cold ensues they become
united the following day into masses six feet broad. The next
two or three weeks are devoted to seal-hunting and seal-killing.
This is the time when the males and females are seeking their
food of small fishes, mollusks, and crustaceans. Frequently no
bay-ice forms, and then the Seals must be sought on the hard
polar-ice. About the 22d to the 24th of March, the Seals resort
to the ice and the females bring forth their young. Later they
seek by preference for this purpose the somewhat firmer bay-ice.
At the time of the birth of the young the males are found with
the females, and sometimes two males to one female. As a rule
the female has but one young, which, if she be not disturbed,
she suckles for about seventeen to eighteen days. The young
develop with extraordinary rapidity, and after three to four
weeks are fat enough to yield a good booty. The whelping
time continues till about the 5th of April; four to five days
later the males leave the "school" ("Stapel") or "shoal," and
depart in a northeasterly direction. The females still remain

---

* Proc. Zoöl. Soc. Lond. 1868, pp. 438, 439; Man. Nat. Hist., etc., Greenland, Mam., p. 67.

for a short time with the young; then they also go off in the direction taken by the males. The young, left to their fate, still remain some days without nourishment, and then also take to the water. If the weather be somewhat favorable, and especially if no snow is falling, immediately follows the "Enterfall", as the destruction of the Seals upon the ice by the seal-killers (Robbenschlägers *) is called.†

The great destruction of Seals in the icy seas about Jan Mayen for many years prior to 1870 began to show its effects so strongly at this time as to raise grave fears for the results. Attention was strongly drawn to the matter in 1871 by Captain Jakob Melsom, of Tönsberg, in an article published that year in Petermann's "Geographische Mittheilungen," entitled "Der Seehundsfang im nordlichen Eismeere." In this memoir he discussed at length the immediate causes that led to the depletion, and suggested a remedy for it which has since been adopted, namely, an international agreement for a "close-season" for the Seals. He traces the decrease in great part to the introduction of steamships into the sealing-fleets, and the too early arrival of vessels at the sealing-grounds. Captain Melsom's paper throws so much light upon the general subject that I deem it of interest to give a translation of the more especially important portions of his memoir.

Says Captain Melsom: "There is good evidence that steam-power is detrimental to the sealing business. Respecting this point I will mention only the fact that steamships have it in their power to reach, nearly every year, the breeding-grounds of the Seals so early that the young are scarcely born before the mothers are killed. The young are then worthless; the real capital, if I may so speak,—the old Seals,—is imprudently expended, and the profits are entirely lost. If in this way great numbers of old female Seals are destroyed without being replaced by a proportionate number of young ones, every one can see what must be the result.‡

"It may be asked why the English, who are still our teachers in this field, have introduced steamships. This I will allow myself to answer. Until the year 1847 the competitors of the

---

* The Germans very appropriately term the butchery of the young Seals upon the ice seal-slaughter ("Robben-schläg"), and the butchers seal-slaughterers ("Robbenschlägers").

† Petermann's Geogr. Mitth., Ergänzungs heft Nr. 26, 1869, p. 81.

‡ "It is well known that the Seal brings forth young only once a year, and only one at a time."

English in seal-hunting in the Arctic Seas were some Danish and German vessels, which certainly were rarely an impediment to these masters of the sea; but now came also the Norwegians, led by Herrn Svend Foyn; and however unskillful they may have been at first,* it was not long before they began to prove troublesome to their old teachers, and as the Norwegians some years later began to make use of the rifle, shooting the full grown male Seal, they by this means—thanks to our good marksmen—were frequently more successful than the English themselves; then it occurred to the latter that by the aid of artificial power they could triumph over the poor Norwegians who had only natural forces at their command; and soon floated colossi with powerful steam-engines and dingy sails upon the waves of the icy seas, terrifying alike the sailing-vessels and the Seals.

"The English have yet another reason besides that already given for the introduction of steam-power, namely, that their voyages to Davis's Strait for Whales absolutely require the use of steam, this enabling them very easily to join in the catch in the northern ice-seas; the ships, returning home at the close of the seal-catching season, discharging their cargoes, etc., then proceeding on the voyage to Davis's Strait, where they arrive in time to engage in whaling; in this way they find employment for their steamships the greater part of the year.

"It is my opinion that the above-mentioned reasons have led the English to employ steam in seal-hunting; they surely saw the hurtfulness of it, but to the stronger belongs the lion's share as long as there is anything to have. Meanwhile the Germans, as well as the Norwegians, grew tired of competing with sailing-vessels against the English steamships, and therefore these nations built steamships in order to obtain an equal footing in this hitherto unequal struggle.

"The English are not at all blind to the fact that their golden time in the Arctic Seas is apparently over, for they well know that now in Norway, as well as in Germany, steamships are built for seal-hunting. I had evidence of this last winter in a conference with one of the before-mentioned [in an earlier part of his paper] English steamship captains; it was not less in-

---

* "I have heard that, as when Foyn for the first time participated in the hunt, the men carried the young Seals on board alive on their backs; later they were conveyed on sledges drawn by four men,—all of which naturally gave great satisfaction to the practical Englishmen."

teresting that it at the same time indicated the business feeling of the English.

"He urged me to try to unite the Norwegian and German shipowners in an agreement that in the future they would not take the young Seals before the 1st of April, till which time the females should remain undisturbed; he would then, on his part, endeavor to bring about a similar agreement on the part of the English shipowners. In passing, I will remark that this idea is a sound one, and that if the close-time should be extended till the 4th, or better still, till the 6th of April, it would be especially favorable for the catch, and it will surely be necessary, sooner or later, to unite upon a preserve-law; but the strange part of the matter is that the English never drew the rein so long as they were the only ones who employed steamships, but only when they feared that they should lose their supremacy.

"As for the rest, it is still a precarious matter how to arrive at the agreement proposed by the Englishman. The Norwegians must first be granted equal privileges with the English in the use of steam-power; I doubt not that the English would then have not less reason to fear this agreement. Perhaps it will result in bringing about the advantage of a close-time, as already proposed by the Englishman.

"There appeared some time since in Dammen's Journal an account of the construction of two steamships built for use in seal-hunting, and it expresses great hope for the continuance of the same. The ice-sea fleet from here (Tönsberg) has considerably increased this year, and next year will be still further augmented. So far as I have thus far been able to determine, the ice-sea fleet from Southern Norway for the year 1872 will consist of eighteen sailing-vessels and eight steamships, besides one sailing-vessel and two steamships owned by foreigners, which ought to be equipped and manned here. I must call this for our small share in a single field a great expansion; indeed, so magnificent that it is time to cry Halt!

"The last four fortunate years have unquestionably aroused so strongly the spirit of speculation that the less fortunate years which preceded them are meanwhile wholly overlooked or quite forgotten. If I mistake not, it is the same with the seal hunting as with the herring fishery—it is periodical.

"The catch of young Seals has considerably diminished during late years; scarcely a doubt prevails as to that. If in spite of this the Norwegians in the last four consecutive years have obtained such favorable results, this good fortune is in part due to the fact that a considerable number of old Seals—owing to the peculiar situation of the ice—have sought refuge on the shores of the neighboring islands, where our accomplished marksmen with their improved rifles have killed great numbers. But to kill in this way year after year the old Seals means ultimately the destruction of the business; and as a consequence of the increasing number of voyages, and the introduction of steam-power into the Norwegian sealing vessels, the hunt has become now more than ever a war of extermination instead of a judicious hunt. But the ice-sea voyages were during a series of years a great blessing to our country and especially to the poorer class of our people; should we not then exert ourselves and do all in our power to preserve the same? Should our ships go continually in the old track, visiting always the usual hunting places, and there continue the already long-waged war of extermination? Is it not high time that we went to the assistance of our less favorably situated countrymen of Northern Norway with one or two of our well-equipped and excellently adapted steam-ships, to aid them in their praiseworthy effort to discover new hunting-grounds, not only in the Kara Sea and on the shores of Nova Zembla, but also eastward of White Island (at the mouth of the Obi) along the coast of Siberia? The rarely traversed stretch between Spitzbergen and Nova Zembla, to the north-ward, should especially be explored with great perseverance, where there is good hope that unknown land and good hunting places may be found, of which wealth one can now scarcely have a presentiment.

"In order to prevent the loss of our ice-sea voyages and with them the capital that now is or may yet be invested in ice-sea ships, I venture to call the attention of my countrymen to the experiment here proposed, as I likewise pray for a hearing in reference to the before-mentioned proposition for a close-time for the Seals, not only on the part of the Norwegians, but also of the Germans and English.

"I have read with great interest Mr. Carl Petersen's communication 'On our Hunting-fields in the Arctic Sea,' ['Über unser Fangsfeld auf dem Eismeer'], and another in the 'Finmarks-post.' It appears from these articles that we are indebted to Consul Finckenhagen, of Hammerfest, for the extension of the hunting-field to Nova Zembla. His vessel, commanded by Captain Carlsen, of Tromsö, took a new course and made a very fortunate trip, since which other hunting expeditions have been made, both from Tromsö and Hammerfest. These enterprising people may not only rejoice in the discovery of new hunting-grounds, but in the accumulation of a considerable amount of scientific information, as Professor Mohn, Director of the Meteorological Institute, of Christiania, can attest.*

"Ought Southern Norway, which furnishes much more capital than our friends in the North, and has in its proud Arctic fleet several steamships, look passively upon the praiseworthy efforts of our less favorably situated countrymen?

"The Swedish Government again equipped last year two ships of war for an expedition into the Arctic waters—naturally for exclusively scientific purposes; for the Swedes, who have no interest in the northern ice-seas, yet sacrifice large sums for the honor of their country and the advancement of knowledge. We can with relatively less expense not only bring honor to our own country, preserve our reputation for seamanship, and serve science, but can indulge the highest hopes of finding rich fields for the employment of our costly Arctic fleet, which perhaps in the old, it may be too quickly despoiled, hunting-fields will be without especially remunerative business."†

From the foregoing it appears that the decrease of the Seals in the Jan Mayen seas, and the improvident slaughter that was yearly waged, had already begun to attract serious attention, as a threatening evil of no small magnitude. Captain Melsom's strong protest seems not to have been without salutary effect. Subsequently the proposition for a close-time for Seals received more and more attention each year, not only in Norway but in Germany and England, with finally the happy result of an international agreement on the part of these countries favorable to the preservation of the Seals.

As regards the general history of the subject it may be of interest to transcribe in the present connection a communication from Mr. Lovenskiold to Mr. Colam, Secretary of the British "Society for the Prevention of Cruelty to Animals", as published by Mr. Frank Buckland in "Land and Water," in the issue of August 28, 1875. This communication is of special interest, not only from its containing a report of a series of resolutions adopted by the shipowners of Southern Norway respecting a close-time, but also much statistical information

---

* Geogr. Mitth., 1870, pp. 194 et seq. ; 1871, Heft i, pp. 35 et seq., Heft iii, pp. 97 et seq., Heft vi, pp. 230 et seq.

† Petermann's Geogr. Mitth., 17 Band, 1871, pp. 341–343.

concerning the sealing business as carried on from the single port of Tönsberg, and especially as relating to the decline of the trade since 1871, foreshadowed in Captain Melsom's above-quoted paper.

Says Mr. Lovenskiold : " I am sorry that I am not very familiar myself with the proceedings in Norway as to seal-fishing, and the protection of that trade. But as I have been honoured by an invitation to attend this meeting of the committee, I should think it the best way of giving the committee some knowledge of those proceedings to give you the substance of some resolutions, carried in a meeting of shipowners the 17th of February of this year [1875]. That meeting was kept at Tourberg [Tönsberg], a town in the southeastern part of Norway, the centre of the seal trade, and of the shipping concerning that trade. The contents of those resolutions is this :—

" 1. A term should be fixed, within which the fishing of Seals in the Arctic Seas should be prohibited.

" 2. The 1st of April would be a suitable term for the beginning of the seal-fishing. The term of the beginning ought not to be fixed later than the 3d of April. The fishing should not be continued after the 5th of June.

" 3. If the protection of the seals shall be of any use, it needs to be enforced by an international treaty, valid for all nations, engaged in shipping and fishing (all sea-faring nations). If such an international treaty can be concluded, we dissuade any protection.

" 4. Any violation by any ship, of the laws concerning the protection of the seals, should be punished with the confiscation of the whole cargo caught by that ship.

" Those resolutions were carried, the three first ones by all votes but two, and the fourth one by all votes but three.

" As it might be of interest for the committee to know the importance of the seal trade, carried on from Norway, I shall give you the following numbers in round figures :

" In the fifteen years, 1860–74, the tonnage of the ships employed in the seal trade from the southern part of Norway, by far the most important, has increased from 5,000 tons in 1860, to 9,000 in 1874. The value of the same ships has increased from £77,500 in 1860 to £175,500 in 1874. In those fifteen years the ships engaged in the seal trade caught together 624,000 young seals, and 376,000 old ones, or at an average per year of 41,600 young seals, and 25,000 old ones. The aggregate value of the seals caught in those fifteen years was £650,000, or at an average per year of £43,350, and the aggregate net gain (surplus) of the trade for the shipowners was £62,200, or at an average per year of £7,150.

" In the last three years from the same part of Norway were employed in the seal trade:—

" 1872. Ten steamers and sixteen sailing-ships, with an aggregate tonnage of about 7,000 tons, and manned with 1,200 sailors.

" 1873. Sixteen steamers and eleven sailing-ships, with an aggregate tonnage of about 8,500, manned with 1,500 sailors.

" 1874. Sixteen steamers and nineteen sailing-ships, with an aggregate tonnage of about 9,000 tons, and manned with 1,600 sailors.

" In these three years all the ships together caught 142,500 young seals, and 128,000 old ones, or at an average per year of 47,500 young seals, and 42,700 old ones.

" The value.of the seals caught in three years, 1872–74, was £16,000, or at an average per year of £60,300 [*sic*]. The steamers engaged in the trade gave in those three years a net gain (surplus) for the shipowners of £9,500; the sailing-ships a loss of £27,500; steamers and sailing-vessels together a loss of £18,000.

"I give you the numbers pursuant to a report in a Norwegian newspaper (*Morgenbladt*) of the 26th April this year. If you compare these numbers concerning the fifteen years (1860–74) and the last three years (1872–74), you will see that the catching of the old seals has considerably increased, but that the catching of the young seals, although the tonnage and the value of the ships employed in the trade has been almost doubled, has only been maintained at the same rate, and that the small surplus of former years has been changed into a considerable loss in the last three years."*

These statistics show the actual occurrence of what Melsom four years before clearly showed must happen under the then prevalent system of indiscriminate slaughter. With the rapid increase of the sealing-fleet was a corresponding decline in the profits of the trade, which soon changed to a considerable annual loss. The relatively small catch of young Seals, and the disproportionate increase in the number of old Seals killed show plainly the state of the business, and expose clearly the ruinous way in which it was prosecuted. As Melsom figuratively ex-

---

* Land and Water (newspaper), August 28, 1875, p. 160.

pressed it, the real capital was being ruinously expended, and the profits had wholly disappeared.

Steps were now, however, promptly taken to remedy the evil, but apparently years must elapse, even with the stringent exercise of every precaution for the due preservation of the Seals, before the golden harvests of the previous decade can again be realized.

In June, 1875, as appears by the British "Law Report" for that year, the British government passed a statute known as the "Seal Fishery Act, 1875,"* which prohibits any British subject from killing or capturing, or attempting to kill or capture, any Seal within the area included between the parallels of 67° and 75° north latitude, and the Greenwich meridians of 5° east longitude and 17° west longitude, under a penalty not exceeding £500 for each offense, one-half to go to the prosecutor. In this act the term "Seal" is defined as meaning "Harp or Saddleback Seal, the Bladder-nosed or Hooded Seal, ·the Ground or Bearded Seal, and the Floe Rat, and includes any animal of the Seal kind which may be specified" by an "Order in Council" under the act. The beginning of the close-time was left to be determined by the Order in Council. The act was made operative by the Queen in Council the 5th of February, 1876, the 3d day of April of every year being fixed as the time when Seals may first be captured.

A few months later a similar act was passed by the Norwegian government, fixing the limits of the protected district by the same boundaries, and prescribing the same date for the beginning of the hunt. The penalty for the violation of the act is 200 to 10,000 crowns. The law received the King's signature the 28th of October, 1876, and was to become operative on the 25th of the following April, upon condition that the other interested governments agree to join with Norway in the enforcement of similar protective legislation.†

---

* The following is a transcript of a portion of the act in question: " . . . The master or person in charge of or any person belonging to any British ship, or any British subject, shall not kill or capture, or attempt to kill or capture any seal within the area mentioned in the schedule to this act, or the part of the area specified in the order, before such day in any year as may be fixed by the order, and the master or person in charge of a British ship shall not permit such ship to be employed in such killing or capturing, or permit any person belonging to such ship to act in breach of this section."

† See Acts of 1876, Nos. 13 and 32. For copies of these acts I am indebted to Professor S. F. Baird, Secretary of the Smithsonian Institution.

I have been unable to determine whether similar legislative action has been taken by Germany, but in all probability such is the case, rendering the protective act a thoroughly international one so far as the three governments chiefly interested are concerned.

4. *Nova Zembla and the Kara Sea.*—Captain Melsom, as above indicated (*anteà*, p. 506) believed, in 1871, that the Kara Sea and the shores of Nova Zembla offered new and profitable hunting-grounds for the Norwegian sealing-fleet, and relates that already several successful voyages had been made to Nova Zembla from Tromsö and Hammerfest. According to Schultz, the Russians, between the years 1830 and 1840, brought "rich cargoes of salmon or trout, of seals and walruses" from Nova Zembla, but he states that later "the product of the fisheries and of the chase diminished; the animals left their usual places of abode and removed to others less accessible. The fishermen consequently ceased going to Novaya-Zemlya, so that in 1850 and 1860 only five vessels sailed for that group of islands.

"The northern island of Novaya-Zemlya is most frequented by fishermen, while those who have strong and well-equipped vessels venture as far north as Matoschkine. The arrangements are made so as to arrive toward the end of June at Novaya-Zemlya, where the fishermen commence their work by hunting the seals and the walrus, and afterward devote themselves to fishing for the common trout, the variety called *Salmo alpinus*, which the Russians call 'golets.'"*

5. *White Sea.*—Many Seals are taken in the White Sea, where they have been hunted by the inhabitants of the neighboring coasts since many hundred years. Now, as formerly, seal-hunting is here mainly prosecuted by the Russians. According to Schultz, the species chiefly hunted is the *Phoca grœnlandica*, which is killed on the ice. The hunt is carried on principally along the eastern shore (which is called the "Winter Coast"), and "in the bays of the Dwina and Mezene, and on the coast of Kanine." During summer, or from May till September, these animals repair to the more remote Arctic Seas, but later make their appearance in the gulfs and bays of the Arctic Coast. They pair on the ice in the White Sea about the beginning of February, especially in the Gulf of the Dwina, at which time the females give

---

* Account of the Fisheries and Seal-hunting in the White Sea, the Arctic Ocean, and the Caspian Sea. Rep. U. S. Com. of Fish and Fisheries, pt. iii, for 1873–74 and 1874–75, originally published (in French) at St. Petersburg in 1873.

birth to their young. The hunt on the "Winter Coast" begins at this time, and continues till the end of March. The chase extends over a district two hundred and thirty miles in length; here numerous huntsmen assemble "from the districts of Archangel, Pinega, and Mezene. The principal place of meeting, and at which generally two thousand huntsmen assemble, is called Kedy, and is located twelve 'versts' (about seven miles) from Cape Voronov. The huntsmen have built at this place about one hundred huts, where there is constant excitement from February till the end of March, while during the rest of the year these huts are deserted.

"About the middle of March, the young phocæ are large enough to leave the ice and swim toward the open sea, whither the old ones do not follow them. They assemble in the Gulf of Mezene, where they rest on the ice and pair. The pieces of ice in the gulf are sheltered from the wind, and are not carried about by the waves, although they melt a little, especially during the rainy periods.

"Numerous societies of huntsmen assemble in the beginning of April at the mouth of the river Kouloï, in order to follow for several weeks the chase of the phocæ on the ice. They use sailing-vessels 22 feet long, with an iron-plated bottom. Every vessel is manned by several huntsmen, is completely equipped, and furnished with provisions and fuel. The huntsmen all leave the shore at the same time; and having reached the floating ice, they draw their vessels on the ice, and there establish a vast encampment. The younger and more active huntsmen are sent out to reconnoiter. Provided with snow-shoes, they hasten in all directions to search for the phocæ. As soon as they observe a flock, they advise the other huntsmen of the fact, and then all run towards the spot, drawing their boats after them. Having arrived within gunshot distance, the most expert are placed in the front rank and commence the chase; for every shot must kill, and not merely wound, lest the cries of the wounded phocæ frighten the whole flock and make them speed away. The animals which are killed are then placed in the boats, and the huntsmen return to the shore—sometimes on the ice, sometimes in the open sea—to deposit there the result of the chase, and bring new provisions to the comrades who had been left there.

"The huntsmen usually receive from their master, provisions and clothing for the whole season, and must give him in return half, or even two-thirds of all the animals which have been killed.

The more hardened and expert a huntsman is, the larger is his share. Every society of twenty huntsmen elects a 'starosta,' (the old one,) whose duty it is to guard the coast and prepare the food, without receiving for this a larger share than the other huntsmen.

"On the western coast of the White Sea, (called the Terski coast,) the phocæ-chase is not as productive as on the eastern coast, because the pieces of ice, driven toward the north, float along the shore. Scarcely more than 15,000 'pouds' (540,000 pounds) of phocæ are caught there every year.

"In these latitudes, the principal meeting-place of the huntsmen is sixteen 'versts' (about nine miles) north of the river Ponoi, and is called Deviataya. Huts are built here, and about five hundred huntsmen assemble, who form themselves into societies. Every society is composed of a master and three huntsmen. While one of the members of the society remains on shore with his sleigh and his reindeer, the other three venture on the pieces of ice to discover the phocæ, which are sleeping there. Every huntsman wears over his clothes a short cloak of reindeer skin, called 'sovik,' and has on his feet large boots lined with fur. At the end of a long strap passed over his shoulder he draws a small boat, weighing 20 kilograms. A game-bag with provisions is attached to his belt. His gun on his shoulder, and having in his hand a long stick, with an iron point, he rapidly and skillfully advances, by means of his snow-shoes, over the vast fields of snow and ice. He acts as guide, and his two comrades follow him in single file, drawing their boat after them. When they have arrived at an expanse of water where phocæ are swimming, two of the huntsmen fire, while the third pushes the boat into the water in order to take up the dead animals, which he hoists into the boat by means of a boat-hook.

"The chase commences early in the morning, and the huntsmen do not return to their hut till evening; a flag hoisted on the shore indicating to them its position."

6. *Caspian Sea.*—Strange as it may seem, the Caspian Sea,—an inland brackish lake, with no natural communication with the oceans, and quite far removed from any other considerable body of water—is the seat of a sealing industry only second in importance to that of the so-called "Greenland" or Jan Mayen Seas. Many more Seals are taken here than Dr. Rink reports the annual catch to be in Greenland, and the number is rather

more than half as large as that taken by the combined sealing fleet of Western Europe in the icy seas about Jan Mayen. The number of skins annually taken during the six years ending with 1872 is given by Schultz as follows: in 1867, 131,723; in 1868, 150,947; in 1869, 128,701; in 1870, 137,030; in 1871, 90,468; in 1872, 156,759, or a yearly average of about 130,000. The species here hunted is the *Phoca caspica* of Nilsson, which by some authors has been regarded merely as a synonym of *Phoca vitulina*. Its habits, however, indicate a species quite distinct. It is said to be from three to six feet long, and to weigh from 72 to 144 pounds. They gather in large herds on the shore, where thousands are sometimes killed in a single hunt.

In 1873 M. Schultz, in his report upon fishing and seal-hunting in the Caspian Sea, described in considerable detail the habits and haunts of the Caspian Seal, and the methods employed for its capture. He says, "The seals love the cold; and, in summer, they seek the deep sea, leaving it in the autumn for their favorite place of abode, the northeastern basin of the Caspian Sea, which is the portion first covered with ice, and where the ice breaks up latest. Numerous herds of seals gather on pieces of floating ice, to rest or to pair. The pairing season lasts from the end of December till January 10. The female every year gives birth to one young, seldom to two. The young have a shining white, silky fur, but after ten days it becomes coarse and turns gray. Then the tender solicitude of the mother ceases, for the little one has to go into the water and swim. Seals that are one year old have gray fur, speckled with black spots.

"The seal is hunted down the western coast of the Caspian Sea, at the mouths of the Volga and the Ural, and in its southern part, especially on the islands of the Gulf of Apchéron.

"The principal meeting-places of seal-hunters are on the seven islands situated north of the Peninsula of Mangyshlak, called the 'Seals' Islands,' on account of the large number of these animals found there. Other islands also abound in phocæ. Thus there have been years when about 40,000 seals were killed on the island of Peshnoï, before the mouths of the Ural, and, in 1846, 1,300 were killed in one night.

"The seals are hunted in three different ways: they are killed with clubs on the islands where they gather; or they are shot with guns; or they are caught in nets. The first-mentioned way is the grandest, and yields the best results.

"The great meeting-place of the huntsmen is Koulali, the largest of the Seal islands, having a length of thirty-five 'versts,' (about twenty miles,) and a breadth of three 'versts,' (about one and two-thirds miles). The hunters, who winter there every year, have built wooden houses, huts, and sheds on this island. The fishing authorities at Astrachan send every year one of their officers to Koulali to superintend the chase and the hunters, where he remains from October till the middle of May. On account of the bustle and noise, the seals have deserted this island for a number of years, and selected for their place of gathering the islands of Sviatoï and Podgornoï.

"In the spring and autumn the seals seek the shore to rest in the sun, one herd arriving after the other. Scarcely has the first settled when a second comes yelling and showing their teeth to drive it away, followed soon by a third, to which it in turn has to lose its place; so that the last herd arriving always drives the first farther back on the coast. The invasion terminates by the arrival of some isolated stragglers.

"Now is the time for the hunters to commence the chase. They carefully observe in what place, and, approximately, in what numbers, the seals have gathered; and then elect as their chief the most experienced and skillful among them. They approach the rookery in boats, either at dusk or during the night, always going against the wind, to conceal their approach.

" After their arrival on shore, the hunters disembark noiselessly, form a line in order to cut off the retreat of the seals, and thus, creeping, advance quite near to the herd, which is sleeping and suspects no danger. On a signal from the chief, the hunters all rise at once and pitilessly attack their unfortunate victims, killing them by a single blow on the snout with the club. The bodies are piled up by means of gaffs, and after a few minutes form a rampart, depriving the survivors of every chance of regaining the sea. The seals howl, groan, bite, and defend themselves, but the hunters, eager for gain, go on killing them without mercy, and soon the whole herd is massacred. It is no infrequent occurrence to see 15,000 dead seals cover the battle-field of a single night.

"After the killing, the dressing of the seals commences, usually about daybreak. The head is cut off, the belly is opened, and the skin is taken off with the thick layer of fat adhering to it. These skins are piled up on the boats, which take them to large sailing vessels, anchored some 'versts' from the shore, on which they are heaped up, each layer being covered with salt. These vessels sail with their cargo to Astrachan, while the hunters return to the coast to carefully clean the battle-field. They bury the bodies and entrails, at some distance, deep in the ground, or throw them into the sea, far from the shore, and carefully obliterate every trace of blood, so that, when another herd of seals arrives, these animals do not see any marks of the slaughter which has taken place; for experience has shown that they never select for their rookery a place from which every trace of the slaughter has not been carefully removed.

"Two hundred seal-hunters, employed by wealthy merchants or fishermen, usually winter on the island of Koulali. Numerous boats, besides, go there every year to participate in the chase. The masters of these boats secure permits* from the fishing authorities and give them to their workmen, who receive their wages in money. . . .

"Another way of hunting the seals is to take them in nets. Immense nets are stretched out, into which the hunters endeavor to chase them by yelling and making a noise. This way of hunting is chiefly employed in the maritime district of the Ural Cossacks and in the Gulf of Sinéyé Mortso, from October till the sea is covered with ice. The nets, called 'okhani,' are 6 'sagenes' (42 feet) deep, and have meshes of seven and a half inches.

"The following is the manner of proceeding: Forty boats join together and elect a chief and an assistant chief. Then the boats sail out to sea with a fair wind, or use their oars, going in a line, thus forming a sort of chain. In every boat there are three nets. The chief, followed by twenty boats, is on the lookout for a herd of seals, which he endeavors to cut off, while his assistant remains with the other half of the fleet at some distance from the shore. When the chief thinks that the time for action has come, he gives the signal by throwing into the sea a bale, to which a flag is fastened. At this signal the boats simultaneously cast their nets, which are all tied together so as to form a wall of meshes, by which the seals are soon completely surrounded. Then the hunters begin to yell and to strike the water with their oars, in order to frighten them. These seek to avoid the danger by plunging, but they rush

* The Russian government derives an average annual income of about $700 from the sale of permits for seal-hunting in the Caspian Sea.

against the barrier of nets, and are caught in the meshes, so that they can be killed without difficulty. This way of hunting is prohibited in those parts of the sea where it injures the fishing or obstructs the first manner of hunting. The chase on the ice is fraught with many dangers, and is, therefore, at present prohibited. The hunters, sitting on little sledges drawn by strong and hardy horses, and provided with food, continue on for several weeks to shoot old seals, and kill young ones while they still have their white and silk-like fur. These hunters brave all dangers; and it has sometimes happened that the south or southwest wind, having detached large masses of ice from the shore, has driven them out into the open sea, where they have floated in all directions, with the adventurous huntsmen on them. These unfortunate hunters usually perish from cold and hunger on these masses of ice, or find their death in the waves."*

7. *North Pacific.*—In the North Pacific the capture of seals for commercial purposes is mainly restricted to the pursuit of the Sea-Elephant (*Macrorhinus angustirostris*), on the coasts of Western Mexico and Lower California. This, although at one time a business of no small importance, was nearly abandoned many years since, and for the best of reasons, namely, the well-nigh complete extinction of the species in consequence of indiscriminate and reckless slaughter. As the history of the subject falls more naturally into the account of that species (to be given later), little further need be said respecting it in the present connection. Many Seals are, of course, annually killed by the natives of the Alaskan, Kamtschatkan, and other coasts of the North Pacific, but I am not aware that sealing is there carried on anywhere, either by the natives or foreign sealing-vessels, to any noteworthy extent.

8. *Antarctic Seas.*—In the southern hemisphere no Seals occur that are the strict representatives of the Greenland and other Seals which, in the northern hemisphere, afford the seal-hunter so lucrative a booty. In the colder south temperate and Antarctic seas is found their commercial representative in that mammoth of the Seal tribe, the Sea-Elephant, or so-called Elephant Seal (*Macrorhinus leoninus*). Here Elephant Seal hunting was for a time prosecuted with great vigor, especially during the early part of the present century. The species was hunted almost exclusively for its oil, and so easily were the animals taken, and

* Rep. U. S. Com. Fish and Fisheries, part iii, 1873–74 and 1874–75, pp. 93, 95.

so indiscriminately and injudiciously were they slaughtered, that in comparatively a few years it became practically exterminated along all those coasts and islands that afforded safe harbors for the vessels engaged in this exceedingly profitable enterprise. At one time abounding on many of the islands off the southern portion of the South American continent, on both the Atlantic and Pacific sides, along favorable stretches of the Patagonian coast, in Terra del Fuego, the Falkland, South Shetland, South Georgian, and other neighboring islands, as well as at the Crozet's, Kerguelen, and Heard's Islands, they are said to be now found in numbers only on the more inaccessible portion of the last-named group of islands. It is difficult, indeed impossible, to give even approximate statistics respecting the numbers of the animals killed, or the amount of Elephant Seal oil obtained. For many years several ships annually obtained partial or complete cargoes from the various localities already mentioned, new stations being sought when the old ones had become exhausted, but the vessels engaged in Sea-Elephant hunting were mostly also engaged in Fur Seal hunting and in whaling, and generally no separate reports of the products of each being given, the statistics of the business are consequently not easily obtainable.

Respecting the history and the present status of Elephant Seal hunting at Kerguelen Land and Heard's Island I quote the following from Dr. J. H. Kidder's recent report on the natural history of Kerguelen Island: "In former years the Kerguelen group of islands was noted as a favorite breeding-place for the sea-elephant (*Macrorhinus leoninus* L.). On this account it has been much frequented by sealers for the last forty years, and resorted to also by whalers as a wintering-place, on account of the great security of Three Island Harbor. The sea-elephants have been so recklessly killed off year after year, no precautions having been taken to secure the preservation of the species, that now they have become very rare. Only a single small schooner, the Roswell King, was working the island during our visit, two others and a bark working Heard's Island, some three hundred miles to the south, where the elephants are still found in considerable numbers. Probably they would long since have abandoned the Kerguelen Islands altogether but for a single inaccessible stretch of coast, Bonfire Beach; where they still 'haul up' every spring (October and November) and breed in considerable numbers. The beach is limited at each end by precipitous cliffs, across which it is quite impossible to transport oil in casks, nor can boats land from the sea, or vessels lie at anchor in the offing, from the fact that the beach is on the west, or windward coast, and exposed to the full violence of the wind. . . .

"The increasing scarcity of the sea-elephant, and consequent uncertainty in hunting it, together with the diminished demand for the oil since the introduction of coal-oil into general use, have caused a great falling off in the business of elephant-hunting. The Crozet Islands, for example, had not been 'worked' for five years, and at Kerguelen there was only one small schooner engaged in this pursuit, two others making Three Island Harbor their headquarters, but spending the 'season' at Heard's Island, three hundred miles to the southward. It may, therefore, be reasonably hoped that these singular animals, but lately far on the way toward extinction, will have an opportunity to increase again in numbers, and that sealers may learn from past experience to carry on their hunting operations with more judgment, sparing breeding females and very young cubs. When the Monongahela visited the Crozet Islands on December 1, they found the sea-elephant very numerous, although left undisturbed for only five seasons."*

At the Falkland Islands, where at the beginning of the present century the Sea-Elephants occurred in great troops, they long since became virtually exterminated, as has been the case at most of the early sealing-grounds. In this work of destruction American vessels have taken a prominent part, and for many years have "maintained a monopoly of the business," most of them sailing from New London, Ct.

As an interesting reminiscence of the palmy days of Sea-Elephant hunting, and as conveying a vivid picture of the scenes and incidents of the business, I quote the following from a recent account by Mr. Charles Lanman, based on the unpublished journals and log-books of some of the chief participants. Says Mr. Lanman:—"But it is of Heard's Island that we desire especially to speak in this paper. It is about eighteen miles long and perhaps six or seven wide; and by right of discovery is an American possession. For many years the merchants of New London cherished the belief that there was land somewhere

---

* Contributions to the Natural History of Kerguelen Island, made in connection with the United States Transit-of-Venus Expedition, 1874-75. Bull. U. S. Nat. Mus., No. 3, pp. 39, 40, 1876.

south of Kerguelen's Island, for in no other way could their captains account for the continuous supply of the sea-elephant on its shores. As long ago as 1849 Captain Thomas Long, then of the Charles Carroll, reported to the owners of his ship that he had seen land from the masthead, while sailing south of Kerguelen's Land; but Captain Head has received the credit of the discovery, although he did not land upon the island. The man who first did this was Captain E. Darwin Rogers. He was on a cruise after sperm whale; his ship was the Corinthian, and he had three tenders ; and his employers were Perkins and Smith —the same Smith heretofore mentioned. Captain Rogers commemorated his success by an onslaught upon the sea-elephants, which he found very numerous on the shore ; and after securing four hundred barrels of oil, improved the first opportunity to inform his employers of what he had done, urging them not only to keep the information secret, but to dispatch another vessel to the newly discovered island. . . . . . The firm purchased a ship at once, Captain Smith took command, and sailed for Heard's Island. With Captain Darwin Rogers as his right-hand man he fully explored the island, named all its headlands and bays and other prominent features, made a map of it, and succeeded in filling all his vessels with oil. Two exploits which he performed with the assistance of his several crews, are worth mentioning:—At one point, which he called the Seal Rookery, they slaughtered five hundred of these animals, and as was afterward found, thereby exterminated the race in that locality ; and they performed the marvelous labor of rolling three thousand barrels of elephant oil a distance of three miles, across a neck of the island, from one hore to another, where their vessel was anchored. The ship which he himself commanded returned in safety to New London with a cargo of oil valued at $130,000, one-half of which was his own property."

Continuing the account he says :—"The number of these animals which annually resort to Heard's Island, coming from unknown regions, is truly immense. In former times the men who hunted them invariably spared all the cubs they met with, but in these latter days the young and old are slaughtered indiscriminately. We can give no figures as to the total yield of elephant oil in this particular locality, but we know that the men who follow the business lead a most fatiguing and wild life, and well deserve the largest profits they can make. While Kerguelen's Land is the place where the ships of the elephant hunters spend the summer months, which season is literally the 'winter of their discontent,' it is upon Heard's Island that the mammoth game is chiefly, if not exclusively, found.   Then it is that the gang of men have the hardihood to build themselves rude cabins upon the island, and there spend the entire winter. Among those who first exiled themselves to this land of fogs and snow and stormy winds, was one Captain Henry Rogers, then serving as first mate, and from his journal, which he kept during this period, we may obtain a realizing sense of the loneliness and hardships of the life to which Americans, for the love of gain, willingly subject themselves in the far-off Indian Ocean.

   " Having taken a glance [in previous portions of the paper not here quoted] at the leading men who identified themselves with the Desolation Islands, and also at the physical peculiarities of those islands, we propose to conclude this paper with a running account of Captain Henry Rogers' adventures during his winter on Heard's Island.

   "He left New London in the brig Zoe, Captain Jas. Rogers, master, Oct. 26, 1856, and arrived at their place of destination February 13, 1857.   For about five weeks after their arrival the crew was kept very busy in rafting to the brig several hundred barrels of oil, which had already been prepared and left over by the crew of a sister vessel, and on the 22d of March, the wintering gang, with Capt. Henry Rogers as their chief leader, proceeded to move their plunder to the shore, and when that work was completed, the brig sailed for the Cape of Good Hope. The gang consisted of twenty-five men, and after building their house, which was merely a square excavation in the ground, covered with boards and made air tight with moss and snow, they proceeded to business.   Those who were expert with the lance did most of the killing; the coopers hammered away at their barrels; and, as occasions demanded, all hands participated in skinning the huge sea-elephants, or cutting off the blubber in pieces of about fifteen pounds each, and then on their back or on rude sledges, transporting it to the trying works, where it was turned into the precious oil.   Not a day was permitted to pass without 'bringing to bay' a little game, and the number of elephants killed ranged from three to as high a figure as forty.   According to the record, if one day out of thirty happened to be bright and pleasant, the men were thankful; for the regularity with which rain followed snow, and the fogs were blown about by high winds, was monotonous beyond conception. . . . Month after month passed away, and there is no cessation in the labors of the elephant hunters. Mist and snow and slaughter, the packing of oil, hard bread and sad beef, fatigue and heavy slumbers—these are the burthen of their song of life."*

METHODS OF CAPTURE, ETC.

The methods employed in capturing Seals vary according to circumstances of locality and other contingencies. In the foregoing pages some of the ways and devices used have been given incidentally at some length in connection with the accounts of certain important sealing districts, but the general subject of Seal capturing claims further and more methodical treatment. Although no very rigid classification of thé methods employed will be attempted, a convenient division may be made under the two heads of "Shore Sealing" and "Ice-Hunting", in accordance with whether the Seals are taken on or near the land, or upon the ice-floes of the high seas. While the former may be carried on partly in boats, the latter requires vessels especially equipped for protracted voyages.

I. SHORE-HUNTING.—The capture of Seals on or near the land is accomplished in various ways, as by the use of nets, the rifle, the sealing-club, the lance, the harpoon, etc. While the use of nets is necessarily restricted to the shore, the club, the rifle, and the lance are the common implements of destruction used also on the ice-floes. The methods employed in shore-hunting vary also with the species pursued and with the season of the year. The capture of Seals by use of the harpoon is mainly practised by the Esquimaux and other northern tribes, and may be termed—

1. *The Greenland method.*—In Greenland about one-sixth of the catch is taken in nets, and the remainder with the harpoon and gun. The Esquimaux method of capturing Seals with the harpoon, has been often described by arctic voyagers, the following account of which, as still practised by the Greenlanders, is here transcribed from Dr. Rink's late work on "Danish Greenland". He says, "The art of catching seals by the harpoon and bladder is still pursued in Greenland exactly in the same way as before Europeans had settled there, without the least

---

* *Forest and Stream* (newspaper), vol. xi, pp. 437, 438, Jan. 2, 1879. A very good account of Sea-Elephant hunting may also be found in Captain Scammon's "Marine Mammals of the North-western Coast of North America," pp. 119–123.

change or improvement; and though some other means of capture have been added, viz, the rifle and the twine-made net, there is some reason to believe that the abolition of the ancient manner of hunting seals would prove fatal to the welfare, if not the existence, of the present race of inhabitants. Still more indispensable to them is the kayak or skin canoe, fitted out especially for this pursuit. It measures upwards of 18 feet in length and about 2 feet in breadth, and weighs about 55 pounds, so that the man on landing can take it in one hand and carry it along with him up the beach. . . . When the kayaker intends to strike a seal with his harpoon, he advances within a distance of about 25 feet from it, then throws the harpoon by means of a piece of wood adapted to support the harpoon while he takes aim with it, and called the 'thrower'. At the same time he loosens the bladder and throws it off likewise. The animal struck dives, carrying away the coiled-up line with great speed; if in this moment the line happens to become entangled with some part of the kayak, or if the bladder is not discharged quick enough, the kayaker in most cases will be capsized without any chance of saving his life by rising again. But if the operation has been entirely successful, the bladder moving on the surface of the water indicates the track of the animal underneath it, and the hunter follows it with the large lance which he throws like the harpoon when the seal appears above the water, repeating the same several times, the lance always disengaging itself and floating on the surface. Finally, when the convulsions of the animal are subsiding, he rows close up to it and kills it with the small hand-spear or knife." *

2. *By means of nets.*—The capture of Seals in nets is mainly limited to the periodical visits of the migratory species to the shore, and occurs chiefly during spring and fall. At some points on the northern shores of Europe, and particularly in the Gulf of Bothnia, the Caspian Sea, and Lake Baikal, sealing nets have been in use for centuries, and are set either from the shore or beneath the ice. Cneiff, in his account of Seal-hunting in East Bothnia,† published originally in 1757, describes,

---

* Danish Greenland, its People and its Products, 1877, pp. 113, 114.

† Bericht vom Seekälberfange in Ostbothnien. Vom Provincialschaffner, Herrn Johann David Kneiff, eingegeben. Der Königl. Schwedischen Akad. der Wissensch. Abhandl., aus der Naturlehre, Haushaltungskunst und Mechanik, auf das Jahr 1757. Aus dem Schwedischen übersetzt, von Abraham Gotthelf Kästner, etc. 19 Band, 1759. pp. 171–186. This is a detailed and very interesting account of Seal-hunting as practised in the Gulf of Bothnia about the middle of the last century.

among other methods, the use of the net. The net, he says, is from sixty to ninety feet in length, and twelve meshes, or about six feet, deep. It is made of linen yarn, spun from good hemp, as coarse as strong sail-yarn, except the lower meshes, which are made from poor hemp, so that when a Seal enters the net and begins to press against it, the lower meshes, if entangled among stones, will easily tear out, and the Seal, feeling the net yield before him, will not turn about and go out. The net is provided with floats of charred fir-wood, pointed at both ends, flattened beneath and rounded above, and about a foot long. These are placed about a foot apart along the upper edge of the net, to which they are firmly bound.* The nets are set in the autumn, from Bartholomew's day till the ice closes, and are used for the capture of what he calls the Bay Sea-Calf, which is doubtless the common Harbor Seal (*Phoca vitulina*). In setting the nets two are commonly fastened together, and are placed near rocks to which the Seals are known to resort. One end of the net is usually fastened by a small cord to a large stone, which is placed on the Seal Rock, the other end of the net being kept in place by an anchor formed of large stones sunk in the water. When the Seal enters the net, thinking to scramble upon the rocks, he immediately thrusts his head through some of the meshes; when he finds the net hanging loosely in the water, he winds himself about in it, believing he is still free, but in turning about to go back he finds his head again through another mesh at the other end of the net, whereupon he thus draws the net around him, and so becomes completely wound up in it, and is held a prisoner by the anchor-stone and line till morning, when he is killed. The nets are placed only where the Seal Rocks lie to the leeward of the land, off rocky points or islands. If the rocks lie to the northward, the nets are laid when the wind is south, and not when it is north, for the Seal seeks the sheltered side where the sea is smooth.

Mr. Lloyd, in his "Game Birds and Wild Fowl of Sweden and Norway" (pp. 420–424), who gives an illustration of the "Stånd-Nät" from Rosted (l. c., p. 420), states that the way of setting the net varies, the net being sometimes placed near a "Skäl-Sten" (Seal Rock), and at other times "across a narrow strait, leading to a bay or inlet of the sea that is resorted to by

---

* A shorter and somewhat earlier description of the Seal net is given by Linné, in his history of his journey through Oeland and Gothland, in 1741. See the German translation, Halle, 1764, p. 221.

seals. Generally," continues Mr. Lloyd, " its innermost end is secured by means of a stout rope to a heavy stone, or to sea-weed, on the 'Skäl-Sten' itself, whilst its outermost end has no other fastening than a small stone of just sufficient weight to keep it in its place, that is, sunk in the deep water beyond. At other times the reverse is the case. The inner end of the net is attached to the 'Skäl-Sten' by a mere thread, whilst its outer extremity is secured to the bottom by a heavy stone. In either case the inner or outer end of the net is left in a measure free, so that when the seal strikes it, the meshes on all sides may more readily collapse about the animal, and the more violently it struggles the more inextricably will it be fixed in the toils.

"The 'Stånd-Nät' is usually set in the evening, and taken up again at a pretty early hour on the following day. If placed near a 'Skäl-Sten' it should be to leeward, because the seal usually mounts the stone on the weather side at night, and in the morning takes to the water in the opposite direction. The chances, therefore, are that in making its plunge into the sea, more especially if its movements be quickened by a blank shot, which is often fired for the purpose, it will be made captive.

" It occasionally happens that the seal is taken in the net of an evening when about to mount the 'Skäl-Sten,' as prior to so doing it is in the frequent habit of making several circuits round the stone for the purpose of ascertaining if all be safe, and should it not observe the net, it runs its head into one or other of the meshes.

" The ' Stånd-Nät,' it should be observed, ought not to be set unless the weather be fine, for if the wind and waves beat on the rock, seals will not take up their night quarters there. To lure these animals into the net, various expedients are resorted to. Bright lights, as is known, greatly excite their curiosity. A fire is therefore made on the shore, or on a rock, in rear of the ' Skäl-Sten,' which has the effect of attracting them to the spot; and as a further inducement, their olfactory nerves are tickled by the fumes of bones and other strong-scented sub-stances, which are cast into the flames. At other times *Kutar*, or seal-cubs, are tied to a line within the net, the cries of which often attract old ones. . . .

" The ' Stånd-Nät ' would appear to be a very destructive engine. We read of as many as fourteen seals having been taken at a single 'haul.' It is chiefly the young ones, however, that are made prisoners. The old ones, let the night be dark as pitch, would seem by scent or otherwise to discern the toils; and even should they get entangled in the meshes, their strength is such, especially in the case of the Grey Seal, that it must be a very strong net to retain them within its folds. Ödman tells us, indeed, that they at times carry away the net altogether. 'A man of my acquaintance,' he goes on to say, 'related to me that he once captured an old seal with portions of ten different nets attached to its body, which was, however, finally secured by the eleventh. On flaying the animal, a part of one of the nets was found to have grown into the skin, and a considerable portion of the others were in a state of decomposition.' When within the folds of the net, the struggles of the seal are most violent, and as it constantly endeavors to 'go ahead', never to retrace its course, it soon becomes so entangled that the captor has difficulty in disengaging it. What with the animal's great exertions, however, in its endeavors to escape, and the want of air, it soon becomes exhausted, and when taken out of the water is often found quite dead."

On the coast of Newfoundland larger nets than those above described are used. Mr. Carroll says: "A seal-net is usually fifty fathoms long and seventeen feet deep. The twine they are made of is about three times the size of salmon-net twine; it will require sixty pounds of such twine to make a seal-net. The net is made on an eight and a half inch card." Each net requires twenty pounds of good cork cut into oval pieces, pointed at each end, seven inches long and two and a half inches wide at the widest part. These are placed one fathom apart on the head rope. The net, with all its attachments, will weigh about two hundred pounds.*

The manner of using these nets, or "seal-frames," is thus described by Mr. Reeks: "Three long nets of strong seal twine are required to construct a frame. One net is firmly secured by anchors parallel with the shore, and at such a distance that the remaining nets, placed one at each end, will just reach the shore, thus forming a kind of oblong figure, the longest net being on the outside. If in the spring, when the Seals migrate from the westward, the net nearest that point is sunk to the bottom; but if in the fall, when the Seals migrate in the reverse direction,—the shores of the island running nearly N. E. and S. W.,—the eastern net is sunk. Two men are required to constantly watch the nets. As soon as a herd of Seals has been

* Seal and Herring Fisheries of Newfoundland, p. 35.

seen to cross the sunken net the top of it is immediately raised to the surface of the water by means of a pulley, and so fastened in that position; the men then commence shouting and firing off guns loaded only with powder, to keep the Seals under water and cause them to 'mesh' in the nets; otherwise they would spring over the nets and escape. When it is seen that no Seals rise to the surface the men launch their boats into the pound and take the Seals from the nets, most of them being drowned, while the others have to be killed.

"As soon as the Seals are got on shore the net is again sunk, and the men, or others employed for the purpose, occupy themselves 'pelting,' or skinning, the Seals until another herd is impounded. In a successful season as many as eighteen hundred Seals have been captured in one of these frames."[*]

In the Caspian Sea the nets, instead of being anchored to the shore, are suspended from boats at a considerable distance from land, as has already been fully described in the account of the Caspian Sea Seal-hunting quoted from Schultz (*anteà*, p. 516). Lloyd also states that in Norway, in winter, when the 'sea is frozen over, the seal-nets are set under the ice. "Small circular holes at stated intervals are first cut in the ice, and afterwards the hauling lines attached to the net are passed, by means of long and forked poles, from the one aperture to the other."[†]

A similar use of nets in Seal-catching prevailed in Lake Baikal a century and a quarter ago. Bell, writing in 1762, describes the process as follows: "The seals are generally caught in winter, by strong nets hung under the ice. The method they use is, to cut many holes in the ice, at certain distances from one another, so that the fishermen can, with long poles, stretch their nets from one hole to another, and thus continue them to any distance. The seals not being able to bear long confinement under the ice, for want of air, seek these holes for relief, and thus entangle themselves in the nets. These creatures, indeed commonly make many holes for themselves, at the setting in of the frost."[‡]

According to Dybowski, nets are still employed for the capture of Seals beneath the ice in Lake Baikal, but apparently in a somewhat different manner. He states that strong nets,

---

[*] Zoölogist, 2d Ser., vol. vi, p. 2542.

[†] The Game Birds and Wild Fowl of Sweden and Norway, p. 423.

[‡] John Bell, Travels from Saint Petersburg in Russia to various parts of Asia, vol. i (Edinburg ed. of 1788), p. 320. (An earlier Glasgow edition was published in 1763.)

made of horse-hair, are inserted through the Seals' breathing-holes in the ice, and that in these nets the Seals, in attempting to reach the surface, become imprisoned.*

Mr. Lloyd describes and figures another kind of net used in the capture of Seals, which he calls the "Ligg-Nät". His description of the "Ligg-Nät", borrowed, as is his figure, from Linné,† is as follows:—"It is attached to two wooden frames, one at each end, which are secured to the bottom of a 'Skäl-Sten.' To the upper bar of the innermost of the frames is fastened a long line reaching to the shore. When one pulls at this line, the net is brought to the surface, but when the line is slackened, it sinks to the bottom. The net, whilst there, is altogether unseen, and the seal, unsuspicious of danger, creeps up, therefore, on to the 'Skäl-Sten'. When the peasant sees that it is asleep, he pulls gently at the line, which brings the net to the surface, and surrounds the stone in the manner of a quadrangular fence. The animal, on awakening from its slumber, casts itself headlong into the water, but cannot extricate itself from the toils before the man, with his harpoon or other implement of destruction, reaches the spot and puts an end to its existence."‡

3. *The Seal-Box.*—Mr. Lloyd also describes another ingenious device for the capture of Seals, used in Norway and Sweden. He says it is called the "*Skäl-Kista*", or Seal-box. "In principle it is the same as the so-called *Watten-Giller*, the expedient commonly adopted to catch rats and mice, viz., a 'balance-board,' placed across a tub of water. It is constructed of logs, and square in form, as seen in the above diagram [referring to a figure of the "Skäl-Kista"], and is sunk in the water up to the letter Y [about the basal third being submerged]. Large stones are afterwards heaped up around and about it, especially at both ends, so as to make it resemble a 'Skäl-Sten' as much as possible. The trap-door T consists of an oblong flat stone, or of plank ends, and swings on an iron bar, the extremities of which rest on the side-walls of the 'Skäl-Kista' itself. To prevent the trap-door T from falling too low there is a spring or stop, so that on the pressure ceasing it at once resumes its horizontal position. This device, as will be readily understood, is covered with sea-

---

* Arch. für Anat. u. Phys., 1873, p. 124.

† Reisen durch Oeland und Gothland, etc. Aus dem schwedischen übersetzt. Halle, 1764, pp. 203, 204, pl. 1, fig. 6.

‡ Game Birds and Wild Fowl, etc., pp. 424, 425.

weed, and when, therefore, the seal, tired of contending with the waves, seeks in all innocence to rest its wearied limbs on what it takes to be a rock, the trap-door swings on its axle, and the yawning gulf beneath presently receives the poor animal; and as the aperture through which it falls is at once closed again, the trap is in readiness to receive others of its comrades who may allow themselves to be similarly beguiled."*

4. *The Seal-Hook.*—In certain parts of the Norwegian coast, and probably elsewhere in Scandinavia, the writer last quoted tells us that Seals are captured by means of barbed hooks, and he depicts the manner of their use. The hooks, he says, quoting from Rosted, should be made of tough iron or steel of at least the thickness of one's finger, with shanks some eighteen inches in length. These are fastened by a half-hitch to a strong horse-hair or hempen line, which is stretched completely around the base of a "Skäl-Sten" or Seal Rock, to which its ends are firmly attached. The hooks are set at low water, and in moderate weather, for in stormy weather the Seals do not usually repair to the rock. At half-ebb of the following tide the rock should be reconnoitered with a telescope. "If any of these animals are then observed to be lying on it, a blank shot (when the boat has approached sufficiently near) should be discharged, which will at once arouse them from their slumbers, and cause them to plunge headlong into the sea, in their progress to which one or more of the company are commonly 'brought up by the run'; for though, when ascending the 'Skäl- Sten,' they are not in the slightest degree impeded by the hooks, which point upwards, and are, moreover, slightly covered with sea-weed, yet in their passage to the water they can hardly pass them unscathed."†

5. *The "Skräckta".*—Mr. Lloyd also describes and figures another ingenious implement adopted in Scandinavia for the destruction of Seals. This consists of a harpoon enclosed in a tube. The tube is made of thick sheet-iron, two feet long and two and a half inches in diameter, with two fixed heads, one at the lower and the other near the upper end. At the bottom is fixed a strong spiral spring, which propels the harpoon, and at the upper end is a projecting trigger, pressure against which serves to discharge the harpoon. Several of these destructive implements are inserted, by the aid of an auger, in a "Skäl-Sten"

---

* Game Birds and Wild Fowl, etc., pp. 425, 426.
† Ibid., pp. 426, 427.

known to be the resort of Seals, and after being set are lightly covered with sea-weed. When the Seal, in creeping up on the rock, comes in contact with the trigger the harpoon is released and becomes lodged in the body of the unlucky animal.*

6. *Other methods.*—In shore-hunting the rifle is often resorted to when other means are unavailing, or the Seals cannot be approached sufficiently near to be dispatched in other ways. The use of the harpoon and bladder, as employed by the Greenlander, has already been described; but the harpoon is also used in other ways, not only by the Esquimaux, but by the inhabitants of Northern Europe, especially, in former times, in Scandinavia. It is employed mainly in winter, when the hunter, usually attired in white, steals upon the Seal while asleep on the ice, or lies in wait for it at its breathing-hole, striking it when it comes to the surface to breathe. The Seal, when struck with the harpoon, is allowed to descend beneath the ice, being held by the line attached to the harpoon. The Seal soon becomes weak from its struggles and is quickly compelled to come to the surface to breathe, when it is easily dispatched and secured.

The seal-club can of course be employed in shore-hunting only when the Seals can be surprised at their favorite landing places, to which, as already detailed, they sometimes repair in herds of thousands. At such points, Schultz tells us, hundreds of seal-hunters congregate by prearrangement and make a combined attack upon the assembled herds of Seals, approaching them stealthily from the sea under cover of darkness, and by cutting off their retreat to the water make a wholesale butchery of the unsuspecting multitude, sometimes destroying, it is said, as many as fifteen thousand Seals in a single night. (See *anteà*, p. 515.)

II. ICE-HUNTING. 1. *In the Gulf of Bothnia.*—The prosecution of sealing voyages in vessels especially equipped for the purpose dates back to at least the beginning of the seventeenth century, but, as already stated, attained no great importance until near the close of that century. A few vessels only, even then, visited the great Arctic sealing-grounds, but in the Baltic sealing voyages appear to have been for a long time prosecuted with considerable regularity. Cneiff has left us a very particular account of the sealing business as carried on about 1750 in the Gulf of Bothnia; it also forms one of the earliest detailed histories of ice-hunting to which I have seen reference.

---

* Game Birds and Wild Fowl, etc., pp. 427, 428.

The species chiefly hunted seems to have been the Gray Seal (*Halichœrus grypus*), this being the only large Seal found in abundance in those waters. According to this writer the hunters were accustomed to make their voyages in open boats, made light and strong, and about fifty feet in length. The keels were shod with iron, and the boats were provided with masts and sails. They were accustomed to start on their voyages about the 25th of February, several boats usually keeping in company, so that if one of them met with an accident the people could be saved by the other boats. Each boat's crew, says Cneiff, numbered eight persons, among whom were a captain or master, a helmsman, and also, in order to have the food quickly prepared, two cooks, the one to provide water, the other, wood, of which they take very little, in order to keep the boat light, the wood taken consisting of a single fir stick about a foot thick and six feet long. Their provisions consisted chiefly of sour bread, to which were added butter, cheese, smoked meat, and salted fish. Brandy was also taken, but the chief drink was the salt or brackish water of the gulf, with which meal was commonly mixed to make it more palatable. Each man also provided himself with two full suits of clothes, so that he could change in case he fell through the ice, the wet clothes being dried by sitting on them in the boat. The most noteworthy article of clothing, says our author, is a skin of calf leather with the hair turned outward. It must be made of the skin of a wholly white calf, so that the Seals may not so readily distinguish the wearer from the ice.

When the time for departure arrived they put into the boat all their various implements and goods and launched it with its wherry from the edge of the firm ice. If there chanced to be a stretch of open water to the southward they sailed through it as far as possible, for in this direction the Gray Seals were most abundant. In case, however, there was new ice along the edge of the old ice, they drew the boat over it, each man pulling by a hair rope fastened to the boat, the captain holding the boat straight by means of a long pole fastened across it while the others drew it. This hard work they were frequently obliged to perform at other times during the journey when meeting with fields of new ice. At such times the heaviest of their things, as the provisions and firewood, were left behind, but never more than an eighth of a mile, when they returned for them with sledges. They drew the boat no farther at one time for fear of the breaking up of the ice, which might separate them from their sledges, and they would thus lose their provisions and other necessaries, they having learned this precaution through such losses. If the new blue ice was too weak to support the boat, they waited till it became stronger, or till a strong favorable wind enabled them to sail through it, in case it was not broken up by the waves. To protect the boat at such times from being cut by the ice, boards were nailed upon it. If the ice was not of great extent the men passed along outside of the boat and broke it through with clubs; if very weak and the wind favorable, they sailed through it without delay or fear. If on the other hand they found the ice strong and comparatively smooth, with not too much snow upon it, they sailed over it, the keel of the boat being protected from wearing by the iron sheathing. In sailing over the ice two men run on the leeward side of the boat and one man on the windward side, who keep the boat steady, while the captain steers it by means of the pole. If after having pursued their journey, by pulling and sailing, for a long time they meet with no Seals on the ice-fields to which the Seals are accustomed to resort, two men from each boat are sent on in advance to search for the Seals. They take with them the wherry, so that they can cross any openings in the ice they may meet with, and also a white dog, which by barking gives them notice if it discovers any Seals.

When at last they arrive at an ice-field on which there are great numbers of Seals, the men hasten with clubs to kill them. The largest of them, says our narrator, are so courageous that they face their pursuers, who, if they do not kill them at once must get out of the way, as the Seals can bite very severely. They leave the young ones till the last, as they are not shy. If there chance to be a great many holes in the ice-floe, so that the Seals can readily get under the ice, the hunters creep stealthily upon them till they get near enough to cut them off from such retreats, and then aim at the largest of the herd. Should there be a great many Seals on a small ice-floe, the men cry like the Seals, and creep toward them on their bellies, often raising their feet and striking them together. But should there be hope of getting only a single shot, they are not permitted to shoot at all, as then, certainly, all the Seals would leave the ice and go directly into the water, save those that were killed. Those Seals that are under the ice are not alarmed by the shooting, and as they come out upon the ice are successively shot, the hunters meantime keeping up their cry in imitation of the Seals, and continuing to strike their feet together as already described. If the hunters have good luck they in this way secure in a single day a large booty.

Another very common way of securing these animals was to watch for them at their breathing-holes, and as they came to the surface for air to transfix them with the harpoon or "seal-iron". The iron being fixed loosely to the shaft the latter is easily detached when the Seal descends again under the ice.   To the iron, however, is fastened a line about six feet in length, which the hunter quickly seizes, and allows the Seal to dance about beneath the ice, the barbed iron preventing its escape.   When the Seal becomes weak and must again obtain air, the line is drawn slowly in as the Seal approaches the air-hole, and finally the Seal is drawn out upon the ice, the hunter being in the meantime aided by his companions.   At other times the young are used as a lure for the capture of the mothers.   For this purpose they employ an iron implement having three barbed hooks, on one of which the young Seal is impaled alive.   The mother hearing its cries approaches it quickly, and immediately embraces it, in the hope to free it, but in so doing presses the other barbed hooks into herself, and both mother and young are drawn out of the water together.

The last method of hunting Seals described by Cneiff, as adopted on these early expeditions, is the following: If they have not already secured a sufficient number of Seals, they seek for them on their return from the south over the ice about the end of spring, when they are then much more surely taken, because they cannot so readily find an opening in the ice through which to escape; then, if attacked, they scramble about over the rough ice in search of openings, during which they are destroyed in such numbers that the sledges are soon loaded with them, and even return the second and third times in case the ice-pack is large and the Seals do not reach open water.   It sometimes happens, on these perilous journeys, that a strong wind breaks up the ice, and the hunter suddenly finds himself on a detached piece of drifting ice, when those who are in the boat must turn to rescue him.   He is fortunate, indeed, if he can bring his slain animals with him; otherwise he must be satisfied to save his life.

Thus it is, says Cneiff, on these dangerous voyages, during

which these poor people are exposed to fierce cold and severe snow-storms, under the open sky, they having no protection save that afforded by the sails of their boat, under which they lie, for it appears that they have not even the comforts of a fire. These perilous journeys occupy commonly two or three months, and sometimes more, according to their success in hunting, remaining out later when they have a long search in finding the Seals, or are late in obtaining a full cargo. Sometimes, however, when very fortunate, the voyage would be completed in five weeks. In case they meet with an abundance of Seals they save only the skin and fat, throwing away the flesh. On their return the products of the voyage are divided equally among the different boats.

It would seem that such exposure and risk would only be undertaken under the incentive of large profits, but on the contrary, after deducting the cost of each man's outfit, and the value of his time if devoted to other pursuits, little is really gained by these arduous and dangerous voyages.

Later in the season (about the end of March) they were accustomed to make a second voyage, this time for the Wikare or Bay Seal (apparently *Phoca vitulina*), for this purpose proceeding northward, with much the same outfit, and in nearly the same manner as on the earlier voyage. These later voyages seem to have been equally beset with danger, fifteen boats, as Cneiff tells us, being lost at one time.*

2. *Off the coast of Newfoundland.*—The season for "ice-hunting" begins at the Newfoundland "sealing-grounds" about the first of March and continues for about two months. The Seals are then on the ice-floes at a considerable distance from land, often several hundred miles. The same vessel, however, sometimes makes two, and on rare occasions three, voyages during the season. Formerly (fifty years ago) vessels engaged in sealing rarely left port before March 17, but more recently have sailed by the first of that month, and sometimes during the last days of February. This, Mr. Carroll claims, is too early, and tends

---

* Abridged from Cneiff's "Bericht vom Seekälberfarge in Ostbothnien." Abhandl. der Köngl. Schwed. Akad. der Wissensch. 19 Band, 1759, pp. 174–183. Lloyd, in his "Game Birds and Wild Fowl of Sweden and Norway" (pp. 433–449) gives a very similar account of what purports to be a history of Seal-hunting in the Gulf of Bothnia in recent times (1866), but his account is little more than a paraphrase of the foregoing, although Cneiff is not cited in this connection. Here and there additional details are given, but in the main Lloyd's account is substantially the same as Cneiff's.

greatly to the detriment of the interests of the sealers themselves, as they thus disturb the Seals at a time when they should be left in peace, or before the "whelping-time" is over. He strongly advocates the prohibition by government of the departure of any vessels for the sealing-grounds before March 5 to 10, and of steamers before the 10th to the 15th of the same month, since otherwise, he observes, "the seal-fishery of Newfoundland may soon, and very soon, dwindle away to such a character that it will not be worth the risk of money to prosecute it."

The vessels employed in the sealing business are "pounded off in the hold," or divided into small compartments to protect the pelts from injury by friction, as well as to preserve the cargo from shifting. The pelts are allowed to thoroughly cool before they are stowed, and are packed "hair to fat to prevent the fat from 'running.'" The owners of sealing-vessels "find all the boats, sealing-gear, powder, shot, and provisions, in consideration of which they are entitled to one half of the seals; the men are entitled to the other half. In steamships the owners find everything required for the prosecution of the voyage, and receive two-thirds of the value of the seals, and the men one-third."*

The voyages are attended with much danger, great hardship, and uncertainty of results, a "good trip" being entirely a matter of chance. Not unfrequently the vessels become "jammed in the ice", and if not crushed in the pack-ice, may be detained for weeks before being able to force their way to the ice-floes which form at this season the grand rendezvous of the Seals. The incidents and dangers ordinarily attending a sealing voyage, as well as the manner of capturing and disposing of the Seals, have been so graphically set forth by Professer Jukes in his entertaining and instructive work entitled "Excursions in Newfoundland", that I transcribe in this connection portions of his account of a sealing cruise participated in by him in March, 1840, in the brigantine "Topaz", Captain Furneaux, of St. John's, Newfoundland. Having, after a week's arduous cruise, fallen in with the Seals and captured a few young ones, he says: "We soon afterwards passed through some loose ice on which the young seals were scattered, and nearly all hands were overboard, slaying, skinning, and hauling. We then got into another lake of water and sent out five punts. The crews

---

* Carroll, Seal and Herring Fishery of Newfoundland, p. 9.

of these joined those already on the ice, and dragging either the whole seals or their 'pelts' to the edge of the water, collected them in the punts, and when one of these was full brought them on board. The cook of the vessel, and my man Simon, with the captain and myself, managed the vessel, circumnavigating the lake and picking up the boats as they put off one after another from the edge of the ice. In this way, when it became too dark to do any more, we found we had got three hundred seals on board, and the deck was one great shambles. When piled in a heap together the young seals looked like so many lambs, and when occasionally, from out of the bloody and dirty mass of carcasses, one poor wretch still alive would lift up its face and begin to flounder about, I could stand it no longer; and arming myself with a hand-spike, I proceeded to knock on the head and put out of their misery all in whom I saw signs of life. After dark we left the lake and got jammèd in a field of ice, with the wind blowing strong from the north-west. The watch was employed in skinning those seals which were brought on board whole, and throwing away the carcass. In skinning, a cut is made through the fat to the flesh, a thickness generally of about three inches, along the whole length of the belly from the throat to the tail. The legs, or 'fippers,' and also the head, are then drawn out from the inside and the skin is laid out flat and entire, with the layer of fat or blubber firmly adhering to it, and the skin in this state is called the 'pelt,' and sometimes the 'sculp.' It is generally about three feet long and two and a half wide, and weighs from thirty to fifty pounds. The carcass when turned out of its warm covering is light and slim, and, except such parts as are preserved for eating, is thrown away."

The next day, continues Mr. Jukes, as soon as it was light, "all hands were overboard on the ice, and the whole of the day was employed in slaughtering young seals in all directions and hauling their pelts to the vessel. The day [March 13] was clear and cold, with a strong north-west wind blowing, and occasionally the vessel made good way through the ice, the men following her and clearing off the seals on each side as we went along. The young seals lie dispersed here and there on the ice, basking in the sun, and often sheltered by the rough blocks and piles of ice, covered with snow. Six or eight may sometimes be seen within a space of twenty yards square. The men, armed with a gaff and a hauling rope slung over their shoulders, disperse about on the ice, and whenever they find a seal strike it a heavy blow in the head, which either stuns the animal or kills it outright. Having killed or at least stunned all they see within a short distance, they skin, or, as they call it, 'sculp' them with a broad clasp-knife, called a sculping-knife, and making two holes along the edge of each side of the skin they lay them one over another, passing the rope through the nose of each pelt and lacing it through the side holes, in such a manner that when pulled tight it draws them into a compact bundle. Fastening the gaff in this bundle, they then put the rope over the shoulder, and haul it away over the ice to the vessel. In this way they bring in bundles of pelts, three, six, or even seven at a time, and sometimes from a distance of two miles. Six pelts, however, is reckoned a very heavy load to drag over the rough and broken ice, leaping from pan to pan, and they generally contrive to keep two or three together to assist each other at bad places, or to pull those out who fall into the water. The ice to-day was in places very slippery and in others broken and treacherous, and as I had not got my boots properly fitted with 'sparable' and 'chisels' I stayed on board and helped the captain and the cook in managing the vessel and whipping in the pelts as they were brought alongside. By twelve o'clock, however, my arms were aching with the work, and on the leeside of the vessel we stood more than knee-deep in warm seal-skins, all blood and fat. Some of the men brought in as many as sixty each in the course of the day, and by night the decks were covered, in many places the full height of the rail. As the men came on board they occasionally snatched a hasty moment to drink a bowl of tea, or eat a piece of biscuit and butter; and as the sweat was dripping from their faces, and their hands and bodies were reeking with blood and fat, and they often spread the butter with their thumbs, and wiped their faces with the backs of their hands, they took both the liquids and the solids mingled with the blood. The deck, of course, when the deck could be seen, was almost as slippery with gore as if it had been ice. Still there was a bustle and excitement in the scene that did not permit the fancy to dwell on the disagreeables, and after a hearty refreshment the men would snatch up their gaffs and hauling ropes, and hurry off in search of new victims: besides, every pelt was worth a dollar! During this time hundreds of old seals were popping up their heads in the small lakes of water and holes among the ice, anxiously looking for their young. Occasionally one would hurry across a pan in search of the snow-white darling she had left, and which she could not recognize in the bloody and broken carcass, stripped of its warm covering, that alone remained of it. I fired several times at these old ones in the afternoon with my rifle from the deck, but without success, as unless the ball hits them on the head, it is a great chance whether it touch any vital part, the body being so thickly clothed with fat. In the evening, however, Captain Furneaux went out on the ice and killed two with his sealing-gun loaded with seal shot. The wind had now sunk to a light air, and the sun set most gloriously, glancing from the golden west across the bright expanse of snow now stained with many a bloody spot and the ensanguined trail which marked the footsteps of the intruders on the peacefulness of the scene. Several vessels came up near us from the south, in the afternoon; but notwithstanding all the slaughter the air as night closed in resounded with the cries of the young seals on every side of us. As the sunlight faded in the west, the quiet moon looked down from the zenith, and a brilliant arch of aurora crossed the heavens nearly from east to west, in a long waving line of glancing light, slowly moving backwards and forwards from north to south across the face of the moon. . . .

" Early in the morning [of the next day, March 14] the crew were out on the ice, and brought in 350 seals. The number hauled in yesterday was 1,380, making the total number now on board upwards of 2,000. After suffering the pelts to lie open on deck for a few hours, in order to get cool, they are stowed away in the hold, being laid one over the other in pairs, each pair having the hair outwards. The hold is divided by stout partitions into several compartments or 'pounds' to prevent too much motion among the seal-skins and keep each in its place. The ballast is heaved entirely out as the pelts are stowed away, and the cargo is trusted to ballast the vessel. In consequence of neglecting to divide the hold into pounds in one of his earlier voyages, Captain Furneaux told us he once lost his vessel. He was detained on his return with 5,000 seals on board, by strong contrary gales which kept him at sea, till by the continued motion and friction his seals began to run to oil. The skins then dashed about from one side of the hold to the other with every roll of the vessel, and he was obliged to run before the wind, which was then blowing from the northwest. The oil spread

from the hold into the cabin and forecastle, floating over everything and forcing the crew to remain on the deck. They got
up some bags of bread, and by putting a pump down through
the oil into the water-casks they managed to get fresh water.
After being in this state some days himself and his crew were
taken out of the vessel by a ship they luckily fell in with, and
carried to St. John's, New Brunswick; but his own vessel, with
her once valuable cargo, and almost all the valuable property
of himself and his crew were necessarily abandoned to the
mercy of the winds and waves, and what became of her was
never known. This was a good practical lesson as to the proper
method of stowing a cargo of seals, and one not likely to be
forgotten. In the present instance, therefore, the pounds were
both numerous and strong." *

In a few days more they completed their cargo and returned
to St. John's with the vessel loaded with between 4,000 and
5,000 Seals. "It was a very good season," Professor Jukes further remarks; "one vessel in two trips brought in eleven thousand Seals, and the total take this year [1840] must have been
considerably upwards of five hundred thousand."† Mr. Reeks
states that in 1866 one vessel, which made two successful trips
to the ice, brought into St. John's harbor 25,000 Seals.‡

To complete the picture here partially drawn of the seal
fishery as pursued by the Newfoundland seal-hunters, I quote
still further from the same author respecting the scenes incident to a sealing voyage of forty years ago. Under date
of March 5, Mr. Jukes writes: "This morning was dark and
foggy, with the wind at southeast. At seven o'clock, after making a tack or two about an open lake and finding no channel,
we dashed into the ice with all sail set, in company with two
other vessels on a north-northwest course. The ice soon got
firmer, thicker and heavier, and we shortly stuck fast. 'Overboard with you! gaffs and pokers!' sung out the captain, and
over went, accordingly, the major part of the crew to the ice.
The pokers were large poles of light wood, six or eight inches
in circumference, and twelve or fifteen feet long. Pounding
with these, or hewing the ice with axes, the men would split
the pans near the bows of the vessel, and then, inserting the
ends of the pokers, use them as large levers, lifting up one side
of the broken piece and depressing the other, and several get-

---

* Excursions in Newfoundland, vol. i, pp. 272–280.

† Ibid., p. 322.

‡ Zoölogist, 2d Ser., vol. vi, 1871, p. 2548.

ting round with their gaffs, they shoved it by main force under the adjoining ice. Smashing, breaking, and pounding the smaller pieces in the course the vessel wished to take, room was afforded for the motion of the larger pans. Laying out great claws on the ice ahead when the wind was light the crew warped the vessel on. If a large, strong pan was met with, the ice-saw was got out. Sometimes, a crowd of men, clinging round the ship's bows and holding on to the bights of ropes suspended there for the purpose, would jump and dance on the ice, bending and breaking it with their weight, shoving it below the vessel, and dragging her on over it with all their force. Up to their knees in water, as one piece after another sank below the cutwater they still held on, hurrahing at every fresh start she made, dancing, jumping, pushing, shoving, hauling, hewing, sawing, till every soul on board was roused into excited exertion. . . . They continued their exertions the whole day, relieved occasionally by small open pools of water; and in the evening we calculated that we had made about fifteen miles. It continued foggy all day, and at night it began to rain. We had seen no vessel since morning—nothing but a dreary expanse of ice and snow stretching away into the misty horizon." The next day "the wind was from the west, and the sky fine and clear. Several vessels were near us, and several more on the horizon. The ice became thicker, stronger, and more compact. We made a few miles in the morning and stuck fast the rest of the day in a very large pan or field of ice, sawing, axing, prising, warping, etc., etc., as yesterday."*

This, in short, was the history of their daily experiences for a week, at the end of which time they first heard the cry of the Seals, and entered upon their work of slaughter.

3. *In the Jan Mayen Seas.*—Seal-hunting in the icy seas about Jan Mayen is conducted under similar conditions and in much the same way as among the ice-floes to the eastward of Newfoundland. Lindeman, in his memoir on the Arctic fisheries as prosecuted from the German seaports, gives a pretty full account of the vessels sailing from the Weser ports, selecting for this purpose the "Hudson", J. H. Westermeyer, commander, as a type, and not only describes her special armature for protection against ice, but her general outfit, including officers and crew, the weapons employed, the commissariat, even to the weekly bill of fare, the "watch", and daily life and duties on

---

* Excurs. in Newfoundland, vol. i, pp. 261–263.

shipboard; and finally gives a history of the voyage she made the year preceding his account, a brief abstract of which may not be here out of place. The captain assembled his crew in January, and on the 21st of February the "Hudson" sailed out of the Weser for her Arctic voyage. The 8th of March found her in N. lat. 71° 18′ and W. long. 3° 8′. On the evening of the 9th they sighted Jan Mayen Island, twenty miles distant to the northwest. On the 14th they encountered heavy winds and a turbulent sea. About the beginning of April she reached the sealing-grounds ("Robbenküste'), the Seals being this year northwestward of Jan Mayen in north latitude 72° and east longitude 2°. Already, numerous vessels were at the place, and on the 14th of April, at 3 p. m., began the slaughter of the young Seals. At eleven o'clock the same evening the "Hudson" had on board 901 young Seals, and on the evening of the following day the number secured reached 2,171. In the course of a few days the crew of the "Hudson" completed their cargo, numbering altogether 5,400 young Seals, which yielded 620 tuns of oil. This, with the skins of the Seals and one Whale ("ein Fisch"), brought 23,983 thalers, gold.

The same writer thus describes the "Seal-coasts" and the hunt. Under the heading "Die Robbenküste, der Robben-schlag," he says:—"The district of the Seal-hunt, if we may so term the butchery of the most patient and submissive of animals, embraces the immensely large area of 6,000 to 8,000 square miles, and though called 'coast' is really no coast, but sea and ice-fields. In this area one comes upon immense herds of Seals, which, according to Yeaman's account, are often twenty to thirty English miles broad. The English call such herds 'Seal's-weddings' or 'Seal-meadows.' The commander, peering through his spy-glass from the 'crow's nest' first discovers the herds of Seals. He shouts the order 'Over all!' The crew costume themselves for the slaughter, their suit consisting of gray linen. Into a leathern belt fastened around the body they stick the skinning-knife. Each provides himself with ropes and a seal-club, the latter implement consisting of a strong stick or shaft, having an iron point, a hammer and hook. Soon the boats are lowered and the men rush into them, and with a loud 'Holulu!' start for the ice. The killing of the Seals upon the ice begins. When the Seals are dead, the skin, together with the fat or blubber, is removed from the body with the skinning-knife. The cabin-boys ('Schiffsjungen') and later all the men draw the skins of the 'dogs,' as the Seals are called in Greenland parlance, with ropes to the ship, where the so-called doctor (or barber) receives them, counting them as soon as they come in at the *Flenssgat*.* The rest of the animal, termed the 'krang', remains on the ice, a booty to the birds and the Polar Bear. The success of the Seal-hunt depends upon quickly taking advantage of the favorable moment. The crew must be constantly quick of hand. Five hundred to six hundred Seals may be killed in a day by a crew of a ship of 180 tons. The difficulty experienced by the men in springing from ice-cake to ice-cake to reach the ship again, is not slight.

"Hunting from boats or vessels is comfortable, and is preferred when there is much open water. They spring from the boats to the ice-floes, kill the Seals in the same way, take them temporarily into the boat and stack them on the first suitable ice-floe. The removal of the skins from the fat is made by the officers on shipboard as opportunity may favor. In this work, following an old Dutch custom, they stand in a row to take 'a little' ("ein 'Lütjer' genommen"),† and occasionally divert themselves with a song. The skins are fastened by hooks to a wooden frame, and the fat is quickly removed and thrown into tubs. The coopers then pack the fat in casks or iron tanks in the lower or middle holds of the vessel. The art of properly removing the fat without injury to the skins is not easy to acquire, and upon this depends in a high degree the value of the skins. I hear that the owner of the 'Albert' has, through a slight modification of the share-money, interested the men in exercising the greatest possible care in removing the skins, in order to secure a good result. The skins are salted, again counted and laid away. By the end of April the proper Seal-hunt is over. Old Seals are rarely to be obtained, they being very watchful; however, the crews of the Norwegian ships are excellent marksmen, and by them many are shot. The value of a young Seal (fat and skin) is 2½ to 3 thalers, while the old ones are worth twice this sum."‡

---

* Opening in the side of a whaling-vessel through which the blubber is taken on board in cutting up a whale.

† "On such occasions the Dutch drunk schnaps from cups. On many Dutch ships it was the custom ('Brauch') or much more the bad custom ('Missbrauch') to take the schnaps-bottle into the boat with them, or hang it by a line from the ship."

‡ Translated from Petermann's Geog. Mitth., Ergänzungsheft No. 26, 1869, pp. 81, 82.

DANGERS AND UNCERTAINTIES OF ICE-HUNTING.—The dangers and uncertainties attending ice-hunting have to some extent been already indicated. The chief hindrances arise from encountering field-ice, within which not only single vessels, but whole fleets, are sometimes held prisoners for weeks, constantly subject to danger from the shifting and grinding of the pack-ice. Among the many dangers to which the ice-hunter is subjected none is greater than that arising from the "rafting" of the ice, which is especially disastrous to steamships. While some vessels, owing to the form of the hull, will "heave out" uninjured, in other cases they will be crushed by the ice passing over them. In general, steamships are said to be in far greater peril when jammed in the ice than sailing-vessels, there being in such cases "no chance whatever" of extricating the former, while the latter usually escape with slight injury. Great danger is said to also arise from large masses of ice being carried by currents against the wind, when, despite every exertion to avert disaster, steamships as well as sailing-vessels are wrecked against the floating islands of ice.

In illustration of the danger from drifting ice I transcribe the following: "In the spring of 1871," says Mr. Carroll, "that splendid new brig, the 'Confederate,' with an experienced captain and seventy-five men, as fine as any country under the sun could produce, left Harbour Grace for the sealing voyage. The brig was driven into Bonavista Bay, jammed in the drift-ice, until it struck the land, seven miles to the westward of Cape Bonavista. There the brig remained for ten days, and not a wag in the water or amongst the ice, the men in anxious waiting for an off-shore wind, when, without any apparent cause, a large flat pan of ice a short distance from the brig moved slowly onwards until it struck the after part of the keel and whipped ten feet of it away. So keen was the cut that it was not observed until the brig began to make water", and the master and men were obliged to abandon her. "Many, in all probability", continues the same writer, "of the steamships at present [1873] engaged in the prosecution of the Seal fishery on the coast of Newfoundland will, without doubt, sooner or later meet with a fate similar to that of the brig 'Confederate.' Sailing vessels will 'heave out' when jammed in the ice and escape uninjured when steamships would be squeezed to atoms."*

Not only are the sealers exposed to dangers from floating ice,

* Seal and Herring Fisheries of Newfoundland, p. 22.

but other risks attend these hardy adventurers. Although the present connection is not the place for an extended history of the disasters incident to the seal-fishery, a single incident in illustration of the danger arising from sudden storms overtaking the seal-hunters when absent from their vessels may here appropriately find place. Scoresby relates, on trustworthy authority, the following that befel the sealing fleet in the Jan Mayen Seas in 1774:

"Fifty-four ships, chiefly Hamburghers, were that year fitted out for the seal-fishery alone, from foreign ports. Most of these, with several English ships, had, in the spring of the year, met together on the borders of the ice, about sixty miles to the eastward of the island of Jan Mayen. On the 29th of March, when the weather was moderate, the whole fleet penetrated within some streams of ice, and sent out their boats in search for seals. While thus engaged, a dreadful storm suddenly arose. So sudden and furious, indeed, was the commencement, and so tremendous and lasting the continuance, that almost all the people who were at a distance from their ships perished." After giving a detailed account of the loss of various ships, as well as boats' crews, he says, "The result of these disasters, when summed up, is dreadful. About 400 foreign seamen, and near 200 British, are said to have been drowned; four or five ships were lost, and scarcely any escaped without damage." *

Although accidents attended with such great fatality are fortunately of rare occurrence, doubtless not a year passes without the loss of numerous vessels and many lives.

Most writers who have given any account of the Seal-fishery refer to the uncertainties of the catch, owing to circumstances wholly beyond the knowledge or control of the sealers. As already stated, a good trip is a matter of chance rather than of foresight or judicious management on the part of the master of the vessel. This uncertainty arises mainly from the unstable character of the ice-floes, which vary in their course with the prevailing direction of the wind and the combined action of the winds and currents. While the Seals. congregate annually on the ice-fields of the same general region, and bring forth their young with surprising regularity as regards season, the place of rendezvous is constantly variable. In like manner the course of the vessel is greatly at the mercy of the elements, or under the control of wholly unforeseen circumstances. The

* Arctic Regions, vol. i, pp. 513–517.

whole matter in question has been thus tersely presented by Mr.
Carroll. "For the last fifty years," says this experienced writer,
"I have been from time to time well and intimately acquainted
with ice-hunting masters; nine-tenths of them when they first
took charge of ice-hunting vessels generally brought into port
what is usually termed 'good saving trips.' It is strange to
say, but not the less true, that the longer a man takes charge
of an ice-hunting vessel the less he knows where to obtain a
trip of old and young seals. In a word, the prosperity of a
sealing voyage, one year with another, depends upon chances,
and I will go farther and say that three-fourths of the heavy trips
of seals' fat that were brought heretofore into port, as well as
the heavy trips of seals' fat brought into port at the present
day, were got also by chance. Spring after spring I have known
ice-hunting vessels to get jammed in the ice, and there kept so
long that the men despaired of obtaining a profitable trip of
seals. Steamships as well as sailing vessels are very often,
owing to gales of wind, obliged to run into the ice for safety,
much against the master's will, and the very place the master
wished above all things to avoid turned out to be the very spot
where what he was after was—plenty of seals."*

SPECIES HUNTED.—The Seals hunted in the North Atlantic
and Arctic waters belong chiefly to four species, namely, the
Harp or Greenland Seal, *Phoca (Pagophilus) grœnlandica*, the
Rough Seal, *Phoca (Pusa) fœtida*, the Harbor Seal (*Phoca vitulina*), and the Hooded Seal (*Cystophora cristata*). The first, by
its numbers, far exceeds in importance all the others together,
and is hence the chief object of pursuit. Two other species, the
Bearded Seal (*Erignathus barbatus*), and the Gray Seal (*Halichœrus grypus*), are also taken when met with, but both are rare
and neither enters largely into the general product of the Seal-
fishery. The Newfoundland Seal-fishery is limited to the capture
of the Greenland, Harbor, and Hooded Seals. The latter is not,
however, a regular object of pursuit, but is taken as opportunity
favors, and some seasons but very few individuals of this species
are met with. The Harbor Seal is taken along the shores, where
it is permanently resident, but comparatively only in small
numbers. The Rough Seal and the Bearded Seal are of considerable importance to the Greenlanders, the former especially,
more than half of the Seals taken by them belonging to this
species.

* Seal and Herring Fisheries of Newfoundland, p. 36.

In addition to the above, the Caspian Seal (*Phoca caspica*) is extensively hunted in the Caspian Sea, and Sea-Elephants on the coast of Lower California and in the Antarctic seas.

ABUNDANCE OF SEALS AT PARTICULAR LOCALITIES.—Respecting the abundance of Seals, particularly at certain localities, and the ease with which they are taken, a few excerpts may here be added to the various incidental references to the subject already made in the general account of Seal-hunting. Mr. H. Y. Hind states that "On March 24, 1857, large ice-fields, driven by the N. and N. W. wind, grounded on the coast of Amherst Island, one of the Magdalen group, and were found to be a vast ' seal meadow.' Not less than 4,000 of these animals, nearly all young, were killed in five days."*

Drs. Koldewey and Pansch, of the German Arctic Expedition of 1869–70, make the following statement :

"The whitish colored young stay on the ice the first few days, and are then killed with clubs by the parties of seal-hunters. . . . The number caught by a single Bremen ship now sometimes amounts to 8 to 10,000 seals; and one may form some idea of the war of destruction waged against these harmless creatures by man, when we hear that of European ships in 1868, five German, five Danish, fifteen Norwegian, and twenty-two British, which were in company in West Greenland, obtained 237,000."†

Mr. Robert Brown states that in the Spitzbergen Sea, the Greenland Seals, at the time of bringing forth of the young, "may be seen literally covering the frozen waste as far as the eye can reach with the aid of a telescope from the 'crow's nest' at the main-royal masthead, and have, on such occasions, been calculated to number upwards of half a million of males and females."‡ It is little wonder that, at such times, but more especially after the young are born and rest helplessly upon the ice, a ship's crew will secure several hundreds in a single day, and quickly fill their vessels with cargoes of ten thousand Seals.

PRODUCTS.

So much has been already said, incidentally, in relation to the products of the Seals and their commercial importance that little need here be added. Of chief importance is the oil, so well known for its valuable properties for illuminating

* Expl. in Labrador, vol. ii, p. 207.
† German Arct. Exped. 1869–70, Eng. ed., 1874, pp. 61, 62.
‡ Proc. Zoöl. Soc. Lon., 1868, p. 418.

purposes and for the lubrication of machinery. The amount annually obtained falls not far short of 90,000 tuns, with a total value of $1,250,000. Next in importance are the skins, which are nearly as valuable as the oil. From very early times they were used for covering trunks, the manufacture of knapsacks, and for many of the uses of ordinary leather. They have been extensively employed, as indeed they are still, for the manufacture of caps, gloves, shoes, and jackets. Of late many have been converted in England into lacquered leather, which is said to be of a superior quality, being beautiful and shining, and of firm texture, and can be furnished at moderate cost. The skins differ in value according to size and color, these varying, of course, with the species and with the age of the animal.

As an article of food, Seals are of the utmost importance to the natives of Greenland, and the northern tribes generally, they deriving from them the greater part of their subsistence. They have been found likewise not unpalatable by our Arctic voyagers, whose sustenance often for long periods has been mainly the flesh of these animals. The Esquimaux and allied tribes of the North are well known to depend upon the Seals not only for their food, but for most of the materials for their boats and sledges, as well as for clothing and the various implements of the chase.

In respect to the character of Seal flesh as food, and the importance of these animals to the Esquimaux, I quote the following from Dr. A. Horner, surgeon to the "Pandora", who, in "Land and Water" for December 18, 1875 (p. 475), thus refers to the general subject:

"From the length of time these people have inhabited this cold country, one naturally expects them to have found some particular food well adapted by its nutritious and heat-giving properties to supply all the wants of such a rigorous climate, and such is found to be the case, for there is no food more delicious to the tastes of the Esquimaux than the flesh of the seal, and especially that of the common seal (*Phoca vitulina*). But it is not only the human inhabitants who find it has such excellent qualities, but all the larger carnivora that are able to prey on them. Seal's meat is so unlike the flesh to which we Europeans are accustomed, that it is not surprising we should have some difficulty at first in making up our minds to taste it; but when once that difficulty is overcome, every one praises its flavour, tenderness, digestibility, juiciness, and decidedly warm-

ing after-effects. Its colour is almost black, from the large
amount of venous blood it contains, except in very young seals,
and is, therefore, very singular looking, and not inviting, while
its flavour is unlike anything else, and cannot be described except
by saying delicious! To suit European palates, there are certain precautions to be taken before it is cooked. It has to be
cut in thin slices, carefully removing any fat or blubber, and
then soaked in salt water for from twelve to twenty-four hours,
to remove the blood, which gives it a slightly fishy flavour. The
blubber has such a strong taste, that it requires an Arctic
winter's appetite to find out how good it is. That of the bearded
seal (*Phoca barbata*) is most relished by epicures. The dain-
tiest morsel of a seal is the liver, which requires no soaking,
but may be eaten as soon as the animal is killed. The heart is
good eating, while the sweetbread and kidneys are not to be
despised.

"The usual mode of cooking seal's meat is to stew it with a
few pieces of fat bacon, when an excellent rich gravy is formed,
or it may be fried with a few pieces of pork, or 'white-man,'
being cut up with the seal, or 'black-man.'

"The Esquimaux make use of every part of the seal, and, it
is said, make an excellent soup by putting its blood and any odd
scraps of meat inside the stomach, heating the contents, and
then devouring tripe, blood and all with the greatest relish.

" For my own part, I would sooner eat seal's meat than mutton or beef, and I am not singular in my liking for it, as several
of the officers on board the Pandora shared the same opinion
as myself. I can confidently recommend it as a dish to be tried
on a cold winter's day to those who are tired of the everlasting
beef and mutton, and are desirous of a change of diet. It is
very fattening, and if eaten every day for several weeks together
is likely to produce rather surprising effects. . . . .

" Seal's meat is a panacea for all complaints among these
primitive people. Our Esquimaux interpreter, 'Joe,' had a
most troublesome cough when we left England, and was convinced he should not get rid of it until he had seal's flesh to eat.
He would not look at any medicine offered to him on board,
but shook his head and said, ' By-and-bye, eat seal, get well.'
His prescription turned out to be a very good one, for he had
not long been feasting on his favourite food before he lost his
cough, and we heard no more of it. For delicate persons, and
especially young ladies and gentlemen who cannot succeed in

making their features sufficiently attractive on chicken and cheesecakes, no diet is likely to succeed so well as delicate cutlets from the loin of a seal.

"For my own part I cannot help thinking that the diminution in the number of seals caught near the principal Danish settlements in Greeland, has a great deal to do with the prevalence of consumption and other diseases among the native inhabitants of those places. Seals are becoming scarcer every year, and, in company with the bison of the North American prairies, will ere long be of the past, and leave the poor Greenlander and Red Indian to follow them."

PREPARATION OF THE PRODUCTS.—The Seals being captured and brought into port, their subsequent treatment as practised by the Newfoundland sealers, may be briefly detailed as follows: After landing and weighing the "pelts," the fat is immediately removed from the skin. This is accomplished by extending the pelt on a table, behind which the skinner stands, holding the skin with the left hand while with a large skinning-knife he removes the fat with the right hand; a good skinner, it is stated, being able to "remove the fat from the skins of four hundred and fifty Harp Seals in ten hours." The skins are then salted and packed, with the flesh side uppermost, and at the end of three weeks are considered cured and fit for shipping. The fat is reduced to oil either by maceration in vats in the sun, or is "rendered" by steam. The latter process is so rapid that at the establishment of John Munn & Co., at Harbour Grace, four thousand "pelts" have been "skinned and rendered into pure Seal oil in twenty-four hours." Although the steam-rendered oil meets with ready sale in consequence of its superior burning qualities and freedom from disagreeable odor, it is less free from smoke than that extracted by the agency of the sun, and for this reason the latter is preferred by the miners. "Formerly every description of Seals' oil was entirely manufactured in wooden vats exposed to the weather," the vats being capable of containing three thousand to four thousand Seals' pelts. When the fat from old Seals is mixed with that from the young, "the oil obtained is somewhat smoky." When drawn off from the tanks, all the oil rendered from the fat of young Seals is sure to come first and is called "pale seal," the other being heavier and darker, and known as "straw color."

From Schultz's minute account of the sealing industry of the Caspian Sea I transcribe the following, as of general interest in

the present connection : " The fat adhering to the skin of the
seal is detached from it, cut into pieces, and melted in caldrons,
after which the oil is poured in barrels. This is the simplest
way of making seal-oil, and the hunters often employ it. But
oil is also manufactured by steam, in establishments built for
this purpose on the left bank of the Volga, opposite Astrachan,
by some rich merchants. Thirty-five 'versts' (about twenty
miles) below Astrachan the Sapojnikow Brothers have built a
steam oil-factory at the 'vataga' (fishing establishment) of Ikri-
annaya. This factory is particularly busy in the spring, when
whole cargoes of seal-fat arrive, which is either boiled immediately in order to extract the oil, or is safely stored away in cellars. These cellars are long, floored, and furnished with four
ventilators and several windows. Large oak-wood tubs, plated
with lead on the inside, and capable of holding 700 'pouds'
(25,200 pounds) of oil each, are placed at intervals in holes dug
in the ground. The oil which runs out of the seal fat piled up
in layers flows into these tubs by way of an inclined plane. The
oil is then poured into barrels. . . . . . Kalmyks are employed
chiefly to detach the fat from the skins. They spread the skin,
with the fur down, on an inclined plank, which they lean against
their breast, in order to have the free use of both their hands.
Then, armed with a two-handled knife, they scrape the fat from
the skin. The oil, which is pure and clear, running down during this operation, flows into a reservoir let into the ground,
holding 400 'pouds' (14,400 pounds,) and forming a cube, each
side of which measures one 'sagene' (7 feet). This work is extremely fatiguing. A strong and experienced Kalmyk can, however, clean 500 or even 700 skins in a single day. The workmen
form associations, sharing their labor and their gain.

"The fat is then melted in large tubs, where it is exposed to
the action of steam. The oil flows through a funnel-shaped apparatus, and, finally, through pipes into immense oak-wood reservoirs. There are three such reservoirs connected by pipes,
and let into the ground, so that the oil from the first flows into
the second, and then into the third, from whence, through cocks,
it passes into casks, which can be shipped as soon as filled. Each
one of these reservoirs has a diameter of 3 'sagenes,' (21 feet,) a
depth of 1 'sagene,' (7 feet,) and can hold 4,800 'pouds' (172,000
pounds) of oil.

"The oil thus extracted forms the first quality. The second
quality is obtained by melting the residue in caldrons, and by

pressing it. The color of this oil is dark-brown. Before the residue is put into the caldrons, capable of holding 200 'pouds' (7,200 pounds) each, it is thrown into a receptacle with an inclined bottom, and the whole mass is stirred violently by means of wooden shovels. This is done in the sunlight, so that the heat may help to melt the mass. This receptacle is joined to the caldron by a large gutter, which is walled up in the furnace. Through this gutter, the residue is led into the caldron, there to melt, which done, the mass is taken out with dippers and cast into a box, which is then pressed. By means of this last operation all the remaining oil contained in the residue is extracted.

"The oil factory of the Sapojnikow Brothers formerly manufactured about 100,000 'pouds' (3,600,000 pounds) of seal-oil, which was sent to Moscow, where it was chiefly used in leather-factories; but during the last fifteen years, this establishment has gone considerably, and other wealthy Astrachan merchants, among them Messrs. Vlasow, Smoline, and Orékhow, have established several factories for the oil.

"The skins of the seals are used for making knapsacks and for covering valises."*

WASTEFUL DESTRUCTION OF SEALS.

There is often a lamentably great and needless waste of Seal-life at the Newfoundland and other sealing-grounds. Mr. Carroll, in 1871, pointedly called the attention of the government authorities to the so-called "panning" process, as a matter calling for statutory regulation. He says, "No greater injury can possibly be done to the seal fishery than that of bulking seals on pans of ice, by crews of ice-hunters. Thousands of seals are killed and bulked, and never seen afterwards. When the men come up with a large number of old and young seals, that cannot get into the water, owing to the ice being in one solid jam, they drive them together, selecting a pan surrounded with rafted ice, on which thousands of seals are placed one over the other, perhaps fifteen deep. A certain number of men is picked out by the ship master to pelt and put on board the bulked seals, whilst others are sent to kill more. It often happens that the men are obliged to go from one to ten miles, before they come up with the seals again, and very often the men pile from five hundred to two thousand in each bulk, which

---

* Rep. U. S. Commis. Fish and Fisheries, pt. iii, 1873–4 and 1874–5, pp. 95, 96.

bulks are from one to two miles apart; care is also taken that flags are stuck up as a guide to direct the men where to find such bulked seals. So uncertain is the weather and precarious the shifting about of the ice, as well as heavy falls of snow and drift, that very often such bulked seals are never seen again by the men that killed and bulked them, as the vessels and steamships are frequently driven by gales of wind far out of sight or reach of them, and frequently wheeled or driven into another spot, where the men again commence killing and bulking as before. In many instances it has happened that the crews of vessels, as well as the crews of steamships, have killed and bulked twice their load. No doubt seals that are bulked are often picked up by the crews of other vessels, but such is the law, that as long as the flags are erected upon the bulks, and the vessel or steamship is in sight, no man can take them, notwithstanding the vessel's or steamship's men that bulked them may be ten miles away from them, whilst another vessel may be driven within a quarter of a mile of the thousands of bulked seals, but owing to the law dare not take them." Sometimes after Seals are bulked heavy gales of wind spring up, driving the vessels or steamships that claim them, as well as any others in the vicinity, twenty or thirty miles from them, and they are thus lost. "Ice-hunting masters make it a rule to have the seals bulked on large flat pans." In this way the skins are damaged by exposure to the weather, being injured by severe frosts, as well as by the sun, so that "between frost and sun thousands of seal skins are rendered valueless." Loss also often happens by the capsizing of the pan of ice on which the skins are piled, and "the seals are never seen afterwards"—this forming the "greatest evil known to ice-hunters." In the spring of 1871, about four miles to the south of Bonavista Cape, there were three pans of ice, marked by flags, on which were piled not less than four thousand seals, but owing to the severity of the weather the men from the shore could not reach them. Owing to the heavy sea and bad weather none of them were ever obtained, as the pans passed over the Flower Rocks upon which the seals were ground to pieces. In the spring of 1872, some five thousand seals, obtained to the westward of Bonavista by the inhabitants of that place, were heaped upon the ice. "There were thirteen flags to be seen in the morning over bulked seals, and when the drift ice struck the land in the evening only six of the flags were visible, the ice having rafted over both flags and seals. Some days after, when the ice moved off from the shore, several bulks of seals were found, but in such a putrid state that they could not be handled. At the lowest calculation," continues Mr. Carroll, " I make bold to state that not less than from ten to twelve thousand pounds currency worth of seals' pelts is lost to the country each sealing voyage [or season], by the present system, carried on by the sealing masters and their crews!" The partial remedy that he suggests is that while no man should have the right to take any Seals of which he is nót the owner as long as the owners watch over them, yet as soon as the proper owners leave them the Seals should be free property to any one who can take them away.*

DECREASE IN THE NUMBER OF SEALS FROM INJUDICIOUS HUNTING.

Formerly so numerous were the Seals commonly hunted in the North Atlantic and Arctic waters (consisting chiefly of the Harp or Greenland Seal), that for many years the annual destruction of hundreds of thousands seemed not in the least to diminish their numbers, and as late as 1873 Mr. Carroll * gave it as his opinion that they were actually on the increase at the Newfoundland sealing-grounds, an opinion concurred in by other authorities. Here, indeed, their number seems unlimited, but it is otherwise in the sealing-districts about Jan Mayen and elsewhere at the various sealing-stations north of the northern coast of Europe. As already detailed (see anteà, pp. 503–510), a marked decline began to be apparent as early as 1865 to 1870, which each succeeding year increased at an alarming rate. Attention was at once directed to the cause, which was evidently overdestruction by the rival sealing-fleets of England, Germany, and Norway, and ruinous and indiscriminate slaughter at improper seasons. The agitation of the matter which followed resulted, as already shown, in the enactment of close-time acts for the protection of the Seals during the period when the young are brought forth. The act on the part of the English came into force in 1876, and soon after similar legislative action was taken by the other interested governments. While a close-time must be favorable to the increase of the Seals, or at least to the maintainance of their present

* Seal and Herring Fisheries of Newfoundland, pp. 32–34.
† Ibid., p. 26.

numbers, too little time has thus far elapsed to show to what extent it may prove beneficial.

The chief victims of the seal-hunter in the Antarctic seas and on the Mexican and Lower Californian coasts—the Sea-Elephants—long since (as previously stated, see *anteà*, pp. 517–522) became practically exterminated on all the islands and coasts where they were formerly hunted, and where at the beginning of the present century they were found in immense troops, and in seemingly exhaustless numbers.

SEALS AND SEAL-HUNTING IN THE OLDEN TIME IN THE GULF OF SAINT LAWRENCE.

This already protracted account of the Seal-fishery may be fittingly closed with the following extract from Charlevoix respecting the Seals of the Gulf of Saint Lawrence, and the Seal-fishery as practised there one hundred and sixty years ago. Charlevoix's account is contained in his letters of travel addressed to the Duchesse of Lesdigiueres, which I give here in the quaint language of the Dodsley translation published 1761. Under date of March 21, 1721, he says: "The sea-wolf owes its name to its cry, which is a sort of howling, for as to its figure it has nothing of the wolf, nor of any known land animal. . . . They never hesitate in this country to place the sea-wolf in the rank of fishes, tho' it is far from being dumb, is brought forth on shore, on which it lives as much as in the water, is covered with hair, in a word, though nothing is wanting to it which constitutes an animal truly amphibious. . . . Thus the war which is carried on against the sea-wolf, though often on shore, and with muskets, is called a fishery; and that carried on against the beaver, though in the water, and with nets, is called hunting.

"The head of the sea-wolf," he continues, "resembles that of a dog; he has four very short legs, especially the hind legs; in every other circumstance he is entirely a fish [il est Poisson]: he rather crawls than walks on his legs; those before are armed with nails, the hind being shaped like fins; his skin is hard, and is covered with a short hair of various colours. There are some entirely white, as they are all when first brought forth; some grow black, and others red, as they grow older, and others again of both colours together.

"The fishermen distinguish several sorts of sea-wolves; the largest weigh two thousand weight, and it is pretended have sharper snouts than the rest. There are some of them which flounce only in water [qui ne sont que fretiller dans l'eau]; our sailors call them *frasseurs*, as they call another sort *nau*, of which I neither know the origin nor meaning. Another sort are called *Grosses têtes*, *Thick-heads*. Some of their young are very alert, and dextrous in breaking the nets spread for them; these are of a greyish colour, and are very gamesome, full of mettle, and as handsome as an animal of this figure can be*; the Indians accustom them to follow them like little dogs, and eat them nevertheless.

"M. Denis [Denys] mentions two sorts of sea-wolves, which he found on the coasts of Acadia; one of them, says he, are so very large, that their young ones are bigger than our largest hogs. He adds that a little while after they are brought forth, the parents lead them to the water, and from time to time conduct them back on shore to suckle them; that this fishery is carried on in the month of February, when the young ones which they are not desirous of catching,† scarce ever go to water; thus on the first alarm the old ones take to flight, making a prodigious noise to advertise their young, that they ought to follow them, which summons they never fail to obey, provided the fishermen do not quickly stop them by a knock on the snout with a stick, which is sufficient to stop them. The number of these animals upon this coast must needs be prodigious, if it is true, what the same author assures us, that eight hundred of these young ones have been taken in one day. . . .

"It is by all agreed that the flesh of the sea-wolf is good eating, but it turns much better to account to make oil of it, which is no very difficult operation. They melt the blubber fat of it over the fire which dissolves into an oil. Oftentimes they content themselves with erecting what they call *charniers*, a name given to large squares of boards or plank, on which is spread the flesh of a number of sea-wolves; here it melts of itself, and the oil runs through a hole contrived for the purpose. This oil when fresh is good for the use of the kitchen, but that of the young ones soon grows rank, and that of the others if kept for any considerable time, becomes too dry [défléche trop]. In this case it is made use of to burn, or in currying leather.

---

* An allusion probably to the *Phoca vitulina*, which is said to be very destructive to nets.

† The original—"lorsque les Petits, ausquels on en veut principalement"—states just the opposite.

It keeps long clear, has no smell, or impurity whatsoever at the bottom of the cask.

"In the infancy of the colony great numbers of the hides of sea-wolves were made use of for muffs. This fashion has long been laid aside, so that the general use they are now put to is the covering of trunks and chests. When tanned, they have almost the same grain with morocco leather; they are not quite so fine, but are less liable to crack, and keep longer quite fresh and look as if new. Very good shoes and boots have been made of them, which let in no water. They also cover seats with them, and the wood wears out before the leather; they tan these hides here with the bark of the oak, and in the dye stuff with which they use black, is mixed a powder made from a certain stone found on the banks of rivers. This is called thunder-stone, or marcasite of the mines.

"The sea wolves couple and bring forth their young on rocks, and sometimes on the ice; their common litter is two, which they often suckle in the water, but oftener on shore; when they would teach them to swim they carry them, say they, on their backs, then throw them off in the water, afterwards taking them up again, and continue this sort of instruction till the young ones are able to swim alone. If this is true, it is an odd sort of fish, and which nature seems not to have instructed in what most sort of land animals do the moment they are brought forth. The sea-wolf has very acute senses, which are his sole means of defense: he is, however, often surprised in spite of all his vigilance, as I have already taken notice; but the most common way of catching them is the following.

"It is the custom of this animal to enter the creeks with the tide; when the fishermen have found out such creeks to which great numbers of sea-wolves resort, they enclose them with stakes and nets, leaving only a small opening for the sea-wolves to enter; as soon as it is high-water they shut this opening, so that when the tide goes out the fishes remain a dry, and are easily dispatched. They also follow them in canoes to the places to which many of them resort, and fire upon them when they raise their heads above water to breathe. If they happen to be no more than wounded they are easily taken; but if killed outright, they immediately sink to the bottom like beavers; but they have large dogs bred to this exercise, which fetch them from the bottom in even seven or eight fathoms of water. Lastly, I have been told, that a sailor having one day surprised a vast herd of them ashore drove them before him to his lodgings with a switch, as he would have done a flock of sheep, and that he with his comrades killed to the number of nine hundred of them. *Sit fides penes autorem.*"*

Subfamily PHOCINÆ *Gray.*

Genus PHOCA, Linné (*emend.*).

Phoca, Linné, Syst. Nat., 1758, i, 37; ibid., 1766, i, 55 (in part).
Pusa, Scopoli, Introd. Hist. Nat., 1777, 490.—Type, *Phoca fœtida.* (See *infra,* under genus *Halichœrus.*)
Calocéphale [*Callocephalus*], F. Cuvier, Mém. du Mus., xi, 1824, 182.—Type, *Phoca vitulina,* Linné.
Calocephalus, F. Cuvier, Dict. Sci. Nat., xxxix, 1826, 544; lix, 1829, 462.
Pagophilus, Gray, "Zoöl. Erebus and Terror, 1844, 3" (subgenus); Cat. Seals Brit. Mus., 1850, 25 (genus).—Type, *Phoca grœnlandica.*
Pagomys, Gray, Proc. Zoöl. Soc. Lond., 1864, 31.—Type, *Phoca fœtida,* Fabricius.
Halicyon, Gray, Proc. Zoöl. Soc. Lond., 1864, 28.—Type, "*Halicyon richardi*" = *Phoca vitulina.*
Haliphilus, Gray, Ann. and Mag. Nat. Hist., xvii, 1866, 446.—Type, *Halichœrus antarcticus,* Peale = *Phoca vitulina.*

Incisors $\frac{3-3}{2-2}$; molars, except the first, 2-rooted and multilobed; facial portion of the skull narrow, elongated, the dorsal outline gradually declining anteriorly; general form of the skull rather flat, depressed, the interorbital region very narrow.

The genus *Phoca,* as here defined, is composed of the smallest species of the family. The three here treated in detail, and the only ones thoroughly known, differ widely in cranial and other osteological characters, and by some writers have been each regarded as the type of a distinct genus, and may be considered as entitled at least to subgeneric rank. The only genus closely allied to *Phoca* (unless *Histriophoca* be excepted, the cranial characters of which are unknown) is *Erignathus,* consisting of a single species, still placed by many writers in the genus *Phoca.* The form of the skull, however, is widely different, the muzzle being broad and short, the frontal region convex and very high, the orbital fossæ and the auditory bullæ very small. *Erignathus* further differs from *Phoca* in having small supra-orbital processes; in the total absence of the acromion process of the scap-

* Journal of a Voyage to North America (Dodsley translation), vol. i, 1761, pp. 222–226. For the original see Journal d'un Voyage fait par ordre du Roi dans l'Amerique septentrinonale (12mo ed., 1744), pp. 211–216.

ula, and in lacking the abrupt eversion of the upper border of the ilia.

The subgenus *Phoca*, consisting, so far as certainly known, of a single species (*Phoca vitulina*), differs from *Pusa* (=*Pagomys*, Gray), and from *Pagophilus*, principally in its generally heavy structure, especially of the skull and dentition, and in the thickness of the body and the shortness of the limbs, particularly of the tibial and radial segments. In *Pusa* and *Pagophilus* the skull is similar in general outlines and proportions, it differing in both from *Phoca* in its generally much slighter structure, very small teeth, flatness of the dorsal aspect of the brain-case, and the slenderness of the muzzle and whole facial region, as well as in the form of the lower jaw. *Pagophilus* differs from *Pusa* in having the posterior nares completely divided into two distinct passages, by the complete ossification of the narial septum, in the broad form of the scapula, and in having only three, instead of four, anchylosed sacral vertebræ—characters possibly of generic rather than subgeneric value. The well-known representatives of these groups are respectively *Phoca fœtida* and *P. grœnlandica*, to which are to be referred also the *Phoca caspica* and the *Phoca sibirica*.*

As already noticed (*anteà*, p. 417) the name *Phoca*, by strict adherence to rules of nomenclature, should be reserved for the *Phoca leonina*, Linné, this being the only Linnean species of *Phoca* left after the removal (in 1824) of *Phoca vitulina* as the type of F. Cuvier's genus *Callocephalus*.† *Pusa* of Scopoli, 1777, with *Phoca fœtida* as the type, however, long antedates *Callocephalus*, and would be strictly the name of the group were *Phoca* set aside. Yet as *Pusa* may be deemed by some as untenable, and as to restrict *Phoca*, on, at best, a slight technicality, to what is now called *Macrorhinus*, would be to subvert all the traditions of nomenclature relating to the generic name of our smaller Phocids, it seems best not to attempt, on so slight a pretext, a change in nomenclature that would doubtless be received with reluctance, if indeed it could be for a long time brought into general use.

Callocephalus has been in more or less general use for the smaller Phocids ever since it was proposed by F. Cuvier in 1824, especially among the earlier French writers, and it has

---

* If the *Phoca nummularis* of Temminck prove to be a valid species its closest affinities are doubtless with *P. vitulina*. On this point see *infra*.

† In the same paper, but eight pages later, *Phoca leonina*, Linné, was made the type of the same author's genus *Macrorhinus*.

been also adopted by several English and German zoölogists. It has been given especial prominence by Dr. Gray and those who have followed his nomenclature. Latterly, however, Gray restricted it to the single species *Phoca vitulina*. Cuvier and his followers placed in it not only *Phoca vitulina, P. fœtida,* and *P. grœnlandica,* but also *P. barbata.*

As early as 1844, Gray removed *Phoca barbata* from the genus *Callocephalus* as the type and sole species of his restricted genus *Phoca.* At the same time *Pagophilus,* with *Phoca grœnlandica* as type, was proposed as a subgenus of *Callocephalus,* but was later raised by him to full generic rank. In 1844 the same author introduced the "genera" *Halicyon* and *Pagomys,* with "*Phoca richardi,*" Gray (=*Phoca vitulina*) as type of the former, while *Phoca fœtida* was chosen as the type of *Pagomys.* In 1866 he bestowed the generic name *Haliphilus* upon the species Gill a few months before had called *Phoca pealei.* This being merely Peale's *Halichœrus antarcticus,* and the same as Gray's *Halicyon richardsi,* the latter had now in use at the same time three generic names for the single species *Phoca vitulina,* namely, *Callocephalus, Halicyon,* and *Haliphilus!*

Pusa, proposed by Scopoli in 1777 as the generic name of a Seal previously figured and described by Müller, is perhaps not without objections, although Scopoli's sole reference being to Müller, the species intended is fixed beyond question, notwithstanding that his diagnosis is a pure absurdity. Müller's figure was copied from a plate published twelve years before by Houttyn, which latter was a copy of a figure published by Albinus in 1756. Albinus's better plate and description were evidently based on an adult female *Phoca fœtida.* The name *Pusa* was overlooked or ignored till Gill in 1872 revived it as a substitute for *Halichœrus.**

PHOCA (PHOCA) VITULINA, *Linné.*

Harbor Seal.

Phoca communis, LINNÉ, Mus. Ad. Fred. i, 1754, 5.
Phoca vitulina, LINNÉ, Syst. Nat., 1758, i, 38; ibid., 1766, i, 56; Faun. Suec., 1761, 2.—MÜLLER, Zool. Dan. Prod., 1776, 1.—SCHREBER, Säuget., iii, [1776?] 333, pl. lxxxiv (figure from Buffon).—ERXLEBEN, Syst. Reg. Anim., 1777, 583.—FABRICIUS, Faun. Grœnl., 1780, 9; Skriv. Naturh.-Selsk., ii, 1791, 98.—GMELIN, Syst. Nat., i, 1788, 63.—KERR, Anim. King., 1792, 123.—EDMONSTON, View of Zetland, ii, 1809, 292.—G.

* For a detailed history of *Pusa,* and a full transcript of Scopoli's diagnosis, see *infra* under genus *Halichœrus.*

560   PHOCA VITULINA—HARBOR SEAL.

CUVIER, Ann. Mus. d'Hist. Nat. Paris, xvii, 1811, 377 (physical and
intellectual faculties).—DESMAREST, Nouv. Dict. d'Hist. Nat., xxv,
1817, 583, pl. xliv, fig. 3; Mam., 1820, 244.—DUVERNOY, Mém. Mus.
d'Hist. Nat. Paris, ix, 1822, 49, 165 (anatomy).—SCORESBY, Voy. to
Greenl., 1823, 416.—E. SABINE, Parry's 1st Voy., Suppl., 1824, cxci.—
HARLAN, Faun. Amer., 1825, 107.—GODMAN, Am. Nat. Hist., i, 1826,
311.—F. CUVIER, Dict. des Sci. Nat., xxxix, 1826, 543.—SCHINZ, Naturg. u. Abild. der Säuget., 1827, 166, pl. lxiv; Synop. Mam., i, 1844,
480.—FLEMING, Hist. Brit. Anim., 1828, 17.—FISCHER, Syn. Mam..
1829, 236.—OWEN, Proc. Com. Zoöl. Soc. Lond., i, 1831, 151 (anatomy).—GRAY, Griffith's Cuv. An. King., v, 1837, 176.—BELL, Brit.
Quad., 1837, 263, figg. (skull and animal); ibid., 2d ed., 1874, 240.—
NILSSON, "Vet. Akad. Handl., 1837, ——; Ill. Fig. till Skand. Faun.
ii, 1840, häft 20;" Arch. für Naturg., 1841, 310; Skand. Faun.,
Daggdj., 1847, 276.—MACGILLIVRAY, Brit. Quad., 1838, 199, pl.
xviii.—BALL, Trans. Roy. Irish Acad., xviii, 1839, pl. iv, figg. 11-13,
pll. v, vi; Sketches of Brit. Seals, 1839, pl. viii, figg. 23-25 (animal),
pl. ix, figg. 26-32 (anatomy).—HAMILTON, Amphib. Carn., 1839, 127,
pll. ii-iv, vi.—RICHARDSON, Zoöl. Beechey's Voy., 1839, 6 (northwest
coast of North America).—BLAINVILLE, Ostéog., Phoca, 1840-1851,
pl. ii (skeleton), pl.v (skull), pl. ix (dentition).—JUKES, Excurs. in
Newfoundland, i, 1842, 309.—THOMPSON, Nat. Hist. Vermont, 1842,
38; ibid., Append., 1853, 13 (Lake Champlain).—WAGNER, Schreber's Säuget., vii, 1846, 26, pl. 84.—'GAIMARD, Voy. en Islande et au
Groënl., 1851, Atlas, pl. xi, figg. 3-5."—"KÖRNER, Skand. Daggdj.,
1855, pl. xi, fig. 1."—GIEBEL, Säuget., 1855, 136.—BLASIUS, Naturg.
Wirb. Deutschl., i, 1857, 248, figg. 136, 137 (skull).—GUNN, Zoölogist, 1864, 9277, 9359.—"HOLMGREN, Skand. Daggdj., 1865, 213."—
GILL, Proc. Essex Inst., v, 1866, 12.—PACKARD, Proc. Bost. Soc. Nat.
Hist., x, 1866, 270 (Labrador).—LLOYD, Game Birds and Wild Fowl
of Sweden and Norway, 1867, 381, colored plate.—ALLEN, Bull. Mus.
Comp. Zoöl., i, 1869, 193 (Massachusetts); ibid., ii, 1870, 25 (comparison with Otariidæ).—REEKS, Zoölogist, 1871, 2541 (Newfoundland).—CORDEAUX, Zoölogist, 1872, 3203 (Lincolnshire coast, England).—ELLIOTT, Cond. of Affairs in Alaska, 1875, 121.—CLARK,
Proc. Zoöl. Soc. Lond., 1874, 556 (Vancouver's Island and California, etc.).—CORNALIA, i, 62 (Mediterranean).—LILLJEBORG, Fauna
öfver Sveriges och Norges Ryggrads., 1874, 672.—COLLET, Bemærkninger til Norges Pattedyrfauna, 1876, 56.—VAN BENEDEN, Ann. du
Mus. Roy. d'Hist. Nat. du Belgique, i, 1877, 19 (geographical distribution, with chart).—RINK, Danish Greenland, its People and its
Products, 1877, 123, 430.—ALSTON, Faun. Scotl., Mam., 1880, 13.

Callocephalus vitulinus, F. CUVIER, Dict. Sci. Nat., xxxix, 1826, 540. —
GRAY, "Zoöl. Erebus and Terror, 1844, 3;" Cat. Osteol. Spec. Brit.
Mus., 1847, 32; Cat. Seals Brit. Mus., 1850, 21, fig. 7; Cat. Seals
and Whales, 1866, 20, fig. 7; Zoölogist, 1872, 3333, 3335 (Brit. Isl.);
Hand-List Seals, 1874, 2, pl. i.—BROWN, Proc. Zoöl. Soc. Lond.,
1868, 340, 411; Man. Nat. Hist., etc., Greenl., 1875, Mam., 41.—
MALM, Göteborgs och Bohusläns Fauna Ryggradsdjuren, 1877, 144.

Phoca variegata, NILSSON, Skand. Faun., i, 1820, 359.

" *Phoca scopulicola,* THIENEMANN, Reisen von Nord. Europa, etc., i, 1824, 59, pl. v (ad. male)" (*apud* Wagner, Blasius and others; referred to *Halichœrus grypus* by Gray).

" *Phoca littorea,* THIENEMANN, Reisen von Nord. Europa, etc., i, 1824, 61, pl. vi (male), pll.vii, viii (skulls)" (*apud auctori var.*).

? *Phoca tigrina,*, LESSON, Man., de Mam., 1827, 206 (="Phoque tigré," Kraschenninikow, Hist. Kamtsch.).

Phoca linnœi, LESSON, Dict. class. d'Hist. Nat., xiii, 1828, 414 (= *P. vitulina,* Linné).

? *Phoca chorisi,* LESSON, Dict. class. d'Hist. Nat., xiii, 1828, 417.—FISCHER, Syn. Mam., 1829, 241 (="Chien de mer de Détroit de Behring," Choris, Voy. Pittoresq., pl. viii).

Phoca canina, PALLAS, Zool. Rosso-Asiat., i, 1831, 114 (at least in part).

? *Phoca largha,* PALLAS, Zool. Rosso-Asiat., i, 1831, 113.

? *Callocephalus largha,* GRAY, Cat. Seals, 1850, 24 (= *P. largha,* Pallas).

? *Pagomys? largha,* GRAY, Cat. Seals and Whales, 1866, 24 (same).

? *Phoca nummularis,* TEMMINCK, Faun. Japon., Mam. Mar., 1842, 3 (=? *P. largha,* Pallas).—WAGNER, Schreber's Säuget., vii, 1846, 24 (same). —VON SCHRENCK, Amur-Lande, i, 1859, 180 (same).—MIDDENDORFF, Sibirish Reise, ii, Th. ii, 1853, 122 (same).

? *Pagomys? nummularis,* GRAY, Proc. Zoöl. Soc. Lond., 1864, 31 (=*Phoca nummularis,* Temminck).

Phoca concolor, DEKAY, New York Zoöl., i, 1842, 53, pl. xviii, fig. 2.

Phoca jubata, HUTCHING, Scenes of Wonder and Curiosity in California, 189, fig.

Halichœrus antarcticus, PEALE, Rep. U. S. Ex. Ex., viii (Mam. and Orn.) 1848, 30, pl. v, (animal) wood-cut (skull).

Lobodon carcinophaga, CASSIN, Rep. U. S. Ex. Ex. (=*Halichœrus antarcticus,* Peale).

Haliphilus antarcticus, GRAY, Ann. & Mag. Nat. Hist., xvii, 1866, 446 (= *Halichœrus antarcticus,* Peale).

Halicyon richardsi GRAY, Proc. Zoöl. Soc. Lond., 1864, 28 (Vancouver's Island); Cat. Seals and Whales, 1866, 30, fig. 9 (skull); Proc. Zoöl. Soc. Lond., 1873, 779 (Japan); Hand-List Seals, 1874, 4. (See CLARK, Proc. Zoöl. Soc. Lond, 1873, 556.)

Phoca pealei, GILL, Proc. Essex Inst. v, 1866, 4, 13 (= *Halichœrus antarcticus,* Peale).—SCAMMON, Marine Mam., 1874, 164, pl. xxii, fig. 1 (="*Phoca pealii,*? Gill)."

Halicyon pealei, GRAY, Suppl. Cat. Seals and Whales, 1871, 2.

Phoca ———, NEWBERRY, Pacif. R. R. Rep. Ex. and Surv., vi, pt. iv, No. 2, 1857, 51.—COOPER, Pacif. R. R. Rep. Ex. and Surv., xii, pt. ii, 1860, 78.

Halicyon? californica, GRAY, Cat. Seals and Whales, 1866, 367 (= "Hair Seal, *Phoca jubata,*" of Hutching).

Phoca fœtida, BARTLETT, Proc. Zoöl. Soc. Lond., 1868, 402 (on the young at birth; a malidentificaton—see SCLATER, ibid., 1871, 701).

Vitulus maris Oceani, RONDELET, De Piscibus, 1554, 458, with fig.—GESNER, Hist. Anim. de Pisci. et Aquat., 1558, 705, (fig.).—ALDROVANDUS, De Piscibus, 1738, 723.

" *Vitulus marinus,* OLAUS MAGNUS, Hist. de Gent. Sept. 1555, 701."

Phoca vulgaris, JONSTON, Hist. Nat. de Piscibus et Cetis, 1649, 221, pl. xliv.

Phoca seu vitulus marinus, RAY, Syn. Quad. 1713, 189.—BRISSON, Reg. Anim.,
    1760, 162.
Phoca, WORM, Mus. Worm., 1655, 289.
Phoca dentibus caninis tectis, LINNÉ, "Syst. Nat., 36"; Fauna Suec., 1746, 4.
Phoca mediæ magnitudinis, STELLER, Nov. Comm. Petrop., ii, 1751, 290.
Spraglet, EGEDE, Gronlands Naturel-Hist., 1741, pl. facing p. 46.
Robbe, Sælhund, PONTOPPIDAN, "Norg. Nat. Hist., ii, 1752, 203, fig."; Nat.
    Hist. von Norwegen, ii, 1753, 237.
Kassigiak, CRANZ, Hist. von Grönl. (zweite Aufl.) i, 1770, 163 (1st ed. 1765,
    not seen); Hist. of Greenl., i, 1767, 123.
Phoque de nôtre Océan, BUFFON, Hist. Nat., xiii, 1765, 333, 339, pl. xlv.
Phoque commune, BUFFON, Hist. Nat., Suppl., vi, 1782, 330, pl. xlvi.—G.
    CUVIER, Oss. Foss., iv, 1823, 278; v, 200.—F. CUVIER, Hist. Nat.
    des Mam., livr. xli, 1824.
Seal, PENNANT, Brit. Zoöl., 1766, 34.
Common Seal, PENNANT, Syn. Quad., 1771, 339; Arctic Zoöl., 2nd ed. i, 1792,
    175.—PARSONS, Phil. Trans., xlvii, 1753, 120, pl. vi.
Leopard Seal, SCAMMON, l. c.
Sœl; Sœlhund; Landsœl; Spragled Sœl, Danish.
Knubbsjäl; Spräcklig Skäl; Wikare Sjœl, Swedish.
Steen-Kobbe, Norwegian.
Seehund; Gemeine Seehund; Gefleckte Seehund; Robbe; Seekalb, German.
Phoque; Phoque commune; Veau-marin; Loup-marin; Chien-marin; French.
*Seal; Common Seal; Harbor Seal; River Seal; Bay Seal; Land Seal; Fresh-
    water Seal; Sea-Calf; Sea-Cat; Sea-Dog*, English authors and Eng-
    lish local names.
Selchie; Selch (Scotland); *Tangfish* (Shetland); *Rawn* (Hebrides).
Native Seal; Ranger; Dotard, Newfoundland sealers.

EXTERNAL CHARACTERS.—Color variable. Above, usually
yellowish-gray, varied with irregular spots of dark brown or
black; beneath, yellowish-white, usually with smaller spots of
dark-brown. Sometimes uniform brownish-yellow above, and
somewhat paler below, entirely without spots; or uniform dark-
gray above, and pale yellowish-white below, everywhere unspotted. Not unfrequently everywhere dark-brown or blackish,
varied with irregular streaks and small spots of yellowish-
brown; the head wholly blackish from the nose to beyond the
eyes; the lips and around the eyes rusty-yellow. Length of
male, 5 to 6 feet; of female, somewhat less. Young at birth
uniform soiled-white or yellowish-white, changing to darker
with the first moult.

The variations in color are almost endless, ranging from uniform yellowish-brown to almost uniform dark-brown, and even
nearly black, with, between these extremes, almost every possible variation, from dark spotting on a light ground to light

spotting on a dark ground. The markings vary in size from very small spots to large, irregular patches and streaks. The more common color is brownish-yellow, varied with spots and

FIG. 43.—Phoca vitulina.

patches of darker, but not unfrequently the general color is blackish, more or less varied with spots, patches and streaks of lighter. The lower surface is generally thickly marked with small oval or roundish spots, smaller and less confluent than those of the upper surface. Specimens from Denmark and the Atlantic coast of North America are indistinguishable from those from Lower California, Washington Territory, and Alaska. Specimens from the Pacific coast present the same wide range of color-variations, and precisely the same phases as those from the shores of the Atlantic.

Captain Scammon gives the weight of two adult females from the Strait of Juan de Fuca, as 56 and 60 pounds respectively. Mr. Michael Carroll gives the weight of adults (sex not stated) as 80 to 100 pounds. Mounted specimens, apparently adult, vary in length from three to five feet. Scammon says that on the Pacific coast it " never exceeds six feet in length," and gives the length of the two above-mentioned females as respectively from "tip of nose to tip of tail" 3 feet 8 inches and 3 feet 10 inches. Mr. Paul Schumacher gives the length of a "female Marbled Seal," sent to the Museum of Comparative Zoölogy from Santa Barbara Island, California, as 6 feet from tip of nose to the end of the hind flippers, which would make the length to the end of the tail about 5 feet 6 inches. Lilljeborg gives the total length to the end of the hind flippers as 5 feet 8 inches (Swedish) or 1740 mm. Bell says, " Length of adult from three to five feet ". Authors generally give the length as from 3 to 6 feet. I find the length of an adult (disarticulated) skeleton to be about 4 feet, or 1225 mm., while Lilljeborg gives the length of the skeleton as 5 feet 1 inch or 1530 mm. In the large series of skins and skulls I have examined very few were marked for sex, and I find nothing explicitly stated by authors in relation to sexual difference in size.

Unlike the *Phoca fœtida*, *P. grœnlandica*, and most other Phocids of the northern waters, the first coat is shed before or soon after birth, but as to the exact time at which it is cast authorities disagree. Mr. Bartlett, in describing a young Seal of this species (wrongly identified at the time as *Phoca fœtida*), born in the Garden of the London Zoölogical Society June 8, 1868, says : " It was born near the edge of the water, and in a few minutes after its birth, by rolling and turning about, was completely divested of the outer covering of *fur* and *hair*, which formed a complete mat, upon which the young animal lay for the hour or two after its birth ".*

---

* Proc. Zoöl. Soc. Lond., 1868, p. 205.

It is sometimes stated that the fœtal coat is retained for four or five days after birth, but other writers affirm that it is shed at the time of birth. Mr. Kumlien, in his MS. notes* on this species, says that the Esquimaux affirm that the "young remain in the white coat but three or four days, differing greatly in this respect from *Pagomys fœtidus*."

DISTINCTIVE CHARACTERS.—The common Harbor Seal, the Ringed Seal, and the Harp Seal, during its earlier stages, are not always certainly distinguishable by color, and are apparently not easy to determine by any other external characters, save one, that have yet been pointed out. The Ringed Seal (*Phoca fœtida*) can always be recognized by the length of the first digit of the manus, which slightly exceeds all the others. When adult, and in the flesh, they must each present well-marked external differences, not only in color but in proportions and form. *P. fœtida* is the smallest of the three, while the Harp Seal (*P. grœnlandica*) is the largest, and when adult, is easily distinguished by coloration alone. *P. vitulina*, judging from the skeleton, is a comparatively robust form, with a large head, broad nose, and rather short limbs. The others are more slender, with a narrower and more pointed nose, and a smaller and more delicately shaped head. By the skull, or by any of the principal bones of the skeleton, particularly of the limbs, they can be easily distinguished, as will be shown by the following rather extended osteological comparisons, with the material for which I am fortunately well provided.

As is well known, *P. vitulina* is easily distinguished from the other species of *Phoca* above named by its heavy dentition, the molars especially being very broad and thick, closely crowded together and set obliquely in the jaw, whereas in both *P. fœtida* and *P. grœnlandica* the teeth are very small, normally implanted, and separated by well-marked diastema. They also

---

* Mr. Ludwig Kumlien, naturalist of the "Howgate Polar Expedition" (1877-'78), kindly placed at my service his report, while in manuscript, on the mammals collected and observed by him in and near Cumberland Sound, from which the extracts given in the following pages as from Mr. Kumlien's "MS. notes" were taken. A year later, and as these pages are passing through the press, his full report has appeared as "Bulletin No. 15" of the United States National Museum, under the title "Contributions to the Natural History of Arctic America, made in connection with the Howgate Polar Expedition, 1877-'78. By Ludwig Kumlien, Naturalist of the Expedition." Washington: Government Printing Office. 1879. 8vo, pp. 179. The account of the Seals occupies pp. 55-64.

have the cingulum smooth, while in *P. vitulina* it is more or less distinctly beaded (sometimes striated) on the anterior portion of the inner side, especially in early life, and on the three anterior teeth. In old age, however, this feature often becomes wholly obliterated. The oblique position of the teeth in the jaw evidently results from their large size, the space for their reception being too short to permit of their standing end to end in the usual manner. Their large size also results in, or necessarily accompanies, a considerable modification of the whole facial portion of the skull, which is greatly thickened and broadened, in comparison with the same part in the other above-mentioned species. Passing to the palatal region, *P. vitulina* and *P. fœtida* present an essential agreement, the posterior nares in both being rather abruptly narrowed posteriorly; the hind border of the palatines is deeply hollowed, and the narial septum is imperfectly developed at its postérior border. In *P. fœtida* it remains wholly unossified behind the palato-maxillary suture, except the buttress-like extensions along the narial roof and floor, and ossification of the septum is carried but little further in *P. vitulina*. In *P. grœnlandica*, however, the septum is fully and even heavily developed to the very end of the squarely truncated hind border of the palatines, dividing vertically the posterior narial opening, which is scarcely at all contracted, into two distinct passages. Its transverse breadth is nearly twice its vertical width, while in *P. vitulina* these dimensions are nearly equal.

The auditory bullæ differ considerably in form in each of the three species here compared. In general form they have in each the outlines of a nearly equilateral triangle, but the sides are set in each at a different angle relatively to the transverse axis of the skull. In *P. fœtida* the anterior border is nearly parallel with the plane of this axis; in *P. vitulina* the two form an acute angle, while in *P. grœnlandica* they form nearly a right angle. The anterior face of the bullæ is nearly plane in *P. fœtida*, strongly hollowed in *P. grœnlandica*, and slightly so in *P. vitulina*. In both *P. fœtida* and *P. grœnlandica* the lateral extension forming the lower border of the meatus auditorius is depressed and swollen or rounded below, forming an abruptly constricted neck to the bulla proper, but in *P. vitulina* it slopes evenly from the highest part of the bulla and terminates in a uniformly tapering triangular point.

The facial portion of the skull, as already intimated, is broad and heavy in *P. vitulina*, to give room and support for the thick strong teeth; in *P. fœtida* it is short and narrow, and uniformly tapering; in *P. grœnlandica* the muzzle is narrow, rather lengthened and attenuated. The teeth of the molar series in *P. vitulina* are relatively about two and a half to three times larger than in either of the other species.

In respect to other characters of the skull, the orbital fossæ are relatively larger in *P. fœtida* than in either of the others, with the inner wall more deeply excavated, and the zygomatic border rather angular (sometimes very markedly so) instead of regularly convex. *P. vitulina* differs still further in the greater development and inward curvature of the malar process of the zygomatic arch. Another striking difference is seen in the general contour of the upper surface of the skull, which in *P. vitulina* is rather sharply convex, with (in old males) the ridges formed for the attachment of the masseter muscles closely approximated along the median line, or sometimes actually meeting to form a low, broad, incipient sagittal crest, while in both *P. grœnlandica* and *P. fœtida* the whole top of the skull is nearly flat, and the ridges for the attachment of the masseter muscles form a thickened line at the edge of the skull where the lateral and dorsal surfaces meet at a rather sharp angle.

The lower jaw in *P. vitulina* is very heavy and short; the symphysis is very short, behind which the rami abruptly bow outward and widely diverge; the rami are very thick, with the axis of expansion nearly vertical, and there is no inward curvature of the inferior border. In *P. grœnlandica* the lower jaw is very slender with a rather long symphysis; the rami are very thin and broad, the inferior borders of which curve inward so as to nearly or quite meet for one-third of the length of the jaw, or nearly as far back as the last molar, while the plane of vertical expansion is very oblique. The lower jaw in *P. fœtida* quite nearly resembles, in general form, that of *P. grœnlandica*. In *P. vitulina* the vertical diameter of the ramus just behind the last molar is only about two and a half times greater than the transverse is at the same point, while in *P. grœnlandica* it is fully four times greater. *P. vitulina* also differs from the others by the abrupt angle formed by the ascending ramus.*

* As will be noticed later (*infra*, p. 573) the lower jaw in *P. vitulina* varies greatly in form and stoutness with age, and probably also with sex. In the foregoing comparison the lower jaw of a very old male *P. vitulina* has been compared with others corresponding in age and sex of *P. grœnlandica* and *P. fœtida*.

Without going into a detailed comparison of the bones of the general skeleton, a few points may be briefly noticed. The scapula has nearly the same general outline in both *P. vitulina* and *P. fœtida*, but differs widely from that of *P. grœnlandica*, mainly through the great development of the posterior upper portion of the blade, which is greatest in *P. vitulina*. In other words the scapula in *P. grœnlandica* is less "sickle-shaped" than in the others owing to the greater development of the pre-scapular portion and the less development of the post-scapular part. In *P. fœtida* the infra-acromial portion is much elongated, so that although the scapula is much smaller than in *P. vitulina*, its length is greater. In *P. fœtida* the length to the breadth* is as 1 to 0.847; in *P. vitulina*, as 1 to 1.155; in *P. grœnlandica* as 1 to 0.908. The width of the post-scapular fossa to the whole breadth is, in *P. fœtida*, as 0.577 to 1; in *P. vitulina* as 0.609 to 1; in *P. grœnlandica* as 0.664 to 1.

The bones of both the fore and the hind limbs vary considerably in size and form with each species, but only the difference in the relative length of the several segments of the limb, compared with its whole length, will be here noted. In both fore and hind limbs the second segment is relatively much shorter in *P. vitulina* than in either *P. fœtida* or *P. grœnlandica;* in *P. grœnlandica* the pes is relatively much lengthened while the manus is of the same length as in *P. vitulina*. The proportionate length of the femur to the tibia varies as follows: in *P. fœtida* the femur to the tibia is as 1 to 2; in *P. vitulina* as 1 to 1.8; in *P. grœnlandica* as 1 to 2.3. The proportionate length of the femur to the pes is as 1 to 2.7 in *P. fœtida*, 1 to 2.4 in *P. vitulina*, and 1 to 2.9 in *P. grœnlandica;* of the femur to the whole limb, respectively, 1 to 5.7; 1 to 4.3; 1 to 6.2. This difference is mainly due to two elements of variation,—the shortness of the tibia in *P. vitulina* and the great length of the pes in *P. grœnlandica*. The ratio of the pes to the whole limb, however, is nearly constant, being as follows: in *P. fœtida*, 1 to 2.14; in *P. vitulina*, 1 to 2.16; in *P. grœnlandica*, 1 to 2.13.

P. vitulina presents another noteworthy point of difference from its allies in the relative shortness of the pelvic bones, which is directly proportionate to that of the tibia and radius, or second limb segments. In both *P. vitulina* and *P. grœnlandica* the length

---

* The supra-scapular epiphysis is in each case omitted from the comparison, and only the scapulæ of adults of comparable ages are employed.

of the pelvis is precisely that of the tibia, being respectively 200 mm. and 255 mm. *P. fœtida* presents a different ratio, due not to the shortness of the pelvis so much as to the great length of the tibia, the tibia measuring 190 mm. and the pelvis 170 mm.

Not only is the dentition exceptionally heavy in *P. vitulina*, but the whole skull is ponderous, in striking contrast with the light thin skull of either of the other species. In other words, *P. vitulina* is a *big-headed*, short-bodied, and short-limbed species. While the linear dimensions of old male skulls fully equal or somewhat exceed the same measurements of equally old male skulls of *P. grœnlandica*, the length of the limbs, and also the entire skeleton, is much less, as shown by the following measurements : *

Species.	Length of the—			
	Skull.	Fore limb.*	Hind limb.	Whole skeleton.
	Millimeters.	*Millimeters.*	*Millimeters.*	*Millimeters.*
Phoca vitulina†	220	499	582	1,225
Phoca grœnlandica†	210	550	677	1,630
Phoca fœtida‡	163	534	450	1,308

* Including scapula.　　† Adult male.　　‡ Adult female.

The fore limb, as well as the total length of the animal, is even actually shorter than in *P. fœtida*, although the latter is a much smaller animal. The ratio of the length of the skull to the length of the whole skeleton in the three species in question is as follows: in *Phoca vitulina*, as 1 to 5.6; in *Phoca grœnlandica*, as 1 to 8; in *Phoca fœtida*, as 1 to 8.6. Measurements (in millimeters) of the principal parts of the skeletons of these three species are presented in the following table, from which it will be seen that the shortness of the caudal vertebræ in *Phoca vitulina* is also a noteworthy point. This is due in part

---

* The largest skull of *P. vitulina* in a series of ten measures 223 mm. in length and 144 mm. in extreme breadth, while the largest skull of *P. grœnlandica* in a series of twelve measures 228 mm. in length and 133 mm. in breadth. No other in the series, however, exceeds a length of 220 mm. or a breadth of 128 mm., old skulls of *P. vitulina* averaging the longer, with the breadth very much greater.

to the small size of these vertebræ, but in part to their reduced number.[*]

Measurements of the principal parts of the skeleton in Phoca vitulina, Phoca grœnlandica, and Phoca fœtida.

Principal parts.	Phoca vitulina.	Phoca grœnlandica.	Phoca fœtida.
Length of the skull	220	210	163
Length of the cervical vertebræ	210	240	200
Length of the dorsal vertebræ	445	510	410
Length of the lumbar vertebræ	216	255	190
Length of the sacral vertebræ[†]	120	100	100
Length of the caudal vertebræ	230	317	245
Length of the scapula	135	152	137
Length of the humerus	114	123	100
Length of the radius	105	130	95
Length of the manus	145	145	128
Length of the pelvis	200	255	170
Length of the femur	112	109	94
Length of the tibia	200	255	190
Length of the pes	270	313	250

[*] The vertebral formulæ of the five species of northern Phocids, of which I have before me several complete skeletons of each, are as follows:

Species.	Cervical vertebræ.	Dorsal vertebræ.	Lumbar vertebræ.	Sacral vertebræ.	Caudal vertebræ.
Phoca vitulina	7	15	5	4	10
Phoca fœtida	7	15	5	4	14
Phoca grœnlandica	7	15	5	3	13
Erignathus barbatus	7	15	5	4	13
Cystophora cristata	7	15	5	4	10

The number of sacral and caudal vertebræ, especially the latter, may be subject to individual variation, as Gerrard (Cat. Bones of Mam. in Brit. Mus., 1862, p. 143) gives 3 sacral and 12 caudal for *P. vitulina*. Lilljeborg also gives only 3, but adds that a very old skeleton has 4. In each of the four skeletons of this species which I have examined 4 vertebræ are firmly anchylosed to form the sacrum; in three of these the caudal series is imperfect. Gerrard gives also only 3 sacral for *Erignathus barbatus* (Ibid., p. 145), while in two skeletons I find 4 in each.

. Lilljeborg gives the length of the tail in *Phoca vitulina* at 69 mm. and in *Phoca fœtida*, male, 150 mm., female, 144.

[†] In two very old skeletons of *Phoca grœnlandica* I find the sacrum to consist of only three anchylosed vertebræ. In both the other species the sacrum consists of four anchylosed vertebræ.

In respect to other features, it may be added that the relative length of the first and second phalanges of the thumb in *P. vitulina* and *P. grœnlandica* is reversed. While the length of the whole digit is nearly the same in the two, the phalanges notably vary, the first phalanx being short and the second long in *P. vitulina*, while in *P. grœnlandica* the first is long and the second short. The relative length of the digits of the manus is nearly the same, in both, *P. fœtida*, as already stated, being easily distinguished from either by its having the first digit decidedly the longest and the others successively shorter.

INDIVIDUAL AND SEXUAL VARIATION.—The wide range of color-variation has been already noted in the general description of the external characters, and this appears to be in great measure independent of either sex or age. After allowing for the thickening of the bones and the development of rugosities for the attachment of muscles, there still remains a considerable range of variation in the skull and other bones that may be considered as purely individual. The variations in the skull are shown to some extent by the subjoined table of measurements, but unfortunately very few of the skulls I have examined have been marked for sex. Those known to be those of aged males are noticeably the largest and heaviest, and the most roughened by tuberosities and incipient crests. The largest and heaviest of all is a very old male skull from Santa Barbara Island, California, but this is nearly paralleled by another from the coast of Massachusetts, also that of a very old male. In adult skulls ranging in length from 210 mm. to 223 mm., the greatest width varies from 124 mm. to 144 mm., with corresponding variations in the dimensions of special parts. The nasal bones vary in both length and width fully twenty-five per cent. of their mean dimensions. There is an equally great amount of variation in the width of the muzzle, and nearly as great in the bones of the palatal region. The form and size of the narial openings are especially subject to variation, as shown in the subjoined table.

In both *Phoca fœtida* and *Phoca grœnlandica* the female skull is much smaller, lighter, and weaker in structure than the male skull, and I believe that corresponding sexual differences in the skull obtain in *Phoca vitulina*, if indeed they are not even still more strongly marked. Among old skulls two well-marked forms occur, differing in the one being much less massive, smaller, and every way slighter than the other, with the facial portion of the skull narrower and the teeth smaller, the lower jaw very much weaker and narrower, and the rami much less bowed outward, scarcely more so than in *P. grœnlandica*. In this slighter form the teeth are so much smaller that occasionally they are placed (in the upper jaw especially) end to end instead of being set obliquely, and even sometimes slightly spaced. Although the skulls are unmarked as to sex, I believe the slighter skulls to be those of females.

The sexual differences in size and cranial characters in this, the *common* Seal of our temperate American and European waters, appear to still remain inadequately investigated, and series of sexed examples seem to be still desiderata in our best collections. The only reference to the subject that I recall are the following incidental observations by Mr. John W. Clark, who, in discussing the assumed distinctive characters of Dr. Gray's *Halicyon richardsi*, says: "The thickening of the lower jaw may be a sexual distinction. A skull, unquestionably of a male, possesses it in a marked degree, while that of a female of apparently about the same age, is slender." [*]

My attention has been forcibly drawn to this matter by a skull (No. 6783, Nat. Mus.) from Plover Bay (Siberian coast of Behring's Straits), which I at first referred unhesitatingly to *Phoca vitulina*, when examined in connection with a large series from both the Atlantic and Pacific coasts of America, but later, when compared again with a smaller series, I thought it might represent a form closely allied to, but still specifically distinct from, *P. vitulina*—probably the so-called *Phoca "nummularis"*. On collating it again with the full series at first examined it seemed undoubtedly to be only an old female of *P. vitulina*. Aside from the general slighter and more delicate structure of the skull, the most notable differences are the smaller, normally implanted, and even slightly spaced molar teeth, the narrowness of the facial portion of the skull, and the corresponding narrowness of the lower jaw and absence of the abrupt outward curvature of the rami at the last molar. In general form the lower jaw is much like that of *P. grœnlandica*, except that the vertical width of the ramus is much less, and the plane of its vertical expansion not nearly so oblique. Other skulls, which are undoubtedly those of *P. vitulina*, so closely resemble this that it is impossible to regard it as otherwise than an exceptionally attenuated female skull of *P. vitulina*. One or two others in the series, also presumed to be female, have the teeth small and implanted in a

---

* Proc. Zoöl. Soc. Lond., 1873, p. 557.

straight line, while in still others the anterior teeth are so implanted, only the posterior two of the series being more or less oblique.

The variations attending increase of age are chiefly the gradual thickening of all parts of the bony framework of the skull, and in the males the development of all the processes for the attachment of muscles, of slight rugosities, an incipient sagittal crest, and a more abrupt outward curvature of the mandibular rami. It is also noteworthy that the teeth are frequently less crowded in the jaw and less oblique in position in the adult and old-age stages than during the earlier periods of development. It would seem hardly necessary to note the varying position with age of the ridges bounding the temporal muscles, since such variation is usually seen in mammals which have these ridges well marked, were it not that a difference in the position of the temporal ridges has been cited by Dr. Gray as a character distinctive of his so-called "*Halicyon richardsi*" as compared with *Phoca vitulina*.*   In very young animals the brain-case is smooth, showing no trace of the temporal ridges; later they are slightly marked and widely diverge; as the age of the animal increases these ridges become stronger and less divergent, and in very aged examples nearly or wholly meet along the median line of the skull, forming a low, broad crest, slightly divided along the middle by a shallow furrow, which may or may not widen posteriorly into a small flat triangular space. †

---

* Hand-List of Seals, etc., 1874, p. 5.

† For additional remarks on individual variation in the characters of the skull see Clark, Proc. Zoöl. Soc. Lond., 1873, pp. 556, 557.

Measurements of nine skulls of Phoca vitulina.

Catalogue number	Locality	Sex	Length	Breadth at mastoid process	Greatest breadth at zygomatic arches	Distance from anterior edge of intermaxillæ to end of pterygoid hamuli	Distance from anterior edge of intermaxillæ to hinder edge of last molar	Distance from anterior edge of intermaxillæ to meatus auditorius	Distance from anterior edge of intermaxillæ to glenoid process	Distance from palato-maxillary suture to end of pterygoid hamuli	Length of alveolar border of maxilla	Width of palatal region at posterior end of maxillæ	Nasal bones, length	Nasal bones, breadth anteriorly	Nasal bones, breadth at fronto-maxillary suture	Breadth of skull at canines	Least breadth of skull interorbitally	Breadth of posterior nares, vertically	Breadth of posterior nares, transversely	Breadth of anterior nares, vertically	Breadth of anterior nares, transversely	Greatest height of skull at auditory bullæ	Length of brain-case	Greatest width of brain-case	Length of lower jaw	Age
*3506	Greenland		170	106	101	86	53	115	110		66	50	42	13	10		12	13	28			74	75	87	104	Young.
*3634	Sable Island, Nova Scotia		221	125	144	122	79	165	157	49	95	62	68	20	13	45	11	20	19	32	32	82	82	98		Very old.
*4713	.....do.....		220	125	137	123	79	165	158	50	95	61	67	21		38	13	20	29	32	32	76	88	97	144	Very old.
†5144	Beverly, Mass.	♂	217	127	139	117	75	163	152	77	80	60	60	20	16	45	15	19	29	30	30	82	85	97	135	Very old.
*14337	St. Paul's Island, Alaska		210			115	76	157	149	48	89	52				38	13	17	28		27	73	82	92	138	Adult.
*6783	Plover Bay, Behring's Straits		208	124	116	113	73	156	144	43	86	51	48	19	12	38	11					75	87	96	138	Rather young.
†3648	"Deception Island"		196			103	70	145	136	39	80	58				36	13	18	28		18	70			130	Young.
*6486	Washington Territory		169	104	99	88	59	117	110	36	63	48	37	13	9	32	13	125	29	18	24	68	72	86	105	Young.
†6157	Santa Barbara Island, Cal	♂	223	130	144	120	76	168	155	42	85	63	61	20	16	48	15	23	28	24	30	82	80	98	142	Very old.

* National Museum, Washington, D. C.　　‡ Type of Peale's *Halichœrus antarcticus*, and of Cassin's *Lobodon carcinophaga*.
† Museum of Comparative Zoölogy, Cambridge, Mass.

GENERAL HISTORY AND SYNONYMY.—The common Seal is mentioned in the earliest works on natural history, having been described and rudely figured by various writers as early as the middle of the sixteenth century, as well as during the seventeenth century. Even down to the time of Linné it was the only species recognized; or, more correctly, all the species known were usually confounded as one species, supposed to be the same as the common Seal of the European coasts. Consequently almost down to the beginning of the present century the "common Seal" was generally supposed to inhabit nearly all the seas of the globe, Buffon, Pennant, Schreber, and others referring to it as an inhabitant of the Southern Hemisphere. Linné distinguished only a single species, even in the later editions of his "Systema Naturæ." As is well known, the smaller species of Seal are with difficulty distinguishable by external characters, particularly during their younger stages. Few, however, are so variable in color as the present, and none has so wide a geographical range. It is hence not surprising that its varying phases should have been made the basis of numerous nominal species. As shown by the above table of synonymy, the species was first introduced into systematic literature by Linné in 1754, under the name *Phoca communis*. He later changed its name to *Phoca vitulina*, which specific designation it has since generally retained. Although the name *vitulina* was, without doubt, based primarily on the animal commonly designated by that name, it originally covered references to other species, but its limitation to the species now under consideration has been so long currently accepted that only needless confusion would result from any hair-splitting device by means of which some later and more strictly applicable name might be substituted. To Fabricius is due the credit of first clearly discriminating the various species of Seals inhabiting the Arctic waters, and by him, in his classic memoir on the Seals of Greenland,* the present species was first described in detail, and its early literary history clearly set forth. Pennant, Schreber, Erxleben, and Gmelin, it is true, had already recognized other species, based, however, mainly on Fabricius's earlier work, "Fauna Grœnlandica," and on Müller's "Prodromus". The latter, so far as the Seals are concerned, rests also on Fabricius's manuscript notes published by Müller. Later, as al-

---

* Udförlig Beskivelse over de Grönlandske Sæle. Skriv. af Naturhistorie-Selskabet, 1ste Bind, 1ste Hefte, 1790, pp. 79–157; 2det Hefte, 1791, pp. 73–170.

ready indicated, numerous writers have contributed to its history.

The first synonym of note was published by Nilsson in 1820, who renamed the species *variegata*, but the name never came into general use, and was soon after abandoned by Nilsson himself, who later adopted *vitulina*. Thienemann, in 1824, described the species as *Phoca littorea*, and also in the same work gave the name *Phoca scopulicola* to an animal referred by some writers to *Phoca vitulina* and by others to *Halichœrus grypus*.* Lesson has furnished his usual quota of synonyms by giving to Linné's *Phoca vitulina* the name *Phoca linnœi*, and by bestowing the names *tigrina* and *chorisi* on the common spotted Seal of the North Pacific, the first being based on the *Phoque tigré* of Kraschenninikow from Kamtschatka, and the other on Choris's figure of his *Chien de mer* of Behring's Straits, neither of which can be positively determined, but may be referred with little doubt to the present species.

The *Phoca canina* of Pallas is generally conceded to be a synonym of *Phoca vitulina*. The same author's *Phoca largha*, applied to a Kamtschatkan Seal, has been the source of more trouble, it being too inadequately described to admit of positive determination. Temminck believed it to be identical with a Seal from Japan, and renamed it *Phoca nummularis*. In his remarks on the Seals of this region he says : "Le troisième Phoque des parages septentrionaux de l'océan pacifique nous est connu d'après trois jeunes individus et d'après un nombre égal de peaux incomplètes d'individus adultes, tous rapportés du Japon par M. M. de Siebold et Bürger. C'est évidemment le deuxième Phoque de Steller, Descr. du Camtsch. p. 107, et l'espèce dont Pallas fait mention en traitant du Phoque commun, l. c. [Zool. Rosso-Asiat.] p. 117, nota 2 ; puis le Phoque, figuré sans le moindre détail descriptif, dans le voyage de Choris, Pl. 8, sous le nom de Phoque du détroit de Behring ; peut-être convient-il également de rapprocher de cette espèce inédite le Phoca largha de Pallas, ibid. p. 113, n°. 43. Quoi qu'il en soit, nous avons cru devoir conférer à ce Phoque le nom qu'il porte, suivant Pallas, l. c. p. 117, chez les Russes, savoir celui de Phoque nummulaire, P h o c a n u m m u l a r i s." There being no evidence to the contrary, it may be assumed, as most subsequent writers have assumed, that Temminck's *Phoca nummu-*

---

* Thienemann's work, "Naturhistorische Bemerkungen gesammelt auf einer Reise im Norden von Europa, vorzüglich in Island in d. Jahrn 1820 bis 1821," I have been unable to see. Giebel and Blasius refer *P. scopulicola* to *P. vitulina*, while Gray assigns it to *H. grypus.*

laris and Pallas's *Phoca largha* are the same. Temminck has given a detailed description of the six skins above mentioned as received from Japan, and also of fragments of the skull removed from the skins of the young individuals. He notes especially the wide range of color-variation presented by his skins, each of which differs considerably from all of the others, the variations being greatest in the adult examples. He gives the length of the largest adult specimen as about five feet, and that of the young as two and a half to three feet. He describes one skin as having exactly the markings represented in Choris's figure, and says it has a close resemblance to certain varieties of the Ringed Seal. The coloration of these specimens, as described, presents nothing incompatible with their reference to either *Phoca fœtida* or *Phoca vitulina*, both of which species occur in the region in question.

The fragments of the skulls are not described in detail, but he says they serve to show that the skull of his *Phoca nummularis* greatly resembles that of the "Phoque à croissant, notamment par la configuration de la région interorbitaire, qui est, par devant, plus large que dans le crâne du Phoque annelé. Quant au système dentaire," he continues, "il n'offre pas la moindre disparité de celui du Phoque à croissant et du Phoque annelé." He states his conclusion as follows : " Ce Phoque est en quelque sorte intermédiaire entre le Phoque à croissant (Phoca oceanica) [=*Phoca grœnlandica*, auct.] et le Phoque annellé, (Phoca hispida, Schreber, Säugth., III, p. 312, n°. 6, Tab. 86 ; Phoca foetida, Müller, Prodr., p. 8 ; Phoca annellata, Nilsson, Skand. Fauna, I, p. 362) ; car il offre beaucoup d'analogie avec le premier par la configuration de son crâne, notamment par celle de la région interorbitaire ainsi par celle de ses dents, tandis qu'il se rapproche davantage du second par son système de coloration."*

Temminck's specimens have also passed under the inspection of Dr. Gray, who says: "This species [Gray's " *Pagomys?* *nummularis*," 1864, his " *Pagomys? Largha*", 1866] is only known from some skins and three fragments of skulls in the Leyden Museum. My excellent friend, Professor Schlegel, the energetic Curator of the Leyden Museum, has most kindly sent to me for examination and comparison the fragments of skulls above referred to : they consist of the face-bone and the lower

* Fauna Japon., Mam. Mar., p. 3.

Misc. Pub. No. 12——37

jaws of three specimens; the most perfect specimen has part of the orbit and the upper part of the brain-case attached to it. They are all from very young specimens, of nearly the same age; and, unfortunately, the most perfect one is without the hinder portion of the palate, so that one cannot make sure that it has the same form of palatine region that is found in *Pagomys;* but the part of the side of the palate that is present, when compared with the same part in *Pagomys*, leads one to think it most likely to be of the same form as in that species.

"The general form and size of the face, and the form of the teeth are those of a skull of *Pagomys fœtidus* of the same age. It only differs from the latter in the lower jaw being rather shorter and broader, in the grinders being larger, thicker, and rather closer together, in the central lobe of the grinders being considerably larger, thicker, and stronger, and in all the lobes of the grinders being more acute. The lower margin of the lower jaw is dilated in front, just as in *Pagomys fœtidus;* but the jaws behind the dilatation diverge more from each other, leaving a wider space between them at the hinder part. The form of the hinder angle of the jaws is very similar in the two species. The orbit is rather smaller and more circular; for in *P. fœtidus* it is rather oblong, being slightly longer than wide. The forehead appears, as far as one can judge by the fragments, to be flatter and broader, and the nose rather shorter." Dr. Gray also adds, in his diagnosis of the species: "The lower jaws short and broad; the grinders thick, with a broad, thick central lobe, and nearly side by side (in the skulls of the young animals)." He also gives comparative measurements of a skull of a young *P. fœtidus* and of *P. nummularis*, but with a good series of young skulls of the former, from the fœtal stage upward, I fail to fully understand his measurements.

"The *Phoca nummularis*," Dr. Gray continues, "has been considered to be identical with *Phoca Largha* of Pallas, from the east shore of Kamtschatka, the *Phoca Chorisii* of Lesson, and the *Phoque tigré* of Kraschennenikow (which has been named *Phoca tigrina* by Lesson), on the strength of their coming from nearly the same district; but I am not aware that specimens of any of the latter species exist to verify the union and determine what are the species described under these names."*

Although neither Temminck nor Gray makes any reference

* Proc. Zoöl. Soc. Lond., 1864, pp. 31, 32; Cat. Seals and Whales, 1866, pp. 24, 25. See also Gray's "Hand-List of Seals," etc., 1874, p. 6.

to *Phoca vitulina* in hiscomparisons of *Phoca nummularis* with other species, the distinctive characters given, so far as they relate to the skull, point decidedly toward *P. vitulina*, especially as respects the dentition, the width of the facial region, etc. The differences in the lower jaw, as compared with that of *P. fœtida* show its closer similarity to that of *P. vitulina*, and the larger, thicker, and more closely set teeth, with their larger and more acute cusps are also differences that point in the same direction. It is to be noted, however, that neither of these writers alludes to the mode of implantation of the teeth, but had they been set obliquely as is *usually* (not always) the case in *P. vitulina*, it is hardly to be supposed that Gray would have failed to so state. As already noted, in *P. vitulina* the size of the teeth, their mode of insertion, and the width of the anterior or facial portion of the skull are subject to considerable variation, the teeth being sometimes set end to end in a straight line, and even with slight spaces between them. Since, however, they are commonly more obliquely and more closely set in young skulls than in adult ones, it seems hardly probable that three young skulls of this species would by chance be found to agree in having the teeth inserted in a straight line, if, indeed, they were all sufficiently intact to show the dentition. The skull from Plover Bay, Behring's Straits (No. 6783, Nat. Mus.), already described (*anteà*, p. 572), seems to agree very closely with the characters given by Gray for Temminck's *Phoca nummularis* (=*"Pagomys? largha,"* Gray, 1866), yet I find in a large series of skulls (more than twenty have been examined) of *Phoca vitulina* a complete gradation between this extremely attenuated example and the very thick, heavy skulls of the oldest males.

While I do not deem it improbable, in view of all the facts of the case, that a species distinct from *Phoca vitulina* but of the same general type of structure, though slenderer, may exist in the North Pacific, and which may be referable to Temminck's *P. nummularis*, I feel disposed to leave the question open and for the present provisionally consider *P. nummularis* as a possible, if not a probable, synonym of *P. vitulina*.

The *Phoca concolor* of DeKay, unquestionably based on the light phase of the common Harbor Seal of our eastern coast, has been referred by nearly all European writers, often with expressions of doubt but frequently with entire positiveness, to *Phoca fœtida*, or the Ringed Seal, in consequence of its light silvery-gray color.

A curious mistake made by Peale in his (suppressed) report
on the Mammals and Birds of the United States Exploring Ex-

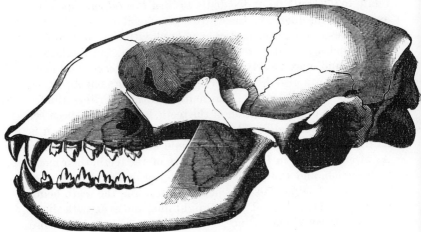

FIG. 44.—"Halichœrus antarcticus", Peale=Phoca vitulina, 4/7 natural size.

pedition under Commodore Wilkes has played an important
rôle in the technical history of the species. A specimen (skin

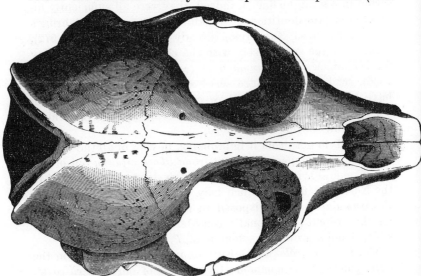

FIG. 45.—"Halichœrus antarcticus", Peale=Phoca vitulina, 4/7 natural size.

and skull,* still extant in the National Museum) was described
under the name *Halichœrus antarcticus,* with the remark, "This

* This skull (No. 3948/3741, Nat. Mus.) is represented in Figs. 44–47, four-
sevenths natural size.

species inhabits the Antarctic Sea. The specimen was obtained on the 10th of March at Desolation Island. It appears from the teeth to be an adult, and is the most perfect specimen brought home by the expedition." Notwithstanding this explicit statement of locality and date of capture, it is undoubtedly erroneous. Dr. Gill, in 1866, first called attention to the probable error, and says, "The '*Halichœrus antarcticus*' of Peale, very erroneously identified with *Lobodon carcinophaga* by Dr. J. E. Gray, is a typical species of *Phoca*, but appears to be identical with a species occurring along the California and Oregon coasts, and consequently there must be some error as to its original habitat in the Antarctic seas. I am happy to add

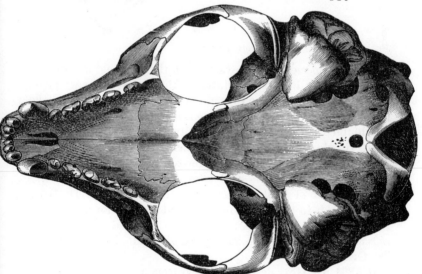

Fig. 46.—"Halichœrus antarcticus", Peale=Phoca vitulina, ½ natural size.

that Mr. Peale himself now doubts the correctness of the labels on the faith of which he gave its habitat, and as a change of name is desirable, I would propose that of *P. Pealii.*"* Having had opportunity of studying the type of the species, and of comparing it with a good series of skulls and skins of the common species of *Phoca* of the Pacific coast of the United States, I have not the slightest doubt as to the correctness of Dr. Gill's interpretation of the case.

Both in 1850 and in 1866 Dr. J. E. Gray included Peale's *H. antarcticus* among the synonyms of *Lobodon carcinophaga*, with

* Proc. Essex Inst., vol. v, 1866, p. 4, footnote.

which, of course, it has very remote affinities.   It seems, however, that he was subsequently not fully satisfied with this alloca-

Fig. 47.—"Halichœrus antarcticus", Peale=Phoca vitulina, ⁴⁄₅ natural size.

tion of the species for he says later in the year last mentioned,
"On rereading Peale's description, I think that it is very probably a new genus, for he says it has six cutting teeth in the
upper jaw, and that the four posterior molar teeth in both jaws
are double-rooted, their crowns many-lobed, the cutting teeth
short, simple and curved; the whiskers flattened, waved on the
edges.   To the animal so characterized the generic name *Haliphilus* may be given."*   Still later, on the basis of Dr. Gill's
determination of Peale's species, as previously noted, Dr. Gray
transferred it to his genus *Halicyon,* and adopted for it the name
Halicyon pealei.†   Cassin, in 1858, also referred Peale's *Halichœrus antarcticus* to *Lobodon carcinophaga,* doubtless following
Dr. Gray.

In 1864, Dr. Gray described a Seal from a skull from Vancouver's Island and a skeleton from Fraser's River, which he referred
not only to a new species but to a new genus, naming it *"Halicyon richardi."* ‡   The validity of the species seems to have
been first called in question by Mr. J. W. Clark in 1873, when
he compared Dr. Gray's type specimen with a skull from San
Francisco, and with others from Newfoundland, and also with
a series of skulls of *Phoca vitulina* from the English coast.   He

---

* Ann. and Mag. Nat. Hist., xvii, 1866, p. 446.
† Suppl. Cat. Seals and Whales, 1871, p. 2.
‡ The species was dedicated by Dr. Gray to " Captain Richard [*lege* Rich-
ards], the Hydrographer of the Admiralty," and ten years later Dr. Sclater
notes the fact that the name should be consequently *richardsi.—Proc. Zoöl.
Soc. Lond.,* 1873, p. 556, footnote.

notes a considerable amount of individual variation in different
skulls of this species, and finds that the characters given by
Dr. Gray for his *Halicyon richardsi* do not hold, and adds, "I
am therefore disposed, so far as present evidence goes, to con-

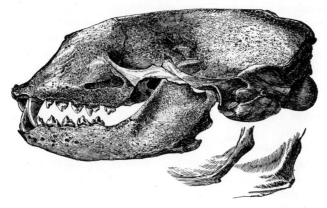

FIG. 48.—"Halicyon richardsi", Gray=Phoca vitulina.*

sider the so-called *Halicyon richardsi* simply a synonym of
Phoca vitulina".† Dr. Gray, however, in his "Hand-List of
Seals" (1874), still retained the species, referring to it three
specimens (two skulls and a skeleton), from, respectively, Fraser's River, Vancouver's Island, and Japan, and stating that
the skin was still unknown. He alludes to Mr. Clark's paper,
and calls attention to several points wherein he believes his
Halicyon richardsi differs from *Phoca vitulina*, and adds that
"though the skulls have some similarity, still there may be a
great difference in the external appearance of the animals." As
regards this point, it has been my good fortune to have access
to a considerable series of both skins and skulls, and I am unable to appreciate any well-marked features in either wherein
the so-called *Halicyon richardsi* differs from the common *Phoca
vitulina* of the North Atlantic. I have had, moreover, specimens
from the vicinity of the original localities whence the species
was described, as well as from points on the Pacific coast as
remote from each other as the Santa Barbara Islands and
Alaska.

---

* From an electrotype of Gray's original figure in Proc. Zoöl. Soc. London,
1864, p. 28. Gray's figure in his "Catalogue of Seals and Whales" (p. 28),
is apparently from the same.

† Proc. Zoöl. Soc. Lond., 1873, p. 556.

Still another synonym, for which we have also to thank Dr. Gray, remains to be considered, namely, his "*Halicyon? californica.*" This is based simply on "The Hair Seal (*Phoca jubata*)" of Hutching s " Scenes of Wonder and Curiosity in California" (p. 189), and is described by Dr. Gray as follows: "A seal without ears, with large, pale rings, which are more or less confluent. Inhab. California."* He quotes parts of Hutching's description, and adds, " It has no affinity to the *Phoca jubata* of the Systematic Catalogue." The extent of Hutching's scientific attainments, so far as the present matter is concerned, may be judged from the italicised portions of the following : " This seal, with which the *coast of California abounds*, is by no means rare, as *almost all the coasts in high southern and northern localities* abound with it."

In recapitulation of the foregoing, it may be noted that of the fourteen distinct synonyms here considered, three (*variegata, linnœi, canina*) are due to the intentional renaming of Linné's *Phoca vitulina;* that three others (*tigrina, chorisi, californica*) are based on the vague descriptions of travellers or unscientific writers; that two (*Lobodon carcinophaga, Phoca jubata*) are the result of malidentifications; and that an erroneous label of locality gave rise to one of the erroneous identifications and led to the introduction of one generic and two additional specific appellations. The four remaining synonyms (*littorea, scopulicola, concolor, richardsi*) have the palliating circumstances of locality and scantiness of material in their favor.

GEOGRAPHICAL DISTRIBUTION.—The Harbor Seal appears to have formerly been much more numerous on portions of our eastern coast than it is at present. Dr. DeKay, writing in 1842, states that the "Common Seal, or Sea Dog," is "now comparatively rare in our [New York] waters," though "formerly very abundant." He adds, "A certain reef of rocks in the harbor of New York is called *Robin's reef*, from the numerous seals which were accustomed to resort there ; *robin* or *robyn* being the name in Dutch for *Seal.* At some seasons, even at the present day, they are very numerous, particularly about the Execution rocks in the Sound; but their visits appear to be very capricious." He further alludes to their capture nearly every year in the Passaic River, in New Jersey, and states that a Seal was taken in a seine in the Chesapeake Bay, near Elko, Maryland, in August, 1824, supposed by Dr. Mitchill, who saw it, to be of this spe-

* Cat. Seals and Whales, 1866, p. 367.

cies.* Although still occasionally appearing on the coast of the
Atlantic States as far southward as North Carolina,† it is of
probably only accidental occurrence south of New Jersey, and
rare south of Massachusetts.

In respect to its occurrence on the New Jersey coast, Dr. C.
C. Abbott, the well-known naturalist of Trenton, N. J., kindly
writes me, in answer to my inquiries on this point, as follows:
"In going over my note-books, I find I have there recorded the
occurrence of seals (*Phoca vitulina*) at Trenton, N. J., as follows:
December, 1861; January, 1864; December, 1866; February,
1870; and December, 1877. In these five instances a single
specimen was killed on the ledge of rocks crossing the river
here and forming the rapids. In December, 1861, three were
seen, and two in February, 1870. A week later one was captured down the river near Bristol, Bucks County, Pennsylvania.
My impression is that in severe winters they are really much
more abundant in the Delaware River than is supposed. Considering how small a chance there is of their being seen when
the river is choked with ice, I am disposed to believe that an
occasional pair or more come up the river, even as high as
Trenton, the head of tide-water, and one hundred and thirty-
eight miles from the ocean.

"On examination of old local histories, I find reference to the
seals as not uncommon along our coast, and as quite frequently
wandering up our rivers in winter. I can find no newspaper
references to the occurrence of seals later than February or
earlier than December, but as historical references to climate,
as well as the memory of aged men still living, show conclusively that our winters are now much milder than they were
even fifty years ago, it is probable that seals did come up the
river earlier in past years.

"In conversation with an old fisherman, now seventy-six
years old, who has always lived at Trenton, and has been a

---

* New York Zoölogy, part i, 1842, pp. 54, 55.
† A recent record of its capture in North Carolina is the following, the
reference, I think, unquestionably relating to the present species:
"SOUTHERN RANGE OF THE SEAL.—The Wilmington, N. C., *Star* of February 28th, mentions the capture, in New River, Onslow County, of a large
female spotted seal, measuring about 7 feet in length, and weighing 250
pounds. This is an interesting note. The species must probably have been
the common harbor seal (*Phoca vitulina*). The same newspaper says one
was reported near Beaufort some time ago."—[W. E. D. SCOTT,] "*Country,*"
vol. i, No. 21, p. 292, March 16, 1878.

good observer, I learn that every winter, years ago, it was expected that one or more seals would be killed ; and that about 1840, two were killed in March, which it was supposed had accompanied a school of herring up the river.

"In my investigations in local archæology I have found, in some of the fresh-water shell-heaps, or rather camp-fire and fishing-village sites along the river, fragments of bones which were at the time identified as those of seals. I did not preserve them as I had no knowledge of their being of interest. They were associated with bones of deer, bear, elk, and large wading birds, and then gave me the impression, which subsequent inquiry has strengthened, that the seal, like many of our large mammals, had disappeared gradually, as the country became more densely settled, and that in pre-European times it was common, at certain seasons, both on the coast and inland."[*]

In later communications (dated January 25 and March 20, 1879) he enclosed to me newspaper slips and notes respecting the capture of eight specimens in New Jersey, mostly near Trenton, during the winter of 1878–'79.

On the coast of Massachusetts they occur in considerable numbers about the mouth of the Ipswich River, where I have sometimes observed half a score in sight at once. They are also to be met with about the islands in Boston harbor, and along the eastern shore of Cape Cod. Captain N. E. Atwood states that they are now and then seen at Provincetown, and that in a shallow bay west of Rainsford Island "many hundreds" may be seen at any time in summer on a ledge of rocks that becomes exposed at low water.[†]

Further northward they become more numerous, particularly on the coast of Maine and the shores of the Gulf of Saint Lawrence, Newfoundland and Labrador, and are also common on the shores of Davis's Strait and in Greenland, where, says Dr. Rink, "it occurs here and there throughout the coast," and is likewise to be met with at all seasons of the year. Mr. Kumlien says it is one of the "rarer species" in the Cumberland waters, but its exact northern limit I have not seen stated.

On the European coasts it is said to occur occasionally in the Mediterranean, and to be not rare on the coast of Spain. It is more frequent on the coasts of France and the British Islands, and thence northward along the Scandinavian peninsula is the

[*] Letter dated Trenton, N. J., Dec. 26, 1878.
[†] See Bull. Mus. Comp. Zoöl., vol. i, p. 193.

commonest species of the family. It also extends northward
and eastward along the Arctic coast of Europe, but late explorers of the Spitzbergen and Jan Mayen Islands do not enu-
merate it among the species there met with. Malmgren states
distinctly that it is not found there,* and it is not mentioned by
Von Heuglin nor by the other German naturalists who have recently visited these islands. From its littoral habits its absence
there might be naturally expected. It is also said by some
writers to occur in the Black and Caspian Seas, and in Lake
Baikal, but the statement is seriously open to doubt, as will be
shown later in connection with the history of the Ringed Seal.

On the Pacific coast of North America it occurs from Southern
California northward to Behring's Strait, where it seems to be an
abundant species. I have examined specimens from the Santa
Barbara Islands, and various intermediate points to Alaska, and
from Plover Bay, on the eastern coast of Siberia. The extent
of its range on the Asiatic coast has not been ascertained. If
it is the species referred to by Pallas under the name *Phoca
canina*, and by Temminck, Von Schrenck, and other German
writers under the name *Phoca nummularis*, as seems probable,
it occurs in Japan and along the Amoor coast of the Ochots
Sea. Von Schrenck speaks of it, on the authority of the natives,
as entering the Amoor River.† The late Dr. Gray referred a
specimen from Japan to his "*Halicyon richardsi*," which, as already shown, is merely a synonym of *Phoca vitulina*. It thus
doubtless ranges southward along the Asiatic coast to points
nearly corresponding in latitude with its southern limit of distribution on the American side of the Pacific.

The Harbor Seal not only frequents the coast of the North
Atlantic and the North Pacific, and some of the larger interior
seas, but ascends all the larger rivers, often to a considerable
distance above tide-water. It even passes up the Saint Lawrence to the Great Lakes, and has been taken in Lake Champlain. DeKay states, on the authority of a Canadian newspaper, that a Seal (in all probability of this species) was taken
in Lake Ontario near Cape Vincent (Jefferson County, New
York) about 1824, and adds that the same paper says that Indian traders report the previous occurrence of Seals in the same
lake, though such instances are rare.‡ Thompson gives two in-

* Wiegm. Arch. für Naturg. 1864, p. 84.
† Reisen im Amoor-Lande, Bd. i, p. 180.
‡ New York Zoölogy, pt. i, 1842, p. 55.

stances of its capture in Lake Champlain; one of the specimens he himself examined, and has published a careful description of it, taken from the animal before it was skinned.*

They are also known to ascend the Columbia River as far as the Dalles (above the Cascades, and about two hundred miles from the sea), as well as the smaller rivers of the Pacific coast, nearly to their sources. Mr. Brown states that "Dog River, a tributary of the Columbia, takes its name from a dog-like animal, probably a Seal, being seen in the lake whence the stream rises."†

HABITS.—The Harbor Seal is the only species of the family known to be at all common on any part of the eastern coast of the United States. Although it has been taken as far south as North Carolina, it is found to be of very rare or accidental occurrence south of New Jersey. Respecting its history here, little has been recorded beyond the fact of its presence. Captain Scammon has given a quite satisfactory account of its habits and distribution as observed by him on the Pacific coast of the United States, but under the supposition that it was a species distinct from the well-known *Phoca vitulina* of the North Atlantic. Owing to its rather southerly distribution, as compared with its more exclusively boreal affines, its biography has been many times written in greater or less detail. Fabricius, as early as 1791, devoted not less than twenty pages to its history, based in part on his acquaintance with it in Greenland,

---

* His record of the capture of these examples is as follows:

"While several persons were skating upon the ice on Lake Champlain, a little south of Burlington, in February, 1810, they discovered a living Seal in a wild state which had found its way through a crack and was crawling upon the ice. They took off their skates, with which they attacked and killed it, and then drew it to the shore. It is said to have been 4½ feet long. It must have reached our lake by way of the Saint Lawrence and Richelieu. . . ."—*Nat. and Civil Hist. of Vermont*, 1842, p. 38.

"Another Seal was killed upon the ice between Burlington and Port Kent on the 23d of February, 1846. Mr. Tabor, of Keeseville, and Messrs Morse and Field, of Peru, were crossing over in sleighs when they discovered it crawling upon the ice, and, attacking it with the butt end of their whips, they succeeded in killing it and brought it on shore at Burlington, where it was purchased by Morton Cole, Esq., and presented to the University of Vermont, where its skin and skeleton are now preserved. . . . At the time the above-mentioned Seal was taken, the lake, with the exception of a few cracks, was entirely covered with ice."—IBID., *Append.*, 1853, p. 13.

†Proc. Zoöl. Soc. Lond., 1868, p. 412, footnote.

and partly on the writings of preceding authors;[*] and much more recently extended accounts of it have been given by Nilsson and Lilljeborg, but unfortunately for English readers the first of these histories is written in Danish and the other in Swedish. It has, however, been noticed quite fully by Bell, Macgillivray, and other British authors, while lesser and more fragmentary accounts of it are abundant. On the New England coast, as elsewhere, it is chiefly observed about rocky islands and shores, at the mouths of rivers and in sheltered bays, where it is always an object of interest. Although ranging far into the Arctic regions, it is everywhere said to be a sedentary or non-migratory species, being resident throughout the year at all points of its extended habitat. Unlike most of the other species, it is strictly confined to the shores, never resorting to the ice-floes, and is consequently never met with far out at sea, nor does it habitually associate with other species. On the coast of Newfoundland, where it is more abundant and better known than at more southerly points, it is said to bring forth its young during the last two weeks of May and the early part of June, resorting for this purpose to the rocky points and outlying ledges along the shore. It is said to be very common along the shores of the Gulf of Saint Lawrence and of Newfoundland in summer, or during the period when the shores are free from ice, but in winter leaves the ice-bound coast for the remoter islands in the open sea. It is at all times watchful, and takes great care to keep out of reach of guns. Still, many are surprised while basking on the rocks, and fall victims to the seal-hunters, while considerable numbers of the young are captured in the seal-nets. They are described as very sagacious, and as possessing great parental affection. Mr. Carroll states that when an old one is found on the rocks with its young it will seize the latter and convey it in its mouth so quickly to the water that there is not time to shoot it; or, if the young one be too large to be thus removed, it will entice it upon its back and plunge with it into the sea. The same writer informs us that this species is a great annoyance to the salmon-fishers, boldly taking the salmon from one end of the net while the fisherman is working at the other end. It is also troublesome in other ways, since whenever the old ones get entangled in

[*] Fabricius appears to have exhaustively presented its literary history, his references to previous authors, in his table of synonymy, occupying nearly four pages.

the strong seal-nets they are able to cut themselves free, a feat it is said no other Seal known in Newfoundland will do.

This species is known to the inhabitants of Newfoundland as the "Native Seal," in consequence of its being the only species found there the whole year. The young are there also called "Rangers," and when two or three years old,—at which age they are believed to bring forth their first young,—receive the name of "Dotards." Here, as well as in Greenland, the skins of this species are more valued than those of any other species, owing to their beautifully variegated markings, and are especially valued for covering trunks and the manufactures of coats, caps, and gloves.* Mr. Brown informs us that the natives of the eastern coast of Greenland prize them highly "as material for the women's breeches," and adds "that no more acceptable present can be given to a Greenland damsel than a skin of the 'Kassigiak,' as this species is there called." The Greenlanders also consider its flesh as "the most palatable of all 'seal-beef'". †

According to Mr. Reeks, the period of gestation is about nine months, the union of the sexes occurring, according to the testimony of the Newfoundlanders, in September. ‡ Only rarely does the female give birth to more than a single young. This agrees with what is stated by Bell and other English authors respecting its season of procreation.

Respecting its general history, I find the following from the pen of Mr. John Cordeaux, who, in writing of this species, as observed by him in British waters, says : "The Seal (*Phoca vitulina*) is not uncommon on that part of the Lincolnshire coast adjoining the Wash. This immense estuary, lying between Lincolnshire and Norfolk, is in great part occupied with large and dangerous sand-banks, intersected by deep but narrow channels. At ebb the sands are uncovered; and at these times, on hot days, numbers of Seals may be found basking and sunning themselves on the hot sands, or rolling and wallowing in the shallow water along the bank. Sometimes a herd of fifteen or twenty of these interesting creatures will collect on some favorite sand-spit; their chief haunts are the Long-sand, near the centre of the Wash; the Knock, along the Lincoln coast; and the Dog's-head sand, near the entrance to Boston Deeps. In the

---

* Carroll, Seal and Herring Fisheries of Newfoundland, 1873, pp. 10, 11.
† Proc. Zoöl. Soc. Lond., 1868, pp. 413, 413.
‡ Reeks (Henry), Zoölogist, 2d ser., vol. vi, 1871, p. 2541.

first week of July, when sailing down the Deeps along the edge
of the Knock, we saw several Seals; some on the bank; others
with their bodies bent like a bow, the head and hind feet only
out of the water. They varied greatly in size, also in color,
hardly any two being marked alike; one had the head and face
dark colored, wearing the color like a mask; in others the upper
parts were light gray; others looked dark above and light below,
and some dark altogether. . . . . The female has one young
one in the year; and as these banks are covered at flood, the
cub, when born, must make an early acquaintance with the
water. In most of the *Phocidæ* the young one is at first covered
with a sort of wool, the second or hairy dress being gradually
acquired; and until this is the case it does not go into the water.
This, however, does not appear to be the case with the common
Seal, for Mr. L. Lloyd says (I believe in his 'Game Birds and
Wild Fowl of Norway and Sweden,' but I have not the book to
refer to) that the cub of the common species, whilst still in its
mother's womb, casts this wooly covering; and when ushered
into the world has acquired its second or proper dress.* If this
is the case, it fully accounts for the cub being able to bear im-
mersion from the hour of its birth. The Seal, if lying undisturbed and at rest, can remain for hours without coming to the
surface."†

I am informed by competent observers that on the coast of
Maine they assemble in a similar manner on sand bars, but take
to the water before they can be closely approached.

Mr. Kumlien (in his MS. notes) observes: "The so called
'Fresh-water Seal' of the whalemen is one of the rarer species
in the waters of Cumberland Sound. They are mostly met with
far up in the fjords, and in the fresh water streams and ponds,
where they go after salmon. They are rather difficult to capture, as at the season when they are commonly met with they
have so little blubber that they sink when shot. . . . The
adult males often engage in severe combats with each other.
I have seen skins so scratched that they were nearly worthless.
In fact, the Eskimo consider a 'Kassiarsoak' (a very large

---

*A statement to this effect is also made by Mr. Carroll, but Mr. Robert
Brown affirms, on the authority of Captain McDonald, that in the Western
Isles of Scotland the young are "born pure white, with curly hair, like the
young of *Pagomys fœtidus*, but within three days of its birth begins to take
dark colors on the snout and tips of the flippers."—*Proc. Zoöl. Soc. Lond.*,
1868, p. 413.

†Zoölogist, 2d ser., vol. vii, 1872, pp. 3203, 3204.

'Kassigiak') as having an almost worthless skin, and seldom
use it except for their skin tents. The skins of the young, on
the contrary, are a great acquisition". He further states that
they do not make an excavation beneath the snow for the reception of the young, like *Phoca fœtida*, "but bring forth later in
the season on the bare ice, fully exposed".

Under the name "Leopard Seal," Captain Scammon has given
a very good account of the habits of this species as observed
by him on the Pacific coast of North America. He speaks of
it as displaying no little sagacity, and considerable boldness,
although exceedingly wary. He says it is "found about outlying rocks, islands, and points, on sand-reefs made bare at low
tide, and is frequently met with in harbors among shipping,
and up rivers more than a hundred miles from the sea. We
have often observed them," he continues, "close to the vessel
when under way, and likewise when at anchor, appearing to
emerge deliberately from the depths below, sometimes only
showing their heads, at other times exposing half of their bodies,
but the instant any move was made on board, they would van-
ish like an apparition under water, and frequently that would be
the last seen of them, or, if seen again, they would be far out
of gun-shot." They come ashore, he observes, "more during
windy weather than in calm, and in the night more than in the
day; and they have been observed to collect in the largest
herds upon the beaches and rocks, near the full and change of
the moon. They delight in basking in the warm sunlight, and
when no isolated rock or shore is at hand, they will crawl upon
any fragment of drift-wood that will float them. Although
gregarious, they do not herd in such large numbers as do nearly
all others of the seal tribe; furthermore, they may be regarded
almost as mutes, in comparison with the noisy Sea Lions. It is
very rarely, however, any sound is uttered by them, but occasionally a quick bark or guttural whining, and sometimes a
peculiar bleating is heard when they are assembled together
about the period of bringing forth their young. At times, when
a number meet in the neighborhood of rocks or reefs distant
from the main-land, they become quite playful, and exhibit much
life in their gambols, leaping out of the water or circling around
upon the surface. . . . Its rapacity in pursuing and devouring the smaller members of the piscatory tribes is quite equal,
in proportion to its size, to that of the Orca. When grappling
with a fish too large to be swallowed whole, it will hold and

handle it between its fore flippers, and, with the united work of its mouth . . . the wriggling prize is demolished and devoured as quickly, and in much the same manner, as a squirrel would eat a bur-covered nut. . . .

"Leopard Seals are very easily captured when on shore, as a single blow with a club upon the head will dispatch them. The Indians about Puget Sound take them in nets made of large hemp-line, using them in the same manner as seines, drawing them around beaches when the rookery is on shore. They are taken by the whites for their oil and skins, but the Indians and Esquimaux make great account of them for food." He adds that the natives of Puget Sound singe them before a fire until the hair is consumed and the skin becomes crisp, when they are cut up and cooked as best suits their taste.*

The apparent fondness of this animal, in common with other species of the family, for music, has been often noted. "During a residence of some years in one of the Hebrides," writes the Rev. Mr. Dunbar, "I had many opportunities of witnessing this peculiarity; and, in fact, could call forth its manifestation at pleasure. In walking along the shore in the calm of a summer afternoon, a few notes from my flute would bring half a score of them within a few yards of me; and there they would swim about, with their heads above water, like so many black dogs, evidently delighted with the sounds. For half an hour, or, indeed, for any length of time I chose, I could fix them to the spot; and when I moved along the water-edge, they would follow me with eagerness, like the Dolphins, who, it is said, attended Arion, as if anxious to prolong the enjoyment. I have frequently witnessed the same effect when out on a boat excursion. The sound of the flute, or of a common fife, blown by one of the boatmen, was no sooner heard, than half a dozen would start up within a few yards, wheeling round us as long as the music played, and disappearing, one after another, when it ceased."†

Although, like other species of the family, evidently attracted by musical sounds, it is perhaps questionable whether they are not as much influenced by curiosity as by any real fondness for music. Any unusual sounds, or unusual movements of any kind, serve to strongly attract them. The writer last quoted states that when he and his pupils were bathing in a small bay where these Seals were abundant they would crowd around them in

*Marine Mam., pp. 165–167.
†Macgillivray's British Quad., pp. 204–205.

numbers, at a distance of a few yards, as though they fancied they were the same species with themselves. "The gambols in the water," he writes, "of my playful companions, and their noise and merriment, seemed, to our imagination, to excite them, and to make them course round us with greater rapidity and animation." Mr. Bell also quotes Mr. Low as referring, in his "Fauna Orcadensis," to their possession of a great deal of curiosity. "If people are passing in boats, they often come quite close up to the boat, and stare at them, following for a long time together; if people are speaking loud, they seem to wonder what may be the matter. The church of Hoy, in Orkney, is situated near a small sandy bay, much frequented by these creatures; and I observed when the bell rang for divine service, all the Seals within hearing swam directly for the shore, and kept looking about them, as if surprised rather than frightened, and in this manner continued to wonder, as long as the bell rung."*

The Harbor Seal yields readily to domestication, and may be easily taught a variety of tricks. In confinement it exhibits great docility, and allows itself to be freely handled without offering resistance or manifesting fear. A specimen exhibited some years since in Boston, included, among its varied accomplishments, performances on the hand-organ. It is said to generally become greatly attached to its keeper, whom it will follow, and whose call it readily obeys. Many accounts have been published of its intelligence and docility in confinement.† Captain Scammon refers to his having had several young ones on ship-board, and says that "in every instance it was but a few weeks before they would follow, if permitted, the one who had especial charge of them, and when left solitary, they would express discontent by a sort of mournful bleating." Of a Leopard Seal, as he terms this species, kept at one time in Woodward's Gardens, in San Francisco, California, he says: "This little favorite has been a resident of that popular and interest-

---

* Bell's British Quad., 1st ed., pp. 265–266.

† Dr. Edmonston seriously records the following: "The young ones are easily domesticated, and display a great deal of sagacity. One in particular became so tame that it lay along the fire among the dogs, bathed in the sea, and returned to the house, but having found the way to the byres, used to steal there unobserved and suck the cows. On this account it was discharged, and sent to its native element."—*A View of the Ancient and Present State of the Zetland Islands, etc.* Vol. ii, 1809, p. 293.

ing resort for over three years, and, although a female, as we were informed, is honored with the title of 'Commodore'. The animal generally makes its appearance close at hand whenever within hearing, if called by name, and when its keeper appears on the lawn, to feed the pelicans, black swans, and other aquatic birds, which are its companions in the artificial pond, the Commodore does not wait to be invited, but, knowing as well as its keeper the meal-hour, is on the watch, and the moment the food-bearer is seen, the little creature—which is not over four feet long—lifts itself out of the water over the curbstones and waddles quickly to its master's side, then holding up its head, with mouth wide open, receives the choice morsels of fish which drop from his hand."*

The food of this species consists largely of fish, but, like other species, it doubtless varies its fare with squids and shrimps. That it aspires to more epicurean tastes is evidenced by its occasional capture of sea-birds. This they ingeniously accomplish by swimming beneath them as they rest upon the water, and seizing them. An eye-witness of this pastime relates an instance as observed by him on the Scottish coast. "While seated on the bents," he writes, "watching a flock of [Herring] Gulls that were fishing in the sea near Donmouth, I was startled by their jerking high in the air, and screaming in an unusual and excited manner. On no previous occasion have I observed such a sensation in a Gull-hood, not even when a Black-head was being pursued, till he disgorged his newly-swallowed fish, by that black-leg, the Skua. The excitement was explained by a Seal [presumably *Phoca vitulina*, this being the only species common at the locality in question] showing above the water with a Herring Gull in his mouth; on his appearing the Gulls became ferocious, and struck furiously at the Seal, who disappeared with the Gull in the water. The Seal speedily reappeared, but on this occasion relinquished his victim on the Gulls renewing their attack. The liberated Gull was so disabled as to be unable to fly, but it had strength enough to hold up its head as it drifted with the tide."†

They are evidently discriminating in their tastes, and not loath to avail themselves of a fine salmon now and then not of their own catching. Their habit of plundering the nets of the fishermen on the coast of Newfoundland has been already

* Marine Mam., pp. 166–167.
† Zoölogist, 2d ser., vol. vi, 1871, p. 2762.

alluded to, but this peculiarity is evidently not confined to the Newfoundland representative of the species, as shown by the following incident related by the writer last quoted. "On a sunny noon in the autumn of 1868," says this observer, "I observed a Seal, not far from the same place, with a salmon in his mouth, which he forced through the meshes of a stake net. The struggling salmon, whose head was in the jaws of the Seal, struck the water violently with his tail, which gleamed like a lustre in the lessening ray. The Seal rose and sunk alternately, keeping seaward to escape Eleys' cartridges from the shore. When above the water he shortened the silver bar, which continued to lash his sides long after its thickest part had disappeared, by rising to his perpendicular, as if to allow the precious metal by its own weight to slip into his crucible. The Seal evidently swallowed above, and masticated below, water—the process lasting about twelve minutes, during which the Seal had travelled a full half-mile."[*]

In their raids upon the nets of the fishermen they become sometimes themselves the victims, being in this way frequently taken along our own coast as well as elsewhere. They are, however, at all times unwelcome visitors. De Kay states that formerly they were taken almost every year in the "fyke-nets" in the Passaic River, greatly to the disgust of the fishermen, the Seals when captured making an obstinate resistance and doing much injury to the nets. Their accidental capture in this way often affords a record of their presence at localities they are not commonly supposed to frequent, as in the Chesapeake Bay, and at even more southerly localities on the eastern coast of the United States.

Owing to the difficulty of capturing this species, and its comparatively small numbers, it is of little commercial importance, although the oil it yields is of excellent quality, and its skins are of special value for articles of dress, and other purposes, in consequence of their beautifully variegated tints. Though not a few are taken in strong Seal-nets, they are usually captured by means of the rifle or heavy sealing gun. On rare occasions they are surprised on shore at so great a distance from water that they are overtaken and killed by a blow on the head with a club. Like other species of the Seal family, the Harbor Seal is very tenacious of life, and must be struck in a vital part by

---

[*] W. Craibe Angus, Zoölogist, 2d ser., vol. vi, 1871, pp. 2762, 2763.

either ball or heavy shot, in order to kill it on the spot. Says Mr. Reeks, "I have been often amused at published accounts of Seals shot in the Thames or elsewhere, but which 'sank immediately'. What Seal or other amphibious animal would not do so if 'tickled' with the greater part of, perhaps, an ounce of No. 5 shot?" He adds that it is only in the spring of the year that this Seal will "float" when killed in the water, but says that he has never seen a Seal "so poor, which, if killed *dead on the spot*, would not have floated from five to ten seconds," or long enough to give "ample time for rowing alongside," supposing the animal to have been killed by shot, and the boat to contain "two hands".* The oil of this species, according to the same writer, sells in Newfoundland for fifty to seventy-five cents a gallon, while the skins are worth one dollar each. Mr. Carroll gives the weight of the skin and blubber of a full-grown individual as ranging from eighty to one hundred pounds, while that of a young one averages, at ten weeks old, thirty to thirty-five pounds. The flesh of the young, the same writer quaintly says, is " as pleasant to the taste as that of any description of salt-water bird."† Its flesh, as already stated, is esteemed by the Greenlanders above that of any other species. Few statistics relating to the capture of this species are available, but the number taken is small in comparison with the "catch" of other species, particularly of the Harp or Greenland Seal. Dr. Rink states that only from one thousand to two thousand are annually taken in Greenland, which is about one to two per cent. of the total catch. They are hunted to a considerable extent, however, wherever they occur in numbers.

The Harbor Seal received this name from its predilection for bays, inlets, estuaries, and fjords, from which habit it is also often termed Bay Seal, and, on the Scandinavian coast, Fjord Seal (Fjordskäl),‡ and also Rock Seal (Steen-Kobbe).

PHOCA (PUSA) FŒTIDA, *Fabricius.*

Ringed Seal.

Phoca, ALBINUS, Acad. Annot., iii, 1756, cap. xv.
Neitsek, CRANZ, Hist. von Grœnl., i, 1765, 164; ibid., English version, 1767, 124.

---

* Zoölogist, 2d ser., vol. vi, 1871, p. 2541.
† Seal and Herring Fisheries of Newfoundland, p. 11.
‡ In Greenland, however, according to Dr. Rink, it is the *Phoca fœtida* that receives this name.

598   PHOCA FŒTIDA—RINGED SEAL

Rough Seal, PENNANT, Synop. Quad., 1771, 341; Hist. Quad., 3d ed., ii, 1792,
        278.
Der rauhe Seehund, SCHREBER, Säugt., iii, [1776?] 312, pl. lxxxvi.
Der graue Seehund, SCHREBER, Säugt., iii, [1776?] 309.
Phoca fœtida, FABRICIUS, Müller's Zool. Dan. Prod., 1776, viii (no descrip-
        tion); Faun. Grœnl., 1780, 13.—DESMAREST, Mam., 1820, 246.—
        RICHARDSON, Parry's 2d Voy., Suppl., 1825, 332.—HARLAN, Faun.
        Amer., 1825, 110.—GODMAN, Am. Nat. Hist., i, 1826, 345.—J. C.
        ROSS, Parry's 3d Voy., App., 18.6, 191; Ross's 2d Voy., App., 1835,
        xix.—FISCHER, Syn. Mam., 1829, 377 (=577).—GRAY, Griffith's
        Cuv. An. King., v, 1827, 178.—BLASIUS, Naturg. Säuget. Deutsch.,
        1857, 251, figg. 138, 139.—BARTLETT, Proc. Zoöl. Soc. Lond., 1868,
        67.—VON MIDDENDORFF, Siberische Reise., iv, 1867, 934.—LILL-
        JEBORG, Fauna öfver Sveriges och Norges Ryggrads., 1874, 682.—
        COLLETT, Bemærk. Norges Pattedyrf., 1876, 57, footnote.—RINK,.
        Danish Greenland, its People and its Products, 1877, 122, 430.
Callocephalus fœtidus, GRAY, Cat. Seals Brit. Mus., 1850, 23; Proc. Zoöl.
        Soc. Lond., 1862, 202 (young, born in confinement).
Pagomys fœtidus, GRAY, Proc. Zoöl. Soc. Lond., 1864, 31; Cat. Seals and
        Whales, 1866, 23; Zoölogist, 1872, 3333, 3335 (British Islands);
        Hand-List Seals, 1874, 6, pl. iii (skull).—KANE, Grinnell Exped.,
        1854, —.—GILL, Proc. Essex Inst., v, 1866, 12. —BROWN, Proc. Zoöl.
        Soc. Lond., 1868, 340, 416; Man. Nat. Hist., etc., Greenland, 1875,.
        Mam., 43.—VAN BENEDEN, Ann. du Mus. Roy. d'Hist. Nat. du Bel-
        gique, i, 1877, 18 (geographical distribution, with chart).—KUM-
        LIEN, Bull. U. S. Nat. Mus., No. 15, 1879, 55 (habits and external
        characters).—"MALMGREN, Öfvers, 1863, 827; Bihang Svenska Exp.,
        1864, 5"; Schwed. Exp. nach Spitz., 1861, 1864, 1868, Passarge's
        German transl., 1869, 78, 180.
Phoca (Pagomys) fœtida, VON HEUGLIN, Reise nach dem Nordpolarmeer, iii,
        1874, 48; ibid., i, 154, 207, 220, 228.
Phoca hispida, SCHREBER, Säugt., iii, [1776?] 312, pl. lxxxvi ("Der rauhe
        Seehund" in text).—ERXLEBEN, Syst. Reg. Anim., 1777, 589.—
        GMELIN, Syst. Nat., i, 1778, 64.—FABRICIUS, Skriv. af Nat. Selsk., ii,
        1791, 74.—KERR, Anim. King., 1792, 125.—DESMAREST, Nouv. Dict.
        d'Hist. Nat., xxv, 1817, 587.—"MELCHIOR, Danske Stats och Norges,
        1834, 234."—HAMILTON, Amphib. Carn.,9   166, pl. viii (from F.
        Cuvier).—GAIMARD, Voy. en Island, etc., Atlas, 1851, pl. x, figg. 1,
        2 (skeleton).—MALMGREN, Öfv. af Kongl. Vetensk.-Akad. Förh.
        Stock., 1863, 143; Arch. für Naturg., 1864, 82.—"HOLMGREN,
        Skand. Däggdj., 1865, 215, fig."—FLOWER, Proc. Zoöl. Soc. Lond.,
        1871, 506 (coast of Norfolk, Engl.; synonymy).—TURNER, Journ.
        Anat. and Phys., iv, 1870, 260; Proc. Roy. Soc. Edinburg, 1869, 70
        (fossil, brick clays, Scotland).—BELL, Hist. Brit. Quad., 2d ed.,
        1874, 247, figg. animal and skull.—ALSTON, Fauna of Scotland,
        Mam., 1880, 14.
Callocephalus hispidus, F. CUVIER, Mém. du Mus., xi, 1824, 189, pl. xii
        (skull); Dict. des Sci. Nat., xxxix, 1826, 547.—LESSON, Man. de
        Mam., 1827, 198.—GRAY, Cat. Seals Br. Mus., 1850, 23.

Phoca vitulina, a, ERXLEBEN, Syst. Reg. Anim., 1777, 587.

Phoca vitulina, β, botnica, GMELIN, Syst. Nat., i, 1778, 63.

Phoca annellata, NILSSON, "Skand. Faun., i, 1820, 362; Illumin. fig. till Skand. Faun., 20:de häft, 1840, pl. xxxviii"; Skan. Fauna., Däggdj., 1847, 283; Arch. für Naturg., 1841, 312.—"THIENEMANN, Reise im Norden Europa, etc., i, 1824, 83,ͺpl. ix (ad. female), pl. x (young), pl. xi (skull), pl. xii (digestive organs)."—SCHINZ, Naturg. u. Abild. der Säugeth., 1827, 167, pl. lxiv.—BALL, Sketches of Irish Seals, fig. 36 (skull, from Thienemann).—SCHINZ, Synop. Mam., i, 1844, 482.—WAGNER, Suppl. Schreber's Säugt., vii, 1846, 29, pl. 84 A.—GAIMARD, Voy. en Island, etc., Atlas, Mam., 1851, pl. xi, fig. 7 (dentition).—GIEBEL, Säuget., 1855, 137.—RADDE, Reisen im Süden vom Ost-Siberien, i, 1862, 296, pl. xiii (animal, skull, dentition, bones of limbs; Lake Baikal).—KARSCH, Hornschuck's Arch. Skand. Beiträge zur Naturgs., ii, 1850, p. 326 (Lake Ladoga).

"Callocephalus annellatus, RÜPPEL, Verzeichness, 167."—"GRAY, Zoöl. Erebus and Terror, 3".

Callocephalus discolor, F. CUVIER, Mém. du Mus., xi, 1824, 186; Dict. Sci. Nat., xxxix, 1826, 545.—LESSON, Man. de Mam., 1827, 198.

Phoca discolor, FISCHER, Syn. Mam., 1829, 237, 375 [=575].—GRAY, Griffith's Cuv. An. King., v, 1827, 177.

Phoca frederici, LESSON, Dict. Class. d'Hist. Nat., xiii, 1828, 416 (=*P. discolor,* F. Cuvier).

Phoca schreberi, LESSON, Dict. Class. d'Hist. Nat., xiii, 1828, 414 (=*P. hispida,* Schreber; *P. fœtida,* Fabricius; *P. annellata,* Nilsson).

Phoca communis var. *octonotata,* KUTORGA, Bull. Soc. Imp. des Nat. de Mosc., 1839, 189, pl. xiii, fig. 1 (animal); pl. xiv, figg. 1–3 (muzzle and mystacial bristle); pl. xv, figg. 1, 2, 5 (bones of forearm and hand); pl. xvi, figg. 1–4 (skull); pl. xviii, fig. 1 (skull).

Phoca communis var. *undulata,* KUTORGA, Bull. Soc. Imp. des Nat. de Mosc., 1839, 191, pl. xiii, fig. 2 (animal); pl. xiv, figg. 4–6 (muzzle, etc.); pl. xv, figg. 3, 4 (bones of fore limb); pl. xvii (skull); pl. xviii, fig. 2 (skull).

" Phoca dimidiata, SCHLEGEL, Mus. Leyd." (*apud* Gray).

Callocephalus dimidiatus, GRAY, Cat. Seals Brit. Mus., 1850, 24.

Callocephalus? *dimidiatus,* GRAY, Cat. Seals and Whales, 1866, 22.

Phoque marbré, F. CUVIER, Hist. Nat. des Mam., livr. ix, 1819 (coast of France,—accidental).

Seal shot near the Orkney Islands, HOME, Phil. Trans., 1822, pl. xxviii (skull).

? Fossil Seal, THOMPSON, Journ. Anat. and Phys., xiii, 1879, 318 (fossil, Scotland).

Fiordsœl, FABRICIUS, l. c.

Stink-Robbe, VON MIDDENDORFF, l. c.

Natsek or Fjord Seal, RINK, l. c.

"Netsick, adults generally; *Tizak,* adult males; *Netsiavik,* young after shedding till one year old; *Ibeen,* young in the white coat, Cumberland Eskimo" (KUMLIEN).

Grä Själ; Grä-Wikare Skäl; Ringlad Skäl; Vikaresjäl, Viksjäl, Swedish.

Steenkobbe, Norwegian.

Der geringelte Seehund; Der rauhe Seehund, German.

Nerpa; Russian.
Ringed Seal ; Marbled Seal; Floe Rat, English.
"Pickaninny Pussy, young, pigeon-English of the whalers " (KUMLIEN).
? *Bodach,* Hebridian.

EXTERNAL CHARACTERS.—Adult, generally blackish-brown above; darkest on the back, lighter on the sides, with large oval, whitish spots; beneath nearly uniform yellowish-white; nose and ring round the eye usually black; mystacial bristles and claws dusky or blackish; pelage rather harsh. Length of the adult male, 5 to 6 feet; female smaller.

The newly-born young are usually white or yellowish-white; the pelage soft and woolly. At the age of about four weeks this gradually gives place to the coarser, more rigid pelage of the adult, and the color changes to dusky, marked sparsely with small blackish spots. Yearlings are often yellowish-white; dusky along the middle of the back, with here and there small spots of blackish.

There is a wide range of individual variation in color, in the newly-born young as well as in the adults, as the following remarks will show.

Three adult specimens from Disco Island, Greenland, present the following variations in color: In No. 8699 (Nat. Mus.) the general color above is yellowish-white, irregularly mottled on the back with oblong spots and streaks of dusky or bluish-black; whole lower parts uniform yellowish-white. In No. 8700 (Nat. Mus.) the dorsal surface is everywhere marbled with light spots having dark centres. There are also patches of dark brown of very irregular outline. The dark-centred light rings are much more distinct than in No. 8699. No. 8698 is yellowish-white marbled with dark brown, the latter tint forming chains of dark-centred light spots. The front part of the head is blackish; the lower parts are uniform yellowish-white. Several yearling specimens, from Cumberland Sound, collected by Mr. Ludwig Kumlien, are whitish or yellowish-white, with small dusky or blackish spots.

Wagner has described a specimen from Labrador as having the back blackish-brown, with a greenish-gray shimmer, and marked with spots of yellowish of varying size, some of them occurring singly, and others joined in pairs into 8-shaped figures; on the sides they form groups of rings, rather symmetrically arranged on the two sides of the body; lower surface pale yellow, with a tinge of olive. A younger specimen, also from

Labrador, is described as duller in color, more grayish, and with coal-black markings.*

Kutorga's variety *octonotata* is described as blackish, darkest along the back, with whitish spots mostly ∞-shaped; lower sur-

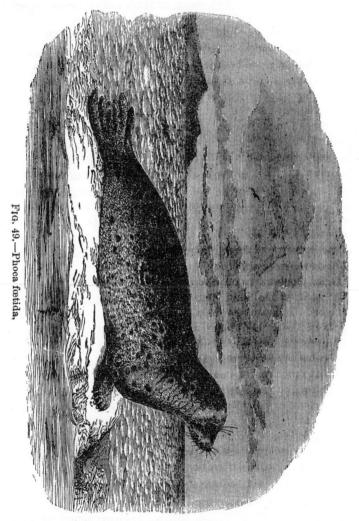

FIG. 49.—Phoca fœtida.

face lighter than the upper; the pelage composed of rather soft, fine hair. The same author's variety, *undulata*, is described as blackish-brown, lighter below and darkest along the

---

* Schreber's Säugt., Th. vii, p. 31.

back, the spots silvery-white, of irregular shape, and rarely in the form of rings.[*]

Nilsson has distinguished a black, a white, and a brownish-gray color-variety. The first is described as dark brownish-black, blacker above and more grayish-brown below, marked everywhere with pale streaks, which sometimes form small whitish oval rings; head and neck with single small whitish spots; nose and eye-rings uniform black; limbs uniform brownish-black. The white variety is described as uniform soiled-white, slightly darker on the middle of the back. The brown variety is said to be uniform brownish-gray; paler below.

Mr. Kumlien states that the new-born young are also "very variable in color; some are pure white; others white on the lower parts, but more or less dusky on the back; others again are fine straw-yellow, with the same dusky variation as in the white ones. The yellow is also variable in the intensity of the shade. Rarely some are found that are quite dusky all over, especially on the head and back; these are generally small and scrawny. The hair," he adds, "is also quite as variable in texture as in color. In some it is fine, long, and woolly (mostly in the pure white examples); in others it is straight or wavy, while some have short and quite hispid hair."

There appears to be also a quite wide range of variation in size; at least the statements of authors indicate that such is the case. It seems probable that in some instances measurements given as those of the adult were really taken from examples not full-grown. Nilsson, in 1837, gave the length of the species as about 3 feet, and later (1847) as 3 to 4 feet. Fabricius says it rarely exceeds 4½ feet in length. Wagner refers to a Labrador specimen as being 4 feet 2 inches long. Capt. J. C. Ross states that the average length, from the nose to the end of the tail, of twenty specimens measured by him, was 55 inches, or 4 feet 7 inches, the hind flipper extending 9 inches beyond the body, thus giving an extreme length of 5 feet 4 inches. He gives the average weight of these same specimens as 199 pounds, and the circumference immediately behind the fore flippers as 49.7 inches.[†] Dr. Rink, however, gives the average weight of seven specimens, "perhaps somewhat below the middle size," as only 84 pounds.[‡] Lilljeborg and Malmgren record much larger dimensions. The former gives the

[*] Bull. Soc. Imp. des Nat. de Moscou, 1839, pp. 189, 191.
[†] Ross's Second Voyage, App., p. xx.
[‡] Danish Greenland, etc., p. 123.

length of the male, from the nose to the end of the tail, as 5 feet 2 inches, or 1,560 mm., and from the nose to the end of the hind flippers as 5 feet 7 inches, or 1,710 mm. The same dimensions of the female he gives as, respectively, 4 feet 6 inches (1,380 mm.) and 4 feet 9 inches (1,470 mm.).* Malmgren states that the largest full-grown individuals he had seen (in the Gulf of Bothnia) attained a length of 5½ Swedish feet from the nose to the end of the tail, and measured nearly 6 feet to the end of the hind flippers. He further states that old or full-grown specimens are rarely taken, and that the measurements usually given are those of specimens one or two years old. He further observes, "Sogar in den finnischen Landseen Ladoga und Pyhäfelkä soll die ganze Länge der alten Individuen nach der Aussage erfahrener Männer beinahe einer Klafter, d. h. sechs Fuss betragen. Doch in diesen Seen wird diese Robbe gewöhnlich im zweiten und dritten Jahre, selten erwachsen, geschossen. Dasselbe ist nach Fabricius auch Grönland der Fall; ich vermuthe daher, dass dies die Ursache ist, wesshalb die Länge des Thieres zu klein angegeben ist, da sie wahrscheinlich nach den ein- oder zweijährigen Individuen, die man am gewöhnlichsten erhält, bestimmt worden ist."† It is certainly evident that the specimens described by Nilsson and Wagner were young. Lilljeborg and Malmgren are the only authors who have apparently examined, or at least described, full-grown specimens.

Mr. Kumlien states that the young are about two feet long when born, and weigh from 4 to 6½ pounds, and that they average 30 inches in length (varying from 23 to 36 inches).

INDIVIDUAL VARIATION AND VARIATIONS DEPENDENT UPON AGE AND SEX.—The Ringed Seal, like the Harbor Seal, varies greatly in color, irrespective of sex and age, both in respect to the ground color and the markings, as has been already shown in the description of the external characters. Like nearly all the Phocids, the young when born are covered with a white or yellowish-white coat of rather soft, woolly hair, which is changed in about four weeks for the sparser, harsher, and darker livery of the adults. The younger animals, however, are grayish or yellowish brown, darker along the middle of the back, and marked irregularly with small dusky spots, the mar-

---

* Fauna Öfver Sveriges och Norges, i, pp. 683, 684.
† Archiv für Naturg., 1864, pp. 83, 84.

bled coloring usually characterizing the adults not being attained till the second or third year. The sexes vary in size, as already noted, the female being considerably the smaller. This difference of size is well shown in the measurements of the skulls given below. Aside from the skull of the female being smaller than that of the male, its structure is weaker, the surface less roughened for the attachment of muscles, the muzzle narrower, the teeth smaller, and the lower jaw much slenderer. In the series of skulls collected by Mr. Kumlien, which were carefully marked for sex, the old males have an average length of about 186 mm., and an average breadth of about 115 mm., while the same dimensions in the old females are respectively 168 mm. and 108 mm.

In general, skulls of the same sex and of corresponding ages vary considerably in details of structure and proportion, but the only purely individual variations worthy of special comment are exhibited in the teeth, which are surprisingly variable in respect to size, and in the number and shape of the accessory cusps. That these variations are not due to age and the accidents of attrition is shown by the fact that they are as well marked when the teeth first cut the gum as at later stages. The last upper molar is especially variable in size and in the prominence of the cusps, the accessory cusps being sometimes well developed and again almost wholly obsolete. The last upper molar has usually only two points, the posterior of which is small, but there is occasionally another still smaller on the anterior inner border of the tooth. Generally the other upper molars have each three cusps, of which the anterior is the smallest, and frequently is wholly obsolete on the second molar when the third, fourth, and fifth molars are each 3-pointed. Frequently, however, all the upper molars, except the first, are 4-pointed, while in a nearly equal percentage of the skulls examined the molars are all only 2-pointed, or all, except the third, which may be 3-pointed. Sometimes the third or fourth upper molar is 3-pointed, while the others are 2-pointed. In one skull all are 3-pointed, *including even the first.*

The lower molars are less subject to variation in respect to the number of points, they being almost invariably 4-pointed, except the first, which is usually 3-pointed. The chief variation I have noticed is that the fifth is sometimes only 3-pointed, like the first, while the first is sometimes 4-pointed, like the others. The size and shape of the cusps vary greatly, being

sometimes thick and short and again slender and very long
These variations are all as strongly marked in the young killed
a few days after birth as in the adult, showing that the variation in the number and shape of the cusps is not due to peculiarities of wearing. In the 3-pointed teeth two accessory cusps
are developed behind the principal one, while in teeth with four
points there is also a small accessory cusp in front of the principal one.   In one instance a first lower molar has two minute
points placed in front of the principal one, there being a
(purely supernumerary) cusp on the outer anterior border of
the tooth. In the males the teeth average (in linear dimensions) about one-eighth larger than in the females; but the size
varies so much in individuals of the same sex that the teeth
are as large in some females as in some males.

Measurements of fourteen skulls of Phoca fœtida.

Measurement	6565	6566	6567	6568	—	8942	7105	16139	16111	6296	16138	6295	6297	16051
Front edge of ramus to last molar.	49	53	42	45	45	…	50	49	47	46	45	44	47	…
Length of lower jaw.	108	109	98	108	108	…	117	120	115	108	102	103	105	…
Greatest width of brain-case.	89	86.5	85	…	84.5	86	89	90	87	90	86	83	81	…
Length of brain-case.	74	70	72	71	72.5	65	71	72	69	70	68	64	66	…
Greatest height of skull at auditory bullæ.	74	69	71	77	72	…	70	70	68	73	69	65	69	…
Breadth of anterior nares, transversely.	23	23.5	21	21	22	22	23.5	23	23	23	22	22	22	…
Breadth of anterior nares, vertically.	21	20	19	19	20	20	21	26	24	21	23	18	22	20
Breadth of posterior nares, transversely.	24	22.5	19	24	24	21	23	25	24	23	26	24	…	…
Breadth of posterior nares, vertically.	16	18	10	15	14	11	11	13	15	10	12	11	13	…
Least breadth of skull interorbitally.	5	6.4	5.5	6	6	6	6	6	7	6	7	5.5	5	4.5
Breadth of skull at canines.	26	28	24	25	25	24	27	29	30	28	27	22	26	25
Nasal bones, breadth at fronto-maxillary suture.	9	7	6	7	6	5	8	9	8	7	8	6	7	5
Nasal bones, breadth anteriorly.	12	12	11	13	10	12	13	13	14	15	22	10	13	23
Nasal bones, length.	41	38	37	37	40	39	43	38	46	35	39	37	30	45
Width of palatal region at posterior end of maxillæ.	46	43	42	46	42	44	44	47	47	45	48	43	48	…
Length of the alveolar border of the maxilla.	65	63	57	64	63	67	67	74	72	69	64	62	50	67
Distance from palato-maxillary suture to end of pterygoid hamuli.	34	38	36	36	37	37	35	40	35	41	32	35	39	…
Distance from anterior edge of inter-maxillæ to glenoid process.	108	104	101	108	109	112	108	120	123	118	110	103	106	113
Distance from anterior edge of inter-maxillæ to meatus auditorius.	117	119	107	116	117	118	117	130	132	124	117	116	118	122
Distance from anterior edge of inter-maxillæ to hinder edge of last molar.	54	54	48	53	51	56	53	58	57	57	54	52	50	50
Distance from anterior edge of inter-maxillæ to end of pterygoid hamuli.	88	90	80	89	88	90	87	96	97	100	82	85	84	…
Greatest breadth at zygomatic arches.	111	106	97	106	100	102	102	113	116	108	108	110	97	104
Breadth at mastoid process.	110	100	102	106	106	102	104	115	112	106	106	102	105	104
Length.	178	167	159	168	171	165	168	186	193	182	168	168	166	175
Sex.							♂	♂		♀	♀	♀		
Locality.	Greenland	do	do	do	do	Unalakleet, Alaska	Plover Bay, Behring's Straits	Gulf of Cumberland (very old)	do	Gulf of Cumberland (quite old)	Gulf of Cumberland (middle aged)	do	do	do
Catalogue number.	6565	6566	6567	6568		8942	7105	16139	16111	6296	16138	6295	6297	16051

DIFFERENTIAL CHARACTERS.—In color the Ringed Seal is not easily distinguishable from certain phases of the Harbor Seal, but it differs from it in its general form, which is much slenderer, with longer limbs and tail, narrower head, and more pointed nose. The Ringed Seal may, however, be distinguished externally from both the Harbor and the Greenland Seals by the form of the manus, in which the first digit is the longest, the others successively slightly decreasing. The cranial characters, and especially the dentition, differ too widely from those of the Harbor Seal to even require a comparison in the present connection, as do also most of the principal bones of the skeleton (see *anteà*, pp. 565–571).

The Ringed Seal differs externally from the Greenland Seal in its smaller size and in the very different coloration of the adults of the two species. When in the " white coat," and in the earlier spotted stages, coloration often fails to be diagnostic, but they may be distinguished by the character of the manus already given. The dentition of these two species, allowing for the difference in size, is quite similar, although the teeth are relatively (as well as absolutely) larger in *Phoca grœnlandica*. In the general form of the skull there is also a close resemblance, although the facial portion is rather more attenuated in *P. grœnlandica*. The form of the palatal region, however, is widely different in the two, the broad shallow posterior nares, completely divided into two separate passages by a bony septum, and the squarely truncate, instead of deeply emarginate, posterior border of the palatine bones serving at a glance to distinguish *P. grœnlandica*.

The relationship of *Phoca fœtida* to the Baikal and Caspian Seals (*Phoca sibirica* and *P. caspica*) is apparently much closer than to any other. The earlier writers, however, as Erxleben, Gmelin, and Pallas, associated them with *Phoca vitulina*, they forming respectively Erxleben's varieties β and γ of this species, and Gmelin's varieties *sibirica* and *caspica*. The Caspian Seal was first recognized as a distinct species by Nilsson in 1837, and called by him *Phoca caspica*. Later its specific distinctness was admitted by Gray (1844), Wagner (1846), and Radde (1862). Nilsson was also the first to make known the fact of its much closer resemblance to *Phoca fœtida* (= *P. annellata*, Nilss.) than to *Phoca vitulina*.[*] Wagner arrived at the

* After detailing its characters, he remarks, " Jeder sieht ein, dass diese Form der *Ph. annellata* viel näher steht, als der *Ph. vitulina*. Doch bildet sie ohne allen Zweifel eine von ersterer bestimmt verschiedene Art : sie ist viel

same conclusion, affirming most emphatically that the Caspian
Seal was in no way closely related to *Phoca vitulina,* but found
its nearest affine in *Phoca fœtida.** The distinctive characters
claimed by these authors for the Caspian Seal, as compared
with *Phoca fœtida,* are larger size, smaller and more widely separated teeth, greater convexity of the cranium, longer, stiffer,.
and more numerous mystacial bristles, and a somewhat different pattern of coloration. The differences claimed by Nilsson
and Wagner were confirmed by Radde† in 1862, who gave a
detailed comparison of the cranial characters of *Phoca caspica*
with those of *Phoca fœtida.* Yet, in face of all this testimony,
we find Mr. Andrew Murray, as late as 1866,‡ affirming that the
Caspian Seal "*is* PHOCA VITULINA," and that the Baikal Seal
is nothing but *Phoca fœtida.*§ It appears, however, that the

---

grösser, anders gefärbt, hat viel stärkeres Barthaar, abstehendere und klei-
nere Zähne, und den Zwischenbalken nach hinten zu abgerundet, wodurch
eine rundliche Uebegangsfläche zwischen Stirn und Schläfengrube entsteht,
wo sich bei *Ph. annellata* stets eine scharfe Kante findet."—*Wiegmann's
Arch. für Naturg.*, 1841, p. 314.

* On this point he says, "Auch aus meiner Vergleichung geht es hervor,
dass *Phoca caspica* keineswegs mit der *Ph. vitulina,* sondern nur mit der
Ph. annellata, in nächste Beziehung treten kann. Als Unterschiede finde
ich, dass die Ringelzeichnung bei *Ph. caspica* minder ausgebildet; ist dafür
sind die Bartschnurren weit zahlreicher, länger und steifer, die Krallen
schwächer und nicht kohlschwarz wie bei *Ph. annellata,* sondern hellbraun
mit weisslichen Spitzen. . . ."—*Schreber's Säugth.*, Theil vii, p. 35.

†Reisen im Süden vom Ost-Sibiriens, vol. i, pp. 296–304.

‡Geograph. Distr. Mam., p. 126.

§ It is perhaps not strange that Mr. Murray should have referred the Seal
of Lake Baikal to *Phoca fœtida,* especially inasmuch as Radde had affirmed
the two to be identical after having compared specimens, but his strange
perversion of the record in the case of the Caspian Seal deserves a passing
notice. He says: "The species in the Caspian [Sea] (*Phoca caspica*) is described as very nearly allied to our common *Phoca vitulina,* and that in Lake
Baikal as equally close to *Phoca fœtida* (*Ph. annellata,* Nilss.), a species
found in the North Atlantic; and but for their geographical position, no
one would think of separating them from these species. In fact, the one *is*
the PHOCA VITULINA, and the other the PHOCA FŒTIDA. Nilsson and Gray
no doubt both consider them distinct, but I do not apprehend that either of
them does so from actual observation [Nilsson characterized *Phoca caspica*
from specimens!], and it is scarcely possible to doubt that the peculiarity
of the locality must have had some influence on their minds. On the other
hand, Pallas, Gmelin, Fischer, [these authors referred *both* to *Phoca vitulina*
as varieties of that species!] and Radde, regard them as belonging to the
two species they resemble, and Radde's personal experiences must outweigh
any foregone conclusion arrived at by others who have not had the advantage of seeing the animals themselves."—*Geogr. Distr. Mam.*, p. 126. That

Caspian Seal* is distinguished not only by well-marked external and cranial characters but by certain strongly-marked peculiarities of habits, coupled with which are to be considered its long-continued geographical isolation and otherwise exceptional conditions of environment.

Radde's views are misrepresented is evident from the following: "*Phoca caspica*," says Radde, "steht in dieser Hinsicht [*i. e.*, allgemeinen Shädelform], wenn ich den einzigen nur vorliegenden Schädel als typischen betracten darf, unbedingt der *Phoca annellata* näher als der gemeinen Robbe" (*Reisen im Süd. vom Ost-Sibiriens*, i, p. 297); and throughout his article takes pains to show how wide are the differences between *Phoca caspica* and *Phoca vitulina*!

* PHOCA (PUSA) CASPICA, *Nilsson.*

Caspian Seal.

Der Seehund, S. G. GMELIN, Reise durch Russl., iii, 1770, 246.
Der caspische Seehund, SCHREBER, Säugth., iii, [1776?] 310.
Phoca vitulina, γ, ERXLEBEN, Syst. Reg. Anim., 1777, 588.
Phoca vitulina, δ, *caspica*, GMELIN, Syst. Nat., i, 1788, 64.—KERR, Anim. King., 1792, 124.
Phoca caspica, NILSSON, "R. Vet. Akad. Handlgr. Stockholm, 1837,—"; Arch. für Naturg., 1841, 313.—SCHINZ, Synop. Mam., i, 1844, 481.—WAGNER, Schreber's Säugth., vii, 1846, 33.—RADDE, Reisen im Süd. vom Ost-Sibiriens, i, 1862, 297–302 (*passim*).
Callocephalus caspicus, GRAY, "Zoöl. Erebus and Terror, 1844, 3"; Cat. Seals Brit. Mus., 1850, 24.
Callocephalus? caspicus, GRAY, Cat. Seals and Whales, 1866, 22.
Phoca canina [var. *caspica*], PALLAS, Zoog. Rosso-Asiat., i, 1831, 116, nota 1 (in part only).

The history of the Caspian Seal, in relation to its literature and synonymy, is briefly as follows: It appears to have been first described in 1770 by S. G. Gmelin in the narrative of his travels in Russia. It was first introduced into systematic zoölogy by Schreber, about six years later, but only under a vernacular name. Schreber's account of the animal was wholly compiled from Gmelin. Erxleben, in 1777, recognized it as a variety of *Phoca vitulina*, without, however, naming it, his citations embracing Gmelin and Schreber, as above. J. F. Gmelin, in 1788, referred it also to *Phoca vitulina*, of which he made it a variety, bestowing upon it the name *caspica*. The next reference to it of importance is made by Nilsson, who, in 1837, dissevered it entirely from *Phoca vitulina*, claimed its specific distinctness, and showed its closer relationship to *P. fœtida* than to *P. vitulina*. Its subsequent history has already been amply detailed. Respecting the habits of the Caspian Seal, Schultz writes as follows:

"These Seals gather in large herds, and, plunging continually into the water, chase scaly fish, of which they eat only the breast, leaving the remainder of the body, with the entrails, to the sea-birds, which are constantly hovering above them. Endowed with a very acute sense of smell, the Seals at times escape the vigilance of their enemies, the fishermen, with the excep-

The Baikal Seal was specifically distinguished much later, it having been referred even by Radde, as late as 1862, to *Phoca annellata* (*=fœtida*). Although it was recognized varietally by Erxleben and Gmelin a century ago, and the question of its specific distinctness raised by Nilsson in 1837, it was first formally separated as a species by Dybowski in 1873,* under the name *Phoca baicalensis.* From three years' observation of the living animal, and from study of the skulls of young and full-grown animals, he reached the conclusion that the Baikal Seal should not be referred to "*Phoca annellata,*" it being easy to distinguish from it at all stages. Dybowski has given excellent figures of the skull, and detailed descriptions of its cranial and external characters, and of its habits, but, unfortunately, makes no comparative references to any other species, nor does he state explicitly in what its distinctive specific characters consist. His very detailed table of measurements, and excellent figures of the skull, however, when compared with *Phoca fœtida*, leave no reason to doubt the specific distinctness of the Baikal Seal from that species. The skull of the Baikal Seal is especially remarkable for its attenuation, and particularly for the length and narrowness of the facial portion. Even the brain-case is narrow, for while its length is the same as in average skulls of *P. fœtida*, its width is less. The orbital fossæ are disproportionately large, whence results a great lateral expansion of the rather slender zygomatic arches, so that the breadth of the skull at the orbits is considerably greater than at the mastoid processes (as 100 to 87), instead of these two dimensions being about equal, as in *P. fœtida.* Dybowski gives the average length of two adult skulls

tion, however, of the young, which, inexperienced as they are, follow the fishing-boats for long distances, and seem to take special pleasure in hearing the fishermen whistle or sing   It is an interesting spectacle to see the young Seals lying on their back, sleeping peaceably while being rocked by the waves, and throwing up from time to time small jets of water by breathing."*

It further appears from the author's detailed account of Seal-hunting in the Caspian Sea (see *anteà*, pp. 514–517) that these animals are preëminently gregarious, and resort, at certain seasons, to favorite localities on the shore in immense herds, to bask in the sun. The pairing season occurs about the beginning of January, and the young are brought forth on the ice. In habits the Caspian Seal thus differs notably from the Ringed Seal, which never resorts to the land in vast herds.

× As translated in Rep. U. S. Com. Fish and Fisheries, pt. iii, 1873–4 and 1874–5, pp. 92, 93.

* Archiv für Anatomie, Physiologie und Weissenschaftliche Medicin, Jahrgang 1873, pp. 109–125, pl. ii, iii.

the same dimensions in exceptionally large old male skulls of *P. foetida* to be 186 mm. and 115 mm. The width at the mastoid processes in the Baikal Seal, however, is only 100 mm.,* against 112 mm. in *P. foetida*. A very strongly marked difference is observable in the relative length of the facial portion of the skull in the two species, this being very much narrower and longer in the Baikal Seal than in the other, this, of course, involving a corresponding narrowness of the nasal bones and the palatal region. The vertical height of the skull is also much less than in *P. foetida*. Without going further into details, it may be sufficient to state that the skull of the Baikal Seal is characterized by great attenuation in every part, with great expansion of the orbits. In dentition and in the general form of the palatal region there is a close agreement with the Ringed Seal, the Baikal Seal being, in a word, a slender form of the *Phoca foetida* type.

It also differs notably in coloration, being apparently never spotted. According to Dr. Dybowski, the adults are silvery-brown ("silberbräunlich") above, and dingy silvery-brown ("schmutzig silberbräun") below; in the younger animals the silver-brown color has a whitish lustre; in the newly-born young the thick, long wool-hair is silvery-white. The length of the full-grown animal is given as 1,300 mm.

While it is pretty clear that the Caspian Seal and the Baikal Seal are both specifically distinct from the Ringed Seal, and that neither of them has any near relationship to *Phoca vitulina*, the points of difference between the two first-named are not so evident. In coloration the Caspian Seal appears to not differ greatly from the Ringed Seal; both consequently differ similarly in color from the Baikal Seal, namely, in being spotted, while the latter is concolor. On the other hand, the Caspian and Baikal Seals agree in being considerably larger than the Ringed Seal, and in the skull being narrower in proportion to its length, with the upper surface more convex. The Baikal Seal, however, appears to be distinguished by the greater attenuation of the facial region coupled with a much greater expansion of the zygomatic arches.

The peculiar features of coloration presented by the Baikal Seal have been given with uniformity since the time of Steller,† who first made them known. Schreber's short description,

---

* Estimated from Dybowski's figure.

† Beschreibung von dem Lande Kamtschatka, 1774, p. 108.

as 192 mm., and their average breadth as 118 mm., which I find
founded on that of Steller, is as follows: "Er ist einfarbig,
silberweiss vom Harren, so gross als der gemeine." He further
says: "Man findet ihm in den beiden sibirischen Landseen
Baikal und Oron, die weit von dem Ocean entfernt sind und
mit demselben durch keinem Fluss Gemeinschaft haben.ᵃ Ob
er vom dem gemeinen wesentlich verschieden sei, ist mir nicht
bekannt." *

Nilsson refers to a specimen supposed to have come from Lake
Baikal as being "Braungrau einfarbig mit flässerer Färbung an
den untern Körpertheilen". Radde says its color is "schön
grau (fast stahlgrau)" on the back, becoming lighter on the
sides, and yellowish-gray beneath. †

"ᵃ Steller, a. o.   ."
* Säugth., iii, 310.—The above is Schreber's account in full.

† PHOCA (PUSA) SIBIRICA, *Allen ex Gmelin.*

Baikal Seal.

The Seal [of Lake Baikal], BELL, Travels from St. Petersb. in Russia to diverse parts of
Asia, "i, 1763, 261"; ibid., i, 1788, 320.
Die virte Sorte Seehunde, STELLER, Beschreibung von dem Lande Kamtschatka, 1774, 108.
Der sibirische Seehund, SCHREBER, Säugth., iii [1776?] 310 (*ex* Steller).
Phoca vitulina, β, ERXLEBEN, Syst. Reg. Anim., 1777, 588 (*ex* Steller et Schreber).
Phoca vitulina, γ, *sibirica,* GMELIN, Syst Nat., i, 1788, 64.—KERR, Anim. King., 1792, 124.
? *Phoca annellata,* NILSSON, Arch. für Naturg., 1841, 312 (in part).
Phoca annellata, RADDE, Reisen im Süd. vom Ost-Sibiriens, i, 1862, 296 (in part,—only the
Baikal specimen).
Phoca baicalensis, DYBOWSKI, Arch. für Anat. u. Phys., 1873, 109, pll. ii, iii (skulls of adult
and young).

This species was apparently first mentioned by Bell (as above cited),
in 1763, who refers to its habits, but gives no account of its characters. It
was quite fully described by Steller in 1774, and was first formally introduced into systematic zoölogy by Schreber about two years later, whose
account is based wholly on Steller's. It was cited as a variety of *Phoca vitulina* by Erxleben in 1777, and named as a variety of that species by Gmelin
in 1788. Nilsson described a skin and an imperfect skull under *Phoca annellata* in 1837, but thought it might prove to be a distinct species, and, moreover, was not certain whether or not his specimen came from Lake Baikal.
Radde, the first naturalist after Steller who described the Baikal Seal,
from an authentic specimen, referred it unhesitatingly to *Phoca annellata*
(=*fœtida*). The single skull on which his observations were based he
stated to be that of a female about three or four years old, but his figures
of it show it to have been much younger, the principal sutures being represented as unobliterated. He found it to be considerably smaller than

From a geographical stand-point there is no *à priori* reason for their identity, they occupying entirely distinct drainage basins, which have had no connection since comparatively remote geological times. Their geographical position, indeed, considered in relation to the present distribution of their nearest allies, as well as to their peculiar environment, is one of the most interesting facts in their history. The Baikal Seal is an inhabitant of a fresh-water lake, while the waters where the other finds a home are only to a slight degree salt. Neither of these remote interior seas has had any recent connection with the Polar Seas, where alone the nearest affines of these Seals are now found. If their oceanic connection was southward (as was most likely that of the Caspian Sea), at the remote

---

either of his skulls of *Phoca fœtida*, but its small size is explainable on the ground of its immaturity. He himself states that his specimen of the Baikal Seal weighed in the flesh only "3¼ Pud" (126 pounds), while the weight of the Baikal Seal, as he says he was informed by the Seal-hunters, ranges from "8 zu 10 Pud" (288 to 360 pounds). Radde's evidently erroneous estimate of the age of his specimen is pointedly noticed by Dr. Dybowski, who, in referring to the fact of young Seals being often mistaken for old ones, adds, "wie es G. Radde gethan hat, der ein 7-8 monatliches junges Thier für ein 3-4 jähriges ausgiebt."[*]

In 1873 Dybowski described the species with admirable fullness, including its external and cranial characters, giving figures of an adult and a young skull, together with a detailed account of its external characters, of not only the adult but young of various ages. He, moreover, was the first to positively claim its specific distinctness, and it is an open question whether his name *baicalensis* ought not to supersede Gmelin's long previously imposed name *sibirica*.

According to Dybowski, these Seals are pretty common in Lake Baikal, but there is rarely opportunity for observing them in summer. The native hunters informed him that they are often seen and shot in the months of July and August on the rocky southwest shore, by lying in wait for them behind rocks. It is during these months that the rutting time occurs, and the young are born in January and February, so that consequently the period of gravidity must be reduced to about six or seven months. The young are said to depend for sustenance exclusively upon the mother's milk for about four months. The lake becomes closed with ice in January, and from that time till the middle of May—a period of about four months—the Seals remain wholly under the ice, but have their breathing-holes through which they obtain air. About the end of March or beginning of April, after the deep snows have become melted by the sun, the hunters seek out these breathing-holes by means of dogs especially trained for the purpose, and capture the Seals in nets placed in the breathing-holes.—*Arch. für Anat. u. Phys.*, 1873, pp. 121-125.—See further Bell's account of their capture about the middle of the last century, already cited (*anteà*, p. 612).

---

[*] Arch. für Anat. u. Phys., 1873, p. 122.

time when these basins formed a part of the great Tertiary sea, of which the Mediterranean and connected interior waters are now the greatly diminished remnants, whence came the stock from which these two allied species of Seal are the descendants? Are we to look for an ancestor in *Phoca fœtida*, or in some allied extinct species, from which came not only these species but also their present northern ally? As shown by the researches of Van Beneden, Seals were abundant in the Pliocene seas of Southern Europe, and among them were forms more or less nearly related to each of the existing types, his genus *Phocanella* being the early representative of the modern *Phoca fœtida*.

GEOGRAPHICAL DISTRIBUTION.—Although the Ringed Seal is a well-known inhabitant of the Arctic Seas, of both hemispheres, the southern limit of its distribution cannot be given with certainty. Wagner* records specimens from Labrador, which is the most southern point on the eastern coast of North America from which it seems to have been reported. It is not enumerated by Jukes or Carroll as among the species hunted by the Newfoundland sealers,† nor is it mentioned by Gilpin‡ as occurring in Nova Scotia. Its occasional presence here and in the Gulf of Saint Lawrence is doubtless to be expected. Further northward, and especially along the shores of Davis's Straits and Greenland, its abundance is well attested. It has also been found as far north as explorers have penetrated, having been met with by Parry as high as latitude 82° 40′. J. C. Ross states that it is

---

* Schreber's Saugt., vii, 1846, p. 31.

† Professor Jukes says four species are known on the coast of Newfoundland, namely, the "Bay Seal" (*Phoca vitulina*), the Harp Seal (*Phoca grœnlandica*), the Hooded Seal (*Cystophora cristata*), and the "Square Flipper" (probably *Halichœrus grypus*). The first he did not see on the ice among the Seals pursued by the sealers. The second is the one that forms the principal object of the chase. The third seems not to be numerous, but occurs occasionally out on the ice-floes with the Harp Seals. The fourth is referred to as very rare, and as being larger than the Hooded Seal. Not one was heard of or seen that season. He supposes it may be the *Phoca barbata.—Excursions in Newfoundland*, vol. i, pp. 308–312.

Carroll states that the species of Seal that are taken on the coast of Newfoundland are the "Square Flipper Seal" (probably *Halichœrus grypus*), the "Hood Seal" (*Cystophora cristata*), the "Harp Seal" (*Phoca grœnlandica*), and the "Dotard" or "Native Seal" (*Phoca vitulina*).—*Seal and Herring Fisheries of Newfoundland*, 1873, p. 10.

‡ The species given by Gilpin as found on the coast of Nova Scotia are the Harbor Seal (*Phoca vitulina*), the Harp Seal (*Phoca grœnlandica*), the Gray Seal (*Halichœrus grypus*), and the Hooded Seal (*Cystophora cristata*).

common on both sides of the Isthmus of Boothia, where it forms
the chief means of subsistence to the inhabitants during eight or
nine months of the year.* It is common in Iceland, and Malmgren and Von Heuglin state it to be numerous at Spitzbergen.
The last-named author gives it as abundant in summer in the
Stor-Fjord and its branches, in Hinlopen Strait, and in the
bays of the northwest coast of Spitzbergen, occurring in great
herds as well as singly, in the open water along the shores and
in the openings in the ice-floes. He states that it is also numerous about Nova Zembla, where great numbers are killed for
their skins and fat.† It is a common species on the coast of
Finland, and further eastward along the Arctic coast of Europe
and doubtless also of Western Asia.‡ It is also a common inhabitant of the Gulf of Bothnia and neighboring waters, and
also of the Ladoga and other interior seas of Finland. It is
said by Blasius to extend southward along the coast of Middle
Europe to North Germany, Ireland, and the British Channel.
Professor Flower has recorded its capture on the coast of Nor-
wich, England, and it undoubtedly occurs at the Orkneys and
the Hebrides, where it is supposed to be represented by the
species known there as "Bodach" or "Old Man". A specimen
was also taken many years since on the coast of France, but here,
as on the shores of the larger British Islands, it can occur as
merely a rare straggler.§ Its fossil remains have been reported

---

* Ross's Sec. Voy., App., 1835, p. xix.
† Reise nach dem Nordpolarmeer, Th. iii, p. 50.
‡ In an account of Professor Nordenskjöld's late Arctic voyage, published
in "Nature" (vol. xxi, p. 40, Nov. 13, 1879), it is stated that *Phoca fœtida*
"was caught in great numbers, and along with fish and various vegetables
forms the main food of the natives" at Cape Serdze (about 120 miles from
Behring's Straits), the point where the "Vega" wintered, this and the Polar
Bear being the only marine mammals seen.
§ Respecting the southern limit of the habitat of this species in Europe,
Professor Flower has the following: "Nilsson speaks of it as being found
on all the Scandinavian coasts, and as having been met with as far south
as the Channel, on the strength of specimens in the Paris Museum from that
locality; but he was unable to find any proofs of its having been met with
on the coast of England. Nor have I been able to discover any positive evidence that it can, at the present day, be reckoned a British species, although
there is little doubt that it must occasionally visit our shores, where its
occurrence would be easily overlooked."—*Proc. Zoöl. Soc. Lond.*, 1871, p. 510.
Collett, contrary to the testimony of Nilsson, excludes it from the mammalian fauna of Norway, and states that he does not know of an authentic
instance of its capture on the Norwegian coast.—*Bemœrkninger til Norges
Pattedyrfauna*, 1876, p. 57, footnote 2.

by Professor Turner as having been found in the brick clays of Scotland. It appears also to be a common species in the North Pacific, there being specimens in the National Museum, unquestionably of this species, from the coast of Alaska, and from Plover Bay, on the Siberian side of Behring's Strait. Its southern limit of distribution along the shores of the North Pacific, on either the American or the Asiatic side, cannot at present be given. Judging from its known distribution in other portions of the Arctic waters, there is no reason to infer its absence from the northern shores of Eastern Asia and Western North America.

GENERAL HISTORY AND NOMENCLATURE.—The earliest notices of this species in systematic works are based on the brief account given by Cranz in 1765, but there appear to be still earlier references to it by Scandinavian writers. As, however, they involve no questions of synonymy, and may in part relate to the Gray Seal (*Halichœrus grypus*), they call for no special remark in the present connection. The "Grå Sial" of Linné's "Fauna Suecica" (1747), however, was referred by Otto Fabricius, in 1791, to *Phoca fœtida*, but recent writers, notably Lilljeborg, have assigned it to *Halichœrus grypus*, but Linné's account seems to be too vague to be positively identified, although it later became the basis of Gmelin's *Phoca vitulina botnica*.

As already' noticed, the early technical history of the species is based on the brief notice of it published in 1765 by the Danish missionary, Cranz, who, in his "Historie von Grönland," referred to it under its native or Eskimo name *Neitsek*. He says it is not very different from the Attarsoak (*Phoca grœnlandica* of systematists) "in size or color, only that the hair is a little browner or a pale white, nor does it lie smooth, but rough, bristly, and intermixed like pig's hair."* Pennant, in 1771, in his "Synopsis of Quadrupeds," called it the Rough Seal, and paraphrased Cranz's description, adding thereto the conjecture: "Perhaps what our *Newfoundland* Seal-hunters call *Square Phipper*". In 1776 it was enumerated in the introduction to Müller's "Zoologiæ Danicæ Prodromus" (p. viii), in a list of Greenland animals supplied by Otto Fabricius after the main body of the work had been printed, where it first receives a systematic name, being there called *Phoca fœtida*. No description is given, but its supposed Icelandic and Greenlandic names are appended, namely, "I. *Utselr*. Gr. *Neitsek, Neitsilek,*" but unfortunately the

---

* English edition, 1767, vol. i, pp. 124, 125.

Seal called Ut-Selur by the Icelanders proves to be *Halichœrus grypus*. It thus happens that the first technical name of the species, as well as some of its earliest vernacular names, relates in part·to the Gray Seal. At about the same time (certainly not earlier) it was described by Schreber in the third part of his "Säugthiere" as Der rauhe Seehund, his description being based entirely on Cranz's and Pennant's. No Latin name is given in the text, but on the plate appears the name *Phoca hispida*. The date of the publication of the fasciculus containing Schreber's description and figure cannot be definitely determined, but contemporary evidence indicates that it must have appeared during the year 1776,* as it is cited by Erxleben in a work published the following year. who adopts *Phoca hispida* for the name of the species. But Erxleben's first reference is to the "Long-necked Seal" of Parsons, whose diagnosis of which Erxleben cites in full. The Long-necked Seal, however, is some indeterminable species of Otary. But all of Erxleben's other references, with one exception (for here "Utselr" is again cited), are pertinent, and his diagnosis is evidently based on the Neitsek of Cranz.

Three years later (1780), Fabricius, in his "Fauna Grœnlandica," gave the first adequate description of the species, under the name *Phoca fœtida*, and quoted *Phoca hispida* as a synonym. Eleven years later (1791), in his celebrated memoir on the Greenland Seals, he reverts to the name *hispida*, conceding it priority, but on what ground is not apparent. The case is thus a peculiar one, and has already received attention at the hands of numerous writers, the matter having been quite recently very fully discussed by Professor Flower.† Although Flower favors the adoption of *hispida*, he admits that "There is nothing either in Schreber's description or figure to identify the species; and it has since been thought (as by A. Wagner in his continuation of Schreber's work, 1846) [*] to refer to a totally distinct animal, viz, *Halichœrus grypus*." He says, further, "Although it may still be a matter of opinion which of these names ought to be

---

* The date on the title-page of the "Dritter Theil" is 1778; the two preceding parts are both dated 1775. The Seals occupy the first pages of the third part.

† Proc. Zoöl. Soc. Lond., 1871, pp. 507–510.

[* Gray, apparently following Wagner, referred, both in 1850 and in 1866, Schreber's *Phoca hispida* to *Halichœrus grypus*, while at the same time he referred Lesson's *Phoca schreberi*, avowedly = *Phoca hispida* Schreber, to his *Callocephalus fœtidus*!]

adopted, it appears to me that, on the whole, preference should be given to *hispida*, on account of priority; for although the earliest descriptions under this name are very meagre and inaccurate, they are avowedly founded on the *Neitsek* of. Cranz, the appellation by which this Seal is known to the Greenlanders to this day, according to Mr. R. Brown,* and are therefore intended for this species, and especially because Fabricius, in 1790 [1791], definitely adopted the name, withdrawing that of *fœtida*. I am further strengthened in this opinion," he continues, "by finding that those eminent Danish naturalists Steenstrup † and Reinhardt‡ both use *hispida* when speaking of this Seal." As regards use, although good authorities have adopted *hispida*, by far the greater number of writers, including equally eminent authorities, among them Lilljeborg and Collett among recent Scandinavian writers, adopt *fœtida*. The question is certainly pretty evenly balanced. Granting, however, that the introduction of the two names was practically simultaneous, and that *fœtida*, as first given, was unaccompanied by a description, while *hispida* had this backing, it is admitted that neither the description nor the figure is of any value in determining what species was intended, and that the Greenland name Neitsek is the only clew to what was meant. Just this clew, backed by the best authority—Fabricius himself—we have also in the case of *fœtida*, while the first real description (in "Fauna Grœnlandica," 1780) of the species was given under this name, and eleven years before the species was recognizably described under the name *hispida* (by Fabricius in 1791). Fabricius gave as his reasons for withdrawing the name *fœtida* and adopting *hispida* that the latter was not only an appropriate name but also the oldest, although he ascribes the name *hispida* to Erxleben. It would seem, however, that he really adopted the name from Pennant, considering Pennant's name "Rough Seal" a strict equivalent of *Phoca hispida*.§ The name *fœtida* appears certainly to be most characteristic.

---

* " 'On the Seals of Greenland,' P. Z. S., 1868, p. 414."

† " 'Melketandsættot hos Remmesælen, Svartsiden, og Fjordsælen (*Phoca barbata*, O. Fabr., *Ph. grönlandica*, O. Fabr., og *Ph. hispida*, Schr.),' Vid. Medd. f. d. Naturh. Forening, 1860. Kjöbh. 1861, s. 251–261."

‡ " 'Om Klapmydsens ufödte Norge og dens Melketandsæt,' Naturh. Foren. Vidensk. Meddelelser, 1864."

§ As being of interest in this connection I submit the following rendering of Fabricius's opening paragraph of his history of the Fiordsæl : "This, next to the Black-side, is the species which is most numerously met with in Greenland. I give to it the Danish name Fiordsæl, because it keeps mostly in the

Another name of considerable prominence in connection with this species is *annellata* of Nilsson, proposed by him in 1820 for Scandinavian representatives of the species, because he did not feel sure of their identity with Greenland examples. This name was adopted later by various writers for a species supposed to be distinct from the *Phoca fœtida* of Greenland, notably among whom are Wagner and Radde, while Giebel held both *fœtida* and *hispida* as synonyms of *annellata*!

The name *discolor*, introduced in 1824 by F. Cuvier as that of a new species, was later abandoned by its author, and never obtained currency except with a few compilers. Lesson, in 1828, characteristically changed it to *frederici*, and at the same time renamed Schreber's *hispida*, calling it *schreberi*.

HABITS, PRODUCTS, AND HUNTING.—The Ringed Seal is preëminently boreal, its home being almost exclusively the icy seas of the Arctic Regions. Its favorite resorts are said to be retired bays and fjords, in which it remains so long as they are filled with firm ice; when this breaks up they betake themselves to the floes, where they bring forth their young. It is essentially a littoral, or rather glacial species, being seldom met with in the open sea. From its abundance in its chosen haunts it is a species well known to Arctic voyagers, and frequent reference is made to it in most of the narratives of Arctic Explorations. These notices are, however, mostly incidental and fragmentary, no one having given a detailed and connected history of the species. I am, therefore, gratified to be able to present, in addition to excerpts from various more or less well-known sources, much fresh information kindly fur-

fjords and rarely goes out to sea. In my Fauna Grönlandica I called it *Phoca fœtida* because it has a stronger stink than the other species. It was previously mentioned under this name, first in my report quoted in Müller's Prodromus (Zool. Dan. Prodr., p. viii). It was then regarded as a new species, as I found it not in Linné; he either did not recognize it or did not distinguish it from the common Seal (*Phoca vitulina*), for at most he only regarded it as a variety of this under the name Grå-Själ (Faun. Suec., p. 2, under Spec. 4). But Pennant, however, gave it as a distinct species, with the name Rough Seal (Syn. Quadr., p. 341, n. 261); afterwards Schreber called it Der rauhe Seehund (Säugth. III. Th. p. 312), and Erxleben *Phoca hispida* (Syst. Regn. Anim. p. 589), which name Gmelin (Syst. Nat., p. 64) has retained. This name is suitable and a very good one for this species on account of its hair, and, although this is also found in the Klapmydsen (*Phoca leonina*, Linné), so are some other characters; wherefore I do not now hesitate to prefer the name *hispida* before *fœtida*, especially as it is the oldest, although the stench is so characteristic."—*Skrifter af Naturhistorie-Selskabat*, 1ste Bind, 2det Hefte, 1791, pp. 74, 75.

nished me by Mr. Ludwig Kumlien, naturalist of the recent
Howgate Polar Expedition. His observations, made chiefly
during several months spent in Cumberland Sound,* are substantially as follows:

"This Seal is very common in all the fjords and bays, from
Hudson's Straits, northward, along Cumberland Island, to the
extreme head of Cumberland Sound; on all the outer islands
about Cape Mercy, and on the west coast of Davis Straits.
I have seen skins from Lake Kennedy that I could not distinguish from those found in Cumberland Sound. This Seal was
never noticed more than a few miles from land; was not met
with in the pack-ice, nor on the Greenland coast, except far up
the fjords. This was in July and August; but I am informed
that they become more common toward autumn, and are found
in considerable numbers some distance from land; they are
less common here, however, than on the west coast. . . . . . .
In the Cumberland waters they are resident and do not migrate
at all unless much disturbed, and then they merely seek a more
secluded locality. On the Greenland coast they appear to migrate up the ice-fjords in summer but to be more generally distributed at other seasons.

"The Netsick, as this species is called by the Cumberland Eskimo, shows a decided predilection for the quiet, still bays and
fjords, seldom venturing far from land. They are the only Seal
caught through the ice in winter, and are consequently the
chief and almost sole dependence of the Eskimo for food, fuel,
light, and clothing. The skins of the adults are made into
summer clothing, while the young are in great demand for under garments and for trousers. Children often have entire suits
made of the skins of the young in the white coat. Such clothing looks very beautiful when new, but they are new but for a
few days, and after this they are repulsive enough. The females were found *enceinte* in the latter part of October, and a
fœtus nearly ready for birth was taken from the uterus January 16. It was 2 feet from the end of the nose to the end of
the hind flippers. It was so doubled in the uterus, however, as
to occupy a space hardly a foot in length; the hind flippers
were turned forward on the tibiæ; the fore flippers hugged the

---

* What Mr. Kumlien's opportunities were for the study of this species
may be inferred from the fact that among the spoils brought with him on
his return are skulls, skins, and skeletons, ranging from the fœtal to the
adult stage, to the number of about fifty specimens.

sides, and the head was bent over on the neck and inclined to one side.

" In a large fjord, known as the greater Kingwak, the tide runs so swiftly at one locality that it never freezes for a space varying from ten to one hundred acres; here the Netsick gather in considerable numbers all winter, and it is a favorite resort for such Eskimo as are fortunate enough to possess a gun. Being but a few miles from our winter harbor, almost daily excursions to these tide rifts were made by our Eskimo hunters. After the 1st of March very few pregnant females were killed at this place, they having by this time chosen the localities for having their young. Those killed after this date were all adult ' Tigak,' or old stinking males.

" It was interesting to observe that the young—yearlings and some two-year-olds, and such as had not yet arrived at maturity—were seldom if ever killed in this open water, but lived in colonies by themselves. When an Eskimo finds a number of *atluks* (breathing-holes) near together he always marks the place by raising little mounds of snow near the holes, for he knows that here is a colony of young animals, which have better skins and meat than the old ones, and are, moreover, much easier to capture. I have counted nearly seventy of these atluks on a space of two acres.

" When a pregnant female has chosen the place where she is to have her young she makes an excavation from six to ten feet in length under the snow, and from three to five feet wide, the height varying with the thickness of the snow covering; the atluk is at one extremity of this excavation, and in such a position that it is always a ready channel of retreat in case of danger.

" The first young were found in the Cumberland waters during the first days of March; still, I have taken a fœtus from the mother in the middle of April. The most profitable time for hunting the young Seal is during the month of April. After this date they have shed so much that the skins are nearly worthless till the hispid hair has got to be of the proper length, when they are considered as the prime article and second only to the young of *Callocephalus vitulinus* in quality.

" The first young one that had begun to shed was taken April 15. I have seen examples that were nearly or quite destitute of the white coat, but still not having the next coat in sight. Such specimens were found to have a very fine coat of the new hair, but so short as not to be perceptible except on close examination, yet showing the exact location and distribution of the dark and light markings; the skin at this time is very black and often much scratched, probably by the mother in trying to make the young one shift for itself.

"I often examined the stomachs of young ones, as well as of adults, but till after they had begun shedding the white coat, and were in all probability twenty-five to thirty days old, I found nothing but the mother's milk. After they begin to shift for th-mselves their food, for a time at least, consists of *Gammari* of different species.

"Before the young begin to shed the white coat they are from 23 to 36 inches from the nose to end of flippers; the average the season through, from a good series of measurements, was about 30 inches. . . . .

"They weigh at birth from four to six and one-half pounds, but the young grow at an astounding rate, becoming exceedingly fat in a few days. The blubber on the young a few days old is almost white and thickly interspersed with blood-vessels; it is not fit to burn.

"There is usually but one young at a birth, still twins are not of rare occurrence, and one instance came under my observation where there were triplets, but they were small, and two of them would probably not have lived had they been born.

"The season for hunting the young at latitude 67° north begins about the middle of March and continues until the latter part of April. The first two weeks of April are the most productive, as later the hair is apt to be very loose, and many even have large bare patches on them. When the season fairly opens the Eskimo hunter leaves the winter encampment with his family and dog team for some favorite resort of this Seal; he soon constructs his snow hut and is as well settled as if it had been his habitation for years; for the Seals he catches bring him and his family food and fuel, and snow to melt for water is always plenty, so that his wants are easily supplied, and he is contented and happy.

"The manner of hunting the young Seal is to allow a dog to run on ahead of the hunter, but having a strong Seal-skin line about his neck, which the Eskimo does not let go of. The dog scents the Seal in its excavation, which could not have been detected from the outside by the eye, and the hunter by a vigorous jump breaks down the cover before the young Seal can reach its atluk, and if he be successful enough to cut off its retreat it becomes an easy prey; otherwise he must use his sealing-hook very quickly or his game is gone. It sometimes happens that the hunter is unfortunate enough to jump the snow down directly over the hole, when he gets a pretty thorough wetting. The women often take part in this kind of sealing and become quite expert. The children begin when they are four or five years old; the teeth and flippers of their first catch are saved as a trophy and worn about the little fellow's neck; this they think will give him good luck when he begins the next year.

"There exists a considerable spirit of rivalry among the mothers as to whose offspring has done the best, size, etc., considered; this runs to such a high pitch that I have known some mothers to *catch* the Seal and then let the child *kill* it, so as to swell the number of his captures.

"Some of the Eskimo hunters, belonging to the 'Florence,' brought as many as seventy at one load. They were kept frozen, and we almost lived on the meat during the season, and learned to like it very much.

"Some of the Hispid Seals pup on the ice, without any covering at all; six instances of this nature came under my observation, and they were all young animals. The young exposed in this manner almost always become the prey of foxes and ravens before they are old enough to take care of themselves.

"As the season advances and the young begin to shed their coats the roof of their *igloo* is often, or perhaps always, broken down, and the mother and young can be seen on sunny days basking in the warm sunshine beside their atluk. The mother will take to the water when the hunter has approached within gunshot, and leave the young one to shift for itself, which generally ends in its staring leisurely at the hunter till suddenly it finds a hook in its side. A stout Seal-skin line is then made fast to its hind flippers and it is let into the atluk; it of course makes desperate efforts to free itself and is very apt to attract the attention of the mother, if she is anywhere in the vicinity. The Eskimo carefully watches the movements of the young one, and as soon as the mother is observed, begins to haul in on the line; the old one follows nearer and nearer to the surface, till, at last, she crosses the hole at the proper depth and the deadly harpoon is planted in her body and she is quickly drawn out. If the mother has seen the hunter approaching the atluk, however, she will not even show herself.

" I have never known of an instance where they have attempted to defend their offspring from man. I once saw a raven trying to kill a young Seal, while the mother was making frantic but very awkward attempts to catch the bird in her mouth.

" When the young first assume the coat of the adults (about the time the ice begins to loosen) they seem possessed of a vast amount of curiosity, and while swimming near the land, as they almost always do, can be lured within gunshot by whistling or singing. They would often play about the schooner, diving underneath and coming up on the opposite side, apparently enjoying it hugely. They delight to swim among the pieces of floating ice in the quiet bays. The young and yearlings of this species are often found together in small bands.

" The adult females will average four feet and a half to the end of the flippers. Such specimens are probably from four to seven years old. The males are a little larger. There is great variation in the skulls, but the sexes can readily be distinguished by the skull alone, the males having a longer and narrower head, with the ridges more prominent.

" It is only the adult males (called ' Tigak,' = Stinker, by the Eskimo) that emit the horribly disagreeable, all-permeating, ever-penetrating odor that has suggested its specific name. It is so strong that one can smell an Eskimo some distance when he has been partaking of the flesh; they say it is more nourishing than the flesh of the females, and that a person can endure great fatigue after eating it. If one of these Tigak comes in contact with any other Seal meat it will become so tainted as to be repulsive to an educated palate; even the atluk of the Tigak can be detected by its odor. [*]

" The food of the adults consists largely of different species of crustaceans, and during winter especially they subsist to a

[* Respecting the fœtid odor emitted by this species, Dr. Rink observes as follows :

" It derives its scientific name from the nauseous smell peculiar to certain older individuals, especially those captured in the interior ice-fjords, which are also on an average perhaps twice as large as those generally occurring off the outer shores. When brought into a hut and cut up on its floor, such a seal emits a smell resembling something between that of assafœtida and onions, almost insupportable to strangers. This peculiarity is not noticeable in the younger specimens or those of a smaller size, such as are generally caught, and at all events the smell does not detract from the utility of the flesh over the whole of Greenland."—Danish Greenland, its People and its Products, p. 123.]

considerable extent upon fish.   I have found in them the remains of *Cottus scorpius, C. grœnlandicus, Gadus agœ* (commonly), and *Liparis vulgaris*.   During the time the adults are
shedding, and for nearly a month previous, I could detect
nothing but a few pebbles in their stomachs; they become poor
at this time, and will sink when shot in the water.

"The milk is thick and rich, and is sometimes eaten by the
natives.   The excrement looks like pale, thickly-clotted blood.

"Albinos are sometimes found, of which the Eskimo tell marvelous stories, one being that when they rise to breathe in their
atluks they come stern first, and in fact they think such animals have their breathing apparatus on the posterior end of the
body.   I imagine this originated from a native once harpooning
an albino in its atluk and finding his harpoon fastened in one
of the hind flippers.   A hairless variety of this Seal is sometimes caught, which the Eskimo call *Okitook*.   I have seen one
such skin; it had a few fine curly hairs scattered over it, but
they were different in texture from the ordinary hair.   I do not
know if the specimen otherwise differed from the ordinary Seal.

"Toward spring, when the sun is shining brightly, these
Seals can be seen in all directions basking on the ice.   Although
to all appearance asleep, they manage to wake up regularly
every few minutes to make sure that there is no danger about.

"At this season it is a favorite method of the Eskimo to
hunt them by crawling flat on his belly toward the Seal, and,
when discovered, to imitate the movements of the animal, and
to advance only when the Seal looks in the opposite direction;
in this manner they often approach so close as to be able to
push them away from their atluks.

"This Seal is of some commercial importance, the Scotch
whalers often buying from the natives during the winter a thousand skins.   These are brought with the blubber, and often cost
the purchaser not over three to seven cents apiece, and this
mostly in tobacco, trinkets, or ship-stores.   To encourage them
to procure more skins, they are furnished with a cheap *breech*-
loading gun and a few hundred cartridges, which they soon
waste, and then their guns are of course worthless.   At the rate
both young and adults are slaughtered at the present day, they
will soon become so scarce that there will not be enough to supply the wants of the natives."*

---

* Copied, with slight verbal changes, from Mr. Kumlien's MS. notes, since
published in "Bull. U. S. Nat. Mus.," No. 15, pp. 55–61.

In addition to the account of the Rough or Ringed Seal given by Mr. Kumlien—which is by far the most important single contribution to its history I have met with—I quote the following. Mr. Robert Brown, in his account of the Mammals of Greenland, says: "They delight to live in retired bays in the neighborhood of the ice of the coasts, and seldom frequent the open sea. In the Greenland and Spitzbergen Seas they chiefly live upon the floes in retired situations at a considerable distance from the margin of the ice. Dr. Wallace observed them for a considerable time in the months of June and July, between N. lat: 76° and 77°, in possession of a large floe, part of which was formed of bay ice, where they had their 'blow-holes' (the *atluk* of the Danes); his ship lay ice-bound for nearly three weeks, at about three miles from this large floe, and hence he had considerable opportunity of observing them. They passed the greater portion of their time apparently asleep beside their holes; and he never saw them all at one time off the ice, unless alarmed by parties from the ship or by the Polar Bear. When the ice slackened away and the sheets of open water formed around the ships, the Seals used to swim near them; and occasionally at these times a few were killed. In the water they are very cautious, swimming near the hunter, gazing on him as if with feelings of curiosity and wonder; but on the ice beside their blow-hole it is almost impossible for the hunter to approach them, so much are they on the alert and so easily alarmed. In Davis's Strait it especially feeds about the base of icebergs and up the ice-fjords. The great ice-fjord of Jakobshavn is a favorite haunt of theirs; the reason for this predilection is apparently that their food is found in such localities in greater abundance. The bergs, even when aground, have a slight motion, stirring up from the bottom the Crustacea and other animals on which the Seals feed; the native, knowing this, frequently endangers his life by venturing too near the iceberg, which not unfrequently topples over upon the eager Seal-hunter." [*]

Dr. Kane thus describes their behavior when basking on the ice. Writing under date of May 20 he says:

"The seal are out upon the ice, one of the most certain signs of summer. They are few in number, and very cautious. We notice that they invariably select an open floe for their hole,

---

[*] Brown, Proc. Zoöl. Soc. Lond., 1868, pp. 414, 415; Man. Nat. Hist., Geol., etc., of Greenland, Mammals, pp. 44, 45.

and that they never leave it more than a few lengths. Their alertness is probably due to their vigilant enemy, the bear. . . . . The first act of a seal, after emerging, is a careful survey of his limited horizon. For this purpose he rises on his fore flippers, and stretches his neck in a manner almost dog-like. This maneuver, even during apparently complete silence, is repeated every few minutes. He next commences with his hind or horizontal flippers and tail a most singular movement, allied to sweeping, brushing nervously as if either to rub something from himself or from beneath him. Then comes a complete series of attitudes, stretching, collapsing, curling, wagging; then a luxurious, basking rest, with his face toward the sun and his tail to his hole. Presently he waddles off about two of his own awkward lengths from his retreat, and begins to roll over and over, pawing in the most ludicrous manner into the empty air, stretching and rubbing his glossy hide like a horse. He then recommences his vigil, basking in the sun with uneasy alertness for hours. At the slightest advance up goes the prying head. One searching glance, and, wheeling on his tail as on a pivot, he is at his hole, and descends head foremost."*

Dr. Richardson describes this species as being less cautious and less active than the Harbor Seal, observing that it is "easily surprised either on land or water, and is moreover a solitary and lazy animal, being wont to lie basking in the sun in place of hunting after its prey, and thus being often found lean from want of nourishment."† They appear, however, to behave quite differently under different circumstances; at least the accounts of authors on these points are more or less at variance. Thus Captain J. C. Ross states: "In the month of May, the Rough Seal, with its young, lie basking in the sun close to holes in the ice, and are at that time very difficult to approach; but the natives imitate both their cry and action so exactly as to deceive the animals until they get sufficiently near to strike them with their spear. Fabricius says it is the most heedless of all the Seals, as well on the ice as in the water. From our experience we would certainly give them a very different character, for none of our sportsmen were ever able to get sufficiently near to shoot them. The natives of Boothia say they

---

* Grinnell Exp., 1854, pp. 375, 376.
† Parry's Second Voyage, App., p. 333.

are not in their prime until the third year, and we never heard them complain of the offensive smell, which their more fastidious brethren in Greenland are said to dislike so extremely."*

Malmgren states that even the young, when lying on the ice, are extremely difficult to kill, for they go immediately into the water on the first view of the hunter, while, on the contrary, he observes, the young of the Gray Seal (*Halichœrus grypus*) has such a terror of the water while it wears its woolly coat that it scrambles out on to the ice as soon as it is thrown into the water.†

The habits of the Ringed Seal, as observed in European waters, seem to agree with what has already been related respecting their life-history in Davis's Strait and Cumberland Sound. Malmgren, for example, states that the females bring forth their young on the western coast of Finland on the ice near the edge of great openings between the 24th of February and the 25th of March, or at the time given by Fabricius and later writers for the same event on the coast of Greenland, and in no respect does their mode of life appear to differ in the icy seas about Spitzbergen from what has already been related.

The Ringed Seal is of far less commercial value than the Harp Seal, but in this respect may be considered as holding the second rank among the northern Phocids. Brown states that "it is chiefly looked upon and taken as a curiosity by the whalers, who consider it of very little commercial importance and call it 'Floe-rat.'" Von Heuglin, however, states that many thousands are annually taken by the sealers for their skins and fat, in the vicinity of Nova Zembla and Spitzbergen. It is of the greatest importance, however, to the Esquimaux and other northern tribes, by whom they are captured for food and clothing. Mr. Brown informs us that it forms, during the latter part of summer and autumn, "the principal article of food in the Danish settlements, and on it the writer of these notes and his companions dined many a time; we even learned to like it and to become quite epicurean connoisseurs in all the qualities, titbits, and dishes of the well-beloved Neitsik! The skin," he continues, "forms the chief material of clothing in North Greenland. All of the οἱ πολλοι dress in Neitsik breeches and jumpers; and we sojourners from a far country soon en-

* Ross's 2d Voy., App., p. xx.
† Arch. für Naturg., 1864, p. 83.

cased ourselves in the somewhat *hispid* but most comfortable nether garments. It is only high dignitaries like 'Herr Inspcktor' that can afford such extravagance as a Kassigiak (*Callocephalus vitulinus*) wardrobe! The Arctic *belles* monopolize them all." *   Rink states that the number annually captured in South Greenland has been calculated at 51,000.†   Capt. J. C. Ross states that the Esquimaux wholly depend upon it for their winter food, and Von Schrenck alludes to the great importance of this animal to the natives of Amoor Land.

Although the methods of capture employed by the Eskimo have already been to some extent described, I transcribe the following from Captain Ross, who says: " . . . when all other animals have retired to a more temperate climate the Seal is sought by the Esquimaux, whose dogs are trained to hunt over the extensive floes of level ice, and to scent out the concealed breathing-holes of the Rough Seal. So soon as one is discovered, a snow wall is built around it, to protect the huntsman from the bitterness of the passing breeze; where, with his spear uplifted, he will sit for hours until his victim rises to breathe, and falls an easy sacrifice to his unerring aim. In this manner a party of thirty hunters killed 150 of these animals during the first two months they remained in our neighborhood; the fishery for ten or twelve miles around was then completely exhausted; so they broke up into various smaller parties and dispersed in various directions." ‡

Dr. Rink states that the Netsek is " stationary throughout the coast" of Danish Greenland. "Only stray individuals of this species", he observes, "emigrate to the main drift-ice of Baffin's Bay in July, and return to the coast when the first bay-ice is forming in September, or occasionally appearing whenever the weather has been stormy. But the chief stock, whose favorite haunts, as has been described, are ice fjords, does not seem to leave the coast at all. It is almost exclusively this seal that is captured as 'utok' and by means of the ice-nets." §

---

* Proc. Zoöl. Soc. Lond., 1868, p. 417 ; Nat. Hist., Geol., etc., Greenland, Mam., p. 45.

† Danish Greenland, etc., p. 123.

‡ Ross's 2d Voy., App., p. xix.

§ Danish Greenland, etc., p. 123.

PHOCA (PAGOPHILUS) GRŒNLANDICA, *Fabricius.*

Harp Seal.

Phoca grœnlandica, FABRICIUS, Müller's Zool. Dan. Prod., 1776, viii; Fauna
Grœnl., 1780, 11; Skriv. Nat. Selsk., i, 1790, 87, pl. xii, fig. 1
(skull).—ERXLEBEN, Syst. Reg. Anim., 1777, 588.—GMELIN, Syst.
Nat., i, 1788, 64.—KERR, Anim. King., 1792, 125.—SHAW, Gen.
Zoöl., i, 1800, 262.—DESMAREST, Nouv. Dict. Sc. Nat., xxv, 1817,
576; Mam., 1820, 245, 376.—"NILSSON, Skand. Faun., i, 1820, 370;
Kongl. Vet. Akad. Handl., Stockholm, 1837, — "; Arch. für Naturg., 1841, 314; Skand. Faun. Däggdj., 1847, 288.—"THIENEMANN,
Reise im Norden von Europa," etc., i, 1824, 104, pl. xiv (ad. male),
pl. xv (ad. female), pl. xvi (male of two years), pl. xvii (male of
one year), pl. xviii (young eight days old), pl. xix (skull), pl. xx
(digestive organs), pl. xxi (attitude in swimming).—RICHARDSON,
Parry's 2d Voy., Suppl., 1825, 336.—HARLAN, Faun. Amer., 1825,
109.—GODMAN, Am. Nat. Hist., i, 1826, 343.—GRAY, Griffith's An.
King., v, 1827, 177, pls. xci, xcii.—J. C. ROSS, Parry's 3d Voy.,
1828, 191; Ross's 2d Voy., Append., 1835, xx.—FISCHER, Synop.
Mam., 1829, 238, 576.—BELL, Brit. Quad., 1837, 269; ibid., 2d ed.,
1874, 252, figg., animal and skull.—MACGILLIVRAY, Brit. Quad.,
1838, 209, pl. xix.—HAMILTON, Amphib. Carniv., 1839, 156, pl.
viii.—BLAINVILLE, Ostéog., Phoca, 1840-1851, pl. v (skull), pl. ix
(dentition).—JUKES, Excurs. in Newfoundland, i, 1842, 309.—
SCHINZ, Synop. Mam., i, 1844, 482.—WAGNER, Schreber's Säugt., vii,
1846, 21, pl. lxxxv A.—GIEBEL, Säugeth., 1855, 136.—BLASIUS, Naturgs. Wirbel. Deutschl., i, 1857, 255, figg. 140, 141.—NORDMANN,
Vit. Medd. Naturh. Forening, 1860 (1861), 25.—MALMGREN, Öfv.
af Kongl. Vet. Akad. Förhl., Stockholm, 1863, 139; Arch. für
Naturg., 1864, 78.—VON MIDDENDORFF, Sibirische Reise, iv, Th.
2, 1867, 934.—FLOWER, Journ. Anat. and Phys., iii, 1868, 270,
fig. 3 (milk-dentition).—QUENNERSTEDT, Kongl. Svens. Vetensk.
Akad. Handl,, vii, No. 3, 1868, 12, pl. i, fig. 1.—"KINBERG, Öfv. af
Kongl. Vet. Akad. Förhl., Stockholm, 1869, 13," (fossil, Sweden).—
MURIE, Proc. Zoöl. Soc. Lond., 1870, 604, pl. xxxii (attitudes and
terrestrial locomotion).—VON HEUGLIN, Petermann's Geogr. Mitth.,
1872, 30.—LILLJEBORG, Fauna öfver Sveriges och Norges Ryggradsjur., 1874, 690.—TURNER, Journ. Anat. and Phys., ix, 1874, 168
(Brit. Isl.).—COLLETT, Bemærk. Norges Pattedyrf., 1876, 58.—RINK,
Danish Greenland, its People and its Products, 1877, 124, 430.—
DAWSON, Canad. Nat., 2d ser., viii, 1877, 340 (fossil, Postpliocene
clays, near Ottawa, Canada).—ALSTON, Faun. Scotland, Mam.,
1880, 14.
Callocephalus grœnlandicus, F. CUVIER, Mém. du Mus., xi, 1824, 186, pl. xii,
fig. 2; Dict. des Sci. Nat., xxxix, 1826, 546.—LESSON, Man. de Mam.,
1827, 197.

Pagophilus grœnlandicus, GRAY, Cat. Seals Brit. Mus., 1850, 25, fig. 8; Cat. Seals and Whales, 1866, 28, fig. 8; Zoölogist, 1872, 3333, 3336 (British Coast, accidental); Hand-List of Seals, 1874, 6, pl. iv.— GILL, Proc. Essex Inst., v, 1866, 12.—PACKARD, Proc. Bost. Soc. Nat. Hist., x, 1866, 271.—BROWN, Proc. Zoöl. Soc. Lond., 1868, 340, 416.—REEKS, Zoölogist, 1871, 2541.—VAN BENEDEN, Ann. du Mus. roy. d'Hist. Nat. du Belgique, i, 1877, 20 (geogr. distr.).— MALM, Göteborgs och Bohusläns Fauna Ryggradsjuren, 1877, 144.

Phoca (*Pagophilus*) *grœnlandica*, VON HEUGLIN, Reisen nach dem Nordpolarmeer, iii, 1874, 51.

Phoca grœnlandica var. *nigra*, KERR, Anim. King., 1792, 125.

Phoca oceanica, LEPECHIN, Act. Acad. Petrop., i, 1777 (1778), 295, pll. vii, viii.—DESMAREST, Mam., 1820, 242, 275.—FISCHER, Synop. Mam., 1829, 238.—HAMILTON, Amphib. Carniv., 1839, 162, pl. viii*.

Callocephalus oceanicus, LESSON, Man. de Mam., 1827, 196.

"*Phoca semilunaris*, BODDAERT, Elen. Anim., 1785, 170."

Phoca albicauda, DESMAREST, Mamm., Suppl., 1822, 839 (locality unknown).

Phoca lagura, G. CUVIER, Oss. foss., 3d ed., v, 1825, 206 (young, "Terra Neuve").—FISCHER, Synop. Mam., 1829, 238 (same).—BLAINVILLE, Ostéogr., Phoca, 1840-1851, pl. ix (dentition).—SCHINZ, Synop. Mam., i, 1844, 483.—GAIMARD, Voy. en Islande, Atlas, 1851, pl. xi, fig. 6 (skull).

Callocephalus lagura, F. CUVIER, Dict. des Sci. Nat., xxxix, 1826, 546.— GRAY, Griffith's An. King., v, 1827, 177.

Phoca mülleri, LESSON, Dict. Class. d'Hist. Nat., xiii, 1828, 413.

Phoca desmaresti, LESSON, Dict. Class. d'Hist. Nat., xiii, 1828, 416 (=*P. albicauda*, Desm.).

Phoca pilayi, LESSON, Dict. Class. d'Hist. Nat., xiii, 1828, 416 (=*P. lagura*, G. Cuv.).

Phoca dorsata, PALLAS, Zoogr. Rosso-Asiat., i, 1831, 112.

? *Phoca albini*, ALEXANDRA, Mem. Acad. Torino, ii, 1850, 141, pl. i-iv (skeleton).

Phoca annellata, GAIMARD, Voy. en Islande, Atlas, 1851, pl. xi, fig. 9.

Swart-süde, EGEDE, Det gamle Grønlands nye Perlustration, 1741, 46, fig.

Blackside Seal, ELLIS, Voy. to Hudson's Bay, 1748, plate facing p. 134.

Attarsoak, CRANZ, Hist. von Grönl., i, 1765, 163.

Vadeselur, OLAFSEN, Reise durch Island, i, 283, ii, 42.

Schwarzseitige Seehund, SCHREBER, Säugt., iii, 310.

Harp Seal, PENNANT, Syn. Quad., 1771, 342; ibid., 1793, ii, 279, pl. xcix.

Phoque à croissant, BUFFON, Hist. Nat., Suppl., vi, 1782, 325.

Harp Seal, SAXBY, Zoölogist, 1864, 9099 (Shetland).—CARROLL, Seal and Herring Fisheries of Newfoundland, 1873, 15.

Grönlandsjäl, Swedish.

Grönlandsäl, Sulryg, Svartside, Norwegian.

Svartside, Danish.

Grönlands-Robbe, Sattel-Robbe, Grönlandische Seehund, German.

Phoque à croissant, French.

Harp Seal, Greenland Seal, Saddleback, Whitecoats (young), English.

Kioluk, Cumberland Eskimo (KUMLIEN).

EXTERNAL CHARACTERS. — *Adult Male.* — General color whitish or yellowish-white, nose and head to behind the eyes black; chin and throat usually with black spots. A broad, lunate spot of black on the sides, extending from the shoulders nearly to the tail, generally broadest anteriorly where the two bands unite on the median line; narrower, and sometimes interrupted, posteriorly, but usually again meeting on the hinder portion of the back, thus forming an ellipsoidal figure. These black bands usually begin over or a little anterior to the shoulders, and extend backward to the end of the tail. There are also generally irregular spots of black on the hind limbs. Length about 5 to 5½ feet, rarely, it is said, attaining the length of 6 feet.

Adult Female.—Similar in general color to the male, but with the black markings indistinct or wholly absent. Size about one-fourth less.

Young.—The new-born young are white or yellowish-white, sometimes pale golden, the pelage soft and woolly. This, after a few weeks, gives place to the coarser, harsher pelage of the adult, and the color becomes pale-gray, darker on the head and lighter below, often with small, dusky spots on the dorsal surface. In the second and third years the general color remains the same, but the spots become larger and darker. In the fourth year, in the males, the spots are still larger, and begin to coalesce; the head becomes black, and the saddle-shaped mark on the sides begins to be clearly distinguishable, but the mature pattern of coloration is said to be not fully developed till the fifth year.

Few Seals* vary so much in color with age as the Harp Seal. This was long since mentioned by Cranz, who says : "All Seals vary annually their colour till they are full grown, but no sort so much as this [the *Attarsoak*], and the Greenlanders vary its name according to its age. They call the fœtus *iblau;* in this state these are white and woolly, whereas the other sorts are smooth and coloured. In the 1st year 't is called *Attarak*, and 't is a cream-colour. In the 2d year *Atteitsiak*, then 't is gray. In the 3d *Aglektok*, painted. In the 4th *Milaktok*, and in the 5th year *Attarsoak*. Then it wears its half-moon, the signal of maturity."†

Fabricius states that it is called during the first year *Atârak*,

---

* Probably parallel variations occur in *Histriophoca fasciata.*
† Hist. of Greenland, English ed., vol. i, 1767, p. 124.

and later in the same summer (after the first moult), *Atàitsiak*, by which name it is also called during the first winter; the second year it is called *Utokáitsiak;* the third year *Aglektók* or *Aglektungoak*, and in the winter *Aglektytsiak;* the fourth year it

FIG. 50.—Phoca grœnlandica.

retains the same name *Agletók*, also varied to *Aglektorsoak*, but after the fourth moult takes the name *Millaktók*. Later it is called *Atârsoak.**

* Skrivter af Naturh.-Selskabet, Bind i, Hefte 1, 1790, pp. 92–94.

Dr. Rink states that at the present day the Greenlanders, as well as the Europeans, divide the "Saddle-backs" into four or five different classes according to their age, but that in familiar language they only distinguish by different names the full-grown animals from the half-grown ones, the latter being called "Blue-sides".*

The young, when first born, are called by the Newfoundland Sealers "White-Coats"; later, during the first moult, "Ragged-Jackets"; when they have attained the black crescentic marks they are termed "Harps", or "Sadlers", and also "Breeding Harps"; the yearlings and two-year-olds are called "Young Harps" or "Turning-Harps", and also "Bedlimers" (or "Bel-lamers", also spelled "Bedlamers"). The older and some recent writers state that the mature pattern of coloration is not attained till the fifth year, while Jukes, Brown, Carroll, and others state that it is acquired in the third or fourth year. There is also a diversity of statement respecting the sexual differences of color in the adults, some writers affirming that the sexes are alike, while others state that the female is without the harp-mark, or has the dark markings of the male only faintly indicated. Mr. Carroll says: "The reason why they are called harp seals, or 'sadlers,' is, the male seal, as well as the female, has a dark stripe on each side from the shoulders to the tail, leaving a muddy white stripe down the back. The male harp seal is very black about the head as well as under the throat. . . . The female harp is of a rusty gray about the head and white under the throat." Both Jukes and Reeks, however, refer to the *absence* of the harp-mark in the female.

Mr. Brown, in his account of the Seals of Greenland, has given a very full account of the changes of color resulting from age and sex, and, in default of a sufficient series of specimens, and of personal experience, I transcribe his observations, as presenting the most explicit and detailed statement available. He says:

"It seems to be almost unknown to most writers on this group that the male and female of the Saddleback are of different colours; this, however, has long been known to the Seal-hunters. *Male.*—The length of the male Saddleback rarely reaches 6 feet, and the most common length is 5 feet, while the female, in general, rarely attains that length. The colour of the male is of a tawny grey, of a lighter or darker shade in different individuals,

* Danish Greenland, its People and its Products, p. 124.

on a slightly straw-coloured or tawny-yellowish ground, having sometimes a tendency to a reddish-brown tint, which latter colour is often seen in both males and females, but especially in the latter, in oval spots on the dorsal aspect. The pectoral and abdominal regions have a dingy or tarnished silvery hue, and are not white, as generally described. But the chief character-istic, at least that which has attracted the most notice, so much as to have been the reason for giving it several names, from the peculiar appearance it was thought to present (*e. g.* 'Harp Seal,' 'Saddleback,' &c.), is the dark marking or band on its dorsal and lateral aspects. This 'saddle-shaped' band commences at the root of the neck posteriorly, and curves downwards and backwards at each side superior to the anterior flippers, reaches downwards to the abdominal region, whence it curves back-wards anteriorly to the posterior flippers, where it gradually dis-appears, reaching further in some individuals than in others. In some this band is broader than in others and more clearly im-pressed, while in many the markings only present an approxi-mation, in the form of an aggregation of spots more or less isolated. The grey colour verges into a darker hue, almost a black tint, on the muzzle and flippers; but I have never seen it white on the forehead as mentioned by Fabricius. The muzzle is more prominent than in any other northern Seal.

"*Female.*—The female is very different in appearance from the male; she is not nearly so large, rarely reaching 5 feet in length, and when fully mature her colour is a dull white or yellowish straw-colour, of a tawny hue on the back, but similar to the male on the pectoral and abdominal regions, only, perhaps, somewhat lighter. In some females I have seen the colour totally differ-ent; it presented a bluish or dark-grey appearance on the back, with peculiar oval markings of a dark colour, apparently im-pressed on a yellowish or reddish-brown ground. These spots are more or less numerous in different individuals. Some Seal-hunters are inclined to think this a different species of Seal from the Saddleback, because the appearance of the skin is often so very different and extremely beautiful when taken out of the water, yet as the females are always found among the immense flocks of the Saddleback, and as hardly two of the latter females are alike, but varying in all stages to the mature female, and on account of there being no males to mate with them, I am inclined to believe with Dr. Wallace that these are only *younger female Saddlebacks.* The muzzle and flippers of the female present the same dark-chestnut appearance as in the male."

In respect to the color of the young, and to changes of color with age, he observes: "(a) The colour after birth is a pure woolly white, which gradually assumes a beautiful yellowish tint when contrasted with the stainless purity of the Arctic snow; they are then called by the sealers 'white-coats,' or 'whitey-coats'; and they retain this colour until they are able to take to the water (when about 14 or 20 days old)." At this time the color "(β) . . . begins to change to that of a dark speckled and then spotted hue; these are denominated 'hares' by the sealers. (γ) This colour gradually changes to a dark bluish colour on the back, while on the breast and belly it is of a dark silvery hue. Young Seals retain this appearance throughout the summer and are termed 'Bluebacks' by the sealers of Spitzbergen, 'Aglektok' by the Greenlanders, Blaa-siden by the Danes. (δ) Thé next stage is called Millaktok by the Greenlanders. The Seal is then approaching to its mature coat, getting more spotted, &c., and the saddle-shape band begins to form. (ε) The last stage (in the male to which these changes refer) is the assumption of the halfmoon-shaped mark on either side, or the 'saddle' as it is called by the northern sealers.

"I consider that about three years are sufficient to complete these changes. This is also the opinion held in Newfoundland, though the Greenland people consider that five years are necessary. I wish, however, to say that these changes do not proceed so regularly as is usually described, some of them not lasting a year, others longer, while, again, several of the changes are gone through in one year; in fact, the coats are always gradually changing, though some of the more prominent ones may be retained a longer and others a shorter time. It would require a very careful and extended study of this animal to decide on this point, which, owing to their migrations, it is impossible to give. After all, these changes and their rapidity vary according to the season and the individual, and really will not admit of other than a general description."*

Dr. Rink gives the weight of a full-grown Saddleback of medium size as 353 pounds, the skin and blubber weighing 116 pounds, the blubber alone in winter amounting to 80 pounds,

* Proc. Zoöl. Soc. Lond., 1868, pp. 417–420; Man. Nat. Hist., Geol. and Phys. of Greenland, etc., 1875, Mam., pp. 45–49.—Compare alse von Heuglin, Reisen nach dem Nordpolarmeer, etc., 1874, pp. 53–54.

while in summer it is scarcely 24 pounds. Carroll, however, gives a much larger weight. He states that " when they are in full flesh the weight of a male Harp Seal varies from 700 to 800 pounds," and that "when prime" the skin and fat alone will weigh 200 pounds, and the same parts of a female 125 pounds. He gives the weight of a Harp Seal when born as 6 to 8 pounds, according to the age of the mother. At fifteen days old he says the "skin and fat will weigh 40 to 45 pounds," and the carcass, after the fat is taken off, about 15 pounds. When thirty days old the weight of the skin and fat he says is only 30 pounds, and at nine months old not more than 40 pounds, but at twelve months is 90 pounds; the young Seals, as stated by other observers, losing much of their fat on being left by their mothers to secure their own food, although the general size continues slowly to increase.

SEXUAL AND INDIVIDUAL VARIATION, AND VARIATIONS DEPENDENT UPON AGE.—The variations in size and color dependent upon sex and age having been already noted, little remains to be said in the present connection beyond a brief reference to the skull. Here the age and sexual variations appear to be strictly parallel with those already described in *Phoca fœtida*. The purely individual variation is also similar in character, except, so far as can be judged from the small series of skulls before me, the variations in the teeth are less marked. The subjoined table of measurements of skulls indicates the range of variation in size and other features. None of the skulls are positively marked for sex, but there seems to be little difficulty in distinguishing the sex by the smaller size, weaker structure, narrower muzzle, and the much weaker dentition in what may be doubtless safely assumed to be the female, since very old skulls, strictly comparable as respects age, vary in just these points. Female skulls appear to rarely exceed, or even quite attain, a length of 200 mm., while old males range from about 210 mm. to 228 mm.

Measurements of twelve skulls of Phoca grœnlandica.

Measurement	*3514	*3503	*3515	*3505	*881	†1146	‡—	‡539	†1628	‡1627	†1094	†1091
Locality	Greenland	...do	...do	...do	Nahant, Mass.		Labrador?	...do	...do	...do	...do	...do
Sex												
Length	215	212	204	168	220	203	220	228	210	212	197	197
Breadth at mastoid processes	118	117	113	108	115	115						
Greatest breadth at zygomatic arches	114	120	113	98.5	117	117	128	133	123	114	114	112
Distance from anterior edge of inter-maxillæ to end of pterygoid hamuli	112	113	110	87	122	113						
Distance from anterior edge of inter-maxillæ to hinder edge of last molar	68	70	64	53	73	67						
Distance from anterior edge of inter-maxillæ to meatus auditorius	150	148	144	115	148	145						
Distance from anterior edge of inter-maxillæ to glenoid process	139	138	133	105	140	134						
Distance from palato-maxillary suture to end of pterygoid hamuli	43		43.5	34								
Length of the alveolar border of the maxillæ	84	87	77	63	90	65						
Width of palatal region at posterior end of maxillæ	50	54	55	45	54	57						
Nasal bones, length	42	38	42	38	45	48						
Nasal bones, breadth anteriorly	15.5	14	14	14	15	16						
Nasal bones, breadth at fronto-maxillary suture	8	6	9	9	9	8						
Breadth of skull at canines	31	33	30	25	34	30						
Least breadth of skull interorbitally	11	7	12.5	10	13	11						
Breadth of posterior nares, vertically	13	12	12	10	14	12						
Breadth of posterior nares, transversely	37	35	35	33	35	32						
Breadth of anterior nares, vertically	30	30	30	22	28	29						
Breadth of anterior nares, transversely	27	26	26	20	30	26						
Greatest height of skull at auditory bullæ	75	79	80	78	78	75						
Length of brain-case	85	85	77	76.5	84	77						
Greatest breadth of brain-case	98	99	93	92	95	96						
Length of lower jaw	130	135	126	104	135	130						
Front edge of ramus to last molar	56	57	56	45	61	56						
Age	Old.	Very old.	Old.	Young.	Very old.	Very old.	Very old.	Very old.	Very old.	Old.	Middleaged.	Middleaged.

* In National Museum, Washington, D. C.   † In Museum of Comparative Zoölogy, Cambridge, Mass.

‡ In Museum of Boston Society of Natural History, Boston, Mass.

GENERAL HISTORY AND NOMENCLATURE.—The Harp Seal,
like the Crested Seal, presents characters, at least in the male
sex, that readily attract the attention of even the casual observer—the one by its "saddle" or "harp-mark" of black on a
light ground, the other by its inflatable hood. Accordingly both
were mentioned by various early writers, but notably by Egede,
Ellis, and Cranz, and the indications they gave of their existence
enter into the technical history of the species, forming as they
do the basis of the first systematic names.   Erxleben described
the species in 1777, under the name *Phoca grœnlandica*, his description being founded mainly on information previously made
public by Cranz.   Fabricius, however, had already designated
the species by this name the previous year, but the only clue
he furnished to the species meant consists merely in his citing
its Icelandic and Greenlandic names.   In 1778 Lepechin described and figured the species under the name *Phoca oceanica*,
between which and *grœnlandica* there is thus almost a question
of priority.*   Although Fabricius in 1790 correctly referred Lepechin's species to *Phoca grœnlandica*, it has since frequently figured in the works of compilers as a distinct species, although
his figures and description† clearly indicate its relationship.

Boddaert, in 1785, added another synonym by renaming the
species *semilunaris*, while Desmarest, in 1822, described what
is believed to have been a young individual of this species
under the name *Phoca albicauda*.   G. Cuvier, in 1825, also described a young specimen as *Phoca lagura*, this name having
for a time considerable currency as that of a veritable species.
Lesson, in 1828, made here his usual contribution of synonyms
by deliberately changing names previously given for those that
better suited his fancy, at his hands the *Phoca grœnlandica* of
authors becoming *Phoca mülleri*, and the two nominal species
previously mentioned as based on young specimens becoming
respectively *Phoca desmaresti* and *Phoca pilayi*.   In 1831 the
species was again intentionally renamed *dorsata* by Pallas, who
quotes as synonyms of *dorsata* both *Phoca grœnlandica* and

---

* Lepechin is usually quoted at 1777, but his paper appears not to have
been published till the following year, thus giving Erxleben's name one
year's priority, and Fabricius's two.

† Lepechin gave the incisive formula as $\frac{4}{4}$,—"In maxilla superiori incisores IV"; "in maxilla inferiori incisores *modo* IV." As suggested by Fabricius nearly a century ago, in the first case "IV" is evidently a lapsus
for VI. (See Fabricius, Skrivter af Naturhistorie-Selskabet, Bd. i, Hf. 1,
1790, p. 97, footnote.)

Phoca oceanica. Gaimard, in 1851, simply through malidentification, referred examples of the present species to *Phoca annellata.*

Of the nine synonyms of this species one is nearly contemporaneous with the tenable name, and under the circumstances of its occurrence was unavoidable; five are due to deliberate, intentional, and needless change of names; two are based on immature examples, and one is the result of malidentification.

GEOGRAPHICAL DISTRIBUTION.—Although the Harp Seal has a circumpolar distribution, it appears not to advance so far northward as the Ringed Seal or the Bearded Seal; yet the icy seas of the North are preëminently its home. It is not found on the Atlantic coast of North America in any numbers south of Newfoundland. A few are taken at the Magdalen Islands, and while on their way to the Grand Banks some must pass very near the Nova Scotia coast. Dr. Gilpin, however, includes it only provisionally among the Seals that visit the shores of that Province. It doubtless occasionally wanders, like the Crested Seal, to points far south of its usual range, as I find a skeleton of this species in the collection of the Museum of Comparative Zoölogy, bearing the legend "Nahant, Mass., L. Agassiz". I have at times felt doubtful about the correctness of the assigned locality, as this seems to be the only proof of the occurrence of this species on the Massachusetts coast. I have, however, recently been informed by Dr. C. C. Abbott, of New Jersey, that a Seal, described to him as being about six feet long, white, with a broad black band along each side of the back, was taken near Trenton, in that State, during the winter of 1878–79. This description can of course refer to no other species than *Phoca grœnlandica*, and as it comes from a wholly trustworthy source it seems to substantiate the occasional occurrence of this species as far south as New Jersey. Von Heuglin gives it as ranging "in den amerikanischen Meeren südwärts bis New York,"* but I know not on what authority.

The Harp Seals are well known to be periodically exceedingly abundant along the shores of Newfoundland, where, during spring, hundreds of thousands are annually killed. In their migrations they pass along the coast of Labrador, and appear with regularity twice a year off the coast of Southern Greenland. Capt. J. C. Ross states that in Baffin's Bay they

---

* Reisen nach dem Nordpolarmeer, p. 56.

keep mostly "to the loose floating floes which constitute what is termed by the whale fishers 'the middle ice' of Baffin's Bay and Davis Straits." He says he never met with them in any part of Prince Regent's Inlet, but states that they are reported by the natives to be very numerous on the west side of the Isthmus of Boothia, but that they are not seen on the east side.* They are well-known visitors to the shores of Iceland, and swarm in the icy seas about Jan Mayen and Spitzbergen. They also occur about Nova Zembla, and Payer refers to their abundance at Franz Josef Land.† They occur in the Kara Sea, and along the Arctic coast of Europe. Malmgren, Lilljeborg, and Collett state that it is of regular occurrence on the coast of Finmark, where it occurs in small numbers from October and November till February. Although reported by Bell and others as having been taken in the Severn, and by Saxby‡ as observed at Baltasound, Shetland, the capture of a specimen in Morecombe Bay, England, reported by Turner§ in 1874, Mr. E. R. Alston says is "the first British specimen that has been properly identified."||

The distribution of this species in the North Pacific is not well known. Pallas (under the name *Phoca dorsata*) records it from Kamtschatka, where its occurrence is also affirmed by Steller. Temminck mentions having examined three skins obtained at Sitka, but adds that it was not observed by "les voyageurs néerlandais" in Japan. In the collections in the National Museum from the North Pacific this species is unrepresented, the species thus far received from there being the following four, namely: *Phoca vitulina, Phoca fœtida, Erignathus barbatus,* and *Histriophoca fascista.*

MIGRATIONS AND BREEDING STATIONS.—The Saddleback, although found at one season or another throughout a wide extent of the Arctic seas, appears to be nowhere resident

---

* Ross's Second Voyage, App., p. xxi.

† New Lands within the Arctic Circle, 1877, p. 266.

‡ Mr. Henry L. Saxby, writing under date of Baltasound, Shetland, March 14, 1864, says, "Several Harp Seals are now to be seen in the deep sheltered voe at Baltasound. This species can scarcely be considered very rare here, but it is said to occur in bad weather, and certainly the present visit forms no exception to the rule, the wind having for some days been blowing heavily from N. E., accompanied by sleet and snow."—*Zoölogist*, vol. xxii, 1864, p. 9090.

§ Journ. Anat. and Phys., vol. ix, 1874, p. 168.

|| Zoöl. Rec., 1874, p. 10; Fauna of Scotland, Mam., 1880, p. 14.

the whole year. Its very extended periodical migrations relate apparently to the selection of suitable conditions for the
production of its young, and occur with great regularity.
Where it spends portions of the year is not well known, while,
on the other hand, it may be found with the utmost certainty
at particular localities during the breeding season. Its most
noted breeding stations are the ice-floes to the eastward of
Newfoundland, and in the vicinity of Jan Mayen, at which localities they appear early in spring in immense herds.

The Seals seen about the shores of Greenland in autumn and
early winter are supposed by most writers to pass the breeding
season in the seas to the eastward of Jan Mayen, but doubtless a very large proportion of the Seals of Hudson's Straits
and neighboring waters to the northward, if not also of Baffin's Bay, really move southward along the Labrador coast to
the Newfoundland waters, since herds of migrating Seals are
regularly observed in autumn to pass in this direction; besides,
it is hard to conceive of any other origin for the immense numbers that resort to the ice-floes off the coast of Southern Labrador and Newfoundland to bring forth their young.

As has been long well known, the Greenland Seal visits the
shores of Greenland both in fall and spring. Dr. Rink states
that "It appears regularly along the southern part of the
coast in September, travelling in herds from south to north, between the islands, and at times resorting to the fjords. . . . .
In October and November the catch is most plentiful; then it
decreases in December, grows scarce in January, and becomes
almost extinct in February."* Mr. Kumlien states that they
"disappear from the Cumberland waters when the ice makes,"
returning again in spring with the appearance of open water.

Their passage southward along the Labrador coast occurs
before the ice forms, and during this journey they are said to
"hug the shore" and freely enter the gulfs and bays. They
appear first in small detachments of half a dozen to a score or
more of individuals; these are soon followed by larger companies, which increase in frequency and numbers; in a few
days they form one continuous procession, filling the sea as far
as the eye can reach. Floating with the Arctic current, their
progress is extremely rapid, and in one short week the whole
multitude has passed. Arriving at the Straits of Belleisle, the
great body are deflected eastward, but many enter the Straits

---

* Danish Greenland, etc., p. 124.

and pass round to the southward of Newfoundland; some, however, spend the winter in the Gulf of Saint Lawrence, where they bring forth their young on the ice in spring.   But the great mass continue onward along the eastern coast of New-foundland as far as Baccalieu Island, at the entrance of Trinity Bay, where they leave the shore for the Grand Banks, at which they arrive about the end of December.   Here they rest for a month, and then turn again northward to seek the ice-floes for breeding stations.   Slowly onward they struggle against the strong current that aided them so much in their southward journey, till they reach the great ice-fields, stretching from the Labrador shore far eastward,—a broad continent of ice.   By the end of February the breeding-sites have been chosen, and the young are born shortly after (generally between the 5th and 10th of March).   Many of the younger Seals (yearlings and two-year-olds), however, still remain on the southern banks. By the end of April the newly-born Seals are strong enough to secure their own food, and in May the numberless multitude resume their northward route, keeping far out at sea to avoid the strong current that courses along the coast.   In May they begin to again arrive on the coast of Southern Greenland, and later visit the more northern shores.

The Seals that resort in such great numbers to the ice-floes east and north of Jan Mayen in spring are believed to come mainly from Greenland, but doubtless a large part really come from the eastward and northward.   Lindeman, speaking of their dispersion after the breeding season, says: "By the end of June they start on their homeward journey to the north and east, the young following; they pass from one outlying point of ice to another, where they lie to rest.   In a single instance they were followed all the way to Spitzbergen, and were here also observed to still pursue an easterly direction.   Whither they go and where they keep themselves till the next spring is certainly a worthy subject of investigation."[*]

As already stated in the general account of Seal-hunting (anteà, pp. 496 et seq.), the Harp Seals assemble early in spring in countless numbers in the vicinity of " the dreary island of Jan Mayen", the ice-floes a little to the eastward and northward of which form their great central rendezvous during the breeding season, and consequently the scene of the grand annual Seal slaughter in the Arctic seas.   Their principal breeding-resort

---

[*] Petermann's Geograph. Mittheil., Ergänzungs Heft Nr. 26, 1869, p.

appears to be a small circular area, having a breadth of about four hundred miles, within which Jan Mayen Island occupies a nearly central point. They are not, however, equally numerous throughout even this limited district, but are most densely massed between the 72d and 73d degrees of north latitude, on or near the 7th meridian east of Greenwich. The exact point, however, varies in different years, in accordance with the varying position of the ice-fields, which is influenced by the prevailing winds and the character of the season. Thus, according to Dr. Wallace, as quoted by Mr. Brown, they were found in 1859 "in considerable numbers not far from Iceland, the most northerly point of which is in N. lat. 66° 44'; this leads me to remark," Mr. Brown continues, "that the Seals are often divided into several bodies or flocks, and may be at a considerable distance from each other, although it is most common to find these smaller flocks on the skirts or at no great distance from the main body." Where the Seals that here congregate in such numbers during the breeding season spend the rest of the year is not well known. Says the writer last quoted, "After the young have begun to take the water in the Spitzbergen sea, they gradually direct their course to the outside streams, where they are often taken in considerable numbers on warm sunny days. When able to provide for themselves, the females gradually leave them and join the males in the north, where they are hunted by the sealers in the months of May and June; and it is especially during the latter month that the females are seen to have joined the males; for at the 'old sealing' (as this is called), in May, it has often been remarked that few or no males are seen in company with the females. Later in the year, in July, there are seen, between the parallels of 76° and 77° N., these flocks of Seals, termed by Scoresby 'Seal's weddings'; and I have found that they were composed of the old males and females and the *bluebacks* [yearlings and two-year-olds], which must have followed the old ones in the north and formed a junction with them some time in June. There is another opinion, that the old females remain and bring their young with them north; but all our facts are against such a theory (*Wallace*).

"These migrations may vary with the temperature of the season, and are influenced by it; it is possible that in the Spitzbergen sea as the winter approaches they keep in advance of it and retreat southward to the limit of the perpetual ice, off the coast of Greenland, somewhere near Iceland, where they spend the winter. We are, however, at a loss regarding the winter habits of these Seals in that region; here no one winters, and there are no inhabitants to note their migrations and ways of life. Different is it, however, on the Greenland shores of Davis's Strait, where in the Danish settlements the Seals form, both with the whites and Eskimo, the staple article of food and commerce, and accordingly their habits and arrival are well known and eagerly watched. The *Atarsoak*, as it is commonly called by the Eskimo, the 'Svartsidede Sælhund' (Black-sided Sealhound) of the Danes, is the most common Seal in all South Greenland. It is equally by this Seal that the Eskimo lives, and the 'Kongl-Grönlandske Handel' makes its commerce. In South Greenland when the Seal generally is talked of, or a good or bad year spoken about, everybody thinks of this Seal; on the other hand, in North Greenland, *Pagomys fœtidus* and *Callocephalus vitulinus* are the most common. These last two species are the only Seals which can be properly said to have their home in Greenland, affecting ice-fjords and rarely going far from the coast. This is not the case with *P. grœnlandicus;* at certain times of the year they completely leave the coast; therefore the Seal-hunting in South Greenland is more dependent upon contingencies than in North Greenland. This Seal arrives regularly in September in companies travelling from the south to north, keeping among the islands; occasionally at this time individuals detach themselves from the drove and go up the inlets. . . ."*

Both Dr. Rink and Mr. Brown believe that it is very improbable that the Seals of South Greenland visit Jan Mayen in the breeding season, deeming it more likely that they resort for this purpose to the southern ice-floes off the Labrador coast. "As to their whereabouts during their absence," however, observes the former, "we are somewhat at a loss for perfect information. There can be no doubt that in spring they retreat to the icefields of the ocean for the purpose of producing their young. It seems most unlikely that the seals from the west coast should have such breeding places to the east of Greenland in the Spitzbergen sea, which would require the whole stock of them to round Cape Farewell at least twice a year. But, considering that just opposite to the west coast extensive

---

* Proc. Zoöl. Soc. Lond., 1868, pp. 421, 422; Man. Nat. Hist., etc., Greenland, Mam., pp. 51, 52.

masses of drift-ice from Baffin's Bay are moving southward throughout the greater part of the year, nothing seems more reasonable to believe than that the seals, having gone their usual beat along the west coast of Greenland put to sea in various latitudes; after which, on crossing Davis Strait, they almost every where will meet with the drift-ice, which they will then follow on its course southward, and on returning they will make the coast of Greenland at some more southerly point, begin their usual migration, and so on."* Mr. Brown, however, adds: "Every one knows when it commences its migration from the south to the north, but nobody knows where the Seal goes to when it disappears off the coast. Between the time they leave the coast in the spring and return in the summer they beget their young; and this seems to be accomplished on the pack-ice a great distance from land, viz, in the Spitzbergen sea. It is at this period that the seal-ships come after them. . . ."

From what has been already stated respecting their passage southward at the beginning of winter along the Labrador coast and the shores of Newfoundland to the Grand Banks, their subsequent movement northward at the beginning of spring to the ice-floes to bring forth their young, and their later migration northward, it seems safe to assume that the Greenland division of these Seals resort mainly in winter to the open waters of the Grand Banks, southeast of Newfoundland, and that after the breeding season they return northward to the Greenland coast; furthermore that the great herds that congregate about Jan Mayen belong mainly to the Arctic waters of the Spitzbergen sea, migrating northward and southward, and more or less westward, with the changes of the season and the position of the ice-fields, and that probably none of the Seals of Baffin's Bay and adjoining waters migrate to the Jan Mayen seas.

As already stated the Harp Seals visit the southern coast of Greenland in May, and appear on the more northern coast in June. "Having visited," says Rink, "the fjords in numerous herds, they again disappear in July and return in September.† Consequently this seal deserts the coast twice a year, and as regularly returns to it in due season, always first making its

* Danish Greenland, etc., 1877, p. 125.
† Mr. Brown says, "This Seal leaves the vicinity of Jakobshavn ice-fjord about the middle of July or beginning of August, and comes back in October very fat. In August and September there are none on that part of the coast."

appearance in the southern, and somewhat later in the northern regions." Why they leave the Greenland coast in August and again visit it in September, and there remain for several months before departing for the south, and where they go during their absence, are questions for which there is as yet no satisfactory answer. It has, however, by some been supposed to relate to the pairing season, which occurs in August, the females on their return in September being found to be with young. Mr. Kumlien states that "a few schools were noticed at different times during September, 1877, and October, 1878, from the islands of the Labrador coast to Cumberland, at times a considerable distance from land. It hence seems probable that many pass this portion of the year at points far to the southward of Greenland."

HABITS.—The Harp Seal is remarkable alike for its abundance and its pelagic and roving habits. Eminently gregarious at all seasons, and doubtless outnumbering all the other species together, it forms the chief basis of the great sealing industry of the northern seas. It is, however, as already shown, nowhere a permanent resident, and during its periodical journeyings traverses a wide breadth of latitude. Although often met with far out at sea, it is never seen far from the floating ice-fields, it generally keeping near the edges of the drifting ice. It appears never to resort to the shores and to be seldom met with on the firm ice. This is doubtless due to the fact that, unlike the Ringed and the Bearded Seals, it never forms for itself an *atluk* or breathing-hole through the ice, and consequently is obliged to keep near the large openings formed by winds or ocean currents. It is generally regarded as less sagacious than most other species, and as submitting, without show of resistance, to the attacks of the sealers.

About the beginning of March they assemble at their favorite breeding stations, selecting for this purpose immense ice-fields far from land. Their best known breeding-grounds are the ice-packs off the eastern coast of Newfoundland, and about the island of Jan Mayen. A few are said to breed on the floating ice in the Gulf of Saint Lawrence, and there are doubtless numerous small outlying colonies in various parts of the North Atlantic and Arctic waters. Mr. Carroll states that off the Newfoundland coast the young are chiefly born between the 5th and 10th of March, or about a week earlier than is the case with the Hooded Seal or the "Square-flipper" (probably *Hali-cœhrus grypus*). According to Lindeman, the young are born much later at the Jan Mayen breeding-grounds, or not till the 23d or 24th of March, the "whelping-time" (as this period is termed in sealing parlance) lasting till about the 5th of April. Only rarely does the female bring forth more than a single young one at a birth. The period of gestation is supposed to be about nine months. If left undisturbed the females are said to suckle the young about fifteen to eighteen days, when the young are so far developed that they are able to take to the water and seek their own nourishment. At this time they begin to shed their white woolly coat, and take on the harsher, grayer pelage that succeeds the fœtal dress.

According to Mr. Carroll, the old "Breeding Harps" are the first to leave the fishing-ground for the purpose of "whelping." In selecting their breeding stations they endeavor to go as far north as they can advance with safety, or until they meet the heavy northern ice, for they know that the more northern the station the more safety there is for the young from the wash of southern storms. Yet, in spite of their delicate instincts, and notwithstanding their great cautiousness, says Lindeman, "it still sometimes happens that heavy northeast storms drive the whole area chosen into the open sea, and the immense mass of young Seals become unfortunately destroyed. I saw many interesting examples," he adds, "of how courageously the mother worked under such an accident in order to bring her young again upon the firm ice, either by trying to swim with it between her fore flippers or by driving it before her and tossing it forward with her nose." Carroll states that all kinds of Seals found about Newfoundland "will at all times endeavor to whelp as near the shore as possible, because instinct teaches them that the nearer the rocks the shallower the water, so that when they abandon their young ones the little creatures will see the bottom so as to enable them to procure their food. When young Seals are whelped near the shore," however, he continues, "and a heavy sea comes on, thousands of them are ground to pieces with the sea against the rocks. I have frequently watched the old female harps bolt up through the ice in a heavy sea and drag their young ones off the ice into the water out of danger. Again, when the ice begins to raft where young Seals are, thousands upon thousands of them are also chopped into piecemeal."

The females take up their stations on the ice very near to each other, the young being thus sometimes born not more than three feet apart; they also all bring forth their young at very nearly the same time. The males accompany the females to the breeding stations, and remain in the vicinity, yet rarely upon the ice, congregating mostly in the open pools between the ice-floes. The mothers leave their young on the ice, to fish in the neighborhood for their own subsistence, but frequently return to the young to suckle them. The young increase rapidly in size, and when three weeks old are said to be nearly half as large as the old ones. At this time they are the fattest and are considered to be in best condition for killing. Later the fat diminishes although the general bulk continues to increase. If undisturbed the old Seals will remain amongst the ice at the breeding-grounds till after the moult, which occurs late in the spring, for the purpose of rubbing off the old hair against the ice.

The annual moulting-time, or "skin-sickness" ("*Hautkrankheit*"), as the Germans expressively term it, is evidently a period of great discomfort, and occurs within four or five weeks after the birth of the young. During this time they rapidly lose their fat, and become more watchful and restless. As Mr. Carroll puts it, about the middle of April the old Seals, and the yearlings and two-year-olds, "mount the ice to scrub themselves". If the day be warm, he adds, "the skin on the back is sure to be sunburnt, so much so, that you can tear it off with your fingers; they will remain on the ice to be killed when once they get sunburnt rather than go in the water. When they do get in the water they will cry with pain and sometimes mount the ice again."

For breeding stations the Seals select "sheet-ice", in which, says Mr. Carroll, they keep holes open through which they may get to their young. A rim soon forms around these holes caused by the freezing of the water forced up by the Seals in passing through them, but they are sure to keep one side of the hole on a level with the water, the side they use in going up and down. They assemble in such numbers that the cry of the vast number of the old and young may be heard to the distance of several miles, particularly if the ear be applied to the ice. The same author states that at the Newfoundland breeding-grounds no wind will break up the "whelping ice" equal to a strong southeast wind; no matter how deep the northern bays

may be, such wind will be sure to break up the ice. It is well
understood, he adds, that the "whereabouts" of the young
Seals depends "on wind and tide".

Mr. Carroll ascribes great sagacity to the Seals in discerning
the character of the weather when they are in danger of being
"embayed". "They are sure to swarm out," he says, "at least
two and sometimes three days before the wind blows in on the
land; they also know when a lake of water is in the sheet or
drift some hundred miles more or less from where they are *by
the reflection of the light through the ice* [!]. . . . When Seals
get embayed and are kept there some number of days and cannot get into the water owing to the ice being jammed, they
begin to travel out in a direct line for the water. Supposing
the water to be fifty miles from them, they know well by scent
where it is, for you will see them stretch out their necks and
sniff; should the ice part in any direction from them they will
at once turn round and avail themselves of it. Much depends
upon the character of the ice they have to travel on as to their
rate of speed; they travel principally by night. I have killed
them with the hair and skin worn off the fore flippers and
bleeding." The same writer states that in cool nights Seals
will travel at an average rate of one mile per hour. Their
speed depends much, however, upon the character of the ice,
on level ice an old Seal being able to outstrip a smart runner
in a distance of sixty yards. They move laboriously, by lifting
themselves off the ice on their fore flippers and drawing up the
hind part of the body, resulting in a "sidelong loping gallop."
In travelling they sometimes become overheated, in which case
the hair becomes loosened and the skin worthless.*

The young Seals are said not to voluntarily enter the water
until at least twelve days old, and that they require four or
five days' practice before they acquire sufficient strength and
proficiency in swimming to enable them to take care of themselves. After they take to the water they congregate by themselves, and when they mount the ice assemble in quite compact
herds.

Professor Jukes refers to a young one that was taken alive
on board his ship as forming a very gentle and interesting pet,
"He lay very quiet on deck, opening and closing his curious
nostrils, . . . and occasionally lifting his fine dark lustrous
eyes as if with wonder at the strange scenes around him." His

* The Seal and Herring Fisheries of Newfoundland, 1873, pp. 24, 25.

short thick fur being dry and clean, gave him a very warm and comfortable appearance. On being patted on the head he drew it in till his face was perpendicular to his body, knitted his brow and closed his eyes and nostrils, thereby assuming a very comical expression of countenance. Although he was fierce when teased, and attempted to bite and scratch, he immediately became quiet on being stroked or patted. They are doubtless easily tamed, and might be made very interesting pets. In the present instance the poor brute was cruelly teased by dogs and men till he became exhausted, and Professor Jukes passed his knife into his heart to end his misery.*

The Harp Seals are stated to swim with great rapidity, propelling themselves with their powerful hind-flippers, one writer estimating their speed when "bolting" under the ice as "at least one hundred miles per hour," and observes that as they pass beneath you you "will observe only a blue shade," even if the water is perfectly clear. Their favorite position when swimming, as affirmed by numerous observers, is on the back or side, in which position they also sleep in the water.

Their social and gregarious instincts seem to be manifested on all occasions; they not only migrate in dense herds, and assemble on the ice in compact bodies, but are rarely met with singly, though occasionally in small groups. As noted on preceding pages, immense herds sometimes fill the sea as far as the eye can reach, or thickly cover the ice over areas of many square miles in extent.

ENEMIES.—Aside from their destruction by man, and not unfrequently by the elements, they find a formidable enemy in the sword-fish, and are extensively preyed upon by sharks. Mr. Carroll, my chief authority on this point, says that when the Seals are floating about on single "pans," he has seen sword-fish, sharks, and other kinds of fish, taking them off. The sword-fish, he says, will get on one side of the pan and press it down to such an angle "that the Seal must slip off among them and be torn to pieces". The Seals appear to have a great terror of these remorseless enemies, for the same authority adds, "I have been on pans of ice when seals mounted the ice to avoid the sword-fish and sharks, and obliged to fire at the monsters to keep them off. A seal will shake with fear, and should a man be on the pan when sword-fish and sharks are after them,

* Excurs. in Newfoundland, vol. i, pp. 283, 284.

they will run between a man's legs for protection".* Doubtless many young Seals not only become the prey of these creatures, but also of the rapacious Orca, so well known to prey upon the young of the Fur Seals.

FOOD.—Like all the Phocids, the Harp Seal is well known to subsist chiefly upon fish, but also in part upon Crustaceans and Mollusks. White-fish and the cod seem to form their chief food off the shores of Newfoundland and Labrador, and from the abundance of the Seals the quantity they consume must be immense. It has even been supposed that the small catch of codfish about the island of Newfoundland is due to the great destruction of these fishes by the Seals, several millions of the latter, it is estimated, spending several months of each year in the vicinity of this island. Allowing only one fish a day to each Seal during the time they stay about the island would require the annual-destruction of several million quintals of these fishes. In their southward migration in autumn along the coast of Labrador, they are said to follow the schools of white-fish, on which to a large extent they are also known to feed. They also follow them into all the bays along the coast in spring. "As long as white-fish are in with the land," says Mr. Carroll, "so sure will seals of every description be there". That they also prey upon the codfish is well proven by Seals being killed with these fish in their mouths, as well as by finding them on the ice to which the Seals have carried them. Mr. Carroll believes that the greater the increase of Seals on the Newfoundland coast the more will the codfish decrease on the same coast. The scarcity of the one thus seems to imply the abundance of the other, so that an abundance of Seals along a coast where cod-fishing is prosecuted is not altogether an unmixed good.

HUNTING AND PRODUCTS.—As so large a part of what has been already said in the general account of the Seal-fishery of the North Atlantic and Arctic waters necessarily relates to the present species, it is scarcely requisite in the present connection to more than recall the leading points of the subject, with the addition of a few details not previously given. As already stated, the sealing-grounds *par excellence* are the ice-floes off the eastern coast of Newfoundland and around Jan Mayen Island,

---

* Seal and Herring Fisheries of Newfoundland, p. 26.

where the present species forms almost the sole object of pursuit. The sealing season lasts for only a few weeks during spring; the enterprise* gives employment during this time to hundreds of vessels and thousands of men, the average annual catch falling little short of a million Seals, valued at about three million of dollars. While the pursuit is mainly carried on in vessels, sailing chiefly from English, German, and Norwegian ports, or from those of Newfoundland and the other British Provinces, many are caught along the shores of the countries periodically visited by these animals, as those of South Greenland, Southern Labrador, Newfoundland, and the Gulf of Saint Lawrence. The pursuit with vessels, and the various incidents connected therewith, have already been detailed, and sufficient allusions have perhaps also already been made to the Greenland method of Seal-hunting (anteà, pp. 522–545).

In consequence of the gregarious habits of the species, and the fact that one-half to two-thirds of those taken are young ones that are not old enough to make any effectual attempt to escape, the success of a sealing voyage depends almost wholly upon the mere matter of luck in discovering the herds. While the old Seals are mostly shot, the young are killed with clubs. In respect to the ease and facility with which they are captured it may be noted that it is not at all unusual, in the height of the season, for the crew of a single small vessel to kill and take on board from five hundred to a thousand in a day. Mr. Brown states: "In 1866 the steamer Camperdown obtained the enormous number of 22,000 Seals in nine days," or an average of 2,500 per day. "It is nothing uncommon," he adds, "for a ship's crew to club or shoot, in one day, as many as from 500 to 800 old Seals, with 2,000 young ones".† Such slaughter is necessarily attended with more or less barbarity, but this seems to be sometimes carried to a needless extreme. The Seals are very tenacious of life, and, in the haste of killing, many are left for a long time half dead, or are even flayed alive. Jukes states that even the young are "sometimes barbarously skinned alive, the body writhing in blood after being stripped of its skin," and they have even been seen to swim away in that state, as when the first blow fails to kill the Seal their hard-hearted murderers "cannot stop to give them a second". "How is it," he adds, "one can steel one's mind to look

---

* For statistics of the Seal-fishery, see anteà, pp. 497–502.
† Man. Nat. Hist., Geol., etc., Greenland, Mam., p. 67, footnote.

on that which to read of, or even think of afterwards, makes one shudder? In the bustle, hurry, and excitement, these things pass as a matter of course, and as if necessary; but they are most horrible, and will not admit of an attempt at palliation."[*] Scoresby[†] and other writers refer to similar heartless proceedings,—as though the necessary suffering attending such a sacrifice of unresisting creatures were not in itself bad enough without the infliction of such needless cruelty. The young Seals not only do not attempt any resistance, but are said to make no effort to move when approached, quietly suffering themselves to be knocked on the head with a club. The old Seals are more wary, and are generally killed with firearms. Scoresby relates that "When the Seals are observed to be making their escape into the water before the boats reach the ice, the sailors give a long-continued shout, on which their victims are deluded by the amazement a sound so unusual produces and frequently delay their retreat until arrested by the blows of their enemies".[‡]

The annual catch of Harp Seals in Greenland is stated by Rink to be 17,500 full-grown "Saddle-backs" and 15,500 "Bluesides", or 33,000 in all. The catch from the Newfoundland ports alone often reaches 500,000, and in the Jan Mayen seas often exceeds 300,000, so that the total annual catch of this species alone doubtless ranges from 800,000 to 900,000.

The commercial products are the oil—used in the lubrication of machinery, in tanning leather, and in miners' lamps—and the skins, which are employed for the manufacture of various kinds of leather and articles of clothing. The skins are said to be mostly sold to English manufacturers, who employ them in the preparation of a superior article of "patent" or lacquered leather. The flesh is esteemed by the Greenlanders as superior to that of their favorite *Neitsek* (*Phoca fœtida*).

GENUS ERIGNATHUS, *Gill.*

Phoca, GRAY, "Zoöl. Erebus and Terror, 1844"; Cat. Seals Brit. Mus., 1850, 27, not of Linné. Type, *Phoca barbata*, Fabricius.
Erignathus, GILL, Proc. Essex Inst., v, 1866, 9. Type, *Phoca barbata*.

Muzzle broad, forehead high, convex; small supraorbital processes. Dental formula as in *Phoca;* teeth small, molars

---

[*] Excurs. in Newfoundland, vol. i, p. 290.
[†] Hist. of the Arct. Reg., vol. i, p. 510.
[‡] Ibid., vol. i, p. 512.

spaced, slightly implanted, early becoming defective by attrition; partly deciduous in old age. Palatal area broad, elliptical, deeply emarginate posteriorly; narial septum incomplete. Lower jaw short, small, the rami outwardly convex. Scapula with no acromion process. Iliac crests not abruptly everted and produced. Middle digits of the manus longest. Mammæ 4.

In respect to the general form of the skull, *Erignathus* differs from *Phoca* in the great height of the skull at the anterior border of the frontals. It differs also in the great breadth, arched form, and elliptical outline of the palate, and in the great depth of the narial fossæ. Although its single species is still commonly placed in the genus *Phoca*, other osteological characters, especially the absence of the acromion process of the scapula and the slight eversion of the iliac border of the pelvis, seem to warrant its separation. Although the animal attains to a large size, the teeth are weak, and in young specimens, or before they have become modified by attrition, are not longer anteroposteriorly, though rather thicker, than in *Phoca fœtida*, and are consequently several times smaller than in *Phoca vitulina*. The first, second, and fifth upper molars are 2-pointed, only the posterior accessory cusp being developed; the third and fourth are 3-pointed, also without an anterior cusp. All the lower molars are 3-pointed, there being an accessory cusp both in front of and behind the principal cusp. Quite early in life the teeth become much worn, and in old age the crowns of the three middle molars become often wholly worn away, leaving only the fangs, and even these sometimes in part disappear. Mr. Kumlien states that "in many adults the teeth can almost be plucked out with the fingers," so slight is their attachment.

ERIGNATHUS BARBATUS (*Fabricius*) *Gill*.

Bearded Seal.

Phoca barbata, FABRICIUS, Müller's Zool. Dan. Prod., 1776, viii; Fauna Grœnl., 1780, 15; Skriv. Nat. Selsk., i, 1790, 139, pl. xiii, fig. 3.—ERXLEBEN, Syst. Reg. Anim., 1777, 590.—GMELIN, Syst. Nat., i, 1788, 65.—KERR, An. King., 1792, 126.—SHAW, Gen. Zoöl., i, 1800.—J. ROSS, Ross's 1st Voy., App., 1819, xli.—DESMAREST, Mam., 1820, 246, 378.—"NILSSON, Skand. Faun., i, 1820, 374 ; K. Vet. Akad. Handl., Stockholm, 1837, —" ; Wiegmann's Arch. für Naturg., 1841, 317 ; Skand. Fauna Däggdj., 1847, 294.—"THIENEMANN, Reise im Norden von Europa, etc., i, 1824, 23, pl. i (ad. female), pl. ii (male of two years), pl. iii (male of one year), pl. iv (skulls)."—RICHARDSON, Parry's 2d Voy., Suppl., 1825, 335.—HARLAN, Faun. Amer., 1825, 111.

—GODMAN, Amer. Nat. Hist., i, 1826, 342.—GRAY, Griffith's Anim.
King., v, 1827, 178; "Zoöl. Erebus and Terror"; Cat. Seals Brit. Mus.,
1850, 27, fig. 9; Cat. Seals and Whales, 1866, 31, fig. 10; Suppl. Cat.
Seals and Whales, 1871, 3; Hand-List Seals, 1874, 8, pl. v.—FIS-
CHER, Syn. Mam., 1829, 240.—J. C. ROSS, Ross's 2d Voy., App., 1835,
xxi.—BELL, Brit. Quad., 1837, 275, fig. of skull (in part only):—
MACGILLIVRAY, Brit. Quad., 1838, 212 (in part only).—HAMILTON,
Amphib. Carn., 1839, 145, pl. "3"—i. e. 5 (in part only); not the
British references, nor the young specimen in Edinb. Mus., *apud*
Brown, Proc. Zoöl. Soc., 1868, 419.—BLAINVILLE, Ostéogr., Phoca,
1840–1851, pl. ix.—TEMMINCK, Faun. Japon., Mam. Marins, 1842,
2 (Japan).—? JUKES, Excurs. in Newfoundland, i, 1842, 312.—WAG-
NER, Schreber's Säugth., vii, 1842, 18.—SCHINZ, Synop. Mam., i, 1844,
481.—VON MIDDENDORFF, Sibir. Reise, ii, 2, 1853, 122.—GIEBEL,
Säugeth., 1855, 134.—VON SCHRENCK, Amur-Lande, i, 1859, 181.—
MALMGREN, Öfv. af Kongl. Vet. Akad. Stockh., 1863, 135; Arch. für
Naturg., 1864, 74.—STEENSTRUP, Vid. Medd. f. d. Naturh. Forening,
1864 (1865), 269.—COLLETT, Bemærk. til Norges Pattedyrf., 1876,
58.—VON MIDDENDORFF, Sibirische Reise, iv, Th. 2, 1867, 934.—
LLOYD, Game Birds and Wild Fowl of Sweden and Norway, 1867,
408, pl., animal.—QUENNERSTEDT, Kongl. Svens. Vetensk. Akad.
Handl., vii, No. 3, 1868, 10.—BROWN, Proc. Zoöl. Soc. Lond., 1868,
340, 424; Man. Nat. Hist., Geol., etc., Greenland, Mam., 1875, 53.—
TORELL and NORDENSKJOLD, Swedischen Exp. nach Spitz. u. Bären
Eiland, etc. (Germ. ed.) 1869, 78 (plate representing a group on the
ice).—LILLJEBORG, Fauna öfver Sveriges och Norges Ryggradsdjur,
i, 1874, 697.—RINK, Danish Greenland, its People and its Products,
1877, 126, 430.—VAN BENEDEN, Ann. du Mus. roy. d'Hist. Nat. du
Belgique, i, 1877, 20 (geogr. distr.).—KUMLIEN, Bull. U. S. Nat.
Mus., No. 15, 1879, 61.

Callocephalus barbatus, F. CUVIER, Mém. du Mus., xi, 1824, 184, pl. xii, fig.
4; Dict. des Sci. Nat., xxxix, 1826, 547.—LESSON, Man. de Mam.,
1827, 198.

Phoca (Callocephalus) barbata, VON HEUGLIN, Reisen nach dem Nordpolarmeer, etc., iii, 1874, 56.

Erignathus barbatus, GILL, Proc. Essex Inst., v, 1866, 12.—? PACKARD, Proc.
Bost. Soc. Nat. Hist., x, 1866, 271.

Phoca leporina, LEPECHIN, Act. Acad. Petrop., i, 1777, 264, pll. viii, ix.—
FABRICIUS, Skriv. Nat. Selsk., i, 2, 1791, 168.—SHAW, Gen. Zoöl., i,
1800, 258.—DESMAREST, Nouv. Dict. Sci. Nat., xxv, 1817, 581; Mam.,
1820, 243, 374.—FISCHER, Synop. Mam., 1829, 237.—HAMILTON,
Amphib. Carniv., 1839, 170, pl. ix.

Callocephalus leporinus, F. CUVIER, Dict. des Sci. Nat., xxxix, 1826, 545.

Phoca lepechini, LESSON, Dict. Class. d'Hist. Nat., xiii, 1828, 415 (= *Phoca
leporina*, Lepechin).

Phoca parsonsi, LESSON, Dict. Class. d'Hist. Nat., xiii, 1828, 414 (=Long-
bodied Seal, Parsons).

Phoca albigena, PALLAS, Zoogr. Rosso-Asiat., i, 1831, 109.

Phoca nautica, PALLAS, Zoogr. Rosso-Asiat., i, 1831, 108.

Phoca naurica, GRAY, Suppl. Cat. Seals and Whales, 1871, 3 (= "*Phoca naurica*
[lege *nautica*] et *Phoca albigena*, Pallas").

Phoca, Vitulus marinus, or *Sea Calf,* PARSONS, Phil. Trans., xlii, 1742–3 (1744), 383, pl. i.
Lachtak, STELLER, Nov. Comm. Acad. Petrop., ii, 1751, 290.
Long-bodied Seal, PARSONS, Phil. Trans., xlvii, 1751–2 (1753), 121.
Uksuk, CRANZ, Hist. von Grœnl., 1765.
Leporine Seal, PENNANT, Syn. Quad., ii, 1793, 277.
Remmesælen, FABRICIUS, 1791, l. c.
Oo-sook, Greenlanders.
Ogjook, Cumberland Esquimaux (KUMLIEN).
Lachtak or *Laktak,* Kamtschatkans.
Storkobbe, Blaakobbe, Havert, Norwegian (VON HEUGLIN).
Hafert-skäl, Storsjäl, Swedish.
Bartrobbe, Bärtige Seehund, German.
? *Square fipper* and *Square flipper,* of Newfoundland Sealers.
Ground Seal, Spitzbergen Sealers (BROWN).
Bearded Seal, Great Seal, English authors.

EXTERNAL CHARACTERS.—Above gray, darker along the middle of the back, the color varying in different individuals. A specimen from Disco Bay, Greenland (Nat. Mus., No. 8697), is gray, varied with black, but without distinct marks or spots. Wagner describes a specimen from Labrador as clear gray above, marbled with large indistinct yellowish spots, among them one on the back of the head more pronounced; sides and whole lower side of the body soiled white. No dark stripe along the back and head. Nilsson describes the color as being a pale gray above, still paler on the sides, and on the belly white; head and neck above, blackish, with a narrow band of the same color along the back. Macgillivray describes a Greenland specimen as having "the fore part of the head brown, the top light yellowish-gray; the hind neck and an obscurely defined space along the back, including the tail, dull brown, the rest dull yellowish-gray". Mr. Kumlien, who has had the opportunity of seeing many specimens in the Arctic regions, says the color is variable, the yellowish-brown being "more or less clouded with lighter or darker markings, irregularly dispersed". The length of adult males is usually given as "about ten feet"; females, rather smaller. I find the total length of an adult female skeleton to be 7 feet 2 inches (2195 mm.).

The young are described as being covered with long, soft, dark gray wool, which, in about two weeks after birth, is replaced by a coat of shorter, more rigid, bluish-gray hair. Nilsson described a specimen supposed to be fœtal as covered with dark gray wool, which is darker on the posterior part of the back and hind feet,* while Lepechin compared the woolly coat of the young (his *Phoca leporina*) to that of *Lepus variabilis.* Mr. Kumlien thus describes a fœtal specimen taken April 28, 1878, near Middliejuacktwack Islands: "Color, uniform grizzly-mouse color, with a tinge of olive-gray. Muzzle, crown, and irregular patches on the back and fore flippers, white. From the nose to the eyes a black line, with another crossing the head behind the eyes, the two forming a perfect cross. Nails, horn-blue, tipped with white. Iris, dark brown. Nose, black. Muzzle, wide, lips full and fleshy, giving the animal a bull-dog expression. Body, long and slender. Beard, pellucid, abundant, white, stout, the bristles becoming shorter toward the nostrils. Hind flippers, large and heavy, looking disproportionate to the size of the body. Hair, rather short, but fine and somewhat woolly, interspersed with another kind, stiff and of a steel-blue color, which I take to be the second coat. The Eskimo are firm in the belief that the Ogjook sheds its first coat within the uterus of the mother. In this case there was certainly an abundance of loose hair in the uterus, but the specimen had been dragged some miles in its envelope over rough ice, besides having been kept three or four days in an Eskimo igloo among a heap of garbage, so that it is not to be wondered at that the hair was loose.

"There was little blubber on the specimen, and this was thickly interspersed with blood-vessels.

"The specimen measured as follows:

	Feet.	Inches.
Extreme length	4	7
Length of head	0	8.25
Width of muzzle	0	4.5
From end of nose to eye	0	3.2
Distance between eyes	0	3.5
Length of fore flipper (to end of nails)	0	7.15
Width of fore flipper	0	4.3
Length of hind flipper	1	0
Greatest expanse of hind flipper	1	1.5 "†

SKULL AND SKELETON.—The principal distinctive osteological features of the Bearded Seal having already been given in connection with the generic diagnosis, little is called for in the present connection, since a detailed account of its osteology does not fall within the scope of the present history of the species.

---

* Archiv. für Naturg., 1841, p. 317.

† MS. notes.

It may be stated, however, in general terms, that the skeleton indicates a general robustness of form, correlating with the rather broad thick head. The relative length of the different limb-segments and vertebral regions is about as in *Phoca græn-landica*, except that the caudal series of vertebræ is much shorter. The bulk of the entire animal, however, must be considerably greater than in *P. grænlandica*. The scapula is long and narrow, the proscapular and postscapular fossæ about equal, the latter not greatly produced at its posterior upper border, as in *Phoca vitulina* and *P. fœtida*. Aside from the absence of the acromion process, it thus differs in its narrow elongated form, and especially in the unusual length of the shaft, from that of either of the three above-named species. *P. grænlandica* presents the opposite extreme, the scapula of which is broad and short.

The exceptional features of the skull are the small size of the orbital fossæ, the rather small size of the auditory bullæ, and the large size of the nasal passages. The general form of the lower jaw is much as in *P. vitulina*, especially resembling it in the lateral convexity of the rami, and in the form of the condylar portion, and in the abruptness of the angle. It is, however, small and weak for the size of the skull, and especially so for the size of the animal.* Perhaps its most striking feature consists in the large process on the hind border just below the condyle, which is twisted over toward the inner edge of the jaw, and has its axis of development in that direction or transverse to the longitudinal axis of the jaw.

In comparison with the skeleton in the above-named species of *Phoca*, the bones of the Bearded Seal are light and porous (less so, however, than in the *Cystophorinæ*); the tuberosities are all rather weakly developed, with a less tendency to anchylosis. To this general laxness of ossification may perhaps be attributed the slight development and consequent lack of abrupt eversion of iliac crests of the pelvis already noted. The subjoined table of linear measurements of the principal bones of the skeleton is taken from that of a quite old female from Cumberland Gulf, collected by Mr. Kumlien.

---

* I find that the lower jaw of a very old male *P. vitulina* just fits an adult female skull of *Erignathus barbatus*, except that the latter is slightly longer.

Measurements of the principal parts of the skeleton in Erignathus barbatus
(♀ *ad.*).

	MM.
Length of the skull	230
Length of the cervical vertebræ	250
Length of the dorsal vertebræ	800
Length of the lumbar vertebræ	390
Length of the sacral vertebræ	175
Length of the caudal vertebræ	350
Length of the scapula	210
Length of the humerus	162
Length of the radius	140
Length of the manus	185
Length of the pelvis	320
Length of the femur	153
Length of the tibia	310
Length of the pes	380
Length of the whole skeleton	2195
Length of the fore limb (exclusive of scapula)	487
Length of the hind limb	843

The series of skulls of this species shows that the female is rather smaller than the male, with a rather weaker osseous structure. While old males have rudimentary but quite strongly marked anteorbital processes, equally old female skulls sometimes show not the slightest trace of them. The largest male skulls do not exceed a length of 260 mm., while one marked as female attains nearly 240 mm. The Bearded Seal, although often stated to be the largest Phocid of the northern seas, has the skull much smaller than either *Halichœrus grypus* or *Cystophora cristata*, while the skeleton of the latter indicates an animal of much greater bulk. Adult female skulls of the Bearded Seal, in fact, scarcely exceed in linear dimensions very large old male skulls of *Phoca vitulina*.

Measurements of nine skulls of Erignathus barbatus.

Catalogue number.	Locality.	Sex.	Age.	Length.	Breadth at mastoid processes.	Greatest breadth at zygomatic arches.	Distance from anterior edge of inter-maxillæ to end of pterygoid hamuli.	Distance from anterior edge of inter-maxillæ to hinder edge of last molar.	Distance from anterior edge of inter-maxillæ to meatus auditorius.	Distance from anterior edge of inter-maxillæ to glenoid process.	Distance from palato-maxillary suture to end of pterygoid hamuli.	Length of the alveolar border of the maxillæ.	Width of palatal region at posterior end of maxillæ.	Nasal bones, length.	Nasal bones, breadth anteriorly.	Nasal bones, breadth at fronto-maxillary suture.	Breadth of skull at canines.	Least breadth of skull, interorbitally.	Breadth of posterior nares, vertically.	Breadth of posterior nares, transversely.	Breadth of anterior nares, vertically.	Breadth of anterior nares, transversely.	Greatest height of skull at auditory bullæ.	Length of brain-case.	Greatest width of brain-case.	Length of lower jaw.	Front edge of ramus to last molar.
6570	Greenland	♂	Old.	237	141	134	128	83	173	163	47	98	70	66	28	16	50	30	31	39	40	37	87	105	114	160	63
6571	do	♂	Old.	257	142	138	129	85	172	164	51	94	67	68	23	15	50	28	28	39	36	34	88	98	113	154	66
6569	do	♂	Rather young.	227	136	137	129	85	170	160	48	96	68	66	22	16	54	27	25	38	33	37	86	98	117	148	68
15685	Cumberland Gulf		Rather young.	...	138	133	...	...	...	...	50	...	65	65	22	15	...	25	30	39	...	...	...	102	112	...	...
7106	Plover Bay, Behring's Sts.	♀	Adult.	208	138	128	116	72	150	141	46	88	55	44	20	14	41	25	28	36	34	34	...	93	118	...	...
7107	do	♀	Adult.	220	...	...	124	80	170	148	51	94	59	52	23	15	43	23	23	38	33	34	...	99	...	...	...
16117	Cumberland Gulf		Old.	230	140	134	134	84	170	163	53	94	68	56	21	14	45	23	28	39	35	33	87	96	115	150	66
16116	do	♀	Old.	238	134	132	132	78	174	156	49	92	64	56	22	15	47	24	21	33	35	33	86	95	112	152	65
6299	do	♀	Adult.	227	141	137	122	81	167	155	51	91	67	58	25	16	50	28	27	37	40	32	92	98	117	150	64

GENERAL HISTORY AND NOMENCLATURE.—The early history of the Bearded Seal is peculiarly involved, owing in part to the vagueness of the early references. The first notice of the species that can be fixed with any degree of certainty is Dr. James Parsons's account, published in 1744, of the "*Phoca, Vitulus marinus*, or Sea-Calf, shewed at Charing-Cross, in Feb., 1742–3", * which, if any reliance can be placed on the figure,—declared by Dr. Parsons to be drawn from life and to have been pronounced by all who saw it to be an excellent likeness,—and the characters given in the text, must be unquestionably the present species. Dr. Parsons says it was a female, and very young, "though Seven Feet and half in Length, having scarcely any Teeth, and having Four Holes regularly placed about the navel, as appears by the Figure, which in time become *Papillæ*." The figure shows the middle finger of the fore-flipper to be the longest, the others regularly decreasing in length on each side, while the hind-flippers terminate squarely, with all the digits of nearly equal size. As this is the only species of Seal found in the northern seas which has four mammæ, and the flippers of the form here indicated, the identity of the species seems beyond question. The size, moreover, corresponds with that of old females of the present species. Its "having scarcely any Teeth" is another strong point in favor of its being the Bearded Seal, since it is well known that in old, or even middle-aged, examples of this species the molar teeth are so much worn down that only the fangs of the greater part of the molars remain, and even these may be in part lacking, while on the other hand no other Seal of this size could show this feature, either from immaturity or attrition due to old age. The identity of Parsons's Long-bodied Seal† with the Bearded Seal (*Phoca barbata*, auct.) was almost universally conceded till 1837, when, from an examination of what was *supposed* to be Dr. Parsons's original specimen, Messrs. Bell and Ball declared it to be *Halichœrus grypus*. Mr. Bell says, "For many years there has been deposited in the British Museum a large specimen of Seal, which has always been considered as the *Phoca barbata*. It was previously in the possession of Mr. Donovan,

---

* Phil. Trans. for the year 1742–1743 (1744), p. 383, pl. i.

† It should be stated that this name was first used by Dr. Parsons for this Seal in a subsequent reference to the same specimen published in 1753 (Phil. Trans., vol. xlvii, p. 121), in which the habitat is given as the coast of Cornwall and the Isle of Wight.

who, I am informed, stated it to be the identical specimen described under the name of 'Long-bodied Seal' by Mr. Parsons, in the forty-seventh volume of the Philosophical Transactions. It has been upon his authority only that *Ph. barbata* has been catalogued as British, and it now proves to be the same species as that lately found on the southern coast of Ireland by Mr. Ball, the *Phoca Gryphus* of Fabricius, *Halichœrus griseus* of Hornschuch and Nilsson. It was by the exhibition of crania of this species by Mr. Ball at the late meeting of the British Association, that Professor Nilsson, who was present, was able to identify it; and a subsequent examination of the specimen in the British Museum led that gentleman to the conclusion that this also is identical with the former." *   Mr. Ball says, " On examining the remains of Donovan's *Ph. barbata,* now in the British Museum, I recognized in it an ill-put-up specimen of our *Halichœrus;* and I presume the stuffer has endeavored to make the specimen correspond with the description of *Ph. barbata* by unduly plumping up the snout and shortening the thumbs, which are evidently pushed in by the wires intended to support the paws." †   Since these announcements the Long-bodied Seal of Parsons has generally been referred to *Halichœrus grypus,* or only doubtfully assigned a place among the synonyms of the Bearded Seal. Inasmuch as the Long-bodied Seal of Parsons forms an important element in the ground-work on which the name *Phoca barbata* reposes, a further inquiry into the question here at issue may be in place.

First, it may be noted, that the identity of Donovan's specimen with that described by Parsons rests on hearsay testimony, namely, a *report* that *he said* it was the same. ‡   Without casting any implication of doubt upon the correct specific determination of Donovan's specimen, as above detailed, it may be observed further that the characters given by Parsons apply to the Bearded Seal and to no other, and, furthermore, that Parsons does not state whether or not his specimen was preserved, nor does he in the original account say where it was captured. In his second notice he gives its habitat as the " Coast of Corn-

---

* History of British Quadrupeds, etc., 1837, pp. 278, 279.

† Ibid., p. 281.

‡ Ball states elsewhere that Donovan's *Phoca barbata* "*seems* to be the individual described by Parsons as the long-bodied seal, and it *appears* to have been on the authority of this specimen that *Phoca barbata* has occupied a place in the British Fauna."—*Trans. Roy. Irish Acad.,* vol. xviii, pt. 1, pp. 90, 91, in a paper read " 12th December, 1836".

wall and the Isle of White" (*lege* Wight?). In view of the admitted uncertainty as to whether Donovan's specimen was the one described by Parsons, and the agreement of Parsons's account and figure (the best figure of any Seal published up to that date), there seems to be no adequate reason for referring Parsons's Long-bodied Seal to any other than the Bearded Seal, with which for three-fourths of a century it was currently associated.

Another early, and in some respects important account of this species, appears to have been given by Cneiff (see *anteà*, p. 530), under the name *Grå Skæl* (or "graue Seekalb" as termed in the German translation of his paper, which is here used). Although Cneiff's (or Kneiff, in the German orthography) Grå Skæl is referred by Lilljeborg and others to *Halichœrus grypus*, its breeding habits seem to forbid its reference to that species, it being said to bring forth its young about the end of February on the ice remote from the land, while *Halichœrus grypus* has its young *in the autumn*, for which purpose *it resorts to the land*, selecting as its breeding haunts rocky shores and small rocky islets. The general habits of the species also better accord with those of the Bearded Seal, especially its forming an *atluk* or breathing-hole through the ice, like *Phoca fœtida*, these two species being the only ones found in the northern seas which have that habit. That it is not *Phoca fœtida* is indicated by its size, which is said to be a full "Klafter" (about $6\frac{1}{4}$ feet) long, and by his comparison of it with the Harbor Seal ("Wikare" or "Meerbusenkalb" = *Phoca vitulina*), from which it appears that the latter is only about half the size of the former.* As respects the color, he says the Gray Sea-Calves are mostly dark gray; many are yellowish; but they are very rarely marked with black and white spots. There is here a closer agreement with the Bearded than with the Gray Seal. There consequently seems to be no reason why Cneiff's Grå Skæl should not be referred to the Bearded Seal, and very strong reasons *against* its reference to *Halichœrus grypus*.†

To resume the early history of the subject, the next notice of the Bearded Seal appears to be Steller's reference, in 1751,

---

* He gives the weight of the graue Seekalb as "18 Lisspfund", and that of the Wikare as "10 Lisspfund ".

† At p. 531, in the account of Cneiff's history of sealing in the Gulf of Bothnia, I gave the species as probably *Halichœrus grypus*, in deference to eminent authority.

to a large Seal occurring in the North Pacific, "quæ magnitudine Taurum superat", and which he says the Kamtschatdales called "Lachtak". He speaks of it as being the·largest of the Seals of those waters. It later formed the "*Phoca maxima,* Steller" of authors, but Steller himself did not originate this phraseology. It is also the basis of the *Phoca lachtak* of Desmarest. These references have been very generally identified with the present species. Cranz, in 1765, referred to it, but without really describing it, under the name Uksuk. Yet the little he had to say about it serves to render it certain that the Uksuk is the *Phoca barbata* of the later systematic writers, it being still known in Greenland under that name (now commonly spelled "Oo-sook"), just as the Pacific representatives of the species are still known under the native name Laktak.

Fabricius was the first to give it a systematic designation, he calling it, in 1776 (in inedited notes in Müller's "Zoologicæ Daniæ Prodromus"), *Phoca barbata,* but the name was unaccompanied by a description. He cites, however, its Icelandic and Greenlandic names "Gramselr" and "Urksuk," by which the species is still known in those countries. Four years later (in his "Fauna Grœnlandica") he fully described the Urksuk of Greenland under the name *Phoca barbata,* when the species became first fairly characterized. In the meantime Erxleben (in 1777) had adopted the name for a large species of Seal, under which designation he cited not only the Uksuk of Cranz, the Gramselr of Iceland, the *Phoca barbata* of Müller's "Prodromus", and the Laktak of Kamtschatka, but also the Long-bodied Seal of Parsons, together with the various names that had been based upon these, either individually or collectively. As he arranged his references chronologically, the first names mentioned are the Long-bodied Seal of Parsons, and the Lachtak, or "*Phoca maxima*", of Steller. His brief diagnosis is evidently based on Cranz's account of the Uksuk.

The name *barbata* is usually ascribed to Müller, 1776, in whose work it first appeared, but rigid constructionists may claim that date as untenable, since no description accompanied the name. In this case it would fall to Erxleben, 1777, who gave of it a brief technical description, and further established it by a full and correct citation of its synonymy.

Almost simultaneously with the appearance of Erxleben's work the species was again indicated by Lepechin (in 1778), under the name *Phoca leporina,* based on the young from the

White Sea. Although he erroneously gave the incisive formula as $\frac{4}{2}$, and based his description and figure on the young still in the white pelage, there has been little doubt among modern writers of its identity with the *Phoca barbata*. For many years, however, the *Phoca leporina* figured in the works of compilers as a distinct species, and became thus a prominent synonym. Lesson, in 1828, renamed it *Phoca lepechini*, at the same time naming the Long-bodied Seal of Parsons *Phoca parsonsi.*

In regard to its general history, it may be added that Fabricius, in 1791, in his monograph of the Greenland Seals, devoted twenty pages to an account of *Phoca barbata*, giving a careful description of its external characters, with detailed measurements, and the first (and a very good) figure of its skull. He adopted for it the Danish vernacular name "Remmsæl," identifying with it Steller's Lacktak, the Icelandic names "Gramselur," "Grœnselr," and "Kampselur," and the Greenlandic names "Urksuk" and "Uksuk," as well as the *Phoca barbata* of Müller's "Prodromus," of Erxleben, of the "Fauna Grœnlandica," and of Gmelin, and also Parsons's Long-bodied Seal, and the subsequent accounts based upon it. The next original information of special importance appears to have been furnished by Thienemann, who, in 1824, in his account of the Seals of Iceland, devoted four plates to its illustration, figuring the adult female, a two-year-old male, a yearling male, and the skull.

In 1831 Pallas introduced two nominal species, referable here, under the names *Phoca nautica* and *Phoca albigena*. With the former he identified the Lacktak of Steller, while he made Lepechin's *Phoca leporina* a synonym of his *Phoca albigena*. These names have been generally referred by subsequent writers, either positively or with reservation, to *Phoca barbata*. Gray, in 1871, separated the Bearded Seals of the North Pacific from those of the North Atlantic as *Phoca "naurica"* (*sic*) apparently wholly on the ground of locality, and referred to this Pallas's *Phoca nautica* and *Phoca albigena*.

Among the more important recent contributors to the history of the species are Malmgren, Von Heuglin, and Collett, the latter, especially, having given a very full account of its habits and distribution on the coast of Norway.

GEOGRAPHICAL DISTRIBUTION.—The present species is circumpolar and extremely boreal in its distribution, and appears to be migratory only as it is forced southward in winter by the extension of the unbroken ice-fields. The southern limit of its range along the Atlantic coast of North America is at present indeterminable. Professor Jukes* gave it doubtfully in his list of the Seals of Newfoundland, supposing it to be the Square Flipper of sealers. Among the many examples of Seals I have had opportunity of examining from Newfoundland, however, I have never met with a specimen of the Bearded Seal. If the Square Flipper of the Newfoundland sealers be really the Bearded Seal, as seems probable, it must be, according to Carroll, of regular occurrence in small numbers about Newfoundland.

Dr. Packard † has attributed it to Labrador, where it undoubtedly occurs, but he gives it on the hypothetical ground that "It is probably the species which is called by the sealers 'Square Flipper,'" and says that adults will "weigh 500 to 600 pounds". Its occurrence in Labrador, however, is apparently established by Wagner, who described a specimen, "das aus Labrador herstammt".‡

Although well known to visit the shores of Greenland, and to range very far north, its limit in this direction still remains undetermined. J. C. Ross states that it approaches the shores of Boothia "only in the summer season," and that in winter it seeks "those parts of the Arctic Ocean which are seldom, if ever, frozen over for any length of time". § Dr. Rink says that it occurs only in small numbers in Greenland, and chiefly at the "northern and southern extremities of the coast." || Mr. Robert Brown's account of its distribution is as follows: "This species has been so often confounded with the Grey Seal (*H. grypus*) and the Saddleback (*P. grœnlandicus*) in different stages and coats, that it is really very difficult to arrive at anything like a true knowledge of its distribution. . . . On the coast of Danish Greenland it is principally caught in the district of Julianshaab a little time before the Klapmyds [*Cystophora cristata*]. It is not, however, confined to South Greenland, but is found at the head of Baffin's Bay and up the sounds of Lancaster, Eclipse, &c., branching off from the latter sea. The Seals seen by the earlier navigators being nearly always referred in

---

* Excurs. in Newfoundland, vol. i, p. 312.
† Proc. Boston Soc. Nat. Hist., vol. x, p. 271.
‡ Schreber's Säugthiere, Band vii, p. 20.
§ Ross's 2d Voy., App., p. xxi.
|| Danish Greenland, etc., p. 126.

their accounts to either *Phoca vitulina* or *P. grœnlandicus*, it is impossible to trace its western range; it is, however, much rarer in the north than in the south of Davis's Strait. Accordingly the natives of the former region are obliged to buy the skin from the natives of the more south of [*sic*] settlements, as it is of the utmost value to them. This Seal comes with the pack-ice round Cape Farewell, and is only found on the coast in the spring. Unlike the other Seals, it has no *atluk*, but depends on broken places in the ice; it is generally found among loose, broken ice and breaking-up floes."*

Mr. Kumlien (MS. notes) says, "This Seal was first noticed a little to the southward of Cape Chidly, and thence northward to our winter harbor, in about lat. 67° N. According to the Eskimo, they are the most common about Cape Mercy, Nugumeute, and the southern Cumberland waters, where they remain all the year, if there is open water. They remain in Cumberland Sound only during the time when there is open water, as they have no *atluk*. On the west coast of Davis Straits they are not rare, but are said by whalemen to diminish in numbers above lat. 75° N. They appear to be more common on the southern shores of the west coast of Davis Straits than on the northern, so that the natives go southward some distance to secure the skins. We noticed them among the pack-ice in Davis Straits in July and August. . . In Cumberland Sound they begin working northward as fast as the floe edge of the ice breaks up, arriving in the vicinity of Annanactook about the latter days of June. In autumn they move southward as fast as the ice makes across the sound, always keeping in open water. They are seldom found in the smaller fjords or bays, but delight in wide expanses of water."

Respecting its southern limit on the coast of Europe, there appears to be no unquestionable record of its capture south of the "North Sea", which locality is given by Gray for various specimens in the British Museum. It was for many years supposed to inhabit the Western Islands of Scotland, and to have occurred at other localities in the British Islands, but on further investigation the species proved to be the Gray Seal. It is consequently omitted from the second edition of Bell's "History of British Quadrupeds". Dr. Gray, writing in 1872, said, "I have never seen a specimen from the coast of Great Britain;

---

* Proc. Zoöl. Soc. Lond., 1868, p. 424; Man. Nat. Hist., Geol., etc., Greenland, 1875, Mam., p. 54.

probably *Halichœrus grypus* was the species taken for it."[*] Its occurrence in Iceland is well attested, and, according to Robert Collett, it is found in small numbers along all the rocky coasts of Norway, from the fjords of Finmark down to latitude 62°,[†] where it occurs all the year. He believes it to have been formerly much more numerous there than it is now. Malmgren also gives it as a rare visitor to the coast of Finmark, and as occurring only late in autumn or winter. He records the capture of one taken near Tromsö about the end of October in 1861. It is stated to be rare about Jan Mayen,[‡] but of frequent occurrence along the ice-fringed shores of West Spitzbergen, where, according to von Heuglin, it is found from July to September, while Malmgren believes it may winter there. Payer gives it as abundant at Franz-Josef Land, where this and the Harp Seal were the only species observed.[§] It has been frequently reported as occurring along the Arctic coast of Europe and Asia. Von Middendorff believes it is this species that the Samoëdes have reported as so abundant at the mouth of the Taimyr River, and as found on the Taimyr Sea. It doubtless not only occurs along the Chatanga to Chatangskij Pogost, but probably reaches the mouth of the Chata.[||]

In respect to its distribution in the North Pacific, Temminck states that its skins are carried to Japan as an article of commerce, and that he has seen an incomplete one brought from that country by Siebold.[¶] He does not state, however, that it inhabits the Japan coast, as some authors have apparently implied. Wagner says, "Das Leidner Museum besitzt Felle von Sitka, aber nicht von Japan."[**] It has not been reported, however, as found at the Fur Seal or Prybilow Islands. There are several specimens in the National Museum collected at Plover Bay, on the Siberian side of Behring's Straits.

Von Schrenck states, on the authority of the natives, that it is common on the southern shore of the Ochots Sea, in the Gulf of Tartary, in the "Amur-Limane," and even in the Amoor River, but adds that the old animals only come into the mouth of the river, while the younger ones go somewhat higher up.

---

[*] Zoölogist, 2d ser., vol. vii, p. 3336.
[†] Bemærkninger til Norges Pattedyrfauna, 1876, p. 58.
[‡] German Arctic Expedition, 1869–70, p. 62.
[§] New Lands within the Arctic Circle, 1877, p. 266.
[||] See von Heuglin, Reisen nach dem Nordpolarmeer, etc., p. 58.
[¶] Fauna Japonica, Mam. Marins, p. 2.
[**] Schreber's Säugthiere, vii, p. 21.

He often saw its skins in Amoor Land, and says they are an article of traffic among the natives as far south as Saghalien Island. But he appears to have seen only the skins in the hands of the natives, and to give its distribution wholly on their testimony. If the limits here assigned, and the locality of Sitka, given by Wagner, be correct, it extends much farther south, along the shores of the North Pacific, than it does on the coast of Europe, but in each case its habitat is bounded by about the same isotherm.

HABITS, PRODUCTS, AND HUNTING.—This large Seal, the largest Phocid of the northern seas, appears to be nowhere abundant, and is usually described as rather solitary, avoiding the company of other species, and as never occurring in large herds like the Harp Seal. Of its habits, as observed on the Atlantic coast of North America, little has been recorded. If the " Square Flipper" Seal of the Newfoundland sealers be this species, of which Mr. Carroll has given some account, it is of not unfrequent occurrence off the shores of that island. From the indications Mr. Carroll gives of its size, the form of the hind flippers (from which it appears to derive its local name), as well as the statement that it has four mammæ, seems to indicate that it can be no other than the present species. As, however, the Gray Seal nearly approaches it in size, and is not enumerated by Mr. Carroll nor Professor Jukes, and apparently is not distinguished by the sealers (although it unquestionably occurs in Newfoundland, as attested by specimens from there in the National Museum), it seems questionable whether the Gray and Bearded Seals are not confounded under the name " Square Flipper ". Since Mr. Carroll's account, however, corresponds in general points so well with the Bearded Seal, I venture to give it provisionally as a part of the history of the species. His account in substance is as follows: The Square Flippers are the largest Seals that are killed on the coast of Newfoundland. They never congregate with any other Seals, and are very scarce, not more than one hundred being taken each sealing voyage throughout the island. Persons who live along the northern bays, and "follow the gun" during the winter and spring, kill a few of them. Many are seen in the Straits of Belleisle, as well as about Saint Paul's Island, in the Gulf of Saint Lawrence. They have their young on the ice about the 20th of March. They are called Square Flippers because the flippers are "square at the top [tip?], thus differing from all [other] species of Seal taken on the coast of Newfoundland." They are very quiet and very fond of their young of which they have never more than one at a time. If seen on the ice they are sure to be killed. The skin and fat of a male Square Flipper, when prime, will weigh "from 7 to 10 cwt.". When in full flesh his weight varies "from 13 to 15 cwt"; the skin and fat of the female when prime, weighs "from 4 to 5 cwt."; the skin and fat of a young square flipper, when sixteen days old, will weigh from 160 to 170 pounds. "The skin of the male and female square flipper is of a cream color, the female has four teats (no other seal known in Newfoundland has more than two). All seals [sic] teats protrude about one inch outside the skin whilst the young is sucking, after which they are drawn in, so as to prevent injury whilst the old seal is crawling on ice or rocks. The oil rendered out of square flippers [sic] fat, old and young, when prime, is considered as pure as the best young Harp oil. Length of an old square flipper, from head to tail, 11 to 12 feet."*

Mr. Kumlien gives the following quite full account of its habits, as observed by him in Cumberland Gulf:

"The Ogjook, as this Seal is termed by the Cumberland Eskimo, delights in basking upon pieces of floating ice, and generally keeps well out at sea. I have never seen any numbers together, but almost always singly. The old males do not seem to agree well, and often have severe battles on the ice-floes when they meet. They use the fore flippers, instead of the teeth, in fighting. . . .

"This seal has the habit of turning a summersault when about to dive, especially when fired at; this peculiarity, which is not shared by any other species that I have seen, is a characteristic by which it may be distinguished at a considerable distance. During May and June they crawl out upon an ice-floe to bask and sleep; at such times they are easily approached by the Eskimo in their kyacks and killed. . . . They dive to great depths after their food, which is almost entirely *Crustacea* and mollusks, including clams of considerable size. . . . In July, during the moulting time, their stomachs contained nothing but stones, some of them nearly of a quarter-pound weight. They seem to eat nothing during the entire time of shedding—probably six weeks. Certain it is they lose all their

---

* The Seal and Herring Fisheries of Newfoundland, together with a condensed History of the Island, 1873, pp. 12, 13.

blubber, and by the middle of July have nothing but 'white-horse',—a tough, white, somewhat cartilaginous substance,—in place of blubber.   At this season they sink when shot.   . . . The young are born upon pieces of floating ice, without any covering of snow. . The season of procreation is during the fore part of May.   After the young have shed their first woolly coat (which they do in a few days), they have a very beautiful steel-blue hair, but generally so clouded over with irregularly dispersed patches of white that its beauty is spoiled.   . . .

"The Ogjook is of great value to the Eskimo, who prize the skins very highly.   All their harnesses, sealing lines, etc., are made from the raw skins; besides this, they make the soles of their boots, and sometimes other portions of their dress, from the skin.   In such localities as the whalemen do not visit, and the natives are obliged to construct skin boats, this seal is in great demand.   It takes fifteen skins for an ominak or skin boat, and these skins require renewing very often.   The skin of the back and belly dries unevenly, so the Eskimo skin the animal along both sides and dry the skin of the upper and lower parts separately.

"It is a prevalent belief among the whalemen that the livers of Seals, and more especially those of this species, are poison-ous; but I am inclined to rate this as imaginative; we ate the livers of all the species we procured without any bad effects." *

The Bearded Seal appears to be a well-known inhabitant of the coast of Norway and of the Arctic islands north of Europe. Collett gives it as occurring in comparatively small numbers along the Norwegian coast, from the fjords of Finmark south-ward to latitude 62°, throughout which range it appears to be resident the whole year.   They make short journeys to the fish-ing-grounds, a few miles off the coast, but for the most part are found constantly at nearly the same localities.   They have, however, their favorite breeding resorts, at which they assem-ble during the breeding season from a considerable area, dis-persing again when the breeding time is over.   One of these breeding-places, and believed to be the most southern one on the Norwegian coast, is at the northwesternmost of the islets of the Froyen group, off Trondhjem Fjord, to which it is sup-posed that nearly all of the individuals found south of latitude 64° resort in the breeding season.   Mr. Collett states that the

---

* Copied from Mr. Kumlien's MS. notes, with slight verbal changes; since published in Bull. U. S. Nat. Mus., No. 15, pp. 61–63.

species is strictly polygamous, the strongest males driving away the younger; yet the number of the females appears to be not much greater than that of the males. He states that one remarkable difference between the individuals found at the southern breeding stations and those living farther north is the different season at which the young are born. He says this occurs in Norway in autumn. The Seals begin to gather at the chosen breeding-place about the middle of September, and, as a rule, the first young are born about the end of that month.* Pairing takes place in the water very soon after the birth of the young, and before the old Seals depart from the breeding-places, which latter event occurs about the end of October. About half of them, however, remain near the outer rocks during winter.†

According to von Heuglin, the female gives birth to her single offspring in February or March, but Malmgren states that he took a ripe fœtus from the mother as late as the 31st of May, and Kumlien says the young are born early in May. Fabricius says late in April or early in May. Carroll, as already noted, says the young are born late in March. As will be noticed later, the exceptional record of an autumnal breeding season for this species given by Collett, on the authority of a correspondent, suggests the possibility that the species really observed was *Halichœrus grypus* or the Gray Seal.‡ The earlier breeding-time given by Carroll, for Newfoundland, may be due to the locality being so far south.

According to Malmgren, they do not frequent open water. As long as the inlets and bays are closed with ice it keeps near openings in the ice, through which it ascends to the surface to rest, but when the fast ice breaks up it keeps among the floating ice near the coast. It does not, however, follow the ice far out to sea, but leaves it and seeks such shores as are skirted with drifting ice. On the coast of Spitzbergen it is rarely met with in summer, owing to the absence of ice, but as soon as the ice again arrives, either from the south or the north, they appear in the bays in great numbers. In Northeast Land, where the inlets are covered with ice till late in August, and where, not far from the land, are many ice-floes, they are common the whole summer. He states that during his stay in Hinlopen

* See next paragraph and *infra*, p. 706.
† Bemærkninger til Norges Pattedyrfauna, pp. 58–60.
‡ See *infra*, p. 706.

Misc. Pub. No. 12——43

Straits some Walrus-hunters shot about sixty of them in the course of two or three days about the beginning of August, and that his harpooners often killed them.* Von Heuglin also refers to its partiality for the neighborhood of ice, and says that on the coast of West Spitzbergen he saw it only in the vicinity of the glaciers that reach the sea. Among the Thousand Islands and in the Stor-Fjord he found it very common, but always singly or in small companies. He states, on the authority of Spörer, that in Nova Zembla it rarely appears on the northern shore of the islands, but commonly visits South Island. He says that, although he saw it there only rarely, it must be sometimes very numerous, as in the course of three days as many as three hundred have been taken by the use of three nets.†

Malmgren states that the Bearded Seal is easily killed when it is in the sea, as it is then not shy, but often comes so stupidly and eagerly about the boat as to be very easily shot. When lying on the ice he describes it as extremely watchful, so that it is impossible to shoot it without using a shooting-screen, such as the Greenlanders employ.‡ Mr. Kumlien, however, states that during May and June, when they crawl out upon an ice-floe to bask and sleep, they are easily approached by the Esquimaux of Cumberland Sound in their kyacks and killed. It is reported to subsist chiefly upon large mollusks and crustaceans. Malmgren records that in the stomachs of all he examined he found large species of *Crangon* and *Hippolyte* (*C. boreas*, *Sabinea septemcarinata*, *Hippolyte polaris*, *H. sowerbyi*, and *H. borealis*), and *Anonyx ampulla* in abundance; occasionally small fishes (*Cottus tricuspis*, Reinh.), and many hundreds of the opercula of species of *Buccinum* and *Natica clausa*, as well as shells of a large *Lamellaria*.§

All writers, from the time of Cranz to the latest observers, testify to its importance to the Esquimaux and other native tribes of the shores it frequents. Its flesh or blubber is said to be more delicate in taste than that of any other species, and to be esteemed as a luxury. Its chief value, however, consists in its skin, which, from its great thickness, is, according to Dr. Rink, "the only one considered fit for making the hunting lines of the kayakers." Von Schrenck speaks of its being used by the na-

* Arch. für Naturg., 1864, pp. 74–75.
† Reisen nach dem Nordpolarmeer in dem Jahren 1870 und 1871, p. 57.
‡ Arch. für Naturg., 1864, p. 77.
§ Ibid., p. 75.

tives of Amoor Land and Saghalien Island for the same purposes as Mr. Kumlien notes in respect to the Esquimaux of
Cumberland Sound. Owing to its scarcity it has no great commercial importance, though sometimes taken by the sealers of
the Spitzbergen sealing-grounds. Rink states that the whole
annual catch of this species in Greenland hardly amounts to
1,000.

GENUS HISTRIOPHOCA, *Gill.*

Histriophoca, GILL, Am. Nat., vii, 1873, 179. (Type, " *Phoca fasciata*, Shaw,
or *P. equestris*, Pallas.")

Cranial characters unknown. Incisors, $\frac{3-3}{2-2}$; C., $\frac{1-1}{1-1}$; M., $\frac{5-5}{5-5}$.
Incisors conical, cylindrical, directed slightly backward. Mo-
lars, except the first, 2-rooted, placed somewhat apart, with
simple crowns directed backward. Sexual differences in color
strongly marked. Males, dark brown, varied with narrow
bands of white. Females, light brown, with the white bands
obsolete.

According to von Schrenck, on whose authority the above
characters are given, the molar teeth, except the first, are 2-rooted,
as in *Phoca* and *Erignathus*, but the crowns resemble those of
the corresponding teeth in *Halichœrus*, being simple and slightly
curved backward. The middle molars (third and fourth) and
sometimes the others, both above and below, show a minute
point or accessory cusp at the base of the principal one, both
in front of it and behind it, but this is a variable feature, not
only as respects the number of teeth thus furnished, but in
some specimens these minute accessory cusps may be wholly
lacking. As the characters of the skull have not been as yet
either figured or described, further comparison with other generic types becomes impracticable.

The genus *Histriophoca* was proposed for the present species
by Dr. Gill in 1873, but has not been fully characterized. Dr.
Gill's diagnosis is as follows: "The structural (and especially
dental) characters of this species, according to Von Schrenck,
indicate a generic distinction from all the familiar forms of the
subfamily *Phocinœ*. The molars, except the first, are two-rooted,
as in the typical *Phocinœ*, but in external form are simply conic,
or have rudimentary cusps, thus resembling *Halichœrus*. This
genus may be called *Histriophoca*." Taking into account the
peculiar pattern of coloration, and the conic, double-rooted

molars, we seem to have a type generically distinct from the ordinary *Phocœ,* and in accordance with this view the genus *Histriophoca* is here provisionally adopted for the *Phoca fasciata* of the early systematic writers.

HISTRIOPHOCA FASCIATA (*Zimm.*), *Gill.*

Ribbon Seal.

Rubbon Seal, PENNANT, Hist. Quad., 1st ed., 1781, ii, 523; 3d ed., ii, 1793, 276, fig. p. 265.

Phoca fasciata, ZIMMERMANN, Geogr. Gesch., iii, 1783, 277 (= "Rubbon Seal," Pennant).—KERR, An. King., 1792, 127 (the same).—SHAW, Gen. Zoöl., i, 1800, 257 (= "Rubbon Seal," Pennant).

Phoca (Otaria?) fasciata, RICHARDSON, Zoöl. Beechey's Voyage, 1830, 6 (= "Ribbon Seal of Pennant, Arct. Zoöl., ii, 165").

Phoca equestris, PALLAS, Zoog. Rosso-Asiat., i, 1831, 111.—VON SCHRENCK, Amur-Lande, i, 1859, 182, pl. ix, fig. 1–3 (animal).

Histriophoca [fasciata], GILL, Amer. Nat., vii, 1873, 179.

Histriophoca fasciata, SCAMMON, Marine Mam., 1874, 140, pl. xxii, fig. 1, 2 (animal, from von Schrenck).

Pagophilus? equestris, GRAY, Suppl. Cat. Seals and Whales, 1871, 2 (in part; includes *Phoca annellata,* Radde!).

Rubbon Seal, PENNANT, l. c.

Ribbon Seal of Alaska, GILL, l. c.

EXTERNAL CHARACTERS.—*Adult male.* General color, dark brown. A narrow yellowish-white band surrounds the neck extending forward to the middle of the head above; another broader yellowish-white band encircles the hinder portion of the body, from which a branch runs forward on each side to the shoulder, the two branches becoming confluent on the median line of the body below, but widely separated above. In other words, the (1) front part of the head, the (2) hind limbs, and the posterior fourth of the body, the (3) top of the neck and the whole anterior half of the back, as well as (4) the fore-limbs and a considerable area at their point of insertion, are dark brown; these four regions being separated by bands of yellowish-white, of variable breadth over different regions of the body. The brown of the anterior part of the dorsal region also extends laterally in the form of a narrow band around the lower part of the neck, where it expands to form a small shield-like spot on the breast. There are also very small spots of brown on the posterior part of the abdominal region.

Adult female.—Uniform pale grayish-yellow or grayish-brown, with the exception of an obscure narrow transverse whitish band across the lower portion of the back. The extremities and the back are darker, with a faint indication of the dark " saddle "-mark seen in the male.

Young.—The young of both sexes are said to resemble the adult female.

Von Schrenck's detailed description, on which the foregoing is mainly based, is substantially as follows : The dark-brown of the head, in the male, is followed by a broad dusky yellowish-gray neck-band, which on the middle line, both above and below, passes forward, but on the sides has the convexity pointing backward. Behind this light neck-band is a broad, long saddle-shaped patch upon the back, which, on the middle line, runs forward in a point, but which extends itself laterally in two narrow bands meeting and expanding on the breast into a pointed spot; posteriorly the dark dorsal patch is also prolonged backward and laterally, but without meeting below. Along the sides of this dorsal area runs a broad, curved, light, soiled yellowish-gray band, with the convexity upward ; these lateral light bands become deflected downward, both anteriorly and posteriorly, and form, by their union, a light band along the belly. Within these light bands anteriorly, on each side, is a large oval dark-brown spot, in which are inserted the anterior extremities. The light ventral area encloses posteriorly two small oval dark-brown spots, and in front of these a third narrower and larger. Behind the dark area on the back is a very broad dorsal cross-band of light yellowish-gray, joining the light bands on the side of the body. Behind this light cross-band the whole posterior part of the body, as well as on the tail and hind limbs, is blackish-brown. As a rule the above-described dark and light color areas are very sharply defined. Sometimes, however, there extends from the dark areas a smaller spot more or less isolated. According to the same writer the color varies considerably in different individuals, one of those he describes having the dark color of a dark grayish-black, and the light markings whitish or straw-yellow. He also states that in the figures given by Siemaschko the light neck-band is deflected backward from the back of the neck to the fore-limbs, leaving the whole breast of the same dark-brown color as the head. Besides this the dark-brown color of the back extends, both posteriorly and anteriorly, to the lower sides of the body, occupying the whole of the ventral surface, with the exception of two light bands which run crosswise around the base of the anterior extremities, and a separate light band that crosses the hinder part of the body. In consequence of the wide departure of the pattern of coloration in Siemaschko's figure from his own examples, von Schrenck is left in doubt as to whether the figure is really a true copy from nature.

The single specimen I have examined (Nat. Mus. No. 9311, Cape Romanzoff, W. H. Dall), a flat skin, lacking the flippers and the facial portion, agrees with von Schrenck's figure in respect to the form and size of the neck-band, but there is a far greater preponderance of light color, which occupies rather more than half the entire surface. Only the posterior sixth of the body is black, and the dark area of the back is very much more restricted, and differs somewhat in outline. In this specimen the breadth of the dark dorsal portion occupies scarcely more than one-third of the whole width of the skin, the light portion on either side nearly equalling it in breadth. It widens over the neck and sends down a lateral branch on each side, the two meeting on the breast. It is contracted over the shoulders, behind which it again expands, and at its posterior border sends down a very narrow branch from the right side to the middle of the belly; its fellow on the opposite side is nearly obsolete, forming merely a broken chain of small dusky spots. There is hence in this example a wide departure from the specimens described by von Schrenck, while the want of symmetry in the two posterior branches of the dorsal spot, and the relatively nearly equal amount of light and dark color, lead one to apprehend a much wider range of individual variation in coloration than von Schrenck apparently suspected, and that after all Siemaschko's figure merely represents a variation in the opposite direction from that here indicated, or an unusual extension of the dark color at the expense of the lighter markings.

SIZE.—Von Schrenck states that this animal is reported to sometimes attain the length of 6½ feet. He gives the length of a full-grown male as 5 feet 6¼ inches (1683 mm.), and that of a full-grown female as 5 feet 3 inches (1600 mm.), based on Wosnessenski's specimens obtained in Kamtschatka, which his hunters informed him were not of the largest size. In other words, it appears to be a Seal of the medium size, or about as large as *Phoca grœnlandica*.

GENERAL HISTORY.—The first account of the present species was published by Pennant, under the name "Rubbon Seal," in the first quarto edition of his "History of Quadrupeds," in 1781 (vol. ii, p. 523). His short description, based wholly on information and a drawing furnished by Dr. Pallas, is as follows: "Seal with very short fine glossy bristly hair, of an uniform color, almost black; marked along the sides, and towards the head and tail, with a stripe of a pale yellow color, exactly resembling a rubbon laid on it by art; words cannot sufficiently convey the idea, the form is therefore engraven on the title of Division III, *Pinnated Quadrupeds*, from a drawing communicated to me by Doctor *Pallas*, who received it from one of the remotest *Kuril* islands.

"Its size is unknown, for Doctor Pallas received only the middle part, which had been cut out of a large skin, so that no description can be given of head, feet, or tail; a shews the part supposed to be next to the head; b that to the tail."*

In Pennant's Arctic Zoölogy (vol. i, 1793, p. 193) there is a shorter but in some respects a more detailed and better account, which I also transcribe. "Rubbon Seal. With very short bristly hair, of an uniform glossy color, almost black: the whole back and sides comprehended within a narrow regular stripe of pale yellow.

"It is to Dr. *Pallas* I owe the knowledge of this species. He received only part of the skin, which seemed to have been the back and sides. The length was four feet, the breadth two feet three; so it must have belonged to a large species. It was taken off the *Kuril* islands."

The markings as represented in Pennant's figure correspond well with those of the animal figured by von Schrenck (presently to be noticed), except that the posterior transverse portion of the band is relatively narrower than in von Schrenck's specimen.

In 1783 Pennant's Rubbon Seal was named *Phoca fasciata* by Zimmermann.† Shaw, without referring to Zimmermann, and probably without knowing that he had named the species, bestowed upon it the same name seventeen years later,‡ and to him the name has been almost universally attributed. The accounts of both these authors were based entirely upon the description given by Pennant, as above quoted, and no further

* Here quoted from the third ed. of Hist. Quad., vol. ii, 1793, p. 277.
† Geograph. Geschict., iii, 277.
‡ Gen. Zoöl., vol. i, 1800, p. 257.

information respecting the species appeared till Pallas in 1831 *
redescribed the species from the original fragment mentioned
by Pennant, and renamed it *Phoca equestris*. In the meantime
the species had been uniformly relegated by authors to the list
of doubtful or inadequately described species. Pallas cites
Pennant's Rubbon Seal as a synonym of his *Phoca equestris*, and
also refers in his description of it to Pennant's figure, but to
neither Zimmermann nor Shaw. He says it is rare in the
Ochots Sea, but is reported to be of frequent occurrence around
the Kurile Islands. His description † adds little of importance
to the information given by Pennant, and apparently relates
to the same specimen.

According to von Schrenck, Hrn. Wosnessenski obtained,
during his residence in Kamtschatka, the first perfect specimens, embracing the old and young of both sexes, thereby es-
tablishing beyond doubt the validity of the species; but this
valuable material remained undescribed until the appearance
of von Schrenck's work on the Mammals of Amoor Land, in
1859.‡   Von Schrenck himself was so fortunate as to also obtain skins of this animal during his journey in Amoor Land,
and to him we are indebted for the first detailed description of
the species, accompanied by excellent colored figures of both
sexes.§   He, however, adopted Pallas's name *Phoca equestris* in
preference to the very appropriate name given half a century
before by Zimmermann, and somewhat later also by Shaw, for
wholly arbitrary reasons.||

---

* Zoogr. Rosso-Asiat., vol. i, 1831, p. 111.

† His description in full is as follows : "*Magnitudine* praecedentes aequasse
vel excessisse videbatur [hence about five feet four inches long from the
nose to the tail, or rather more], pellis enim portio c solo dorso exsecta
quatuor fere dodrantum latitudinem et sex ad septem dodrantum longitudinem habebat. *Color* totius brunneus, seu fuscus, cum brunnei tinctura,
uniformis. *Pili* breves, laevigati, rigidi ut in Ph. canina (=*Phoca vitulina*).
Insula lata alba, ut *amiciss*. *Pennant* delineavit, antice angulo versus
cervicem coëuns, per latera introrsum arcuata, postice transversa trabe
connexa, totum dorsi discum includit.—Optandum, ut haec singularis
species perfectius innotescat."

‡ Von Schrenck alludes to a very brief and unimportant reference to the
species by Siemaschko, in a work published in the Russian language in 1851.

§ Reisen und Forschungen in Amur-Lande, i, 1859, pp. 182–188, pl. ix.

|| He appears not to have known of Zimmermann's reference to the spe-
cies, but speaks of Shaw's name as "eine Bezeichnung, die jedoch gegen-
wärtig gegen den uhrsprünglichen, vom Entdecker selbst stammenden und
nur durch das verzögerte Erscheinen der Zoographia Rosso-Asiatica später
bekannt gewordenen Namen *Ph. equestris* zurücktreten muss."—L. c., p. 182.

As von Schrenck figured only the external characters, we have still to regret the absence of illustrations of the skull and dentition.

Gray, in 1866,* referred, without question, Shaw's *Phoca fasciata* to *Phoca fœtida*, and also, doubtfully, the *Phoca equestris* of both Pallas and von Schrenck—a rather strange proceeding, in view of von Schrenck's excellent description of the species and striking figure. In 1871,† however, he raised it to the rank of a species, under the name "*Pagophilus? equestris*", referring to it, however, Radde's *Phoca annellata*, an entirely different animal.

The next reference to the species I am able to find is Dr. Gill's account,‡ already cited, in which he mentions two skins in the collection of the Smithsonian Institution, collected by Mr. Dall at Cape Romanzoff, cites von Schrenck's account of its dental and other characters, and proposes for it the generic name *Histriophoca*.

Captain Scammon's account of the species,§ published in 1874, completes, so far as known to me, the written history of the species. Captain Scammon gives a figure of the animal, apparently copied from von Schrenck's.

GEOGRAPHICAL DISTRIBUTION.—According to Pallas, the present species occurs around the Kurile Islands and in the Ochots Sea. Von Schrenck states that Hr. Wosnessenski obtained specimens that were killed on the eastern coast of Kamtschatka, and that he himself saw skins of examples killed on the southern coast of the Ochots Sea, where, however, the species seems to be of rare occurrence. He further states that it occurs also in the Gulf of Tartary, between the island of Saghalien and the mainland, but apparently not to the southward of that island, the southern point of which (in latitude 46° N.) he believes to be the southern limit of its distribution. Mr. Dall secured specimens taken at Cape Romanzoff. Captain Scammon states, "It is found upon the coast of Alaska, bordering on Behring Sea, and the natives of Ounalaska recognize it as an occasional visitor to the Aleutian Islands. . . . . The Russian traders, who formerly visited Cape Romanzoff,

---

* Cat. Seals and Whales, p. 23.
† Suppl. Cat. Seals and Whales, p. 2.
‡ Amer. Nat., vol. vii, 1873, pp. 178, 179.
§ Marine Mam., 1874, p. 140, pl. xxi.

from St. Michael's, Norton Sound, frequently brought back the skins of the male *Histriophoca,* which were used for covering trunks and for other ornamental purposes." This writer also states that he "observed a herd of Seals upon the beaches at Point Reyes, California," in April, 1852, which, "without close examination, answered to the description given by Gill" of the present species. Probably, however, a "close examination" would have shown them to be different, as no examples are yet known from the Californian coast, and the locality is far beyond the probable limits of its habitat. Its known range may, therefore, be given as Behring's Sea southward—on the American coast to the Aleutian Islands, and on the Asiatic coast to the island of Saghalien.

HABITS.—Almost nothing appears to have been as yet recorded respecting the habits of the Ribbon Seal. Von Schrenck gives us no information of importance, and we search equally in vain for information elsewhere. All of the four specimens obtained by Wosnessenski were taken on the eastern coast of Kamtschatka, at the mouth of the Kamtschatka River, about the end of March. According to the reports of hunters, it very rarely appears at this locality so early in the season, being not often met with there before the early part of May. The natives use its skins, in common with those of other species, for covering their snow-shoes.

GENUS HALICHŒRUS, *Nilsson.*

Halichœrus, NILSSON, "Faun. Skand., i, 1820, 377." Type, *Halichœrus griseus,* Nilsson = *Phoca grypus,* Fabricius.

Pusa, GILL (ex "Scopoli, 1777"), Johnson's New Univ. Cycl., iii, 1877, 1226 (not *Pusa,* Scopoli, which was based on *Phoca fœtida*).

Dental formula as in *Phoca.* Molars conical, as broad as long, with very small accessory cusps when young, all single-rooted, except the last lower and two last upper ones. Facial portion of the skull greatly developed, forming nearly half the length of the skull, and very broad—broader at the base of the zygomatic process of the maxillary than the brain-case. Interorbital bridge thick, high; orbital fossæ large; brain-case very small, forming less than one-third of the length of the skull, instead of nearly one-half, as in *Phoca, Erignathus, Cystophora,* etc. Strongly developed sagittal and occipital crests in old age in the males.

Halichœrus forms (except possibly *Monachus*) the most strongly marked generic type among the *Phocinæ*, and in view of the striking peculiarities of the skull it is not surprising that Dr. J. E. Gray should have allotted it a subfamily (or "tribal") rank,[*] but why he should have associated it with the Walrusses seems hard to conceive. While the dental formula is the same as in the other genera of *Phocinæ*, the teeth depart widely in their simple conical cylindrical form from what is met with in the other genera, as well as in being mostly single-rooted. The proportions of the skull are almost the reverse of what is met with in the other genera. The preorbital portion forms nearly half the length of the skull, and has a proportionally remarkable breadth, the width of the skull at the base of the zygomatic process of the maxillary considerably exceeding the greatest width of the brain-case, instead of being only about half as wide, as is the case in *Phoca*. The brain-case is disproportionately small, being scarcely longer than in *Phoca grœnlandica* or *P. vitulina*, although the total length of the skull is one-third greater, while the breadth of the brain-case is actually less! The opening of the anterior nares is simply immense, in comparison with any other representative of the subfamily *Phocinæ*, being even larger than in *Cystophora*. The interorbital region, correlatively with the nasal passages, is also greatly thickened. In old age, at least in the males, the sagittal crest is greatly developed (15 mm. high in a specimen before me), as are also the occipital ridges. The postorbital region thus strikingly recalls the highly developed crests of the Otaries. In respect to the general skeleton I am unable to speak, my material being limited to two skulls and a few skins.

GENERAL HISTORY AND DISCUSSION OF THE "GENUS PUSA" OF SCOPOLI.—The genus *Halichœrus*, distinguished by Nilsson in 1820, has been until recently without a synonym. In 1877, however, Dr. Gill revived the name *Pusa*[†] of Scopoli,

---

[*] Suppl. Cat. Seals and Whales, 1871, p. 3.

[†] Concerning the etymology, signification, and early use of this singular word the following may be of interest. According to Houttuyn (Natuurlyke Historie, etc., Deel i, Stuk ii, 1761, p. 15) and Müller (Natursystem, Theil i, 1773, p. 199) *Pusa* is simply the Greenlandic word for Seal. The first use of the word by European authors seems to have been by Anderson (1746), and soon after by Cranz, who, however, spelled it *Pua*, and gave it as the Greenlandic equivalent of the Latin *Phoca* (Historie von Grœnl.,

1777,* supposing it to have been based on the *Phoca grypus* of Fabricius. Dr. Gill does not appear to have anywhere given reasons for this interpretation. In Johnson's "Cyclopedia," as above cited, he simply calls the Gray Seal "*Pusa* (*Halichœrus*) *grypus*", which is doubtless to be interpreted as *Pusa* (= *Halichœrus*) *grypus*. Dr. Coues, however, has had occasion to consider *Pusa* in relation to its use by Oken, in 1816, as a generic designation for the Sea Otter. In referring to this point Dr. Coues observes: "*Pusa* had, however, already been used by another writer in connection with a genus of Seals now commonly known as *Halichœrus*, but in such a peculiar way as to raise one of those technical questions of synonymy which authors interpret differently, in absence of fixed rule. Scopoli based his *Pusa* upon a figure of Salomon [*lege* Philipp Ludwig Statius] Müller's, recognizable with certainty as *Halichœrus*, and gave characters utterly irreconcilable with those of this animal. This is the whole case. Now it may be argued that there being no such animal whatever as Scopoli says his *Pusa* was, his name drops out of the system, and *Pusa* of Oken, virtually an entirely new term, is tenable for something else, namely, for the Sea Otter. On the other hand, Scopoli's quotations show ex-

---

1765, p. 161). The same form of the word is used by Schreber (Säugth., Theil iii, p. 285). Erxleben (Syst. Reg. Anim., 1777, p. 586) gives *Purse* and *Kassigiak* as the Greenlandic names of *Phoca vitulina*. Fabricius, in 1790 (Skrivter af Naturhistorie-Selskabet, Bd. i, Hefte 1, 1790, p. 90 and foot-note 30), gives *Puirse* in his text as one of the Greenlandic names of the Harp Seal, and in a footnote gives a further account of the word. He says: *Pua*, as written by Cranz, and after him by Schreber, is erroneous, this word meaning a lung. But *Pûse*, or *Pûese*, as Professor Glahn (Anmærkninger til Cranzes Hist., p. 150) corrected it, is not wholly right. Likewise incorrect is Anderson's *Pusa* in his "Efterr. om Strat-Davis, § LV". It is from here that Scopoli learned the name *Pusa* as he has used it for his supposed new genus of animals, which, however, is nothing more than a species of Seal (see Beschäft. Berl. Ges. Naturs., IV, B., p. 464).

It thus appears that the name *Pusa*, with its various orthographic forms, was originally simply a generic term for Seals in general, the Greenlandic equivalent of the Latin *Phoca*, the English *Seal*, etc. In view of this is it improbable that the pigeon-term "*Pussy*," said to be commonly employed by the northern sailors and sealers of various nationalities for young Seals in the white coat, may not be a corruption of the Greenlandic *Pusa*?

* See Johnson's New Universal Cyclopedia, vol. iii, 1877, p. 1226. He also employed the name in the same sense in 1876 in his anonymous "List of the Principal Useful or Injurious Mammals" of North America. For an account of the last-named publication see *anteà*, p. 22.

actly what he meant, in spite of his inept diagnosis; his name
Pusa therefore holds, and cannot be subsequently used by
Oken in a different connection."*

An examination of the case, however, shows that Müller's
plate is *not* "recognizable with certainty" as that of any particular species of Seal. Scopoli's diagnosis is simply an absurdity, as the subjoined transcript† sufficiently shows, his
reference to Müller's description and plate affording the only
real basis for his genus *Pusa*. As already stated, the figure
cannot be positively referred to any particular species of Seal.
The description given by Müller ‡ records few characters that
are not applicable to any species of Earless Seal. Those which
are not thus applicable appear to relate to *Phoca grypus*, Fabricius, and I so at first interpreted the description, but later I
found it necessary to go further back in the history of the subject. The plates of Müller's work, so far as the mammals are
concerned, prove on collation to be very close copies of those
given by Houttuyn (with the exception of three that appear to
be here for the first time published) twelve years earlier, if, indeed, some of them were not actually printed from the same
etchings. Müller says (l. c., p. 201), " Der Professor Albinus in
Leiden zergliederte den 24. Februar, 1748. in Gegenwart des
Herrn Houttuyns einen Seehund, welcher Tab. XI. fig. 6. ab-

---

* Fur-bearing Animals, 1877, p. 337.

† "PVSA. Scop. Pedes antici unguiculati, postici connati in pinnam sexlobam, ad quorum originem superne exit pinna lanceolata, horizontalis.
Dentes incisores quatuor, canini supra sex, infra quatuor. Auriculæ nul-
læ. Pili breves.
"Descriptionem & iconem dedit Cl. MÜLLERVS S. N. *Tom.* I. *Tab.* XI. *fig.*
6."—*Introdvctio ad Historiam Natvralem sistens Genera Lapidvm, Plantarvm, et
Animalivm hactenvs detecta,*" etc., 1777, p. 490, genus 433. For this transcript I am indebted to Dr. Edward J. Nolan, secretary of the Academy of
Natural Sciences of Philadelphia, the library of this Academy containing
the only copy of the work known to me as existing in this country. (Since
the above was written a copy of Scopoli's above-named work has been received at the library of Harvard College.)

‡ Des Ritters Carl von Linné Königlich Schwedischen Leibarztes, &c.,
&c., vollständiges Natursystem nach der zwölften lateinischen Ausgabe
und nach Anleitung des holländischen Houttuynischen Werks mit einer
ausführlichen Erklärung ausgefertiget von Philipp Ludwig Statius Müller
Prof. der Naturgeschichte zu Erlang und Mitglied der Röm. Kais. Akademie
der Naturforscher, &c. Erster Theil. Von den säugenden Thieren. Mit 32
Kupfern. Nürnburg, bey Gabriel Nicolaus Raspe, 1773. Eine andere Art
eines Seehundes, pp. 201, 202, pl. xi, fig. 6.

gebildert ist." . . . Referring now to Houttuyn,* we find this statement: "*Fig.* 6, [pl. xi] is die van een Zee-Hond, welken de Hooggeleerde Heer ALBINUS, in 't Jaar 1748, den 24 February, te Leiden, op de Vertoonplaats der Ontleedkunde, in myn by- zyn heeft laaten openen" (l. c., p. 16). Later (l. c., pp. 28, 29) he gives a description of the specimen here referred to as dis- sected in his presence by Professor Albinus, where he says, "De Heer ALBINUS heeft in den Zee-Hond, hier voor in *Fig.* 6 ["pl. xi, fig. 6," in the margin] afgebeeld, onder anderen, het volgende opgemeckt," citing at this point, in a footnote, "*An- not. Acad.* Libr. III, Cap. XV." Before turning to Albinus's account it may be well to state that Müller's and Houttuyn's plates here cited are identical, even to the notation, and that Müller's description is merely a slightly abridged translation of Houttuyn's account.†

On referring to Albinus, we find not only a very full and lucid account of the external and some other characters of the specimen Houttuyn saw him dissect, but also the original of both Houttuyn's and Müller's figures! Albinus's figure differs from the others only in being much more finely executed. But besides the figure copied by Houttuyn, Albinus gives several detail figures, which demonstrate that the specimen could not have been *Halichœrus grypus*. Albinus's description shows him to have been not only one of the most accomplished anatomists

---

* Natuurlyke Historie of uitvoerige Beschryving der Dieren, Planten, en Mineraalen, Volgens het Samenstel van den Heer Linnæus. Met naauwkeu- rige Afbeeldingen. Eerste Deels, Tweede Stuk. Vervolg der Zoogende Die- ren. Te Amsterdam. By F. Houttuyn, M D CC LXI.

† Since writing the above I have met with a reference to Scopoli's *Pusa* by Hermann, in his elaborate account of the Monk Seal of the Mediter- ranean, in which he criticises severely Scopoli's absurd diagnosis, and sug- gests explanations of some of Scopoli's erroneous characters. As Hermann (Beschäftigungen der Berlinischen Gesellschaft Naturforschender Freunde, 4 Band, 1779, p. 464, footnote) intimates, his "Pedes . . . . postici connati in pinnam sexlobam" is based on a very stupid misunderstanding of Müller's figure, in which only the upper edge of the left hind flipper is seen above the right one. Although the shading renders the figure per- fectly intelligible, Scopoli evidently counted this upper edge of the left hind flipper as the sixth lobe of a single appendage, the whole forming his six-lobed "pinna". If we may suppose the transposition of two words ("incisores" and "canini") by typographical error in Scopoli's dental for- mula, the rendering would be correct, namely, Dentes *canini* quatuor, *inci- sores* supra sex, infra quatuor. But this we fear is lenient judgment, al- though it would seem that Scopoli must have known better than to delib- erately ascribe *ten canines* to any mammal.

of his time, but a well-trained observer; his description alone
would show beyond doubt that the species was not *Halichœrus*.
Houttuyn's account seems to have been in part based on that
of Albinus, but includes statements that lead one to suppose it
may have been in part based on his own original notes. To
show that the species was not *Halichœrus grypus*, but beyond
doubt *Phoca fœtida*, I quote portions of the description given
by Albinus:* "Delecta ad me est phoca, capta in vicino mari,
[North Sea] longa pedes sex & dimidium, ab ore ad extremos
pedes postiores usque: ex quo reliquae dimensiones in figura
subjecta (*a*) invenientur; in quo forma bestiae cum cura expressa.
Venter autem plenior, eo quod gravida erat, embryonem con-
tinens longitudinis pedalis. . . . Labium superius ab utraque
parte nasi in magnum globum modice protuberans; in quibus
pilorum species, similitudine felis, quadrupedumque aliarum
multarum. . . . Latiores quam crassiores, ab utraque parte
plani, marginibus rotundis. Per marginum longitudinem veluti
serrati, eminentiis ovatis per sinus lunatos distinctis. Respond-
ent margines sibi invicem, sinus sinibus eminentiae eminentiis:
itaque tanquam per intervalla constricti, quadam nodorum inter-
mediis locis specie. A principio margines recti. . . . Dentes
in maxilla superiore sex continui in parte priore, quorum medii
quatuor minores. Inde, sed modico interjecto intervallo, quod
canini inferiores subeant, canini, adunci, maximi omnium.
Post hos in lateribus maxillares ab utraque parte quinque,
parvi, veluti tricuspides, mucrone medio majore. Infra maxillares totidem, & canini, sed primores tantummodo quatuor,
ostendentibus praesepiolis, è quibus excussi fuerant. . . .
In ventre post umbilicum mammarum notae geminae, foramina
referentes, extremi digiti auricularis capacia. . . . Pili,
quibus corium tectum, breves, tenues, laeves, à capite directi
ad caudam, pedesque extremos. . . . Color iis ad fulvum
vergens, maculis fuscis toto corpore crebris. In ventre & pec-
tore color pallidior. Cauda & pedes postici toti fusci, sine
maculis, praeterquam ad digitorum exortum, ubi in exteriore
parte maculae fulvae, parvae, paucaeque. Fusci quoque pedes
antici, sed tamen extrinsecus aliquantum maculosiores. Omnes
ex interiore parte sine maculis fusci, pilis mollioribus.

---

* B. S. Albini Academicarum Annotationum Liber Tertius. Continet ana-
tomica, physiologica, pathologica, zoographica. Leidae. Apud J. & H.
Verbeek, Bibliopolas. cɪɔɪɔcclvi [1756]. Caput XV. *De phoca.* Liber
iii, pp. 64–71.

"Embryo masculus, pilosus quidem, sed tam subtiliter, ut facile existimaretur depilis. Color pallidior, nec nisi in dorso maculosus. Jam spectabilis mystax, & superciliis respondentes sylvulae. Digiti pedum distincti. Ungues visendi . . . " (l. c., pp. 64–71).

The plate (pl. vi, Libr. iii) accompanying Albinus's memoir gives (fig. 1) a side view of the animal; a half front view (fig. 2) of the head, with the mouth wide-open, displaying the dentition; a view (fig. 3) of the posterior end of the body from below, showing the genital opening, the tail, and hind flippers; a diagram (fig. 4) of the genito-anal orifice; a claw (fig. 5) of one of the anterior digits, and (fig. 6) one of the mystacial bristles of natural size. Even in the large figure the tricuspid character of the molar teeth is seen, while in the enlarged view of the head this is still more distinctly shown. In this the five molars, both above and below, of the right side, are represented as small, distinctly three-pointed, the middle point the longest, while the teeth are separated by slight intervals, the dentition thus in every respect agreeing unmistakably with that of *Phoca fœtida*. The large figure shows also the first claw of the fore limb to be the largest, another distinctive character of *Phoca fœtida*. It consequently follows that if *Pusa* is tenable in a generic sense it must be held for *Phoca fœtida*, in place of *Pagomys* of much later date, by those who would, generically, separate *Phoca fœtida* from the other Seals. The condition of the fœtus also points to *Phoca fœtida*, which has its young early in March.

Houttuyn's description, and consequently Müller's, to which Scopoli refers, is merely a loose abridged version of that given by Albinus, in which they omit to state that the length given includes the outstretched hind flippers. They also describe the molar teeth simply as being pretty sharp ("de Kiezen zelfs eenigermaate scherp," *Houttuyn;* "die Backenzähne ziemlich scharf," *Müller*), and speak of the fœtus as being nearly naked ("en was nog byna kaal," *Houttuyn;* "fast kahl," *Müller*), but in no other point is there any noteworthy discrepancy. Albinus's account of the fœtus shows it to have been nearly mature, and the date of the dissection being given by Houttuyn as the 24th of February, is, as already noted, further proof that the species was not *Halichœrus grypus*.

It is barely possible that the specimen figured and described may have been *Phoca grœnlandica ;* the large size alone favors this view, but if the animal were measured along the curvature of the body instead of in a straight line between the two extremities, the dimensions given would not be too large for a fullgrown female *P. fœtida.* The figures show it could not have been *Phoca vitulina,* while *Halichœrus grypus* is entirely out of the question; for the phrase "parvi, veluti tricuspides, mucrone medio majore" cannot be applied to the large, conical, single-pointed molars of *Halichœrus,* even if we had not the figures to show the small size and tricuspid character of the teeth.

HALICHŒRUS GRYPUS (*Fabricius*) *Nilsson.*

Gray Seal.

? *Grå Siäl,* LINNÉ, Fauna Suecica, 1746, 4.—CNEIFF, "Svenska Vid. Acad. Handl., xix, 1757, 171."

Ut-Selur, Wetrar-Selur, OLAFSEN, Reise durch Island, i, 1774, 260, 281.

Der Graue Seekalb, KNEIFF, Abhandl. Köngl. Schwed. Akad. Wissen., xix, 1759, 171.

Phoca grypus, FABRICIUS, Skriv. af Naturh.-Selsk., i, 2, 1791, 167, pl. xiii, fig. 4 (skull).—HALLGRIMSSON, Krøyer's Naturh.-Tidsskrift, ii, 1838-'39, 91 (Iceland).

Halichœrus grypus, NILSSON, "Kongl. Vet. Akad. Handl. Stockholm, 1837, ——"; Arch. für Naturg., 1841, 318; "Illum. Fig. till Skand. Faun., ii, 1837, pl. xxxiv; text, 1840, 'i, 20'"; Skand. Fauna, Daggdjuren, 1847, 299.—WAGNER, Schreber's Säugth., vii, 1846, 12.—SCHINZ, Synop. Mam., i, 1844, 483.—GRAY, Cat. Seals Brit. Mus., 1850, 30, fig. 10; Cat. Seals and Whales, 1866, 34, fig. —; Zoölogist, 1872, 3333, 3335; Hand-List of Seals, 1874, 9, fig. 4, pl. vii (skull, juv.).— HORNSCHUCH and SCHILLING, Arch. fur Naturg., 1851, 21.—GIEBEL, Säugeth., 1855, 133.—BLASIUS, Naturg. Säugeth. Deutschl., 1857, 256, figg. 142-144.—NORDMANN, Vid. Medd. f. d. natur. Foren., 1860 (1861), 307.—MALMGREN, Öfv. af Kongl. Vet. Akad. Stockh., 1863, 135; Arch. für Naturg., 1864, 74.—"HOLMGREN, Skand. Daggdjuren, 1865, 220, fig."—GILL, Proc. Essex Inst., v, 1866, 12.—BROWN, Proc. Zoöl. Soc. Lond., 1868, 340, 426; Man. Nat. Hist., Geol., etc., Greenland, Mam., 1875, 54.—TURNER, Journ. Anat. and Phys., iv, 1870, 270 (coast of Scotland); ibid., vii, 1873, 273 (abnormal dentition); Trans. Roy. Soc. Edinb., xxvii, 275.—LILLJEBORG, Fauna öfver Sveriges och Norges Ryggradsjur, 1874, 709.—VAN BENEDEN, Ann. du Mus. d'Hist. Nat. du Belgique, i, 1877, 18 (geogr. distr.).

Pusa (Halichœrus) grypus, GILL, Johnson's New Univ. Cycl, iii, 1877, 1226.

Phoca gryphus, LICHTENSTEIN, Abhand. d. Berlin Akad., 1822-23 (1825), Phys. Kl., 1.—FISCHER, Syn. Mam., 1829, 239.—MACGILLIVRAY, Brit. Quad., 1838, 214.—BLAINVILLE, Ostéogr., Phoca, 1840-1851, pl. ix (dentition).

Misc. Pub. No. 12——44

Halichœrus gryphus, BALL, Trans. Roy. Irish Acad., xviii, 1838, 89, pl. i
(animal, female and two young), pl. ii (skull, female), pl. iii (lower
jaw and teeth).—BELL, Brit. Quad., 1837, 278 (figg. skull and animal); ibid., 1874, 262.—REINHARDT, Krøyer's Naturhist.-Tidsskr.,
iv, 1843, 313; Isis, 1845, 702.—REEKS, Zoölogist, 1871, 2549 (Newfoundland).—GRAY, Ann. and Mag. Nat. Hist., 4th ser., ix, 1872,
322 (occurrence in Brit. Isl.); ibid., xiv, 1874, 96 (Cornwall).—
MALM, Goteborgs och Bohusläns Fauna Ryggradsjuren, 1877, 145.

Halichœrus griseus, Nilsson, Skand. Faun., i, 1820, 377.—HORNSCHUCH, Isis,
1824, 810.—LESSON, Man. de Mam., 1827, 205.—HAMILTON, Amphib.
Carniv., 1839, 174, pl. x.—SELBY, Ann. Nat. Hist., vi, 1841, 462
(Farn Islands).

Phoca halichœrus, THIENEMANN, "Reise in Norden Europa's, i, 1824, 142."

Phoca scopulicola, THIENEMANN, "Reise in Norden Europa's, i, 1824, 59, pl.
v (adult male)." (Iceland.)—FISCHER, Syn. Mam., 1829, 237 (from
Thienemann and Lesson).

Callocephalus scopulicolus, Lesson, Man. de Mam., 1820, 199 (=*Phoca scopulicola,* Thienemann).

Phoca thienemanni, Lesson, Dict. class. d'Hist. Nat., xiii, 1828, 414 (=*Phoca
scopulicola,* Thienemann).

Halichœrus macrorhynchus, HORNSCHUCH and SCHILLING, Arch. für Naturgesch., 1851, 22.

Halichœrus pachyrhynchus, HORNSCHUCH and SCHILLING, Arch. für Naturgesch., 1851, 22.

Phoca barbata, EDMONSTONE, View of the Zetland Islands, ii, 1809, 294.—
SELBY, Zoölog. Journ., ii, 1826, 465.—FLEMING, Hist. Brit. Anim.,
1828, 18.—JENYNS, Brit. Vert., 1835, 16.—MACGILLIVRAY, Brit.
Quad., 1838, 212 (in part).—HAMILTON, Amphib. Carnivora, 1839,
145, pl. "3", iv (in part only—the biographical matter relating to
British localities).—BELL, Hist. Brit. Quad., 1837, 274 (the British
references only).

Seal from the South Seas, HOME, Phil. Trans., 1822, pl. xxvii (skull).

Gray Seal, English; *Graue Seehund,* German; *Grå Siäl, Grå Skäl,* Swedish;
Krumsnude de Sœl, Danish; *Tapvaist,* Hebridian; *Haaf-fish,* Orca-
dian; *Ut-Selur, Wetrar-Selur,* Icelandic.

EXTERNAL CHARACTERS.—Color of the adults silver-gray,
ash-gray, or dusky-gray, with obscurely defined spots of dusky
or blackish, the general color varying in different individuals
from nearly uniform silvery, or yellowish-white, to dusky or even
black, the lighter examples with or without blackish spots.
The young are at first white or yellowish-white, but soon become dingy-yellow, blotched irregularly with blackish-gray,
and later acquire still darker tints. The pelage in the young
is soft and woolly, in the adults short and rigid and rather
sparse, the hairs flattened, adpressed, often recurved at the
tips. The mystacial bristles are abundant, large, stiff, flattened, and waved or crenulated on both margins. Fore feet

with the first and second toes longest and subequal; hind feet deeply emarginate, the outer toes forming long lappets; nails of all the digits well developed, but those of the fore feet much larger than those of the hind feet. Length of the adult male about 8 feet, rarely 9 feet; of the adult female, about 6½ to 7 feet. Females smaller and lighter colored than the males.

FIG. 51.—Halichœrus grypus.

A specimen in the National Museum, probably an adult male, from Sable Island, is silvery-gray, with spots of black and white, the latter confined mainly to the sides of the body and neck. A female from Sealand is whitish-gray, with large obscure spots of darker and touches of dusky. Mr. Ball says that the color varies greatly in different individuals, and that of the many specimens he had seen he did not remember "that any two were precisely similar". He describes an adult female as appearing of a uniform silvery-gray when seen from the front, but when viewed from the rear seemed of a sooty-brown color, while the spots or blotches were only distinctly visible from a side view. He says, "The very young females are generally of a dull yellowish white, with rather long hair, which falls off in about six weeks after birth, and gives place to a shorter and more shining coat of a warm, dingy yellow, variously blotched with blackish gray; the whole becoming gradually more dull, the blotching more indistinct, and a general dark shade spreading on the back as the animals advance in age." He describes a young male which "has long yellowish hair slightly tinged with brownish black on the back; is black on the nose, chin, and cheeks, and on the palms of the forefeet."* He gives the length of a skeleton of "a very aged female" as "seven feet two inches".† Selby, who observed the species at the Farn Islands, gives the length of the full-grown male as eight feet and the color as dark gray, or nearly black, and says the female is smaller and greenish-white, sparsely spotted with darker; the young as yellowish-white, changing to gray at the first moult.‡ Hallgrimsson makes the same observations in relation to the Utselur of Iceland, which he identifies with the *Phoca grypus* of Fabricius, stating that the males are not only larger than the females, but are blackish-gray, or sometimes wholly black, while the females are lighter colored; and adds that the new-born young are covered with a white woolly coat.§ Nilsson describes a young female, about four feet long, taken in August, as silver-gray marbled or irregularly spotted above, on the sides and on the limbs with black, most numerously on the sides and limbs; below white, with scattered spots of black. Another young female, about three and a half feet long, killed in July, as pale ash-gray above, varied with blackish or dusky spots; the sides, limbs, and under parts white. Another young female, about

---

*Trans. Roy. Irish Acad., vol. xviii, 1837, p. 90.
†Bell's Hist. Brit. Quad., 1837, p. 283.
‡Ann. Nat. Hist., vol. vi, 1841, p. —.
§Isis, 1841, p. 291.

four and a half feet long, killed July 20, was dark-gray above, along the back still darker or blackish-gray, and paler on the sides; back and sides with irregular spots of black of various sizes; nose and limbs brownish-gray, unspotted.*

Edmonstone† gives the weight of this seal as "45 stone of 14 pounds each" (= 630 pounds), and its length as "10 to 12 feet."

SKULL.—The two skulls before me indicate great variations resulting from age, especially in the thickening of the bones and the development of heavy sagittal and occipital crests. The old skull (No. 4717, Nat. Mus., Sable Isl., N. S.), is presumably that of a very old male, and differs from any which I have seen figured in its large size and greatly produced crests. The teeth in the young, especially in the lower jaw,‡ have slight but distinct accessory cusps, which become wholly obliterated later in life. In the old skull already mentioned the crowns are much worn, and the roots are very thick and strong. The strongly marked distinctive features of the skull have already been noticed (anteà, p. 683). In all probability the sexual differences are strongly marked, especially in weaker structure and slighter crests in the female.§ To judge by Ball's figure (l. c., pl. ii) of the skull said to be that of a very aged female, they may be wholly lacking. I subjoin the following measurements of the two above-mentioned skulls:

---

*Apud Wagner, Schreber's Säugthiere, Band vii, 1846, pp. 15, 16.

†A view of the ancient and present state of the Zetland Islands, etc., vol. ii, 1809, p. 294.

‡See Ball, Trans. Roy. Irish Acad., xviii, pl. iii.

§ I regret especially in this connection my inability to consult Hornschuch and Schilling's "Kurze Notizen über die in der Ostsee Vorkommenden Arten der Gattung Halichœrus, Nilsson." Greifswald, 1850.

Measurements of two skulls of Halichœrus grypus.

	Young	Very old
Age.		
Length of lower jaw.	137	217
Greatest width of brain-case.	92	100
Length of brain-case.	80	98
Greatest height of skull at auditory bullæ.	80	97
Breadth of anterior nares, transversely.	...	...
Breadth of anterior nares, vertically.	...	...
Breadth of posterior nares, transversely.	29	36
Breadth of posterior nares, vertically.	21	38
Least breadth of skull interorbitally.	19	35
Breadth of skull at canines.	40	90
Nasal bones, breadth at fronto-maxillary suture.	17	...
Nasal bones, breadth anteriorly.	27	...
Nasal bones, length.	47	...
Width of palatal region at posterior end of maxillæ.	54	78
Length of alveolar border of maxillæ.	78	152
Distance from palato-maxillary suture to end of pterygoid hamuli.	51	78
Distance from anterior edge of intermaxillæ to glenoid process.	140	237
Distance from anterior edge of intermaxillæ to meatus auditorius.	149	250
Distance from anterior edge of intermaxillæ to hinder edge of last molar.	73	...
Distance from anterior edge of intermaxillæ to end of pterygoid hamuli.	120	198
Greatest breadth at zygomatic arches.	120	198
Breadth at mastoid processes.	124	158
Length.	208	320
Sex.	...	...
Locality.	Greenland	Sable Island, N. S.
Catalogue number.	6573	4717

GEOGRAPHICAL DISTRIBUTION.—The Gray Seal appears to
be not only one of the least abundant of the northern Phocids,
but also to be restricted to a rather narrow range. It is wholly
confined to the North Atlantic, and even here is found only
within comparatively narrow limits. On the American coast
it occurs as far southward as Sable Island, Nova Scotia, where
its presence is attested by specimens in the National Museum,
collected there by Mr. P. S. Dodd. This, however, is the southernmost point at which it is known to occur. Mr. Reeks says,
"It is comparatively rare in the Straits of Labrador and Belle
Isle, although very few seasons pass without a few being captured either on the ice or in the 'seal frames.'"* Beyond this
point to the northward it has been recorded by Mr. Brown as
probably occurring on the coast of Greenland. He says, "In
1861, a little south of Disco Island, we killed a Seal the skull
of which proved it to be of this species; and again this summer [1867] I saw a number of skins in Egedesminde and other
settlements about Disco Bay, which appeared to be of this species. Though the natives do not seem to have any name for it,
the Danish traders with whom I talked were of opinion that
the *Graskäl*, with which they were acquainted as an inhabitant
of the Cattegat, occasionally visited south and the more southerly northern portions of Greenland with the herds of *Atak*
(*P. grœnlandicus*). The skull to which I refer, though carefully
examined at the time, was afterwards accidentally destroyed
by a young Polar Bear which formed one of our ship's company on that northern voyage; therefore, though perfectly convinced of its being entitled to be classed as a member of the
Greenland fauna, I am not in a position to assert this with
more confidence than as being a very strong probability. It
should be carefully looked for among the herds of *P. grœnlandicus* when they arrive on the coast."† It is not, however,
given by Dr. Rink as an inhabitant of Greenland, nor was it
obtained by Mr. Kumlien during his recent sojourn in Cumberland Sound. I find, in short, no evidence of its occurrence
on the North American or Greenland coasts other than that
already given. Its occurrence in Iceland, however, is abundantly substantiated, and it is also rather common along the
shores of Northern Europe. Nilsson states that it has been

---

* Zoölogist, 2d ser., vol. vi, 1871, p. 2549.

† Proc. Zoöl. Soc. Lond., 1868, p. 427; Man. Nat. Hist. Greenland, etc.,
1875, Mammals, p. 55.

long known to inhabit all the seas that border Scandinavia, in the East Sea as well as in the Sound, in the Cattegat, and in the North Sea; the same statement being also made by Blasius and other later authorities. Collett gives it as found sparingly along the whole coast of Norway, from latitude 58° to 70°. It is not mentioned by von Heuglin as an inhabitant of Spitzbergen, Jan Mayen, and Nova Zembla, while Malmgren distinctly states that it does not reach Spitzbergen. He says there is some reason to believe it occurs in small numbers on the coast of Finmark, where it was observed by Lilljeborg (at Tromsö) in 1848.

Mr. Ball and others are authority for its common occurrence on the southern coast of Ireland, and it has for a long time been known as an inhabitant of the Orkneys, the northern coast of Scotland, the Hebrides, and the Farn Islands. Gray states that it has been found in various parts of the Irish Sea and St. George's Channel; that he has heard of it in the Isle of Man, and believes that it occurs as far south as Land's End and the Scilly Isles.* He also states that there is little doubt of its presence on the north coast of Cornwall, and that he had been informed that many Seals of very large size haunt the caverns on the coast of Plymouth.† Bell‡ refers to its capture in the Isle of Wight, and says living specimens have been received by the Zoölogical Society from the coast of Wales.

To summarize the foregoing, it may be stated that the Gray Seal ranges from Nova Scotia and the British Islands northward to Greenland and Finmark, but is absent from the islands of the Arctic Ocean.

GENERAL HISTORY AND NOMENCLATURE.—The earliest notice of the Gray Seal that requires attention in the present connection, if not the earliest that can be with certainty identified, was given by Cneiff in his account of the Seal fishery of the Gulf of Bothnia, published about the middle of the last century.§ On this account is based "Der graue Seehund" of Schreber.

*Ann. and Mag. Nat. Hist., 4th ser., vol. ix, 1872, p. 322.
†Ibid., vol. xiv, 1874, p. 96.
‡ Hist. Brit. Quad., 1874, p. 265.
§ Bericht vom Seekälberfange in Ostbothnien. Vom Provincialschaffner, Herrn Johann David Kneiff eingegeben. < Der Königl. Schwedischen Akademie der Wissenchaften, Abhandlungen, etc., auf das Jahr 1757. Aus dem Schwedischen übersetzt, von Abraham Gotthelf Käftner, Band xix, 1759, pp. 171–186.

The Grå Siäl of Linné's "Fauna Suecica" also appears to relate to the present species. Olafsen, in his "Reise durch Island," published in 1772, repeatedly refers to it under the names "Ut-Selur" and "Wetrar-Selur", giving quite a full account of its habits. Although he omits all mention of its size and external characters, his description of its habits, particularly of its resorting to low islands and rocky shores in November, to bring forth its young, seems to identify his Ut-Selur with the present species. Furthermore, the Ut-Selur of the Icelanders has been determined by Hallgrimsson to be the *Phoca grypus* of Fabricius, which the latter referred doubtfully to his *Phoca hispida*.

Although Schreber's "Der grosse Seehund" is referred by nearly all writers to the Bearded Seal, it is compounded of two species, the diagnosis being based on the "Utsuk" of Cranz, and hence on the Bearded Seal, while his account of its habits is derived from Olafsen, and relates entirely to the Ut-Selur of that author, and consequently to *Halichœrus grypus*.

The Gray Seal received its first systematic name at the hands of Fabricius in 1791, who briefly referred to it under the name *Phoca grypus*,* and gave a good figure of its skull.† Fabricius's name appears to have for a long time escaped the notice of subsequent writers, it being conspicuously absent from the works of compilers down to about 1835. In the meantime the species was again brought to light by Nilsson, who, in 1820, renamed it *Halichœrus griseus*. Lesson, in 1827, also mentioned it under this name, as did Fischer in 1829. Lichtenstein, in 1822, seems to be the first to recall the Fabrician name, modified, however, to *gryphus*. Thienemann described the species in 1824 as an inhabitant of Iceland, under the name *Phoca halichœrus*, and at the same time added a nominal species based on the young, which he designated as *Phoca scopulicola*. In 1850 Hornschuch and Schilling, after an examination of a series of fifty skulls collected in the East Sea, arrived at the conclusion

---

* By many writers, as shown in the above table of synonymy, this name is rendered *gryphus*, but, as first pointed out by Nilsson in 1827, and since restated by several German authors, the correct orthography is *grypus*, or crooked-nosed, the Danish name, as given by Fabricius, being "Krumsnudede Sæl. (See also on this point Lilljeborg, Fauna Sveriges och Norges, i, 1874, p. 711, footnote.)

† The original skull (lacking the lower jaw) figured by Fabricius is doubtless still extant in the Museum of Lund, for as late as 1841 Hallgrimsson stated that through the kindness of Professor Reinhardt he had had the opportunity of comparing with it a skull of the Ut-Selur he obtained in Iceland.

that the genus *Halichœrus* was there represented by three species, they describing as new *H. macrorhynchus* and *H. pachyrhynchus*, but these have not been accepted as valid by subsequent writers.*

As will be noticed later more in detail, the Gray and the Bearded Seals have often been confounded, especially by British authors; consequently all the references to the Bearded Seal as a British species really relate to *Halichœrus grypus.*†

---

*I know of Hornschuch and Schilling's brochure only through citations by authors. I quote the following abstract of the paper from Prof. Andreas Wagner's "Bericht über die Leistungen in der Naturgeschichte der Säugethiere während des Jahres 1850": "Ihre 3 Arten sind folgende: 1) *H. macrorhynchus*, die langschnauzige Meerrobbe (H. grypus s. griseus Nilss.); Rücken aschgrau, ins Grünliche schiessend, schwach silberartig schillernd und mit wenigen kleinen graubraunen Flecken bestreut. Schädel in allen seinen Theilen s e h r gestreckt und sein oberer Umriss bildet eine in der Mitte stark gesenkte Linie; die Eckzähne stark. Obwohl Zeichnung und Farbe bedeutend variiren, so behalten sie doch immer einen eigenthümlichen Typus, welches auch bei den folgenden Arten der Fall ist. —2) *H. Grypus* Fabr., die krummnasige Meerrobbe; Rükken weisgrau, stark ins Grünlichblaue ziehend mit starkem Silberschiller und vielen grösseren und kleineren, unregelmässigen, mehr oder minder ineinander verfliessenden schwarzen Flecken. Schädel kurz und ziemlich hoch, sein oberer Umriss bildet eine bogenförmige Linie; die Eckzähne sind schwächer als bei der folgenden Art.—3) *H. pachyrhynchus*, die dickschnauzige Meerrobbe; Rükken silb.rweiss, ins Grünlichblaugraue schiessend, glänzend, mit kleinen und mässig grossen, länglichen, schwarzbraunen, unregelmässigen Flecken. Schädel ziemlich kurz, viel weniger gestreckt als bei H. macrorhynchus, sein oberer Umriss bildet eine beinahe gerade, bis zum Anfange der Nasenbeine sich etwas erhebende, dann sich stark senkende Linie; die Zähne stärker als bei H. macrorhynchus."— Troschel's *Arch. für Naturg.*, 1851, ii, p. 29.

†Mr. Selby admits that much of this confusion, at least so far as regards the "Great Seal of the Farn Islands," is due to his erroneously referring it, in 1826, to the *Phoca barbata*, but to Mr. Selby is also due the credit of later (in 1841) making known the affinities of the "Great Seal" of the Farn Islands. After alluding to the fact that the large Seal of the Northumberland coast was referred to *Phoca barbata* by both Jenyns and Bell on his authority, and stating the reasons that led to his erroneous determination, he says: " . . . having requested the person who at present rents these [Farn] islands to send me the heads of any Seals he might be fortunate enough to kill, at the usual time of his visiting the island to which they retire to calve, (which they do about the 10th or 15th of November,) I have had an opportunity of examining three heads, which I received in a fresh state about six weeks ago, one being that of an adult female, the other two belonging to younger animals, all of which upon examination proved to belong to *Halichœrus griseus*, agreeing in every essential character with Mr. Bell's description of that animal, and with the drawings given me by Mr. Ball; and as no other

HABITS.—Respecting the Gray Seal as an American animal little or nothing seems to have been written. As an inhabitant of Ireland, the Hebrides, the Orkneys, the North and Baltic Seas, and Iceland, its history is better known. As will be noticed later, however, there are discrepant accounts respecting important points. Hallgrimsson has given a very interesting notice of these animals as observed in Iceland by A. Thorlacius, a trustworthy merchant and experienced hunter, of Stikkjisholm, Iceland, whose letter about them, as given by Hallgrimsson, may be rendered as follows:

"The Utsel is here very common in the Bredebugt, and especially on the coast of Westland. When full-grown it is four or five ells [8 to 10 feet] long; the male is probably still larger, and is always larger than the female. Its food consists partly of various kinds of fishes, as haddock, flounders, catfish (*Cottus*), etc., and partly of crustaceans and other lower animals, as starfishes, etc., especially in winter, when the fishes mostly seek the deep water. The animals here named I have myself seen them eat, as they chance to bring them to the surface of the water. Although this species of Seal occurs here in large numbers, only a few fully grown ones are taken, because they are not so easily killed here as the younger ones are, their strong skulls being not easily penetrated by bullets, and there are also very few expert marksmen here. Besides, they are very shy and watchful. Three weeks before the beginning of winter * [about October 1], the full-grown Utselur begin to come about the rocks and islets near the land, where they bring forth their young. They choose especially such rocks as are not covered by the spring-tides, and also the lower islands that have not too precipitous shores. Here the females have their young about fourteen days before the commencement of winter [about the second week of October]. The young are thickly covered

species of Seal has hitherto been recognized or met with by those who for a long series of years have been in the habit of seeing and taking these animals in this particular locality, I have now scarcely a doubt but that the whole of the colony that has so long inhabited the Farn Islands belongs to this species."—*Ann. and Mag. Nat. Hist.*, vol. vi, 1841, p. 463.

A little earlier than this (in 1837) Mr. Ball determined the large Seal of the Irish coast, till then also supposed to be *Phoca barbata*, to be *Halichœrus grypus* (Trans. Roy. Irish Acad., vol. xviii, 1837, pp. 89-98), since which time *Phoca barbata*, auct., has generally been excluded from the British Fauna.

* Hallgrimsson says in a note of explanation, "According to the Icelandic division of the year this falls between the 19th and 26th of October."

with soft, whitish-yellow woolly hair; this it gradually loses,
and does not enter the water till the moult is wholly completed,
at which time it is four or five weeks old. During the time the
young are lying upon the dry land they do not leave their
places, but every tide their mothers crawl up to them to suckle
them. Sometimes the females leave their young so near the
sea that the waves reach them, and by the spring-tides they
are swept along and carried helpless from one rock to another,
for while the milk-hair is worn the young Seal is able to swim
but little and is still less able to dive. In this condition it is
called by the Icelanders Sjovelkjingur (Sea-rover); such un-
fortunates are weak and emaciated, while those that have
remained undisturbed are fat and well conditioned. These are
called Volselr. The young is fattest when it is 'half ready',
that is to say when it has lost the milk-hair from the head and
feet; but later it becomes poorer, because the mother then
allows it to get hungry, in order to induce it to leave its resting
place and go into the sea. This happens about the end of the
third week of winter (middle of November) or a little later;
consequently the young are found to be best for killing when
three weeks old. The Utsel is blackish-gray; some are entirely
pure black, especially the males; the females are somewhat
lighter. It has a long nose and a big head, which in the old
males appears as if it were angular. These have a fierce aspect
and are very irritable and quarrelsome. They often fight with
each other on the shore, and bite so powerfully that they retire
from the conflict bleeding and mangled. They are also dangerous to the men who hunt them on the shore if they approach
carelessly, which they therefore must always do from the side.

"Respecting the age to which this species of Seal attains I
can say nothing that can be positively relied upon; yet they
apparently live to be very old. But I know with certainty that
the period of pregnancy continues for nine months."*

Mr. Selby has given a very interesting account of his observations on the Gray Seal as observed at the Farn Islands,†
based on his own frequent visits to these islands, and also on
"the long experience of a respectable individual, now upwards

* Isis, 1841, pp. 291, 292, originally published in Krøyer's Naturhist.-
Tidsskrift, Band ii, Heft i, 1837, pp. 97, 98.

† "Observations on the Great Seal of the Farn Islands, showing it to be the
Halichœrus griseus, Nilss., and not the *Phoca barbata*." By P. J. Selby, Esq.,
F. L. S., &c., &c. <Annals and Magazine of Natural History, vol. vi,
1841, pp. 462–466.

of eighty years of age, who succeeded his father, and continued to rent these islands till within the last eight or ten years.
From his account," continues Mr. Selby, "it appears that these
Seals were much more abundant some forty or fifty [now
eighty or ninety] years ago than they are now, which he partly
attributes to the great destruction he himself committed among
them (having been a first-rate Seal-hunter), and to the annoyance they have since been subjected to by the erection of the
present outer lighthouse, which is built upon an island to
which they were in the habit of retiring to rest during the
recess of the tide.

"In the year 1772, this old gentleman informs me that he
killed seventy-two young Seals, all of this species, and once
also killed fourteen old ones, in one day, upon the Crimston
Rock, the small island upon which they mostly calve, an event
that takes place, as I have previously observed, in the month
of November; and as the rutting season begins about the last
week in February or first week in March, it would appear that
the period of gestation of the *Halichœrus griseus* is about eight
and a half or nine months. The young when first calved are
nearly three feet in length, and grow very rapidly till they
quit the rock and are able to follow their dams to the water,
which is generally about a fortnight after birth; when first
calved they are covered with a longish soft woolly hair, of a
yellowish white or cream-colour, which gives place before they
quit the rock to a shorter hair of a grisly hue. If an opportunity offers, the young are sometimes tethered by a rope and
kept upon the rock a week or two beyond the usual time, in
order to get them of as large a size and as fat as possible before
they are slaughtered; but this must not be persisted in too long,
otherwise the dams are apt to forsake or refuse to suckle them
at the stated times of tide. The food of *Halichœrus* consists
entirely of fish, not restricted, it is supposed, to any particular
species, though they show a great predilection for the *Cyclo-
stoma lumpus* (Lump-sucker), particularly to the female, which
there goes by the name of the *Hush*. . . . They swim with
great strength and rapidity, and are frequently submerged for
two or three minutes, during which they make great progress,
and re-appear many gunshots distant from the place where they
went down, and they seem to delight and sport in the rapid
and heavy currents which exist among the Islands. They show
great curiosity in gazing at anything strange, and will remain

stationary for minutes together, with the head and neck out of
the water, staring at a boatman or any other object that at-
tracts their attention. This curiosity, in parts where they were
not often disturbed, procured me frequent shots with the rifle;
for when I observed them basking upon the rocks, twenty or
thirty in a herd, during the ebb of tide, I used to land at some
distance and make all haste to the point where they were assembled; and though I might not get within shot before they
took to the sea, I was sure of some of them re-appearing quite
within distance after their first plunge into the water. In this
way I have killed several, but never had the good luck to secure the carcass; for even though some of them floated a short
time after death, which, however, is rarely the case, they were
certain to be swept away and buried in the heavy stream which
runs past the point I have mentioned, and where the Seals
were generally assembled, before the boat could come round
and reach them. . . .

"The Great Seal seldom wanders to any great distance from
the Farn Islands, as it is only seen occasionally as far north as Berwick Bay, and off Dunstanborough and Coquet Island to the
south. It also seems jealous of the presence of any other species within its peculiar precincts, as the Common Seal, *Phoca
vitulina*, is scarcely ever seen within its territory, though small
herds frequent the coast of the main land nearly opposite, upon
the bar of Budle Bay, and at Holy Island."

Mr. James Wilson, in his "Notes regarding the distinctive
habits of the Scotch Phocæ or Seals," as observed by him at
the Western Islands, says, on the authority of Mr. Archibald
M'Neill, "The largest of these . . . . is known by the native name of *Tapvaist*, and although it associates occasionally
with the other kinds, yet it differs in many respects in its habits. I presume it to be the species usually designated by our
British writers as the Great Seal, or *Phoca barbata*." Although
it "is observed occasionally," he continues, "on shore with individuals of other kinds, ... it may be characterized as
being of solitary habits, and as frequenting the most remote
and undisturbed situations. It is neither so lively nor so
watchful as the common seal, nor is it so easily alarmed. . . .
One of the most characteristic and distinctive traits in its history is derived from its period of production, viz. the end of
September or commencement of October,—while that of the
common seal is usually the beginning of June. . . . The

young of the *Tapvaist* or great seal . . . . is invariably whelped above water-mark, and, it is said, during spring tides. They remain in a helpless condition on the rocks, for several weeks, before they can swim, and during this time they cast most of their long hair." *

A much earlier account, relating to this locality and species, has been given by Martin in his "Description of the Western Islands of Scotland," published in 1716. "On the western coast of this island [Harris] lies the rock Eousmil, about a quarter of a mile in circumference, and is still famous for the yearly fishing of seals there, in the end of October. This rock belongs to the farmers of the next adjacent lands. . . . . These farmers man their boats with a competent number fit for the business, and they always embark with a contrary wind, for their security against being driven away by the ocean, and likewise to prevent them from being discovered by the seals, who are apt to smell the scent of them, and presently run to sea. When this crew is quietly landed, they surround the passes, and then the signal for the general attack is given from the boat, and so they beat them down with big staves. The seals at this onset make towards the sea with all speed, and often force their passage over the necks of the stoutest assailants, who aim always at the forehead of the seals, giving many blows before they are killed; and if they are not hit exactly on the front, they contract a lump on their forehead, which makes them look very fierce;† and if they get hold of the staff with their teeth, they carry it along to sea with them. Those that are in the boat shoot at them as they run to sea, but few are catched that way. The natives told me that several of the biggest seals lose their lives by endeavouring to save their young ones, whom they tumble before them towards the sea. I was told, also, that three hundred and twenty seals, young and old, have been killed at one time in this place. The reason for attacking them in October is, because in the beginning of this month the seals bring forth their young on the ocean side; but these on the east side, who are of lesser stature [*Phoca vitulina*, probably], bring forth their young in the middle of June."‡

---

* Mag. Zool. and Botany, vol. i, 1837, pp. 540, 541.

† This seems to point to the Hooded Seal as being possibly involved in the account here quoted, although it evidently relates mainly to the Gray Seal.

‡ Pinkerton's Collection of Voyages and Travels, vol. iii, pp. 594, 595.

The same writer alludes as follows to the uses and supposed medicinal qualities of the flesh of these Seals:

"The natives [of the Island of Harris] salt the seals with the ashes of burnt sea-ware, and say they are good food: the vulgar eat them commonly in the spring-time, with a long pointed stick instead of a fork, to prevent the strong smell which their hands would otherwise have for several hours after. The flesh and broth of fresh young seals is by experience known to be pectoral; the meat is astringent, and used as an effectual remedy against diarrhea and dysenteria: the liver of a seal being dried and pulverised, and afterwards a little of it drunk with milk, aquavitæ, or red wine, is also good against fluxes. . . . The seal, though esteemed fit only for the vulgar, is also eaten by persons of distinction, though under a different name, to wit, ham: this I have been assured of by good hands, and thus we see that the generality of men are as much led by fancy as judgment in their palates, as well as in other things. The popish vulgar, in the islands to the southward from this, eat these seals in Lent instead of fish."*

Edmonstone, in his account of the Zetland Isles (vol. ii, 1809, p. 294), also refers at some length to this species under the names *Phoca barbata* and "Haaf Fish," stating, among other things, that the "young are brought forth in the months of September, October, and November."

The habits of the Gray Seal in the Gulf of Bothnia, where it was formerly very abundant, appear to be very different from what they are described to be at other localities, especially in respect to the season of reproduction. Although I have met no recent account of its habits as observed there, Lilljeborg and other Scandinavian writers quote Cneiff's account of the Gray Seal of the Baltic as referring without question to *Halichœrus grypus*. The Gray Seal, according to Cneiff, was so numerous about the middle of the last century as to occur in herds of several hundreds, and was regularly hunted for its fat and skins.† He describes it as somewhat migratory, leaving the Baltic at the approach of winter for the more northerly

---

* Pinkerton's Collection of Voyages and Travels, vol. iii, pp. 595,596.

† For a full abstract of Cneiff's account of the Bothnian Seal-fishery see *anteà*, pp. 530–534. As already stated (*anteà*, p. 664), there is reason to doubt whether this account does not relate to *Erignathus barbatus* rather than to *Halichœrus grypus*, in respect to the season of giving birth to the young and breeding habits, it agreeing with the former and not with the latter.

parts of the Bothnian Gulf. He says the female has its young *about the end of February on the ice,* but also says that they breed on the rocks when there is opportunity. The young does not at first enter the water, unless forced into it by the breaking up of the ice, the female suckling it upon the ice. During the first eight days after its birth it is wholly white, but after this the hair begins to fall, first on the head and fore feet, which at the end of fourteen days are blackish gray. As the Gray Seal cannot continue under the ice in winter without frequently coming to the surface for air, it has therein various small breathing-holes, which are so small at the top that it can only thrust its head through or even merely the nose, but they are wider below and perfectly round, being easily made so by the fore feet. They also have larger openings through which they ascend to the surface to repose, or, during the breeding season, to suckle the young. When the ice breaks up, before the young are strong enough to go south, as sometimes happens, and while they are still congregated in large herds at the breeding-places, they seek out the largest and soundest pieces of ice, on which they and their young can remain in greatest safety. At such times they all wish to get on the same piece, where perchance there may not be room for all; they therefore begin to fight with each other, biting and bullying, so that after the strife one may see large wounds on their bodies. It is worthy of remark that the Gray Seals and the Wikare or Harbor Seals do not associate together.

The Gray Seals, he continues, begin to lose their old hair about the 25th of March, which they rub off against the ice. At about this time the old Gray Seals with their young, which are no longer suckled, return to the East Sea. The Gray Seals stay in winter in the Gulf of Bothnia, probably because they find there thick ice which is not so liable to be broken up by strong winds as in other seas; consequently they can there bring their young into the world with greater safety. The only migration noticeable appears to support this opinion, for it takes place as soon as the young are large enough to obtain their own food. It is also noteworthy that the young of the Gray Seal know how to take a straight line for the East Sea, going so directly as to cross stony ground or a point of land, if it lies in their way; consequently the return often costs them their life. *

* Königl. Schwed. Akad. der Wissen., xix Band, 1759, pp. 172–174.
Misc. Pub. No. 12——45

In many respects Cneiff's above-cited account recalls the habits of the Ringed Seal, but his description of it indicates its weight to be about twice that of the Wikare or Harbor Seal, which seems to preclude the supposition of its being the Ringed Seal. Besides this, the Gray Seal is well known to be an abundant species in the Baltic, where the Bearded Seal, with which Cneiff's account seems in most particulars to agree, is not reported to occur. The breeding season is here distinctly affirmed to occur about the end of February,* while in Iceland and in the Western Islands of Scotland it occurs in October and November. At the last-named localities the species resorts to outlying rocks for its breeding-sites, while in the Bothnian Gulf the young are brought forth on the ice. The large herds here met with, in contrast with the small parties seen elsewhere, is also a noteworthy discrepancy not easily explained.

With one exception, this is the only Seal that is known to bring forth its young in the autumn. Collett states that the Bearded Seal breeds in October on the coast of Norway, and this again is the only instance known to me of the Bearded Seal having been reported as breeding at any other time than very early in spring. Although both the Gray and the Bearded Seals have about the same range on the Norwegian coast, it may seem rash to question the report of so trustworthy a naturalist as Mr. Collett, yet, if I rightly understand his remarks, his information touching this point is given at second hand, and it therefore seems possible that his correspondent may have mistaken the Gray Seal for the Bearded Seal. If such be the case, the breeding of the Gray Seal in the Bothnian Gulf in spring may be regarded as exceptional in the history of the species, while the reported breeding of the Bearded Seal in autumn would be the result of a malidentification of the species.†

---

* Lichtenstein has described a young example, still in the white coat, taken on the Pommeranian coast, March 28, 1821, which seems to confirm Cneiff's account of the breeding of this species early in spring.—*Abhandl. der Berlin Akad.*, 1822–23 (1825), p. 1.

† In order not to do Mr. Collett injustice I quote the following: After referring to what he believes to be the southernmost breeding station of the Bearded Seal on the coast of Norway, namely, on some rocky islets off Trondhjem Fjord, he says: "Af de udførlige Meddelelser, som jeg efterhaanden har modtaget af disse Øers Eier, Hr. Borthen, fremgaar det, at *Phoca barbata* i sit Levesæt og Yngleforholde i flere vigtige Henseender skiller sig fra de af vore øvrige nordiske Sæler, hvorom vi have nogen Kundskab.

GENUS MONACHUS, *Fleming.*

Monachus, FLEMING, Phil. Zoöl., ii, 1822, 187.
Pelage [*Pelagius*], F. CUVIER, Mém. du Mus., xi, 1824, 193, 196, pl. xiii.
    Type, *Phoca monachus,* Hermann.
Pelagius, F. CUVIER, Dict. Sci. Nat., xxxix, 1826, 550.
Heliophoca, GRAY, Ann. and Mag. Nat. Hist., xiii, 1854, pl. xiii (young).

" Muzzle rather elongate, broad, hairy, with a slight groove between the nostrils; whiskers small, quite smooth, flat, tapering. Fore feet short; fingers gradually shorter to the inner one; claws 5, flat, truncate. Hind feet hairy between the toes; claws very small; hair short, adpressed, with very little or no under-fur. Skull depressed; nose rather depressed, rather elongate, longer than the length of the zygomatic arch; palate angularly notched behind. Cutting-teeth $\frac{4}{4}$, large, notched within, the middle upper much smaller, placed behind the intermediate ones. Canines large, conical, sharp-edged. Grinders $\frac{5-5}{5-5}$, large, crowded, placed obliquely with regard to the central palatine line; crown large, conical, with several small conic rhombic tubercles. Lower jaw angulated in front below, with diverging branches, the lower edge of the branches rounded, simple. The grinders, except the two first in both jaws, are implanted by two roots; their crown is short, compressed, conical, with a cingillum [*sic*] strongly developed on their inner side, and developing a small anterior and posterior accessory cusp; the upper jaw is much less deep than in *Halichœrus;* the canines are relatively large, and the nasal bones are much shorter."—*Gray.*

The genus *Monachus* was placed by Gray, as late as 1866 and previously, in the subfamily *Stenorhynchinœ.* In 1871 he raised it to the rank of a distinct " tribe" (*Monachina*) or subfamily. The single species usually referred to it—the Mediterranean Seal, *Phoca monachus,* Hermann—has even been placed by Wagner, Giebel, and other German writers in the genus *Leptonyx,*

En af de mærkeligste af disse Afvigelser er Tidspunktet for dens Yngletid. Denne indtræffer nemlig i Norge om Høsten, efterat Individerne i Midten af September have samlet sig paa de bestemte Ynglepladse; umiddelbart derefter foregaar Ungens Kastning, i Regelen i den sidste Uge af September."— *Bemærkninger til Norges Pattedyrfauna,* 1876, p. 59. See also Lilljeborg's " Fauna öfver Sveriges och Norges Ryggradsdjur," 1874, p. 702, where the same is given in substance, and also Lilljeborg's note (note 2, p. 702) in reference to Fabricius's and Malmgren's observations on the breeding season of *Phoca barbata.*

nearly equivalent to the subfamily *Stenorhynchinæ* of Gray and other recent writers. Gill, in 1866, transferred it to the *Phocinæ*. Its introduction into the North American fauna rests on the provisional assignment of the Seal of the West Indian waters to this genus.

MONACHUS? TROPICALIS, *Gray.*

West Indian Seal.

Seal, DAMPIER, Voy. round the World, ii, 2, 3d ed., 1705, 23.
Cystophora antillarum, GRAY, Proc. Zoöl. Soc. Lond., 1849, 93 (in part only).
Phoca tropicalis, GRAY, Cat. Seals Brit. Mus., 1850, 28.
Monachus tropicalis, GRAY, Cat. Seals and Whales, 1866, 20; Hand List of
       Seals, 1874, 11.
—— *wilkianus*, GOSSE, Naturalist's Sojourn in Jamaica, 1851, 307.
Pedro Seal, GOSSE, l. c.; *Jamaica Seal*, GRAY, l. c.

CHARACTERS.*—Incisors $\frac{2-2}{2-2}$; canines $\frac{1-1}{1-1}$; molars $\frac{5-5}{5-5}$, five-lobed, conical, rugose at base. Soles and palms naked. Anterior digits with well-developed nails; posterior digits with the nails rudimentary. Mystacial bristles long, flexible, smooth. Color intense uniform black, or black varied with gray. Pelage very short, stiff, closely appressed. Length of adult male, about ten feet.

Although the existence of Seals in the West Indian waters has been known for two centuries, a most tantalizing uncertainty still prevails in respect to their characters and affinities. I had hoped to be able in the present connection to clear up some of these doubts, but as my efforts to obtain specimens have thus far proved fruitless, I have to content myself with giving a transcript of what has already been written about them, with such critical remarks as the case suggests.

So far as known to me, Dampier was the first to record the existence of Seals in the Caribbean Sea, but he gives no description of them, his reference consisting of an account of a sealing voyage made to the Alacrane Reef in 1675, and incidents relating thereto. His account, however, shows that at that time they were so abundant at that locality as to be sought there for their oil, and where, in fact, for some years previously, the sealing business had been an industry of considerable commercial importance.

DAMPIER'S ACCOUNT, 1675.—In describing the "Alacrane"

---

* Compiled from Hill and Gosse.

Islands,* under the marginal date "*An.* 1675," he says, "Here are many Seals: they come up to sun themselves only on two or three of the Islands, I don't know whether exactly of the same kind with those in colder Climates, as I have noted in my former Book, they always live where there is plenty of Fish.

"To the North of these Islands lyes a long ledge of Rocks, bending like a Bow; it seems to be 10 or 12 Yards wide, and about 4 Leagues long, and 3 Leagues distant from the Island. They are above Water, all joyning very close to one another, except at one or two Places, where are small Passages about nine or ten Yards wide; 'twas through one of these that Providence directed us in the Night; for the next Morning we saw the Riff about half a Mile to the North of us, and right against us was a small Gap by which we had come in hither, but coming to view it more nearly with our Boat, we did not care to venture out that way again. . . . There we Anchored and lay three or four days, and visited most of them, and found plenty of such Creatures, as I have already described.

"Though here was great store cf such good Food and we like to want, yet we did neither salt any, nor spend of it fresh to save our Stock. I found them all but one Man averse to it, but I did heartily wish them of another mind, because I dreaded wanting before the end of the Voyage; a hazard which we needed not to run, there being such plenty of Fowls and Seals (especially of the latter), that the Spaniards do often come hither to make Oyl of their Fat; upon which account it has been visited by English-men from *Jamaica*, particularly by Capt. *Long:* who, having the Command of a small Bark, came hither purposely to make Seal Oyl, and anchored on the North side of one of the sandy Islands, the most convenient Place, for his design: ———— Having got ashore his Cask to put his Oyl in, and set up a Tent for lodging himself and his Goods, he began to kill the Seal, and had not wrought above three or four days before a fierce North-wind blew his Bark ashore. By good fortune she was not damnified: but his Company being but small, and so despairing of setting her afloat again, they

---

* "The *Alacranes* are 5 or 6 low sandy Islands, lying in the Lat. of about 23 d. North, and distant from the Coast of *Jucatan* about 25 Leagues; the biggest is not above a Mile or two in Circuit. They are distant from one another 2 or 3 Miles, not lying in a Line, but scattering here and there, with good Channels of 20 or 30 Fathom Water, for a Ship to pass between."— DAMPIER, *Two Voyages to Campeachey*, etc., in his *Voyage round the World*, vol. ii, part 2, 3d ed., 1705, p. 23.

fell to contriving how to get away; a very difficult task to accomplish, for it was 24 or 25 Leagues to the nearest Place of the Main, and above 100 Leagues to *Trist*, which was the next English Settlement. But contrary to their expectation, instead of that, Capt. *Long* bid them follow their Work of Seal-killing and making Oyl; assuring them that he would undertake at his own peril to carry them safe to *Trist*. This though it went much against the grain, yet at last he so far prevailled by fair Words, that they were contented to go on with their Seal-killing, till they had filled all their Cask." The narrative continues that by a to them lucky accident "two *New-England* Ketches going down to *Trist*, ran on the backside of the Riff, where they struck on the Rocks, and were bulged". Captain Long and his crew assisted them to unlade their goods and bring them ashore, in requital for which they helped him to launch his own vessel, "and lading his Oyl, and so they went merrily away for *Trist*." Captain Dampier adds, "The whole of this Relation I had from Captain *Long* himself." *

How long the capture of Seals for commercial purposes continued after this date, or whether it was ever carried on at other points in these waters, I have no means of determining.† Owing to the limited area to which they were restricted, and consequently their necessarily small numbers, it is evident that they could not long have survived in force under such vigorous persecution.

HILL'S AND GOSSE'S ACCOUNTS, 1843, 1851.—A description of this Seal (and the first one, so far as I can learn) was published, according to Mr. Gosse, by Mr. Richard Hill, in the "Jamaica Almanack for 1843." Mr. Gosse, in 1851, in his work entitled "A Naturalist's Sojourn in Jamaica," republished Mr. Hill's account, and added thereto further remarks on the species, based largely on information communicated by Mr. Hill. As Mr. Hill's description is nearly inaccessible, while Mr. Gosse's book is by no means easy of access, I here transcribe the whole account as given at length by Mr. Gosse, under the heading "The Pedro Seal": "In the Jamaica Almanack for 1843, Mr. Hill published a Memoir on a Seal inhab-

---

* Ibid., pp. 25–28.

† Olafsen, in his "Reise durch Island," p. 284, refers to the Great Seal of the Antilles, and cites "Joh. Sam. Hallen's Natur-Geschichte der Thiere, p. 593 und 581," as containing a further account. Hallen's work being inaccessible to me, I am unable to state what information may be there found.

iting the Pedro Kays, a reef of rocks, lying off the south coast of Jamaica. As it appears to be a species unknown to naturalists, and as the publication in which it was described had only a transient and local interest, I transcribe the Memoir at length, adding to it such particulars of the natural history of the animal as have since been communicated to me by my friend.

" 'The differences which exist in the crania of the *Phocidœ*, and other discrepancies of structure which have been remarked as distinguishing the several genera into which the family is divided, would appear to make the Seal from the Pedro Shoal more allied to the *Ph. vitulina* of Linn. (*Calocephalus*, Fr. Cuv.) than to any of which we have detailed accounts, although very different from all.* The shoulders, legs, and thighs are concealed within the body, and the hand is extremely flattened and fin-like. The cranium is large, high, and convex:—there are ten molar teeth, and two canines in the upper jaw, and the same number in the lower; these, with *four incisors*, above and below, make in all thirty-two teeth. They are *five-lobed* and conical, and they terminate in a base of extremely rough enamel. The teeth are so disposed that when the mouth is closed there is no interspace above or below them, the points of the upper teeth filling the depressed intervals of the lower ones. Having no external auricle, and ears with foramina so small as to be hardly perceptible, the species belongs to the *Inauriculata* of Peron, or the earless division of Seals. The nostrils are narrow fissures, which appear like two slits in the nose, and are frequently and rapidly closed. The small orifices of the ears are in a similar manner rapidly opened and shut. The lips are full and fleshy, and covered with numerous strong bristles, very flexible, of a black hue with transverse bars of grey. The colour of the body is an intense and uniform black. The hair is short and stiff, and extremely and curiously close. The close bristly covering prevails everywhere except on the palms of the flippers, which are bare. The fore paw has much more the form of a foot than of a hand, the first finger answering to the thumb being the longest. There are nails only on

---

* "From Mr. Hill's description it appears to have the incisors and nailless hind feet of *Stenorhynchus*, with the molars of *Calocephalus*. The data are perhaps not sufficient to warrant the formation of a new genus, but I may be permitted to propose the trivial name of *Wilkianus* for the species, in honour of George Wilkie, Esq., to whose courtesy I am indebted for the skin of an adult specimen, probably of the same kind, shot by himself."

the fore paws, those of the hinder being rudimentary.   The
eyes are large, black, and full, and the irides crimson.

" ' When the specimen from which these notes were made
first arrived it was very lively, and so sensible to the slightest
touch, that however lightly the hand might be placed on the
fur, it felt the contact, and moved rapidly away, jerking the
whole body forwards.   When left unmolested it was playful.
It ploughed the water with the nose, and snorted as it drew
the head out.   It grunted like a pig, and barked, growled, and
snarled, like a dog.   It was fond of turning upon the back and
lying dozing.   In this posture it slept and basked in the sun.
It refused all food, and lived four months without eating.
Symptoms of dulness only appeared in the last month, when it
was found to be labouring under some disease of the head;
and when it died it was discovered to have become totally
blind, the dark pupil of the eye having disappeared, together
with the crimson colour of the iris.   It was surprisingly fat, notwithstanding its long fast.   The fat was four inches thick, and
yielded four gallons of oil.   It was a male, but the organs of
generation were not externally perceptible.   This organization
is accordant with the peculiarities of the Seal tribe : in the female the teats are concealed in the skin, and the lacteal fulness
swells with the rotundity of the body, so that the animal does
not suffer pain or inconvenience when crawling on land; and
the bifid termination of the tongue, another peculiarity, is an
adaptation which enables the young of the *Phocidæ* to seize
the nipple under comparatively difficult circumstances, attendant on lactation.   The occipital aperture, which remains for a
long time unossified in this tribe of animals, being still open,
though reduced to a very small orifice,—this Seal may be considered to have been only just full grown.   The unworn sharpness of the teeth indicated the same fact.

" ' The measurements of this specimen were as follows :

	Feet.	Inches.
Total length along the back from snout to tip of tail	4	2
From snout to insertion of fore paw	1	6
From insertion of fore paw to hind paw	2	10
Circumference of body near fore paws	3	2
Circumference at hind paws	1	6
Breadth of back at fore paws	1	0
From one fore paw to the other, extended	2	6
Length of fore paw	0	10
Length of hind paw	0	11

	Feet.	Inches.
Breadth of head across ears, measured horizontally	0	7
Length of head	0	9
Breadth of nose	0	4½
Length of tail	0	3

" ' The Kays frequented by these Seals are situated at about a degree south from this Island, and form portions of an extensive and dangerous line of rocks on a shoal about 100 miles long, the two extremities of soundings touching nearly the 77th and 79th meridians of W. longitude. These banks rise precipitously from the deep ocean, with reefs formed, like the usual rocks in these seas, of coral, with an accumulation of shells and calcareous sand. The depth of water varies from 7 to 17 fathoms. A scanty vegetation covers the principal group of islands, which are what are properly called the Pedro Kays. The detached islets about 90 miles apart, known as the Portland and Rattlesnake Rocks, are nearly the eastern and western extremities of the bank. This shoal has always been visited as an excellent and inexhaustible fishing ground; and, probably from the variety and abundance of its aquatic animals and marine productions, it received from the Spaniards the name of Vivero, a word equally designating a warren or fish-pond. The principal supply of turtles for the Kingston Market is derived from these shoals, and the rocks are numerously tenanted by sea-birds.'

"In the spring of 1846 George Wilkie, Esq., paid a visit to these Kays and succeeded in obtaining a larger specimen of the Seal. Some notes with which he kindly furnished me, through the medium of Mr. Hill, of the peculiarities of the different islets, depict natural difficulties in the access to Seal Kay, sufficient to account for the meagreness of the information about Seals, possessed by the host of egg-gatherers, who annually resort to those rocks and shoals. Seal Kay lies about three miles to leeward of the principal group. It is about two acres in extent, and rises to twenty feet in height, but is entirely destitute of all terrestrial vegetation. Address, in landing, requires to be combined with strength, hardihood, and perseverance; and frequently before a footing can be obtained, the Seals, the objects of attraction, have escaped to the waters, and continue to avoid the shore as long as intruders remain upon the island. 'When Mr. Wilkie's party first landed in their late visit, they surprised some five Seals on shore. They immediately succeeded in heading a "Bull," or Male Seal, both big and burly, and killed him. He proved to be an aged patriarch, with teeth nearly worn to the stumps, and a hide gashed and seamed with scars, got in many a fierce fight;—and about ten feet in length.

"'In the scramble which the Seal makes to regain the water, nothing is to be remarked but the violence and impatience with which he jerks his body forward; but when he plunges from the shore into the sea, it is no small treat to see the suddenness with which the uncouth animal, so unwieldy and helpless on land, becomes gracefully alert in the ocean. The command with which he strikes through the water, the velocity with which he cleaves the flood, the ease with which he winds the mazes of the rocks, and dashes forward into the hidden recesses of the deep, are beautifully interesting in a creature looking so essentially a quadruped. When the boat is afloat again, the Seals come trooping out to reconnoitre. At a depth of about three feet they paddle about, gazing up through the clear liquid with an expression of countenance beaming with curiosity and intelligence. They dodge around the boat, occasionally ascending to the surface, to renew their inspirations of air, and to look upon their island home, to ascertain whether they may return thither and be at rest.

"'A grown-up cub about four feet long had been taken by the people. One Seal was observed more persevering in her watchfulness and assiduity to regain the shore, than the rest. This was conjectured to be the dam of the slaughtered young one. The maternal instinct did not exhibit any stronger emotion than this anxious vigilance. The young one was sufficiently grown to be no longer dependent on the mother. Had it been still sucking, there was enough to show that the parental passion would have merged fearlessness into fury, and inquietude for the safety of its young, into unsparing vengeance for its fate.

"'Without doing more than referring to Weddell's observation, that the jaw of the Seals he describes was so powerful in the agonies of death as to grind stones into powder, it seemed, from the condition of the teeth of some eight that were taken during the time Mr. Wilkie's party were on the Pedros, that their strength is exercised in more laborious work than crushing the bones of fishes. The opinion that the more experienced fishermen expressed was, that they fed as generally on molluscous animals as on fish, and that their teeth suffered much wear and tear in the work of breaking shells. Yet it is remarkable that the contents of the stomachs of those killed gave them no insight into the nature of their food:—they were invariably empty.

"'I must not omit to mention that our friends had one opportunity of closely observing the progression of the Seal when ascending the beach. The advance was by zigzag movements. It was evident that the ground was first gripped by one fore flipper, then by the other, that the body advanced first to the right, then to the left, as one or the other flipper took its hold of the earth, and helped the body onward. They seemed to delight in basking in the sun, and to huddle together, and grunt out their pleasure in each others' company.'

"The skin of one of the specimens obtained in this expedition Mr. Wilkie kindly presented to me; a courtesy the value of which was enhanced by the fact of its being one of the chief of the *opima spolia*, a sort of trophy of his own exploits. It is now in the British Museum. As the skull was not preserved, the actual identity of the species with the smaller specimen, described by Mr. Hill, cannot with certainity be established; and there seems a little discrepancy in the proportions, as will be seen by comparing the admeasurements of Mr. Hill's, already given, with the following, which were taken from Mr. Wilkie's specimen:—

	Feet.	Inches.
Length from nose to tip of tail	6	6
Circumference at fore paws	3	4
Length of fore paw	0	11½
Length of hind paw	0	10¾
Length of tail	0	2

"The fur is of a nearly uniform dirty ash-gray, black at the base, and gray at the tips of the hairs; it is slightly mottled on the belly; it is very close and stiff, and not more than one-fourth of an inch long. The *vibrissæ* or whiskers are from an inch to an inch and three-quarters long; white, with one on each side brown."*

GRAY'S ACCOUNTS, 1849–1874.—Mr. Gosse, as appears from the foregoing, transmitted a specimen to the British Museum,—

---

* A Naturalist's Sojourn in Jamaica. By Philip Henry Gosse, A. L. S., &c. Assisted by Richard Hill, Esq., Cor. M. Z. S. Lond., Mem. Counc. Roy. Soc. Agric. of Jamaica. London: Longman, Brown, Green, and Longmans, 1851, pp. 107–114.

the skin last described in the above transcript, and, so far as Mr. Gosse's narrative goes to show, the only one he ever saw. Dr. Gray, however, in 1849, described another "skin and skull" from "the West Indies", which he later stated were sent home direct by Mr. Gosse from Jamaica.   Dr. Gray's description of these specimens is as follows: "We have lately received from the West Indies the skin and skull of a seal which evidently belongs to the same genus as the crested seal of the northern hemisphere.   The skull, or rather the teeth, when compared with those of the Greenland specimens, induce me to believe that it is distinct from them.   It chiefly differs in the form of the outer upper cutting teeth and canines.   In *all* the specimens, both old and young, from the North Sea, the outer upper cutting teeth and the canines are narrow and compressed.   In the West Indian skull, which is that of a very young specimen, the outer upper cutting teeth and the canines are broad, strongly keeled on each side and longitudinally plaited within. In this skull the 4th grinder has only a single root, and the 5th grinder has two; the crowns of the teeth are plaited and tubercular like those of the North Sea specimens.   The face is rather broader than in a skull of the northern kind of nearly the same size.   This species may be called *Cystophora antillarum*.

" We have received an imperfect skin of a seal from Jamaica, which was brought home by Mr. Gosse.   It is unfortunately without any bones.   The whiskers are short, thick, white, cylindrical, regularly tapering, and without any appearance of a wave or twist.   In this character it most agrees with *Phoca barbata*." *

The following year he redescribed these specimens, claiming Jamaica as the habitat of his *Cystophora antillarum*, and stating that the specimens on which it was based were from "Mr. Gosse's collection", as follows:

"2. Phoca tropicalis. Jamaica Seal.
"Grey-brown; hair very short, strap-shaped, closely adpressed, black with a slight grey tip; whiskers short, thick, cylindrical, regularly tapering, without any appearance of wave or twist; fingers gradually shorter.
"Inhab. Jamaica.
"a. Skin imperfect, without skull.
"Skin referred to in description of Cystophora antillarum, *Gray, Proc. Zoöl. Soc.*, 1849, 93."†

_____
* Proc. Zoöl. Soc. Lond., 1849, p. 93.
† Cat. Seals Brit. Mus., 1850, p. 28.

"2. CYSTOPHORA ANTILLARUM. WEST INDIAN HOODED SEAL.

"Skull, face broad. The outer upper cutting teeth and the canines broad, strongly keeled on each side and longitudinally plaited within. Fur grey brown, lips and beneath yellow.

"Cystophora antillarum, *Gray, Proc. Zoöl. Soc.*, 1849, 93.

" Inhab. West Indies.

"*a.* Stuffed specimen. West Indies, Jamaica. Mr. Gosse's collection.

"*b.* Skull of a very young specimen. The face is broader than the skull of *C. cristata* of the same size. The crowns of the teeth are plaited and tubercular. The 4th grinder has only a single root, the 5th has two. West Indies, Jamaica. Mr. Gosse's collection.

"Specimen described, *Gray, Proc. Zoöl. Soc.*, 1849, 93." *

These descriptions were repeated, *verbatim*, by Gray in 1866, in his " Catalogue of Seals and Whales in the British Museum" (pp. 20, 43), without additional remark, but in his account of the genus *Monachus* (ibid., p. 18) he says : "As the other sub-tropical Seal, *Phoca tropicalis* (Gray, Cat. Seals B. M. 28), from Jamaica, described from an imperfect skin without a skull, has similar small smooth whiskers [as *Monachus albiventer*], it may very probably, when its skull has been examined, be found to belong to this genus, which will then prove to be a subtropical form of the family."† It appears, however, that in the mean-time a figure of the *Cystophora antillarum* had appeared in an " inedited" plate in the " Zoölogy of the Erebus and Terror,"‡ at least such is here so cited by Gray, but whether the skull or the external characters are represented is not stated.

It will be noticed that in the above descriptions no measure-ments are given, nor any indication of size, nor is the structure of the hind feet referred to, not so much as to state whether they are or are not provided with nails. The only further infor-mation Dr. Gray has vouchsafed to us, based on his own obser-vations, is contained in his " Hand List of Seals," etc., pub-lished in 1874, where he says under *Cystophora antillarum* (p. 18): "Animal, stuffed, young male? 1005 *a.* Skull, young,

---

* Cat. Seals Brit. Mus., 1850, p. 38.

† These remarks appeared originally twelve years before in his description of his genus *Heliophoca* (Proc. Zoöl. Soc. Lond., 1854, p. 44), and hence the comparison of the whiskers was first made with his *Heliophoca atlantica*, which he later (1866) referred to *Monachus albiventer*. *Heliophoca* was charac-terized as having the " cutting teeth 6/4," but this seems to have been a typo-graphical error for 4/4, as with this correction the whole description of the genus *Heliophoca* was in 1866 introduced under *Monachus*.

‡ On this, as on many other occasions, I have to lament the absence from all the libraries of this vicinity of the part of this work treating of the mam-mals.

broken. Muzzle rather dilated . . . This skull is exceedingly like that of the young *C. cristata*." In 1866, under "Additions and Corrections" in his " Catalogue of Seals and Whales"
(p. 367), he cites for the first time " Hill's Jamaica Almanack,
1843", and adds as a synonym " The Pedro Seal (Phoca Wilkianus), *Gosse, Nat Sojourn in Jamaica*, 307, 308", and quotes
the descriptions of specimens there given.

GILL ON THE WEST INDIAN SEALS, 1866.—Dr. Gill in 1866
(but before he had seen Gray's "Catalogue of Seals and
Whales" of that date) thus referred to the West Indian Seals:
"The relations of the Jamaican Seal, rejoicing in the two names,
Phoca tropicalis, Gray, and ——?! *Wilkianus*, Gosse (1851), are
very uncertain. Mr. Gosse obtained a single skin. The exact
origin of the *Cystophora antillarum* was not mentioned in the
original description, and its West Indian habitat requires confirmation."* Dr. Gray,† a little later, in referring to Dr. Gill's
above-quoted remarks, reaffirmed that the specimens of both
his species were obtained in Jamaica by Mr. Gosse.

ANALYSIS AND DISCUSSION OF THE FOREGOING.—From the
foregoing,—the only information at present accessible on the
subject,—what conclusions may be drawn respecting the number of species and affinities of the West Indian Seals? Are
there two species or one, and what is their relationship? In
the first place, it may be noted that Gray's *Phoca tropicalis* and
Gosse's Pedro Seal (—— *wilkianus*), the latter named specifically, but referred to no particular genus, are one and the same
thing, the former being based on Mr. Gosse's specimen. That
such a species exists is beyond question, while, as will be noticed fully later, its generic affinities seem to be with *Monachus*,
to which genus Dr. Gray finally referred it. Secondly, it is to
be noted that the specimens made known by Hill and Gosse,
and all their observations respecting the Jamaican Seals, relate to this type, and in no way suggest the genus *Cystophora*.
In the third place, no one can doubt but that the specimens
on which Gray based his *Cystophora antillarum* were correctly
referred to *Cystophora*. Every point of the description renders this evident, while Dr. Gray himself says, in his last reference to the species, twenty-five years after it was first described, "This skull is exceedingly like that of the young *C.*

---

* Proc. Essex Institute, vol. v, No. 1, April, 1866, p. 4, footnote.
† Ann. and Mag. Nat. Hist., xvii, 1866, p. 445.

cristata."* Finally, did the specimens on which *Cystophora antillarum* was based come from the West Indies? Dr. Gray says, in his first reference to them in 1849, they were "lately received from the West Indies." In the next paragraph he says, "We have received an imperfect skin of a Seal from Jamaica, which was brought home by Mr. Gosse," certainly implying that the specimens mentioned just before were not from Mr. Gosse, and probably not from Jamaica. On the other hand, Mr. Gosse's account of the Pedro Seal indicates that Mr. Gosse himself never even saw other specimens of this Seal than the skin he sent to the British Museum, his whole account of the species, aside from a description of this skin, being avowedly given at second hand. Yet Dr. Gray the next year, in redescribing these specimens, made the skin received from Mr. Gosse the basis of his *Phoca tropicalis*, while the "skin and skull" on which *Cystophora antillarum* became now exclusively based received a definite locality and history, namely, "West Indies, Jamaica, Mr. Gosse's Collection." † It was a year later when Mr. Gosse published his account of the Pedro Seal, and if these specimens, alleged to have been received from Jamaica through him, had related to the Pedro Seal, or to any other Jamaican Seal, it is probable that he would not have failed to refer to them in treating of the Seals of Jamaica. Dr. Gray, however, doubtless firmly believed in their Jamaican origin, for he not only gives for them the habitat and history above quoted in all his subsequent notices of *Cystophora antillarum*, but in 1866, in replying to Dr. Gill's remark that its West Indian habitat required confirmation, says he (Gill) overlooked "the fact that they were both [*Phoca tropicalis* and *Cystophora antillarum*] collected in Jamaica and sent home direct from the island by Mr. Gosse." ‡

In regard to the occurrence of the genus *Cystophora* in Jamaican waters, there are, in the present state of our knowledge of the subject, only two alternatives, one of which implies the acceptance of Gray's alleged origin of his specimens of *C. antillarum* as valid, while the other assumes an accidental error of locality; since its presence or absence there, so far as we now know, turns upon this point. In favor of the latter alternative is the pretty strong inference, derivable from the gen-

* Hand List of Seals. etc., 1874, p. 18.
† Cat. Seals, p. 38.
‡ Ann. and Mag. Nat. Hist., xvii, 1866, p. 145.

eral history of the case, that Mr. Gosse obtained or saw only a single specimen of Seal in Jamaica,—namely, the skin forming the basis of Gray's *Phoca tropicalis*.* Furthermore, the genus *Cystophora*, as now known, is a subarctic type, the occurrence of which within the tropics seems at least very improbable. Agreeing, therefore, with Dr. Gill that the West Indian habitat of *Cystophora antillarum* even still "requires confirmation," I can recognize in the present connection only a single species of West Indian Seal,—namely *Phoca tropicalis*, Gray, 1850,= *Monachus tropicalis*, Gray, 1874.

AFFINITIES OF THE JAMAICAN OR PEDRO SEAL.—In respect to the characters and affinities of the Jamaican Seal, we have nothing of importance beyond the information furnished by Messrs. Hill and Gosse. The incisive formula of $\frac{2-2}{2-2}$ shows that it is neither *Cystophora* nor *Macrorhinus*, nor even a typical member of the subfamily *Phocinæ*. In this respect it agrees with *Monachus* and with the Stenorhyncine Seals, with which types it also agrees in the rudimentary condition of the nails on the hind feet. It agrees with *Monachus* in the structure of

---

* After transmitting this article to the printer it seemed to me desirable to settle, if possible, the question of the West Indian origin of the specimens on which Dr. Gray based his *Cystophora antillarum*, and I accordingly addressed a letter of inquiry on the subject to Mr. P. H. Gosse. He not only promptly replied, but in a subsequent letter kindly gave me permission to publish his letter, the greater part of which I here transcribe:

"SANDHURST, TORQUAY, *Jan.* 18, 1880.

" MY DEAR SIR: In reply to your inquiry about West Indian Seals, I may say, with certainty, (notwithstanding the length of time that has elapsed,) that Dr. J. E. Gray was in error, in supposing that more than *one species* was actually delivered to the Brit. Museum, from Jamaica, *by me.* This was the skin mentioned in my ' Nat. Soj. Jam.', p. 314. . . . .

"The Seal of the Pedro Kays is certainly not a *Cystophora.* I know nothing of this; nor of any other Seal from the Antilles than the species I have described in 'N. S. J.'

"Believe me, my dear sir, very truly yours,

"P. H. GOSSE."

This makes it evident that Dr. Gray was mistaken in his statement that the specimens of his *Cystophora antillarum* were "collected in Jamaica and sent home direct from the island by Mr. Gosse." I will now add that I believe it safe to refer *Cystophora antillarum* to the well-known *Cystophora cristata*, even Gray himself having stated that the young skull on which it was mainly based "is exceedingly like that of the young *C. cristata*," and to assume that the supposition of its West Indian origin was wholly a mistake.

the mystacial bristles, and in the palms being bare, in which characters it differs from any of the *Stenorhynchinæ*. It appears to differ from *Monachus* in no essential character, except in the structure of the teeth, which seem to agree better with those of *Lobodon*. Mr. Hill describes the molars as "five-lobed and conical," and as "terminating in a base of extremely rough enamel." In *Monachus* the molars are very thick, broad, and conical, with a small accessory cusp before and behind the principal one, and a roughened cingulum. As in all other characters the agreement is closer with *Monachus* than with any of the Antarctic genera, I accept provisionally Gray's reference of the species to *Monachus*, especially as Mr. Hill's description of the dentition is, on the whole, rather vague. Besides this, *Monachus* is the only subtropical genus of the family, unless the Jamaican Seal prove to be a distinct generic type.

GEOGRAPHICAL DISTRIBUTION.—Respecting the present geographical distribution of the West Indian Seal, I am indebted for valuable information to Mr. R. W. Kemp, who, under date of "Key West, Fla., April 29, 1878," wrote me as follows: "Some two or three years ago there were two seen near Cape Florida. It was supposed that they had strayed from some of the Bahama Islands, as there are some few to be found in that vicinity. I am informed by reliable parties that Seals are to be found in great numbers at the Anina Islands, situated between the Isle of Pines and Yucatan. One of my informants says that as he was sailing about the islands fishing and wrecking, he and his party discovered a number of Seals on one of them, and went on shore to kill some, merely 'for fun'. On nearing the shore the Seals all got into the water. They then hid themselves in the shrubbery along the beach, and in about ten or fifteen minutes the Seals came on the beach again. The men, armed with axes, sprang upon them, the Seals trying to get into the water again. Two of them were killed, and another one, as one of the men came up to him, turned around and barked furiously at him, which frightened the poor man so badly (he having never seen one before, and knowing nothing of their habits) that he almost fainted. The Seals are said to be very easily killed or captured alive. They yield a great deal of oil. The skins are very large, but not easy to cure, on account of their fatty substance." In a later letter he refers to their great rarity on the Florida coast, where he says they occur "only once or twice in a life-time", but alludes to their comparative abundance on the coast of Yucatan, and their occasional occurrence at the Bahama Islands.

Mr. L. F. de Pourtalès also informs me that there is a rock on Salt Key Bank, near the Bahamas, called "Dog Rock", presumably from its having been formerly frequented by the Seals. Also, that his pilot, in 1868–69, told him he had himself killed Seals among the rocky islets of Salt Key Bank.

I learn from Mr. S. W. Garman, who accompanied Mr. Agassiz during his dredging expedition in the Caribbean Sea, in the United States Coast Survey steamer "Blake," during the winter of 1877–78, that the Seal of those waters is well known to the wreckers and turtle-hunters of that region, and that they often kill them for their oil. He also informs me that these animals had also been frequently seen and killed by one of the officers of the "Blake," especially about the Isle of Pines, south of Cuba, and at the Alacranes, where, as already noted, they occurred in such abundance at the time of Dampier's visit in 1676 as to be extensively hunted for their oil. They are also known to the whalers who visit these waters.

The specimens described by Messrs. Hill and Gosse were taken at the Pedro Kays, off the southern coast of Jamaica, where thirty years ago they appear to have occurred in considerable numbers.

On a "Chart of the Environs of Jamaica," published in 1774,* as well as on later maps of this region, are indicated some islets off the Mosquito Coast, in about latitude 12° 40′, which bear the name "Seal Kays," doubtless in reference to the presence there of these animals.

It therefore appears that the habitat of the West Indian Seal extends from the northern coast of Yucatan northward to the southern point of Florida, eastward to the Bahamas and Jamaica, and southward along the Central American coast to about latitude 12°. Although known to have been once abundant at some of these localities, it appears to have now wellnigh reached extinction, and is doubtless to be found at only a few of the least frequented islets in various portions of the area above indicated. Being still well known to many of the wreckers and turtle-hunters, it seems strange that it should

---

* History of Jamaica, vol. i, facing title-page. The work is anonymous, but the authorship is attributed to Edward Long.

have so long remained almost unknown to naturalists. The only specimen extant in any museum seems to be the imperfect skin transmitted by Mr. Gosse to the British Museum thirty years ago. Consequently respecting none of the Pinnipeds, at least of the northern hemisphere, is information still so desirable.

Subfamily CYSTOPHORINÆ, *Gray*.

Genus CYSTOPHORA, *Nilsson*.

Cystophora, Nillson, "Skand. Fauna, i, 1820, 382." Type, *Cystophora borealis*, Nilsson = *Phoca cristata*, Erxleben.

Stemmatope [*Stemmatopus*], F. Cuvier, Mém. du Mus., xi, 1824, 196. Type, *Phoca cristata*, Erxleben.

Stemmatopus, F. Cuvier, Dict. Sci. Nat., xxxix, 1826, 551; ibid., lix, 1829, 464.

Mirounga, Gray, Griffith's An. King., v, 1827, 463 (in part).

Incisors $\frac{2-2}{1-1}$. Molar teeth with small, plaited crowns, a distinct neck, and very thick, swollen roots, all simple-rooted except the fifth upper, which is double-rooted, as is also sometimes the fourth upper. Nasal bones rather short, small; palatal surface broad, flat; hind border of palatines concave. Nasal passages deep, broad, nearly divided posteriorly by a long septum; interorbital region broad; muzzle narrow, rather produced. Auditory bullæ greatly swollen, the anterior border nearly straight (convex in youth). Brain-case short, broad; prominent occipital crests and well-developed anteorbital processes in the adult males; also, with a large inflatable sac on the nose, which is absent in the females and in the young males. Digits of both fore and hind limbs armed with large powerful claws. Outer digits of pes but little longer than the middle ones.

Cystophora agrees closely with *Macrorhinus* in the form and general character of the teeth; also in neither do the intermaxillaries rise in front to meet the nasals, as is the case throughout the *Phocinæ*. *Cystophora* differs from *Macrorhinus* in the form of the basisphenoid and basioccipital bones, which in *Cystophora* constitute a broad flat interbullar space, this region in *Macrorhinus* being narrow and deeply hollowed. Also in the form of the hind feet, which in *Cystophora* are only slightly emarginate on the distal border, whereas in *Macrorhinus* the hind feet are deeply forked, in consequence of the outer digits far exceeding in length the middle ones. The nails are also rudimentary in the last-named genus, while in *Cystophora* they form strong, well-developed claws. A further difference in external characters consists in the form of the nasal appendage of the adult males, the large inflatable sac met with in *Cystophora* being represented in *Macrorhinus* by a long flexible proboscis, resulting in a widely different physiognomical expression.

In respect to general form, *Macrorhinus* is heavily developed anteriorly, all the bones of the fore limbs being especially massive, while those of the hind limbs are rather weak, and the feet small. The scapula is very large and broad, the width at the widest part being equal to the length. The acromion process is strongly developed, and the crest placed very near the posterior border, two-thirds of the width of the scapula being in front of the crest. While the length of the skeleton (adult males being compared in each case) in *Macrorhinus leoninus* is twice that of *Cystophora cristata*, and the bulk of the whole animal must be many times greater, the hind limb is only a little larger than in the latter (for detailed measurements see *infra*, pp. 733 and 750). While the humerus and radius are each twice as long in *Macrorhinus* as the corresponding parts in *Cystophora*, the tibia is scarcely a third longer, while the relative length of the pelvis in the two is as 5 to 4!

Cystophora, so far as is certainly known, is represented by only a single species, which is restricted to the colder parts of the North Atlantic. A second species has been attributed to the Caribbean Sea, but, as already shown (*anteà*, p. 720), there seems to be reason for believing the locality to have been wrongly assigned.

CYSTOPHORA CRISTATA (*Erxl.*), *Nilss.*

Hooded Seal.

Phoca leonina, LINNÉ, Syst. Nat., 1766 (in part,—only the reference to Ellis's "Seal with a Cawl"; not *Phoca leonina*, Linné, 1758).—FABRICIUS, Müller's Zool Dan. Prod., 1776, viii ; Faun. Grœnl., 1780, 7 (excluding part of the references; not *Phoca leonina*, Linné, 1758).—WALLACE, Proc. Roy. Phys. Soc. Edinb., 1862, 393.

Klamüts, EGEDE, Det gamle Grönlands Nye Perls., etc., 1741, pl. facing p. 46.

Seal with a Cawl, ELLIS, Voyage to Hudson's Bay, etc., 1748, pl. facing p. 134.

Neitsersoak, CRANZ, Historie von Grönland, 1765.

Hooded Seal, PENNANT, Synop. Quad., 1771, 342 (based on Egede and Cranz).

Klappmüze, SCHREBER, Säugt., iii, 312 (based on the foregoing).

Phoca cristata ERXLEBEN, Syst. Reg. Anim., 1777, 590 (based exclusively on Egede, Ellis, Cranz, Pennant, and Schreber, as above).—GMELIN, Syst. Nat., i, 1778, 64.—FABRICIUS, Skrivter af Naturh.-Selskabet, i, Hefte 2, 1791, 120 (in part only).—KERR, Anim. King., 1792, 126.— DESMAREST, Nouv. Dict. des Sci. Nat., xxv, 1817, 580; Mam., 1820, 241, 371.—DEKAY, Ann. New York Lyc. Nat. Hist., i, 1824, 94, pl. vii.— LUDLOW and KING, ibid., 99 (anatomy).—HARLAN, Faun. Amer., 1825, 106.—GODMAN, Am. Nat. Hist., i, 1826, 336.—LESSON, Dict. Class. d'Hist. Nat., xiii, 1828, 412.—FISCHER, Syn. Mam., 1829, 241.— VON BAER, Bull. Acad. Imp. des Sci. de St. Pétersb., iii, 1838, 350.— HAMILTON, Amphib. Carniv., 1839, 197, pl. xiv (from DeKay).— BLAINVILLE, Ostéogr., Phoca, 1840-1851, pl. v (skull). — SCHINZ, Synop. Mam., i, 1844, 485.—GERVAIS, Zool. et Pal. Français, 1848-1852, 139, pl. xlii (animal, skull, and dentition. Coast of France, accidental); ibid., 1859, 270.

Cystophora cristata, NILSSON, "K. Vet. Akad. Handl. Stockholm, 1837, —"; Wiegmann's Arch. für Naturg., 1841, 326; "Illum. Fig. till Skand. Fauna, 20:de häftet, 1840"; Skand. Faun. Däggdj., 1847, 312.— WAGNER, Schreber's Säugt., vii, 1842, 48.—GRAY, Proc. Zoöl. Soc. Lond., 1849, 91 (variation in dentition); Cat. Seals Brit. Mus., 1850, 36, fig. 13; Cat. Seals and Whales, 1866, 41, fig. 14; Zoölogist, 1872, 3334, 3338 (distribution); Hand List Seals, 1874, 17, fig. 11, pl. xiii (skull, juv.).—GIEBEL, Säugeth., 1855, 142.—BLASIUS, Naturg. Wirbel. Deutschl., i, 1857, 258, figg. 145-147.—MALMGREN, Öfv, af Kongl. Vet.-Akad. Förh. Stockholm, 1863, 134; Arch. für Naturg., 1864, 72.—REINHARDT, Vidensk. Meddel. Natur. Foran., 1864 (1865), 248, 277 (milk-dentition).—GÜNTHER, Zoöl. Rec., ii, 1865, 38 (abstract of Reinhardt's paper on milk-dentition).—GILL, Proc. Essex Inst., v, 1866, 13.—PACKARD, Proc. Bost. Soc. Nat. Hist., x, 1866, 271 (Labrador).—QUENNERSTEDT, K. Svens. Veten.-Akad. Handl., Bd. vii, 1868, No. 3, 25.—BROWN, Proc. Zoöl. Soc. Lond., 1868, 435; Man. Nat. Hist., etc., Greenland, 1875, Mam., 64.—ALLEN, Bull. Mus. Comp. Zoöl., i, 1869, 193 (Massachusetts).—REEKS, Zoölogist, 1871, 2548 (Newfoundland).—WALKER, Scottish Nat., ii, 1873 (St. Andrews, Scotland).—MOORE, Proc. Liverpool Soc., xxvii, 1873, xiii (England).—COBBOLD, Proc. Zoöl. Soc. London, 1873, 741 (same).— VON HEUGLIN, Reisen nach dem Nordpolarmeer, iii, 1874, 66.—LILLJEBORG, Fauna öfv. Sveriges och Norges Ryggradsdjur, 1874, 721.— VAN BENEDEN, Ann. du Mus. roy. d'Hist. Nat. du Belgique, pt. i, 1877, 17 (geogr. distr., with chart).—RINK, Danish Greenland, its People and its Products, 1877, 126, 430.—SCHULTZ. Rep. U. S. Comm. Fish and Fisheries, pt. iii, 1873-74 and 1874-75, 53.

Stemmatopus cristatus, F. CUVIER, Mém. du Mus., xi, 1824, 196, pl. xiii; Dict. des Sci. Nat., xxxi, 1826, 551.—LESSON, Man. de Mam., 1827, 201.— DEKAY, Zoöl. N. Y., i, 1842, 55, pl. xv, fig. 1.—JUKES, Excurs. in Newfoundland, i, 1842, 319.

Mirounga cristata, GRAY, Griffith's An. King., v, 1827, 463.

Phoca cucullata, BODDAERT, "Elen. Anim., 1785, 107."

Phoca mitrata, "Milbert MS.", G. CUVIER, Oss. foss., 3me ed., v, 1825, 210, pl. xviii, fig. 3.—F. CUVIER, Dents de Mam., 1825, 122, pll. xxxviii B, xxxix.—FISCHER, Syn. Mam., 1829, 241:—HAMILTON, Amphib. Carniv., 1839, 204, pl. xv ("from Dict. des Sci. Nat.").

Stemmatopus mitratus, GRAY, "Brooke's Cat. Mus., 1826, 36."

Cystophora borealis, NILSSON, "Skand. Faun., 1820, 383."

Phoca leucopla, THIENEMANN, "Reise im Norden von Europa, etc., 1824, 102, pl. xiii (young); Bull. Sci. Nat., 1825, v, 261."—FISCHER, Syn. Mam., 1829, 337.

Phoca isidorei, LESSON, Rev. Zool., 1843, 256 (Isle d'Oleron, France, accidental).

? *Cystophora antillarum*, GRAY, Proc. Zoöl. Soc. Lond., 1849, 93 (excluding the skin from Jamaica received from Mr. Gosse); "Zoöl. Erebus and Terror, t. ined.," *apud* Gray; Ann. and Mag. Nat. Hist., 1850, 58; Cat. Seals Brit. Mus., 1850, 38; Cat. Seals and Whales, 1866, 43; Hand List Seals, etc., 1874, 18.

"*A Seal new to the British Shores*, CLARKE, 1847, 4to. fig. of animal & skull."

Hood Seal, CARROLL, Seal and Herring Fisheries of Newfoundland, 1873, 13.

Klapmydsen, REINHARDT, Vidensk. Meddel. fra den Naturh. Foren., i, Kjöbenhavn, 1864 (1865), 248 (milk-dentition).

Neitsersoak (♂), *Nesaursalik* (♀), *Kakortak* (young), Greenlandic.

Blåssjälen, Blass-Skäl, Klapmyts, Klappmysta, Swedish.

Tevyak, Russian.

Klapmyds, Danish.

Klappmütze, Blaserobbe, German.

Phoque à capuchon, French.

Hooded Seal, Crested Seal, Bladdernose, English.

EXTERNAL CHARACTERS.—Color above bluish-black, lighter on the sides and ventrally, thickly varied with small irregular spots of whitish; head and limbs nearly uniform black. Sometimes the light grayish-white tint prevails, varied with spots of dark-brown or blackish. Length of full-grown male about 7½ to 8 feet; of full-grown female about 7 feet. The young are born white, with a soft woolly pelage, but this is soon changed for the harsh, stiff covering of the adults, and the color changes to a uniform brownish, or more or less silvery-gray, lighter on the sides, and whitish below.

DeKay describes the male as having the " Head small in proportion to the body, with a moveable muscular bag on its summit, extending from the muzzle to about five inches behind the eyes, and in certain positions nearly covering the internal canthi. This sac is twelve inches long, and, when fully distended, nine inches high, covered with short hairs, and with slight transverse wrinkles. The nostrils are round, each two inches in diameter, and pierced in the anterior part of this hood. When the hood or nasal sac is not inflated, the septum nasi can be distinctly felt, elevated into a ridge about six inches high. . . . Nasal sac bright brown or rufous." *

---

* New York Zoölogy, pt. i, pp. 55, 56.

Nearly all writers, from Fabricius to the present time, speak of the hood as a sexual character.   Mr. Carroll, whose familiarity with the species should render him an authority on this

FIG. 52.—Cystophora cristata.

point, says distinctly: " The female of this species has no hood on the head."*   Mr. Brown, however, observes: "It is asserted by the sealers that this bladder is a sexual mark, and is not found on the female".   But he adds: "I do not think there is any just ground for this belief"; yet he presents no reasons for

---

* Seal and Herring Fisheries of Newfoundland, p. 14.

this statement.* The testimony on this point is too explicit, however, to be set aside on the mere ground of opinion.

Respecting the size of this species, DeKay gives the total length of the male specimen described by him as 90.5 inches, of

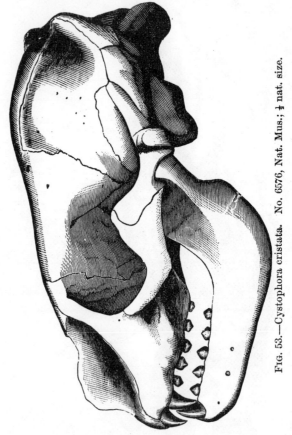

FIG. 53.—Cystophora cristata. No. 6576, Nat. Mus.; ¼ nat. size.

which the tail formed 6.5 inches. He says its weight was 500 to 600 pounds (*loc. cit.*, p. 56). Carroll says the length of the Hooded Seal, "from nose to tail", is 6 to 8 feet, and that the weight of the male, "when in full flesh", varies from 800 to 900 pounds, of which the skin and fat form 350 to 450 pounds, while the skin and fat of the female, "when prime", will weigh only 150 to 200 pounds. This indicates that the female is much smaller than the male (*loc. cit.*, p. 14). Quennerstedt gives the

---

* Proc. Zoöl. Soc. Lond., 1868, p. 436, footnote; Manual Nat. Hist., Geol., etc., Greenland, Mam., 1875, p. 64, footnote.

following measurements : Male, from the point of the nose to the end of the tail, 2190 mm. (7 feet 3 inches, Swedish measure) ; from the point of the nose to the end of the hind extremities 2400 mm.

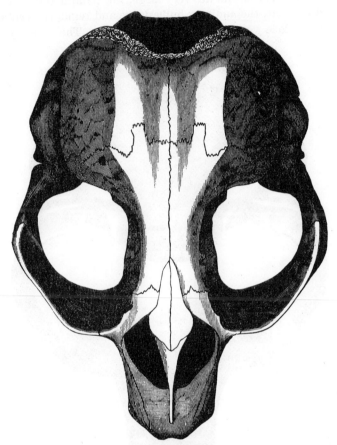

FIG. 54.—Cystophora cristata.   No. 6576, Nat. Mus ; ¼ nat. size.

(8 Swedish feet); length of the fore limb 480 mm.; of the hind limb 420 mm.; of the tail 165 mm.   Female, from the point of the nose to the end of the tail, 1980 mm. (6½ Swedish feet.*) He gives the length of two young ones a few days old as respectively 3 feet 4 inches and 3 feet 9 inches.   I find the length of a skeleton of an adult male, from the point of the nose to the end of the tail, to be 1970 mm.

* K. Sv. Vet.-Akad. Handl., Band vii, No. 3, 1868, pp. 25, 26.

SKELETON AND SKULL.—The skull of the Crested Seal (Figg. 53–56, from a rather young example) is nearly quadrate in general form, with the anteorbital portion abruptly and greatly narrowed. The orbits are very large, the palatal region very broad and flat, the auditory bullæ large and regularly swollen, the brain-case very short and very broad. The interorbital

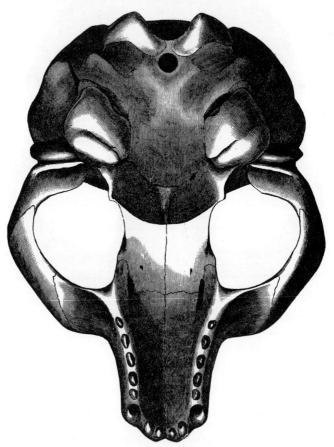

FIG. 55.—Cystophora cristata. No. 6576, Nat. Mus.; ½ nat. size.

region is broad, regularly narrowing anteriorly. The teeth are remarkable for their small plaited (not lobed) crowns and the very large size of the roots, which are not only long but greatly swollen. Usually only the last (5th) upper molar is 2-rooted, and in this the double character of the root is only indicated by the deep longitudinal grooves on each side. Not unfrequently, however, the fourth upper molar is either distinctly double-rooted, or incipiently so, the two fangs being connate and more or less completely united. It sometimes happens that the 4th upper molar is double-rooted on one side and single-rooted on the other. In old age the bones of the skull become greatly

FIG. 56.—Cystophora cristata. No. 6576, Nat. Mus.; ½ nat. size.

thickened, the surface assumes a more or less rugose character, with incipient sagittal and well-developed occipital crests. The sutures, even in old age, remain open nearly throughout the skull.

The usual sexual variations are observable in the smaller size, weaker structure, and smoother surface of the bones in the females. An adult female skull gives a length of 220 mm. against a length of 275 mm. in an old male. Detailed measurements are given in the subjoined table.

Measurements of six skulls of Cystophora cristata.

Measurement	*6577	*6574	*6576	†1084	†1145	†1085
Catalogue number.	*6577	*6574	*6576	†1084	†1145	†1085
Locality.	Greenland	do	do	Newfoundland	do	do
Sex.				♂	♂	♀
Length.	235	202	202	264	275	220
Breadth at mastoid processes.	153	137	144	163	173	147
Greatest breadth at zygomatic arches.	162	152	160	210	220	165
Distance from anterior edge of intermaxillæ to pterygoid hamuli.	132	122	125	165	180	132
Distance from anterior edge of intermaxillæ to hinder edge of last molar.	67	68	66	90	91	74
Distance from anterior edge of intermaxillæ to meatus auditorius.	170	155	150	200	205	165
Distance from anterior edge of intermaxillæ to glenoid process.	153	144	145	188	196	150
Distance from palato-maxillary suture to end of pterygoid hamuli.	65	56	52	68	72	55
Length of the alveolar border of the maxillæ.		80	74	68	123	95
Width of palatal region at posterior end of maxillæ.	66	60	60	72	75	65
Nasal bones, length.	47.5	45	47	59	70	42
Nasal bones, width anteriorly.	19	21	22	30	33	20
Nasal bones, width at fronto-maxillary suture.	10	11	10	17	17	14
Breadth of skull at canines.	44	42.5	43	62	72	45
Least breadth of skull, interorbitally.	30	24	30	40	51	27
Breadth of posterior nares, vertically.	20	16	20	25	30	20
Breadth of posterior nares, transversely.	45	43	47	57	58	47
Breadth of anterior nares, vertically.	42	36	42	70	60	42
Breadth of anterior nares, transversely.	40	38	40	55	52	38
Greatest height of skull at auditory bullæ.	102			115	120	100
Length of brain-case.	91	81	88	86	90	80
Greatest width of brain-case.	115	114	114	120	138	120
Length of lower jaw.	153	144	147	170	185	145
Front edge of ramus to last molar.					75	57
Age.	Adult.	Rather young.	Rather young.	Very old.	Very old.	Old.

* National Museum, Washington, D. C.

† Museum of Comparative Zoölogy, Cambridge, Mass.

In respect to the general skeleton, it may be noted that the bones are very thick in proportion to their length, and of a rather light open structure. The vertebræ are short and broad, especially their centra, the apophysial elements being also thick and short. The scapula is very large, with the crest medial and the acromian process well developed. In general outline and proportions it bears a close resemblance to that of *Phoca grœnlandica*, and is consequently entirely unlike the long, narrow, lunate scapula of *Erignathus barbatus*. The femur is short and stout, being one-third shorter but much thicker than in the last-named species. The same stoutness of form characterizes the humerus and the other limb-bones. While the humerus is of the same length as in the Bearded Seal, it is much thicker, besides differing much in other features. There is thus a strongly marked difference in the relative length of the upper segments of the fore and hind limbs in the two species. The pelvis has the same general form as in the species of the restricted genus *Phoca*. The following measurements are from a disarticulated skeleton of an adult male of apparently medium size:

Measurements of the principal parts of the skeleton in Cystophora cristata (♂ ad.).

	MM.
Length of the skull	265
Length of the cervical vertebræ	275
Length of the dorsal vertebræ	630
Length of the lumbar vertebræ	320
Length of the sacral vertebræ	190
Length of the caudal vertebræ	290
Length of the scapula	213
Length of the humerus	160
Length of the radius	155
Length of the manus	215
Length of the pelvis	318
Length of the femur	127
Length of the tibia	315
Length of the pes	410
Length of the whole skeleton	1,970
Length of the fore limb (excluding scapula)	530
Length of the hind limb	852

GEOGRAPHICAL DISTRIBUTION AND MIGRATIONS. — The Hooded or Crested Seal is restricted to the colder parts of the North Atlantic and to portions of the Arctic Sea. It ranges from Greenland eastward to Spitzbergen and along the Arctic coast of Europe, but is rarely found south of Southern Norway and Newfoundland.   As is the case with other pelagic species, stragglers are sometimes met with far to the southward of the usual range of the species.   On the North American coast it appears to be of uncommon occurrence south of the point already mentioned, as it is said by Gilpin* to be "a rare visitor to the shores of Nova Scotia."   Like the Harp Seal, it appears also to be regularly migratory, but owing to its much smaller numbers and less commercial importance, its movements are not so well known.   Carroll states that it visits the coast of Newfoundland at the same time as the Harp Seal, or about the 25th of February, the time, however, varying with the state of the weather.   He further states that Hooded Seals always keep to the eastward of the Harp Seals, amongst the heavy ice; also that they are quite numerous in spring in the Gulf of Saint Lawrence, where "many of them are killed by persons who reside on St. Paul's Island."†   Dr. Packard states that it "is not uncommonly, during the spring, killed in considerable numbers by the sealers" along the coast of Labrador.‡   Rink says, "It is only occasionally found along the greater part of the coast [of Greenland], but visits the very limited tract between 60° and 61° N. lat., in great numbers, most probably in coming from and returning to the east side of Greenland.   The first time it visits us is from about May 20 till the end of June, during which it yields a very lucrative catch."§   Robert Brown observes, "With regard to the favourite localities of this species of Seal Cranz and the much more accurate Fabricius disagree—the former affirming that they are found mostly on great ice islands where they sleep in an unguarded manner, while the latter states that they delight in the high seas, visiting the land in April, May, and June.   This appears contradictory and confusing; but in reality both authors are right, though not in an exclusive sense."   Again he says : "This Seal is not common anywhere.   On the shores of Greenland it is chiefly found beside large fields of ice, and comes to the coast, as was remarked by Fabricius long ago, at certain times of the year.   They are chiefly found in South Greenland, though it is erroneous to say that they are exclusively confined to that section.   I have seen them not uncommonly about Disco Bay, and have killed them in Melville Bay,

* Proc. and Trans. Nova Scotia Inst. Nat. Sci., vol. iii, pt. 4, p. 884.
† Seal and Herring Fisheries of Newfoundland, pp. 13, 14.
‡ Proc. Bost. Soc. Nat. Hist., vol. x, p. 271.
§ Danish Greenland, etc., 1877, p. 126.

in the most northerly portion of Baffin's Bay. They are principally killed in the district of Julianshaab, and then almost solely in the most southern part, on the outtermost islands, from about the 20th of May to the last of June; but in this short time they supply a great portion of the food of the natives and form a third of the colony's yearly production. In the beginning of July the Klapmyds leaves, but returns in August, when it is much emaciated. Then begins what the Danes in Greenland call the *maigre Klapmydse fangst*, or the 'lean-Klapmyds-catching', which lasts from three to four weeks. Very seldom is a Klapmyds to be got at other places, and especially at other times. The natives call a Klapmyds found single up a fjord by the name of *Nerimartont*, the meaning of which is 'gone after food'. They regularly frequent some small islands not far from Julianshaab, where a good number are caught. After this they go further north, but are lost sight of, and it is not known where they go to (Rink, *l. c.*). Those seen in North Greenland are mere stragglers, wandering from the herd, and are not a continuation of the migrating flocks. Johannes (a very knowing man of Jakobshavn) informed me that generally about the 12th of July a few are killed in Jakobshavn Bay (lat. 69° 13' N.). It is more pelagic in its habits than the other Seals, with the exception of the Saddleback."[*]

I conclude the account of the geographical distribution of the Hooded Seal in Baffin's Bay with the following from Mr. Kumlien's account:

"The bladder-nose appears to be very rare in the upper Cumberland waters. One specimen was procured at Annanactook in autumn, the only one I saw. The Eskimo had no name for it, and said they had not seen it before. I afterward learned that they are occasionally taken about the Kikkerton Islands in spring and autumn. I found their remains in the old kitchenmiddens at Kingwah. A good many individuals were noticed among the pack-ice in Davis's Straits in July."[†]

On the European coast this species is said to be of not very common occurrence on the northern coast of Norway, but more to the southward only stragglers appear to have been met with.[‡] In March and April, according to Malmgren, they are

---

[*] Proc. Zoöl. Soc. Lond., 1868, pp. 436, 437; Man. Nat. Hist., etc., Greenland, Mam., pp. 65, 66.　　　　[†] Bull. U. S. Nat. Mus., No. 15, 1879, p. 64.

[‡] Says Blasius, writing in 1857, "An den südlichen Küstenländern der Nordsee hat man sie bis jetzt noch nicht gesehen."—*Naturgesch. der Säugeth. Deutschlands*, p. 260.

seen about Jan Mayen, and they are said to occur on the coast of Finmark, and at the mouth of the White Sea. Von Baer* and Schultz also state that it is rarely found not only in the White Sea, but along the Timanschen and Mourman coasts. Von Heuglin says it appears to be found in the Spitzbergen waters only on the western coast of these islands,† and states that they are not known to occur at Nova Zembla. He gives its principal range as lying more to the westward, around Iceland and Greenland.

It thus appears that the range of the Crested Seal is restricted mainly to the Arctic waters of the North Atlantic, from Spitzbergen westward to Greenland and Baffin's Bay, and thence southward to Newfoundland. Stragglers have been captured, however, far to the southward of these limits, on both sides of the Atlantic. Thus Gray observes:

"A young specimen has been taken in the River Orwell; at the mouth of the Thames; and at the Island of Oleron, west coast of France, but I greatly doubt if it had not escaped from some ship coming from North America; there is no doubt of the determination of the species. The one caught on the River Orwell, 29th June, 1847, is in the Museum of Ipswich, and was described by Mr. W. B. Clarke, on the 14th August, 1847, in 4to, with a figure of the Seal and skull. The one taken on the Isle d'Oleron is in the Paris Museum, and is figured, with the skull, in Gervais, Zool. and Paléont. Franc., t. 42, and is called *Phoca Isidorei*, by Lesson, in the Rev. Zool., 1843, 256. The young is very like that of *Pagophilus grœnlandicus*, but is immediately known from it by being hairy between the nostrils, and by the grinders being only plated and not lobed on the surface." ‡

Its capture has occurred a few times on the coast of the United States, as far from its usual range even as on the European coast. A large Seal is occasionally seen on the coast of

---

*Bull. Acad. Imp. des Sci. de St. Pétersb., iii, 1838, p. 350.

† Malmgren, writing some years earlier, says that in recent times it has not been observed with certainty at Spitzbergen, though reported as occurring there by Martens and Scoresby. Possibly, he says, during its summer wanderings it may extend to the latitude of Spitzbergen. During Torell's first journey to Spitzbergen a young individual was killed in the vicinity of Bear Island. He says it is only exceptionally taken by the Seal hunters about Jan Mayen, only a comparatively small number being captured.—*Arch. für Naturgesch.*, 1864, p. 72.

‡ J. E. Gray, Zoölogist, 2d ser., vol. vii, 1872, p. 3338.

Massachusetts, which has been supposed to be the Crested Seal, but just what this large Seal is remains still to be determined.* DeKay, in 1824, recorded † the capture of a male example of this species in a small creek that empties into Long Island Sound at East Chester, about fifteen miles from New York City. Twenty years ‡ later he refers to this as the first and only known instance of its occurrence within the limits of the State of New York, where, he says, "it can only be regarded as a rare and accidental visitor." Professor Cope, however, has recorded its capture in the Chesapeake Bay, where he says it has twice occurred.§ The first specimen was recorded in 1865 ¶ as "some species of *Cystophora*", taken near Cambridge, Maryland, on an arm of the Chesapeake Bay, eighteen miles from salt water, by Mr. Daniel M. Henry". The specimen, it is said, "measured 6¾ feet, and weighed, when living, about 330 lbs.". Although Professor Cope adds, "Whether this species is the *C. cristata* or *antillarum*, can not be determined, owing to the imperfection of extant descriptions", there is no reason for doubting that it was really the Crested Seal, a conclusion to which Professor Cope seems to have later arrived. Although Gray's suggestion anent the English specimen naturally arises, namely, transportation from the north in some ship, it seems more probable that they were really wanderers from the usual home of the species.

---

* In my " Catalogue of the Mammals of Massachusetts," I refer to this large Seal as follows, supposing it to be the Hooded Seal : "From accounts I have received from residents along the coast of a seal of very large size observed by them, and occasionally captured, I am led to think this species is not of unfrequent occurrence on the Massachusetts coast. Mr. C. W. Bennett informs me of one taken some years since in the Providence River, a few miles below Providence, which he saw shortly after. From his very particular account of it I cannot doubt that it was of this species. Mr. C. J. Maynard also informs me that a number of specimens have been taken at Ipswich within the past few years, that have weighed from seven hundred to nine hundred pounds. It seems to be most frequent in winter, when it apparently migrates from the north."—*Bull. Mus. Comp. Zoöl.*, vol. i, No. 8, 1869, pp. 193, 194. This identification was made almost solely on the ground of size, taken in connection with the fact that the species had been taken in Long Island Sound near New York City. The question, however, may fairly be raised whether the large Seals more or less frequently seen on the coast of New England are not really the Gray Seal (*Halichœrus grypus*).

† Ann. New York Lyceum Nat. Sci., vol. i, 1824, p. 94.

‡ New York Zoöl., pt. i, p. 56.

§ New Topog. Atlas of Maryland, 1873, p. 16.

¶ Proc. Acad. Nat. Sci. Phila., 1865, p. 273.

Misc. Pub. No. 12——47

GENERAL HISTORY AND NOMENCLATURE.—The first references to the Hooded Seal, or at least the earliest that have any importance in relation to the technical history of the species, are the brief allusions to it made by Egede in 1741, by Ellis in 1748, and by Cranz in 1765, the first two of whom gave each, in addition to their textual notices, a grotesquely rude figure of the animal, although they show clearly that no other species than the present could have been intended. On these references and figures are based Pennant's "Hooded Seal," and Schreber's· "Klappmüze". Primarily they are the basis also of Erxleben's *Phoca cristata* (1777), the first tenable specific designation of the species. Eleven years earlier (1766), however, Fabricius had called the species *Phoca leonina*, apparently confounding it with the *Phoca leonina* of Linné, 1758, which has reference entirely to the Sea Elephant of the southern seas. In 1766 Linné also partly confounded the two species by citing as synonyms of his *Phoca leonina* the Klapmüts of Egede, Ellis's "Seal with a Cawl," etc. Although Fabricius retained for this species the name *Phoca leonina*, in his "Fauna Grœnlandica," published in 1780, he abandoned it in 1791 for *Phoca cristata*, but at the same time kept up the confusion of this species with the southern Sea Elephant by citing the references to that species as synonyms. With slight exceptions, the name *cristata* has since prevailed as the designation of the species, although Boddaert renamed it *cucullata* in 1785, and Nilsson in 1820 applied to it the name *borealis*. Milbert labelled a specimen he sent from New York to the Paris Museum *Phoca mitrata*, which name was published by G. Cuvier in 1825, and subsequently came into some prominence, through the labors of compilers, as that of a supposed second species of Crested Seal. Thienemann, in 1824, described the young under the name *leucopla*, and Lesson, in 1843, added *isidorei*, based, as already noticed, on a specimen captured on the coast of France.

The chief stumbling-block in the technical history of this species has been Cuvier's *Phoca mitrata*. To show how imperfectly the Hooded or Crested Seal was known by a prominent writer on the Pinnipeds as late as 1839, and also to indicate the confusion that arose from the *Phoca mitrata*, I quote a few passages from Dr. Robert Hamilton's "Amphibious Carnivora" (pp. 197, 204-206). He begins by saying that "It is with considerable hesitation we place the Crested Seal in the same genus with the Mitrata". He correctly gives for a "Plate of the Cristata" a copy of that published by DeKay in the "Annals of the New York Lyceum of Natural History," of the New York specimen, but has to regret that DeKay's description was inaccessible to him, and so falls back upon "the accurate Fabricius" for the chief part of his account of the species, adding thereto a few anatomical details from Drs. Ludlow and King, based, like DeKay's figure, on the specimen taken near New York. Of the *Phoca mitrata* he says: "The designation of *Mitred* Seal appears to have been first applied by Camper, and a cranium with this label was found·in his museum, in 1811, by Baron Cuvier. This specimen was supposed [doubtless correctly] to have been procured in the Northern Ocean. Soon after making this observation Cuvier received from Mr. Milbert, of New York, a young animal of this genus, from which a skeleton was prepared, and which was found perfectly to correspond with Camper's specimen. The locality of its capture was not indicated. It has probably been from these materials that the plate in the Pl. de Dict. des Scien. Nat., of which ours is a copy, has been prepared, though this is not expressly stated. The learned author of the work [F. Cuvier] here referred to has certainly been unfortunate in making this animal identical with the Crested Seal." After quoting from Cuvier's and Blainville's accounts of this specimen he concludes by stating: "The dimensions, the habits, and even the locality of this singular species seems to be nearly unknown," and quotes "as the only gleanings we have detected" a part of Cranz's account of his " Neitsersoak," which he (Cranz) says is also called "Clapmutz," and hence is merely Cranz's account of the Crested Seal!

All the names, both vernacular and technical, that have been applied to this species relate to the peculiar inflatable hood of the male. Respecting the vernacular designations, Mr. Brown gives the following: "Popular names.—'*Bladdernose*' or, shortly '*Bladder*' (of northern sealers, Spitzbergen sea [also of Newfoundland]); *Klappmysta* (Swedish); *Klakkekal, Kabbutskobbe* (Northern Norse); *Kiknebb* (Finnish); *Avjor, Fatte-Nuorjo,* and *Oaado* (Lapp); *Klapmyds* (Danish; hence Egede, Greenl., p. 46: the word *Klapmyssen,* used by him on page 62 of the same work, Engl. trans., and supposed by some commentators to be another name, means only *the* Klapmyds, according to the Danish orthography): *Klapmütze* (German, hence Cranz, Greenl., i, p. 125: I have also occasionally heard the English sealers call it by this name, apparently learnt from the Dutch and German sailors). All of these words mean the 'Seal with a *cap* on,' and are derived from the Dutch, who style the frontal appendage of this species a *mutz* or cap, hence the Scotch *mutch*. This prominent characteristic of the Seal is also commemorated in various popular names certain writers have applied to it, such as *Blas-Skäl* (Bladder-Seal) by Nilsson (Skand. Faun., i, p. 312), [hence *Blase-Robbe* by various German writers,] *Hooded Seal* by Pennant (Synopsis, p. 342), *Seal with a caul* by Ellis (Hudson Bay, p. 134), in the French vernacular *Phoque à capuchon*, and in the sealers' name of *Bladdernose*, ♂ *Neitersoak*, ♀ *Nesaursalik* (Greenland), and *Kakortak* (when two years old)." *

HABITS.—As already noted in the account of the geographical distribution of this species, it is, like the Harp Seal, pelagic and migratory, preferring the drift ice of the "high seas" to the vicinity of land, and seems rarely if ever to resort to rocky islands or shores. It brings forth its young on the ice, remote from the land, in March, a week or ten days later than the Harp Seal, with which it appears only rarely to associate, although the two species are often found on neighboring ice-floes.† It is commonly described as the most courageous and combative of the Phocids, often turning fiercely upon its pursuers. Dr. Rink states that its pursuit is hazardous to a man in a frail kayak, and that its destruction is facilitated by the use of the rifle, the hunter first shooting it from the ice-floes and afterward dispatching it with the harpoon from the kayak. Although it will pursue a man and bite him, Brown states that "as long as the memory of the oldest inhabitant of South Greenland extends, only one man in the district of Julianshaab (where they are chiefly captured) has been killed by the bite of the Klapmyds, though not unfrequently the harpoon and line have been broken." Various writers speak of the difficulty of killing it with the seal-club, and state that it is hard to kill with the sealing-gun unless hit on the back of the neck behind

---

* Proc. Zool. Soc. Lond., 1868, pp. 435–436; Man. Nat. Hist. Greenland, etc., Mam., p. 64.

† Says Brown, "It is affirmed, curiously enough, that the *Bladdernose* and the *Saddleback* are rarely or ever [*sic*] found together; they are said to disagree. At all events the latter is generally found on the inside of the pack, while the former is on the outside."—*Proc. Zoöl. Soc. Lond.*, 1868, p. 437; *Man. Nat. Hist. Greenland*, etc., *Mam.*, p. 65. Jukes and Carroll, from entirely independent observations, make substantially the same statement.

the hood, the inflatable "hood" of the male affording no small degree of protection from the effects of the club, or even the ordinary heavy seal-shot. Mr. Carroll says that no matter how large the gun, or how heavy the shot you fire at him, you will not kill him, even if within the length of the gun, unless he rises in the water so that you shoot him in the throat, or he turns the side of his head toward you.

The Hooded Seal is described as very active when in the water. It swims very low, with only the top of the head above the surface. During the rutting season the males wage fierce battles for the possession of the females, the noise of which may be heard miles away. At times the sexes are said to live apart, but associate in families during the breeding season. Their affection for each other, and especially for their young, is represented as very strong, both parents remaining by them with such persistency that the whole family are easily killed. It often happens, says Carroll, that if the female or young one be killed the male will mount the ice and take the dead one in his mouth and bring it into the water, in which act he is very often himself killed. The female is reported to be far less fierce than the male, but even she will allow herself to be killed before she will abandon her young one. Jukes represents the young of this species which he had on shipboard as tamer and more gentle than the young of the Harp Seal, and that when teased it did not offer to scratch and bite so much as did the young Harps.

FOOD.—The food of this species doubtless consists chiefly of fishes of different species. Malmgren supposed it to subsist mainly on those of large size. That it also feeds upon squids, and probably on other mollusks, is evinced by their remains having been found in their stomachs, as well as "the beaks of large cuttle fish."*

HUNTING AND PRODUCTS.—This species, owing to its scarcity, is of relatively small commercial importance, yet many are taken every year by the Newfoundland and Jan Mayen sealers; generally no separate estimates, however, are given of the number taken. Dr. Rink states that the average annual catch in Greenland is 3,000. The flesh is greatly esteemed by the Greenlanders.

The Hooded Seal is usually taken on the ice, but Mr. Reeks states that many are also shot in the spring of the year by the

---

* Jukes, Excurs. in Newfoundland, vol. i, p. 312.

settlers along the coast of Newfoundland. As already stated,
the hood of the male affords such a protection to its owner as
to render the animal so provided very hard to kill with the
ordinary seal-club, or even with a heavy load of shot; and they
are, furthermore, "at times very savage, and it requires great
dexterity on the part of the seal-hunters to keep from being
bitten".

GENUS MACRORHINUS, *F. Cuvier.*

Macrorhine [*Macrorhinus*], F. CUVIER, Mém. du Mus., xi, 1824, 200, pl. xiii.
     Type and sole species, *Phoca proboscidea*, Péron.
Macrorhinus, F. CUVIER, Dict. Sci. Nat., xxxix, 1826, 552; ibid, lix, 1829, 464.
Macrorhyna, GRAY, Griffith's An. King., i, 1827, 180 ("misprint", Gray).
Mirounga, GRAY, Griffith's An. King., v, 1827, 179 (in part).
Rhinophora, WAGLER, Nat. Syst. Amph., 1830, 27.
Cystophora, NILSSON (in part), 1837, and of various later authors.
Morunga, GRAY, List Ost. Spec. Brit. Mus., 1847, 33.

Dental characters as in *Cystophora.* Basisphenoid and basioccipital bones deeply arched and the interbullar space narrow.
Hind feet deeply bilobed, with the claws rudimentary. Males
with an elongated tubular proboscis.

The genus *Macrorhinus* was founded by F. Cuvier in 1824 on
the Proboscis Seal (*Phoca proboscidea,* Péron) or the Elephant
Seal of the Southern Seas, which, until within the last fifteen
years, was supposed to be sole representative of the genus. In
1866 a second species was described by Dr. Gill from the coast
of California, thus adding this remarkable genus to the North
American fauna. As already shown (*anteà,* p. 723), *Macrorhinus* agrees closely with *Cystophora* in dental and cranial
characters, the skull being an exaggerated form of that of
Cystophora, while it differs from the latter in the form of the
nasal appendage of the males, in the form of the hind limbs,
and in the relative size of the fore and hind limbs.

A prominent synonym of *Macrorhinus* is Gray's barbarous
term *Mirounga,* proposed in 1827 for both this species and *Cystophora cristata,* but changed in 1847 to *Morunga* and restricted
to the Elephant Seal of the Southern Seas. Although uniformly
adopted by Gray, and by those who follow Gray as authority,
it is clearly antedated, even in its first form, by three years by
Macrorhinus of F. Cuvier, although Gray's incorrect citation of
F. Cuvier at 1827 makes them apparently synchronous.*

* It may be here noted that Gray persistently gives the date of the eleventh
volume of the "Mémoirs du Museum" containing F. Cuvier's paper, "De
quelque espèces de Phoques et des groupes génériques entre lesquels ils se

MACRORHINUS ANGUSTIROSTRIS, *Gill.*

Californian Sea Elephant.

Macrorhinus angustirostris, GILL, Proc. Essex Inst., v, 1866, 13; Proc. Chicago
    Acad. Sci., i, 1866, 33.—SCAMMON, Proc. Acad. Nat. Sci. Phila.,
    1869, 63; Marine Mam., 1874, 115, pl. xx, figg. 1, 2.
Morunga angustirostris, GRAY, Suppl. Cat. Seals and Whales, 1871, 5.
Sea Elephant, SCAMMON, J. Ross Browne's Resources of the Pacific Slope
    [App.], 129; Overland Monthly, iii, 112–117, Nov. 1870.
Elefante marino, of Mexicans and old Californians.
Elephant Seal; Sea Elephant, English.

EXTERNAL CHARACTERS.—Color light dull yellowish-brown,
varied with gray, rather darker on the back, more yellowish
below. Hair very harsh and stiff. The new coat is said to have
a slightly bluish cast. Mystacial bristles black, in four to six
rows; the longest five to seven inches long; flattened, with
waved or beaded edges. A group of bristles over each eye, the
largest nearly as thick and as long as any of the mystacial
bristles, and two or three on each side of the face, midway between the nose and eye. Extremity of hind flippers deeply
emarginate, hairy, without nails. Fore flippers armed with
strong nails; web deeply notched between the fourth and fifth
digits, slightly so between the third and fourth, and a slight in-
dentation between the second and third. (Description based
on three examples from Santa Barbara Island.)

"The sexes vary much in size, the male being frequently triple
the bulk of the female; the oldest of the former will average
fourteen to sixteen feet; the largest we have ever seen measured
twenty-two feet from tip to tip." "The adult females average
ten feet in length between extremities." *Scammon.* "Round
the under side of the neck, in the oldest males, the animal appears to undergo a change with age; the hair falls off, the skin
thickens and becomes wrinkled—the furrows crossing each other,
producing a checkered surface—and sometimes the throat is
more or less marked with white spots. Its proboscis extends

---

partagent", as 1827, instead of 1824, the correct date. He also cites F. Cuvier's article on "Les Phoques" in the "Dictionnaire des Sciences Naturelles" as published in tome lix, 1829, although it appeared originally in
tome xxix, 1826. The result is a postdating by three years of F. Cuvier's
generic and specific names, favoring, as it happens (accidentally or otherwise), several of Gray's names published in "Griffith's Animal Kingdom"
in 1827.

from opposite the angle of the mouth forward (in the larger males) about fifteen inches, when the creature is in a state of quietude, and the upper surface appears ridgy; but when the

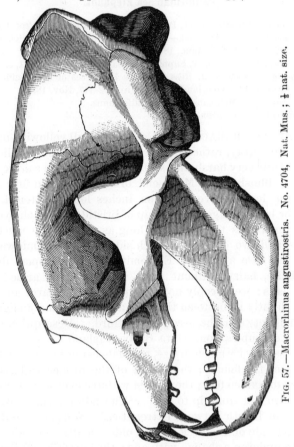

FIG. 57.—*Macrorhinus angustirostris.* No. 4704, Nat. Mus. ; ⅛ nat. size.

animal makes an excited respiration, the trunk becomes elongated, and the ridges nearly disappear." The females "are destitute of the proboscis, the nose being like that of the common seal, but projecting more over the mouth." *Scammon.*

Captain Scammon gives the following measurements, in feet and inches, of two large females taken on the coast of Lower California:

	"No. 1.		No. 2.	
"Length from tip to tip	9	0	10	0
Round the body behind the fore flippers	5	10	5	9
Length of tail	0	2	0	2¼

	No. 1.	No. 2.
Breadth of tail at root	0 2	0 2½
Length of posterior flippers	1 7	1 7
Expansion of posterior flippers	1 8	1 8
Length of fore flippers	1 5	1 2
Width of fore flippers	0 6	0 6
Round extremity of body at root of tail	1 6	1 7

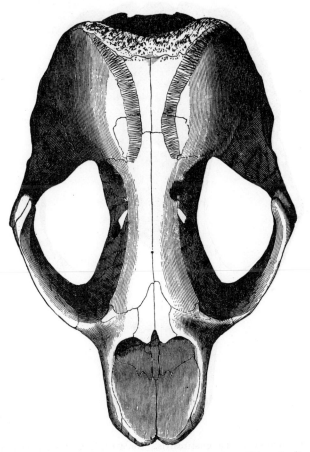

FIG. 58.—Macrorhinus angustirostris. No. 4704, Nat. Mus.; ⅛ nat. size.

	No. 1.	No. 2.
From tip of nose to corner of mouth	0 7	0 8
Opening of mouth	0 4½	0 4½
From tip of nose to eye	0 8	0 9
From tip of nose to fore flippers	2 7	3 0
Length of fissure between the eyelids	0 0	0 1¾ "

Two salted skins, male and female, obtained from Santa Barbara, by Professor O. C. Marsh, which I had an opportunity of
examining as they were unpacked at Professor Ward's establishment, measured respectively from tip of nose to end of tail
5 feet and 4 feet 10 inches.   One of these specimens now meas-

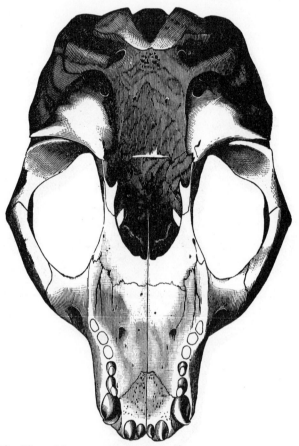

FIG. 59.—Macrorhinus angustirostris.   No. 4704, Nat. Mus.; ⅛ nat. size.

ures, mounted, 6 feet 4 inches.   They were young individuals,
probably yearlings.   In neither was the proboscis at all developed.   Captain Scammon gives the length of a "new-born
pup" as 4 feet.

SKULL.—The skull differs from that of *Cystophora cristata*
as already described, but its general contour is essentially the

same, of which it may be considered as a magnified type. It is somewhat arched above, the region of greatest convexity being at the orbits. The facial portion is abruptly narrowed and somewhat produced, the width of the skull at the base of the malar process of the maxillaries being nearly twice as great as

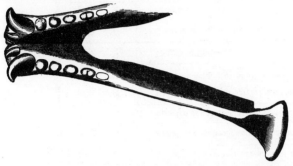

FIG. 60.—Macrorhinus angustirostris. No. 4704, Nat. Mus.; ⅓ nat. size.

at the canines. The following measurements of two examples will serve to indicate its general proportions. One is marked as female, but whether either is full-grown is not known. For purposes of comparison the same table contains measurements of two old male skulls of the Sea Elephant of the Southern Seas.

Measurements of two skulls each of Macrorhinus angustirostris and M. leoninus.

Catalogue number.	Locality.	Sex.	Length.	Breadth at mastoid processes.	Greatest breadth at zygomatic arches.	Distance from anterior edge of inter-maxillæ to pterygoid hamuli.	Distance from anterior edge of inter-maxillæ to hinder edge of last molar.	Distance from anterior edge of inter-maxillæ to glenoid process.	Distance from palato-maxillary suture to end of pterygoid hamuli.	Length of the alveolar border of the maxillæ.	Width of palatal region at posterior end of maxillæ.	Nasal bones, length.	Nasal bones, width anteriorly.	Nasal bones, width at fronto-maxillary suture.	Breadth of skull at canines.	Least breadth of skull interorbitally.	Breadth of posterior nares, vertically.	Breadth of posterior nares, transversely.	Breadth of anterior nares, vertically.	Breadth of anterior nares, transversely.	Greatest height of skull at occipital condyles.	Length of brain-case.	Greatest width of brain-case.	Length of lower jaw.	Front edge of ramus to last molar.	Age.
*15270	Lower California	……	343	184	214	198	108	244	85	134	83	52	38	23	81	39	……	……	60	57	134	123	153	340	82	……
*4704	St. Bartholomew's Bay, Lower California	♀	320	180	200	195	98	228	85	122	85	48	29	21	72	40	52	49	56	52	112	119	146	320	78	……
†1178	Hurd's Island, Indian Ocean.	♂	480	290	360	340	130	365	173	190	170	65	60	32	125	82	55	70	……	……	200	128	213	345	120	Old.
†1179	…do………	♂	510	310	365	330	140	……	145	200	170	65	80	40	160	98	65	67	……	……	190	124	220	350	120	Old.

* *Macrorhinus angustirostris.* Specimens in National Museum, Washington, D.C. No. 4704 is the original type of the species.

† *Macrorhinus leoninus.* Specimens in Museum of Comparative Zoölogy, Cambridge, Mass.

In respect to the skeleton I have no account to render, my material being restricted to the before-mentioned skins and skulls. As a contribution to the history of the genus, however, I append measurements of the principal parts of an old male skeleton of the Southern Sea Elephant, one of two nearly complete skeletons in the Museum of Comparative Zoölogy, obtained by Captain Edwin Church at Hurd's* Island, and received from Mr. R. H. Chapell, of New London.

Measurements of the principal parts of the skeleton in Macrorhinus leoninus (δ ad.).

	MM.
Length of the skull	480
Length of the cervical vertebræ	570
Length of the dorsal vertebræ	1,690
Length of the lumbar vertebræ	670
Length of the sacral vertebræ	250
Length of the caudal vertebræ (approximate)	680
Length of the scapula	330
Length of the humerus	335
Length of the radius	310
Length of the manus (approximate)	550
Length of the pelvis	390
Length of the femur	200
Length of the tibia	415
Length of the pes (approximate)	450
Length of the os penis	335
Length of the whole skeleton	4,340
Length of the fore limb (excluding scapula)	1,195
Length of the hind limb	1,065

Peters† states that the skull of a male, apparently an old animal, from Kerguelen Island, measures 50 centimetres, and the vertebral column 370 centimetres,—substantially the same as the above. The skull is intermediate in length between the two of which measurements have already been given (*anteà*, p. 748); the length of the vertebral column given by Peters is a little (16 centimetres) less, but this may be due to differences in the mode of measurement, as I have included the probable length of the intervertebral cartilage.

COMPARISON WITH THE SOUTHERN SEA ELEPHANT.—So far as can be determined by descriptions, the Northern and

---

* This name I find variously spelled by different writers, to wit: "Herd's," "Heard's," and "Hurd's."

† Monatsb. der K. P. Akad. der Wissensch. zu Berlin, 1877, p. 393, footnote.

the Southern Sea Elephants* differ very little in size, colòr, or
other external features. Captain Scammon gives the average
length of the full-grown male of the northern species as twelve
to fourteen feet, and says that the largest he ever measured had
a length of twenty-two feet "from tip to tip". Péron gives the
length of the southern species as twenty to twenty-five, and
even thirty feet, with a circumference of fifteen to eighteen
feet. Anson gives the length as twelve to twenty feet, and the
circumference as eight to fifteen feet. Pernety records the total
length as twenty-five feet. Scammon gives the length of the
young of the northern species, at birth, as four feet; and Péron
gives four or five feet as the length of the young at birth for
the southern species. The skeletons of the two old males of the
southern species, already mentioned, allowing for the intervertebral cartilages that have disappeared in maceration, measure respectively not over fifteen and sixteen feet, adding to
which the length of the hind flipper and the proboscis gives a
total length, from "tip to tip," of about twenty-one to twenty-
two feet. From the foregoing we may infer that the usual
difference in size between the two species is not great, the

---

* It is here assumed that the Sea Elephants of the Southern Hemisphere
are all referable to a single species, the *Phoca leonina* of Linné, 1758, based
on the Sea Lion of Lord Anson, which was renamed *Phoca elephantina* by
Molina, in 1782, and again renamed *Phoca proboscidea* by Péron, in 1816, and
of which *Phoca byroni* of Desmarest, and also *Phoca ansoni* of the same author (the latter species in part only), and the *Mirounga patagonica* of Gray
are synonyms. I am aware, however, that Peters has recently proposed
the recognition of four species, namely, *Cystophora leonina* (= Anson's Sea
Lion), *C. falklandica* (= Pernety's Sea Lion), *C. proboscidea* (ex Péron), and
C. kerguelensis (the Sea Elephant of Kerguelen Island). He seems not, however, to have arrived at this course by an examination of an extensive suite
of specimens from various localities, as he refers in this connection to only
a single old male example from Kerguelen Island. He seems to have been
influenced merely by the varying statements in respect to size and some
other features given by Pernety, Anson, and Péron. His entire prešenta-
tion of the case is as follows: "Pernety gibt von seinem Seelöwen eine lange
Mähne, eine Totallänge von 25 Fuss und einen Durchmesser der Basis der
Eckzähne von 3 Zoll an. Pérons See-Elephanten sollen bis 30 Fuss lang und
von blaugrauer Farbe sein. Vielleicht sind alle diese Arten verschieden
und es würde dann der Name *C. leonina* L. bloss dem Anson'schen Seelöwen
zu belassen sein, während die *C. falklandica*, wie man die von Pernety benennen könnte, die *C. proboscidea* Péron, die *C. angustirostris* Gill der nörd-
lichen Hemisphäre und die von Kerguelenland besonderen Arten angehören
würden. Für den letzteren Fall schlage ich vor, diese Art *kerguelensis* zu
benennen" (Monatsb. d. K. P. Akad. Wissensch. zu Berlin. (1875, p. 394,
footnote).

southern species on the whole appearing to be somewhat the larger of the two.

In respect to color, Eaton says of the Sea Elephant of Kerguelen Island: "Some examples are uniformly reddish brown; others are pale, blotched and spotted with darker grey".* Peters describes an old example from the same locality as dark brown, the hairs being white at the base, and brown or blackish-brown at the tips. Péron gives the color as sometimes grayish, sometimes bluish-gray, more rarely brownish-black. Lizars† has described the color of a female as dark olive brown, shading to a yellowish-bay upon the belly. Scammon gives the color of California examples as light brown, but adds that the new coat has a bluish cast. It thus appears that the two species present a striking similarity of coloration.

I have before me two skulls of the northern species, both of which are those of adult females, and also four skulls of the southern species, but of these two are those of adult males, and the others those of animals probably less than three months old. Of the two young skulls, one is from Kerguelen Island (Kidder), and the other from the "Cape of Good Hope" (Layard). The adult skulls being of different sexes are not comparable. They differ greatly in size (see measurements, *anteà*, p. 748), and necessarily in other features, but to what extent these differences are merely sexual cannot be stated.

From the foregoing it would seem that the northern and southern Sea Elephants, though presumably distinct, are closely allied, as well in structural characters as in habits. In respect to geographical distribution, I am not aware that the southern species has been found north of about the 35th degree of south latitude (the island of Juan Fernandez), or the northern species south of about the 24th degree of north latitude. It may consequently be safely assumed that the two forms have been long isolated, and that the southern is an offshoot from northern stock, since the only other known species of the *Cystophorinæ* is also northern.

GEOGRAPHICAL DISTRIBUTION.—The Sea Elephant seems to have been formerly very abundant on the coast of California and Western Mexico, whence it became long since nearly

---

* Proc. Roy. Soc. London, vol. xxiii, 1875, p. 502. See also Philosoph. Transact., vol. clxviii, 1879, p. 96.

† Hamilton's Marine Amphibia, p. 211.

extirpated. Captain Scammon, in writing (about 1852) of Cedros Island, off the coast of Lower California, says: "Seals and Sea Elephants once basked upon the shores of this isolated spot in vast numbers, and in years past its surrounding shores teemed with sealers, Sea Elephant and Sea Otter hunters; the remains of their rude stone-houses are still to be seen in many convenient places, which were once the habitations of these hardy men."* A few Sea Elephants are still found at Santa Barbara Island, where they are reported, however, to be nearly extinct. Whether or not they still occur elsewhere along the California coast I am without means of determining, although it is probable that a small remnant still exists at other points, where scarcely more than a quarter of a century ago vessels were freighted with their oil. Neither is it possible to determine with certainty the limits of their former range. Captain Scammon, who doubtless obtained his information from trustworthy sources, states that it extended from Cape Lazaro, latitude 24° 46′ north, to Point Reyes, in latitude 38°, or for a distance of about two hundred miles. As already stated (anteà, p. 290), Dampier, in 1686, met with Seals on the islands off the western coast of Mexico, as far south as latitude 21° to 23°, but of what species his record unfortunately fails to show. They were doubtless either Sea Elephants or Sea Lions (*Zalophus californianus*), and may have included both. This rather implies its former extension, two hundred years ago, considerably to the southward of the limit assigned by Captain Scammon, on probably traditional reports current among the residents of this part of the coast at the time of his visit there in 1852.

GENERAL HISTORY.—The California Sea Elephant was first described by Dr. Gill, in 1866, from a skull of a female in the Museum of the Smithsonian Institution, received from Saint Bartholomew's Bay, Lower California. Its external characters were first made known by Captain C. M. Scammon in 1869, and the species was redescribed by him in 1874, with detailed measurements of two adult females and a newly-born pup. This is all that has thus far appeared relating to its technical history. Captain Scammon, as early as 1854, gave some account of the habits of this species, under the name Sea Elephant, and earlier incidental references to it doubtless occur in the narratives

---

* In J. Ross Browne's "Resources of the Pacific Slope" [App.], p. 129.

of travellers. Dr. Gill observes, in his paper already cited, "For a long time, the fact that a species of the genus *Macrorhinus* or Elephant-Seal inhabits the coast of Western North America has been well-known. But, on account of the want of opportunity for comparison of specimens, the relations of the species have not been understood". I fail to find, however, in any technical account of the Sea Elephant, any previous notice of their occurrence on the coast of North America. Dr. Gill compares his specimen of the skull with the figure of the skull of *Morunga elephantina* (=*Macrorhinus leoninus*) published by Dr. Gray in the "Zoölogy of the Erebus and Terror", and says if that "represents an equally old female, the present species must be very distinct". He adds, "I do not know the size of the original of that gentleman's figure. Some of the differences, however, cannot be the effect of age, and there can exist no doubt that the present form is at least distinct from those described by the Cuviers, Blainville, and Gray. In allusion to the peculiarly narrowed and produced snout of the female, the name *Macrorhinus angustirostris* is conferred upon it." *

HABITS.—We are indebted to Captain Scammon, who has fortunately had favorable opportunities for observation, for everything of importance that has thus far been recorded respecting the habits of the Sea Elephant of California. "The habits of these huge beasts," he tells us,† "when on shore, or loitering about the foaming breakers, are in many respects like those of the Leopard Seals [*Phoca vitulina*]. Our observations on the Sea Elephants of California go to show that they have been found in much larger numbers from February to June than during other months of the year; but more or less were at all times found on shore upon their favorite beaches, which were about the islands of Santa Barbara, Cerros, Guadalupe, San Bonitos, Natividad, San Roque, and Asuncion, and some

* "On a new species of the genus *Macrorhinus*." Proc. Chicago Acad., i, 1866, pp. 33–34.
† Marine Mammalia, 1874, pp. 117–119. See also Proc. Acad. Nat. Sci. Phila., 1869, pp. 63–65, where the account here quoted was first published. See further J. Ross Browne's Resources of the Pacific Coast [Append.], p. 129, where the same author has also given a short account of its habits as observed at Cedros (or Cerros) Island in 1852. Also an article entitled "Sea-Elephant Hunting," in the "Overland Monthly" magazine, iii, pp. 112–117, Nov., 1870.

of the most inaccessible points on the main-land between Asuncion and Cerros. When coming up out of the water, they
were generally first seen near the line of surf; then crawling up
by degrees, frequently reclining as if to sleep; again, moving
up or along the shore, appearing not content with their last
resting-place. In this manner they would ascend the ravines,
or 'low-downs,' half a mile or more, congregating by hundreds.
They are not so active on land as the seals; but, when excited
to inordinate exertion, their motions are quick—the whole body
quivering with their crawling, semi-vaulting gait, and the animal at such times manifesting great fatigue. Notwithstanding
their unwieldiness, we have sometimes found them on broken
and elevated ground, fifty or sixty feet above the sea.

"The principal seasons of their coming on shore, are, when
they are about to shed their coats, when the females bring forth
their young (which is one at a time, rarely two), and the mating season. These seasons for 'hauling up' are more marked
in southern latitudes. The different periods are known among
the hunters as the 'pupping cow,' 'brown cow,' 'bull and cow,'
and 'March bull' seasons;* but on the California coast, either
from the influence of climate or some other cause, we have noticed young pups with their mothers at quite the opposite
months. The continual hunting of the animals may possibly
have driven them to irregularities. The time of gestation is
supposed to be about three-fourths of the year. The most
marked season we could discover was that of the adult males,
which shed their coats later than the younger ones and the females. Still, among a herd of the largest of those fully matured (at Santa Barbara Island, in June, 1852), we found several cows and their young, the latter apparently but a few days
old.

" When the Sea Elephants come on shore for the purpose of
'shedding', if not disturbed they remain out of water until the
old hair falls off. By the time this change comes about, the
animal is supposed to lose half its fat; indeed, it sometimes
becomes very thin, and is then called a 'slim-skin'.

"In the stomach of the Sea Elephant a few pebbles are
found, which has given rise to the saying that 'they take in
ballast before going down' (returning to the sea). On warm

---

* Referring to the habits of the Southern Sea Elephant (*Macrorhinus leo-
ninus*), as he had " learned from ship masters who have taken Seals about
Kerguelen's Land, the Crozets, and Hurd's Island." See Proc. Acad. Nat.
Sci. Phila., 1869, p. 64.

and sunny days we have watched them come up singly on smooth beaches, and burrow in the dry sand, throwing over their backs the lose particles that collect about their fore limbs, and nearly covering themselves from view; but when not disturbed, the animals follow their gregarious propensity, and collect in large herds." "The largest number I ever found in one herd," he states in another connection, "was one hundred and sixty-five, which lay promiscuously along the beach or up the ravine near by."

Nothing further respecting the breeding habits or sexual relations of the species appears to have been as yet recorded, but they may be presumed to be similar to those of the Sea Elephant of the Antarctic Seas.*

CHASE AND PRODUCTS.—The mode of capturing Sea Elephants on the coast of California has been described by Captain Scammon, who "in 1852, when the 'gold fever' raged", was compelled by force of circumstances "to take command of a brig bound on a 'sealing, Sea-elephant and whaling voyage' or abandon sea life—at least temporarily". He says: "The sailors get between the herd and the water; then, raising all possible noise by shouting, and at the same time flourishing clubs, guns, and lances, the party advance slowly toward the rookery, when the animals will retreat, appearing in a state of great alarm. Occasionally an overgrown male will give battle, or attempt to escape; but a musket-ball through the brain dispatches it; or some one checks its progress by thrusting a lance into the roof of its mouth, which causes it to settle on its haunches, when two men with heavy oaken clubs give the creature repeated blows about the head, until it is stunned or killed. After securing those that are disposed to show resistance, the party rush on the main body. The onslaught creates such a panic among these peculiar creatures, that, losing all control of their actions, they climb, roll, and tumble over each other, when pre-

---

* The Sea Elephants appear to be exceptional among the *Phocidæ* in the great disparity of size between the sexes, in which, as well as in their breeding habits, they closely resemble the Otaries. Although, unlike the latter, they have not the power of using the hind limbs in locomotion on land, and are hence unable to walk, they manage to crawl to a considerable distance from the sea,—according to Scammon, a "half a mile or more". The habits of the Southern Sea Elephant (*Macrorhinus leoninus*) were long since described by Anson and Pernety, and later by Péron, but their accounts seem in some respects to be tinged with romance. According to these writers the males fight desperately for the possession of the females.

vented from farther retreat by the projecting cliffs. We recollect in one instance, where sixty-five were captured, that several were found showing no signs of having been either clubbed or lanced, but were smothered by numbers of their kind heaped upon them. The whole flock, when attacked, manifested alarm by their peculiar roar, the sound of which, among the largest males, is nearly as loud as the lowing of an ox, but more prolonged in one strain, accompanied by a rattling noise in the throat. The quantity of blood in this species of the seal tribe is supposed to be double that contained in an ox, in proportion to its size.

"After the capture the flaying begins. First, with a large knife, the skin is ripped along the upper side of the body the whole length, and then cut down as far as is practicable, without rolling it over; then the coating of fat that lies between the skin and flesh—which may be from one to seven inches in thickness, according to the size and condition of the animal—is cut into 'horse-pieces', about eight inches wide, and twelve to fifteen long, and a puncture is made in each piece sufficiently large to pass a rope through. After flensing the upper portion of the body, it is rolled over, and cut all around as above described. Then the 'horse-pieces' are strung on a raft-rope (a rope three fathoms long, with an eye-splice in one end), and taken to the edge of the surf; a long line is made fast to it, the end of which is thrown to a boat lying just outside the breakers; they are then hauled through the rollers and towed to the vessel, where the oil is tried out by boiling the blubber, or fat, in large pots set in a brick furnace for the purpose. The oil produced is superior to whale oil for lubricating purposes. Owing to the continual pursuit of the animals, they have become nearly if not quite extinct on the California coast, or the few remaining have fled to some unknown point for security." * He also states that a fat bull, eighteen feet long, taken by the brig "Mary Helen", in 1852, yielded two hundred and ten gallons of oil.

---

* Marine Mammalia, pp. 118, 119.

APPENDIX.

A.—MATERIAL EXAMINED.*

FAMILY ODOBÆNIDÆ.

ODOBÆNUS ROSMARUS.

List† of specimens examined.

Catalogue number.	Sex.	Locality.	In collection of—	Nature of specimen.	Age.
7156	♀	North Greenland	National Museum.	Skull	Old.
4645	♀	Greenland	....do	....do	Very old.
9570	♀	......do	....do	....do	Young.
	♂	......do	....do	Skin	Adult.
1860	♀	Sable Island	....do	Skull	Old.
	♂	Greenland	Boston Soc.N.Hist.	....do	Young.
1720	♀	......do	Mus. Comp. Zoöl	....do	Old.
1721	♂	......do	....do	Skeleton	Old.
	♂	......do	....do	Skull	Old.
....	....	......do	....do	....do	Young.
	♂	Davis's Straits	Prof. H. A. Ward	Skin and skull	Adult.
	♂	......do	....do	....do	Adult.
	♂	......do	....do	....do	Young.
	....	Greenland	Amherst College	Skin	Young.

*Only that relating to North American species is recorded in the subjoined tables.
† In part only; quite a number of skulls have been examined in different museums that are not here recorded.

ODOBÆNUS OBESUS.

List of specimens examined.

Catalogue number.	Sex.	Locality.	In collection of—	Nature of specimen.	Age.
11746	♂	Walrus Island, Alaska	National Museum.	Skull.............	Very old.
9475	♂	......do	...do	....do	Very old.
14395	♂	......do	...do	....do	Very old.
14396	♂	......do	...do	....do	Young.
14397	♂	......do	...do	....do	Very old.
6780	♂	North Pacific	...do	....do	Very old.
7889	♂	......do	...do	....do	Very old.
	♀	Walrus Island, Alaska	... do	Skeleton	Very old.
	♂	.....do	...do	Skin	Very old.
	♂	North Pacific	Prof. H. A. Ward ..	Skin and skull ..	Young.
	♂	Walrus Island, Alaska....	Boston Soc.N.Hist.	Skeleton	Very old.
	♂	......do	Mus. Comp. Zoöl ..	....do	Very old.
	♂	......do	...do	Skin	Very old.
	♂	......do	...do	Skull.............	Very old.
	♂	......do	...do	....do	Very old.
	♂	......do	...do	....do	Very old.
...		......do	...do	....do	Young.
	♂	North Pacific	...do	....do	Very old.
....		......do	Boston Soc.N.Hist.	....do	Young.
...		......do	...do	....do	Young.

FAMILY OTARIIDÆ.

EUMETOPIAS STELLERI.

List of specimens examined.

Catalogue number.	Sex.	Locality.	In collection of—	Collected by-	Nature of specimen.
2920	♂	St. Paul's Island, Alaska .	Museum of Comparative Zoölogy.	Charles Bryant..	Skin and skeleton.
2921	♂	......do	...do	....do	Do.
11675	♂	......do	National Museum.	....do	Skin.
......	♂	......do	...do ....,.	....do	Do.
8166	♂	St. George's Island, Alaska	....do	W. H. Dall	Skull.
8162	♀	......do	...do	....do	Do.
8163	♀	......do	...do	....do	Do.
15359	♂	Unalashka..................	...do	....do	Do.
4703	O ♂	San Francisco, Cal.........	....do	Dr. W. O. Ayres.	Do.
4702	♂	......do	...do	....do	Do.
6906	♂	Monterey, Cal	....do	Dr. C. S. Canfield	Do.
3631	♂	......do	...do	A. S. Taylor	Do.
4701	♂	Farallone Island, Cal	....do	Dr. W. O. Ayres.	Do.
13217	♂	......do	...do	C. M. Scammon..	Do.
1767	♂	......do	Mus. Comp. Zoöl ..	F. Bierstadt.....	Do.

ZALOPHUS CALIFORNIANUS.

List of specimens examined.

Catalogue number of skull.	Corresponding number of skin.	Original number.	Sex.	Locality.	In collection of—	Collected by—	Nature of specimen.	Age.
			♂	California	National Museum.		Skull......	Very old.
261			♂	...do	...do		....do	Very old.
14506			♂	Santa Barbara Island, Cal.	...do	Capt. C. M. Scammon.	....do	Rather young.
14507			♀	...do	...do	....do	....do	Adult.
15254			♂	San Nicolas Island, Cal.	...do	P. Schumacher.	....do	Middle-aged.
15255			♂	...do	...do	....do	....do	Middle-aged.
15660			O	California.	...do	....do	....do	Few days old.
15661			O	...do	...do	....do	....do	Few days old.
6159			♀	Santa Barbara Island, Cal.	Museum of Comparative Zoölogy.	....do	Skeleton ..	Old.
6160			O	...do	...do	....do	....do	About four months.
6158	5839		O	...do	...do	....do	Skull and skin.	Fœtal.
6156			O	...do	...do	....do	...do	A few months.
6153		5	O	...do	...do	....do	....do	A few months.
6154	5876	3	O	...do	...do	....do	Skull......	A few months.
6155		4	O	...do	...do	....do	....do	A few months.
6146	5786		♂	...do	...do	....do	Skull and skin.	Old.
6150	5787		♀	...do	...do	....do	...do	Old.
6156	5788	10	♀	...do	...do	....do	....do	Adult.
6152	5677	11	♀	...do	...do	....do	....do	Adult.
6147	5785	3	♂	...do	...do	....do	....do	Nearly adult.
6148	5789	8	♂	...do	...do	....do	....do	Rather young.
6149	5678		♀	...do	...do	....do	....do	Middle-aged.
6161			♂	...do	...do	....do	....do	Rather young.
6162			♂	...do	...do	....do	....do	Two or three years.
6163			♂	...do	....do	....do	....do	Two or three years.
6164			♂	...do	...do	...do	...do	
6165			♀	...do	...do	...do	....do	Adult.
6166			♀	...do	...do	...do	....do	Adult.
1132			♂	San Diego, Cal.	...do	Hassler Expedition.	Skull......	Young.

CALLORHINUS URSINUS.

List of specimens examined.

Catalogue number.	Sex.	Locality.	In collection of—	Collected by—	Nature of specimen.
11726	♂	Saint Paul's Island, Alaska	National Museum.	Capt. C. Bryant .	Skull.
11695	♂	......do	....do	....do	Skull.
11715	♂	......do	....do	....do	Skull.
11701	♂	......do	....do	....do	Skull.
11733	♂	......do	....do	....do	Skull.
11698	♂	......do	....do	....do	Skull.
11689	♂	......do	....do	....do	Skull.
7109	♀	......do	....do	Colonel Bulkley.	Skull.
14508	♀	......do	....do	....do	Skull.
12737	♀	......do	....do	H. W. Elliott....	Skull.
12738	♀	......do	....do	....do	Skull.
12739	♀	......do	....do	....do	Skull.
8415	♀	Aleutian Islands.............	....do	Dr. T. T. Minor .	Skull.
8825	O	......do	....do	....do	Skull.
11270	♀	......do	....do	....do	Skull.
9080	♀	Straits of Juan de Fuca...	....do	Capt. C. M. Scam-mon.	Skull.
9079	♀	......do	do	....do	Skull.
1839	♂	......do	....do	....do	Skull.
6536	♀	Puget Sound.................	....do	J. G. Swan......	Skull.
6537	♀	......do	....do	....do	Skull.
6558	♀	......do	....do	....do	Skull.
6911	♀	......do	....do	....do	Skull.
	♂	Saint Paul's Island, Alaska	....do	H. W. Elliott....	Skin.
	♀	......do	....do	....do	Skin.
	♀	......do	....do	....do	Skin.
	O	......do	....do	....do	Skin.
	O	......do	....do	....do	Skin.
	O	......do	....do	....do	Skin.
2922	♂	......do	Mus. Comp. Zoöl ..	Capt. C. Bryant .	Skin.
2923	♂	......do	....do	....do	Skin.
2924	♀	......do	....do	....do	Skin.
2925	♀	......do	....do	....do	Skin.
2926	O	......do	....do	....do	Skin.
2927	O	......do	....do	....do	Skin.
......	♂	......do	....do	....do	Skeleton.
1088	♂	......do	....do	....do	Skeleton.
1086	♀	......do	....do	....do	Skeleton.
1087	♀	......do	....do	....do	Skeleton.
1149	O	......do	....do	....do	Skull.
1150	O	......do	....do	....do	Skull.
1787	♂	......do	....do	Alaska Com. Co.	Skeleton.
1788	♂	......do	....do	....do	Skeleton.
1789	♂	......do	....do	....do	Skeleton.
1784	♀	......do	....do	....do	Skeleton.
1785	♀	......do	....do	....do	Skeleton.
1786	♂	......do	....do	....do	Skeleton.
5325	O	......do	....do	C. J. McIntyre ..	Skin.

FAMILY PHOCIDÆ.

PHOCA VITULINA.

List of specimens examined.

Catalogue number.	Sex.	Locality.	In collection of—	Collected by—	Nature of specimen.
3506	♂	Greenland	National Museum.	....	Skull. *
3634	....	Sable Island, N. S..........	....do	P. S. Dodd.......	Skull.†
4716	♀	......do	....do	...do	Skull.
4713	♂	......do	...do	...do	Skull.‡
4714	....	......do	...do	...do	Skull.
3506	....	......do	...do	...do	Skull.
5285	....	England	Mus. Comp. Zoöl ..	....	Skeleton.
1788	....	Penekese Island, Mass....	...do	....	Skeleton.
1143	....	Nahant, Mass...........	....do	L. Agassiz.......	Skeleton.
1142	....	Massachusetts..........	...do	....	Skeleton.
5144	....	Beverly, Mass	....do	W. R. Cabot.....	Skeleton.
6157	....	Santa Barbara Island, Cal.	...do	P. Schumacher..	Skull. §
12284	....	Aleutian Islands..........	National Museum.	W. H. Dall	Skull. ‡
11742	....	St. Paul's Island, Alaska..	....do	Charles Bryant..	Skull.
14337	♂	......do	...do	H. W. Elliott....	Skull.
9480	....	Kanai Island, Alaska	...do	F. Bischoff......	Skull.
6783	♂	Plover Bay, Behring's Sts.	....do	Col. Buckley	Skull.
6559	....	Washington Territory....	....do	J. G. Swan.......	Skull,
6485	....	......do	....do	Lt. G. W. White.	Skull.
6486	....	......do	....do	...do	Skull.
6535	....	......do	....do	S. Jones........	Skull.
9081	....	Straits of Juan de Fuca...	....do	Capt. C. M. Scammon.	Skull.
3931	....	San Francisco, Cal	....do	U. S. Ex. Ex....	Skull. ‡
3648	....	"Deception Island"	....do	...do	Skull. ‖
3741	....	......do	....do	...do	Skin. ‖
7778	....	Alaska	....do	W. H. Dall.......	Skin.
7779	....	......do	....do	...do	Skin.
7789	....	Washington Territory	...do	Lt. G. W. White.	Skin.
7781	....	...do	...do	...do	Skin.
9516	....	Kanai, Alaska	...do	W. H. Dall.......	Skin.
8201	....	Sealand, Denmark	...do	Copenhagen Mus.	Skin.
12043	....	....	...do	....	Skin.
5852	....	Sable Island, N. S.........	...do	P. S Dodd.......	Skin.
5853	....	......do	...do	...do	Skin.
3742	....	California..............	...do	....	Skin.
7782	....	Washington Territory	...do	....	Skin.
3320	....	......do	...do	Dr. J. S. Newberry	Skin.
5680	....	Santa Barbara Island, Cal.	Mus. Comp. Zoöl ..	P. Schumacher ..	M't'd skin.
5783	....	......do	...do	...do	M't'd skin.
5784	....	......do	....do	...do	M't'd skin.
......	....	Orleans, Mass..........	....do	J. A. Allen	M't'd skin.

* Adult. † Old. ‡ Young. § Very old.

‖ The type of "*Halichœrus antarcticus*, Peale," and "*Lobodon carcinophaga*, Cassin."

PHOCA FŒTIDA.

List of specimens examined.

Catalogue number.	Sex.	Locality.	In collection of—	Collected by—	Nature of specimen.
6560	....	Greenland	National Museum.	Copenhagen Mus.	Skull.
6567	....	......do	....do	....do	Skull.
6568	....	......do	....do	....do	Skull.
6565	....	......do	....do	....do	Skull.
3504	....	......do	....do	S. Sternberg.....	Skull.
3503	....	......do	....do	....do	Skull.
8698	....	Disco Bay, Greenland.....	....do	Copenhagen Mus.	Skull.
8699	....	......do	....do	... do ...:......	Skull.
8700	....	......do	....do	....do	Skull.
8942	....	Unalakleet, Alaska	....do	W. H. Dall......	Skull.
7105	...	Plover Bay, Behring's Sts.	....do	....do	Skull.
6785	...	......do	....do	....do	Skull.
6297	♀	Gulf of Cumberland	Mus. Comp. Zoöl.	L. Kumlien......	Skeleton.*
6295	♀	......do	....do	....do	Skull.*
6296	♂	......do	....do	...do	Skull.*
6298	♀	......do	....do	....do	Skeleton.†
16106	....	......do	National Museum.	....do	Skeleton.*
16077	♀	......do	....do	....do	Skeleton.‡
16061	..	......do	....do	....do	Skeleton.§
16033	....	......do	....do	....do	Skeleton.‖
16034	....	......do	....do	....do	Skeleton.‖
16050	....	......do	....do	....do	Skeleton.‖
16098	♀	......do	....do	....do	Skeleton.§
16138	♂	......do	....do	....do	Skull.*
16139	♂	......do	....do	....do	Skull.*
16111	♂	......do	....do	....do	Skull.*
16023	♂	......do	....do	....do	Skull.‡
16051	♀	......do	....do	....do	Skull.*
16058	....	......do	....do	....do	Skull.†
16135	....	......do	....do	....do	Skull.†
16137	....	......do	....do	....do	Skull.†
16082	....	......do	....do	....do	Skull.†
16101	....	......do	....do	....do	Skull.†
16134	....	......do	....do	....do	Skull.†
16068	....	......do	....do	....do	Skull.†
16066	....	......do	....do	....do	Skull.†
16136	....	......do	....do	....do	Skull.†
16134	..	......do	....do	....do	Skull.‖
16055	....	......do	....do	....do	Skull.‖

*Adult. †Young in white coat. ‡Young. §Few days old. ‖Fœtal.

PHOCA GRŒNLANDICA.

List of specimens examined.

Catalogue number.	Sex.	Locality.	In collection of—	Collected by—	Nature of specimen.
3514	....	Disco Bay, Greenland.....	National Museum.	Copenhagen Mus.	Skull.
3515	....	......do	....do	....do	Skull.
3805	....	......do	....do	....do	Skull.
881	....	(?)	....do	(?)	Skull.
8762	....	Disco Bay, Greenland.....	....do	Copenhagen Mus.	Skin.
4582	....	Hudson's Bay?..............	....do	(?)	Skin.*
12039	....	St. John's, Newfoundland .	....do	J. M. Harvey....	Skin.*
12040	....	......do	....do	....do	Skin.*
5791	....	Newfoundland.............	Mus. Comp. Zoöl..	....	Skin.*
5139	....	"Massachusetts"..........	....do	....	Skeleton.
1146	....	"Nahant, Mass."..........	....do	"L. Agassiz" ...	Skeleton.

*In the white coat.

ERIGNATHUS BARBATUS.

List of specimens examined.

Catalogue number.	Sex.	Locality.	In collection of—	Collected by—	Nature of specimen.
6569	♂	Greenland	National Museum.	Copenhagen Mus.	Skull.
6570	♂	......do	....do	....do	Skull.
6571	♂	......do	....do	....do	Skull.
7103	....	Plover Bay, Behring's Straits.	....do	Col. Buckley....	Skull.
7104	♀	......do	....do	....do	Skull.
7106	♀	......do	....do	....do	Skull.
7107	♀	......do	....do	....do	Skull.
15685	....	Cumberland Gulf..........	....do	W. A. Mintzer ..	Skull.
5697	....	Disco Bay, Greenland.....	....do	Copenhagen Mus.	Skin.
16116	♀	Cumberland Gulf.........	....do	L. Kumlien	Skeleton.
16117	....	......do	....do	....do	Skull.
16112	....	......do	....do	....do	Skeleton.*
6299	♀	......do	Mus. Comp. Zoöl ..	....do	Skull.

*Fœtal.

HISTRIOPHOCA FASCIATA.

List of specimens examined.

Catalogue number.	Sex.	Locality.	In collection of—	Collected by—	Nature of specimen.
9311	....	Cape Romanzoff	National Museum.	W. H. Dall	Skin.

HALICHŒRUS GRYPUS.

List of specimens examined.

Catalogue number.	Sex.	Locality.	In collection of—	Collected by—	Nature of specimen.
4717	♂	Sable Island, Nova Scotia	National Museum.	P. S. Dodd	Skull.*
6593	....	Greenland	....do	Copenhagen Mus.	Skull.†
14011	....	St. John's, Newfoundland .	....do	J. M. Harvey ...	Skull.†
5857	....	Sable Island, Nova Scotia .	....do	P. S. Dodd	Skin.‡
8694	♀	Sealand, Denmark	....do	Copenhagen Mus.	Skin.

    * Very old.   † Young.   ‡ Adult.

CYSTOPHORA CRISTATA.

List of specimens examined.

Catalogue number.	Sex.	Locality.	In collection of—	Collected by—	Nature of specimen.
6577	....	Greenland	National Museum.	Copenhagen Mus.	Skull.
6576	....	......do	....do	....do	Skull.
6574	....	......do	....do	....do	Skull.
14012	....	Newfoundland	....do	M. Harvey......	Skull.*
14013	....	......do	....do	....do	Skull.*
1083	....	......do	Mus. Comp. Zoöl ..	Michael Carroll.	Skull.
1084	....	......do	....do	....do	Skeleton.
1085	....	......do	....do	....do	Skeleton.
5790	♀	......do	....do	....do.	Mounted.
8696	....	Disco Bay, Greenland.....	National Museum.	Copenhagen Mus.	Skin.
16022	♀	Gulf of Cumberland	....do	L. Kumlien	Skull.†

    * Young.   † "Two years old."

MACRORHINUS ANGUSTIROSTRIS.

List of specimens examined.

Catalogue number.	Sex.	Locality.	In collection of—	Collected by—	Nature of specimen.
4704	....	San Bartholomew's Bay, Lower California.	National Museum.	Dr. W. O. Ayres.	Skull.
13526	....	Coast of Lower California .	....do	C. M. Scammon .	Skull.
......	....	......do	....do	....	Skin.
......	....	......do	Prof. O. C. Marsh .	....	Skin.
......	....	......do	......do	....	Skin.

B.—ADDITIONS AND CORRECTIONS.

FAMILY ODOBÆNIDÆ.

ODOBÆNUS ROSMARUS.

To the bibliographical references to this species already given (*anteà*, pp. 23-26) add the following:

Rostunger oder *Rossmer*, OLAFSEN, Reise durch Island, i, 1774, 189, § 525 (uses of the tusks); ii, 118, § 1861 (occurrence in Iceland).
Trichechus rosmarus, E. SABINE, Parry's First Voy., Suppl., 1824, cxci.—RICHARDSON, Parry's Second Voy., Suppl., 1825, 337.—J. C. ROSS, Parry's Third Voy., 1826, 192 (distribution—Spitzbergen, Walden Island, etc.).—SCHINZ, Syn. Mam., i, 1844, 487.—LLOYD, Game Birds and Wild Fowl of Sweden and Norway, 1867, 444.—FEILDEN, Nares's Voy. to the Polar Sea, 1875-76, ii, 1878, 196.—ALSTON, Proc. Nat. Hist. Soc. Glasgow, 1879, 97 (Outer Hebrides); Fauna of Scotland, Mam., 1880, 15 (Western Scotland).
Odobænus rosmarus, QUENNERSTEDT, Kongl. Svenska Vetenskaps-Akademiens Handlingar, vii, No. 3, 1868, 10.

SIZE AND EXTERNAL APPEARANCE.—To the remarks already given the following may be added:

Dr. E. Sabine gives the length of a young male, from the point of the nose to the end of the hind flippers, as 10 feet 3 inches, and the weight as 1,384 pounds.—*Parry's First Voyage*, Suppl., 1824, p. cxci.

" One of the largest Walruses we saw was killed on the ice near Shannon, on the 27th of August, 1869, by Dr. Copeland. It measured 9 feet 11 inches. . . . It [the Walrus] is from 9 feet 6 inches to 16 feet 6 inches long, weighs about 20 cwt., and its skin is 3½ inches thick (a sort of massive coat of mail), with a head of infinite ugliness, rather large eyes, and tusks sometimes 30 inches long (of a sort of ivory), which help the creature to obtain his food (chiefly mussels) from the bottom of the sea, and, together with the breast-fins, help him to climb on to the floating ice to a place of rest [compare *anteà*, p. 137].

Round his jaws are long eat-like bristles, as thick as a large darning-needle. Demoniacal as his appearance is his voice is as bad—a jerking imitative scream, lowing and puffing, often repeated, and in which it seems to delight."—*German Arctic Expedition*, 1869–70, p. 479.

GEOGRAPHICAL DISTRIBUTION.—To the remarks on this subject add the following:

Nova Zembla and Northern Coast of Europe.—According to Alexander Schultz, Walruses are caught "on the coasts of Novaya Zemblya and the islands of Vaïgatch and Kalonyen."

He also states that "About a dozen sailing-vessels devote themselves habitually to hunting the Walrus from Cape Kanine to the mouth of the Kara".—*Rep. U. S. Commis. Fish and Fisheries*, Part iii, 1876, pp. 53, 56.

Franz-Josef Land.—Payer reports Walruses as seen on two occasions near the coast of Franz-Josef Land.—*New Lands within the Arctic Circle*, p. 266.

Abundance in Wolstenholme Sound.—"Two floe-pieces two or three feet thick, and each covering an area of about half a mile, were black with the large ungainly creatures", in Wolstenholme Sound, August, 1871.—*Narrative of the "Polaris" North Polar Expedition*, 1876, p. 72.

Spitzbergen, etc.—Mr. J. C. Ross, in Parry's "Third Voyage" (1824, p. 192), says that "*Trichechus rosmarus*" is "Very uncommon along the western coast of Spitzbergen and the Low Islands of Phipps; but none seen to the northward of Walden Island."

Iceland.—Respecting the former occurrence of the Walrus in Iceland Olafsen observes "Dahingegen hat man hier [Ost-Island] mehr Rostunge als an andern Orten, inbesondere kam 1708 eine ungewöhnliche Menge davon nach den Ost-fiorden."—*Reise durch Island* (German translation), Theil ii, p. 118, §1861.

SUPPOSED PRESENCE OF WALRUSES IN THE ANTARCTIC SEAS.—In the last footnote to page 176 reference is made to Mr. R. Brown's belief that "It is not unlikely that it [the Walrus] may even be found in the Antarctic regions", in relation to which I there observe: "This idea I have not seen elsewhere revived since the early part of the present century". I find, however, that Schinz, in his "Synopsis Mammalium" published as late as 1844 (vol. i, p. 487), says : " Es [*Trichechus rosmarus*] bewohnt die Polarmeere *beide Pole,* man ist aber noch im Zweifel, ob das Wallross des Südpolarmeers dieselbe Art mit denen der arctischen Meere sein ".

THE WALRUS A FORMIDABLE ANTAGONIST.—To the previous remarks on this subject (*anteà,* pp. 107–133, *passim*), add the following : " In the summer of 1869 a boat excursion to Cape Wynn with difficulty escaped the destruction of their craft. Another time they were followed by a herd and succeeded in reaching the shore of an island, where, though only for a short time, they were blockaded in. The longer you live in Arctic regions the less can you persuade yourself to attack these creatures in their own element, unless forced by pressing circumstances, i. e., want of either food or of oil, and then it is advisable, if in boats, to provide oneself with cartridges."— *German Arctic Expedition,* 1869–70, p. 481.

CURIOSITY AND FEARLESSNESS OF THE WALRUS.—"One peculiarity [of the Walrus], which under some circumstances may be very dangerous, is its curiosity. Should one of these monsters see a boat, it raises itself astonished above the surface, utters at once a cry of alarm, swimming towards it as quickly as possible. This call brings up others, awakens the sleepers which the boat had carefully avoided, and in a short time the small vessel is followed by a number of these monsters, blustering in apparent or real fury in all their hideousness. The creatures may possibly be only actuated by curiosity, but their manner of showing it is unfortunately so ill-chosen that one feels obliged to act on the defensive. The bellowing, jerking, and diving herd is now but a short distance from the boat. The first shot strikes, and this inflames their wrath, and now begins a wild fight in which some of the black sphynxes are struck with axes on the flappers, with which they threaten to overturn the boat. Others of the men defend themselves with a spear or with the blade of an oar."—*German Arctic Expedition,* 1869–70, p. 481.

LOCOMOTION; USE OF THE TUSKS IN CLIMBING.—Captain Sir Edward Belcher, in "The last of the Arctic Voyages" (vol. i, p. 93), thus describes the Walrus's manner of mounting an ice-floe: "But here, within a few feet, deliberately did I watch the progress of the animal in effecting its purpose. In the first place, the tail and fins, exerting their full power in the water, gave such an impetus, that it projected about one-third of the body of the animal on to the floe. It then dug its tusks with such terrific force into the ice that I feared for its brain, and, leech-like, hauled itself forward by the enormous muscular power of the neck, repeating the operation until it was secure. The force with which the tusks were struck into the ice appeared not only sufficient to break them, but the concussion was so heavy that I was surprised that any brain could bear it. Can any one then be surprised, when they are informed, that they 'die hard,' even when shot through the brain?"

FIGURES OF THE WALRUS.—In Lloyd's "Game Birds and Wild Fowl of Sweden and Norway, together with an Account of the Seals and Salt-Water Fishes of those Countries" (London, 1867)—a work which was inaccessible to me till after the Monograph of the Walruses was put in type—occur two admirable plates, illustrative of the Walrus (facing pp. 444 and 457), the first in tint, the other plain. The colored plate (drawn by Körner) gives a full side view of one individual, and a side and front view respectively of the heads and front parts of two others. The other (drawn by Wolf) represents an encounter between a Walrus and a Polar Bear. In this illustration the Walrus is in the attitude of walking, with the hind feet turned forward and the fore feet bent backward. This is one of the most characteristic and truthful representations I have yet met with.

In the "German Arctic Expedition, 1869–70", there is also a figure (wood-cut, p. 369) of a young Walrus,—a side view, with the hind flippers turned backward.

ODOBÆNUS OBESUS.

DESTRUCTION OF THE PACIFIC WALRUS.—Attention has already been called (see *anteà*, pp. 185–186) to the rapid diminution of the Pacific Walrus, and to the alarm the natives have of late years felt respecting the disappearance of their chief means of support. The following (here copied from the Boston *Daily Advertiser* of October 4, 1879) shows how speedily their fears were realized:

"A letter from E. F. Nye, barque Mt. Wollaston, off Cape

Lisburne, Arctic ocean, written to the New Bedford Standard, and dated August 2 [1879], says:—

"'This season up to the present time has been a successful one. Fifty-one whales have been taken by the fleet, against thirty-two at the same time last year, and the whales have run large, averaging about 100 barrels of oil, and say 80,000 pounds of whalebone in all; also about 11,000 walrus against 12,000 last year; the walrus making less oil than usual, as fewer females are killed and a larger proportion of male walrus than in years past. . . . The trading vessels have about 6000 pounds of whalebone and a small quantity of ivory compared with former years; about half the fleet are in this vicinity, the other half are all over to Cape Seege and the western walrus-ing, destroying them by the thousands; about 11,000 have been taken and 30,000 or 40,000 destroyed this year. Another year or perhaps two years will finish them,—there will hardly be one left, and I advise all natural history societies and museums to get a specimen while they can. Fully one-third of the population south of St. Lawrence bay perished the past winter for want of food, and half the natives of St. Lawrence Island died; one village of 200 inhabitants all died except-ing one man. Mothers took their starving children to the burying-grounds, stripped the clothing from their little emaci-ated bodies, and then strangled them or let the intense cold end their misery. It is heart-rending to hear them tell how they suffered. Captain Cogan has taken very few walrus; he says that for every one hundred walrus taken a family is starved, and I concur in his opinion. I should like to see a stop put to this business of killing the walrus, and so would most of those engaged in it. Almost every one says that it is starving the natives, and if one of our whalers should be wrecked on the coast in the fall, the crew must perish.'"

FAMILY OTARIIDÆ.

OTARIES AT THE GALAPAGOS ISLANDS.—To the footnote to page 211, the following may be added:

An early reference to the occurrence of the Southern Fur Seal at these islands is given by Pennant (Arct. Zoöl., i, 1792, p. 199; see also Hist. Quad., third ed., ii, 1793, p. 282) on the authority of Woodes Rogers (Voy., p. 136, 265). On referring to Callander's account of Rogers's voyage, the only one to which I have access, I find the following, which seems to have been generally overlooked by later writers on these animals: "Seals haunt some of these islands [Galapagos], but not so numerous, nor their fur so good as at *Juan Fernandez.*" He refers to a very large one that repeatedly attacked him and adds, "This amphibious beast was as big as a large bear."*

In the footnote to page 211 it is stated that the occurrence of Eared Seals at the Galapagos Islands seems not to be generally known. While there is no reference to their occurrence at these islands by Gray and other leading writers on the group, I find that Mr. Salvin, in discussing the probable habitat of the *Xema furcata* (Ibis, 1875, p. 497), alludes to the fact incidentally, as follows : "Still even were its Arctic character established, it [*Xema furcata*] may yet be an inhabitant of the Galapagos Islands, where an *Otaria* belonging to a northern species exists and formerly abounded". "Northern", however, should read *southern*, for the *Otaria* is *O. jubata*, and consequently is not a *northern* type.

FOSSIL OTARIES.—Add at page 217 the following: Professor McCoy has recently described and figured† fossil remains of Eared Seals from the Pliocene of Victoria, under the name *Arctocephalus williamsi*, sp. nov. The skull figured, which he refers to as an "old male skull", bears a close resemblance to the skull of a female of *Zalophus lobatus*, from which, judging from his description and figures, it does not very materially differ.

CAPTURE OF SEA LIONS FOR MENAGERIES.—The following interesting account of the capture of Sea Lions alive for menageries is from the "Illustrated Guide and Catalogue of Woodward's Gardens," of San Francisco, California (San Francisco, 1880, pp. 50–52), where it is credited to the "Santa Barbara Press":

"Nearly all the live seal, sea lions and sea elephants that have been furnished Woodward's Gardens, in San Francisco, and that have been sent to the Old World and the Eastern States during the last fifteen years, have been captured from the Santa Barbara Islands, across the channel from this city.

"Every year there are more or less of these animals captured

---

* Callander's Voyages, vol. iii, 1768, p. 307.
† Prod. Palæont. Victoria, decade v, p. 7, pll. xli and xliv.

on the islands, for the purpose of supplying menageries in the Eastern States, and parties engaged in the business always come to Santa Barbara to secure men for the purpose, who have had years of experience in capturing them.

"The mode of capturing these animals is simple, yet very exciting, and while it is not considered much of a trick to cage an ordinary sized seal, it is a big contract to capture a bull that weighs 1,500 pounds or more, without seriously injuring the animal.

"Three or four expert vaqueros usually approach the animals that are out on the rocks near the beach, select, perhaps from a hundred or more, the big bull which usually starts for the water, and when the animal arrives at a convenient place on the sand, if possible three riatas are thrown simultaneously, one over the animal's neck, one over either of his front flippers, and one over his rear flippers, making a spread eagle of him instantly. The riata that holds the rear flippers takes away the motive power of the animal, and while his other front flipper is 'lassoed', the riatas are all fastened to the rocks or trees near by, or held by the engaged, while the large box—which has already been made—without the cover, is brought and carefully stood on end close behind the animal, unobserved, and, with a man on top of it is dropped suddenly over the sea lion as he lies stretched at full length on the sand. Small ropes are worked under the box and the animal, and lashed to the top of the box, and at a given signal the riatas are loosened, and the animal is free to move around in his cage at will. The cages are made out of strong fence boards, firmly nailed to the scantling in the corners and on the sides. They are about four feet high, four and a half feet wide, and from eight to fourteen feet in length, and always made before the animal is captured. After the animals are caged, several strong ropes are made fast around the cage and blocks hitched into these ropes, and the cage, with the animal, is drawn through the water to the schooner, near by, and hoisted on board. Fish and water are given the captured, but they often go ten to twenty days without eating."

ZALOPHUS CALIFORNIANUS.

PERIOD OF GESTATION, ETC.—Under date of June 2, 1880, Mr. Frank J. Thompson, superintendent of the Cincinnati Zoölogical Garden, wrote me that a second Californian Sea

Lion had just been born at the Garden. In respect to the period of gestation, etc., Mr. Thompson later kindly gave me the following interesting particulars: "The mother arrived in the Garden on July 2, 1879, and was seen in copulation with the male several times between July 10 and 15. The young one was born May 31, 1880, making the period of gestation about ten and a half months. It was evidently the first calf, and therefore, as is generally the case, the period of gestation was a little lengthened. The youngster is just fourteen days old this morning [June 14], but does not as yet show the least desire to go into the water. He will follow his mother to the edge of the water and there quietly remain while she takes her bath. We had the mother in our possession thirty days before she ate, and as she must have been captured twenty-five or thirty days previous, she was without food for some fifty or sixty days. She was shipped from San Francisco, California, by rail, in a simple wicker basket, and I do not believe she had a drop of water *in transitu.*"

CALLORHINUS URSINUS.

BREEDING OFF THE COAST OF WASHINGTON TERRITORY.—
In a letter from Mr. James G. Swan, Field Assistant of the United States Commission of Fish and Fisheries, dated Neah Bay, Washington Territory, July 17, 1880, kindly communicated by Dr. Coues, contains the following respecting the breeding of the Fur Seal off the coast of Washington Territory:

"Several new facts and theories have been developed by my investigations about Fur Seals this season. The fact that they do have pups in the open ocean off the entrance to Fuca Strait, is well established by evidence of every one of the sealing captains, the Indians, and my own personal observations. Doctor Power says the fact does not admit of dispute. The theory of the captains is, first, that this fact proves conclusively that these Seals do not go to Behring's Sea to have their young, and hence they argue that they do not go there at all, but 'haul out' for purposes of reproduction on some undiscovered islands in the North Pacific, or go at once to the coast of Japan or Siberia where they are known to abound. It seems as preposterous to my mind to suppose that all the Fur Seals of the North Pacific go to the Pribylov Islands as to suppose that all the salmon go to the Columbia and Frazer's River or to the Yakon.

"The question is one of interest, and I have suggested to Professor Baird his having blank forms of questions furnished the captains of all the vessels engaged in sealing, for them to fill out with their observations during the season or during their voyages. These blanks could be sent to the custom-houses at San Francisco, Port Townsend, and Victoria, and given to the captains, with their other papers, when they clear on their sealing voyages, with instructions to fill them out and return them to the custom-house at the end of their voyage.

"A series of such observations, made during several successive seasons, would enable us to ascertain definitely the facts about the Fur Seal, whose habits are but little known except at the rookeries."

Prof. D. S. Jordan, the well-known ichthyologist, to whom the letter was addressed, adds: "I may remark that I saw a live Fur Seal pup June 1 [1880], at Cape Flattery, taken from an old seal just killed, showing that the time of bringing them forth was just at hand."

These observations, aside from the judicious suggestions made by Mr. Swan, are of special interest as confirming those made some years ago by Captain Bryant, and already briefly recorded (*anteà*, p.   ) in this work. They seem to show that at least a certain number of Fur Seals repair to secluded places suited to their needs as far south as the latitude of Cape Flattery, to bring forth their young.

FAMILY PHOCIDÆ

EXTINCT SPECIES.—Prof. A. Leith Adams, in a paper " On Remains of Mastodon and other Vertebrata of the Miocene Beds of the Maltese Islands" (Quar. Journ. Geol. Soc., vol. xxxv, part 3, August, 1879, p. 524, pl. xxv, figg. 1, 2), has described and figured four teeth and a portion of the left ramus of a seal from the calcareous sandstone of Gozo, Malta, under the name *Phoca rugosidens* (*Phoca rugosidens*, Owen apud Adams). They indicate a species of about the size of *Monachus albiventer*, with which in respect to the character of the teeth the species may be compared. Mr. Adams says that canine teeth of large size, referable to the *Phocidæ*, are of common occurrence in the sand bed, and are also somewhat plentiful in the nodule-seams of the calcareous sandstone.

INDEX.

[The figures in black-faced type refer to the descriptions of the families, genera, and species.]

A.

Aglektok, 632, 633.
Aglektorsoak, 633.
Aglektungoak, 633.
Aglektytsiak, 633.
Alactherium, 12, 14.
    cretsii, 13.
Amphibia, 11.
Arctocéphales, 210.
Arctocephalina, 11, 187, 188, 189, 414.
Arctocephalus, 11, 190, 191, 192, 209,
    210, 225, 231, 275, 312.
    antarcticus, 190, 198, 201, 202, 205,
      207, 212, 222.
    argentatus, 199, 202.
    australis, 197, 199, 202, 205, 207,
      210, 211, 222, 224, 226, 330, 333.
    brevipes, 204, 213.
    californianus, 197, 199, 233, 249,
      314, 338.
    cinereus, 191, 197, 198, 199, 213,
      217, 294, 372.
    delalandi, 190, 191, 196, 197, 213.
    elegans, 204, 213.
    eulophus, 199, 200, 202, 204, 216,
      316.
    falklandicus, 191, 196, 197, 199,
      200, 202, 204, 205, 207, 208, 210,
      213, 228, 372.
    forsteri, 199, 204, 205, 213, 372.
    gazella, 204, 213.
    gilliespii, 197, 276.
    grayi, 211.
    grayii, 199, 200, 202, 204.
    hookeri, 191, 197, 202, 209.
    lobatus, 191, 196, 197, 209, 217,
      293, 294.
    monteriensis, 190, 191, 197, 199,
      233, 249, 293, 314, 337.
    nigrescens, 191, 197, 199, 205, 211,
      222.
    nivosus, 198, 199, 202, 203, 213.

Arctocephalus—Continued.
    philippii, 199, 202, 204.
    pusillus, 203, 204.
    schisthyperoës, 198, 202, 203, 213.
    schistuperus, 198.
    williamsi, 770.
    ulloæ, 191.
    ursinus, 197, 200, 204, 312.
Arctophoca, 191, 192, 206, 210.
    argentata, 192, 199, 204, 206.
    elegans, 202, 213.
    falklandica, 192.
    gazella, 202, 213.
    nigrescens, 192, 204.
    philippii, 192, 199, 206.
Atàitsiak, 633.
Atak, 425.
Atârak, 631.
Atarpek, 433.
Atarpiąk, 429, 432, 433.
Atârsoak, 633.
Attarak, 632.
Attarsoak, 425, 631, 632.
Atteitsiak, 632.
Auvekæjak, 432.
Avjor, 739.

B.

Bartrobbe, 657.
Bedlamers, 634.
Bedlimers, 634.
Bellamers, 634.
Beluga catodon, 147.
Bête à la grande dent, 26, 82.
Bodach, 600.
Bos marinus, 26, 82.
Blaakobbe, 657.
Blase-Robbe, 740.
Bladdernose, 726, 740.
Blase-Seehund, 427.
Blandrush, 425.

NATURAL SCIENCES IN AMERICA

An Arno Press Collection

Allen, J[oel] A[saph]. **The American Bisons,** Living and Extinct. 1876

Allen, Joel Asaph. **History of the North American Pinnipeds:** A Monograph of the Walruses, Sea-Lions, Sea-Bears and Seals of North America. 1880

American Natural History Studies: The Bairdian Period. 1974

American Ornithological Bibliography. 1974

Anker, Jean. **Bird Books and Bird Art.** 1938

Audubon, John James and John Bachman. **The Quadrupeds of North America.** Three vols. 1854

Baird, Spencer F[ullerton]. **Mammals of North America.** 1859

Baird, S[pencer] F[ullerton], T[homas] M. Brewer and R[obert] Ridgway. **A History of North American Birds:** Land Birds. Three vols., 1874

Baird, Spencer F[ullerton], John Cassin and George N. Lawrence. **The Birds of North America.** 1860. Two vols. in one.

Baird, S[pencer] F[ullerton], T[homas] M. Brewer, and R[obert] Ridgway. **The Water Birds of North America.** 1884. Two vols. in one.

Barton, Benjamin Smith. **Notes on the Animals of North America.** Edited, with an Introduction by Keir B. Sterling. 1792

Bendire, Charles [Emil]. **Life Histories of North American Birds** With Special Reference to Their Breeding Habits and Eggs. 1892/1895. Two vols. in one.

Bonaparte, Charles Lucian [Jules Laurent]. **American Ornithology:** Or The Natural History of Birds Inhabiting the United States, Not Given by Wilson. 1825/1828/1833. Four vols. in one.

Cameron, Jenks. **The Bureau of Biological Survey:** Its History, Activities, and Organization. 1929

Caton, John Dean. **The Antelope and Deer of America:** A Comprehensive Scientific Treatise Upon the Natural History, Including the Characteristics, Habits, Affinities, and Capacity for Domestication of the Antilocapra and Cervidae of North America. 1877

Contributions to American Systematics. 1974

Contributions to the Bibliographical Literature of American Mammals. 1974

Contributions to the History of American Natural History. 1974

Contributions to the History of American Ornithology. 1974

Cooper, J[ames] G[raham]. **Ornithology.** Volume I, Land Birds. 1870

Cope, E[dward] D[rinker]. **The Origin of the Fittest:** Essays on Evolution and **The Primary Factors of Organic Evolution.** 1887/1896. Two vols. in one.

Coues, Elliott. **Birds of the Colorado Valley.** 1878

Coues, Elliott. **Birds of the Northwest.** 1874

Coues, Elliott. **Key To North American Birds.** Two vols. 1903

Early Nineteenth-Century Studies and Surveys. 1974

Emmons, Ebenezer. **American Geology:** Containing a Statement of the Principles of the Science. 1855. Two vols. in one.

Fauna Americana. 1825-1826

Fisher, A[lbert] K[enrick]. **The Hawks and Owls of the United States in Their Relation to Agriculture.** 1893

Godman, John D. **American Natural History:** Part I — Mastology and **Rambles of a Naturalist.** 1826-28/1833. Three vols. in one.

Gregory, William King. **Evolution Emerging:** A Survey of Changing Patterns from Primeval Life to Man. Two vols. 1951

Hay, Oliver Perry. **Bibliography and Catalogue of the Fossil Vertebrata of North America.** 1902

Heilprin, Angelo. **The Geographical and Geological Distribution of Animals.** 1887

Hitchcock, Edward. **A Report on the Sandstone of the Connecticut Valley,** Especially Its Fossil Footmarks. 1858

Hubbs, Carl L., editor. **Zoogeography.** 1958

[Kessel, Edward L., editor]. **A Century of Progress in the Natural Sciences:** 1853-1953. 1955

Leidy, Joseph. **The Extinct Mammalian Fauna of Dakota and Nebraska,** Including an Account of Some Allied Forms from Other Localities, Together with a Synopsis of the Mammalian Remains of North America. 1869

Lyon, Marcus Ward, Jr. **Mammals of Indiana.** 1936

Matthew, W[illiam] D[iller]. **Climate and Evolution.** 1915

Mayr, Ernst, editor. **The Species Problem.** 1957

Mearns, Edgar Alexander. **Mammals of the Mexican Boundary of the United States.** Part I: Families Didelphiidae to Muridae. 1907

Merriam, Clinton Hart. **The Mammals of the Adirondack Region,** Northeastern New York. 1884

Nuttall, Thomas. **A Manual of the Ornithology of the United States and of Canada.** Two vols. 1832-1834

Nuttall Ornithological Club. **Bulletin of the Nuttall Ornithological Club:** A Quarterly Journal of Ornithology. 1876-1883. Eight vols. in three.

[Pennant, Thomas]. **Arctic Zoology. 1784-1787.** Two vols. in one.

Richardson, John. **Fauna Boreali-Americana;** Or the Zoology of the Northern Parts of British America, Containing Descriptions of the Objects of Natural History Collected on the Late Northern Land Expeditions Under Command of Captain Sir John Franklin, R. N. Part I: Quadrupeds. 1829

Richardson, John and William Swainson. **Fauna Boreali-Americana:** Or the Zoology of the Northern Parts of British America, Containing Descriptions of the Objects of Natural History Collected by the Late Northern Land Expeditions Under Command of Captain Sir John Franklin, R. N. Part II: The Birds. 1831

Ridgway, Robert. **Ornithology.** 1877

Selected Works By Eighteenth-Century Naturalists and Travellers. 1974

Selected Works in Nineteenth-Century North American Paleontology. 1974

Selected Works of Clinton Hart Merriam. 1974

Selected Works of Joel Asaph Allen. 1974

Selections From the Literature of American Biogeography. 1974

Seton, Ernest Thompson. **Life-Histories of Northern Animals: An Account of the Mammals of Manitoba.** Two vols. 1909

Sterling, Keir Brooks. **Last of the Naturalists:** The Career of C. Hart Merriam. 1974

Vieillot, L. P. **Histoire Naturelle Des Oiseaux de L'Amerique Septentrionale,** Contenant Un Grand Nombre D'Especes Decrites ou Figurees Pour La Premiere Fois. 1807. Two vols. in one.

Wilson, Scott B., assisted by A. H. Evans. **Aves Hawaiienses:** The Birds of the Sandwich Islands. 1890-99

Wood, Casey A., editor. **An Introduction to the Literature of Vertebrate Zoology.** 1931

Zimmer, John Todd. **Catalogue of the Edward E. Ayer Ornithological Library.** 1926